the Astronomy Place

YOUR CONVENIENT ONLINE ACCESS TO THE MOST POPULAR ASTRONOMY STUDENT WEBSITE AVAILABLE

With your purchase of a new copy of ***The Essential Cosmic Perspective,* Third Edition,** you should have received a Student Access Kit for **the Astronomy Place.** The kit contains instructions and a code for you to access this dynamic website. Your Student Access Kit looks like this:

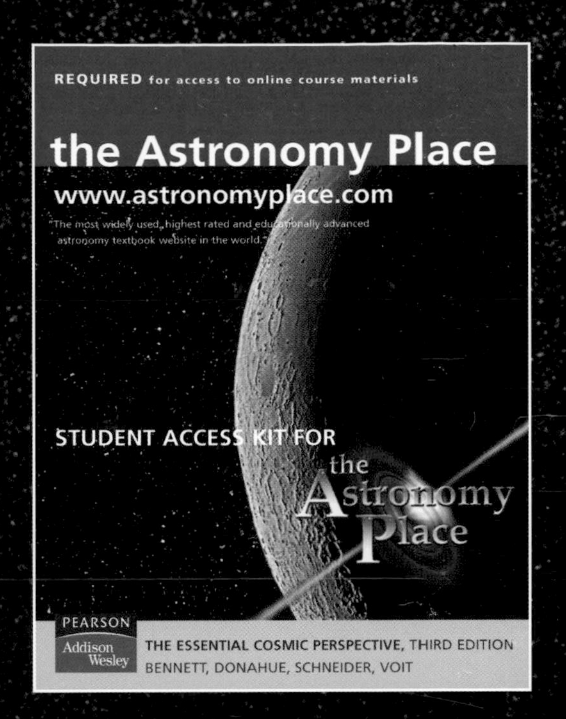

DON'T THROW YOUR ACCESS KIT AWAY!

If you did not purchase a new textbook or cannot locate the Student Access Kit and would like to access the wealth of AstronomyPlace resources, you may purchase your subscription online with a major credit card. Go to www.astronomyplace.com, click on the textbook cover for ***The Essential Cosmic Perspective,* Third Edition,** click Buy Now and follow the on-screen instructions.

WHAT IS Astronomy Place?

Astronomy Place is a dynamic website that features award-winning self-paced animated and interactive tutorials—each designed specifically to help you master key concepts throughout in the course. The site also features narrated and animated movies, chapter-specific quizzes, summaries and overviews, self-test quizzes, flashcards for reviewing, weblinks and more.

To log in to Astronomy Place
After you register using the instructions in the Student Access Kit or purchase access online, simply go to www.astronomyplace.com, click on your book cover and type your Login Name and Password (that you created during registration).

Minimum System Requirements
Windows: 266 MHz; Windows NT, 2000, or XP
Macintosh: 233 MHz; OS 9.2 or higher
Both:

- 64 RAM installed
- 800 x 600 screen resolution
- Browsers: PC: Internet Explorer 5.0 or higher, Netscape Communicator 6.2 or higher; Mac: Internet Explorer 5.0 or higher or Safari 1.2
- Plug Ins: Shockwave Player 8, Flash Player 7, QuickTime 6.0
- Internet connection with 56K modem

Joining an online class (if available)
An online class may be available to you for the Astronomy Place website associated with this textbook. If your instructor chooses to include this as part of your coursework, your instructor will provide you with a *Class ID.*

To participate in an online class, (after you have logged in to Astronomy Place for ***The Essential Cosmic Perspective,* Third Edition**), click Join a Class and follow the on-screen instructions, providing the Class ID when asked. From then on, whenever you log in to this site with your login name and password, you will have access to this online class.

Technical Support
M-F 9am – 6pm Eastern (US & Canada)
http://supportform.pearsoned.com

WHAT STUDENTS ARE SAYING...

"These tutorials greatly increased my knowledge and confidence ... I like ... that you learn something and then you are able to apply it through questions and visual sequences."
—James Tomlinson

"... one of the most awesome educational programs I have ever logged on to. I am pleased with the understanding the phases of the moon that this program has provided with me with."
—David Richardson

"[After the tutorial on the Astronomy Place] I was amazed how well I understood phases of the moon. ... The Light and Spectroscopy tutorial was especially helpful for understanding absorption and emission line spectrums. I had a firm grasp of those concepts after I did the tutorial."
—Cassalyn David

The Essential COSMIC PERSPECTIVE

THIRD EDITION

Jeffrey Bennett
University of Colorado at Boulder

Megan Donahue
Michigan State University

Nicholas Schneider
University of Colorado at Boulder

Mark Voit
Michigan State University

San Francisco Boston New York
Capetown Hong Kong London Madrid Mexico City
Montreal Munich Paris Singapore Sydney Tokyo Toronto

Senior Executive Editor: *Adam Black, Ph.D.*
Developmental Editor: *Elisa Adams*
Assistant Editor: *Stacie Kent*
Managing Producer: *Claire Masson*
Senior Marketing Manager: *Christy Lawrence*
Production Supervisor: *Shannon Tozier*
Production Management: *Elm Street Publishing Services, Inc.*
Composition: *Thompson Type*
Illustrators: *John and Judy Waller, Scientific Illustrators*
Manufacturing Manager: *Pam Augspurger*
Text Designer: *Andrew Ogus and Mark Ong*
Cover Designer: *Hespenheide Design*
Prepress Services: *H & S Graphics*
Text Printer and Binder: *VonHoffmann Press*
Cover Printer: *Phoenix Color*
Cover Credit: Global Cluster: NASA/STScI; Moon: Eckhard Slawik/Photo Researchers, Inc.

CIP data is on file at the Library of Congress

ISBN: 0-8053-8934-2

1 2 3 4 5 6 7 8 9 10—VHC—07 06 05 04

www.aw-bc.com/physics

We shall not cease from exploration
And the end of all our exploring
Will be to arrive where we started
And know the place for the first time.

T. S. Eliot

DEDICATION

To all who have ever wondered about the mysteries of the universe. We hope this book will answer some of your questions—and that it will also raise new questions in your mind that will keep you curious and interested in the ongoing human adventure of astronomy.

And, especially, to the members of the "baby boom" that has occurred among the authors and editors during the writing of this book: Michaela, Emily, Rachel, Sebastian, Elizabeth, Nathan, Grant, Georgia, Brooke, Brian, and Angela. The study of the universe begins at birth, and we hope that you will grow up in a world with far less poverty, hatred, and war so that all people will have the opportunity to contemplate the mysteries of the universe into which they are born.

Brief Contents

PART I

DEVELOPING PERSPECTIVE

PART II

KEY CONCEPTS FOR ASTRONOMY

PART III

LEARNING FROM OTHER WORLDS

PART IV

STARS

PART V

GALAXIES AND BEYOND

PART VI

LIFE ON EARTH AND BEYOND

Detailed Contents

PART II

KEY CONCEPTS FOR ASTRONOMY

PART III

LEARNING FROM OTHER WORLDS

PART IV
STARS

Preface

We humans have gazed into the sky for countless generations. We have wondered how our lives are connected to the Sun, Moon, planets, and stars that adorn the heavens. Today, through the science of astronomy, we know that these connections go far deeper than our ancestors ever imagined. This book tells the story of modern astronomy and the new perspective, *The Essential Cosmic Perspective*, that it gives us on ourselves and our planet.

This book grew out of our experience teaching astronomy to both college students and the general public over the past 25 years. During this time, a flood of new discoveries fueled a revolution in our understanding of the cosmos but had little impact on the basic organization and approach of most astronomy textbooks. We felt the time had come to rethink how to organize and teach the major concepts in astronomy to reflect this renaissance in understanding. This book is the result.

WHO IS THIS BOOK FOR?

The Essential Cosmic Perspective is designed as a textbook for college courses in introductory astronomy, but is suitable for anyone who is curious about the universe. We assume no prior knowledge of astronomy or physics, and the book is especially suited to students who do not intend to major in mathematics or science.

We have tailored *The Essential Cosmic Perspective* to one-semester survey courses in astronomy by carefully selecting the most important topics in astronomy and presenting them with only as much depth as can be realistically learned in one semester. This book may also be used for two-semester astronomy sequences, though instructors of such courses may wish to consider the more comprehensive version of this book, titled *The Cosmic Perspective*.

ABOUT THE THIRD EDITION

If you are familiar with prior editions of *The Essential Cosmic Perspective*, this third edition might at first be unrecognizable. We have completely rewritten the book from beginning to end, in order to make it much easier for instructors to teach from and students to learn from in a fast-paced one-semester course. Nevertheless, our guiding philosophies, which we outline below, remain the same. Here, briefly, is a list of the major changes you'll find in this third edition:

- Streamlined Presentation: To make it more realistic for this book to be covered in one semester, we have streamlined the presentation significantly. The book now has only 18 chapters, and we have significantly shortened the word count by focusing on only the most important concepts for a one-term course.
- New Layout: We have given the book a more student-friendly layout by using one text column rather than two. This makes the book more open and easier to read, and also gives the students a wide margin in which to make notes that will aid them in studying. We have also added highlighted essential points. These signposts (easy to recognize with their blue font) help students to find the relevant discussion in the text quickly and easily when they need to review a particular topic.
- Chapter Structure Centered on Learning Goals: We now begin each chapter with a clearly spelled-out list of Learning Goals that take the form of questions. These learning goals are repeated as the subsection headings throughout the book. This makes it much easier for students to know exactly what they are learning and why, and also aids studying by making it easier to find topics in the book. Each chapter now concludes with a summary that offers brief answers to the Learning Goal questions plus thumbnail images to remind students of key figures.
- Enhanced Media Tie-ins: In addition to the book itself, we have developed more of our popular and critically acclaimed multimedia tutorials, and have built a new library of Interactive Figure study aids. (These are key figures from each chapter that are better understood through interactive versions for reasons of geometry, scale, time evolution, or multiple representation.) These, along with a new gradebook and a wealth of other resources are available on the Astronomy Place Web site (www.astronomyplace.com)—the most pedagogically advanced, highest student-rated and most widely used astronomy textbook study Web site in the world. Please note that these features are tied directly to this textbook and no other, and have been developed with the active involvement of the authors. Thus, for example, you'll find the same pedagogical philosophy and the same jargon in the web tutorials as you'll find in the book; you'll also notice consistency in the illustration styles. Icons throughout the book point to key media tutorials and exercises. A new Media Explorations section summarizes the media resources for each chapter and provides media-related activities to make it easier for professors to utilize these popular resources in their course.

THEMES OF *THE ESSENTIAL COSMIC PERSPECTIVE*

The Essential Cosmic Perspective offers a broad survey of modern understanding of the cosmos and of how we have built that understanding. Such a survey can be presented in a number of different ways. We have chosen to interweave a few key themes throughout the book, each selected to help make the subject more appealing to students who may never have taken any formal science courses and who may begin the course with little understanding of how science works. We built our book around the following five key themes:

- *Theme 1: We are a part of the universe and thus can learn about our origins by studying the universe.* This is the overarching theme of *The Essential Cosmic Perspective*, as we continually emphasize that learning about the universe helps us understand ourselves. Studying the intimate connections between human life and the cosmos gives students a reason to care about astronomy and also deepens their appreciation of the unique and fragile nature of our planet and its life.
- *Theme 2: The universe is comprehensible through scientific principles that can be understood by anyone.* The universe is comprehensible because the same physical laws appear to be at work in every aspect, on every scale, and in every age of the universe. Moreover, while the laws generally have been discovered by professional scientists, their fundamental features can be understood by anyone. Students can learn enough in one term of astronomy to comprehend the basic reasons for many phenomena they see around them—phenomena ranging from seasonal changes and phases of the Moon to the most esoteric astronomical images that appear in the news.
- *Theme 3: Science is not a body of facts but rather a process through which we seek to understand the world around us.* Many students assume that science is just a laundry list of facts. The long history of astronomy can show them that science is a process through which we learn about our universe—a process that is not always a straight line to the truth. That is why our ideas about the cosmos sometimes change as we learn more, as they did dramatically when we first recognized that Earth is a planet going around the Sun rather than the center of the universe. In this book, we continually emphasize the nature of science so that students can understand how and why modern theories have gained acceptance and why these theories may still change in the future.
- *Theme 4: A course in astronomy is the beginning of a lifelong learning experience.* Building upon the prior themes, we emphasize that what students learn in their astronomy course is not an end but a beginning. By remembering a few key physical principles and understanding the nature of science, students can follow astronomical developments for the rest of their lives. We therefore seek to motivate students enough that they will continue to participate in the ongoing human adventure of astronomical discovery.
- *Theme 5: Astronomy affects each of us personally with the new perspectives it offers.* We all conduct the daily business of our lives with reference to some "world view"—a set of personal beliefs about our place and purpose in the universe that we have developed through a combination of schooling, religious training, and personal thought. This world view shapes our beliefs and many of our actions. Although astronomy does not mandate a particular set of beliefs, it does provide perspectives on the architecture of the universe that can influence how we view ourselves and our world, and these perspectives can potentially affect our behavior. For example, someone who believes Earth to be at the center of the universe might treat our planet quite differently from someone who views it as a tiny and fragile world in the vast cosmos. In many respects, the role of astronomy in shaping world views may represent the deepest connection between the universe and the everyday lives of humans.

PEDAGOGICAL PRINCIPLES OF *THE ESSENTIAL COSMIC PERSPECTIVE*

No matter how an astronomy course is taught, it is very important to present material according to a clear set of pedagogical principles. The following list briefly summarizes the major pedagogical principles that we apply throughout this book. (The Instructor's Guide describes these principles in more detail.)

- *Stay focused on the big picture.* Astronomy is filled with interesting facts and details, but they are meaningless unless they fit into a big-picture view of the universe. We therefore take care to stay focused on the big picture (essentially the themes discussed above) at all times. A major benefit of this approach is that although students may forget individual facts and details after the course is over, the big picture framework should stay with them for life.
- *Always provide context first.* We all learn new material more easily when we understand why we are learning it. In essence, this is simply the idea that it is easier to get somewhere when you know where you are going. We therefore being the book (chapter 1) with a broad overview of modern understanding of the cosmos, so that students can understand what they will be learning in the rest of the book. We maintain this "context first" approach throughout the book by always telling students what they will be learning, and why, before diving into the details.
- *Make the material relevant.* It's human nature to be more interested in subjects that seem relevant to our lives. Fortunately, astronomy is filled with ideas that touch each of us personally. For example, the study of our solar system helps us better understand and appreciate our planet

Earth, and the study of stars and galaxies helps us learn how we have come to exist. By emphasizing our personal connections to the cosmos, we make the material more meaningful, inspiring students to put in the effort necessary to learn it.

- *Emphasize conceptual understanding over stamp collecting of facts.* If we are not careful, astronomy can appear to be an overwhelming collection of facts that are easily forgotten when the course ends. We therefore emphasize a few key conceptual ideas that we use over and over again. For example, the laws of conservation of energy and conservation of angular momentum (introduced in Section 4.3) reappear throughout the book, and we find that the wide variety of features found on the terrestrial planets can be understood through just a few basic geological processes (see Section 7.1). Research shows that, long after the course is over, students are far more likely to retain such conceptual learning than individual facts or details.

- *Proceed from the more familiar and concrete to the less familiar and abstract.* It's well known that children learn best by starting with concrete ideas and then generalizing to abstractions later. In fact, the same is true for many adults. We therefore always try to "build bridges to the familiar"—that is, to begin with concrete or familiar ideas and then gradually draw more general principles from them.

- *Use plain language.* Surveys have found that the number of new terms in many introductory astronomy books is larger than the number of words taught in many first courses in foreign language. In essence, this means the books are teaching astronomy in what looks to students like a foreign language! Clearly, it is much easier for students to understand key astronomical concepts if they are explained in plain English without resorting to unnecessary jargon. We have gone to great lengths to eliminate jargon as much as possible or, at minimum, to replace standard jargon with terms that are easier to remember in the context of the subject matter.

- *Recognize and address student misconceptions.* Students do not arrive as blank slates. Most students enter our courses not only lacking the knowledge we hope to teach but often holding misconceptions about astronomical ideas. Therefore, to teach correct ideas, we must also help students recognize the paradoxes in their prior misconceptions. We address this issue in a number of ways, the most obvious being the presence of many Common Misconceptions boxes. These summarize commonly held misconceptions and explain why they cannot be correct.

The Topical (Part) Structure of *The Essential Cosmic Perspective*

The Essential Cosmic Perspective is organized into six broad topical areas (the six Parts are in the table of contents), each approached in a distinctive way designed to help maintain the focus on the themes previously mentioned. Here, we summarize the guiding philosophy through which we have approached each topic. We also highlight a few of the major changes in the third edition and list a few of the key new figures that illustrate the general improvements we have made in this new edition. (The Instructor's Guide describes the topical structure in much more detail.)

Part I Developing Perspective (Chapters 1–3)

Guiding Philosophy: Introduce the big picture, the process of science, and the historical context of astronomy.

The basic goal of these chapters is to give students a big picture overview and context for the rest of the book and to be sure they develop an appreciation for the process of science and how science has developed through history. Chapter 1 offers an overview of our modern understanding of the cosmos, thereby giving students perspective on the entire universe. Chapter 2 provides an introduction to basic sky phenomena, including seasons and phases of the Moon, and a perspective on how phenomena we experience every day are tied to the broader cosmos. Chapter 3 discusses the nature of science, offering a historical perspective on the development of science and giving students perspective on how science works and how it differs from nonscience.

New for the third edition Users of the second edition will want to be aware of the following organizational changes we have made to Part I for the third edition:

1. We have enhanced our coverage of the seasons, and moved this coverage and our discussion of precession to Chapter 2 (from Chapter 1 in the previous edition).
2. The 10-page tour of the solar system that formerly appeared in Chapter 1 now appears in chapter 6.
3. We have expanded and completely revised both the discussion of the Greek role in the historical development of science (Section 3.2) and the discussion of the nature of science (Section 3.4).

Key Figures We have added many new figures in this edition, and revised many others. To better understand our pedagogical approach to Part I, we encourage you to look at some of the following figures: 1.1, 1.2, 1.10, 1.15, 1.16, 2.13, 2.20, 2.28, 3.21, 3.23

Part II Key Concepts for Astronomy (Chapters 4–5)

Guiding Philosophy: Bridges to the familiar.

These two chapters lay the groundwork for understanding astronomy through what is sometimes called the universality of physics—the idea that a few key principles governing matter, energy, light, and motion explain both the phenomena of our daily lives and the mysteries of the cosmos. Chapter 4 covers the laws of motion, the crucial conservation laws of angular momentum and energy, and the universal law of gravitation. Chapter 5 covers the nature of light and matter, the formation of spectra, and telescopes.

New for the third edition Users of the second edition will want to be aware of the following organizational changes we have made to Part II for the third edition:

1. We have eliminated the second edition's Chapter 4 (matter and energy), instead integrating its content into the new Chapters 4 and 5.
2. Section 4.3 on conservation laws in astronomy is new to this edition, introducing students to these two very important laws in a single, concise section.

Key Figures To better understand our pedagogical approach to Part II, we encourage you to look at some of the following figures: 4.3, 4.5, 4.15, 5.2, 5.6, 5.22

Part III Learning From Other Worlds (Chapters 6–9)

Guiding Philosophy: Learning about Earth by learning about the solar system.

This set of chapters begins with a broad overview of the solar system and discussion of solar system formation in Chapter 6; this chapter also includes discussion of extrasolar planets. The next three chapters then focus respectively on the terrestrial planets, the jovian planets, and the small bodies of the solar system. Note that Part III is essentially independent of Parts IV and V, and thus can be covered either before or after them.

New for the third edition Users of the second edition will want to be aware of the following organizational changes we have made to Part III for the third edition:

1. The overview of our solar system that begins Chapter 6 now includes the 10-page solar system tour that previously appeared in Chapter 1.
2. We have eliminated the second edition's chapter 11 (Planet Earth), instead integrating its content into the new Chapters 7 and 18.
3. Chapter 7 now proceeds in a planet-by-planet order through its discussion of the terrestrial planets, though our emphasis remains on a discussion of how and why the planets differ from one another. Section 7.1 uses Earth to illustrate the basic principles of geology and atmospheres, which we then apply to the other terrestrial planets in the next three sections. We then conclude in Section 7.5 by returning to Earth to see how we can understand its unique features and the human impact in light of what we have learned from other worlds.

Key Figures To better understand our pedagogical approach to Part III, we encourage you to look at some of the following figures: 6.1, 6.15, 6.27, 7.1, 7.15, 7.27, 7.31, 7.36, 7.38, 8.1, 8.26, 8.30, 9.8, 9.9, 9.20

Part IV Stars (Chapters 10–13)

Guiding Philosophy: We are intimately connected to the stars.

These are our chapters on stars and stellar lifecycles. Chapter 10 covers the Sun in depth, so that it can serve as our concrete model for building an understanding of other stars. Chapter 11 describes the general properties of other stars, how we measure these properties, and how we classify stars with the HR diagram. Chapter 12 covers stellar evolution, tracing the birth-to-death lives of both low- and high-mass stars. Chapter 13 covers the end points of stellar evolution: white dwarfs, neutron stars, and black holes.

New for the third edition Users of the second edition will want to be aware of the following organizational changes we have made to Part IV for the third edition:

1. We have reorganized Chapter 10 on the Sun by moving all the material on solar activity to a single, consolidated section (10.3).
2. In both Chapters 11 and 12, we have restructured the narrative so that students learn all the important ideas of stellar properties (in Chapter 11) or stellar evolution (in Chapter 12) before we show these ideas on the HR diagram. This approach recognizes that the HR diagram is conceptually difficult for many students, and that students can better understand it by first learning the conceptual ideas and then seeing how these ideas are summarized on the diagram.

Key Figures To better understand our pedagogical approach to Part IV, you might in particular wish to look at some of the following figures: 10.3, 10.6, 10.14, 11.8, 11.10, 12.3, 12.5, 12.10, 12.20, 13.9

Part V Galaxies and Beyond (Chapters 14–17)

Guiding Philosophy: Present galaxy evolution in a way that parallels the teaching of stellar evolution, and integrate cosmological ideas in the places where they most naturally arise.

These chapters cover galaxies and cosmology. Chapter 14 presents the Milky Way as a paradigm for galaxies in much the same way that Chapter 10 uses the Sun as a paradigm for stars. Chapter 15 presents the variety of galaxies, how we determine key parameters such as galactic distances and age, and current understanding of galaxy evolution. Chapter 16 focuses on dark matter and dark energy and their role in the fate of the universe. Chapter 17 covers the theory of the Big Bang.

New for the third edition Users of the second edition will want to be aware of the following organizational changes we have made to Part V for the third edition:

1. Chapter 14 on the Milky Way now includes a discussion of the evolutionary history of our galaxy.
2. We have reorganized Chapter 15 so that the discussion of cosmic distances and ages is better integrated into the context of galaxy evolution.
3. We have rewritten Chapter 16 to include much more discussion of dark energy and its possible relevance to the fate of the universe. In particular, please note our new Section 16.1, which explains the meaning of the terms dark matter and dark energy so that students have appropriate context for the subsequent discussion.

Key Figures To better understand our pedagogical approach to Part V, we encourage you to look at some of the following figures: 14.1, 14.3, 14.18, 14.20, 15.1, 15.16, 15.19, 15.29, 16.2, 16.13, 16.14, 16.15, 17.3

Part VI Life on Earth and Beyond (Chapter 18)

Guiding Philosophy: The study of life on Earth helps us understand the search for life in the universe.

This Part consists of a single chapter. It may be considered optional, to be used as time allows. Those who wish to teach a more detailed course on astrobiology may wish to consider the text *Life in the Universe*, by Bennett, Shostak, and Jakosky.

New for the third edition Users of the second edition will want to be aware of the following organizational changes we have made to Part VI for the third edition:

1. This chapter is quite different from the comparable chapter in the second edition, and in particular now includes a discussion of life on Earth, which formerly appeared in Chapter 11.
2. We have increased emphasis on how understanding astronomy gives us insight into the challenge of interstellar travel, largely designed so that students will be able to critically evaluate reports of UFOs.

Key Figures To better understand our pedagogical approach to Part VI, we encourage you to look at some of the following figures: 18.3, 18.4, 18.7, 18.15

PEDAGOGICAL FEATURES OF *THE ESSENTIAL COSMIC PERSPECTIVE*

Alongside the main narrative, *The Essential Cosmic Perspective* includes a number of pedagogical devices designed to enhance student learning. Here is a brief summary, beginning with features new to the third edition.

NEW **Learning Goals** Presented as key questions at the start of each chapter and used as subsection titles in the book, these goals help students focus their attention on the most important concepts.

NEW **Chapter Summary** The end-of-chapter summary offers concise answers to the learning goal questions, helping reinforce student understanding of key concepts from the chapter.

NEW **Highlighted Essential Points** These blue text callouts call attention to key points and helps students find the relevant discussion in the text.

NEW **Key Concept Figures** Dozens of new figures have been added and many more have been improved so that nearly every key concept described in the text is now accompanied by a figure that summarizes it visually. Thus, students can get an overview of all the key chapter concepts by studying the illustrations (with their captions), then go back to read the chapter in detail.

NEW **Wavelength/Observatory Icons** For astronomical photographs (or art that might be confused with photographs), simple icons identify the wavelength band of the photo or identify the figure as an art piece or computer simulation. Along with the wavelength icon for photos, another icon indicates whether the image came from ground-based or space-based observations.

NEW **Interactive Figures** In each chapter, key figures that students often struggle with because of geometry/perspective, scale, time evolution, or tying together multiple representations, are now supplemented with interactive versions online.

NEW **Media Explorations** Each chapter ends with a section or page of Media Explorations that highlight some of the many award-winning interactive tutorials and other media resources available to aid students in studying the chapter material and provide suggestions for how to easily incorporate these into your course. These sections include suggested Web projects designed for independent research.

- **Astronomy Place Interactive Tutorials** Specific Lessons from within the highly-acclaimed interactive tutorials on www.astronomyplace.com are referenced above specific section titles so students know where to find additional, self-paced help.
- **The Big Picture** Every chapter narrative ends with this feature. It helps students put what they've learned in the chapter into the context of the overall goal of gaining a new perspective on ourselves and our planet.
- **End-of-Chapter Questions** Each chapter includes an extensive set of exercises that can be used for study, discussion, or assignment. NEW in this edition: We have added Math Help and Exercises at the end of most problem sets. These point to relevant mathematical topics that we have included on the Astronomy Place Web site, and thus are for the benefit of those who wish to delve into some of the mathematical details behind the concepts covered in this book.

- **Think About It** This feature, which appears throughout the book as short questions integrated into the narrative, gives students the opportunity to reflect on important new concepts. It also serves as an excellent starting point for classroom discussions.
- **Common Misconceptions** These boxes address and correct popularly held but incorrect ideas related to the chapter material.
- **Special Topic Boxes** These boxes contain supplementary discussion topics related to the chapter material but not prerequisite to the continuing discussion.
- **Cross-References** When a concept is covered in greater detail elsewhere in the book, we include a cross reference, in brackets, to the relevant section (e.g., [Section 5.2]).
- **Glossary** A detailed glossary makes it easy for students to look up important terms.
- **Appendixes** The appendixes include a number of useful references and tables, including key constants (Appendix A), key formulas (Appendix B), key mathematical skills (Appendix C), and numerous data tables and star charts.

RESOURCES AND SUPPLEMENTS FOR *THE ESSENTIAL COSMIC PERSPECTIVE*

The Essential Cosmic Perspective is much more than just a textbook. It is a complete package of resources designed to help both teachers and students. Here is a brief summary of the available resources and supplements.

- FREE with All New Books: **Astronomy Place (www.astronomyplace.com)**. The Astronomy Place Web site offers a complete library of critically acclaimed Interactive Tutorials in addition to a wealth of study resources for students, a gradebook, course management tools, and many other resources for teachers. With more than 70,000 users per month, this is the most popular astronomy textbook Web site available to students. A subscription to the site is included free with every new book. Look for your personal access kit with your new book. If you did not receive an access kit with your book, you may purchase access online at www.astronomyplace.com. Among the many resources at the Astronomy Place, you'll find:
    - **Critically acclaimed interactive tutorials** 22 in-depth Interactive Tutorials are now available, providing more than 60 individual tutorial lessons and hundreds of interactive tools and animations, each focused on a key concept. The text includes icons in section headers to point to specific online tutorial lessons, plus suggested tutorial activities in the Media Explorations sections at the end of each chapter.
    - **Online, multiple-choice chapter quizzes** New for the third edition, the Astronomy Place now has three quizzes for each chapter in the book. One quiz focuses on basic definitions and ideas, one on more conceptual questions, and the third on identification and understanding of photos and important figures from the text.
    - New **Interactive Figures** Key figures that students often struggle with because of geometry/perspective, scale, time evolution or tying together multiple representations, are now cited in the figure caption and an interactive version provided online.
    - **A powerful gradebook** Students can use the tutorials and quizzes for self-study. However, if you want to make this material part of the grade, a new gradebook allows you to track your students completion of the tutorials and their Tutorial Exercise and quiz grades.
    - **And much more** Animated movies, flash cards, study resources for individual chapters, and many other useful study aids can be found at the Astronomy Place Web site.
- FREE with All New Books: ***Voyager: Skygazer, College Edition.*** Based on Voyager III, one of the world's most popular planetarium programs, *SkyGazer* makes it easy for students to learn constellations and explore the wonders of the sky through interactive exercises. The *Skygazer* CD is packaged free with all new copies of this book. It comes preloaded with 75 demos, and suggested activities appear in the Media Explorations section at the end of each chapter in the book.
- FREE with New Books: **The Addition Wesley Astronomy Tutor Center**. This center provides one-on-one tutoring by qualified college instructors in any of four ways: phone, fax, email, and the Internet during evening and weekend hours. Tutor center instructors will answer questions and provide help with examples and exercises from the text. Tutor center registration is free with new books only when the professor orders books with the special tutor center package. (Professors: Contact your local Addison Wesley sales representative if you wish to order this package.) Otherwise, it can be purchased separately. See www.aw-bc.com/tutorcenter for more information.
- **Astronomy Media Workbook** (ISBN 0-8053-8755-2). This student supplement offers an extensive set of printed activities and more in-depth projects suitable for labs or homework assignments that use the Astronomy Place Web site tutorials and Skygazer software.

Several additional supplements are available for instructors only. Contact your local Addison Wesley sales representative to find out more about the following supplements:

- **The Essential Cosmic Lecture Launcher CD** (ISBN 0-8053-8951-2). This CD provides a wealth of presentation tools to help prepare course lectures. It includes a set of PowerPoint® slides for every chapter in the textbook, a comprehensive collection of more than 1,000 high-resolution figures from the book and other astronomical sources, and a library of more than 250 interactive applets and simulations, all built specifically to enhance the same pedagogy presented in the text.

- **Instructor's Guide** (ISBN 0-8053-8943-1). This guide contains a detailed overview of the text, sample syllabi for courses of different emphasis and duration, suggestions on teaching strategies, answers or discussion points for all Think About It questions in the text, solutions to end-of-chapter problems, and a detailed reference guide summarizing media resources available for every chapter and section in the book.
- **Carl Sagan's Cosmos (DVD or Video)**. The Best of Cosmos and the complete, revised, enhanced, and updated Cosmos series are available free to qualified adopters of *The Essential Cosmic Perspective*.
- **Test Bank**. Available in both computerized (ISBN 0-8053-8950-4) or printed (ISBN 0-8053-8949-0) form, the Test Bank contains a broad set of multiple-choice, true/false, and free-response questions for each chapter, including the questions from the online quizzes.
- **Transparency Acetates** (ISBN 0-8053-8944-X). For those who use overhead projectors in lectures, this set contains more than 100 images from the text.

ACKNOWLEDGMENTS

A textbook may carry author names, but it is the result of hard work by a long list of committed individuals. We could not possibly list everyone who has helped, but we would like to call attention to a few people who have played particularly important roles. First, we thank our editors and friends at Addison Wesley who have stuck with us through thick and thin, including Adam Black, Linda Davis, Stacie Kent, Christy Lawrence, Stacy Treco, Liana Allday, Vivian McDougal, and Shannon Tozier. Special thanks to our past and present production teams, especially Mary Douglas and Brandi Nelson; our art and design team, Blakeley Kim, Mark Ong, Judy Waller, John Waller, and Angel Chavez; our supplements team, including Tom Fleming and Stacy Palen (work on the Cosmic Lecture Launcher CD), Jonathan Williams, Dave Brain, and Kelly Cline (work on the Test Bank), and Michael LoPresto (Media Workbook author); and our Web team, led by Claire Masson, Jim Dove, and Ian Shakeshaft.

We've also been fortunate to have an outstanding group of reviewers whose extensive comments and suggestions helped us shape the book. We thank all those who have reviewed drafts of the book in various stages, including:

Christopher M. Anderson, *University of Wisconsin*
Peter S. Anderson, *Oakland Community College*
Keith Ashman, *University of Missouri, Kansas City*
John Beaver, *University of Wisconsin at Fox Valley*
Timothy C. Beers, *Michigan State University*
Priscilla J. Benson, *Wellesley College*
David Brain, *University of California Berkeley Space Sciences Laboratory*
David Branch, *University of Oklahoma*
Jean P. Brodie, *UCO/Lick Observatory, University of California, Santa Cruz*
James Brooks, *Florida State University*
Daniel Bruton, *Stephen F. Austin State University*
Eric Carlson, *Wake Forest University*
Supriya Chakrabarti, *Boston University*
Dipak Chowdhury, *Indiana University—Purdue University at Fort Wayne*
Chris Churchill, *New Mexico State University*
Josh Colwell, *University of Colorado*
Kevin Crosby, *Carthage College*
Christopher Crow, *Indiana University Purdue University, Fort Wayne*
John M. Dickey, *University of Minnesota*
Robert Egler, *North Carolina State University at Raleigh*
David Falk, *Los Angeles Valley College*
Robert A. Fesen, *Dartmouth College*
Douglas Franklin, *Western Illinois University*
Sidney Freudenstein, *Metropolitan State College of Denver*
Martin Gaskell, *University of Nebraska*
Richard Gelderman, *Western Kentucky University*
David Graff, *U.S. Merchant Marine Academy*
Richard Gray, *Appalachian State University*
Kevin Grazier, *Jet Propulsion Laboratory*
Alan Greer, *Gonzaga University*
David Griffiths, *Oregon State University*
David Grinspoon, *University of Colorado*
Bruce Gronich, *University of Texas, El Paso*
Jim Hamm, *Big Bend Community College*
Charles Hartley, *Hartwick College*
Joe Heafner, *Catawba Valley Community College*
Richard Holland, *Southern Illinois University, Carbondale*
Joseph Howard, *Salisbury University*
Richard Ignace, *University of Wisconsin*
Bruce Jakosky, *University of Colorado*
Adam Johnston, *Weber State University*
Lauren Jones, *Gettysburg College*
William Keel, *University of Alabama*
Julia Kennefick, *University of Arkansas*
Steve Kipp, *University of Minnesota, Mankato*
Kurtis Koll, *Cameron University*
John Kormendy, *University of Texas, Austin*
Eric Korpela, *University of California, Berkeley*
Kristine Larsen, *Central Connecticut State University*
Ana Marie Larson, *University of Washington*
Larry Lebofsky, *University of Arizona*
Nancy Levenson, *University of Kentucky*
Patrick Lestrade, *Mississippi State University*
David M. Lind, *Florida State University*
Michael LoPresto, *Henry Ford Community College*

William R. Luebke, *Modesto Junior College*
Marie Machacek, *Massachusetts Institute of Technology*
Steven Majewski, *University of Virginia*
Phil Matheson, *Salt Lake Community College*
Marles McCurdy, *Tarrant County College*
Stacy McGaugh, *University of Maryland*
Barry Metz, *Delaware County Community College*
William Millar, *Grand Rapids Community College*
Dinah Moche, *Queensborough Community College of City University, New York*
Zdzislaw E. Musielak, *University of Texas, Arlington*
Gerald H. Newsom, *Ohio State University*
Brian Oetiker, *Sam Houston State University*
John P. Oliver, *University of Florida*
Russell L. Palma, *Sam Houston State University*
Bryan Penprase, *Pomona College*
Charles Peterson, *University of Missouri, Columbia*
Jorge Piekarewicz, *Florida State University*
Harrison B. Prosper, *Florida State University*
Monica Ramirez, *Aims College, Colorado*
Christina Reeves-Shull, *Richland College*
Elizabeth Roettger, *DePaul University*
Roy Rubins, *University of Texas, Arlington*
Rex Saffer, *Villanova University*
John Safko, *University of South Carolina*
James A. Scarborough, *Delta State University*
Ann Schmiedekamp, *Pennsylvania State University, Abington*
Joslyn Schoemer, *Denver Museum of Nature and Science*
James Schombert, *University of Oregon*
Gregory Seab, *University of New Orleans*
Larry Sessions, *Metropolitan State College of Denver*
Paul Sipiera, *William Harper Rainey College*
Michael Skrutskie, *University of Virginia*
Mark H. Slovak, *Louisiana State University*
Norma Small-Warren, *Howard University*
Dale Smith, *Bowling Green State University*
John Spencer, *Lowell Observatory*
Darryl Stanford, *City College of San Francisco*
John Stolar, *West Chester University*
Jack Sulentic, *University of Alabama*
C. Sean Sutton, *Mount Holyoke College*
Beverley A. P. Taylor, *Miami University*
Brett Taylor, *Radford University*
Donald M. Terndrup, *Ohio State University*
Frank Timmes, *School of Art Institute of Chicago*
David Trott, *Metro State College*
Nicole Vogt, *New Mexico State University*
Darryl Walke, *Rariton Valley Community College*
Fred Walter, *State University of New York, Stony Brook*
James Webb, *Florida International University*
Mark Whittle, *University of Virginia*
Paul J. Wiita, *Georgia State University*
Jonathan Williams, *University of Florida*
J. Wayne Wooten, *Pensacola Junior College*
Scott Yager, *Brevard College*
Arthur Young, *San Diego State University*
Min S. Yun, *University of Massachusetts, Amherst*
Dennis Zaritsky, *University of California, Santa Cruz*
Robert L. Zimmerman, *University of Oregon*

Historical Accuracy Reviewer Owen Gingerich, *Harvard—Smithsonian*

In addition, we thank the following colleagues who helped us clarify technical points or checked the accuracy of technical discussions in the book:

Thomas Ayres, *University of Colorado*
Cecilia Barnbaum, *Valdosta State University*
Rick Binzel, *Massachusetts Institute of Technology*
Howard Bond, *Space Telescope Science Institute*
David Brain, *University of California Berkeley Space Sciences Laboratory*
Humberto Campins, *University of Florida*
Robin Canup, *Southwest Research Institute*
Kelly Cline, *Carroll College*
Josh Colwell, *University of Colorado*
Mark Dickinson, *Space Telescope Science Institute*
Jim Dove, *Metropolitan State College of Denver*
Harry Ferguson, *Space Telescope Science Institute*
Andrew Hamilton, *University of Colorado*
Todd Henry, *Georgia State University*
Dave Jewitt, *University of Hawaii*
Hal Levison, *Southwest Research Institute*
Mario Livio, *Space Telescope Science Institute*
Mark Marley, *New Mexico State University*
Kevin McLin, *University of Colorado, Boulder*
Rachel Osten, *University of Colorado, Boulder*
Bob Pappalardo, *University of Colorado*
Michael Shara, *American Museum of Natural History*
Glen Stewart, *University of Colorado*
John Stolar, *West Chester University*
Dave Tholen, *University of Hawaii*
Nick Thomas, *MPI/Lindau (Germany)*
Dimitri Veras, *University of Colorado*
John Weiss, *University of Colorado, Boulder*
Don Yeomans, *Jet Propulsion Laboratory*

Finally, we thank the many people who have greatly influenced our outlook on education and our perspective on the universe over the years, including Tom Ayres, Fran Bagenal, Forrest Boley, Robert A. Brown, George Dulk, Erica Ellingson, Katy Garmany, Jeff Goldstein, David Grinspoon, Robin Heyden, Don Hunten, Bruce Jakosky, Joan Marsh, Catherine McCord, Dick McCray, Dee Mook, Cheri Morrow, Charlie Pellerin, Carl Sagan, Mike Shull, John Spencer, and John Stocke.

Jeff Bennett
Megan Donahue
Nick Schneider
Mark Voit

About the Authors

JEFFREY BENNETT

Jeffrey Bennett received a B.A. in biophysics from the University of California, San Diego (1981) and a Ph.D. in astrophysics from the University of Colorado, Boulder (1987). He currently spends most of his time as a teacher, speaker, and writer. He has taught extensively at all levels, including having founded and run a science summer school for elementary and middle school children. At the college level, he has taught more than fifty classes in subjects ranging from astronomy, physics, and mathematics, to education. He served two years as a visiting senior scientist at NASA headquarters, where he helped create numerous programs for science education. He also proposed the idea for and helped develop the *Voyage* Scale Model Solar System, which opened in 2001 on the National Mall in Washington, D.C. (He is pictured here with the model Sun.) In addition to this astronomy textbook, he has written college-level textbooks in astrobiology, mathematics, and statistics, and a book for the general public, *On the Cosmic Horizon* (Addison Wesley, 2001). He also recently completed his first children's book, *Max Goes to the Moon* (Big Kid Science, 2003). When not working, he enjoys participating in masters swimming and in the daily adventures of life with his wife Lisa, his children Grant and Brooke, and his dog, Max. You can read more about his projects on his personal Web site, www.jeffreybennett.com.

MEGAN DONAHUE

Megan Donahue is an associate professor in the Department of Physics and Astronomy of Michigan State University. Her current research is mainly on clusters of galaxies: their contents—dark matter, hot gas, galaxies, active galactic nuclei—and what they reveal about the contents of the universe and how galaxies form and evolve. She grew up on a farm in Nebraska and received a bachelor's degree in physics from MIT, where she began her research career as an X-ray astronomer. She has a Ph.D. in astrophysics from the University of Colorado, for a thesis on theory and optical observations of intergalactic and intracluster gas. That thesis won the 1993 Trumpler Award from the Astronomical Society for the Pacific for an outstanding astrophysics doctoral dissertation in North America. She continued post-doctoral research in optical and X-ray observations as a Carnegie Fellow at Carnegie Observatories in Pasadena, California, and later as an STScI Institute Fellow at Space Telescope. Megan was a staff astronomer at the Space Telescope Science Institute until 2003, when she joined the MSU faculty. Megan is married to Mark Voit, who is also a frequent collaborator of hers on many projects, including this textbook and the raising three children, Michaela, Sebastian, and Angela. Between the births of Sebastian and Angela, Megan qualified for and ran the Boston Marathon. She hopes to run another one soon.

NICHOLAS SCHNEIDER

Nicholas Schneider is an associate professor in the Department of Astrophysical and Planetary Sciences at the University of Colorado and a researcher in the Laboratory for Atmospheric and Space Physics. He received his B.A. in physics and astronomy from Dartmouth College in 1979 and his Ph.D. in planetary science from the University of Arizona in 1988. In 1991, he received the National Science Foundation's Presidential Young Investigator Award. His research interests include planetary atmospheres and planetary astronomy, with a focus on the odd case of Jupiter's moon Io. He enjoys teaching at all levels and is active in efforts to improve undergraduate astronomy education. Off the job, he enjoys exploring the outdoors with his family and figuring out how things work.

MARK VOIT

Mark Voit is an associate professor in the Department of Physics and Astronomy at Michigan State University. He earned his A.B. in astrophysical sciences at Princeton University and his Ph.D. in astrophysics at the University of Colorado in 1990. He continued his studies at the California Institute of Technology, where he was a research fellow in theoretical astrophysics, then moved on to Johns Hopkins University as a Hubble Fellow. Before coming to Michigan State, Mark worked in the Office of Public Outreach at the Space Telescope, where he developed museum exhibitions about the Hubble Space Telescope and was the scientist behind NASA's HubbleSite. His research interests range from interstellar processes in our own galaxy to the clustering of galaxies in the early universe. He is married to co-author Megan Donahue, and they try to play outdoors with their three children whenever possible, enjoying hiking, camping, running, and orienteering. Mark is also author of the popular book *Hubble Space Telescope: New Views of the Universe*.

How to Succeed in Your Astronomy Course

USING THIS BOOK

Each chapter in the book is designed to make it easy for you to study effectively and efficiently. To get the most out of each chapter, you might wish to use the following study plan:

- Begin by reading the Learning Goals to make sure you know what you will be learning about in each chapter.
- Before reading in depth, start by skimming the chapter, focusing only on the illustrations. Study each illustration and read the captions so that you will get an overview of the key chapter concepts.
- Next, read the chapter narrative. Try to answer the Think About It questions as you go along, but you may save the other boxed features (Common Misconceptions and Special Topics) to read later.
- After reading the chapter once, go back through and read the boxed material. Also look for the tutorial icons that tell you when there is a relevant Web-based tutorial on the Astronomy Place (www.astronomyplace.com). If you are having difficulty with a concept, be sure you try the tutorial.
- Study the Chapter Summary by first trying to answer the Learning Goals questions for yourself, then checking your understanding against the answers given in the summary.
- Check your understanding by trying the online quizzes at www.astronomyplace.com. Do the visual and basic quizzes first. Once you clear up any difficulties you have with these, try the conceptual quiz.

THE KEY TO SUCCESS: STUDY TIME

The single most important key to success in any college course is to spend enough time studying. A general rule of thumb for college classes is that you should expect to study about 2 to 3 hours per week *outside* of class for each unit of credit. For example, based on this rule of thumb, a student taking 15 credit hours should expect to spend 30 to 45 hours each week studying outside of class. Combined with time in class, this works out to a total of 45 to 60 hours spent on academic work—not much more than the time a typical job requires, and you get to choose your own hours. Of course, if you are working while you attend school, you will need to budget your time carefully.

As a rough guideline, your studying time in astronomy might be divided as shown in the table at the top of p. xxi. If you find that you are spending fewer hours than these guidelines suggest, you can probably improve your grade by studying more. If you are spending more hours than these guidelines suggest, you may be studying inefficiently; in that case, you should talk to your instructor about how to study more effectively.

GENERAL STRATEGIES FOR STUDYING

- Don't miss class. Listening to lectures and participating in discussions is much more effective than reading someone else's notes. Active participation will help you retain what you are learning.
- As you read, make notes to remind yourself of ideas you'll want to review in more detail later. The best way to do this is to make notes in the margins of the book. If you want to mark text for later review, don't high-light—underline! Using a pen or pencil to underline material requires greater care than highlighting and therefore helps keep you alert as you study. Be careful to underline selectively—it won't help you later if you've underlined everything.
- Budget your time effectively. One or 2 hours each day is more effective, and far less painful, than studying all night before homework is due or before exams.
- If a concept gives you trouble, do additional reading or studying beyond what has been assigned. And if you still have trouble, ask for help: You surely can find friends, colleagues, or teachers who will be glad to help you learn.
- Working together with friends can be valuable in helping you understand difficult concepts. However, be sure that you learn *with* your friends and do not become dependent on them.
- Be sure that any work you turn in is of *collegiate quality*: neat and easy to read, well organized, and demonstrating mastery of the subject matter. Although it takes extra effort to make your work look this good, the effort will help you solidify your learning and is also good practice for the expectations that future professors and employers will have.

If Your Course Is	*Times for Reading the Assigned Text (per week)*	*Times for Homework Assignments (per week)*	*Times for Review and Test Preparation (average per week)*	*Total Study Time (per week)*
3 credits	2 to 4 hours	2 to 3 hours	2 hours	6 to 9 hours
4 credits	3 to 5 hours	2 to 4 hours	3 hours	8 to 12 hours
5 credits	3 to 5 hours	3 to 6 hours	4 hours	10 to 15 hours

PREPARING FOR EXAMS

- Study the review questions, and rework problems and other assignments; try additional questions to be sure you understand the concepts. Study your performance on assignments, quizzes, or exams from earlier in the term.
- Study the relevant online tutorials and chapter quizzes available at www.astronomyplace.com.
- Study your notes from lectures and discussions. Pay attention to what your instructor expects you to know for an exam.
- Reread the relevant sections in the textbook, paying special attention to notes you have made on the pages.
- Study individually *before* joining a study group with friends. Study groups are effective only if every individual comes prepared to contribute.
- Don't stay up too late before an exam. Don't eat a big meal within an hour of the exam (thinking is more difficult when blood is being diverted to the digestive system).
- Try to relax before and during the exam. If you have studied effectively, you are capable of doing well. Staying relaxed will help you think clearly.

1
Our Place in the Universe

LEARNING GOALS

1.1 OUR MODERN VIEW OF THE UNIVERSE
- What is our place in the universe?
- How did we come to be?
- How can we know what the universe was like in the past?
- Can we see the entire universe?

1.2 THE SCALE OF THE UNIVERSE
- How big is Earth compared to our solar system?
- How far away are the stars?
- How big is the Milky Way Galaxy?
- How big is the universe?
- How do our lifetimes compare to the age of the universe?

1.3 SPACESHIP EARTH
- How is Earth moving in our solar system?
- How is our solar system moving in the Milky Way Galaxy?
- How do galaxies move within the universe?
- Are we ever sitting still?

Far from city lights on a clear night, you can gaze upward at a sky filled with stars. Lie back and watch for a few hours, and you will observe the stars marching steadily across the sky. Confronted by the seemingly infinite heavens, you might wonder how Earth and the universe came to be. If you do, you will be sharing an experience common to humans around the world today and in thousands of generations past.

Modern science offers answers to many of our fundamental questions about the universe and our place within it. We now know the basic content and scale of the universe. We know the age of Earth and the approximate age of the universe. And, although much remains to be discovered, we are rapidly learning how the simple ingredients of the early universe developed into the incredible diversity of life on Earth.

In this first chapter, we will survey the content and history of the universe, the scale of the universe, and the motions of Earth. We'll develop a "big picture" perspective on our place in the universe that will provide a base on which we can build a deeper understanding in the rest of the book.

1.1 OUR MODERN VIEW OF THE UNIVERSE

If you observe the sky carefully, you can see why most of our ancestors believed that the heavens revolved about a stationary Earth. The Sun, Moon, planets, and stars appear to circle around our sky each day, and we cannot feel the constant motion of Earth as it rotates on its axis and orbits the Sun. Thus, it seems quite natural to assume that we live in an Earth-centered, or *geocentric,* universe.

Nevertheless, we now know that Earth is a planet orbiting a rather average star in a vast universe. The historical path to this knowledge was long and complex. In later chapters, we'll see that many ancient beliefs made a lot of sense in their day and changed only when people were confronted by strong evidence to the contrary. We'll also see how the process of science has enabled us to acquire this evidence and thereby discover that we are connected to the stars in ways our ancestors never imagined. First, however, it's useful to have at least a general picture of the universe as we know it today.

• What is our place in the universe?

We can describe our place in the universe with what we might call our "cosmic address," illustrated in Figure 1.1. Earth is a planet in our **solar system**, which consists of the Sun and all the objects that orbit it: nine planets and their moons, the chunks of rock we call asteroids, the balls of ice we call comets, and countless tiny particles of interplanetary dust.

Our Sun is a star, just like the stars we see in our night sky. The Sun and all the stars we can see with the naked eye make up only a small part of a huge, disk-shaped collection of stars called the **Milky Way Galaxy**. A galaxy is a great island of stars in space, containing from a few hundred million to a trillion or more stars. The Milky Way Galaxy is relatively large, containing more than 100 billion stars. Our solar system is

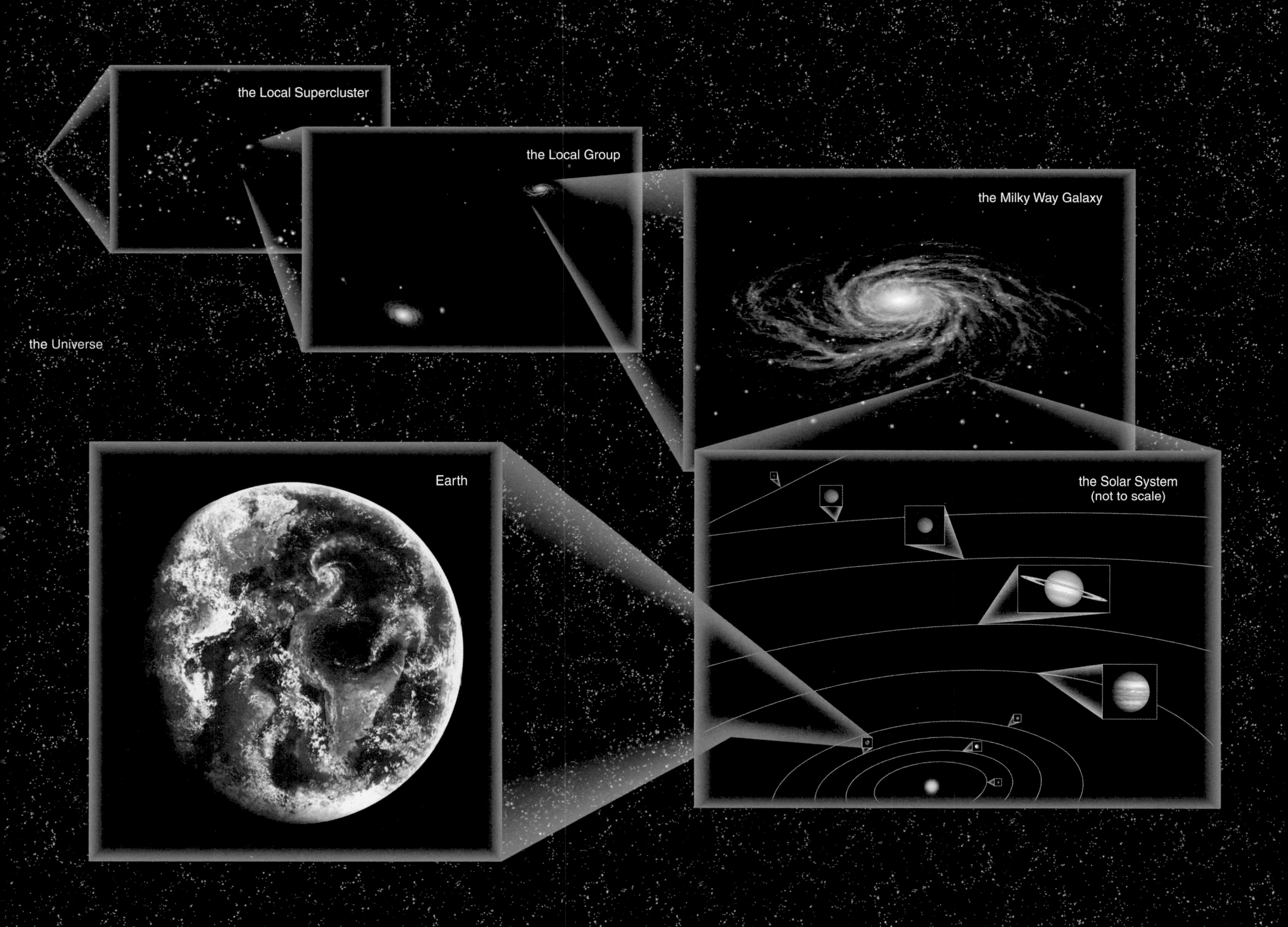

Figure 1.1

This painting illustrates our cosmic address.

located a little over halfway from the galactic center to the edge of the galactic disk.

Billions of other galaxies are scattered through space. Some galaxies are fairly isolated, but many others are found in groups. Our Milky Way, for example, is one of the two largest galaxies among about 40 galaxies in the **Local Group**. Groups of galaxies with more than a few dozen members are often called **galaxy clusters**.

We live on one planet orbiting one star among more than 100 billion stars in the Milky Way Galaxy, which in turn is one of billions of galaxies in the universe.

On a very large scale, the universe appears frothlike, with galaxies and galaxy clusters loosely arranged in giant chains and sheets. The galaxies and galaxy clusters are more tightly packed in some places than in others, forming giant structures called **superclusters**. The supercluster to which our Local Group belongs is called, not surprisingly, the **Local Supercluster**. Between the vast groupings of galaxies lie huge voids containing few, if any, galaxies. Finally, the **universe** is the sum total of all matter and energy, encompassing the superclusters and voids and everything within them.

THINK ABOUT IT Some people think that our tiny physical size in the vast universe makes us insignificant. Others think that our ability to learn about the wonders of the universe gives us significance despite our small size. What do *you* think?

• How did we come to be?

According to modern science, we humans are newcomers in an old universe. We'll devote much of the rest of this textbook to studying the scientific evidence that backs up this idea. To help prepare you for this study, let's look at a quick overview of the scientific history of the universe, as summarized in Figure 1.2.

The Big Bang As we'll discuss in Section 1.3, telescopic observations of distant galaxies show that the entire universe is *expanding*. That is, average distances between galaxies are increasing with time. If the universe is expanding, everything must have been closer together in the past. From the observed rate of expansion, astronomers estimate that the expansion started about 14 billion years ago. We call this beginning the **Big Bang**. The three cubes in the upper left corner of Figure 1.2 represent the expansion through time of a small piece of the entire universe.

The rate at which galaxies are moving apart suggests that the universe was born about 14 billion years ago, in the event we call the Big Bang.

The universe as a whole has continued to expand ever since the Big Bang, but on smaller size scales the force of gravity has drawn matter together. Structures such as galaxies and clusters of galaxies occupy regions where gravity has won out against the overall expansion. That is, while the universe as a whole continues to expand, individual galaxies and their contents do *not* expand. This idea is illustrated by the expanding cube in Figure 1.2. Notice that as the cube as a whole grows larger, the matter within it tends to clump into galaxies and galaxy clusters. Most galaxies, including our own Milky Way, probably formed within a few billion years after the Big Bang.

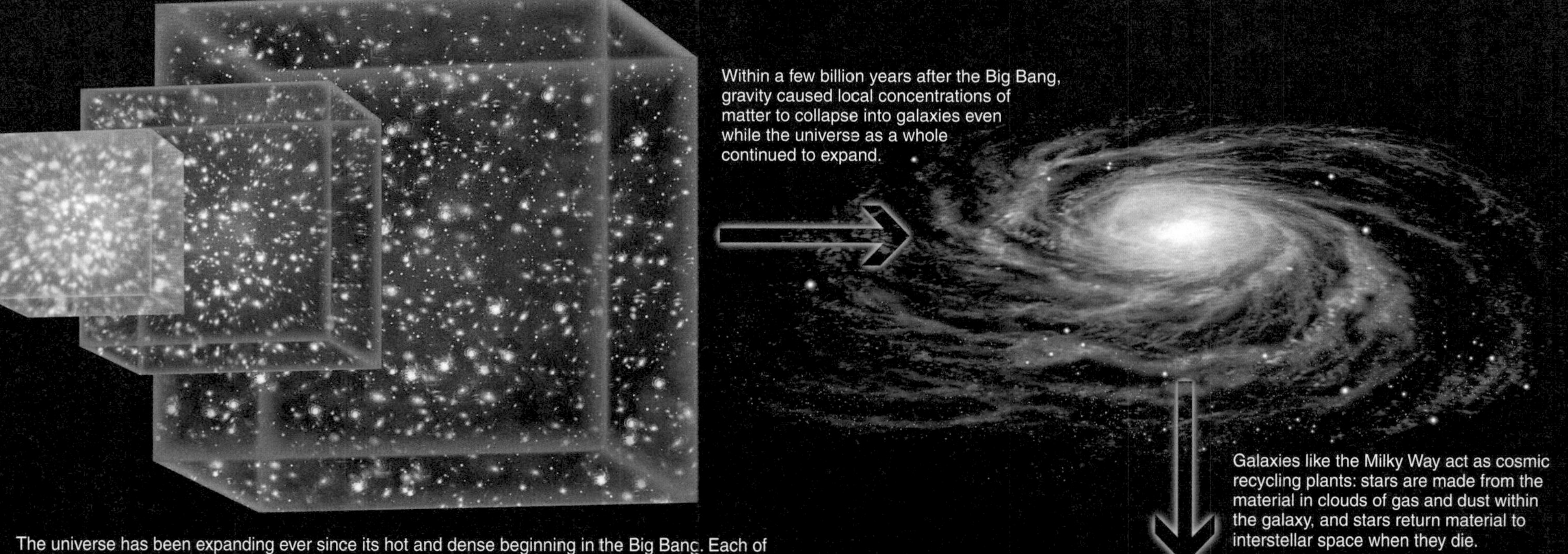

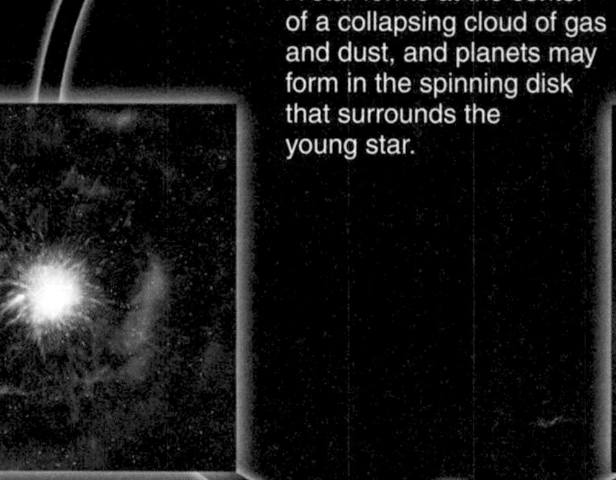

Figure 1.2

Our cosmic origins: This sequence of paintings illustrates current scientific understanding of the history of our universe. It shows how matter produced in the Big Bang changed and came to be incorporated in planet Earth, where it has become part of all living things, including us.

Stellar Lives and Galactic Recycling Within galaxies, gravity drives the collapse of clouds of gas and dust to form stars and planets. Stars are not living organisms, but they nonetheless go through "life cycles." After their births in giant clouds of gas and dust, stars shine for millions or billions of years. The energy that makes stars shine comes from **nuclear fusion**, the process in which lightweight atomic nuclei smash together and stick (or fuse) to make heavier nuclei. Nuclear fusion occurs deep in a star's core throughout its life. A star "dies" when it finally exhausts all its usable fuel for fusion.

Stars are born in interstellar clouds, produce energy and new elements through nuclear fusion, and release those new elements to interstellar space when they die.

In its final death throes, a star blows much of its content back out into space. In particular, massive stars die in titanic explosions (called *supernovae*). The returned matter mixes with other matter floating between the stars in the galaxy, eventually becoming part of new clouds of gas and dust from which new generations of stars can be born. Thus, galaxies function as cosmic recycling plants, recycling material expelled from dying stars into new generations of stars and planets. This cycle is

BASIC ASTRONOMICAL OBJECTS, UNITS, AND MOTIONS

This box summarizes a few key astronomical definitions introduced in this chapter and used throughout the book.

Basic Astronomical Objects

star Our Sun and other ordinary stars are large, glowing balls of gas that generate heat and light through nuclear fusion in their cores.

planet A moderately large object that orbits a star. Planets may be rocky, icy, or gaseous in composition, and they shine primarily by reflecting light from their star. (Astronomers sometimes disagree about what counts as a planet, because there are no official minimum or maximum sizes.)

moon (or satellite) An object that orbits a planet. The term *satellite* is also used more generally to refer to any object orbiting another object.

asteroid A relatively small and rocky object that orbits a star. Asteroids are sometimes called *minor planets* because they orbit much like planets but are smaller than anything we consider to be a true planet.

comet A relatively small and icy object that orbits a star.

Collections of Astronomical Objects

solar system Our solar system consists of the Sun and all the material that orbits it, including the planets. The term *solar system* technically refers only to our own star system (because *solar* means "of the Sun"), but it is sometimes applied to other star systems.

star system A star (sometimes more than one star) and any planets and other materials that orbit it. (Roughly half of all star systems contain two or more stars.)

galaxy A great island of stars in space, containing from a few hundred million to a trillion or more stars, all held together by gravity and orbiting a common center.

cluster (or group) of galaxies A collection of galaxies bound together by gravity. Small collections (up to a few dozen galaxies) are generally called *groups*, while larger collections are called *clusters*.

supercluster A gigantic region of space where many individual galaxies and many groups and clusters of galaxies are packed more closely together than elsewhere in the universe.

universe (or cosmos) The sum total of all matter and energy—that is, all galaxies and everything between them.

observable universe The portion of the entire universe that can be seen from Earth, at least in principle. The observable universe is probably only a tiny portion of the entire universe.

Astronomical Distance Units

astronomical unit (AU) The average distance between Earth and the Sun, which is about 150 million kilometers. (More technically, 1 AU is the length of the semimajor axis of Earth's orbit.)

light-year The distance that light can travel in 1 year, which is about 9.46 trillion kilometers.

Terms Relating to Motion

rotation The spinning of an object around its axis. For example, Earth rotates once each day around its axis, which is an imaginary line connecting the North Pole to the South Pole through the center of Earth.

revolution (orbit) The orbital motion of one object around another. For example, Earth revolves (orbits) around the Sun once each year.

expansion (of the universe) We say that the universe is expanding because the average distance between galaxies is increasing with time. Note that while the universe as a whole is expanding, individual galaxies and their contents (as well as groups and clusters of galaxies) are *not* expanding.

illustrated in the lower right of Figure 1.2. Our own solar system is a product of many generations of such recycling.

Stars Manufacture the Elements of Earth and Life The recycling of stellar material is connected to our existence in an even deeper way. By studying stars of different ages, we have learned that the early universe contained only the simplest chemical elements: hydrogen and helium (and a trace of lithium). We and Earth are made primarily of other elements, such as carbon, nitrogen, oxygen, and iron. Where did these other elements come from? Apparently, all the elements besides hydrogen and helium were manufactured by stars, either through the nuclear fusion that makes them shine or through nuclear reactions accompanying the explosions that end their lives.

We are "star stuff"—made of material that was manufactured in stars from the simple elements born in the Big Bang.

The recycling of matter in the Milky Way Galaxy had already been taking place for several billion years by the time our solar system formed, about 4.6 billion years ago. By that time, stars had converted about 2% of the original hydrogen and helium into heavier elements. Thus, the cloud that gave birth to our solar system was made of about 98% hydrogen and helium and 2% of everything else. This 2% may sound small, but it was more than enough to make the small rocky planets of our solar system, including Earth. On Earth, some of these elements became the raw ingredients of life, ultimately blossoming into the great diversity of life on Earth today.

In summary, most of the material from which we and our planet are made was created inside stars that died before the birth of our Sun. We are intimately connected to the stars because we are products of stars. In the words of astronomer Carl Sagan (1934–1996), we are "star stuff."

• How can we know what the universe was like in the past?

You may wonder how we can claim to know anything about what the universe was like in the distant past. The answer is that we can actually see into the past by studying light from distant stars and galaxies.

Light travels extremely fast by earthly standards. The speed of light is 300,000 kilometers per second. At this speed it would be possible to circle Earth nearly eight times in just 1 second. Nevertheless, even light takes a substantial amount of time to travel the vast distances in space. For example, light takes about 1 second to reach Earth from the Moon, and about 8 minutes to reach Earth from the Sun. Light from the stars takes many years to reach us, so we measure distances to the stars in units called **light-years**. One light-year is the distance that light can travel in 1 year—about 10 trillion kilometers, or 6 trillion miles. (You can calculate this distance by multiplying the speed of light by the number of seconds in one year.) Note that a light-year is a unit of *distance,* not time.

Because light takes time to travel through space, we are led to a remarkable fact:

The farther away we look in distance, the further back we look in time.

For example, the brightest star in the night sky, Sirius, is about 8 light-years from our solar system. This means it takes light from Sirius about 8 years to reach us. Thus, when we look at Sirius, we are seeing it as it was about 8 years ago.

THE MEANING OF A LIGHT-YEAR

A recent advertisement illustrated a common misconception by claiming "It will be light-years before anyone builds a better product." Notice that this advertisement misuses the term light-years to imply time, when it is in fact a unit of distance. If you are unsure whether the term light-years is being used correctly, try testing the statement by remembering that 1 light-year is approximately 10 trillion kilometers, or 6 trillion miles. The advertisement then reads "It will be 6 trillion miles before anyone builds a better product," which clearly does not make sense.

Figure 1.3

The Great Galaxy in Andromeda is about 2.5 million light-years away, so this photo captures light that traveled through space for 2.5 million years to reach us. Thus, we see the galaxy as it looked about 2.5 million years ago.

Light takes time to travel the vast distances in space. When we look deep into space, we also look far into the past.

The effect is more dramatic at greater distances. The Great Galaxy in Andromeda (also known as the Andromeda Galaxy and M31) lies about 2.5 million light-years from Earth. Figure 1.3 is therefore a picture of how this galaxy looked about 2.5 million years ago, when early humans were first walking on Earth. We see more distant galaxies as they were even further back into the past.

It's also amazing to realize that any "snapshot" of a distant galaxy is a picture of both space and time. For example, because the Andromeda Galaxy is about 100,000 light-years in diameter, the light we see from the far side of the galaxy must have left on its journey to us some 100,000 years before the light from the near side. Thus, in addition to showing us how the Andromeda Galaxy looked more than 2 million years ago, the single photograph in Figure 1.3 shows different parts of the galaxy spread over a time period of 100,000 years. When we study the universe, it is impossible to separate space and time.

THINK ABOUT IT Suppose that, at this very moment, students are studying astronomy on a planet somewhere in the Andromeda Galaxy. What would they see as they look from afar at our Milky Way? Could they know that we exist here on Earth? Explain.

• Can we see the entire universe?

The fact that looking deep into space means looking far back in time allows us to observe how the universe has changed through time. Remember that we think the universe is about 14 billion years old. If we look at a galaxy that is 7 billion light-years away, we see it as it looked 7 billion years ago—which means we see it as it was when the universe was only half its current age. If we look at a galaxy that is 12 billion light-years away, we see it as it was 12 billion years ago, when the universe was only 2 billion years old. Thus, simply by looking to great distances, we can see what parts of the universe looked like when the universe was younger. The remarkable photo that opens this chapter (p. 1), taken by the Hubble Space Telescope, shows some galaxies that are more than 12 billion light-years away. (To learn more about this photo, see Web Project 4 on p. 23)

Because the universe is about 14 billion years old, we cannot observe light coming from anything more than 14 billion light-years away.

Even with the most powerful telescopes imaginable, there's a limit to how far we can see into space. This limit is imposed by the age of the universe (Figure 1.4). In a universe that is 14 billion years old, we cannot possibly see anything more than 14 billion light-years away. If we wanted to look more than 14 billion light-years away—say, to a distance of 15 billion light-years—we'd be trying to look to a time before the universe existed, when there's nothing to see. Thus, our **observable universe**—the portion of the entire universe that we can potentially observe—consists only of objects that lie within 14 billion light-years of Earth. This fact does not put any limit on the size of the *entire* universe, which may be far larger than our observable universe. We simply have no hope of seeing or studying anything beyond the bounds of our observable universe.

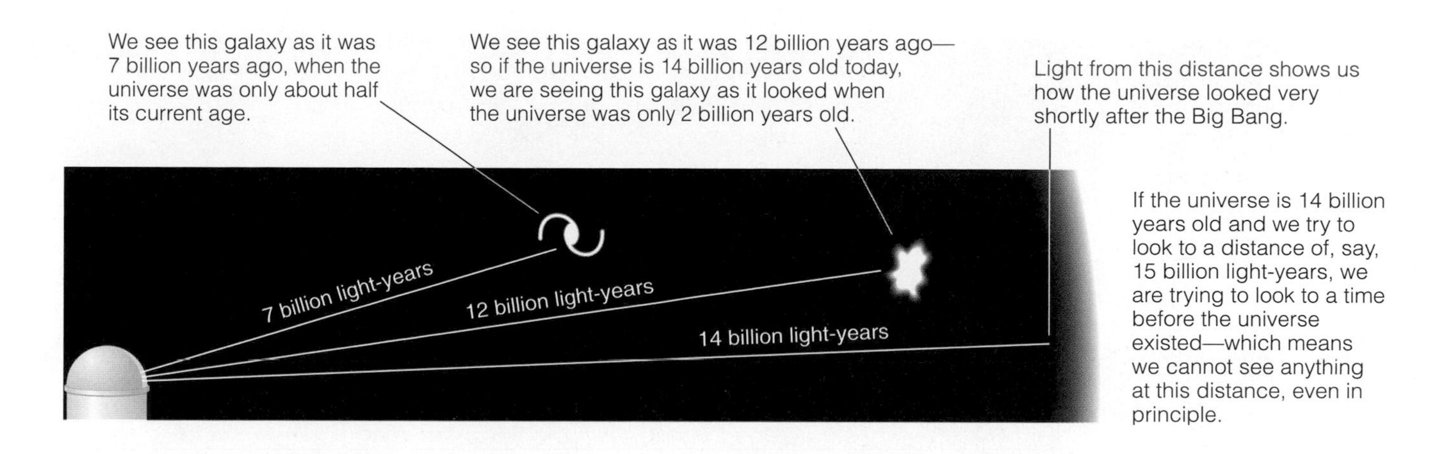

Figure 1.4

Powerful telescopes allow us to observe how the universe has changed with time.

Scale of the Universe Tutorial, Lessons 1–3

1.2 THE SCALE OF THE UNIVERSE

The numbers in our description of the size and age of the universe probably have little meaning for you—after all, they are literally astronomical. In this section, we will try to give those figures some meaning.

Virtual Tour of the Solar System

• How big is Earth compared to our solar system?

One of the best ways to develop perspective on cosmic sizes and distances is to imagine our solar system shrunk down to a scale on which you could walk through it. The Voyage scale model solar system in Washington, D.C. (Figure 1.5) makes such a walk possible. The Voyage model shows the Sun and the planets, and the distances between them, at *one ten-billionth* their actual sizes and distances. Figure1.6a shows the planets at their scaled sizes, and Figure 1.6b shows a map of the locations of the planets in the Voyage model. If you study the figures carefully, you'll notice several key features of our solar system:

- The Sun and planets are tiny compared to the distances between them. On the Voyage scale, the Sun is about the size of a large grapefruit, while the planets range in size from dust-speck-size Pluto to marble-size Jupiter. Earth is about the size of the ball point in a pen and is located about 15 meters (16.5 yards) from the Sun.
- The planets fall into two clear groups by distance. The four inner planets—Mercury, Venus, Earth, and Mars—all lie within just a few steps of the Sun on this scale. The outer planets—Jupiter, Saturn, Uranus, Neptune, and Pluto—are much more widely separated. For example, Jupiter is more than three times as far as Mars from the Sun.
- One of the most striking features of the solar system is its emptiness (that's why we call it *space!*). To show the complete orbits of the planets around the Sun, the Voyage model would require an area measuring over a kilometer (0.6 mile) on a side, which is larger than most college campuses. Spread over this large area, only the grapefruit-size Sun, the nine planets, and a few moons would be big enough to notice with your eyes.

Figure 1.5

This photo shows the pedestals housing the Sun (the gold sphere on the nearest pedestal) and the inner planets in the Voyage scale model solar system (Washington, D.C.). The building at the left is the National Air and Space Museum. (The model planets are encased in the sidewalk-facing disks visible at about eye level on the planet pedestals.)

a This painting shows the planets and Sun at *one ten-billionth* of their actual sizes. (Distances are *not* to scale in this painting.)

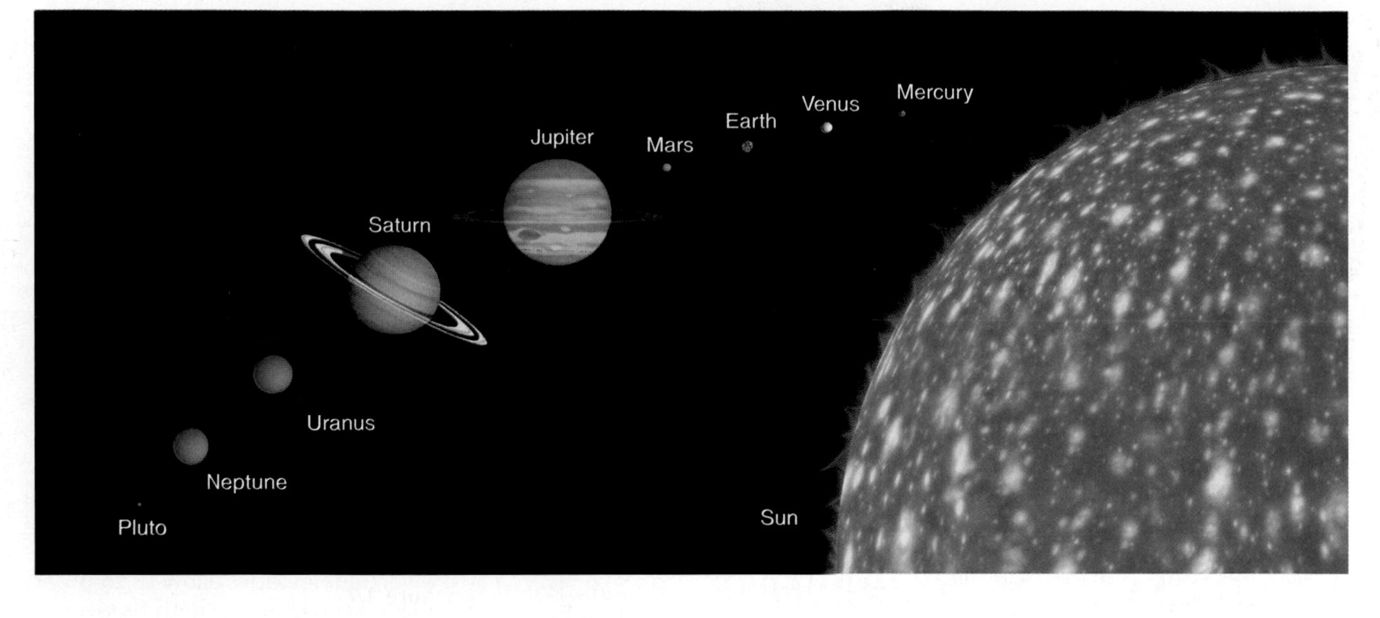

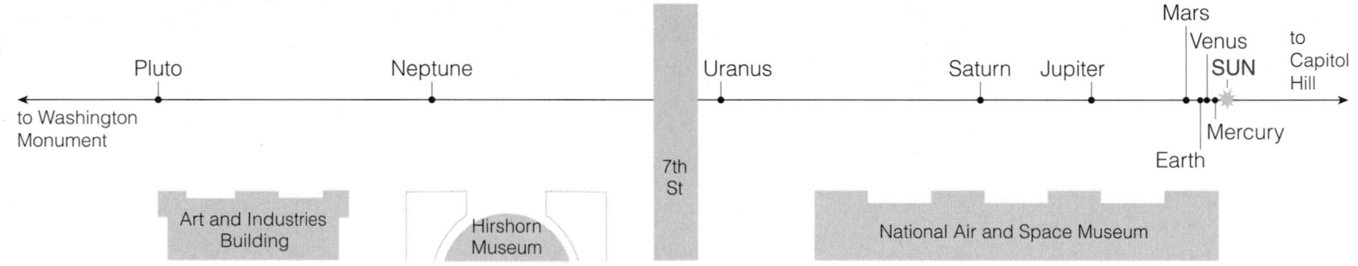

b This map shows the locations of the Sun and planets in the Voyage model.

Figure 1.6

The Voyage scale model represents the solar system at *one ten-billionth* of its actual size. Planets are shown in a line for easy comparison of distances; in reality, each planet orbits the Sun independently and a perfect alignment essentially never occurs.

THINK ABOUT IT Earth is the only place in our solar system—and the only place we yet know of in the universe—with conditions suitable for human life. How does visualizing the Earth to scale affect your perspective on human existence? How does it affect your perspective on our planet? Explain.

On a scale in which the Sun is the size of a grapefruit, Earth is the size of a ball point from a pen, orbiting the Sun at a distance of 15 meters.

Seeing our solar system to scale also helps us put space exploration into perspective. The Moon is the farthest point to which humans have traveled. Twelve astronauts walked on the Moon between 1969 and 1972 as part of the Apollo program (Figure 1.7). On the 1-to-10-billion Voyage scale, the Moon is a barely visible speck about two thumb-widths (4 centimeters) from Earth (see Figure 2.18). Thus, if you place the model Earth in the center of your palm, the farthest place humans have ever reached also lies within your palm. Sending humans to Mars—a dream of many people—will require a journey some 200 times as long.

Our robotic spacecraft have gone much farther, visiting every planet in our solar system except Pluto. These trips stretch the limits of modern technology, and it can take years to reach the outer planets. For example, while in the Voyage model you can walk from Earth to Pluto in only about 10 minutes, a spacecraft now being readied for the journey will take a decade to reach Pluto. Our solar system is a vast place with many worlds that we have barely begun to explore.

• How far away are the stars?

Imagine that you start at the Voyage model Sun in Washington, D.C. You walk the roughly 600-meter distance to Pluto (just over $\frac{1}{3}$ mile) and then decide to keep going to find the nearest star besides the Sun. How far would you have to go?

Amazingly, you would need to walk to California. That is, on the same scale that allows you to walk from the Sun to Pluto in minutes, even the nearest stars would be more than 4,000 kilometers (2,500 miles) away. If this answer seems hard to believe, you can check it for yourself. A light-year is about 10 trillion kilometers, which becomes 1,000 kilometers on the 1-to-10-billion scale (because 10 trillion ÷ 10 billion = 1,000). The nearest star system to our own, a three-star system called Alpha Centauri (Figure 1.8), is about 4.4 light-years away. That distance becomes about 4,400 kilometers (2,700 miles) on the 1-to-10-billion scale, or roughly equivalent to the distance across the United States.

On the same scale on which Pluto is just a few minutes walk from the Sun or Earth, the distance to the nearest stars is equivalent to the distance across the United States.

The tremendous distances to the stars give us some perspective on the technological challenge of astronomy. For example, because the largest star of the Alpha Centauri system is roughly the same size and brightness as our Sun, viewing it in the night sky is somewhat like being in Washington, D.C., and seeing a very bright grapefruit in San Francisco (neglecting the problems introduced by the curvature of the Earth). It may seem remarkable that we can see this star at all, but the blackness of the night sky allows the naked eye to see it as a faint dot of light. It looks much brighter through powerful telescopes, but we still cannot see any features of the star's surface.

Now, consider the difficulty of seeing *planets* orbiting nearby stars. It is equivalent to looking from Washington, D.C., and trying to see ball points or marbles orbiting grapefruits in California or beyond. You probably won't be surprised to learn that we have not yet seen such planets directly. Indeed, the bigger surprise may be that we *have* discovered more than 100 planets around other stars through indirect techniques [Section 6.5].

Our examination of stellar distances also offers a sobering lesson about the possibility of travel to the stars. Although science fiction shows like *Star Trek* and *Star Wars* make interstellar travel seem easy, the reality is far different. Consider the *Voyager 2* spacecraft. Launched in 1977, *Voyager 2* flew by Jupiter in 1979, Saturn in 1981, Uranus in 1986, and Neptune in 1989. (Its trajectory did not take it near Pluto.) *Voyager 2* is now bound for the stars at a speed of close to 50,000 kilometers per hour—about 100 times as fast as a speeding bullet. Even at this speed, *Voyager 2* would take about 100,000 years to reach Alpha Centauri if it were headed in that direction (which it's not). Convenient interstellar travel remains well beyond our present technology.

• How big is the Milky Way Galaxy?

The vast separation between our solar system and Alpha Centauri is typical of the separations among star systems here in the outskirts of the Milky Way Galaxy. Thus, the 1-to-10-billion scale is useless for thinking about distances beyond the nearest stars, because more distant stars would not fit on Earth on this scale. Visualizing the entire galaxy requires a new scale.

Figure 1.7

The Moon is the most distant place ever visited by humans, yet on the 1-to-10-billion scale of the Voyage model it is only about 4 centimeters (1.5 inches) from Earth. This famous photograph from the first Moon landing (Apollo 11 in July 1969) shows astronaut Buzz Aldrin, with Neil Armstrong reflected in his visor. Armstrong was the first to step onto the Moon's surface, saying, "That's one small step for [a] man, one giant leap for mankind."

Figure 1.8

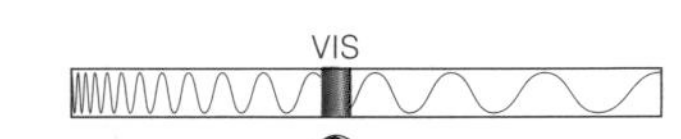

This photograph and diagram show the constellation Centaurus, which is visible from tropical and southern latitudes. Alpha Centauri's real distance is about 4.4 light-years, which becomes 4,400 kilometers (2,700 miles) on the 1-to-10-billion Voyage scale.

COMMON MISCONCEPTIONS

CONFUSING VERY DIFFERENT THINGS

Most people are familiar with the terms solar system and galaxy, but people sometimes mix them up. Notice how incredibly different our solar system is from our galaxy. Our solar system is a single star system consisting of our Sun and the various objects that orbit it, including Earth and eight other planets. Our galaxy is a collection of more than 100 billion star systems—so many that it would take thousands of years just to count them. Thus, confusing the terms solar system and galaxy means making a mistake by a factor of 100 billion—a fairly big mistake!

Let's further reduce our solar system scale by a factor of 1 billion (making it a scale of 1 to 10^{19}). On this new scale, each light-year becomes 1 millimeter, and the 100,000-light-year diameter of the Milky Way Galaxy becomes 100 meters, or about the length of a football field. Visualize a football field with a scale model of our galaxy centered over midfield. Our entire solar system is a microscopic dot located around the 20-yard line. The 4.4-light-year separation between our solar system and Alpha Centauri becomes just 4.4 millimeters on this scale—smaller than the width of your little finger. If you stood at the position of our solar system in this model, millions of star systems would lie within reach of your arms.

It would take thousands of years just to count out loud the number of stars in the Milky Way Galaxy.

Another way to put the galaxy into perspective is to consider its number of stars—more than 100 billion. Imagine that tonight you are having difficulty falling asleep (perhaps because you are contemplating the scale of the universe). Instead of counting sheep, you decide to count stars. If you are able to count about one star each second, on average, how long would it take you to count 100 billion stars in the Milky Way? Clearly, the answer is 100 billion (10^{11}) seconds, but how long is that? Amazingly, 100 billion seconds turns out to be more than 3,000 years. (You can confirm this by dividing 100 billion by the number of seconds in 1 year.) Thus, you would need thousands of years just to *count* the stars in the Milky Way Galaxy, and this assumes you never take a break—no sleeping, no eating, and absolutely no dying!

• How big is the universe?

As incredible as the scale of our galaxy may seem, the Milky Way is only one of at least 100 billion galaxies in the observable universe. Just as it would take thousands of years to count the stars in the Milky Way, it would take thousands of years to count all the galaxies.

Think for a moment about the total number of stars in all these galaxies. If we assume 100 billion stars per galaxy, the total number of

Jan. 1 The Big Bang

Feb. The Milky Way forms.

Sept. 3 Earth forms.

Sept. 22 Early life on Earth

stars in the observable universe is roughly 100 billion × 100 billion, or 10,000,000,000,000,000,000,000 (10^{22}).

Roughly speaking, there are as many stars in the observable universe as there are grains of sand on all the beaches on Earth.

How big is this number? Visit a beach. Run your hands through the fine-grained sand. Imagine counting each tiny grain of sand as it slips through your fingers. Then imagine counting every grain of sand on the beach and continuing on to count *every* grain of dry sand on *every* beach on Earth. If you could actually complete this task, you would find that, roughly speaking, the number of grains of sand is about the same as the number of stars in the observable universe (Figure 1.9).

Figure 1.9

The number of stars in the observable universe is about the same as the number of grains of dry sand on all the beaches on Earth.

THINK ABOUT IT Contemplate the fact that there may be as many stars in the observable universe as grains of sand on all the beaches on Earth and that each star is a potential sun for a system of planets. With so many possible homes for life, do you think it is conceivable that life exists only on Earth? Why or why not?

• How do our lifetimes compare to the age of the universe?

Now that we have developed some perspective on the scale of space, we can do the same for the scale of time. Imagine the entire history of the universe, from the Big Bang to the present, compressed into a single year. We can represent this history with a *cosmic calendar,* on which the Big Bang takes place at the first instant of January 1 and the present day is just before the stroke of midnight on December 31 (Figure 1.10). For a universe that is about 14 billion years old, each month on the cosmic calendar represents a little more than 1 billion years.

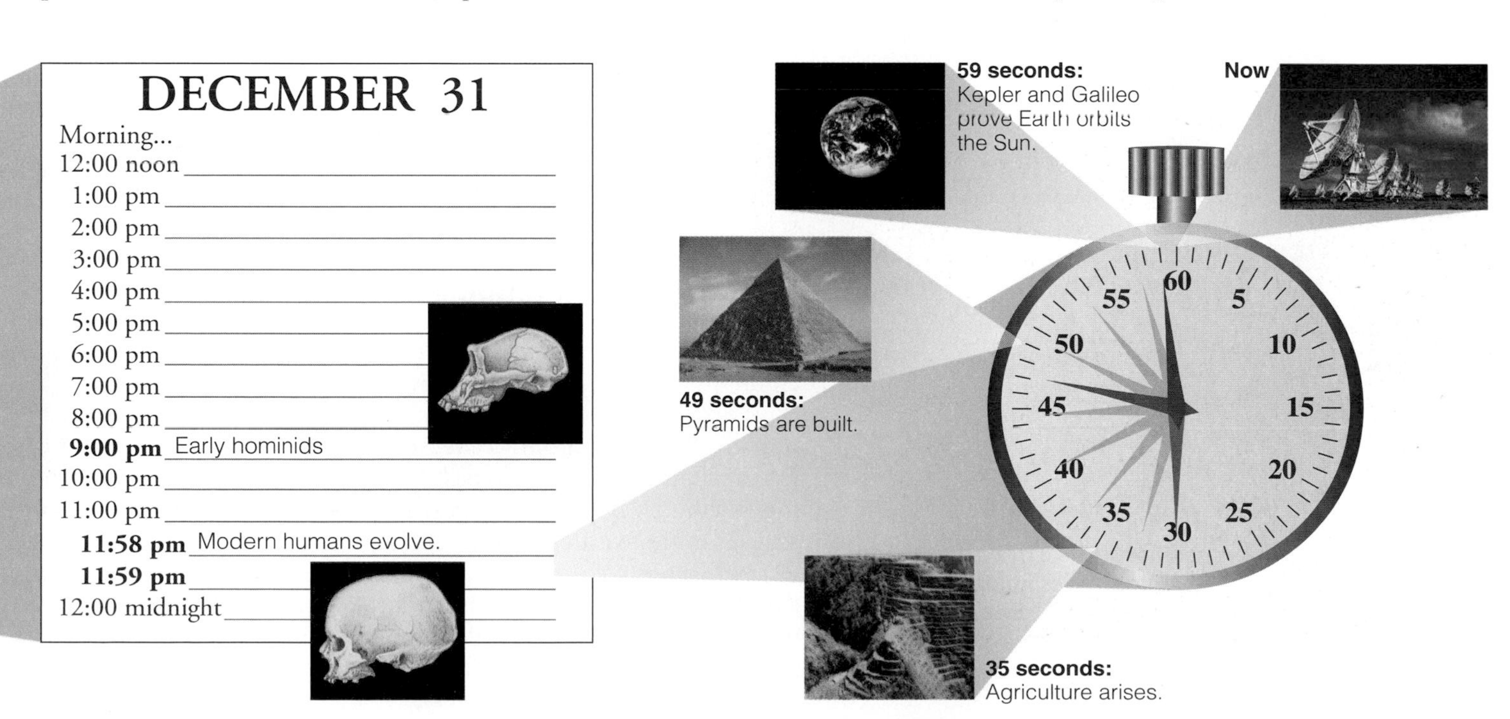

Figure 1.10

The cosmic calendar compresses the 14-billion-year history of the universe into 1 year, so that each month represents a little more than 1 billion years (more precisely, 1.17 billion years). This cosmic calendar is adapted from a version created by Carl Sagan.

On this time scale, the Milky Way Galaxy probably formed sometime in February. Many generations of stars lived and died in the subsequent cosmic months, enriching the galaxy with the "star stuff" from which we and our planet are made.

Our solar system and our planet did not form until early September on this scale, or 4.6 billion years ago in real time. By late September, life on Earth was flourishing. However, for most of Earth's history, living organisms remained relatively primitive and microscopic in size. On the scale of the cosmic calendar, recognizable animals became prominent only in mid-December. Early dinosaurs appeared on the day after Christmas. Then, in a cosmic instant, the dinosaurs disappeared forever—probably due to the impact of an asteroid or a comet [Section 9.4]. In real time, the death of the dinosaurs occurred some 65 million years ago, but on the cosmic calendar it was only yesterday. With the dinosaurs gone, small furry mammals inherited Earth. Some 60 million years later, or around 9 P.M. on December 31 of the cosmic calendar, early hominids (human ancestors) began to walk upright.

If we imagine the 14-billion-year history of the universe compressed into one year, a human lifetime lasts only a fraction of a second.

Perhaps the most astonishing thing about the cosmic calendar is that the entire history of human civilization falls into just the last half-minute. The ancient Egyptians built the pyramids only about 11 seconds ago on this scale. About 1 second ago, Kepler and Galileo proved that Earth orbits the Sun rather than vice versa. The average college student was born about 0.05 second ago, around 11:59:59.95 P.M. on the cosmic calendar. On the scale of cosmic time, the human species is the youngest of infants, and a human lifetime is a mere blink of an eye.

1.3 SPACESHIP EARTH

Wherever you are as you read this book, you probably have the feeling that you're "just sitting here." Nothing could be further from the truth. In fact, you are being spun in circles as Earth rotates, you are racing around the Sun in Earth's orbit, and you are careening through the cosmos in the Milky Way Galaxy. In the words of noted inventor and philosopher R. Buckminster Fuller (1895–1983), you are a traveler on *spaceship Earth*. In this section, we'll take a brief look at the motion of spaceship Earth through the universe.

• How is Earth moving in our solar system?

The most basic motions of Earth are its **rotation** (spin) and its **orbit** (sometimes called *revolution*) around the Sun.

Earth rotates once each day around its axis, an imaginary line connecting the North Pole to the South Pole through the center of Earth. Earth rotates from west to east (counterclockwise as viewed from above the North Pole), which is why the Sun and stars appear to rise in the east and set in the west each day.

Although we do not feel any obvious effects from Earth's rotation, the speed of rotation is substantial (Figure 1.11). Unless you live very near the North or South Poles, you are whirling around Earth's axis at a speed of more than 1,000 kilometers per hour (600 miles per hour)—faster than most airplanes travel.

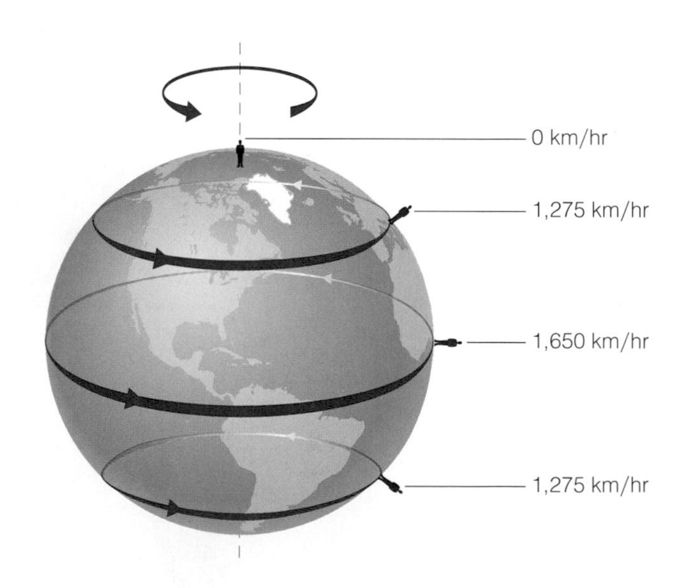

Figure 1.11

As Earth rotates, your speed around Earth's axis depends on your location. Almost everyone lives in places where rotation whirls you around the axis at a speed of more than 1,000 km/hr. Notice that you are always rotating from west to east—which is why the Sun rises in the east and sets in the west.

At the same time Earth is rotating, it is also orbiting around the Sun. It takes 1 year to complete each orbit. Again, while we don't feel the effects of this motion, the speed is quite impressive. The orbit is not quite a perfect circle, which makes Earth's orbital speed vary slightly [Section 3.3], but at all times we are racing around the Sun at a speed in excess of 100,000 kilometers per hour (60,000 miles per hour).

Earth rotates on its axis once each day and orbits the Sun once each year.

Earth's orbital path defines a flat plane that we call the **ecliptic plane** (Figure 1.12). Earth's axis is tilted by $23\frac{1}{2}°$ from a line *perpendicular* to the ecliptic plane. This **axis tilt** happens to be oriented so that it points almost directly at a star called Polaris, or the North Star. Keep in mind that the idea of axis tilt makes sense only in relation to the ecliptic plane. That is, the idea of "tilt" by itself has no meaning in space, where there is no absolute up or down. In space, "up" and "down" mean only away from the center of Earth (or another planet) and toward the center of Earth, respectively.

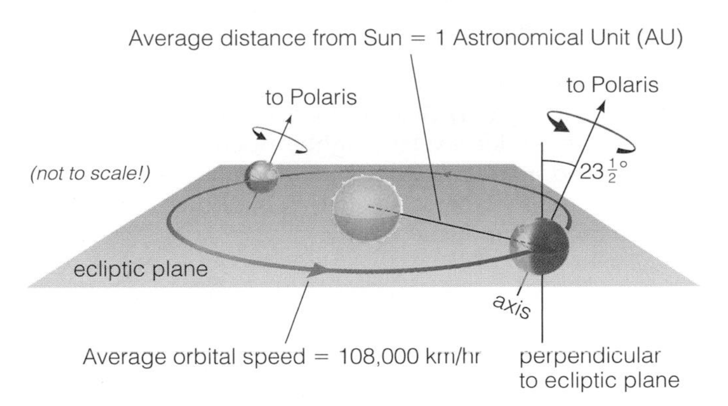

Figure 1.12

Earth orbits the Sun once each year, at an average distance of 1 AU, or about 150 million kilometers. Earth's orbit defines the ecliptic plane, and Earth's rotation axis is tilted $23\frac{1}{2}°$ from a line *perpendicular* to this plane. Notice that Earth both rotates and orbits counterclockwise as viewed from above the North Pole.

THINK ABOUT IT If there is no up or down in space, why do you think nearly all globes have the North Pole on top and the South Pole on the bottom? Would it be equally correct to have the South Pole on top or to turn the globe sideways? Explain.

One astronomical unit (AU) is Earth's average distance from the Sun, and is about 150 million kilometers.

As you study Figure 1.12, notice that Earth orbits the Sun in the same direction that it rotates on its axis (counterclockwise as viewed from above the North Pole). This is not a coincidence but a consequence of the way our planet was born. As we'll discuss in Chapter 6, Earth and the other planets were born in a spinning disk of gas that surrounded our Sun when it was young, and Earth rotates and orbits in the same direction as the disk was spinning. Notice also that we give a special name to the average distance between the Sun and Earth. This distance is called an **astronomical unit**, or **AU**, and it is about 150 million kilometers (93 million miles).

• How is our solar system moving in the Milky Way Galaxy?

Rotation and orbit are only a small part of the travels of spaceship Earth. In fact, our entire solar system is on a great journey within the Milky Way Galaxy.

Our Local Solar Neighborhood Let's begin with the motion of our solar system relative to nearby stars in what we call our *local solar neighborhood,* the region of the Sun and nearby stars. The box in Figure 1.13 shows that stars within the local solar neighborhood move essentially at random relative to one another. These stars (or the stars of any other small region of the galaxy) generally move quite fast relative to one another. For example, we are moving relative to nearby stars at an average speed of about 70,000 kilometers per hour (40,000 miles per hour), about three times as fast as the Space Station orbits Earth. Given these high speeds, why don't we see nearby stars racing around our sky?

Figure 1.13

This painting illustrates the motion of our solar system within the Milky Way Galaxy. The "zoom in" box shows that stars in our local solar neighborhood move essentially at random relative to one another. But at the same, the entire galaxy is rotating, so that all these stars orbit around the center of the galaxy.

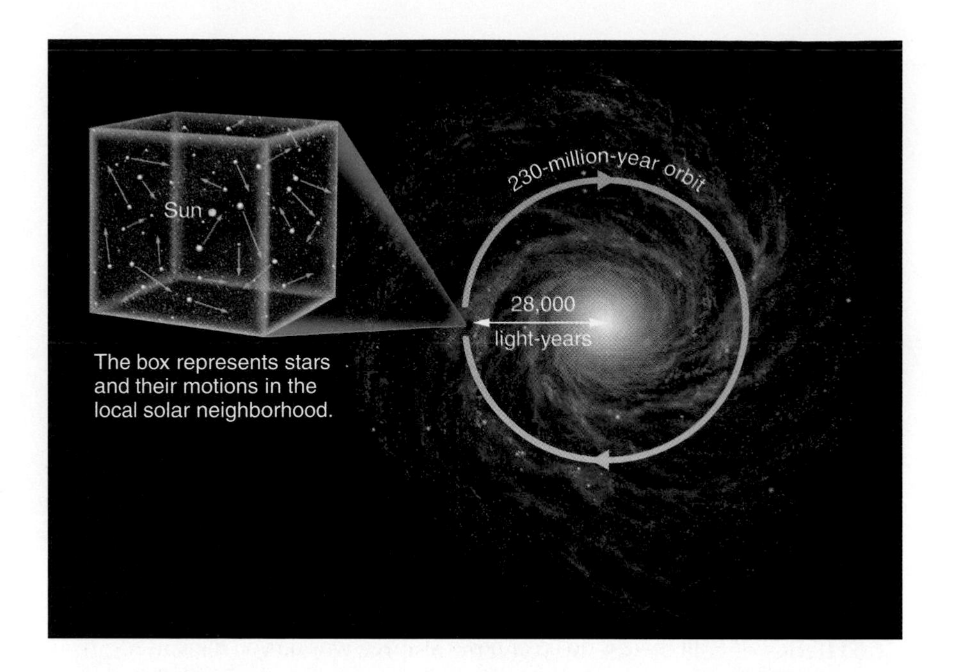

Stars in our local solar neighborhood move in essentially random directions relative to each other.

The answer lies in their vast distances from us. You've probably noticed that a distant airplane appears to move through your sky more slowly than one flying close overhead. If we extend this idea to the stars, we find that even at speeds of 70,000 kilometers per hour, stellar motions would be noticeable to the naked eye only if we watched them for thousands of years. That is why the patterns in the constellations seem to remain fixed. Nevertheless, in 10,000 years the constellations will be noticeably different from those we see today. In 500,000 years they will be unrecognizable. If you could watch a time-lapse movie made over millions of years, you *would* see stars racing across our sky.

THINK ABOUT IT Despite the chaos of motion in the local solar neighborhood over millions and billions of years, collisions between star systems are extremely rare. Explain why. (*Hint:* Consider the sizes of star systems, such as the solar system, relative to the distances between them.)

Galactic Rotation If you look closely at leaves floating in a stream, their motions relative to one another might appear random, just like the motions of stars in the local solar neighborhood. As you widen your view, you see that all the leaves are being carried in the same general direction by the downstream current. In the same way, as we widen our view beyond the local solar neighborhood, the seemingly random motions of its stars give way to a simpler and even faster motion: The entire Milky Way Galaxy is rotating. Our solar system, located about 28,000 light-years from the galactic center, completes one orbit of the galaxy in about 230 million years. Even if you could watch from outside our galaxy, this motion would be unnoticeable to your naked eye. However, if you calculate the speed of our solar system as we orbit the center of the galaxy, you will find that it is close to 800,000 kilometers (500,000 miles) per hour.

The Sun and other stars in our neighborhood orbit the center of the galaxy every 230 million years, because the entire galaxy is rotating.

Careful study of the galaxy's rotation reveals one of the greatest mysteries in science—one that we will study in depth in Chapter 16. Stars at different distances from the galactic center orbit at different speeds, and we can learn how mass is distributed in the galaxy by measuring these speeds. Such studies have turned up an enormous surprise: It seems that the stars in the disk of the galaxy represent only the "tip of the iceberg" compared to the mass of the entire galaxy (Figure 1.14). That is, most of the mass of the galaxy seems to be located outside the visible disk, in what we call the galaxy's *halo*. We don't know the nature of this mass. Because we have not detected any light coming from it, we call it **dark matter**. Studies of other galaxies suggest that they also are made mostly of dark matter. In fact, most of the mass in the universe seems to be made of this mysterious dark matter, but we do not yet know what it is.

• How do galaxies move within the universe?

The billions of galaxies in the universe also move relative to one another. Within the Local Group (see Figure 1.1), some of the galaxies move toward us, some move away from us, and two small galaxies (known as the Large and Small Magellanic Clouds) apparently orbit our Milky Way Galaxy. Again, the speeds are enormous by earthly standards. For example, the Milky Way is moving toward the Great Galaxy in Andromeda at about 300,000 kilometers per hour (180,000 miles per hour). Despite the high speed, we needn't worry about a collision anytime soon. Even if the

Figure 1.14

This painting shows an edge-on view of the Milky Way Galaxy. Most visible stars reside within the galaxy's thin *disk*, which runs horizontally across the page in this figure. Careful study of galactic rotation suggests that most of the mass lies in the galactic *halo*—a large, spherical region that surrounds and encompasses the disk. Because this mass emits no light that we have detected, we call it dark matter.

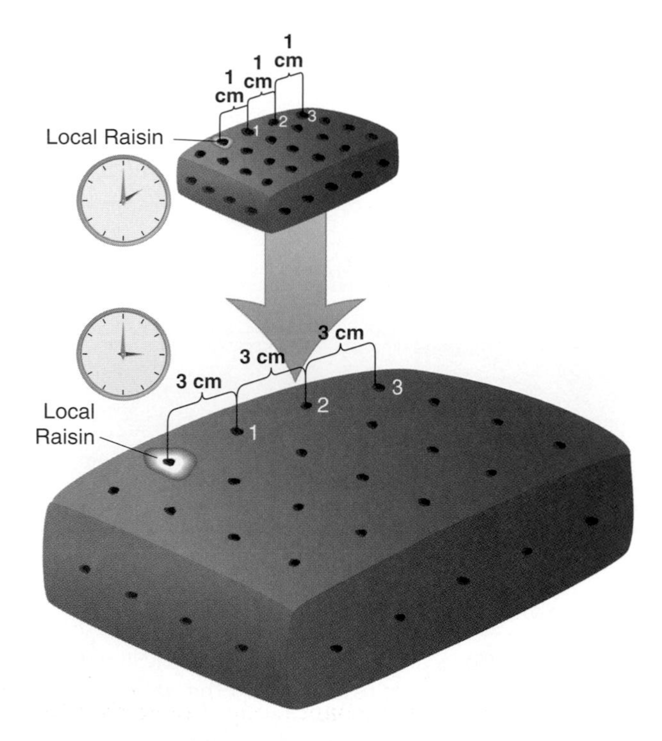

Distances and Speeds As Seen from the Local Raisin

Raisin Number	*Distance Before Baking*	*Distance After Baking (1 hour later)*	*Speed*
1	1 cm	3 cm	2 cm/hr
2	2 cm	6 cm	4 cm/hr
3	3 cm	9 cm	6 cm/hr
⋮	⋮	⋮	⋮

Figure 1.15 Interactive Figure

An expanding raisin cake illustrates basic principles of the expansion of the universe. From the outside, the raisin cake appears to expand uniformly. From the inside, anyone living in one of the raisins would find that all other raisins are moving away as the cake expands, with more distant raisins moving away faster. This analogy shows why the fact that more distant galaxies move away from us faster than nearer ones implies that our universe is expanding.

Milky Way and Andromeda Galaxies are approaching each other head-on (which they might not be), it will be nearly 10 billion years before any collision begins.

When we look outside the Local Group, however, we find two astonishing facts that were first recognized in the 1920s by Edwin Hubble, for whom the Hubble Space Telescope was named:

1. Virtually every galaxy outside the Local Group is moving *away* from us.
2. The more distant the galaxy, the faster it appears to be racing away.

These facts might make it sound like we suffer from a cosmic case of chicken pox, but there is a much more natural explanation: *The entire universe is expanding.* We'll save details about this expansion for later in the book (Chapter 15), but you can understand the basic idea by thinking about a raisin cake baking in an oven.

Imagine that you make a raisin cake in which the distance between adjacent raisins is 1 centimeter. You place the cake in the oven, where it expands as it bakes. After 1 hour, you remove the cake, which has expanded so that the distance between adjacent raisins has increased to 3 centimeters (Figure 1.15). The expansion of the cake seems fairly obvious. But what would you see if you lived *in* the cake, as we live in the universe?

Pick any raisin (it doesn't matter which one), call it the Local Raisin, and identify it in the pictures of the cake both before and after baking. Figure 1.15 shows one possible choice for the Local Raisin, with three nearby raisins labeled. The accompanying table summarizes what you would see if you lived within the Local Raisin. Notice, for example, that Raisin 1 starts out at a distance of 1 centimeter before baking and ends up at a distance of 3 centimeters after baking, which means it moves a distance of 2 centimeters away from the Local Raisin during the hour of baking. Hence, its speed as seen from the Local Raisin is 2 centimeters per hour. Raisin 2 moves from a distance of 2 centimeters before baking to a distance of 6 centimeters after baking, which means it moves a distance of 4 centimeters away from the Local Raisin during the hour. Hence, its speed is 4 centimeters per hour, or twice as fast as the speed of Raisin 1. Generalizing, the fact that the cake is expanding means that all raisins are moving away from the Local Raisin, with more distant raisins moving away faster.

Distant galaxies are all moving away from us, with more distant ones moving faster, telling us that we live in an expanding universe.

Hubble's discovery that galaxies are moving in much the same way as the raisins in the cake, with most moving away from us and more distant ones moving away faster, implies that the universe in which we live is expanding much like the raisin cake. If you now imagine the Local Raisin as representing our Local Group of galaxies and the other raisins as representing more distant galaxies or clusters of galaxies, you have a basic picture of the expansion of the universe. Like the expanding dough between the raisins in the cake, *space* itself is growing between galaxies. More distant galaxies move away from us faster because they are carried along with this expansion like the raisins in the expanding cake. Many billions of light-years away, we see galaxies moving away from us at speeds approaching the speed of light.

There's one important distinction between the raisin cake and the universe: Because a cake is small in size, it has a center and edges that we can see. In contrast, we do not think the universe has a true center or edges. Anyone living in any galaxy would see other galaxies moving away, with more distant ones moving faster, and would have an observable universe extending 14 billion light-years in all directions from their location. No place can claim to be any more "central" than any other place.

• Are we ever sitting still?

We and our planet are constantly on the move through the universe, and at surprisingly high speeds.

As we have seen, we are never truly sitting still. Figure 1.16 summarizes the motions we have covered. We spin around Earth's axis at more than 1,000 km/hr, while our planet orbits the Sun at more than 100,000 km/hr. Our solar system moves among the stars of the local solar neighborhood at typical speeds of 70,000 km/hr, while also orbiting the center of the Milky Way Galaxy at a speed of about 800,000 km/hr. Our galaxy moves among the other galaxies of the Local Group, while all other galaxies move away from us at speeds that grow greater with distance in our expanding universe. Spaceship Earth is carrying us on a remarkable journey!

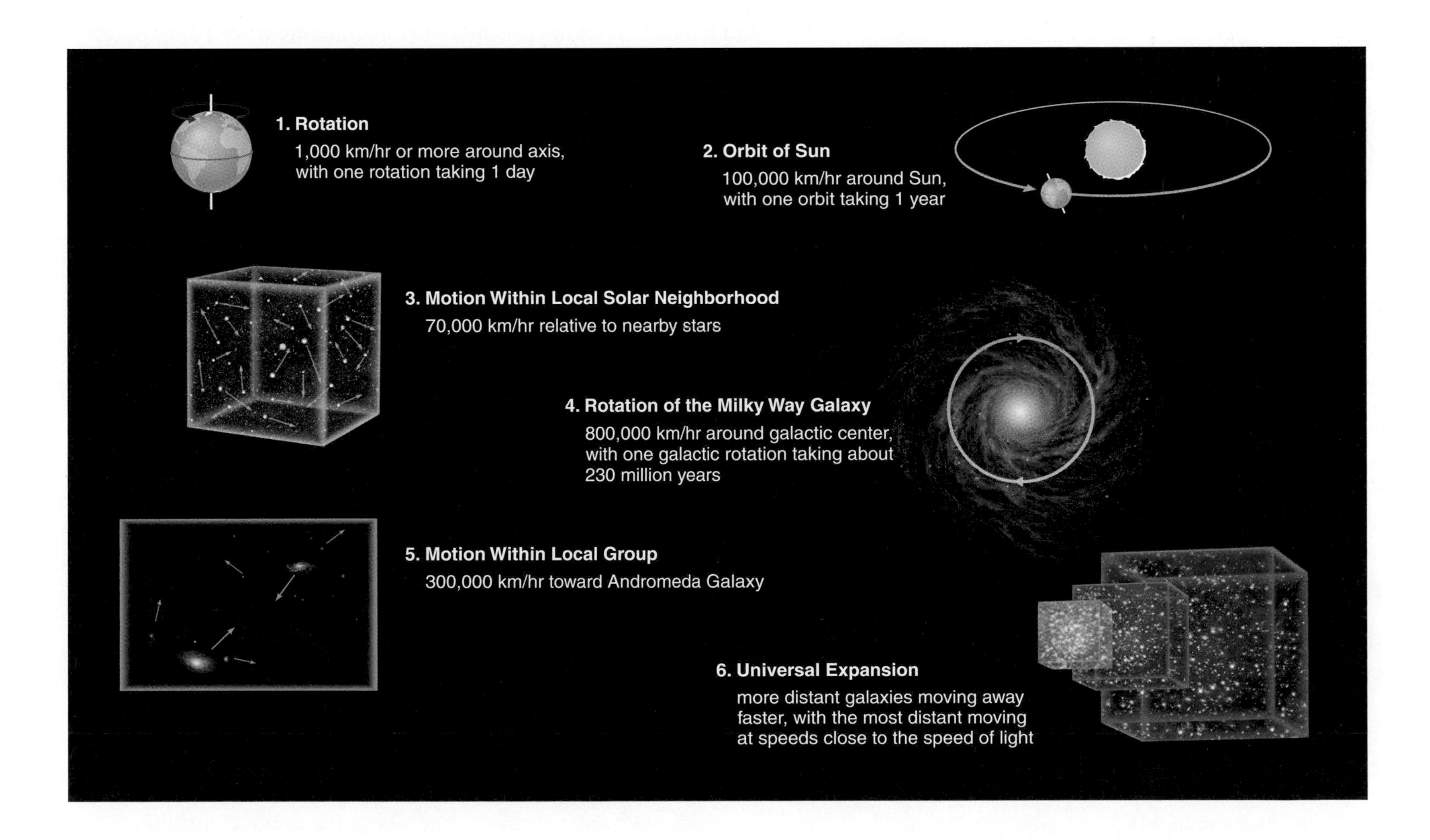

Figure 1.16

This figure summarizes the basic motions of Earth in the universe, along with their associated speeds.

THE BIG PICTURE
Putting Chapter 1 into Context

In this first chapter, we developed a broad overview of our place in the universe. As we consider the universe in more depth in the rest of the book, remember the following "big picture" ideas:

- Earth is not the center of the universe but instead is a planet orbiting a rather ordinary star in the Milky Way Galaxy. The Milky Way Galaxy, in turn, is one of billions of galaxies in our observable universe.
- We are "star stuff." The atoms from which we are made began as hydrogen and helium in the Big Bang and were later fused into heavier elements by massive stars. When these stars died, they released these atoms into space, where our galaxy recycled them into new stars and planets. Our solar system formed from such recycled matter some 4.6 billion years ago.
- Cosmic distances are literally astronomical, but we can put them in perspective with the aid of scale models and other scaling techniques. When you think about these enormous scales, don't forget that every star is a sun and every planet is a unique world.
- We are latecomers on the scale of cosmic time. The universe was already more than half its current age when our solar system formed, and it took billions of years more before humans arrived on the scene.
- All of us are being carried through the cosmos on spaceship Earth. Although we cannot feel this motion in our everyday lives, the associated speeds are surprisingly high. Learning about the motions of spaceship Earth gives us a new perspective on the cosmos and helps us understand its nature and history.

1.1 OUR MODERN VIEW OF THE UNIVERSE

- **What is our place in the universe?**

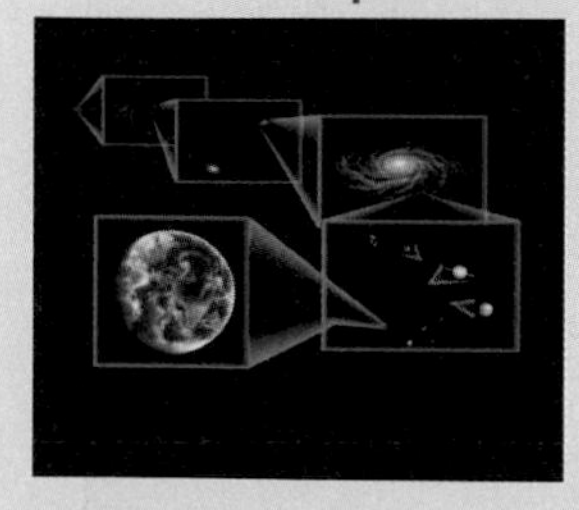

Earth is a planet orbiting the Sun. Our Sun is one of more than 100 billion stars in the **Milky Way Galaxy**. Our galaxy is one of about 40 galaxies in the **Local Group**. The Local Group is one small part of the **Local Supercluster**, which is one small part of the **universe**.

- **How did we come to be?**

The universe began in the **Big Bang** and has been expanding ever since, except in localized regions where gravity has caused matter to collapse into galaxies and stars. The Big Bang essentially produced only two chemical elements: hydrogen and helium. The rest have been produced by stars and recycled within galaxies from one generation of stars to the next, which is why we are "star stuff."

- **How can we know what the universe was like in the past?**

We can actually see into the past by studying light from distant stars and galaxies. Light takes time to travel through space, so the farther away we look in distance, the further back we look in time. Thus, when we look billions of **light-**

years away, we see pieces of the universe as they were billions of years ago.

- **Can we see the entire universe?**

No. The age of the universe limits the extent of our **observable universe**. For a universe that is 14 billion years old, our observable universe extends to a distance of 14 billion light-years. If we tried to look beyond that distance, we'd be trying to look to a time before the universe existed.

1.2 THE SCALE OF THE UNIVERSE

- **How big is Earth compared to our solar system?**

On a scale of 1 to 10 billion, the Sun is about the size of a grapefruit. Planets are much smaller, with Earth the size of a ball point and Jupiter the size of a marble on this scale. The distances between planets are huge compared to their sizes, with Earth orbiting 15 meters from the Sun on this scale.

- **How far away are the stars?**

On the 1-to-10-billion scale, it is possible to walk from the Sun to Pluto in just a few minutes. On the same scale, the nearest stars besides the Sun are thousands of kilometers away.

- **How big is the Milky Way Galaxy?**

On a scale where the Milky Way is the size of a football field, the distance to the nearest star would be only about 4 millimeters. There are so many stars in our galaxy that it would take thousands of years just to count them.

- **How big is the universe?**

The observable universe contains at least 100 billion galaxies. The total number of stars in the observable universe is roughly the same as the number of grains of dry sand on all the beaches on Earth.

- **How do our lifetimes compare to the age of the universe?**

On a cosmic calendar that compresses the history of the universe into 1 year, human civilization is just a few seconds old, and a human lifetime lasts only a fraction of a second.

1.3 SPACESHIP EARTH

- **How is Earth moving in our solar system?**

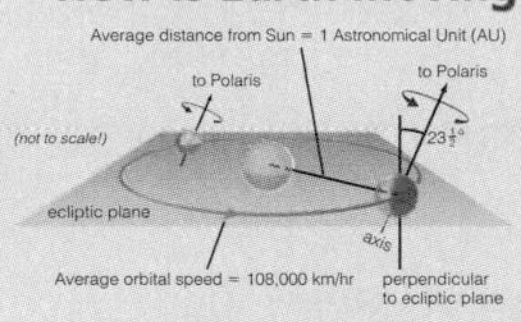

Earth **rotates** on its axis once each day and **orbits** the Sun once each year. Earth orbits at an average distance from the Sun of 1 **AU** and with an **axis tilt** of $23\frac{1}{2}°$ to a line perpendicular to the **ecliptic plane**.

- **How is our solar system moving in the Milky Way Galaxy?**

We move seemingly randomly relative to other stars in our local solar neighborhood. The speeds are substantial by earthly standards, but the stars are so far away that their motion is undetectable to the naked eye. Our Sun and other stars in our neighborhood orbit the center of the galaxy every 230 million years, because the entire galaxy is rotating.

- **How do galaxies move within the universe?**

Galaxies move essentially at random within the Local Group, but all galaxies beyond the Local Group are moving away from us. More distant galaxies are moving faster, which tells us that we live in an expanding universe.

- **Are we ever sitting still?**

We are never truly sitting still. We spin around Earth's axis and orbit the Sun. Our **solar system** moves among the stars of the local solar neighborhood, while orbiting the center of the Milky Way Galaxy. Our galaxy moves among the other galaxies of the Local Group, while all other galaxies move away from us in our expanding universe.

EXERCISES AND PROBLEMS

REVIEW QUESTIONS

1. What do we mean by a *geocentric* universe? In broad terms, contrast a geocentric view of the universe with our modern view of the universe.
2. Briefly describe the major levels of structure (such as planet, star, galaxy) in the universe.
3. What do we mean when we say that the universe is *expanding*? Why does an expanding universe suggest that the universe began in what we call the *Big Bang*?
4. Briefly explain what Carl Sagan meant when he said that we are "star stuff."
5. How fast does light travel? What is a *light-year*?
6. Explain the statement *The farther away we look in distance, the further back we look in time.*
7. What do we mean by the *observable universe*? Is it the same thing as the entire universe?
8. Suppose we use a scale on which the Sun is the size of a large grapefruit (one ten-billionth of its actual size). How big and how far away is Earth? How big and how far away are the other planets in our solar system? How far away are the nearest stars besides the Sun?
9. Describe at least two ways to put the scale of the Milky Way Galaxy into perspective. Then describe at least one way to put the size of the observable universe into perspective.
10. Imagine describing the cosmic calendar to a friend. In your own words, give your friend a feel for how the human race fits into the scale of time.
11. Define *astronomical unit*, *ecliptic plane*, and *axis tilt*. Explain how each is related to Earth's rotation and/or orbit.

12. What is the shape of the Milky Way Galaxy? Where is our solar system located within the galaxy? How does our solar system move within the galaxy?
13. Distinguish between our galaxy's *disk* and *halo*. Where do most visible stars reside? Where does the mysterious *dark matter* seem to reside?
14. What key observations by Edwin Hubble lead us to conclude that the universe is expanding? Briefly explain how these observations imply that the universe is expanding.

DOES IT MAKE SENSE?

Decide whether each statement makes sense and explain why it does or does not.

Example: I walked east from our base camp at the North Pole.

Solution: The statement does not make sense because *east* has no meaning at the North Pole—all directions are south from the North Pole.

15. Our solar system is bigger than some galaxies.
16. The universe is about 14 billion light-years old.
17. It will take me light-years to complete this homework assignment!
18. Someday we may build spaceships capable of traveling at a speed of 1 light-minute per hour.
19. Astronomers recently discovered a moon that does not orbit a planet.
20. NASA plans soon to launch a spaceship that will leave the Milky Way Galaxy to take a photograph of the galaxy from the outside.
21. The observable universe is the same size today as it was a few billion years ago.
22. Photographs of distant galaxies show them as they were when they were much younger than they are today.
23. At a nearby park, I built a scale model of our solar system in which I used a basketball to represent Earth.
24. Because nearly all galaxies are moving away from us, we must be located at the center of the universe.

PROBLEMS

(Quantitative problems are marked with an asterisk.)

25. *Raisin Cake Universe.* Suppose that all the raisins in a cake are 1 centimeter apart before baking and 4 centimeters apart after baking.
    a. Draw diagrams to represent the cake before and after baking.
    b. Identify one raisin as the Local Raisin on your diagrams. Construct a table showing the distances and speeds of other raisins as seen from the Local Raisin.
    c. Briefly explain how your expanding cake is similar to the expansion of the universe.
26. *Scaling the Local Group of Galaxies.* Both the Milky Way Galaxy and the Great Galaxy in Andromeda (M31) have a diameter of about 100,000 light-years. The distance between the two galaxies is about 2.5 million light-years.
    a. Using a scale on which 1 centimeter represents 100,000 light-years, draw a sketch showing both galaxies and the distance between them to scale.
    b. How does the separation between galaxies compare to the separation between stars? Based on your answer, discuss the likelihood of galactic collisions in comparison to the likelihood of stellar collisions.

*27. *Distances by Light.* Just as a light-year is the distance that light can travel in 1 year, we define a light-second as the distance that light can travel in 1 second, a light-minute as the distance that light can travel in 1 minute, and so on. Calculate the distance in kilometers represented by each of the following: 1 light-second; 1 light-minute; 1 light-hour; 1 light-day.

*28. *Driving to the Planets (and Stars).* Imagine that you could drive your car at a constant speed of 100 km/hr (62 mi/hr), even in space. (In reality, the law of gravity would make driving through space at a constant speed all but impossible.)
    a. How long would it take to drive all around the Earth? Assume you can drive across both land and ocean. (*Hint:* Use Earth's circumference of approximately 40,000 kilometers.)
    b. Suppose you started driving from the Sun. How long would it take to reach Earth? How long would it take to reach Pluto? (You can find planetary distance data in Appendix E.)
    c. How long would it take to drive the 4.4 light-years to Alpha Centauri? (*Hint:* Remember that 1 light-year is approximately 10 trillion kilometers.)
    d. Suppose you wanted to reach Alpha Centauri in only 100 years. How fast would you have to go? Compare this speed to the speeds of our fastest current spacecraft (around 50,000 km/hr).

DISCUSSION QUESTIONS

29. *Vast Orbs.* Although many people may have gleaned hints of the vast size of the cosmos, Dutch astronomer Christiaan Huygens may have been the first person to truly understand both the large sizes of other planets and the great distances to other stars. This is what he wrote, way back in about 1690: "How vast those Orbs must be, and how inconsiderable this Earth, the Theatre upon which all our mighty Designs, all our Navigations, and all our Wars are transacted, is when compared to them. A very fit consideration, and matter of Reflection, for those Kings and Princes who sacrifice the Lives of so many People, only to flatter their Ambition in being Masters of some pitiful corner of this small Spot." What do you think he meant by this statement? Explain clearly.
30. *Infant Species.* In the last few tenths of a second before midnight on December 31 of the cosmic calendar, we have developed an incredible civilization and learned a great deal about the universe, but we also have developed technology through which we could destroy ourselves. The midnight bell is striking, and the choice for the future is ours. How far into the next cosmic year do you think our civilization will survive? Defend your opinion.
31. *A Human Adventure.* Astronomical discoveries clearly are important to science, but are they also important to our personal lives? Defend your opinion.

MATH HELP AND EXERCISES

For additional help and practice with mathematical concepts applicable to this chapter, the Astronomy Place web site has the following resources with detailed explanations, worked examples and practice problems:

- How far is a light-year?

For a complete list of media resources available, go to **www.astronomyplace.com** and choose Chapter 1 from the pull-down menu.

ASTRONOMY PLACE WEB TUTORIALS

Tutorial Review of Key Concepts

Use the interactive **Tutorial** at **www.astronomyplace.com** to review key concepts from this chapter.

Scale of the Universe Tutorial

Lesson 1 Distances Scales: The Solar System
Lesson 2 Distances Scales: Stars and Galaxies
Lesson 3 Powers of 10

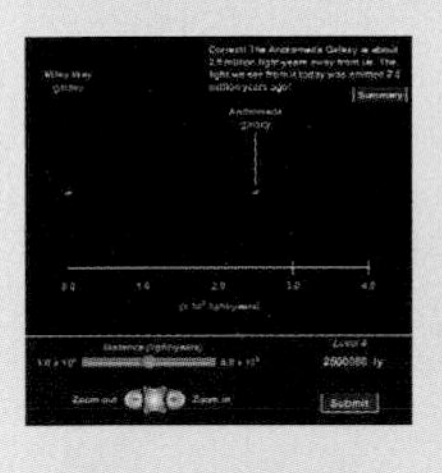

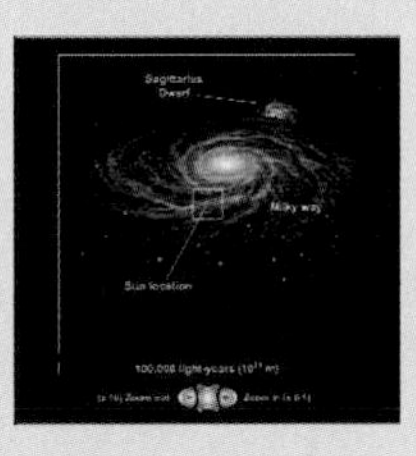

Supplementary Tutorial Exercises

Use the interactive **Tutorial Lessons** to explore the following questions.

Scale of the Universe Tutorial, Lesson 2

1. Could we use spaceships like those that we use to explore our solar system to explore planets around other stars? Why or why not?
2. How does the distance between galaxies in the Local Group compare to the sizes of the galaxies?
3. Why are collisions between galaxies more likely than collisions between stars within galaxies?

Scale of the Universe Tutorial, Lesson 3

1. Why are powers of 10 useful for describing and comparing distances in the universe?
2. If you begin from the scale of the observable universe, by how many powers of 10 must you zoom in to see the orbits of the planets in our solar system?
3. How many powers of 10 separate the scale on which you can see all the planetary orbits from the scale on which you can see individual people?

EXPLORING THE SKY AND SOLAR SYSTEM

Of the many activities available on the *Voyager: SkyGazer* **CD-ROM** accompanying your book, use the following files to observe key phenomena covered in this chapter.

Go to the **File: Basics** folder for the following demonstrations.

1. Chicago 10000AD
2. Dragging the Sky

Go to the **Explore** menu for the following demonstrations.

1. Solar Neighborhood
2. Paths of the Planets

MOVIES

Check out the following narrated and animated short documentaries available on **www.astronomyplace.com** for a helpful review of key ideas covered in this chapter.

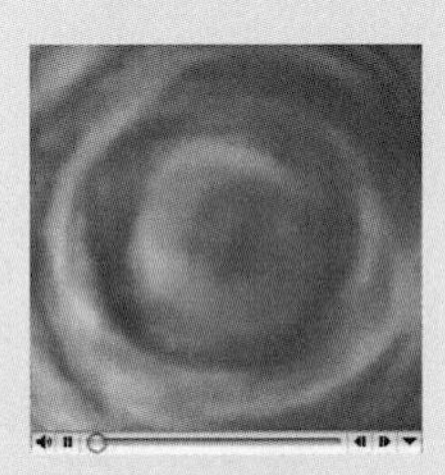

From the Big Bang to Galaxies Movie

WEB PROJECTS

Take advantage of the useful Web links on **www.astronomyplace.com** to assist you with the following projects.

1. *Astronomy on the Web.* The Web contains a vast amount of astronomical information. Starting from the links on the textbook Web site, spend at least an hour exploring astronomy on the Web. Write two or three paragraphs summarizing what you learned from your Web surfing. What was your favorite astronomical Web site, and why?
2. *Tour Report.* Take the virtual tour of the Voyage scale model solar system on the text Web site (**www.astronomyplace.com**). After completing it, imagine that a friend asks you the following questions. Answer each question in one paragraph.
   a. Is the Sun really much bigger than Earth?
   b. Is it true that the Sun uses nuclear energy?
   c. Would it be much harder to send humans to Mars than to the Moon?
   d. In elementary school, I heard that Neptune is farther from the Sun than Pluto. Is this true?
   e. I read that Pluto is not really a planet. What's the story?
   f. Why didn't they have any stars besides the Sun in the scale model?
   g. What was the most interesting thing you learned during your tour?
3. *NASA Missions.* Visit the NASA Web site to learn about upcoming astronomy missions. Write a one-page summary of the mission you feel is most likely to give us new astronomical information during the time you are enrolled in your astronomy course.
4. *The Hubble Ultra Deep Field.* The photo that opens this chapter is called the Hubble Ultra Deep Field. Find the photo on the Hubble Space Telescope web site, and learn how it was taken, what it shows, and what we've learned from it. Write a short report on your findings.

2
Discovering the Universe for Yourself

LEARNING GOALS

2.1 PATTERNS IN THE NIGHT SKY
- What are constellations?
- How do we locate objects in the sky?
- Why do stars rise and set?
- Why don't we see the same constellations throughout the year?

2.2 THE REASON FOR SEASONS
- What causes the seasons?
- How do we mark the progression of the seasons?
- Does the orientation of Earth's axis change with time?

2.3 THE MOON, OUR CONSTANT COMPANION
- Why do we see phases of the Moon?
- What causes eclipses?

2.4 THE ANCIENT MYSTERY OF THE PLANETS
- What was once so mysterious about the movement of planets in our sky?
- Why did the ancient Greeks reject the real explanation for planetary motion?

This is an exciting time in the history of astronomy. A new generation of telescopes is probing the depths of the universe. Increasingly sophisticated space probes are collecting new data about the planets and other objects in our solar system. Rapid advances in computing technology are allowing scientists to analyze the vast amount of new data and to model the processes that occur in planets, stars, galaxies, and the universe.

One goal of this book is to help *you* share in the ongoing adventure of astronomical discovery. One of the best ways to become a part of this adventure is to do what other humans have done for thousands of generations: Go outside, observe the sky around you, and contemplate the awe-inspiring universe of which you are a part. In this chapter, we'll discuss a few key ideas that will help you understand what you see in the sky.

2.1 PATTERNS IN THE NIGHT SKY

Imagine yourself far from city lights, gazing up at the sky on a clear, moonless night. More than 2,000 stars may be visible to your naked eye, along with the whitish band of light that we call the *Milky Way* (Figure 2.1). The stars you see at any particular time of night vary with your location on Earth and the time of year, but the overall appearance of the night sky is much the same today as it has been for many thousands of years. In this section, we'll discuss how we make sense of major features of the night sky.

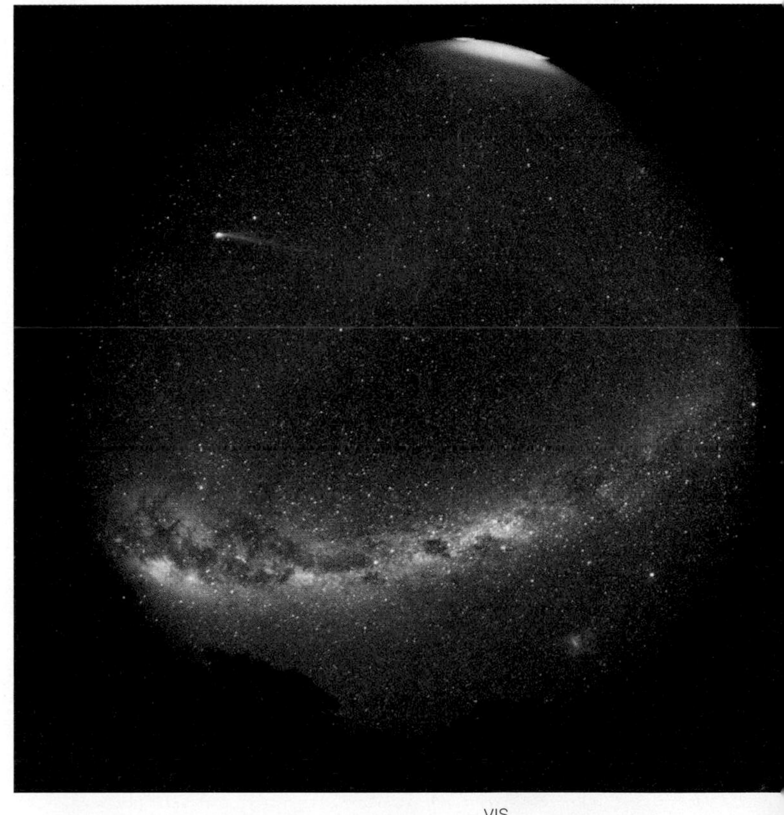

Figure 2.1

A "fish-eye" photograph of the Australian night sky. The Milky Way is the prominent whitish band, with dark lanes running through it. Comet Hyakutake is also visible (near the upper left) in this 1996 photo.

• What are constellations?

As you gaze upward at the night sky, your mind may group some of the stars into patterns that look like familiar shapes or objects. People of nearly every culture have seen and named such patterns. We usually refer to them as constellations, but to astronomers the term has a more precise meaning: a **constellation** is a *region* of the sky with well-defined borders. Just as every spot of land in the United States is part of some state, every point in the sky belongs to some constellation. For example, Figure 2.2 shows the borders of the constellation Orion and several of its neighbors. The familiar patterns of stars merely help us locate these constellations.

A constellation is a region of the sky with well-defined borders.

The official borders of the constellations were set in 1928 by members of the International Astronomical Union (IAU), an association of astronomers from around the world. The IAU divided the sky into 88 constellations (see Appendix I), choosing constellation names that were familiar to European and American astronomers. Most constellations visible in the Northern Hemisphere have names that can be traced back to civilizations of the ancient Middle East. Most of the official names of the constellations visible from the Southern Hemisphere were given by seventeenth-century European explorers.

Figure 2.2

Red lines mark official borders of several constellations near Orion. Yellow lines connect recognizable patterns of stars within constellations. Sirius, Procyon, and Betelgeuse form a pattern that spans several constellations and is called the Winter Triangle. It is easy to see on clear winter evenings.

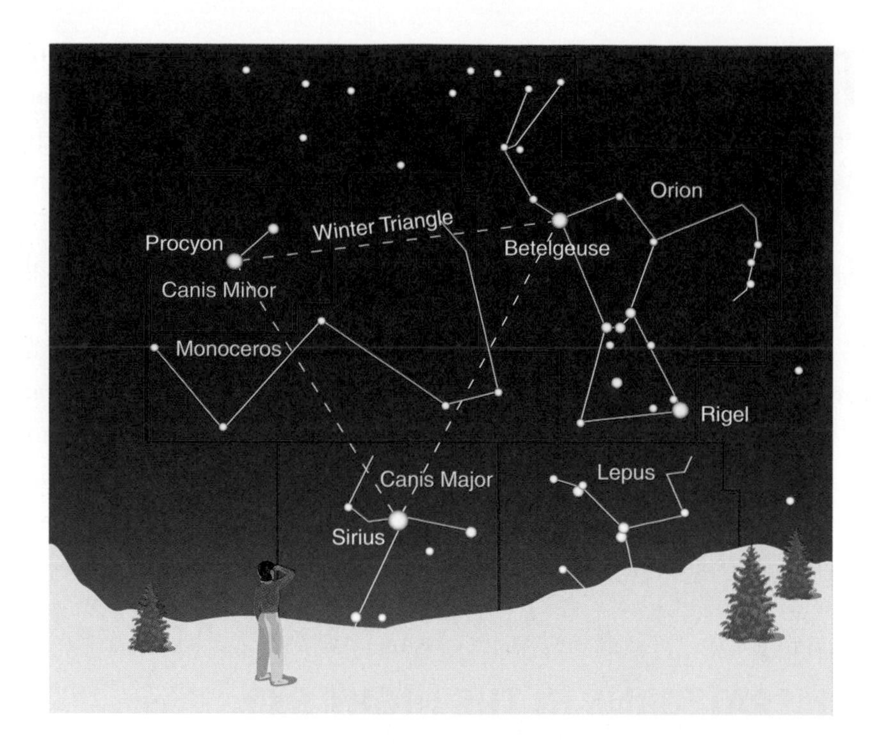

The Celestial Sphere The stars in a particular constellation appear to lie close to one another but may be quite far apart in reality, because they may lie at very different distances from Earth (Figure 2.3). This illusion occurs because we lack depth perception when we look into space, a consequence of the fact that the stars are so far away [Section 1.2]. The ancient Greeks mistook this illusion for reality, imagining the stars to lie on a great **celestial sphere** that surrounds the Earth.

All stars appear to lie on a *celestial sphere*, but in reality they lie at different distances from Earth.

We now know that Earth does not really lie in the center of a giant ball of stars, but we can still use the idea of a celestial sphere to help us understand and map the sky. Just as a globe shows us the layout of Earth's surface, a model of the celestial sphere shows how the stars appear to be arranged in the sky. A typical model of the celestial sphere (Figure 2.4) shows patterns of bright stars, borders of constellations, and four special points and circles:

- The **north celestial pole** is the point directly over Earth's North Pole.
- The **south celestial pole** is the point directly over Earth's South Pole.
- The **celestial equator**, which is a projection of Earth's equator into space, makes a complete circle around the celestial sphere.
- The **ecliptic** is the path the Sun follows as it appears to circle around the celestial sphere once each year.

Notice that the ecliptic crosses the celestial equator at a $23\frac{1}{2}°$ angle, because that is the tilt of Earth's axis. In fact, the ecliptic on the celestial sphere is the projection into space of Earth's orbital plane (called the *ecliptic plane* [Section 1.3]).

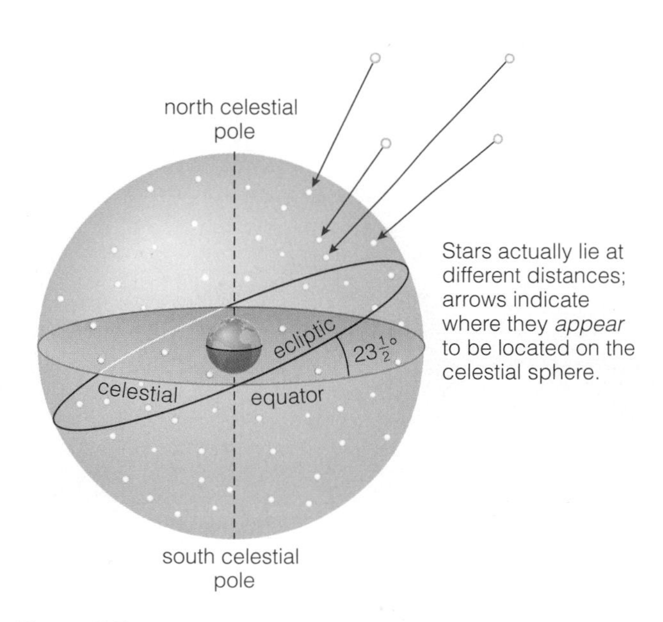

Figure 2.3

Stars appear to lie on a great celestial sphere that surrounds the Earth, but this is an illusion created by our lack of depth perception when we look into space.

The Milky Way The band of light that we call the Milky Way can also be shown on the celestial sphere, since it circles all the way around the sky, passing through more than a dozen constellations. Although the

Milky Way can be difficult to see near bright city lights, it is quite prominent in dark skies.

The "Milky Way" in the night sky is actually our view into the disk of our Milky Way Galaxy.

The Milky Way in the night sky and the Milky Way Galaxy are closely related. You'll understand why if you remember that we are looking at our galaxy from within its disk (see Figures 1.1, 1.14). No matter what direction we look within the disk of the galaxy, our view takes in countless stars and vast clouds of interstellar gas and dust. The stars and bright clouds make a narrow band in our sky because the disk of the galaxy is fairly thin. The dark lanes running through the Milky Way occur in places where we are looking at particularly dark and dense interstellar clouds. These dark clouds hide the light of stars that lie behind them. If you look closely, you'll notice that the Milky Way's width is different in different directions. It looks widest in the direction of the constellation Sagittarius, because that is the direction in which we are looking toward the galaxy's wide, central bulge.

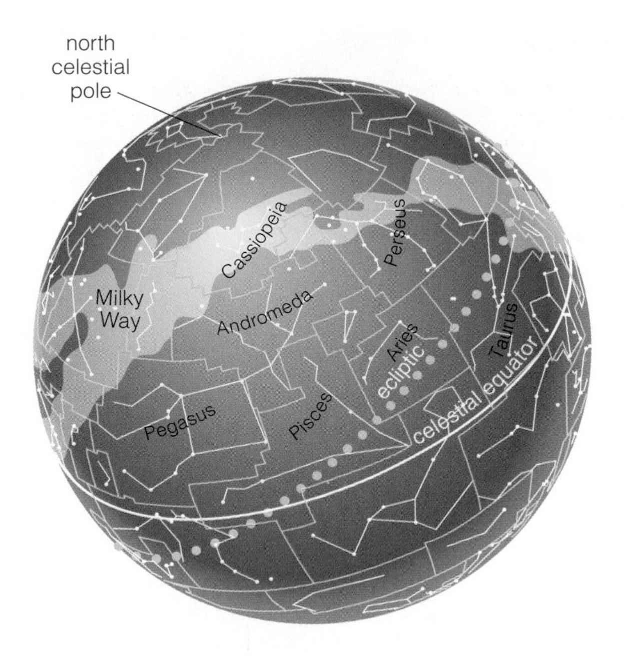

Figure 2.4

A model of the celestial sphere shows the patterns of the stars, the borders of the 88 official constellations, the ecliptic, and the celestial equator and poles. Because the celestial sphere represents the view from Earth, we imagine Earth to reside in the center of the sphere.

• How do we locate objects in the sky?

The constellations always stay the same on the celestial sphere, at least on the time scale of human lifetimes. But what we actually see when we look into the sky depends on our location on Earth and varies from hour to hour and night to night. The reason for these changes is that we see only half of the celestial sphere at any particular moment from any particular location, while the other half is blocked from view by the ground. The first step in making sense of our local view is to have a way to describe the locations of the stars and other objects that we see.

The Local Sky Picture yourself standing in a flat, open field. The sky appears to take the shape of a dome, making it easy to understand why people of many ancient cultures believed we live on a flat Earth under a great dome that encompasses the world. Today, we use the idea of a dome to define the **local sky**—the sky as seen from wherever you happen to be standing.

We use several key features of the local sky to serve as references for locating objects (Figure 2.5). The boundary between Earth and sky defines the **horizon**. The point directly overhead is the **zenith**. The **meridian** is an imaginary half-circle stretching from the horizon due south, through the zenith, to the horizon due north.

We pinpoint an object in the local sky by stating its altitude above the horizon and direction along the horizon.

We can pinpoint the position of any object in the local sky by stating its **direction** along the horizon and its **altitude** above the horizon. For example, Figure 2.5 shows a person pointing to a star located in the direction of southeast at an altitude of 60°. (The zenith has an altitude of 90° but no direction, because it is straight overhead.)

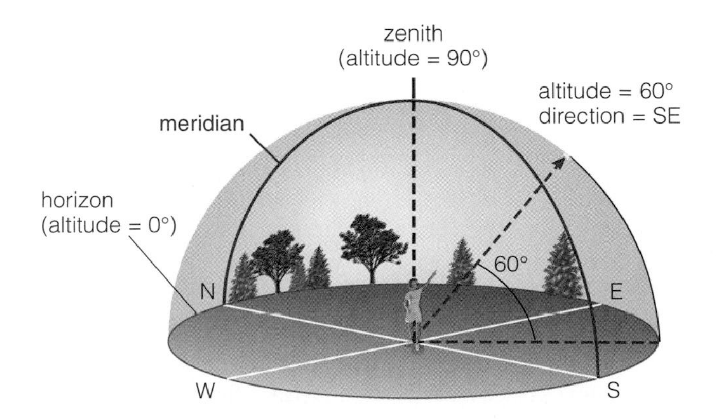

Figure 2.5

From any place on Earth, the local sky looks like a dome (hemisphere). This diagram shows key reference points in the local sky. It also shows how we describe any position in the local sky by its altitude and direction.

Angular Sizes and Distances In addition to stating an object's location in our sky, we often want to describe its size or how far it is from other objects in the sky. For example, to describe the pattern of stars in a particular constellation, we might wish to say how large the pattern appears in the sky and where individual stars are located relative to others. It's easy to do this, as long as we remember that we can describe only *angular* sizes and distances in the sky. Our lack of depth perception on the

a The angular size of the Moon is about $\frac{1}{2}$° (which is also the angular size of the Sun).

b The angular distance between the two "pointer stars" of the Big Dipper (which point to the North Star, Polaris; see Figure 2.11a) is about 5°.

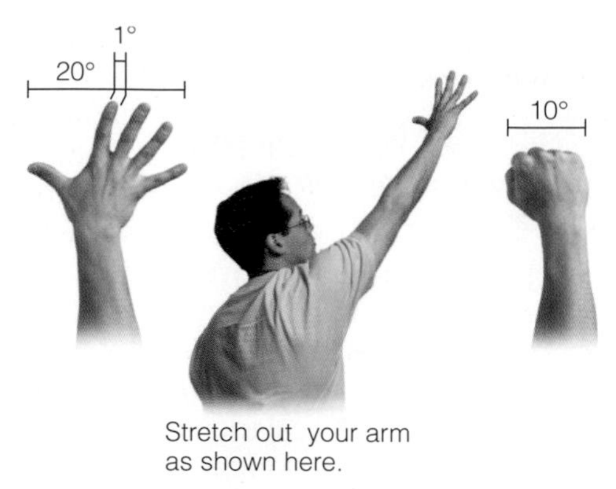

c You can estimate angular sizes or distances with your outstretched hand.

Figure 2.6

We measure *angular sizes* or *angular distances,* rather than actual sizes or distances, when we look at objects in the sky.

celestial sphere means we have no way to judge the true sizes or separations of the objects we see.

The **angular size** of an object is the angle it appears to span in your field of view. For example, the angular sizes of the Sun and Moon are each about $\frac{1}{2}$° (Figure 2.6a). Notice that angular size does not by itself tell us much about an object's true size, because angular size also depends on distance: The farther away an object is, the smaller it will look in angular size. For example, the Sun is about 400 times larger in diameter than the Moon, but appears to be the same size in our sky because it is also about 400 times farther away.

The farther away an object is, the smaller its angular size in our sky.

The **angular distance** between a pair of objects in the sky is simply the angle that appears to separate them. For example, the angular distance between the "pointer stars" at the end of the Big Dipper's bowl is about 5° (Figure 2.6b). You can use your outstretched hand to make rough estimates of angles in the sky (Figure 2.6c).

For more precise astronomical measurements, we subdivide each degree into 60 **arcminutes** and subdivide each arcminute into 60 **arcseconds**. We abbreviate arcminutes with the symbol ′ and arcseconds with the symbol ″. For example, we read 35°27′15″ as "35 degrees, 27 arcminutes, 15 arcseconds."

COMMON MISCONCEPTIONS

THE MOON ILLUSION

You've probably noticed that the full moon appears to be larger when it is near the horizon than when it is high in your sky. However, this apparent size change is an illusion. If you measure the angular size of the full moon on a particular night, you'll find it is about the same whether it is near the horizon or high in the sky. The Moon's angular size in the sky depends only on its true size and its distance from Earth. Although this distance varies over the course of the Moon's monthly orbit, it does not change enough to cause a noticeable effect on a single night. You can confirm that the Moon's angular size remains the same by measuring it. You may also be able to make the illusion go away by viewing the Moon upside down between your legs when it is on the horizon.

THINK ABOUT IT Children often try to describe the sizes of objects in the sky (such as the Moon or an airplane) in inches or miles, or by holding their fingers apart and saying, "It was THIS big." Can we really describe objects in the sky in this way? Why or why not?

• Why do stars rise and set?

Now that we understand how to locate objects in the sky, we can consider how and why they appear to move through the local sky. The daily rise and set of the Sun, Moon, planets, and stars occurs because of Earth's daily rotation. If we could view this motion from "outside" the celestial sphere, it would appear very simple (Figure 2.7): Every object that we see on the celestial sphere appears to make a daily circle around Earth. However, the motion can look a little more complex in the local sky.

In the local sky, the horizon essentially cuts the celestial sphere in half, as shown in Figure 2.8 for a typical location in the Northern Hemisphere. As Earth rotates—making the celestial sphere appear to circle around us—we see stars and other objects on the celestial sphere move across our local sky. Different stars follow different paths through the sky, depending on where they are located on the celestial sphere. If you study Figure 2.8 carefully, you'll notice the following key facts about the paths of stars through the sky (the same ideas apply to the Sun, Moon, and planets):

- Stars relatively near the north celestial pole remain perpetually above the horizon. They never rise or set but instead make daily *counter-clockwise* circles around the north celestial pole. We say that such stars are **circumpolar**.
- Stars relatively near the south celestial pole never rise above the horizon at all.
- All other stars have daily circles that are partly above the horizon and partly below it. Because Earth rotates from west to east (counter-clockwise as viewed from above the North Pole), the stars appear to rise in the east and set in the west.

Stars appear to circle around us, going from east to west through the sky, because of Earth's rotation.

The photo that opens this chapter (p. 24) shows a beautiful time-exposure photograph that helps us see the daily paths of stars. It was taken at Arches National Park in Utah. The star paths that make complete circles belong to circumpolar stars, and the north celestial pole lies at the center of these circles. The circles grow larger for stars farther from the north celestial pole. If they are large enough, the circles cross the horizon, so that the stars rise in the east and set in the west. The same ideas apply in the Southern Hemisphere, except that circumpolar stars are those near the south celestial pole.

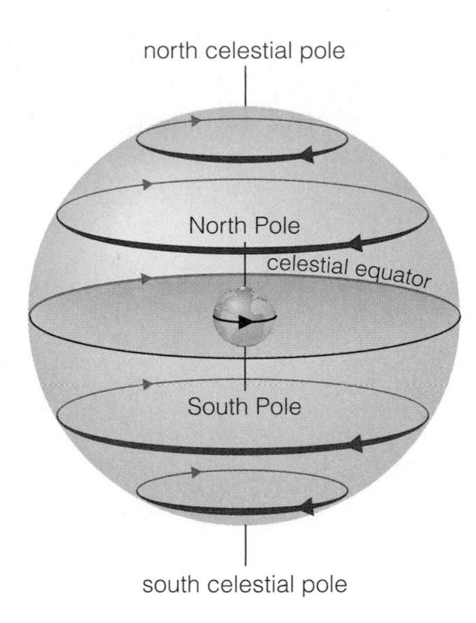

Figure 2.7

Earth rotates from west to east (black arrow), making the celestial sphere *appear* to rotate around us from east to west (red arrows).

• Why don't we see the same constellations throughout the year?

The daily circling of the sky is not the only way in which you can observe changes in the locations of stars. If you travel far north or south, you'll see a different set of constellations than you see at home. And even if you stay in one place, you'll see different constellations at different times of year. Let's explore why.

Variation with Latitude To understand why the visible constellations vary with north or south travel, we must first review how we locate points on Earth (Figure 2.9). **Latitude** is defined to be 0° at the equator and increases northward and southward. The North Pole and South Pole have latitude 90°N and 90°S, respectively. Note that "lines of latitude" are actually circles running parallel to the equator. **Longitude** measures east-west position, so "lines of longitude" are semicircles extending from the North Pole to the South Pole. By international treaty, the line of longitude passing through Greenwich, England, is defined to be longitude 0°. This line is called the **prime meridian**. Stating a latitude and a longitude pinpoints a location on Earth. For example, Figure 2.9 shows that Rome

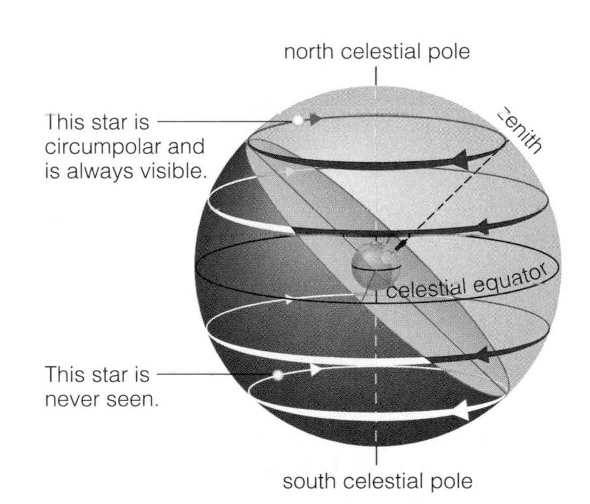

Figure 2.8

This diagram shows the local sky for a typical location in the Northern Hemisphere. The local horizon slices through half the celestial sphere, which is why the local sky looks like a dome. Because the slice is at an angle to the equator, the simple daily circles of stars are tilted in the local sky. Note: It may be easier to follow the star paths in the local sky if you rotate the page so that the zenith points up.

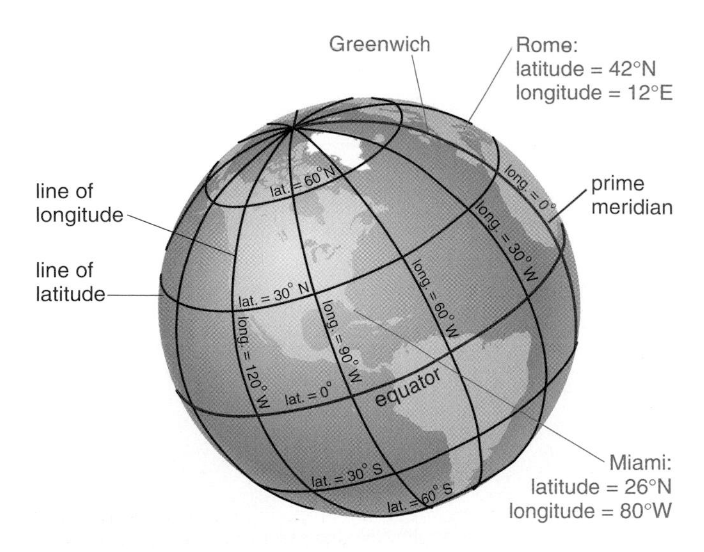

Figure 2.9

We can locate any place on Earth's surface by its latitude and longitude. Latitude measures angular distance north or south of the equator. Longitude measures angular distance east or west of the prime meridian, which passes through Greenwich, England.

lies at about 42°N latitude and 12°E longitude and that Miami lies at about 26°N latitude and 80°W longitude.

You can see why visible constellations vary with latitude by looking again at Figure 2.8. If we move the person standing on Earth to a different latitude, it shifts the position of her horizon and zenith on the celestial sphere. For example, Figure 2.10 shows how the local sky appears relative to the celestial sphere for the latitudes of the North Pole (90°N) and Sydney, Australia (34°S).

The constellations you see depend on your latitude, but not on your longitude.

Note that although the sky varies with latitude, it does *not* vary with longitude. For example, Charleston (South Carolina) and San Diego are at about the same latitude, so people in both cities see the same set of constellations at night. (However, because they are in different time zones, night comes to Charleston about 3 hours before it comes to San Diego.)

If you study diagrams like those in Figure 2.8 and 2.10 carefully, you can discover many interesting facts about the sky. For example, notice that all stars visible at the North Pole are circumpolar, and you can only see stars that lie on the northern half of the celestial sphere. These facts explain why the Sun remains above the horizon for 6 months at the North Pole: The Sun lies on the northern half of the celestial sphere for half of each year (see Figure 2.4), so during these 6 months it must circle the North polar sky daily just like a circumpolar star.

The altitude of the celestial pole in your sky is equal to your latitude.

The diagrams also show a fact that is very important to navigation: *The altitude of the celestial pole in your sky is equal to your latitude.* For example, if the north celestial pole appears in your sky at an altitude of 40° above your north horizon, your latitude is 40°N. Similarly, if the south celestial pole appears in your sky at an altitude of 34° above your south horizon, your latitude is 34°S.

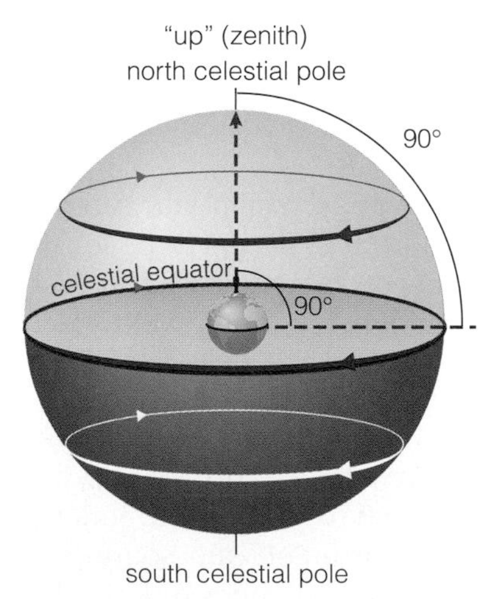

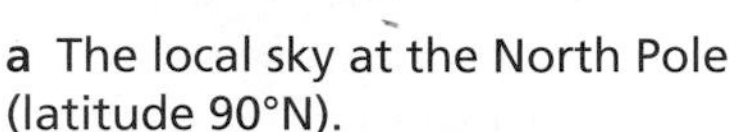

a The local sky at the North Pole (latitude 90°N).

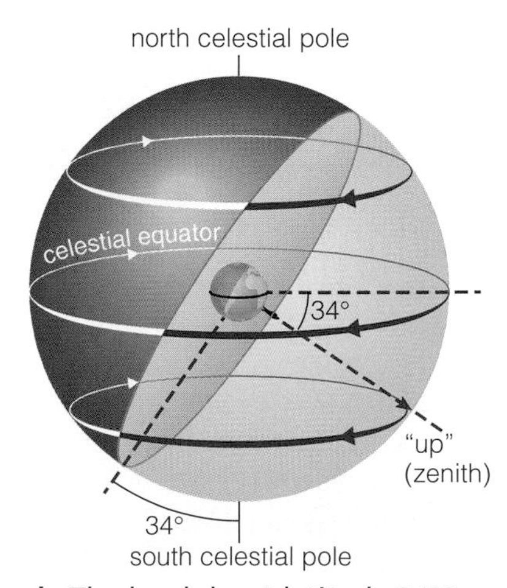

b The local sky at latitude 34°S.

Figure 2.10 Interactive Figure

The sky varies with latitude. Notice that the altitude of the celestial pole that is visible in your sky is always equal to your latitude.

COMMON MISCONCEPTIONS

WHAT MAKES THE NORTH STAR SPECIAL?

Most people are aware that the North Star, Polaris, is a special star. Contrary to a relatively common belief, however, it is *not* the brightest star in the sky. More than 50 other stars are either considerably brighter or comparable in brightness. Polaris is special because it is so close to the north celestial pole. This position makes it very useful in navigation, because it closely marks the direction of due north and because its altitude in your sky is nearly equal to your latitude.

Thus, you can determine your latitude just by finding the celestial pole in your sky. Finding the north celestial pole is fairly easy, because it lies very close to the star Polaris, also known as the North Star (Figure 2.11a). In the Southern Hemisphere, you can find the south celestial pole with the aid of the Southern Cross (Figure 2.11b).

THINK ABOUT IT Answer for your latitude: Where is the north (or south) celestial pole in your sky? Where should you look to see circumpolar stars? What portion of the celestial sphere is never visible in your sky?

SPECIAL TOPIC: HOW LONG IS A DAY?

We usually think of a day as the time it takes for Earth to rotate once, and this is indeed the time it takes for stars to make one full circuit through our sky (for example, from the moment a star is highest in our sky one day until it is highest again the next day). However, this rotation period is *not* exactly 24 hours. Instead, Earth's rotation period is about 23 hours and 56 minutes (more precisely $23^h56^m4.09^s$)—or about 4 minutes short of 24 hours. What's going on?

We define two different types of day. Earth's rotation period of about 23 hours and 56 minutes is called a **sidereal day**. *Sidereal* (pronounced *sy-dear-ee-al*) means "related to the stars," so the name sidereal day comes from the fact that it is the time it takes stars to make a full circuit through our sky. Our 24-hour day, which we call a **solar day**, is the average time it takes the Sun to make one circuit through the sky.

A simple demonstration shows why the solar day is about 4 minutes longer than the sidereal day (see figure below). Set an object on a table to represent the Sun, and stand a few steps away to represent Earth. Point at the Sun and imagine that you also happen to be pointing toward a distant star that lies in the same direction. If you rotate (counterclockwise) while standing in place, you'll again be pointing at both the Sun and the star after one full rotation (figure a). However, because Earth orbits the Sun at the same time that it rotates, you can make the demonstration more realistic by taking a couple of steps around the Sun (counterclockwise) while you rotate (figure b). After one full rotation, you will again be pointing in the direction of the distant star, so this represents a sidereal day. But it does not represent a solar day, because you will not yet be pointing back at the Sun. If you wish to point again at the Sun, you need to make up for your orbital motion by making slightly more than one full rotation. This "extra" bit of rotation makes a solar day longer than a sidereal day.

The only problem with this demonstration is that it exaggerates Earth's daily orbital motion. In reality, Earth moves about 1° per day around its orbit (because it makes a full 360° orbit in 1 year, or about 365 days). Because a single rotation means rotating 360°, Earth must actually rotate about 361° with each solar day (figure c). The extra 1° rotation takes about $\frac{1}{360}$ of Earth's rotation period, which is about 4 minutes.

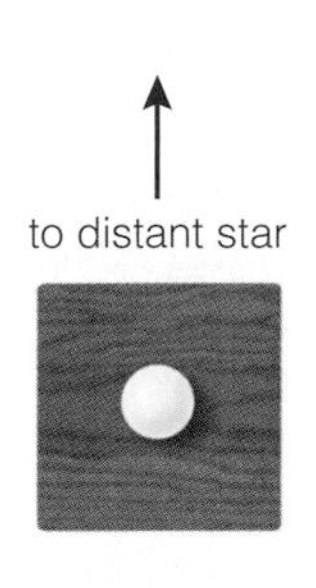

a One full rotation represents a sidereal day and means you are again pointing to the same distant star.

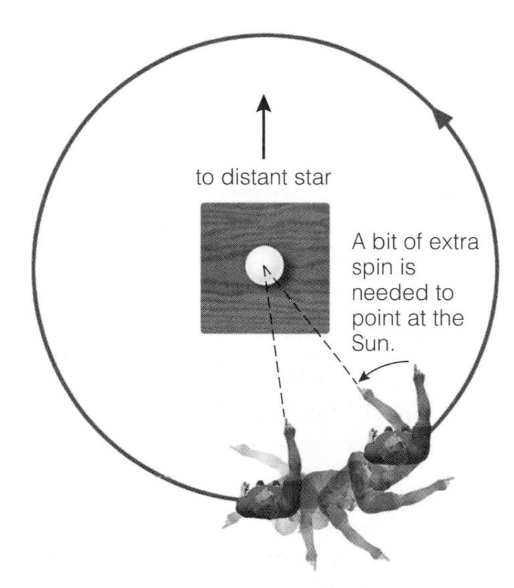

b While you are "orbiting" the Sun, one rotation still returns you to pointing at a distant star, but you need slightly more than one full rotation to return to pointing at the Sun.

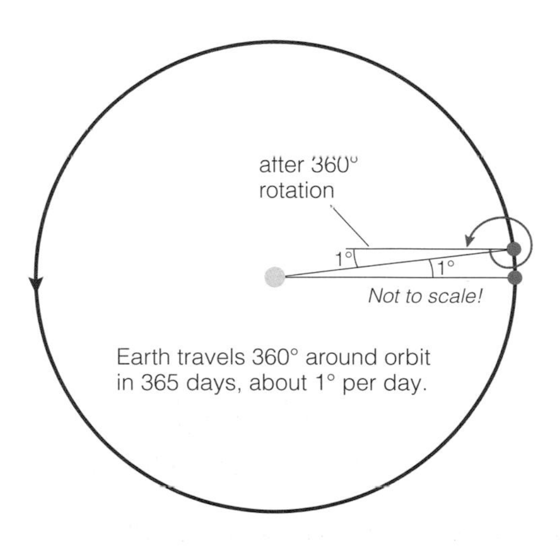

c Earth travels about 1° per day around its orbit, so a solar day requires about 361° of rotation.

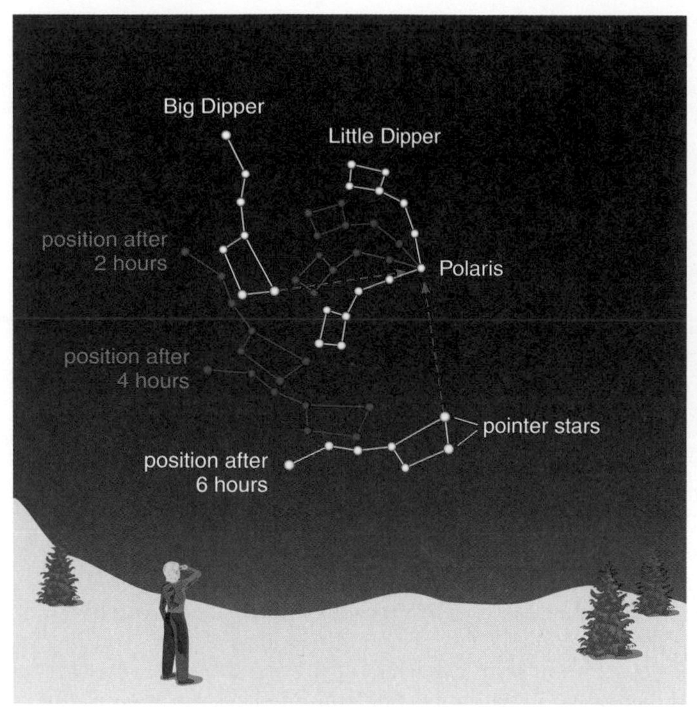

a In the Northern Hemisphere, the pointer stars of the Big Dipper point to the North Star, Polaris, which lies within 1° of the north celestial pole. Note that the sky appears to turn *counterclockwise* around the north celestial pole.

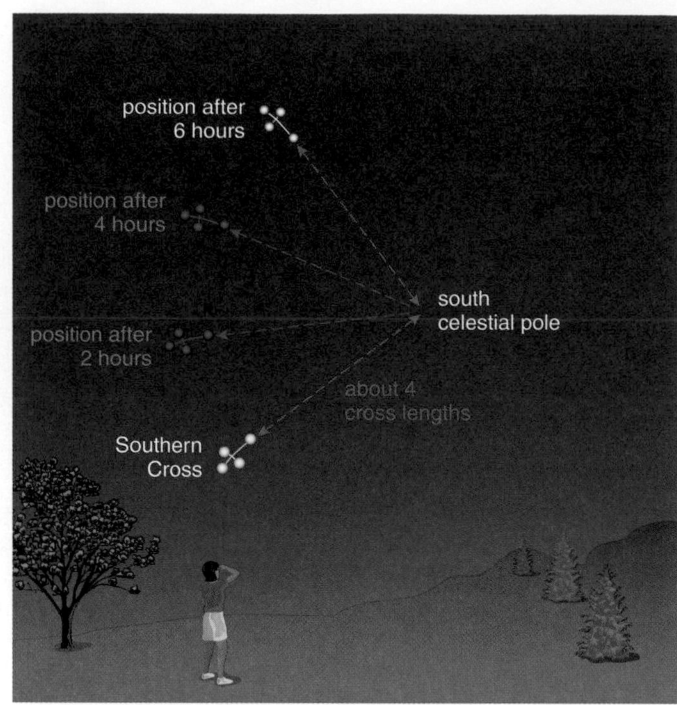

b In the Southern Hemisphere, the Southern Cross points to the south celestial pole, which is not marked by any bright star. The sky appears to turn *clockwise* around the south celestial pole.

Figure 2.11

You can determine your latitude by measuring the altitude of the celestial pole in your sky.

COMMON MISCONCEPTIONS

STARS IN THE DAYTIME

Because we don't see stars in the daytime, some people believe that the stars vanish in the daytime and "come out" at night. In fact, the stars are always present. The reason your eyes cannot see stars in the daytime is that their dim light is overwhelmed by the bright daytime sky. You *can* see bright stars in the daytime with the aid of a telescope, and you may see stars in the daytime if you are fortunate enough to observe a total eclipse of the Sun. Astronauts can also see stars in the daytime. Above Earth's atmosphere, where no air is present to scatter sunlight through the sky, the Sun is a bright disk against a dark sky filled with stars. (However, because the Sun is so bright, astronauts must block its light and allow their eyes to adapt to darkness if they wish to *see* the stars.)

Variation with Time of Year The night sky changes through the year because of Earth's changing position in its orbit around the Sun. Figure 2.12 shows how this works. As we orbit the Sun over the course of a year, the Sun *appears* to move against the background of the distant stars in the constellations. We don't see the Sun and the stars at the same time, but if we could we'd notice the Sun gradually moving eastward along the ecliptic, completing one circuit each year. The constellations along the ecliptic are called the constellations of the **zodiac**. (Tradition places 12 constellations along the zodiac, but the official borders include a wide swath of a thirteenth constellation, Ophiuchus.)

The constellations visible at a particular time of night change as we orbit the Sun.

The Sun's apparent location along the ecliptic determines which constellations we see at night. For example, Figure 2.12 shows that the Sun appears to be in Leo in late August. We therefore cannot see Leo in late August, because it moves with the Sun through the daytime sky. However, we can see Aquarius all night long, since it is opposite Leo on the celestial sphere. Six months later, in February, we see Leo at night while Aquarius is above the horizon only in the daytime.

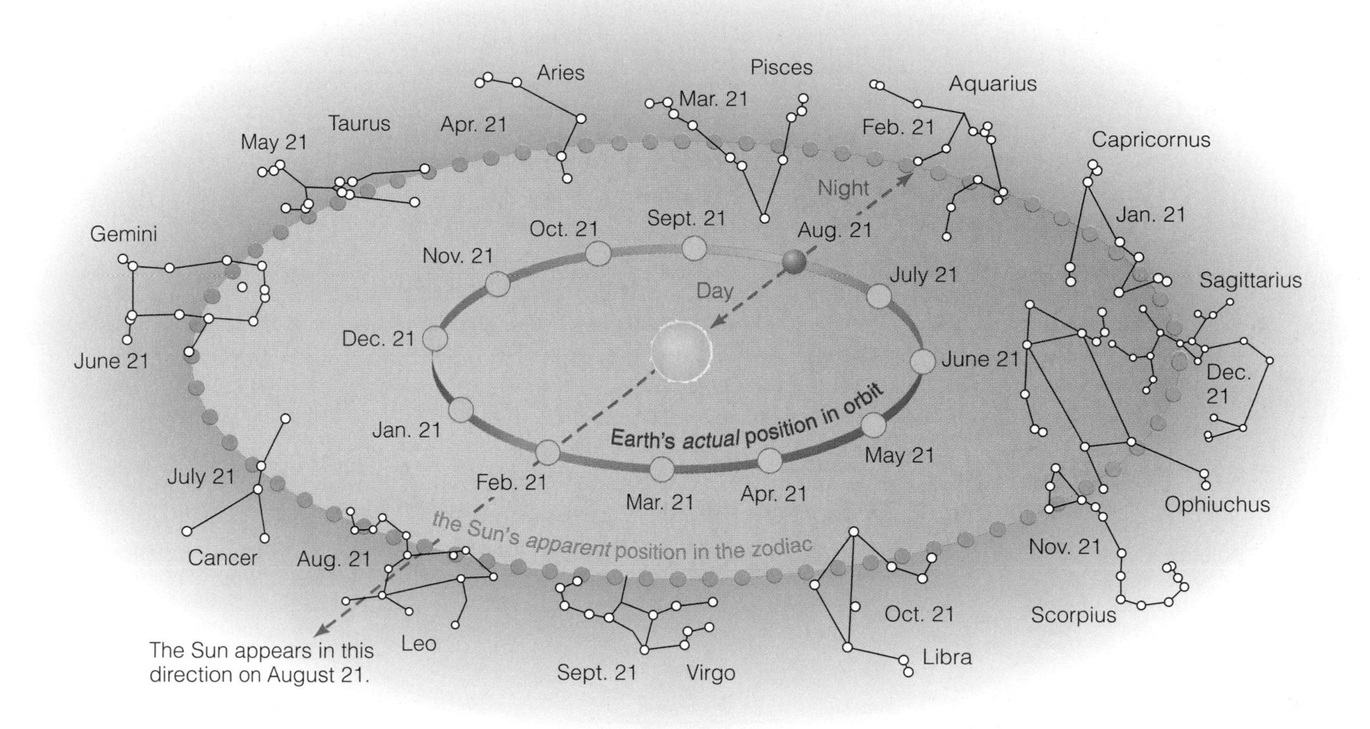

Figure 2.12 Interactive Figure

This diagram shows why the Sun appears to move steadily eastward along the ecliptic, through the constellations of the zodiac. As Earth orbits the Sun, we see the Sun against the background of different zodiac constellations at different times of year. For example, on August 21 the Sun appears to be in Leo.

Seasons Tutorial, Lessons 1–3

2.2 THE REASON FOR SEASONS

We have seen how Earth's rotation makes the sky appear to circle us daily and how the night sky changes as Earth orbits the Sun each year. The combination of Earth's rotation and orbit also leads to the progression of the seasons. In this section, we'll explore the reason for seasons.

• What causes the seasons?

You know that we have seasonal changes, such as longer and warmer days in summer and shorter and cooler days in winter. But why do the seasons occur? The answer is that the tilt of Earth's axis causes sunlight to fall differently on Earth at different times of year. Figure 2.13 shows the idea.

Earth's axis points in the same direction in space all year round, which means its orientation *relative to the Sun* changes as Earth orbits the Sun.

Notice that Earth's axis remains pointed in the same direction in space (toward Polaris) throughout the year [Section 1.3]. However, because Earth orbits the Sun, the orientation of the axis *relative to the Sun* changes over the course of a year. For example, look at the left side of Figure 2.13. The Northern Hemisphere is tipped toward the Sun, making it summer there, while the Southern Hemisphere is tipped away from the Sun, making it winter there. Half an orbit later (the right side of Figure 2.13), the situation is reversed. The Northern Hemisphere is in winter because it is tipped away from the Sun, while the Southern Hemisphere is in summer because it

COMMON MISCONCEPTIONS

THE CAUSE OF SEASONS

When asked what causes the seasons, many people mistakenly answer that the seasons are caused by variations in Earth's distance from the Sun. By realizing that the Northern and Southern Hemispheres experience opposite seasons, you'll see that Earth's varying distance from the Sun *cannot* be the cause of the seasons. If it were, both hemispheres would have summer at the same time. The real cause of the seasons is Earth's axis tilt, which causes the Northern and Southern Hemispheres to alternately receive more or less direct sunlight.

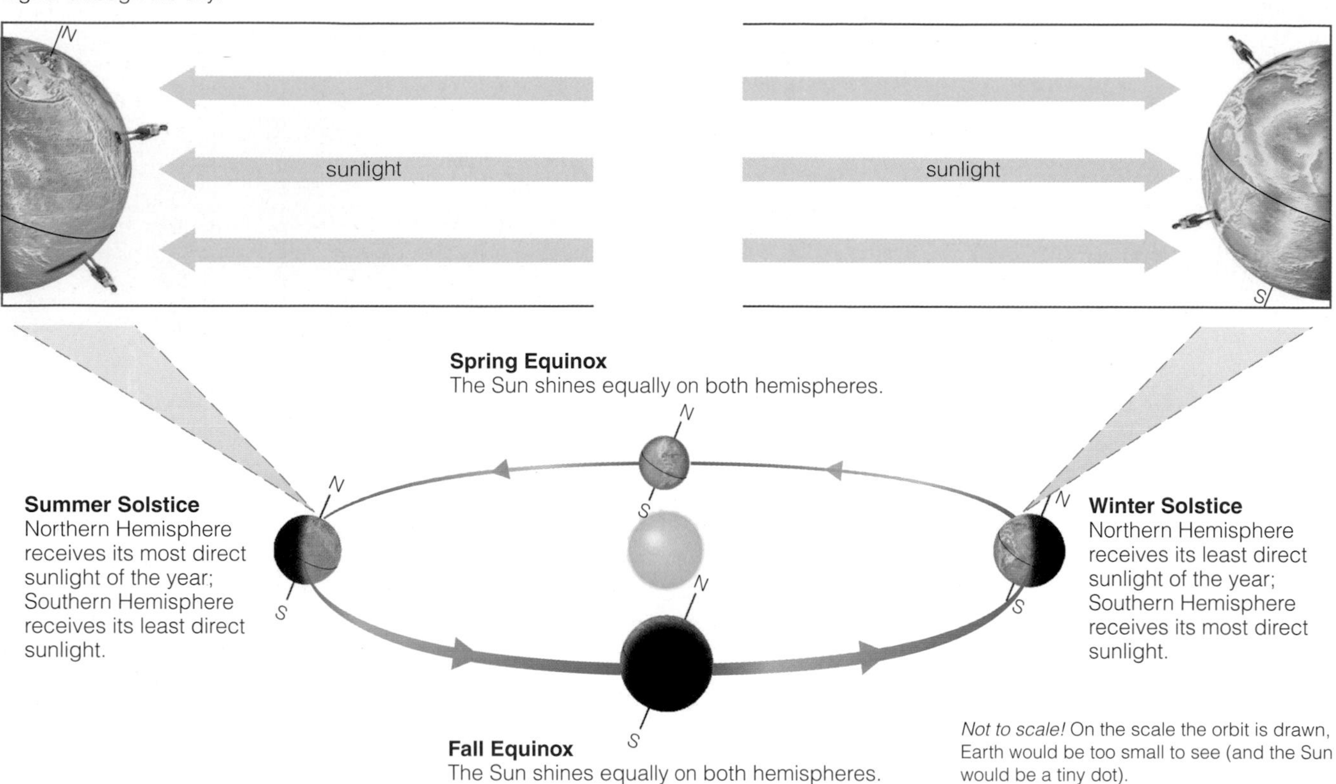

Figure 2.13 Interactive Figure

Earth's seasons are caused by the tilt of its axis. Notice that the axis points in the same direction (toward Polaris) throughout the year, which means the Northern Hemisphere is tipped toward the Sun on one side of the orbit and away from the Sun on the other side of the orbit. The same is true for the Southern Hemisphere, but on opposite sides of the orbit.

is tipped toward the Sun. Thus, the two hemispheres experience opposite seasons.

We can also see why the days are longer and warmer in summer and shorter and cooler in winter. Sunlight strikes the summer hemisphere at a steeper angle than it strikes the winter hemisphere. The steeper angle means sunlight is more concentrated, which is why summer tends to be warmer than winter. The steeper angle also means the Sun follows a longer and higher path through the summer sky, which is why the days are long and midday shadows are short. You can see the annual change in how high the Sun rises in your sky by observing the Sun's position at the same time each day (Figure 2.14).

Summer occurs in your hemisphere when sunlight hits it more directly, and winter occurs when the sunlight is less direct.

Notice that the seasons on Earth are caused only by the axis tilt and *not* by any change in Earth's distance from the Sun. Although Earth's orbital distance varies over the course of each year, the variation is fairly small: Earth is only about 3% farther from the Sun at its farthest point than at its nearest. The difference in the strength of sunlight due to this small change in distance is easily overwhelmed by the effects caused by the $23\frac{1}{2}°$ axis tilt. (*Note:* Some people also misinterpret diagrams like Figure 2.13, getting the mistaken impression that orbital position can make one hemisphere closer to the Sun. The mistake happens because figures like this are not drawn to scale. If the figure were drawn to scale, Earth would be microscopic, showing that there can be no practical difference in the distances of the two hemispheres from the Sun.)

THINK ABOUT IT Jupiter has an axis tilt of about 3°, small enough to be insignificant. Saturn has an axis tilt of about 27°, or slightly greater than that of Earth. Both planets have nearly circular orbits around the Sun. Do you expect Jupiter to have seasons? Do you expect Saturn to have seasons? Explain.

• How do we mark the progression of the seasons?

Today, we use a calendar to mark the progression of the seasons. In ancient times, however, people tracked the seasons by observing the Sun's changing position in our sky. Let's explore how the Sun's changing path allows us to mark the changing of the seasons.

Solstices and Equinoxes To help us mark the changing seasons, we define four special moments in the year, each of which corresponds to one of the four special positions in Earth's orbit shown in Figure 2.13.

We mark the progression of the seasons with four special moments during the year: the spring and fall equinoxes and the summer and winter solstices.

The **summer solstice**, which occurs around June 21 each year, is the moment when the Northern Hemisphere receives its most direct sunlight (and the Southern Hemisphere gets its least direct sunlight). The **winter solstice**, which occurs around December 21, is the moment when the Northern Hemisphere receives its least direct sunlight. The **spring equinox** (or *vernal equinox*) occurs around March 21, at the moment when the Northern Hemisphere goes from being tipped slightly away from the Sun to being tipped slightly toward the Sun. The **fall equinox** (or *autumnal equinox*) occurs around September 22 and marks the opposite change, when the Northern Hemisphere first starts to be tipped away from the Sun. Note that the two equinoxes are the times when the Sun shines equally on both hemispheres.

The exact dates and times of the solstices and equinoxes vary from year to year but stay within a couple of days of the dates given here. In fact, our modern calendar is designed so that the solstices and equinoxes occur around the same dates each year. The pattern of leap years was set so that the average length of the calendar year is the same as the actual time between one spring equinox and the next. We generally add a day to make leap year every fourth year, because the actual length of a year is about $365\frac{1}{4}$ days. However, because the precise length of the year is about 11 minutes short of $365\frac{1}{4}$ days, we must occasionally skip a leap year in order to keep the calendar synchronized with the seasons over many centuries. Leap year is skipped when a century changes (for example, in years 1700, 1800, 1900) *unless* the century year is divisible by 400. Thus, 2000 was a leap year because it is divisible by 400 ($2{,}000 \div 400 = 5$), but 2100 will *not* be a leap year.

Identifying the Solstices and Equinoxes Ancient people recognized the days on which the solstices and equinoxes occur by observing the Sun in the sky. Many ancient structures were used for this purpose, including Stonehenge in England and the Sun Dagger in New Mexico [Section 3.1].

The Sun rises precisely due east and sets precisely due west *only* on the days of the spring and fall equinoxes.

The equinoxes occur on the only two days of the year that the Sun rises precisely due east and sets precisely due west (Figure 2.15). These are also the only two days when the Sun is above and below the horizon for precisely 12 hours each. The summer solstice occurs on the

Figure 2.14

This composite photograph shows images of the Sun at 8- to 10-day intervals over the course of a year, always from the same place and at the same time of day (technically, at the same "mean solar time"). This photo looks east, so north is to the left and south is to the right. The three bright streaks show the path of the Sun's rise on three particular dates. Notice the dramatic change in the Sun's altitude and sunrise path over the course of the year. (The "figure 8" shape (called an *analemma*) occurs because of a combination of Earth's axis tilt and Earth's varying speed as it orbits the Sun.)

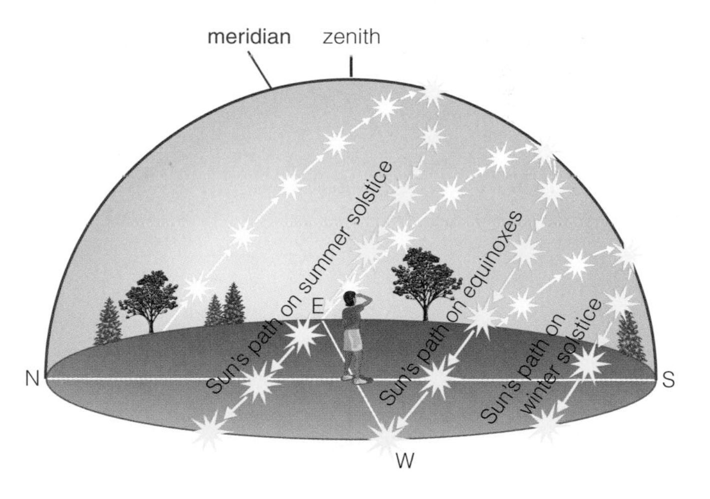

Figure 2.15

This diagram shows the Sun's path on the solstices and equinoxes for the Northern Hemisphere sky (latitude 40°N). Notice that the Sun rises exactly due east and sets exactly due west only on the equinoxes. The summer solstice occurs on the day that the Sun rises and sets farthest to the north and reaches its highest point in the sky. The winter solstice occurs on the day that the Sun rises and sets farthest to the south and traces its lowest path through the sky.

day that the Sun follows its longest and highest path through the Northern Hemisphere sky (and its shortest and lowest path through the Southern Hemisphere sky). It is therefore the day that the Sun rises and sets farther to the north than on any other day of the year, and on which the noon Sun reaches its highest point in the Northern Hemisphere sky. The opposite is true on the day of the winter solstice, when the Sun rises and sets farthest to the south and the noon Sun is lower in the Northern Hemisphere sky than on any other day of the year.

First Days of Seasons We usually say that each equinox and solstice marks the first day of a season. For example, the day of the summer solstice is usually said to be the "first day of summer." Notice, however, that the summer solstice occurs when the Northern Hemisphere has its *maximum* tilt toward the Sun. Thus, you might wonder why we consider the summer solstice to be the beginning rather than the midpoint of summer.

In part, the answer is that the choice of the summer solstice as the "first" day of summer is somewhat arbitrary. However, the choice makes sense in at least two ways. First, it was much easier for ancient people to identify the days on which the Sun reached extreme positions in the sky—such as reaching its highest point on the summer solstice—than other days in between. Second, we usually think of the seasons in terms of weather, and the solstices and equinoxes correspond quite well with the beginnings of seasonal weather patterns. For example, although it is around the summer solstice that the Sun's path through the Northern Hemisphere sky is longest and highest, it is *not* usually the warmest time of the year. Instead, the warmest days tend to come one to two months later. To understand why, think about what happens when you heat a pot of cold soup. Even though you may have the stove turned on high from the start, it takes a while for the soup to warm up. In the same way, it takes some time for sunlight to heat the ground and oceans from the cold of winter to the warmth of summer. Thus, "midsummer" in terms of weather comes in late July and early August, which makes the summer solstice a pretty good choice for the "first day of summer." For similar reasons, the winter solstice is a good choice for the first day of winter, and the spring and fall equinoxes are good choices for the first days of those seasons.

Seasons Around the World The names of the solstices and equinoxes reflect the northern seasons, which can make things sound strange when we talk about seasons in the Southern Hemisphere. For example, on the *summer* solstice it is *winter* in the Southern Hemisphere. This apparent injustice to people in the Southern Hemisphere arose because the solstices and equinoxes were named long ago by people living in the Northern Hemisphere. A similar injustice is inflicted on people living in equatorial regions. If you study Figure 2.13 carefully, you'll see that Earth's equator gets its most direct sunlight on the two equinoxes and its least direct sunlight on the solstices. Thus, people living near the equator don't experience four seasons in the same way as people living at mid-latitudes. Instead, equatorial regions generally have rainy and dry seasons, with the rainy seasons coming when the Sun is higher in the sky.

At very high latitudes, the Sun becomes circumpolar in summer, remaining above the horizon all day long.

In addition, seasonal variations around the times of the solstices are more extreme at high latitudes. For example, Alaska has much longer summer days and much longer winter nights than Florida. In fact,

COMMON MISCONCEPTIONS

HIGH NOON

When is the Sun directly overhead in your sky? Many people answer "at noon." It's true that the Sun reaches its *highest* point each day when it crosses the meridian, giving us the term "high noon" (though the meridian crossing is rarely at precisely 12:00). However, unless you live in the Tropics (between latitudes $23\frac{1}{2}$°S and $23\frac{1}{2}$°N), the Sun is *never* directly overhead. In fact, any time you can see the Sun as you walk around, you can be sure it is *not* at your zenith. Unless you are lying down, seeing an object at the zenith requires tilting your head back into a very uncomfortable position.

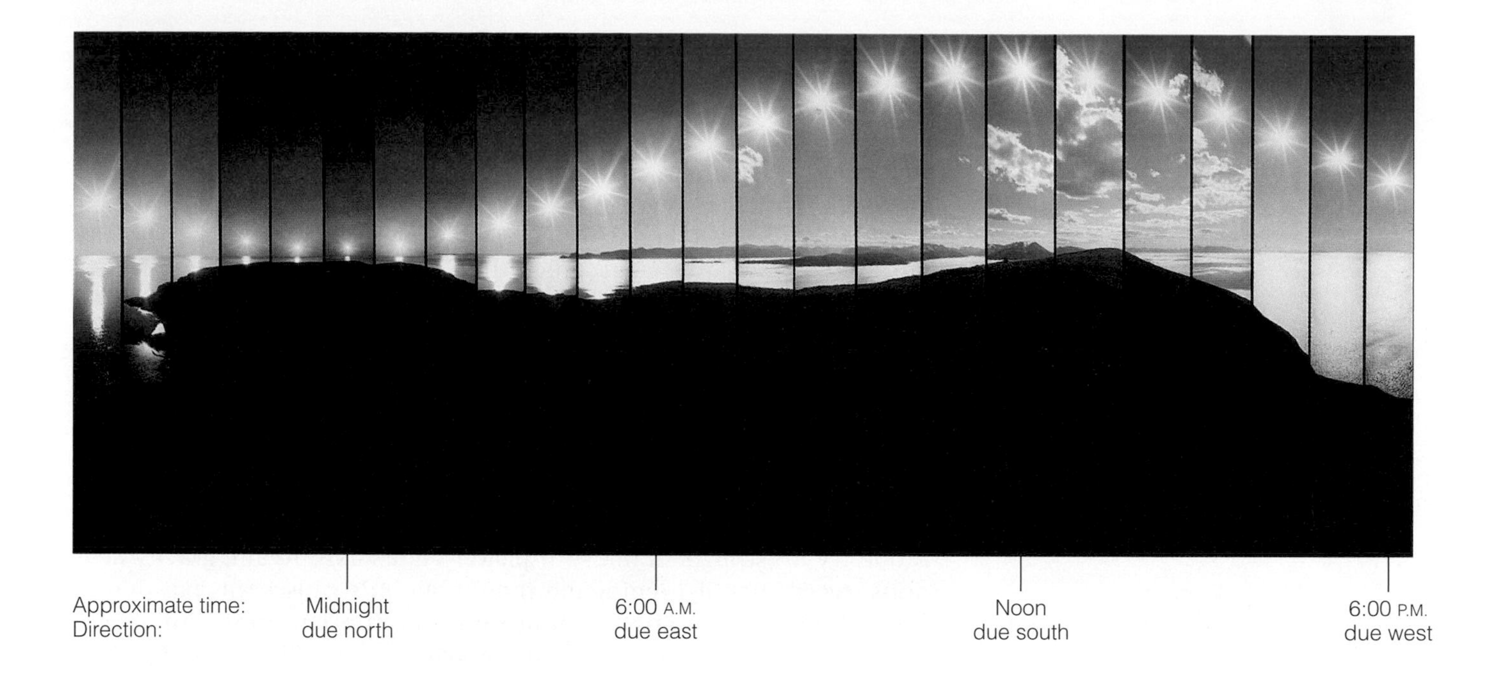

Figure 2.16

This sequence of photos shows the progression of the Sun all the way around the horizon on the summer solstice at the Arctic Circle. Notice that the Sun does not set but instead skims the northern horizon at midnight. It then gradually rises higher, reaching its highest point at noon, when it appears due south.

the Sun becomes circumpolar at very high latitudes (within the *Arctic* and *Antarctic Circles*) in the summer. The Sun never sets during these summer days in what we call the *land of the midnight Sun* (Figure 2.16). Of course, the name "land of noon darkness" would be more appropriate in the winter, when the Sun never rises above the horizon at these high latitudes.

• Does the orientation of Earth's axis change with time?

As we've seen, the seasons are caused by the $23\frac{1}{2}°$ tilt of Earth's axis. So you may be wondering if this tilt ever changes. In fact, the *amount* of the tilt hardly changes at all—to our knowledge, Earth's axis tilt has remained close to $23\frac{1}{2}°$ since shortly after our planet was born, more than 4 billion years ago. However, the *orientation* of the axis tilt gradually changes.

The change in the way Earth's axis points in space arises from something called **precession**. Precession occurs with many rotating objects. You can see it easily by spinning a top (Figure 2.17a). As the top spins rapidly, you'll notice that its axis also sweeps out a circle at a somewhat slower rate. We say that the top's axis *precesses*. Earth's axis precesses in much the same way, but far more slowly (Figure 2.17b).

The tilt of Earth's axis remains close to $23\frac{1}{2}°$, but the direction the axis points in space changes slowly with the 26,000-year cycle of precession.

Each cycle of Earth's precession takes about 26,000 years. Today, the axis points toward Polaris, making it our North Star. However, because of precession, Polaris has been a good North Star for only a few centuries, and a few centuries from now it will no longer be a good marker for due north. Some 13,000 years from now, the axis will point toward the star Vega (within a few degrees), making Vega the North Star at that time. In 26,000 years, the axis will again point the way it does now. Note that the axis does not point near any bright star during most of each cycle.

Because the amount of the axis tilt remains about the same at all times, precession does not affect the pattern of the seasons. However, be-

SUN SIGNS

You probably know your astrological "sign." When astrology began a few thousand years ago, your sign was supposed to represent the constellation in which the Sun appeared on your birth date. However, this is no longer the case for most people. For example, if your birthday is the spring equinox, March 21, a newspaper horoscope will show that your sign is Aries, but the Sun appears in Pisces on that date. In fact, because of precession, your astrological sign generally corresponds to the constellation in which the Sun *would have appeared* on your birth date if you had lived about 2,000 years ago. The astrological signs are based on the positions of the Sun among the stars as described by the Greek scientist Ptolemy, who lived and worked in the second century A.D.

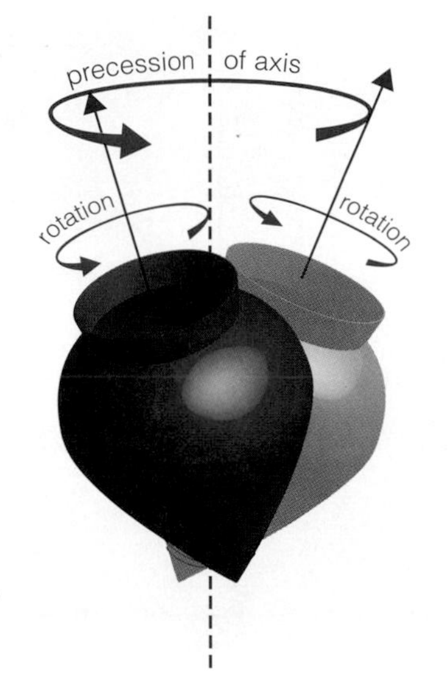

a A spinning top wobbles, or precesses, more slowly than it spins.

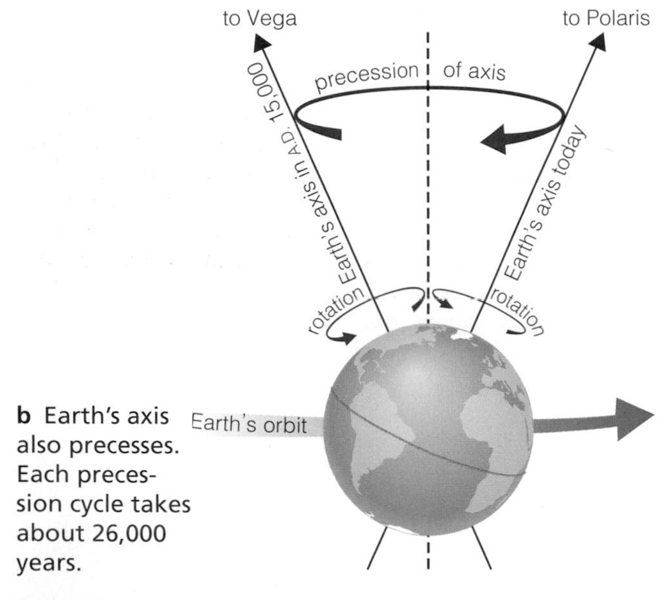

b Earth's axis also precesses. Each precession cycle takes about 26,000 years.

Figure 2.17 Interactive Figure

Precession affects the orientation of a spinning object's axis, but not the amount of its tilt.

cause the solstices and equinoxes correspond to points in Earth's orbit that depend on the direction the axis points in space, their positions in the orbit gradually shift with the cycle of precession. As a result, the constellations associated with the solstices and equinoxes gradually change. For example, a couple of thousand years ago the Sun appeared in the constellation Cancer on the day of the summer solstice, but it now appears in Gemini on that day. That explains something you can see on any world map: The latitude at which the Sun is directly overhead on the summer solstice ($23\frac{1}{2}°$ N) is called the Tropic of Cancer, telling us that it got its name back when the Sun used to appear in Cancer on the summer solstice.

Precession arises from the effect of gravity on a tilted, spinning object. A spinning top precesses because Earth's gravity tries to pull over its tilted spin axis. Gravity does not succeed in pulling it over (at least until friction slows the rate of spin) but instead causes the axis to precess. The spinning Earth precesses because gravitational tugs from the Sun and Moon try to "straighten out" our planet's tilted axis. Again, gravity does not succeed in straightening out the tilt but only causes the axis to precess. It's important to note that precession can occur only with objects that are *not* perfect spheres, because gravity needs some type of bulge to tug on in order to change the spin axis. The top is clearly not spherical. Earth is close to spherical but is slightly fatter around the equator. Thus, the gravitational tugs from the Sun and Moon actually pull on Earth's excess equatorial mass.

THINK ABOUT IT You can observe and experiment with precession using an inexpensive toy *gyroscope*. Find a gyroscope and see how it works. How is its motion similar to that of the precessing Earth? How is it different? (*Hint:* Compare the time scales for the cycles of precession.)

astronomyplace.com **Phases of the Moon Tutorial, Lessons 1–3**

2.3 THE MOON, OUR CONSTANT COMPANION

Now that we have discussed the basic patterns of motion in our sky and the pattern of the seasons, we turn our attention to the Moon. Aside from the Sun, the Moon is the brightest and most noticeable object in our sky.

The Moon is our constant companion in space, orbiting Earth about once every $27\frac{1}{3}$ days. Its average distance from Earth is about 380,000 kilometers (240,000 miles), but the precise distance varies because the orbit is not quite a perfect circle. Figure 2.18 shows the Moon's orbit on the same scale we used for the model solar system in Section 1.2. Remember that on this scale, the Sun is about the size of a large grapefruit and is located about 15 meters from Earth. Thus, the entire orbit of the Moon would fit easily inside the Sun, and for practical purposes we can consider the Earth and Moon to share the same orbit around the Sun.

Like all objects in space, the Moon appears to reside on the celestial sphere. Earth's daily rotation makes the Moon appear to rise in the east and set in the west each day. In addition, because of its orbit around Earth, the Moon appears to move eastward from night to night through the constellations of the zodiac. Each circuit through the constellations takes the same $27\frac{1}{3}$ days that the Moon takes to orbit Earth. If you do the math, you'll see that this means the Moon moves relative to the stars by about $\frac{1}{2}°$—its own angular size—each hour. You can notice this gradual motion in just a few hours, by checking the Moon's position in comparison to bright stars near it in the sky.

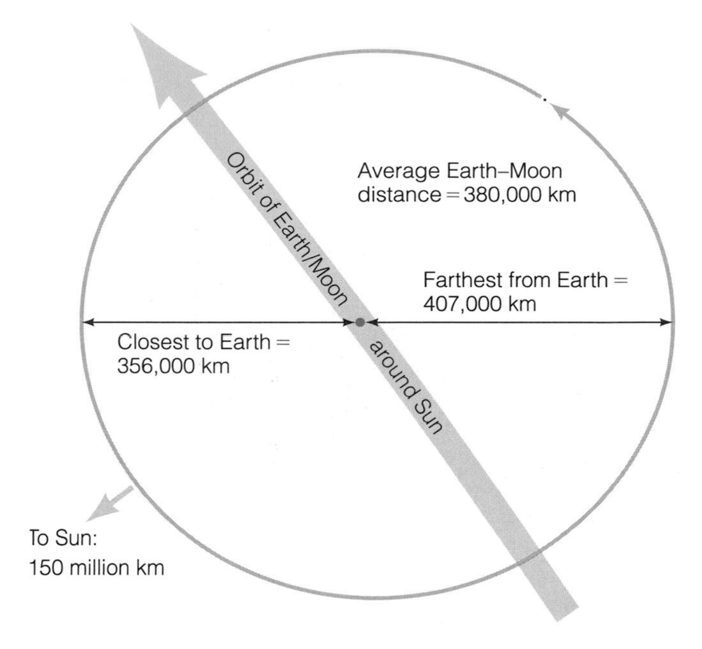

Figure 2.18

The Moon's orbit around Earth, shown on the same 1-to-10-billion scale discussed in Section 1.2 (see Figure 1.6). The sizes of Earth and the Moon are also shown on the same scale. The segment shown of our orbit around the Sun looks nearly straight because the distance to the Sun is so great in comparison with the size of the Moon's orbit.

• Why do we see phases of the Moon?

As the Moon moves through the sky, both its appearance and the time at which it rises and sets change with the cycle of **lunar phases**. The phase of the Moon on any given day depends on its position relative to the Sun as it orbits Earth.

The phase of the Moon depends on its position relative to the Sun as it orbits Earth.

The easiest way to understand the lunar phases is with the simple demonstration illustrated in Figure 2.19. Take a ball outside on a sunny day. (If it's dark or cloudy, you can use a flashlight instead of the Sun; put the flashlight on a table a few meters away and turn it on to shine on you.) Hold the ball at arm's length to represent the Moon while your head represents Earth. Slowly spin around (counterclockwise), so that the ball goes around you just like the Moon orbits Earth. As you turn, you'll see the ball go through phases just like the Moon. If you think about what's happening, you'll realize that the phases of the ball result from just two basic facts:

1. Half the ball always faces the Sun (or flashlight) and therefore is bright, while the other half faces away from the Sun and therefore is dark.
2. As you look at the ball at different positions in its "orbit" around your head, you see different combinations of its bright and dark faces.

For example, when you hold the ball directly opposite the Sun, you see only the bright portion of the ball, so this represents the "full" phase. When you hold the ball at its "first quarter" position, half the face you see is dark and the other half is bright.

We see lunar phases for the same reason. Half the Moon is always illuminated by the Sun, but the amount of this illuminated half that we see from Earth depends on the Moon's position in its orbit. The photographs in Figure 2.19 show how the phases look. Each complete cycle of phases, from one new moon to the next, takes about $29\frac{1}{2}$ days—hence the origin of the word *month* (think "moonth"). This is slightly longer than the Moon's actual orbital period because of Earth's motion around the Sun during the time the Moon is orbiting around Earth.

The Moon's phase affects not only its appearance, but also its rise and set times.

Notice that the different phases rise, reach their highest points, and set at particular times. For example, because full moon occurs when the Moon is opposite the Sun in the sky, the full moon must rise around sunset, reach its highest point in the sky at midnight, and set around sunrise. Similarly, because first-quarter moon occurs when the Moon is about 90° east of the Sun in our sky, it must rise around noon, reach its highest point around 6pm, and set around midnight. Figure 2.19 lists the approximate rise, highest point, and set times for each phase. (The exact times vary with latitude, time of year, and other factors.)

MOON IN THE DAYTIME AND STARS ON THE MOON

In traditions and stories, night is so closely associated with the Moon that many people mistakenly believe that the Moon is visible only in the nighttime sky. In fact, the Moon is above the horizon as often in the daytime as at night, though it is easily visible only when its light is not drowned out by sunlight. For example, a first-quarter moon is easy to spot in the late afternoon as it rises through the eastern sky, and a third-quarter moon is visible in the morning as it heads toward the western horizon.

Another misconception appears in illustrations that show a star in the dark portion of the crescent moon. A star in the dark portion appears to be in front of the Moon, which is impossible because the Moon is much closer to us than is any star.

Figure 2.19 Interactive Figure

A simple demonstration explains the phases of the Moon. The woman's head represents Earth and the ball represents the Moon. If she turns slowly in a circle, so that the ball "orbits" her just like the Moon orbits Earth, she'll see the ball go through phases just like the Moon. Notice that the half of the ball (or Moon) facing the Sun is always illuminated while the half facing away is always dark, but the woman sees different combinations of the illuminated and dark portions depending on the ball's (or Moon's) location in its orbit. (The new moon photo shows blue sky, because a new moon is always close to the Sun in the sky and hence hidden from view by the bright light of the Sun.)

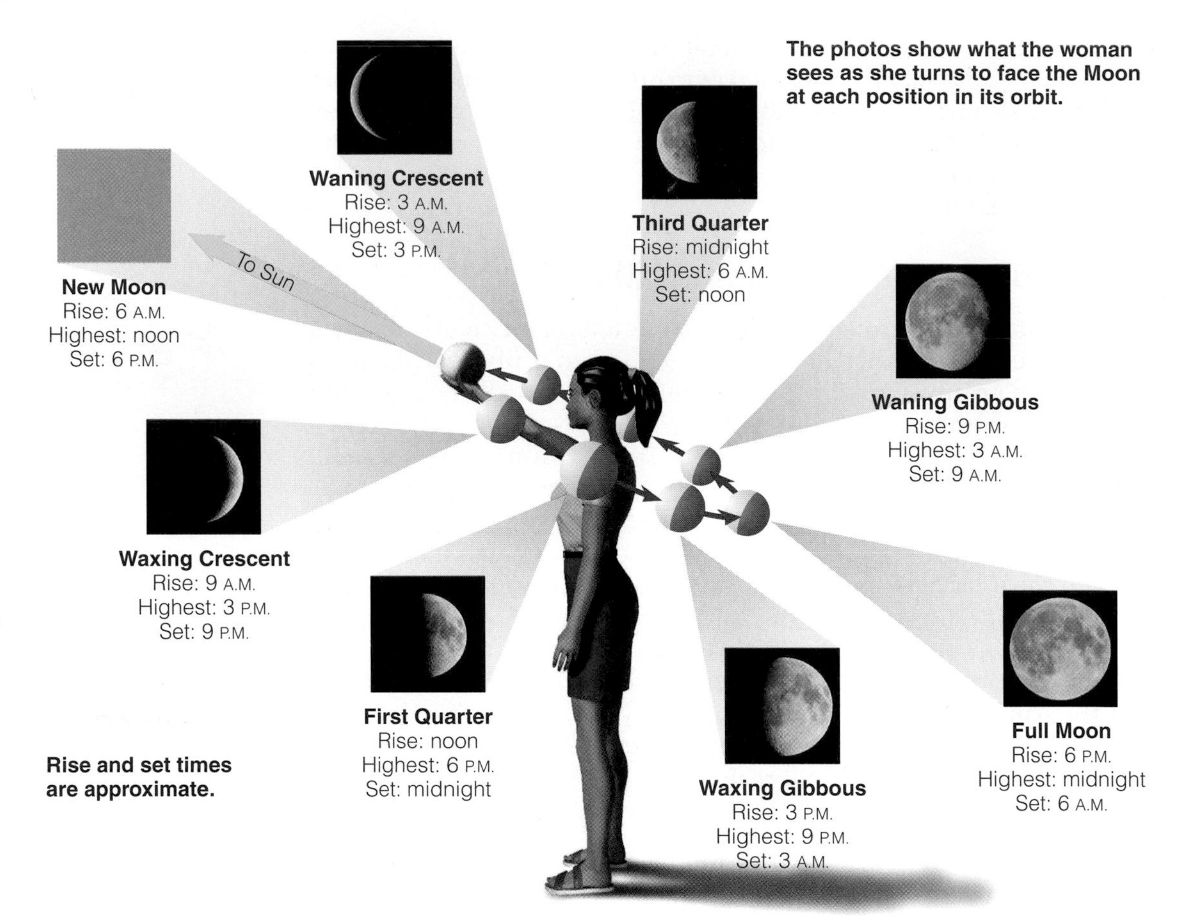

THINK ABOUT IT Suppose you go outside in the morning and notice that the visible face of the Moon is half light and half dark. Is this a first-quarter or third-quarter moon? How do you know?

Eclipses Tutorial, Lessons 1–3

• What causes eclipses?

The phases of the Moon are the most obvious effect of the Moon's orbit around Earth. Occasionally, however, we see something much more dramatic. The Moon and Earth cast shadows in sunlight, and these shadows can create **eclipses** when the Sun, Earth, and Moon fall into a straight line. Eclipses come in two basic types:

- A **lunar eclipse** occurs when Earth lies directly between the Sun and Moon, so that Earth's shadow falls on the Moon.
- A **solar eclipse** occurs when the Moon lies directly between the Sun and Earth, so that the Moon's shadow falls on the Earth. People living within the area covered by the Moon's shadow will see the Sun blocked or partially blocked from view.

To understand when and why eclipses occur, we must investigate the Moon's orbit of Earth a little more deeply.

THE "DARK SIDE" OF THE MOON

Although we see many *phases* of the Moon, we do not see many *faces.* In fact, from Earth we always see (nearly) the same face of the Moon (for reasons we'll discuss in Chapter 4). This leads to a common point of confusion. The term *dark side of the Moon* really should be used to mean the night side—that is, the side facing away from the Sun. Unfortunately, *dark side* traditionally meant what would better be called the *far side*—the face that never can be seen from Earth. Many people still refer to the far side as the "dark side," even though this side is not necessarily dark. For example, during new moon the far side faces the Sun and hence is completely sunlit. The only time the far side is completely dark is at full moon, when it faces away from both the Sun and Earth.

Conditions for Eclipses Look once more at Figure 2.19. The figure makes it look like the Sun, Earth, and Moon line up with every new and full moon. Thus, if this figure told the whole story of the Moon's orbit, we would have both a lunar and a solar eclipse every month. But we don't.

We see a lunar eclipse when Earth's shadow falls on the Moon, and a solar eclipse when the Moon blocks our view of the Sun.

The missing piece of the story in Figure 2.19 is that the Moon's orbit is slightly inclined (by about 5°) to the ecliptic plane (the plane of Earth's orbit around the Sun). To visualize this inclination, imagine the ecliptic plane as the surface of a pond, as shown in Figure 2.20. Because of the inclination of its orbit, the Moon spends most of its time either above or below this surface. It crosses *through* this surface only twice during each orbit: once coming out and once going back in. The two points in each orbit at which the Moon crosses the surface are called the **nodes** of the Moon's orbit.

As you study Figure 2.20, notice that the nodes are aligned the same way (diagonally on the page) throughout the year. As a result, the nodes lie in a straight line with the Sun and Earth only about twice each year. (For reasons we'll discuss shortly, it is not *exactly* twice each year.) Thus, we find the following two conditions that must be met for an eclipse to occur:

1. The phase of the Moon must be full (for a lunar eclipse) or new (for a solar eclipse), since those are the only phases at which the Sun, Earth, and Moon can lie in a straight line.

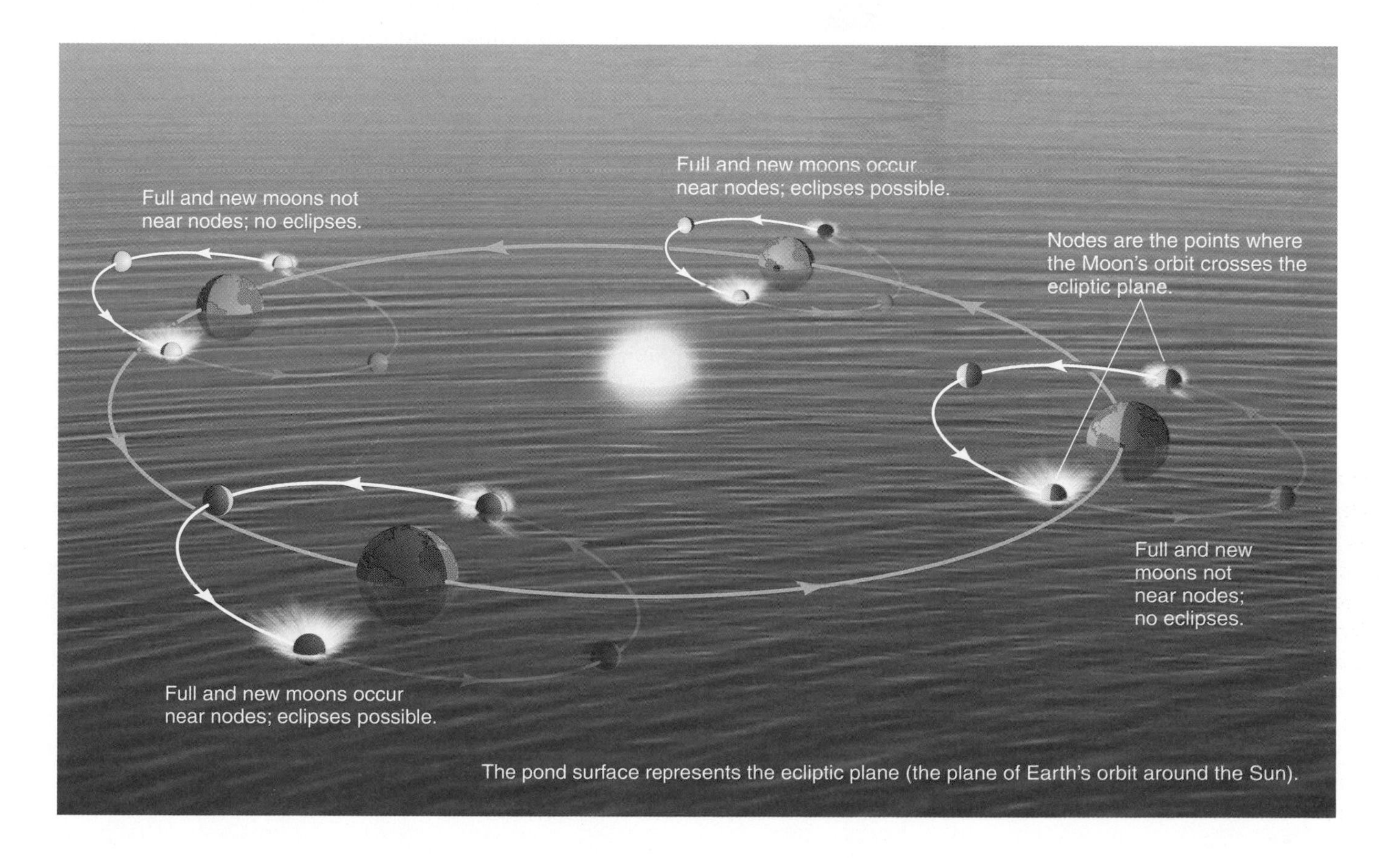

Figure 2.20

This illustration represents the ecliptic plane as the surface of a pond. The Moon's orbit is slightly tilted (by about 5°) to the ecliptic plane, so the Moon spends half of each orbit above the plane (the pond surface) and half below it. The points at which the orbit crosses the ecliptic plane are the *nodes* of the Moon's orbit. Eclipses occur only when the Moon both is at a node (passing through the pond surface) *and* has a phase of either new moon (for a solar eclipse) or full moon (for a lunar eclipse)—as is the case with the lower left and top right orbits shown.

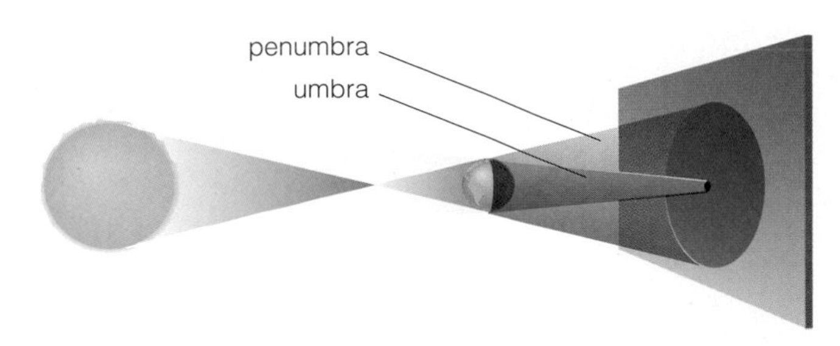

Figure 2.21

The shadow cast by an object in sunlight. Sunlight is fully blocked in the umbra and partially blocked in the penumbra.

2. The new or full moon must occur during one of the periods when the nodes of the Moon's orbit are aligned with the Sun and Earth, since those are the only times when the new and full moons lie in the ecliptic plane.

We see an eclipse only when a full or new moon occurs at one of the points where the Moon's orbit crosses the ecliptic plane.

Although there are two basic types of eclipse—lunar and solar—each of these types can look different depending on precisely how the shadows fall. The shadow of the Moon or Earth consists of two distinct regions: a central **umbra**, where sunlight is completely blocked, and a surrounding **penumbra**, where sunlight is only partially blocked (Figure 2.21). Thus, an umbral shadow is totally dark, while a penumbral shadow is only slightly darker than no shadow. Let's see how this affects eclipses.

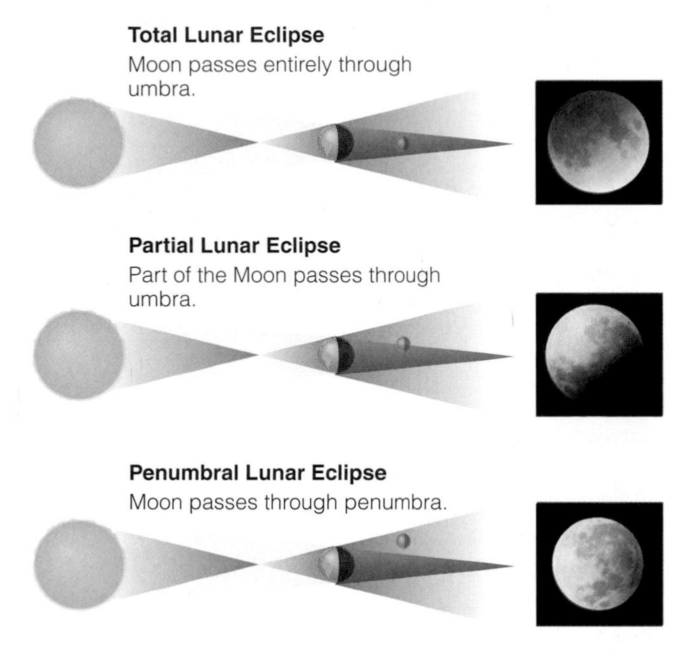

Figure 2.22 Interactive Figure

The three types of lunar eclipse.

Lunar Eclipses A lunar eclipse begins at the moment when the Moon's orbit first carries it into Earth's penumbra. After that, we will see one of three types of lunar eclipse (Figure 2.22). If the Sun, Earth, and Moon are nearly perfectly aligned, the Moon will pass through Earth's umbra, and we will see a **total lunar eclipse**. If the alignment is somewhat less perfect, only part of the full moon will pass through the umbra (with the rest in the penumbra), and we will see a **partial lunar eclipse**. If the Moon passes *only* through Earth's penumbra, we will see a **penumbral lunar eclipse**.

Penumbral eclipses are the most common type of lunar eclipse, but they are not visually impressive because the full moon darkens only slightly. Earth's umbral shadow clearly darkens part of the Moon's face during a partial lunar eclipse, and the curvature of this shadow demonstrates that Earth is round. A total lunar eclipse is particularly spectacular because the Moon becomes dark and eerily red during **totality**—the time during which the Moon is entirely engulfed in the umbra. Totality typically lasts about an hour. The Moon becomes dark because it is in shadow, and red because Earth's atmosphere bends some of the red light from the Sun toward the Moon.

Solar Eclipses We can also see three types of solar eclipse (Figure 2.23). If a solar eclipse occurs when the Moon is relatively close to Earth in its orbit, the Moon's umbra touches a small area of Earth's surface (no more than about 270 kilometers in diameter). Anyone within this area will see a **total solar eclipse**.

A total solar eclipse is visible only within the small area where the Moon's umbral shadow touches Earth's surface.

Surrounding the region of totality is a much larger area (typically about 7,000 kilometers in diameter) that falls within the Moon's penumbral shadow. Anyone within this region will see a **partial solar eclipse**, in which only part of the Sun is blocked from view. If the eclipse occurs when the Moon is relatively far from Earth, the umbra may not reach Earth's surface at all. In that case, anyone in the small region of Earth directly behind the umbra will see an **annular eclipse**, in which a ring of sunlight surrounds the disk of the Moon. (Again, anyone in the surrounding penumbral shadow will see a partial solar eclipse.)

The combination of Earth's rotation and the orbital motion of the Moon causes the Moon's umbral and penumbral shadows to race across the face of Earth at a typical speed of about 1,700 kilometers per hour. As a result, the umbral shadow traces a narrow path across Earth, and totality never lasts more than a few minutes in any particular place.

A total solar eclipse is a spectacular sight. It begins when the disk of the Moon first appears to touch the Sun. Over the next couple of hours, the Moon appears to take a larger and larger "bite" out of the Sun. As totality approaches, the sky darkens and temperatures fall. Birds head back to their nests, and crickets begin their nighttime chirping. During the few minutes of totality, the Moon completely blocks the normally visible disk of the Sun, allowing the faint *corona* to be seen (Figure 2.24). The surrounding sky takes on a twilight glow, and planets and bright stars become visible in the daytime. As totality ends, the Sun slowly emerges from behind the Moon over the next couple of hours. However, because your eyes have adapted to the darkness, totality appears to end far more abruptly than it began.

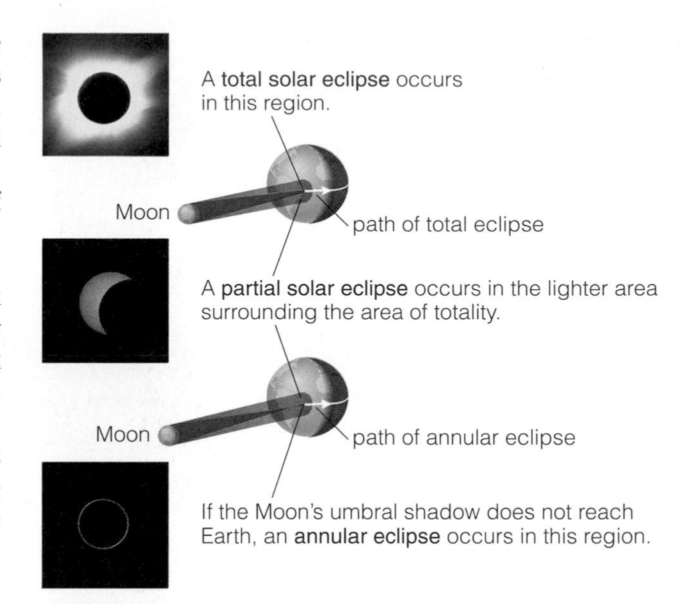

Figure 2.23 Interactive Figure

The three types of solar eclipse. The diagrams show the Moon's shadow falling on Earth; note the dark central umbra surrounded by the much lighter penumbra.

Predicting Eclipses

Few phenomena have so inspired and humbled humans throughout the ages as eclipses. For many cultures, eclipses were mystical events associated with fate or the gods, and countless stories and legends surround them. Much of the mystery of eclipses probably stems from the relative difficulty of predicting them.

Look again at Figure 2.20. The two periods each year when the nodes of the Moon's orbit are nearly aligned with the Sun are called **eclipse seasons**. Each eclipse season lasts a few weeks, so some type of lunar eclipse occurs during each eclipse season's full moon, and some type of solar eclipse occurs during its new moon.

If Figure 2.20 told the whole story, eclipse seasons would occur every 6 months, and predicting eclipses would be easy. For example, if eclipse seasons always occurred in January and July, eclipses would always occur on the dates of new and full moons in those months. But the figure does not show one important thing about the Moon's orbit: The nodes slowly move around the orbit. As a result, eclipse seasons occur slightly less than 6 months apart (about 173 days apart) and do not recur in the same months year after year.

The general pattern of eclipses repeats with the approximately 18-year saros cycle.

The combination of the changing dates of eclipse seasons and the $29\frac{1}{2}$-day cycle of lunar phases makes eclipses recur in a cycle of about 18 years $11\frac{1}{3}$ days. This cycle is called the **saros cycle**. Astronomers in many ancient cultures identified the saros cycle and thus could predict *when* eclipses would occur. However, the saros cycle does not account for all the complications involved in predicting eclipses. If a solar eclipse occurred today, the one that would occur 18 years $11\frac{1}{3}$ days from now would not be visible from the same places on Earth and might not be of the same type. For example, one might be total and the other only partial. No ancient culture achieved the ability to predict eclipses in every detail.

Today, we can predict eclipses because we know the precise details of the orbits of Earth and the Moon. Many astronomical software packages can do the necessary calculations. Table 2.1 lists upcoming total lunar eclipses, and Figure 2.25 shows paths of totality for upcoming total solar eclipses.

Figure 2.24

This multiple-exposure photograph shows the progression of a total solar eclipse. Totality (central image) lasts only a few minutes, during which time we can see the faint corona around the outline of the Sun. This photo was taken July 22, 1990, in La Paz, Mexico. The foreground church was photographed at a different time of day.

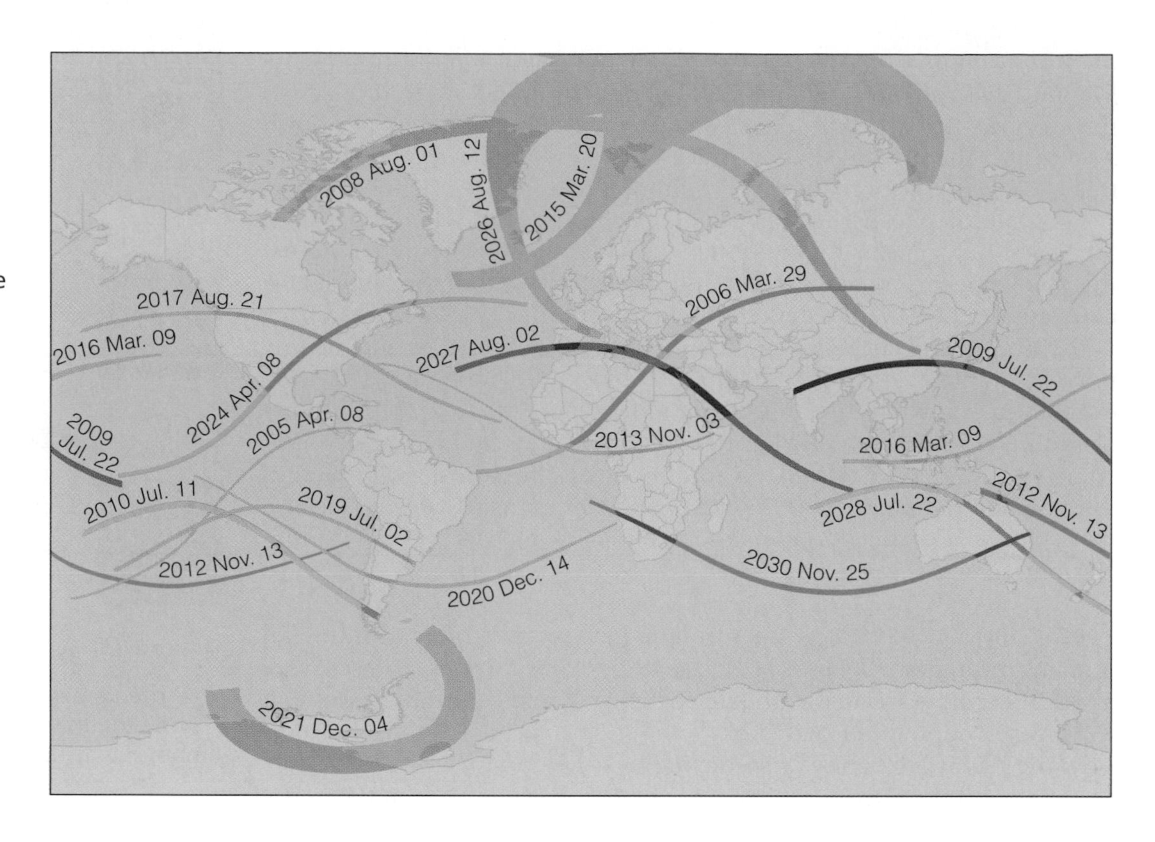

Figure 2.25

This map shows the paths of totality for solar eclipses from 2004 to 2030. Paths of the same color represent eclipses occurring in successive saros cycles, separated by 18 years 11 days. For example, the 2024 eclipse occurs 18 years 11 days after the 2006 eclipse (both shown in green).

TABLE 2.1
Total Lunar Eclipses through 2010

Date
October 28, 2004
March 3, 2007
August 28, 2007
February 21, 2008
December 21, 2010

2.4 THE ANCIENT MYSTERY OF THE PLANETS

We've now covered the appearance and motion of the stars, Sun, and Moon in the sky. That leaves us with the planets yet to discuss. As we'll soon see, planetary motion posed an ancient mystery that played a critical role in the development of modern civilization.

Five planets are easy to find with the naked eye: Mercury, Venus, Mars, Jupiter, and Saturn. Mercury is visible only infrequently, and then only just after sunset or just before sunrise because it is so close to the Sun. Venus often shines brightly in the early evening in the west or before dawn in the east. If you see a very bright "star" in the early evening or early morning it is probably Venus. Jupiter, when it is visible at night, is the brightest object in the sky besides the Moon and Venus. Mars is often recognizable by its red color, though you should check a star chart to make sure you aren't looking at a bright red star. Saturn is also easy to see with the naked eye, but because many stars are just as bright as Saturn, it helps to know where to look. (It also helps to know that planets tend not to twinkle as much as stars.) Sometimes several planets may appear close together in the sky, offering a particularly beautiful sight (Figure 2.26).

• What was once so mysterious about the movement of planets in our sky?

Ancient people carefully observed the movements of the planets among the stars, and you can observe the same motions for yourself. On any particular night, the planets move through the sky just like stars, rising in the east and setting in the west. However, if you observe planetary positions among the stars over a period of weeks or months, you'll notice that the planets wander slowly through the constellations of the zodiac. (The word *planet* comes from the Greek for "wandering star.") By itself, this wandering might not be too surprising. After all, the Sun and Moon also move among the constellations. But the planets move in what seems to be a very strange way.

Unlike the Sun and Moon, which move steadily eastward relative to the stars, the planets vary substantially in both speed and brightness as they move among the stars. Moreover, while the planets *usually* move eastward relative to the stars, they occasionally reverse course completely, moving westward rather than eastward through the zodiac. A period during which a planet appears to move westward relative to the stars is called a period of **apparent retrograde motion** (*retrograde* means "backward"). These periods may last from a few weeks to a few months, depending on the planet (Figure 2.27).

Ancient astronomers could easily "explain" the daily paths of the stars through the sky by imagining that the celestial sphere was real and that it really rotated around Earth each day. But the apparent retrograde motion of the planets posed a far greater mystery: What could cause the planets sometimes to go backward? As we'll discuss in Chapter 3, the ancient Greeks came up with some very clever ways to explain the occasional backward motion of the planets, despite being wedded to the incorrect idea of an Earth-centered universe. However, the Greek explanation was quite complex, and ultimately wrong.

In contrast, apparent retrograde motion has a simple explanation in a Sun-centered solar system. You can demonstrate it for yourself with the help of a friend (Figure 2.28a). Pick a spot in an open field to represent the Sun. You can represent Earth, walking counterclockwise around the Sun, while your friend represents a more distant planet (such as Mars, Jupiter, or Saturn) by walking counterclockwise around the Sun at a greater distance. Your friend should walk more slowly than you, because more distant planets orbit the Sun more slowly. As you walk, watch how

Figure 2.26

This photograph shows a rare planetary grouping in which all five planets that are easily visible to the naked eye appeared close together in the sky. It was taken near Chatsworth, New Jersey, just after sunset on April 23, 2002. The next such close grouping of these five planets in our sky will not occur until September 2040.

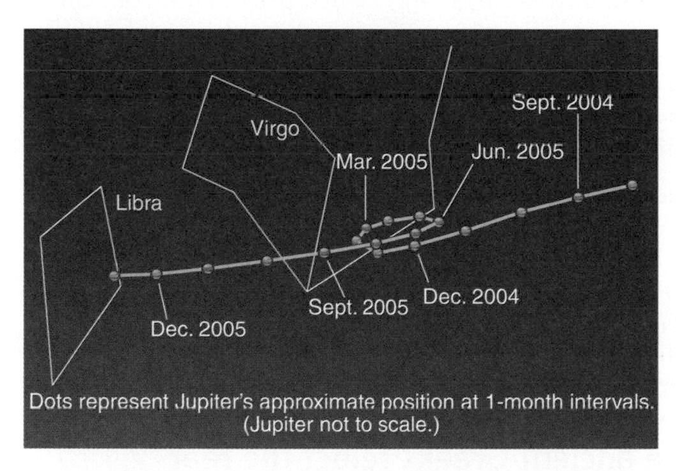

a This diagram shows Jupiter's approximate position among the stars in our sky during 2004–2005. Notice the backward path between February and June 2005.

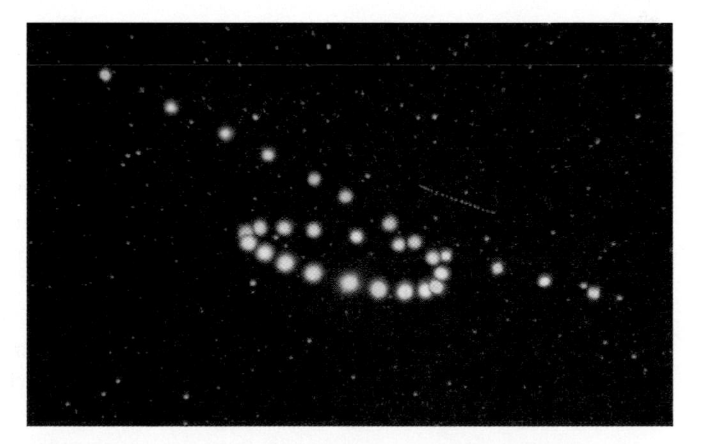

b This composite photograph shows Mars between early June and late November 2003. The progression goes from right to left, with apparent retrograde motion in the middle. The 29 individual photos were taken at 5- to 8-day intervals. (The white dots in a line just right of center are the planet Uranus, which by coincidence was in the same part of the sky.)

Figure 2.27

Apparent retrograde motion.

Figure 2.28 Interactive Figure

Apparent retrograde motion—the occasional "backward" motion of the planets relative to the stars—has a simple explanation in a Sun-centered solar system.

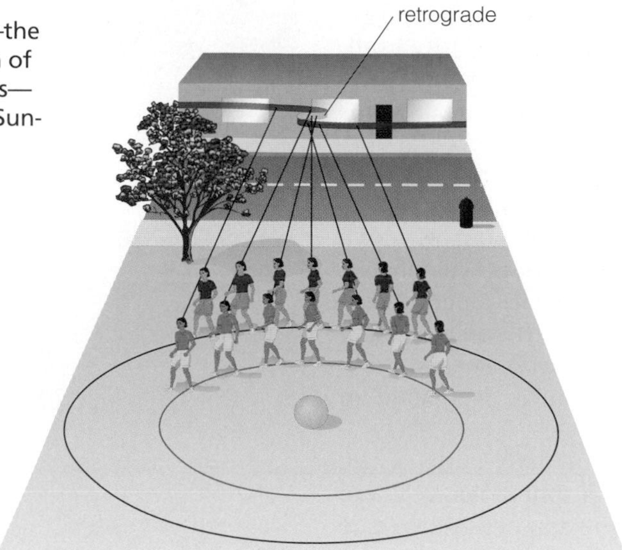

a This demonstration shows why planets sometimes seem to go backward relative to distant stars. Watch how your friend (in red) usually appears to you (in blue) to move forward against the background of the building in the distance but appears to move backward as you catch up to and pass him or her in your "orbit."

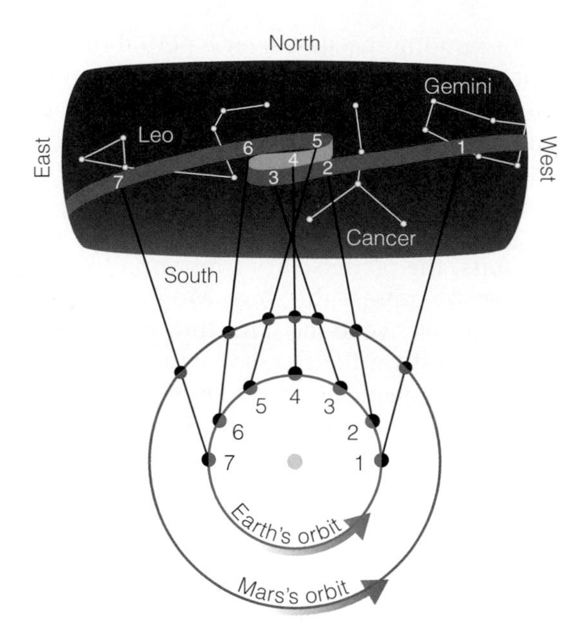

b This diagram shows how the idea from the demonstration applies to a planet. Follow the lines of sight from Earth to Mars in numerical order. Notice that Mars appears to move westward relative to the distant stars as Earth passes it by in its orbit (roughly from points 3 to 5 in the diagram).

your friend appears to move relative to buildings or trees in the distance. Although both of you always walk the same way around the Sun, your friend will appear to move backward against the background during the part of your "orbit" at which you catch up to and pass him or her. To understand the apparent retrograde motions of Mercury and Venus, which are closer to the Sun than is Earth, simply switch places with your friend and repeat the demonstration.

A planet appears to move backward relative to the stars during the period when Earth passes it by in its orbit.

This demonstration applies directly to the planets. For example, because Mars takes about 2 years to orbit the Sun (actually 1.88 years), it covers about half its orbit during the 1 year in which Earth makes a complete orbit. If you trace lines of sight from Earth to Mars from different points in their orbits, you will see that the line of sight usually moves eastward relative to the stars but moves westward during the time when Earth is passing Mars in its orbit (Figure 2.28b). Like your friend in the demonstration, Mars never actually changes direction. It only *appears* to change direction from our perspective on Earth.

• Why did the ancient Greeks reject the real explanation for planetary motion?

If the apparent retrograde motion of the planets is so readily explained by recognizing that Earth orbits the Sun, why wasn't this idea accepted in ancient times? In fact, the idea that Earth goes around the Sun was suggested as early as 260 B.C. by the Greek astronomer Aristarchus. No one knows why Aristarchus proposed a Sun-centered solar system, but the fact that it explains planetary motion so naturally probably played a role. Nevertheless, Aristarchus's contemporaries rejected his idea, and the Sun-centered solar system did not gain wide acceptance until almost 2,000 years later.

Although there were many reasons for the Greek reluctance to abandon the idea of an Earth-centered universe, perhaps the most prominent involved their inability to detect something called **stellar parallax**. Extend your arm and hold up one finger. If you keep your finger still and alternately close your left eye and right eye, your finger will appear to jump back and forth against the background. This apparent shifting, called *parallax,* occurs because your two eyes view your finger from opposite sides of your nose. If you move your finger closer to your face, the parallax increases. If you look at a distant tree or flagpole instead of your finger, you may not notice any parallax at all. Thus, parallax depends on distance, with nearer objects exhibiting greater parallax than more distant objects.

If you now imagine that your two eyes represent Earth at opposite sides of its orbit around the Sun and that your finger represents a relatively nearby star, you have the idea of stellar parallax. That is, because we view the stars from different places in our orbit at different times of year, nearby stars should *appear* to shift back and forth against the background of more distant stars (Figure 2.29).

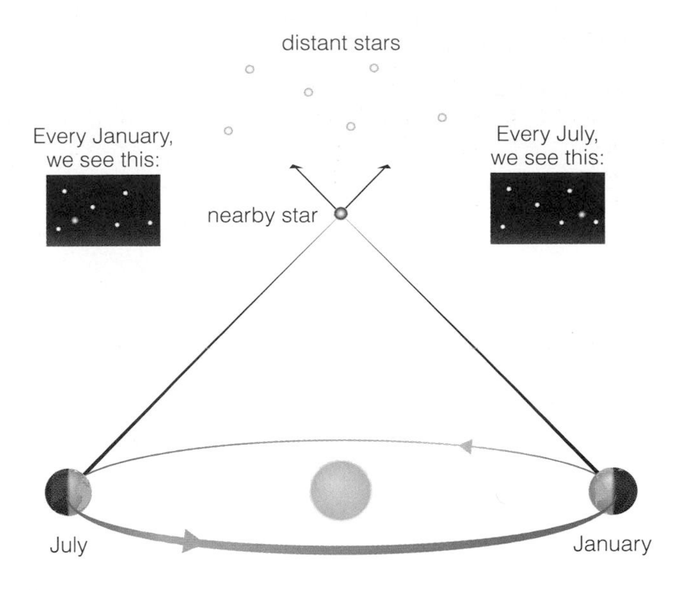

Figure 2.29

Stellar parallax is an apparent shift in the position of a nearby star as we look at it from different places in Earth's orbit. This figure is greatly exaggerated; in reality, the amount of shift is far too small to detect with the naked eye.

The Greeks knew that stellar parallax should occur if Earth orbits the Sun, but they could not detect it.

Because the Greeks believed that all stars lay on the same celestial sphere, they actually expected to see stellar parallax in a slightly different way. If Earth orbited the Sun, they reasoned that at different times of year we would be closer to different parts of the celestial sphere and thus would notice changes in the angular separations of stars. However, no matter how hard they searched, they could find no sign of stellar parallax. They concluded that one of the following must be true:

1. Earth orbits the Sun but the stars are so far away that stellar parallax is undetectable to the naked eye.
2. There is no stellar parallax because Earth remains stationary at the center of the universe.

Unfortunately, with notable exceptions such as Aristarchus, the Greeks rejected the correct answer (1) because they could not imagine that the stars could be *that* far away. Today, we can detect stellar parallax with the aid of telescopes, providing direct proof that Earth really does orbit the Sun. Careful measurements of stellar parallax also provide the most reliable means of measuring distances to nearby stars [Section 11.1].

THINK ABOUT IT How far apart are opposite sides of Earth's orbit? How far away are the nearest stars? Describe the challenge of detecting stellar parallax. It may help to visualize Earth's orbit and the distance to the stars on the 1-to-10-billion scale used in Chapter 1.

The ancient mystery of the planets drove much of the historical debate over Earth's place in the universe. In many ways, the modern technological society we take for granted today can be traced directly back to the scientific revolution that began because of the quest to explain the strange wandering of the planets among the stars in our sky. We will turn our attention to this revolution in the next chapter.

THE BIG PICTURE

Putting Chapter 2 into Context

In this chapter, we surveyed the phenomena of our sky. Keep the following "big picture" ideas in mind as you continue your study of astronomy:

- You can enhance your enjoyment of learning astronomy by observing the sky. The more you learn about the appearance and apparent motions of the sky, the more you will appreciate what you can see in the universe.
- From our vantage point on Earth, it is convenient to imagine that we are at the center of a great celestial sphere—even though we really are on a planet orbiting a star in a vast universe. We can then understand what we see in the local sky by thinking about how the celestial sphere appears from our latitude.
- Most of the phenomena of the sky are relatively easy to observe and understand. The more complex phenomena—particularly eclipses and apparent retrograde motion of the planets—challenged our ancestors for thousands of years and helped drive the development of science and technology.

2.1 PATTERNS IN THE NIGHT SKY

- **What are constellations?**

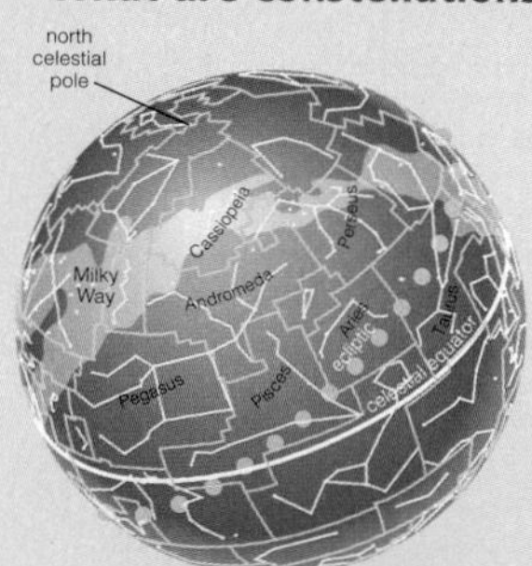

Constellations are regions of the sky with well-defined borders, drawn so that we can map them on the **celestial sphere.** A model of the celestial sphere shows these borders and patterns of stars within them, the **north** and **south celestial poles,** the **celestial equator,** and the **ecliptic.**

- **How do we locate objects in the sky?**

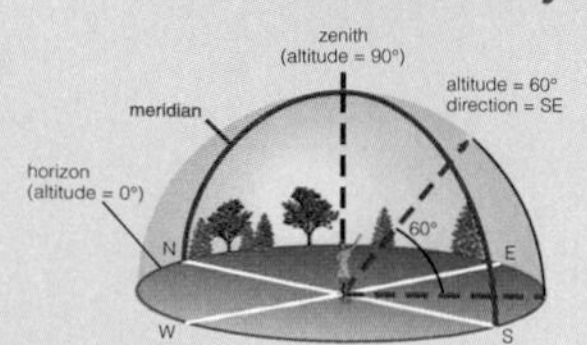

We can pinpoint any object in the **local sky** by stating its **altitude** (in degrees) above the horizon and its **direction** along the horizon. The **horizon** is the boundary between Earth and sky, and the **zenith** is the point directly overhead. The **meridian** runs from due south to due north through the zenith.

- **Why do stars rise and set?**

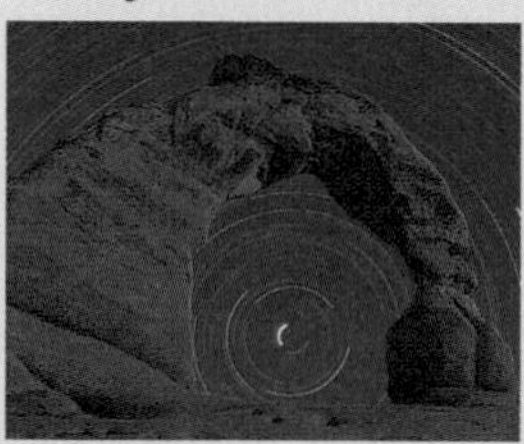

Earth's rotation makes stars appear to circle around Earth each day. A star whose complete circle lies above our horizon is said to be **circumpolar.** Other stars have circles that cross the horizon, making them rise in the east and set in the west each day.

- **Why don't we see the same constellations at all times?**

The visible constellations vary with time of year because our night sky lies in different directions in space as we orbit the Sun. The constellations vary with latitude because your latitude determines the orientation of your horizon relative to the celestial sphere. The sky does not vary with longitude.

2.2 THE REASON FOR SEASONS

- **What causes the seasons?**

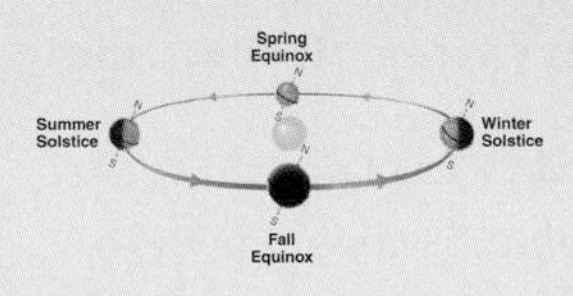

The tilt of Earth's axis causes the seasons. The axis points in the same direction (toward Polaris) throughout the year. Thus, as Earth orbits the Sun, sunlight hits different parts of Earth more directly at different times of year.

- **How do we mark the progression of the seasons?**

The **summer** and **winter solstices** are the times during the year when the Northern Hemisphere gets its most and least direct sunlight, respectively. The **spring** and **fall equinoxes** are the two times when both hemispheres get equally direct sunlight.

- **Does the orientation of Earth's axis change with time?**

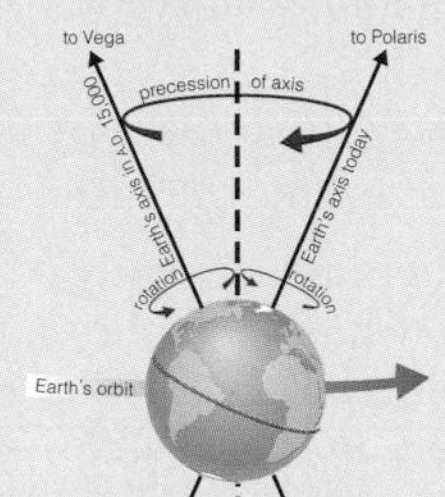

Earth's 26,000-year cycle of **precession** changes the orientation of the axis in space, although the tilt remains about $23\frac{1}{2}°$. The changing orientation of the axis does not affect the pattern of seasons, but it changes the identity of any "north star" and shifts the locations of the solstices and equinoxes in Earth's orbit.

2.3 THE MOON, OUR CONSTANT COMPANION

- **Why do we see phases of the Moon?**

The **phase** of the Moon depends on its position relative to the Sun as it orbits Earth. The half of the Moon facing the Sun is always illuminated and the other half dark, but from Earth we see varying combinations of the illuminated and dark faces.

- **What causes eclipses?**

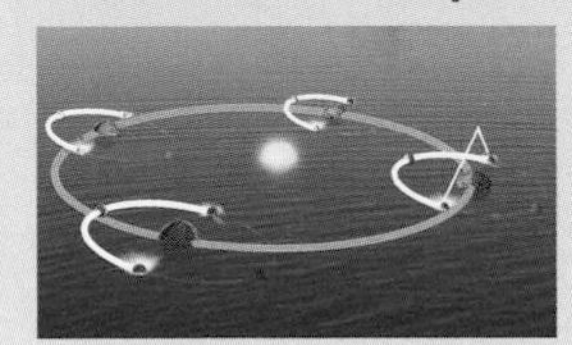

We see a **lunar eclipse** when Earth's shadow falls on the Moon, and a **solar eclipse** when the Moon blocks our view of the Sun. We do not see an eclipse at every new and full moon because the Moon's orbit is slightly inclined to the ecliptic plane. Eclipses come in different types, depending on where the dark **umbral** and lighter **penumbral** shadows fall.

2.4 THE ANCIENT MYSTERY OF THE PLANETS

- **What was once so mysterious about the movement of planets in our sky?**

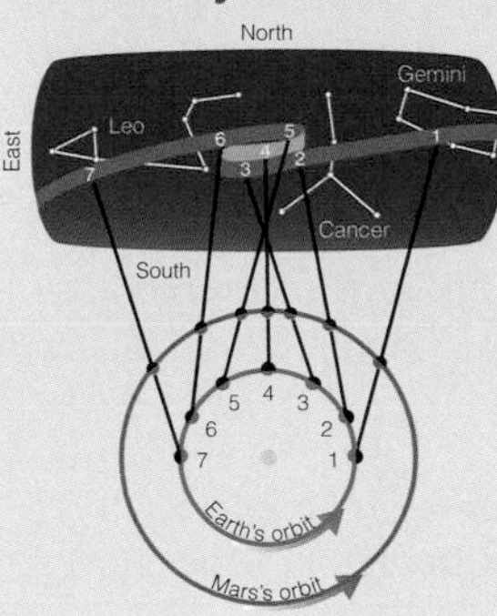

Planets generally drift eastward relative to the stars over the course of the year, but for weeks or months they reverse course in their periods of **apparent retrograde motion.** This motion occurs when Earth passes by (or is passed by) another planet in its orbit, but it posed a major mystery to ancient people who assumed Earth to be at the center of the universe.

- **Why did the ancient Greeks reject the real explanation for planetary motion?**

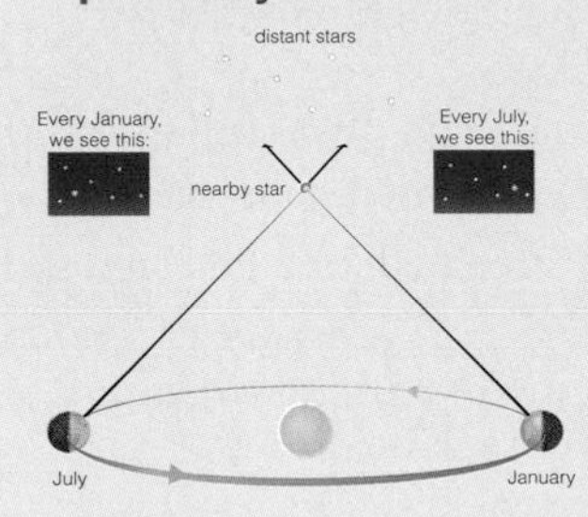

The Greeks rejected the idea that Earth goes around the Sun in part because they could not detect **stellar parallax**—slight apparent shifts in stellar positions over the course of the year. To most Greeks, it seemed unlikely that the stars could be so far away as to make parallax undetectable to the naked eye, even though that is, in fact, the case.

EXERCISES AND PROBLEMS

REVIEW QUESTIONS

1. Suppose you were making a model of the celestial sphere with a ball. Briefly list all the things you would need to mark on your celestial sphere, and how you would decide where to put them on the sphere.
2. Why does the local sky look like a dome? Define *horizon*, *zenith*, and *meridian*. How do we describe the location of an object in the local sky?
3. Explain why we can measure only *angular sizes* and *angular distances* for objects in the sky. What are *arcminutes* and *arcseconds*?
4. What are *circumpolar stars*? Are more stars circumpolar at the North Pole or in the United States? Explain.
5. What are *latitude* and *longitude*? Does the sky vary with latitude? Does it vary with longitude? Explain.
6. What is the *zodiac*, and why do we see different parts of it at different times of year?

7. Suppose Earth's axis had no tilt. Would we still have seasons? Why or why not?
8. Briefly describe what is special about the summer and winter solstices and the spring and fall equinoxes.
9. What is precession, and how does it affect the sky that we see from Earth?
10. Briefly describe the Moon's cycle of *phases*. Can you ever see a full moon at noon? Explain.
11. Suppose you lived on the Sun (and could ignore the heat). Would you still see the Moon go through phases as it orbits Earth? Why or why not?
12. Why don't we see an *eclipse* at every new and full moon? Describe the conditions that must be met for us to see a *solar* or *lunar eclipse.*
13. What do we mean by the *apparent retrograde motion* of the planets? Why was it difficult for ancient astronomers to explain but easy for us to explain?
14. What is *stellar parallax*? Briefly describe the role it played in making ancient astronomers believe in an Earth-centered universe.

DOES IT MAKE SENSE?

Decide whether the statement makes sense and explain why it does or does not. (For an example, see Chapter 1, "Does It Make Sense?")

15. If you had a very fast spaceship, you could travel to the celestial sphere in about a month.
16. The constellation Orion didn't exist when my grandfather was a child.
17. When I looked into the dark fissure of the Milky Way with my binoculars, I saw what must have been a cluster of distant galaxies.
18. Last night the Moon was so big that it stretched for a mile across the sky.
19. I live in the United States, and during my first trip to Argentina I saw many constellations that I'd never seen before.
20. Last night I saw Jupiter right in the middle of the Big Dipper. (*Hint:* Is the Big Dipper part of the zodiac?)
21. Last night I saw Mars move westward through the sky in its apparent retrograde motion. (*Hint:* How long does it take to notice apparent retrograde motion?)
22. Although all the known stars appear to rise in the east and set in the west, we might someday discover a star that will appear to rise in the west and set in the east.
23. If Earth's orbit were a perfect circle, we would not have seasons.
24. Because of precession, someday it will be summer everywhere on Earth at the same time.

PROBLEMS

(Quantitative problems are marked with an asterisk.)

25. *New Planet.* Suppose we discover a planet in another solar system that has a circular orbit and an axis tilt of 35°. Would you expect this planet to have seasons? If so, would you expect them to be more or less extreme than the seasons on Earth? If not, why not?
26. *Your View.*
    a. Find your latitude and longitude, and state the source of your information.
    b. Describe the altitude and direction in your sky at which the north or south celestial pole appears.
    c. Is Polaris a circumpolar star in your sky? Explain.
27. *View from the Moon.* Suppose you lived on the Moon, in which case you would see Earth going through phases in your sky. Assume you live near the center of the face that looks toward Earth.
    a. Suppose you see a full Earth in your sky. What phase of the Moon would we see here on Earth? Explain.
    b. Suppose that people on Earth see a full moon. What phase would you see for Earth? Explain.
    c. Suppose that people on Earth see a waxing gibbous moon. What phase would you see for Earth? Explain.
    d. Suppose that people on Earth are viewing a total lunar eclipse. What would you see from your home on the Moon? Explain.
28. *A Farther Moon.* Suppose the distance to the Moon were twice its actual value. Would it still be possible to have a total solar eclipse? Why or why not?
29. *A Smaller Earth.* Suppose Earth were smaller. Would solar eclipses be any different? If so, how? What about lunar eclipses? Explain.
30. *Observing Planetary Motion.* Find out what planets are currently visible in your evening sky. At least once a week, observe the planets and draw a diagram showing the position of each visible planet relative to stars in a zodiac constellation. From week to week, note how the planets are moving relative to the stars. Can you see any of the apparently "erratic" features of planetary motion? Explain.
31. *A Connecticut Yankee.* Find the book *A Connecticut Yankee in King Arthur's Court,* by Mark Twain. Read the portion that deals with the Connecticut Yankee's prediction of an eclipse (or read the entire book). In a one- to two-page essay, summarize the episode and how it helps the Connecticut Yankee gain power.

*32. There are 360° in a full circle.
    a. How many arcminutes are in a full circle?
    b. How many arcseconds are in a full circle?
    c. The Moon's angular size is about $\frac{1}{2}$°. What is this in arcminutes? In arcseconds?

DISCUSSION QUESTIONS

33. *Earth-Centered Language.* Many common phrases reflect the ancient Earth-centered view of our universe. For example, the phrase "the Sun rises each day" implies that the Sun is really moving over Earth. We know that the Sun only *appears* to rise as the rotation of Earth carries us to a place where we can see the Sun in our sky. Identify other common phrases that imply an Earth-centered viewpoint.
34. *Flat Earth Society.* Believe it or not, there is an organization called the Flat Earth Society. Its members hold that Earth is flat and that all indications to the contrary (such as pictures of Earth from space) are fabrications made as part of a conspiracy to hide the truth from the public. Discuss the evidence for a round Earth and how you can check it for yourself. In light of the evidence, is it possible that the Flat Earth Society is correct? Defend your opinion.

For a complete list of media resources available, go to **www.astronomyplace.com** and choose Chapter 2 from the pull-down menu.

ASTRONOMY PLACE WEB TUTORIALS

Tutorial Review of Key Concepts

Use the interactive **Tutorials** at **www.astronomyplace.com** to review key concepts from this chapter.

Seasons Tutorial

Lesson 1 Factors Affecting Seasonal Changes
Lesson 2 The Solstices and Equinoxes
Lesson 3 The Sun's Position in the Sky

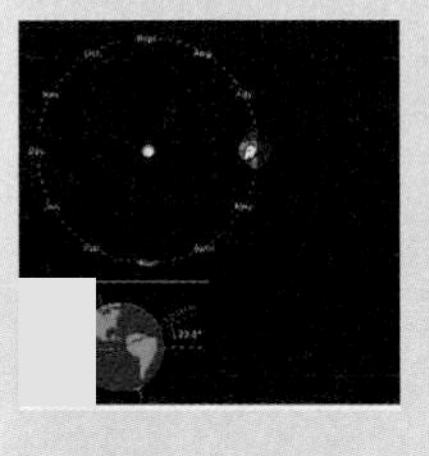
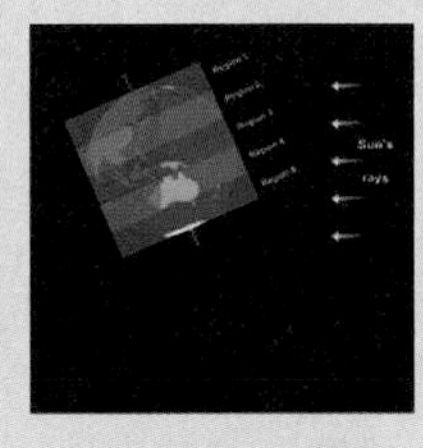
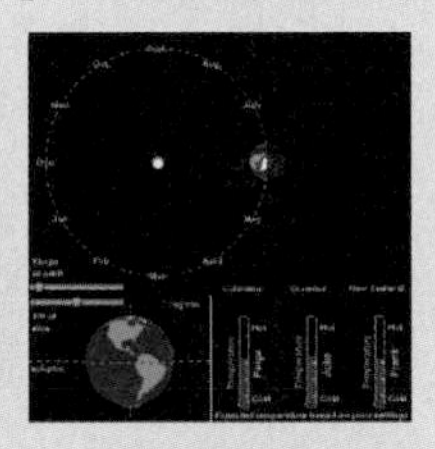

Phases of the Moon Tutorial

Lesson 1 The Causes of Lunar Phases
Lesson 2 Time of Day and Horizons
Lesson 3 When the Moon Rises and Sets

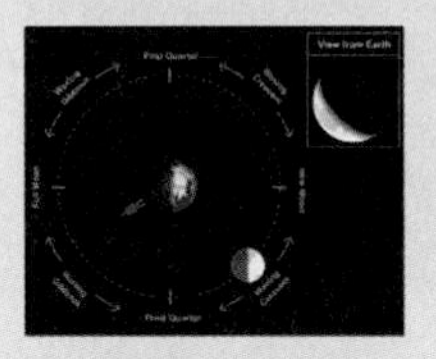
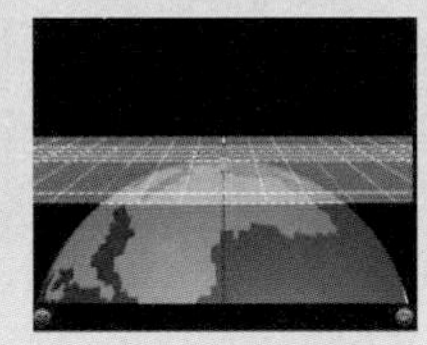
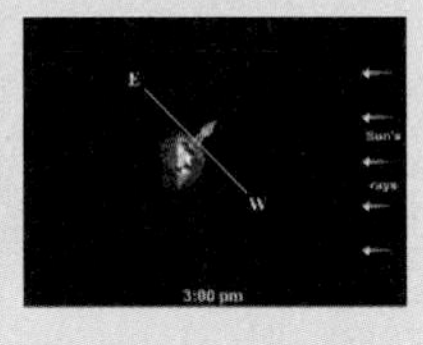

Eclipses Tutorial

Lesson 1 Why and When Do Eclipses Occur?
Lesson 2 Types of Solar Eclipses
Lesson 3 Lunar Eclipses

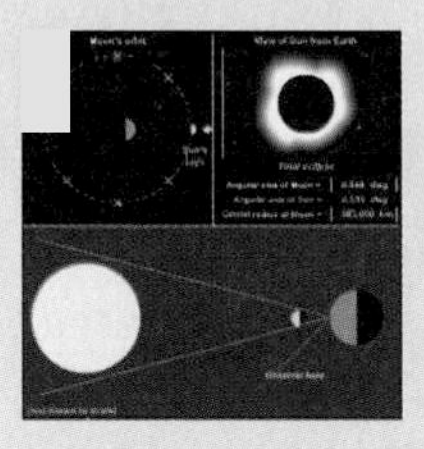
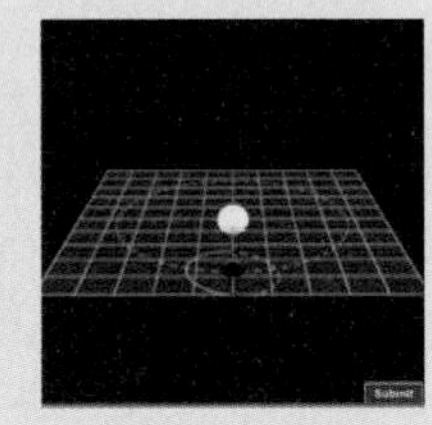
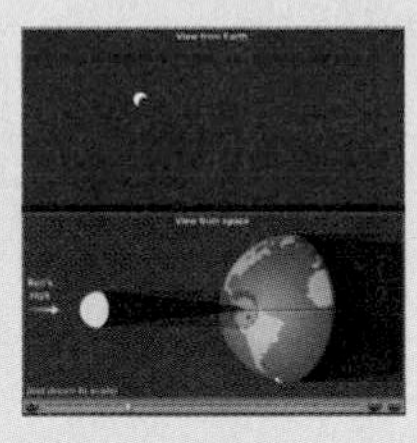

Supplementary Tutorial Exercises

Use the interactive **Tutorial Lessons** to explore the following questions.

Seasons Tutorial, Lesson 1

1. What factors affect a planet's surface temperature?
2. Why does the tilt of Earth rather than Earth's orbit have a greater effect on Earth's seasons?
3. Why is the equator always hot and why are the poles always cold despite the passage of the seasons?

Phases of the Moon Tutorial, Lesson 3

1. When does a new moon rise and set?
2. In which phase does the Moon set just after the Sun?
3. What factors influence our view of the Moon from Earth?

Eclipses Tutorial, Lesson 1

1. What happens during a solar eclipse?
2. What happens during a lunar eclipse?
3. What role does the Moon's orbit play in the appearance of a solar eclipse?

EXPLORING THE SKY AND SOLAR SYSTEM

Of the many activities available on the ***Voyager: SkyGazer* CD-ROM** accompanying your book, use the following files to observe key phenomena covered in this chapter.

Go to the **File: Basics** folder for the following demonstrations.

1. Wide Field Milky Way
2. Eclipse 1991–1992 Views
3. Winter Sky

Go to the **File: Demo** folder for the following demonstrations.

1. Russian Midnight Sun
2. Earth Orbiting the Moon
3. Mars in Retrograde

Go to the **Explore** menu for the following demonstrations.

1. Shadows on Earth
2. Phases of the Planets

WEB PROJECTS

Take advantage of the useful Web links on **www.astronomyplace.com** to assist you with the following projects.

1. *Sky Information.* Search the Web for sources of daily information about sky phenomena (such as lunar phases, times of sunrise and sunset, or dates of equinoxes and solstices). Identify and briefly describe your favorite source.
2. *Constellations.* Search the Web for information about the constellations and their mythology. Write a short report about one or more constellations.
3. *Upcoming Eclipse.* Find information about an upcoming solar or lunar eclipse that you might have a chance to witness. Write a short report about how you could best witness the eclipse, including any necessary travel to a viewing site, and what you can expect to see. Bonus: Describe how you could photograph the eclipse.

3
The Science of Astronomy

LEARNING GOALS

3.1 THE ANCIENT ROOTS OF SCIENCE
- In what ways do all humans employ scientific thinking?
- How did astronomical observations benefit ancient societies?
- What did ancient civilizations achieve in astronomy?

3.2 ANCIENT GREEK SCIENCE
- Why does modern science trace its roots to the Greeks?
- How did the Greeks explain planetary motion?
- How did Islamic scientists preserve and extend Greek science?

3.3 THE COPERNICAN REVOLUTION
- How did Copernicus, Tycho, and Kepler challenge the Earth-centered idea?
- What are Kepler's three laws of planetary motion?
- How did Galileo solidify the Copernican revolution?

3.4 THE NATURE OF SCIENCE
- How can we distinguish science from nonscience?
- What is a scientific theory?

Today we know that Earth is a planet orbiting a rather ordinary star, in a galaxy of more than a hundred billion stars, in an incredibly vast universe. We know that Earth, along with the entire cosmos, is in constant motion. We know that, on the scale of cosmic time, human civilization has existed for only the briefest moment. How did we manage to learn these things?

It wasn't easy. Astronomy is the oldest of the sciences, with roots extending as far back as recorded history allows us to see. But the most impressive advances in knowledge have come in just the past few centuries.

In this chapter, we will trace how modern astronomy grew from its roots in ancient observations, including those of the Greeks. We'll pay special attention to the unfolding of the Copernican revolution, which laid the foundation for nearly all of modern science. Finally, we'll explore the nature of modern science and the scientific method.

3.1 THE ANCIENT ROOTS OF SCIENCE

A common stereotype holds that scientists walk around in white lab coats and somehow think differently than other people. In reality, scientific thinking is a fundamental part of human nature. In this section, we will trace the roots of science to experiences common to nearly all people and nearly all cultures.

• In what ways do all humans employ scientific thinking?

Scientific thinking comes naturally to us. By about a year of age, a baby notices that objects fall to the ground when she drops them. She lets go of a ball—it falls. She pushes a plate of food from her high chair—it falls too. She continues to drop all kinds of objects, and they all plummet to Earth. Through powers of observation, the baby learns about the physical world, finding that things fall when they are unsupported. Eventually, she becomes so certain of this fact that, to her parents' delight, she no longer needs to test it continually.

One day somebody gives the baby a helium balloon. She releases it, and to her surprise it rises to the ceiling! Her understanding of nature must be revised. She now knows that the principle "all things fall" does not represent the whole truth, although it still serves her quite well in most situations. It will be years before she learns enough about the atmosphere, the force of gravity, and the concept of density to understand *why* the balloon rises when most other objects fall. For now, she is delighted to observe something new and unexpected.

Scientific thinking is based on everyday ideas of observation and trial-and-error experiments.

The baby's experience with falling objects and balloons exemplifies scientific thinking. In essence, it is a way of learning about nature through careful observation and trial-and-error experiments. Rather than thinking differently than other people, modern scientists simply are trained

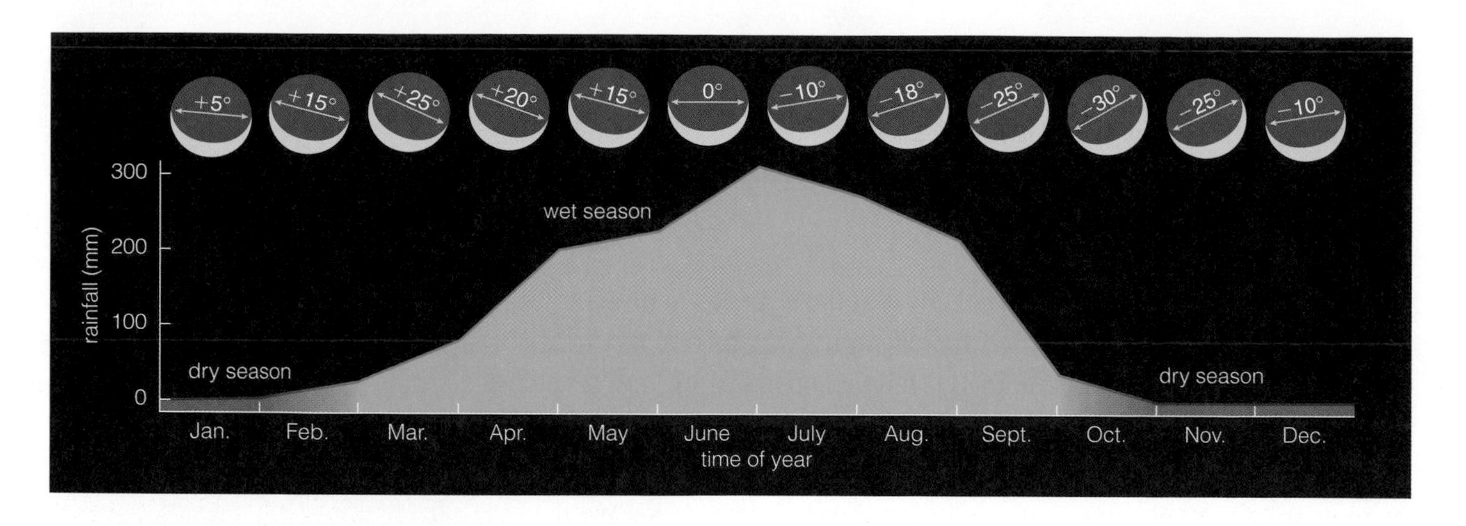

Figure 3.1

The roots of science lie in careful observation of the world around us. This diagram shows how central Africans used the orientation of the waxing crescent Moon to predict the rainfall. The graph depicts the annual rainfall pattern in central Nigeria, and the Moon diagrams show the way in which the angle of the "horns" of a waxing crescent moon (relative to the western horizon) changes during the year. (Adapted from *Ancient Astronomers* by Anthony F. Aveni.)

to organize everyday thinking in a way that makes it easier for them to share their discoveries and employ their collective wisdom. This type of clear and organized thinking is at the heart of science.

THINK ABOUT IT Describe a few cases where you have learned by trial and error while cooking, participating in sports, fixing something, learning on the job, or in any other situation.

Just as learning to communicate through language, art, or music is a gradual process for a child, the development of science has been a gradual process for humanity. Science in its modern form requires painstaking attention to detail, relentless testing of each piece of information to ensure its reliability, and a willingness to give up old beliefs that are not consistent with observed facts about the physical world. For professional scientists, these demands are the "hard work" part of the job. At heart, professional scientists are like the baby with the balloon, delighted by the unexpected and motivated by those rare moments when they—and all of us—learn something new about the universe.

• How did astronomical observations benefit ancient societies?

We will discuss modern science shortly, but first we will explore how it arose from the observations of ancient peoples. Our exploration begins in central Africa, where people of many indigenous societies predict the weather with reasonable accuracy by making careful observations of the Moon. The Moon begins its monthly cycle as a crescent in the western sky just after sunset. Through long traditions of sky watching, central African societies learned that the orientation of the crescent "horns" relative to the horizon is closely tied to rainfall patterns (Figure 3.1). No one knows when central Africans first developed the ability to predict weather using the lunar crescent, but the tradition may date back thousands of years.

Why did ancient people make such careful and detailed observations of the sky? In part, it was probably to satisfy their inherent curiosity. But astronomy also played a practical role for them. They used the changing positions of the Sun and stars to keep track of the time and seasons, crucial skills for people who depended on agriculture. Some cultures even learned

TABLE 3.1 *The Seven Days of the Week and the Astronomical Objects They Honor*

In English, the correspondence between celestial objects and days is clear only if we look at the names of the objects used by the Teutonic tribes that lived in the region of modern-day Germany. French and Spanish are included because they make some of the correspondences easier to see.

	Teutonic			
Object	*Name*	*English*	*French*	*Spanish*
Sun	Sun	Sunday	dimanche	domingo
Moon	Moon	Monday	lundi	lunes
Mars	Tiw	Tuesday	mardi	martes
Mercury	Woden	Wednesday	mercredi	miércoles
Jupiter	Thor	Thursday	jeudi	jueves
Venus	Fria	Friday	vendredi	viernes
Saturn	Saturn	Saturday	samedi	sábado

to navigate by the Sun and stars. Many people developed ideas and stories to explain what they saw in the heavens.

Ancient people used observations of the sky to keep track of the time and seasons and as an aid in navigation.

Modern measures of time come directly from ancient observations of motion in the sky. The length of our day is the time it takes the Sun to make one full circuit of the sky. The length of a month comes from the Moon's cycle of phases [Section 2.3], and our year is based on the cycle of the seasons [Section 2.2]. The seven days of the week were named after the seven naked-eye objects that appear to move among the constellations: the Sun, the Moon, and the five planets recognized in ancient times (Table 3.1).

• What did ancient civilizations achieve in astronomy?

Nearly all ancient civilizations practiced astronomy at some level. Many built remarkable structures for observing the sky. Let's explore a few of the many ways that ancient societies studied the sky.

Determining the Time of Day In the daytime, ancient peoples could tell time by observing the Sun's path through the sky. Many cultures probably used the shadows cast by sticks as simple sundials. The ancient Egyptians built huge obelisks, often inscribed or decorated in homage to the Sun, that probably also served as simple clocks (Figure 3.2). At night, ancient people could estimate the time from the position and phase of the Moon (see Figure 2.19) or by observing the constellations visible at a particular time of night (see Figure 2.12).

We can trace the origins of our modern clock to ancient Egypt, some 4,000 years ago. The Egyptians divided the daylight and darkness into 12 equal parts, much as we break the 24-hour day into 12 hours each of a.m. and p.m. (The abbreviations *a.m.* and *p.m.* stand for the Latin terms *ante meridiem* and *post meridiem*, respectively, which mean "before the middle of the day" and "after the middle of the day.")

Figure 3.2

This ancient Egyptian obelisk, which stands 83 feet tall and weighs 331 tons, resides in St. Peter's Square at the Vatican in Rome. It is one of 21 surviving obelisks from ancient Egypt, most of which are now scattered around the world. Shadows cast by the obelisks may have been used to tell time.

Marking the Seasons Many ancient cultures built structures to help them mark the seasons. One of the oldest standing human-made structures served such a purpose: Stonehenge in southern England (Figure 3.3). Stonehenge served both as an astronomical device for keeping track of the seasons and as a social and religious gathering place.

One of the most spectacular structures used to mark the seasons was the Templo Mayor in the Aztec city of Tenochtitlán, located on the site of modern-day Mexico City (Figure 3.4). Twin temples stood on top of a flat-topped, 150-foot-high pyramid. From the vantage point of a royal observer watching from the opposite side of the plaza, the Sun rose directly through the notch between the temples on the equinoxes. Like Stonehenge, the Templo Mayor served important social and religious functions in addition to its astronomical role.

Many ancient cultures aligned their buildings with the cardinal directions (north, south, east, and west), enabling them to mark the rising and setting of the Sun relative to the building orientation. Some cultures created monuments with a special astronomical purpose. For example, someone among the ancient Anasazi people carved a spiral known as the Sun Dagger on a vertical cliff face near the top of a butte in Chaco Canyon, New Mexico (Figure 3.5). The Sun's rays form a dagger of sunlight that

Figure 3.3

The remains of Stonehenge today. It was built in stages from about 2750 B.C. to about 1550 B.C.

Figure 3.4

This scale model shows the Templo Mayor and the surrounding plaza as they are thought to have looked before Aztec civilization was destroyed by the Spanish Conquistadors. The structure was used to help mark the seasons.

Figure 3.5

The Sun Dagger, a summer solstice marker built by the Anasazi. It is an arrangement of rocks that shape the Sun's light into a dagger, with a spiral carved on a rock face behind them. The dagger pierces the center of the carved spiral only at noon on the summer solstice.

Figure 3.6

The Moon rising between two stones of the 4,000-year-old sacred stone circle at Callanish, Scotland (on the Isle of Lewis in the Scottish Hebrides). The full moon rises in this position only once every 18.6 years.

pierces the center of the carved spiral only once each year—at noon on the summer solstice.

Marking Lunar Cycles Many ancient civilizations paid particular attention to lunar phases, often using them as the basis for calendars. The months on lunar calendars generally have either 29 or 30 days, chosen to make the average agree with the approximately $29\frac{1}{2}$-day lunar cycle. Thus, a 12-month lunar calendar has only 354 or 355 days, or about 11 days fewer than a calendar based on the Sun. Such a calendar is still used in the Muslim religion. That is why the month-long fast of Ramadan (the ninth month) begins about 11 days earlier with each subsequent year.

Other lunar calendars take advantage of the fact that 19 years is almost precisely 235 lunar months. That is, every 19 years we get the same lunar phases on about the same dates. (This 19-year cycle is known as the *Metonic cycle*.) A lunar calendar can therefore be kept roughly synchronized to the seasons by adding a thirteenth month to 7 of every 19 years (making exactly 235 months in each 19-year period), ensuring that "new year" comes on approximately the same date every nineteenth year. The Jewish calendar uses this idea, adding a thirteenth month in the third, sixth, eighth, eleventh, fourteenth, seventeenth, and nineteenth years of each 19-year cycle. This also explains why the date of Easter changes from year to year: The New Testament ties the date of Easter to the Jewish festival of Passover, the date of which is set by the Jewish lunar calendar.

Remarkable ancient achievements included accurate calendars, eclipse prediction, navigational tools, and elaborate structures for astronomical observations.

Some ancient cultures learned to predict eclipses by recognizing the 18-year saros cycle [Section 2.3]. In the Middle East, the ancient Babylonians achieved remarkable success in predicting eclipses more than 2,500 years ago. The Mayans of Central America also appear to have been experts at eclipse prediction, but we know few details about their accomplishments because the Spanish Conquistadors burned most Mayan writings.

The complexity of the Moon's orbit leads to other long-term patterns in the Moon's appearance. For example, the full moon rises at its most southerly point along the eastern horizon only once every 18.6 years, a phenomenon that can be observed from the 4,000-year-old sacred stone circle at Callanish, Scotland (Figure 3.6).

Observing Planets and Stars Many ancient cultures also made careful observations of planets and stars. The Chinese, for example, began keeping remarkably detailed records of astronomical observations at least 5,000 years ago.

Other people built elaborate structures designed to observe planets or stars. The Mayan observatory at Chichén Itzá had windows strategically placed for observations of Venus (Figure 3.7). In the Nazca desert of Peru, we find hundreds of lines and patterns etched in the sand, many aligned to point to places where bright stars or the Sun rose at particular times of year. The patterns, many of which are large figures of animals, may represent constellations (Figure 3.8).

THINK ABOUT IT Animal figures like that in Figure 3.8 show up clearly only when seen from above. As a result, some UFO enthusiasts argue that the patterns must have been created by aliens. What do you think of this argument? Defend your opinion.

Figure 3.7

The ruins of the Mayan observatory at Chichén Itzá.

Figure 3.8

Hundreds of lines and patterns are etched in the sand of the Nazca desert in Peru. This aerial photo shows a large etched figure of a hummingbird.

Structures for astronomical observation also were popular in North America. Lodges built by the Pawnee people in Kansas featured strategically placed holes for observing the passage of constellations that figured prominently in their folklore. In the northern plains of the United States, Native American Medicine Wheels had "spokes" aligned with the rising and setting of bright stars, as well as with the rising and setting of the Sun on the equinoxes and solstices (Figure 3.9).

Figure 3.9

The Big Horn Medicine Wheel in Wyoming. The 28 "spokes" radiating out from the center probably relate to the Moon's cycle of phases (with 28 rather than 29 because there is no spoke for the day of the new moon).

Perhaps the people most dependent on knowledge of the stars were the Polynesians, who lived and traveled among the many islands of the mid- and South Pacific. Because the next island in a journey usually was too distant to be seen, poor navigation meant becoming lost at sea. As a result, the most esteemed position in Polynesian culture was that of the Navigator, a person who had acquired the detailed knowledge necessary to navigate great distances among the islands. The Navigators employed a combination of detailed knowledge of astronomy and equally impressive knowledge of the patterns of waves and swells around different islands (Figure 3.10).

3.2 ANCIENT GREEK SCIENCE

Before a structure such as Stonehenge or a Medicine Wheel could be built, careful observations had to be made and repeated over and over to ensure their validity. Careful, repeatable observations also underlie modern science. Thus, elements of modern science were present in many early human cultures. If the circumstances of history had been different, almost any culture might have been the first to develop what we consider modern science. In the end, however, history takes only one of countless possible paths. The path that led to modern science emerged from the ancient civilizations of the Mediterranean and the Middle East—and especially from ancient Greece.

Greece began to rise as a center of power around 500 B.C. In about 330 B.C., Alexander the Great (356–323 B.C.) began a series of conquests that expanded the Greek empire throughout the Middle East, absorbing

Figure 3.10

A traditional Polynesian navigational instrument.

Figure 3.11

These renderings show an artist's reconstruction, based on scholarly research, of how the Great Hall and a scroll room of the Library of Alexandria might have looked.

a The Great Hall.

b A scroll room.

the former empires of Egypt and Mesopotamia. Alexander helped spread Greek culture because he was more than just a military leader. He also had a keen interest in science and education, perhaps because his personal tutor had been the great philosopher Aristotle (384–322 B.C.).

Alexander encouraged the pursuit of knowledge and respect for foreign cultures. On the Nile delta in Egypt, he founded the city of Alexandria, which soon became a center of world culture. The heart of Alexandria was a great library and research center that opened in about 300 B.C. (Figure 3.11). It remained the world's preeminent center of research for some 700 years, until it was finally destroyed in the fifth century A.D. At its peak, it may have held more than a half million books, handwritten on papyrus scrolls. Most of these books were ultimately burned, their wisdom lost forever.

THINK ABOUT IT Estimate the number of books you're likely to read in your life, and compare this number to the half million books once housed in the Library of Alexandria. Can you think of other ways to put into perspective the loss of ancient wisdom resulting from the destruction of the Library of Alexandria?

COMMON MISCONCEPTIONS

COLUMBUS AND A FLAT EARTH

A widespread myth holds that Columbus proved Earth to be round rather than flat. In fact, knowledge of the round Earth predated Columbus by nearly 2,000 years. However, it is probably true that most people in Columbus's day believed Earth to be flat, largely because of the poor state of education. The vast majority of the public was illiterate and unaware of the scholarly evidence for a spherical Earth. Interestingly, Columbus's primary argument with other scholars concerned the *distance* from Europe to Asia going westward—and it was Columbus who was wrong. He underestimated the true distance and as a result was woefully unprepared for the voyage to Asia he thought he was undertaking. Indeed, his voyages would almost certainly have ended in disaster had it not been for the presence of the Americas, which offered a safe landing well to the east of Asia.

• Why does modern science trace its roots to the Greeks?

The ancient Greeks stand out as the first people who tried to explain the motions of astronomical objects in ways that relied on logic and geometry, without resorting to the supernatural. They sought to understand the architecture of the universe by constructing models of nature. The idea of modeling is central to modern science.

Scientific models differ somewhat from the models you may be familiar with in everyday life. In our daily lives, we tend to think of models as miniature physical representations, such as model cars or airplanes. In contrast, a scientific **model** is a conceptual representation whose purpose is to explain and predict observed phenomena. For example, a model of Earth's climate uses logic and mathematics to represent what we know about how the climate works. Its purpose is to explain and predict climate changes, such as the changes that may occur with global warming.

The Greeks developed models of nature that aimed to explain and predict observed phenomena.

Just as a model airplane does not faithfully represent every aspect of a real airplane, a scientific model may not fully explain all our observations of nature. Nevertheless, even the failings of a scientific model can be useful, because they often point the way toward building a better model.

The Greek **geocentric model** of the cosmos, so named because it placed a spherical Earth at the center of the universe, sought to explain the motions of the Sun, Moon, planets, and stars. It may seem primitive from our modern perspective, but it was among the greatest intellectual accomplishments of ancient times.

We do not know precisely when the Greeks first thought that Earth is round, but this idea was already being taught by about 500 B.C. by the famous mathematician Pythagoras (c. 560–480 B.C.). More than a century later, Aristotle cited observations of Earth's curved shadow on the Moon during lunar eclipses as evidence for a spherical Earth.

Figure 3.12

This model represents the Greek idea of the heavenly spheres (c. 400 B.C.). Earth is a sphere that rests in the center. The Moon, the Sun, and the planets each have their own spheres. The outermost sphere holds the stars.

• How did the Greeks explain planetary motion?

The Greek geocentric model was based both on observations and on a strong belief that the heavens must be geometrically "perfect." Plato (428–348 B.C.) asserted that all heavenly objects must move in perfect circles and must reside on huge, perfect spheres encircling Earth. Aristotle echoed and amplified Plato's ideas, arguing forcefully in favor of an Earth-centered universe.

The Greeks thereby came to imagine that Earth is surrounded by a series of nested spheres holding the Moon, Sun, planets, and stars (Figure 3.12). The difficulty with this model was that it made it hard to explain the apparent retrograde motion of the planets [Section 2.4]. How could the planets sometimes go backward in our sky if they were moving in perfect circles?

The Greeks solved this problem by using multiple invisible spheres to explain the motion of each planet. By carefully choosing the sizes, rotation axes, and rotation speeds for the invisible spheres, the Greeks were able to have every sphere turn in a perfect circle even while their combined action reproduced the observed motions of planets in our sky. The idea was refined for centuries, reaching its culmination with the work of Claudius Ptolemy (c. A.D. 100–170; pronounced *tol-e-mee*). Building on the work of other Greeks (most notably Apollonius and Hipparchus), Ptolemy created a quantitative model that could be used to predict planetary positions in the sky. We refer to Ptolemy's model as the **Ptolemaic model** to distinguish it from earlier geocentric models.

The Ptolemaic model had each planet move on a small circle whose center moves around Earth on a larger circle.

The essence of the Ptolemaic model was that each planet moves on a small circle whose center moves around Earth on a larger circle (Figure 3.13). (The small circle is called an *epicycle*, and the larger circle is called a *deferent*.) A planet following this circle-upon-circle motion traces a loop as seen from Earth, with the backward portion of the loop mimicking apparent retrograde motion. However, to make his model agree well with observations, Ptolemy had to include a number of other complexities. For example, his model had many even smaller circles that

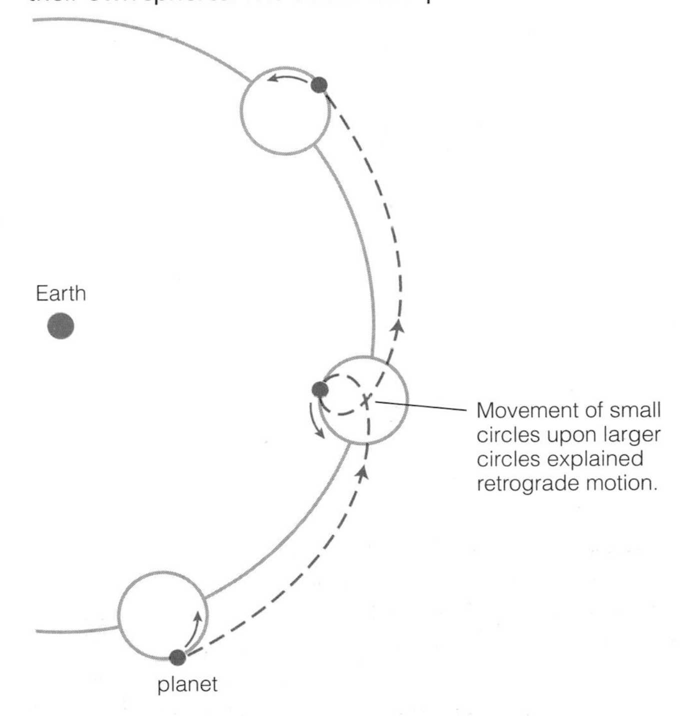

Figure 3.13 Interactive Figure

The Ptolemaic model explained apparent retrograde motion by supposing that each planet moved around Earth on a small circle that turned upon a larger circle. As shown by the dashed curve, this makes the planet trace a loop as seen from Earth. The backward portion of the loop represents apparent retrograde motion.

moved on the original set of small circles, and he positioned some of the large circles slightly off-center from Earth.

Ptolemy's model could correctly forecast future planetary positions to within a few degrees of arc—which is about the angular extent of your hand held at arm's length against the sky. This was considered quite accurate at the time. Indeed, his model generally worked so well that it remained in use for the next 1,500 years. When his book describing the model was translated by Arabic scholars around A.D. 800, they gave it the title *Almagest,* derived from words meaning "the greatest compilation."

• How did Islamic scientists preserve and extend Greek science?

Much of Greek knowledge was lost with the destruction of the Library of Alexandria. That which survived was preserved primarily thanks to the rise of a new center of intellectual inquiry in Baghdad (in present-day Iraq). While European civilization fell into the period of intellectual decline known as the Dark Ages, scholars of the new religion of Islam sought knowledge of mathematics and astronomy in hopes of better understanding the wisdom of Allah. During the eighth and ninth centuries A.D., scholars working in the Muslim empire centered in Baghdad translated and thereby saved many ancient Greek works.

Islamic scholars preserved and extended ancient Greek scholarship, and their work helped ignite the European Renaissance.

Around A.D. 800, the Islamic leader Al-Mamun (A.D. 786–833) established a "House of Wisdom" in Baghdad with a mission much like that of the destroyed Library of Alexandria. Founded in a spirit of openness and tolerance, the House of Wisdom employed Jews, Christians, and Muslims, all working together in

SPECIAL TOPIC: ERATOSTHENES MEASURES EARTH

In a remarkable feat, the Greek astronomer and geographer Eratosthenes estimated the size of Earth in about 240 B.C. He did it by comparing the altitude of the Sun on the summer solstice in the Egyptian cities of Syene (modern-day Aswan) and Alexandria.

Eratosthenes knew that the Sun passed directly overhead in Syene on the summer solstice. He also knew that in the city of Alexandria to the north the Sun came within only 7° of the zenith on the summer solstice. He therefore reasoned that Alexandria must be 7° of latitude to the north of Syene (see figure). Because 7° is $\frac{7}{360}$ of a circle, he concluded that the north-south distance between Alexandria and Syene must be $\frac{7}{360}$ of the circumference of Earth.

Eratosthenes estimated the north-south distance between Syene and Alexandria to be 5,000 stadia (the *stadium* was a Greek unit of distance). Thus, he concluded that:

$$\frac{7}{360} \times \text{circumference of Earth} = 5000 \text{ stadia}$$

From this he found Earth's circumference to be about 250,000 stadia.

Today, we don't know exactly what distance a stadium meant to Eratosthenes. Based on the actual sizes of Greek stadiums, it must have been about 1/6 kilometer. Thus, Eratosthenes estimated the circumference of the Earth to be about $\frac{250{,}000}{6} = 42{,}000$ kilometers—remarkably close to the modern value of just over 40,000 kilometers.

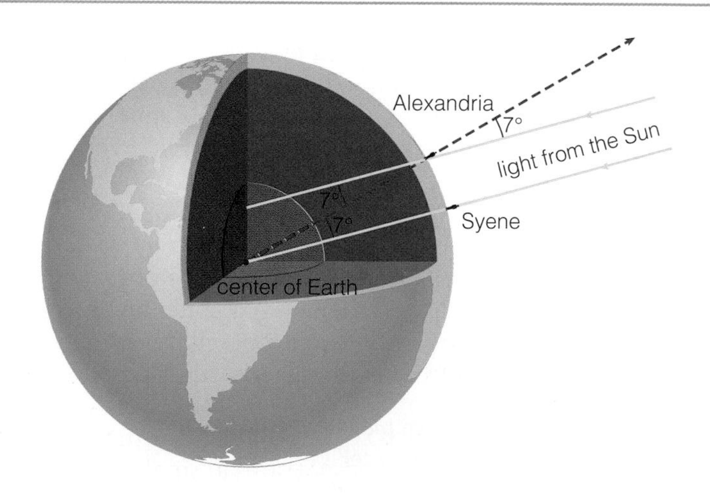

At noon on the summer solstice, the Sun appears at the zenith in Syene but 7° shy of the zenith in Alexandria. Thus, 7° of latitude, which corresponds to a distance of $\frac{7}{360}$ of Earth's circumference, must separate the two cities.

scholarly pursuits. Using the translated Greek scientific manuscripts as building blocks, these scholars developed algebra and many new instruments and techniques for astronomical observation. Most of the official names of constellations and stars come from Arabic because of the work of the scholars at Baghdad. If you look at a star chart, you will see that the names of many bright stars begin with *al* (e.g., Aldebaran, Algol), which simply means "the" in Arabic.

The Islamic world of the Middle Ages was in frequent contact with Hindu scholars from India, who in turn brought knowledge of ideas and discoveries from China. Hence, the intellectual center in Baghdad achieved a synthesis of the surviving work of the ancient Greeks and that of the Indians and the Chinese. The accumulated knowledge of the Arabs spread throughout the Byzantine empire (the eastern part of the former Roman empire). When the Byzantine capital of Constantinople (modern-day Istanbul) fell to the Turks in 1453, many Eastern scholars headed west to Europe, carrying with them the knowledge that helped ignite the European Renaissance.

3.3 THE COPERNICAN REVOLUTION

Modern science, as we generally think of it, arose during the European Renaissance. Within a half-century after the fall of Constantinople, Polish scientist Nicholas Copernicus began the work that ultimately overturned the Earth-centered, Ptolemaic model. Over the next century and a half, philosophers and scientists (who were often one and the same) debated and tested his radical view of the cosmos. Ultimately, the new ideas introduced by Copernicus fundamentally changed the way we perceive our place in the universe. This dramatic change, known as the **Copernican revolution**, spurred the development of virtually all modern science and technology.

• How did Copernicus, Tycho, and Kepler challenge the Earth-centered idea?

Recall that the idea of a Sun-centered solar system was suggested by Aristarchus [Section 2.4] more than 1,700 years before Copernicus was born. However, few people took the idea seriously before the time of the Copernican revolution. Let's examine how three key individuals—Copernicus, Tycho Brahe, and Kepler—challenged the prevailing dogma that held that our planet must be the center of the universe.

Copernicus Nicholas Copernicus was born in Torún, Poland, on February 19, 1473. His family was wealthy and he received a first-class education in mathematics, medicine, and law. He began studying astronomy in his late teens. By that time, tables of planetary motion based on the Ptolemaic model were noticeably inaccurate.

In his quest for a better way to predict planetary positions, Copernicus adopted Aristarchus's Sun-centered idea. He was probably motivated in large part by the much simpler explanation for apparent retrograde motion offered by a Sun-centered system (see Figure 2.28). As he worked out the mathematical details of his model, Copernicus also discovered simple geometric relationships that allowed him to calculate each planet's

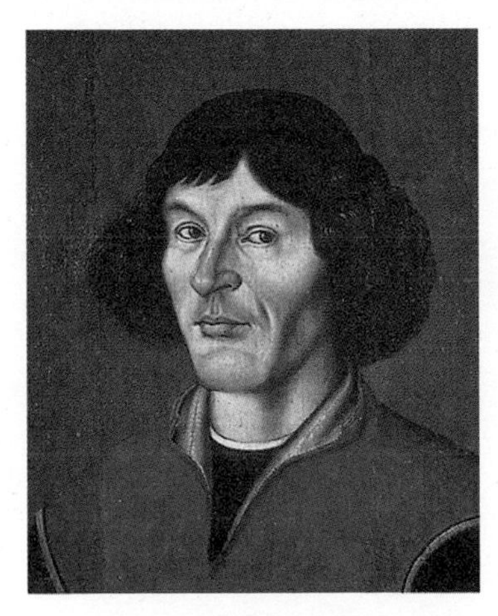

Copernicus (1473–1543)

orbital period around the Sun and its relative distance from the Sun in terms of Earth–Sun distance. The success of his model in providing a geometric layout for the solar system further convinced him that the Sun-centered idea must be correct.

Copernicus was hesitant to publish his work, fearing that his suggestion that Earth moved would be considered absurd. Nevertheless, he discussed his system with other scholars, generating great interest. At the urging of some of these scholars, including some high-ranking officials of the Catholic Church, he finally agreed to publish his work. Copernicus saw the first printed copy of his book, *De Revolutionibus Orbium Caelestium* ("Concerning the Revolutions of the Heavenly Spheres"), on the day he died—May 24, 1543.

Copernicus's Sun-centered model had the right general ideas, but its predictions were not substantially better than those of Ptolemy's Earth-centered model.

Despite having the correct idea of a Sun-centered system, Copernicus's model did not make substantially better predictions of planetary positions than Ptolemy's model. The problem was that Copernicus still believed that heavenly motion must occur in perfect circles. Because the true orbits of the planets are *not* circles, Copernicus found it necessary to add circles upon circles to his system, just as in the Ptolemaic system. As a result, his complete model was no more accurate and no less complex than the Ptolemaic model, and the Sun-centered idea won relatively few converts in the 50 years after it was published.

Tycho Brahe (1546–1601)

Tycho Part of the difficulty faced by astronomers who sought to improve either the Ptolemaic or the Copernican system was a lack of quality data. The telescope had not yet been invented, and existing naked-eye observations were not very accurate. In the late 1500s, Danish nobleman Tycho Brahe (1546–1601), usually known simply as Tycho (pronounced tie-koe), set about correcting this problem.

When Tycho was a young boy, his family discouraged his interest in astronomy. He therefore followed his passion in secret, learning the constellations from a miniature model of a celestial sphere that he kept hidden. As he grew older, Tycho was often arrogant about both his noble birth and his learned abilities. At age 20, he fought a duel with another student over which of them was the better mathematician. Part of Tycho's nose was cut off, and he designed a replacement piece made of silver and gold.

In 1563, Tycho decided to observe a widely anticipated alignment of Jupiter and Saturn. To his surprise, the alignment occurred nearly two days later than Copernicus had predicted. Resolving to improve the state of astronomical prediction, he set about compiling careful observations of stellar and planetary positions in the sky.

Tycho's fame grew after he observed what he called a *nova,* meaning "new star," in 1572 and proved that it was much farther away than the Moon. (Today, we know that Tycho saw a *supernova*—the explosion of a distant star [Section 12.3].) In 1577, Tycho observed a comet and proved that it too lay in the realm of the heavens. Others, including Aristotle, had argued that comets were phenomena of Earth's atmosphere. King Frederick II of Denmark decided to sponsor Tycho's ongoing work, providing him with money to build an unparalleled observatory for naked-eye observations (Figure 3.14). After Frederick II died in 1588, Tycho moved to Prague, where his work was supported by German emperor Rudolf II.

Tycho's accurate, naked-eye observations provided the data needed to improve the Copernican system.

Over a period of three decades, Tycho and his assistants compiled naked-eye observations accurate to within less than 1 arcminute—less than the thickness of a fingernail viewed at arm's length. Despite the quality of his observations, Tycho never succeeded in coming up with a satisfying explanation for planetary motion. He was convinced that the *planets* must orbit the Sun, but his inability to detect stellar parallax [Section 2.4] led him to conclude that Earth must remain stationary. Thus, he advocated a model in which the Sun orbits Earth while all other planets orbit the Sun. Few people took this model seriously.

Figure 3.14

Tycho Brahe in his naked-eye observatory, which worked much like a giant protractor. He could sit and observe a planet through the rectangular hole in the wall as an assistant used a sliding marker to measure the angle on the protractor.

Kepler Tycho failed to explain the motions of the planets satisfactorily, but he succeeded in finding someone who could: In 1600, he hired the young German astronomer Johannes Kepler (1571–1630). Kepler and Tycho had a strained relationship, but Tycho recognized the talent of his young apprentice. In 1601, as he lay on his deathbed, Tycho begged Kepler to find a system that would make sense of the observations so "that it may not appear I have lived in vain."

Kepler was deeply religious and believed that understanding the geometry of the heavens would bring him closer to God. Like Copernicus, he believed that Earth and the other planets traveled around the Sun in circular orbits. He worked diligently to match circular motions to Tycho's data.

Kepler labored with particular intensity to find an orbit for Mars, which posed the greatest difficulties in matching the data to a circular orbit. After years of calculation, Kepler found a circular orbit for Mars that *almost* matched Tycho's observations—the largest discrepancies between his model predictions and Tycho's data were about 8 arcminutes.

Kepler surely was tempted to ignore these discrepancies and attribute them to errors by Tycho. After all, 8 arcminutes is barely one-fourth the angular diameter of the full moon. But Kepler trusted Tycho's careful work. The small discrepancies finally led Kepler to abandon the idea of circular orbits—and to find the correct solution to the ancient riddle of planetary motion. About this event, Kepler wrote:

> *If I had believed that we could ignore these eight minutes [of arc], I would have patched up my hypothesis accordingly. But, since it was not permissible to ignore, those eight minutes pointed the road to a complete reformation in astronomy.*

By using elliptical orbits, Kepler created a Sun-centered model that predicted planetary positions with outstanding accuracy.

Kepler's key discovery was that planetary orbits are not circles but instead are a special type of oval called an **ellipse**. You probably know how to draw a circle by putting a pencil on the end of a string, tacking the string to a board, and pulling the pencil around (Figure 3.15a). Drawing an ellipse is similar, except that you must stretch the string around *two* tacks (Figure 3.15b). The locations of the two tacks are called the **foci** (singular, **focus**) of the ellipse. By altering the distance between the two foci while keeping the same length of string, you can draw ellipses of varying **eccentricity**, a quantity that describes how much an ellipse deviates from a perfect circle (Figure 3.15c). A circle is an ellipse with zero eccentricity, and greater eccentricity means a more elongated ellipse.

Kepler's decision to trust the data over his preconceived beliefs marked an important transition point in the history of science. Once he abandoned perfect circles in favor of ellipses, Kepler soon came up with a

Johannes Kepler (1571–1630)

Figure 3.15 Interactive Figure

An ellipse is a special type of oval. These diagrams show how an ellipse differs from a circle and how different ellipses vary in their eccentricity.

a Drawing a circle with a string of fixed length.
b Drawing an ellipse with a string of fixed length.
c *Eccentricity* describes how much an ellipse deviates from a perfect circle.

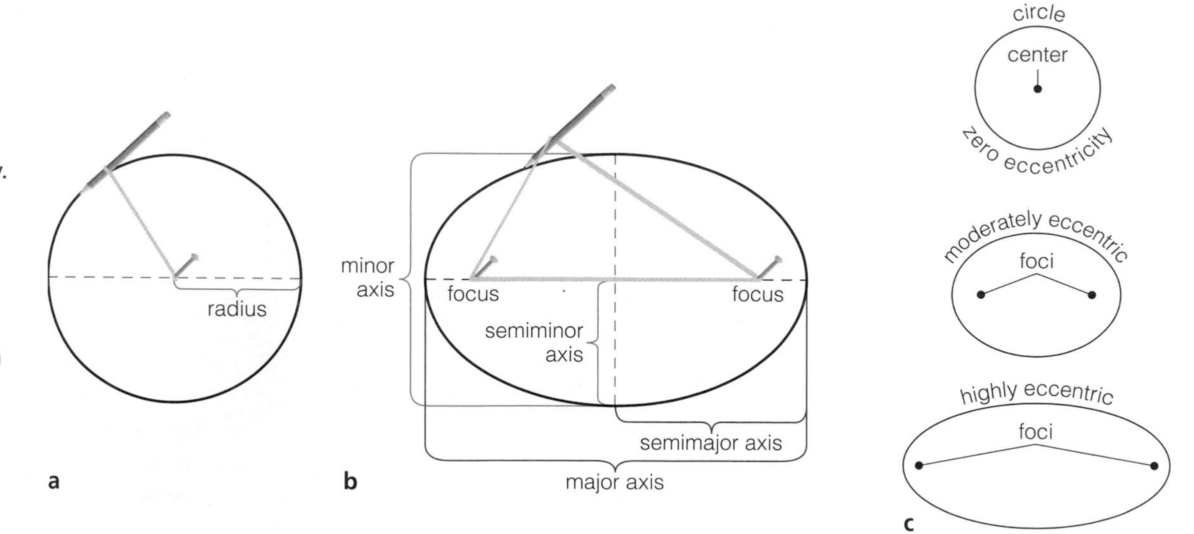

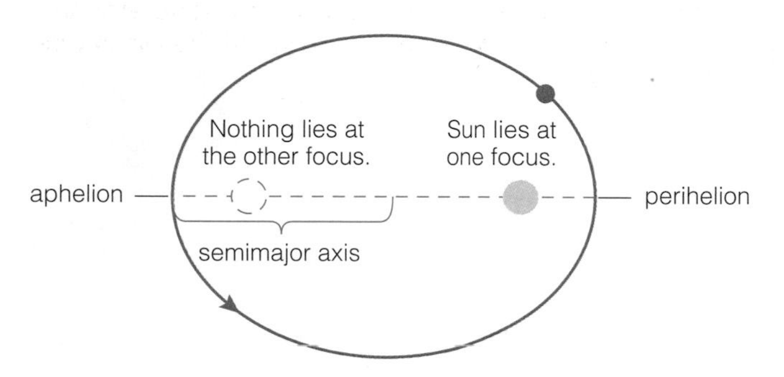

Figure 3.16 Interactive Figure

Kepler's first law: The orbit of each planet about the Sun is an ellipse with the Sun at one focus. (The eccentricity shown here is exaggerated compared to the actual eccentricities of the planets.)

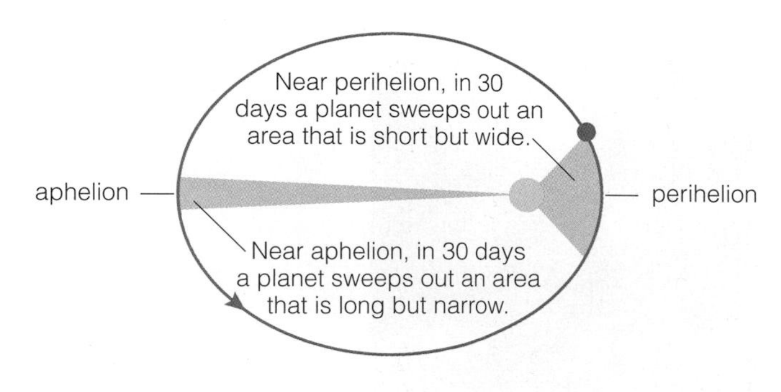

Figure 3.17 Interactive Figure

Kepler's second law: As a planet moves around its orbit, it sweeps out equal areas (the shaded regions) in equal times.

model that could predict planetary positions with far greater accuracy than Ptolemy's Earth-centered model. Kepler's model withstood the test of time and became accepted not only as a model of nature but as a deep, underlying truth about planetary motion.

astronomyplace.com **Orbits and Kepler's Laws Tutorial, Lessons 2–4**

• What are Kepler's three laws of planetary motion?

Kepler summarized his discoveries with three simple laws that we now call **Kepler's laws of planetary motion**. He published the first two laws in 1610 and the third in 1618.

Kepler's first law: The orbit of each planet about the Sun is an ellipse with the Sun at one focus.

Kepler's first law tells us that the orbit of each planet about the Sun is an ellipse with the Sun at one focus (Figure 3.16). (There is nothing at the other focus.) In essence, this law tells us that a planet's distance from the Sun varies during its orbit. It is closest at the point called **perihelion** (from the Greek for "near the Sun") and farthest at the point called **aphelion** (from the Greek for "away from the Sun"). The *average* of a planet's perihelion and aphelion distances is called its **semimajor axis**. We will refer to this simply as the planet's average distance from the Sun.

Kepler's second law: As a planet moves around its orbit, it sweeps out equal areas in equal times.

Kepler's second law states that as a planet moves around its orbit, it sweeps out equal areas in equal times. As shown in Figure 3.17, this means the planet moves a greater distance when it is near perihelion than it does in the same amount of time near aphelion. That is, the planet travels faster when it is nearer to the Sun and slower when it is farther from the Sun.

Kepler's third law: More distant planets orbit the Sun at slower average speeds, obeying the precise mathematical relationship $p^2 = a^3$.

Kepler's third law tells us that more distant planets orbit the Sun at slower average speeds, obeying a precise mathematical relationship. The relationship is written $p^2 = a^3$, where p is the planet's orbital period in years and a is its average distance

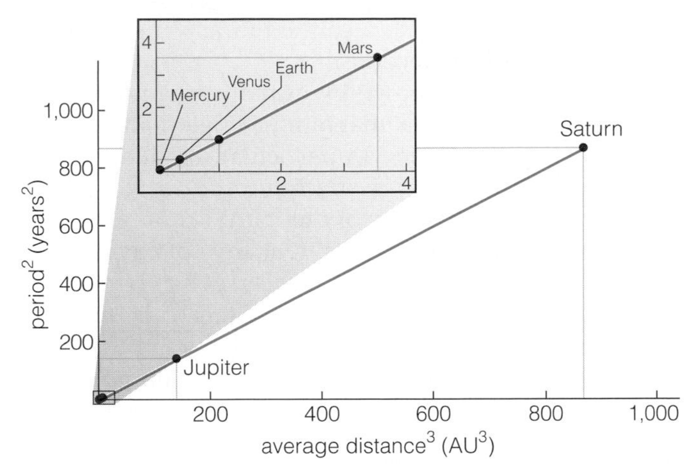

a The precise statement of Kepler's third law is $p^2 = a^3$, where *p* is a planet's orbital period in years and *a* is its average distance from the Sun in AU. The graph shows this relationship for the planets known in Kepler's time. The straight line tells us that p^2 (plotted along the vertical axis) is equal to a^3 (plotted along the horizontal axis).

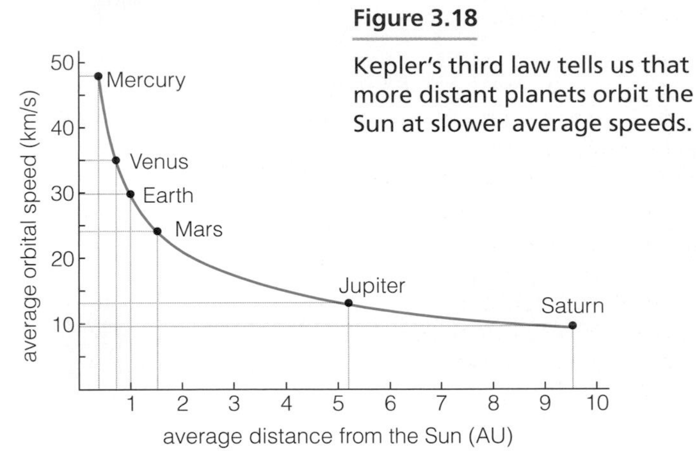

Figure 3.18

Kepler's third law tells us that more distant planets orbit the Sun at slower average speeds.

b Because Kepler's third law relates a planet's orbital distance to its orbital time (period), we can use it to calculate a planet's average orbital speed. This graph shows the result—more distant planets orbit the Sun more slowly. (Kepler knew the form of the relationship but could not determine speeds in km/s because the numerical value of the astronomical unit was not yet known.)

from the Sun in astronomical units. Figure 3.18a shows the $p^2 = a^3$ law graphically. Notice that the square of each planet's orbital period (p^2) is indeed equal to the cube of its average distance from the Sun (a^3). Figure 3.18b plots the actual periods and average distances, confirming that more distant planets orbit the Sun more slowly.

THINK ABOUT IT Suppose a comet has an orbit that brings it quite close to the Sun at its perihelion and beyond Mars at its aphelion, but with an average distance (semimajor axis) of 1 AU. According to Kepler's laws, how long would the comet take to complete each orbit of the Sun? Would it spend most of its time close to the Sun, far from the Sun, or somewhere in between? Explain.

• How did Galileo solidify the Copernican revolution?

The success of Kepler's laws in matching Tycho's data provided strong evidence in favor of Copernicus's placement of the Sun, rather than Earth, at the center of the solar system. Nevertheless, many scientists still voiced reasonable objections to the Copernican view. There were three basic objections, all rooted in the 2,000-year-old beliefs of Aristotle and other ancient Greeks.

- First, Aristotle had held that Earth could not be moving because, if it were, objects such as birds, falling stones, and clouds would be left behind as Earth moved along its way.
- Second, the idea of noncircular orbits contradicted Aristotle's claim that the heavens—the realm of the Sun, Moon, planets, and stars—must be perfect and unchanging.
- Third, as the ancients had argued, stellar parallax ought to be detectable if Earth orbits the Sun [Section 2.4].

Galileo (1564–1642)

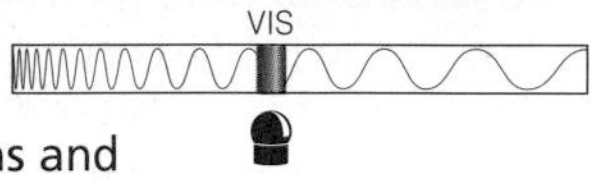

Figure 3.19

The shadows cast by mountains and crater rims near the dividing line between the light and dark portions of the lunar face prove that the Moon's surface is not perfectly smooth.

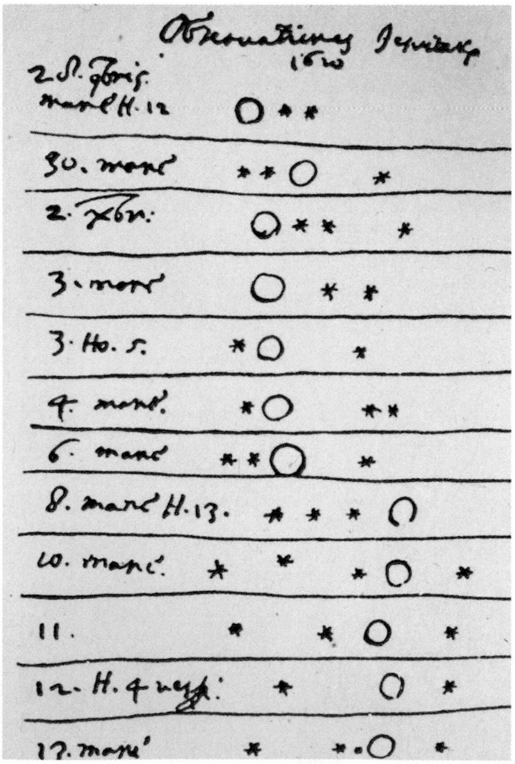

Figure 3.20

A page from Galileo's notebook written in 1610. His sketches show four "stars" near Jupiter (the circle) but in different positions at different times (and sometimes hidden from view). Galileo soon realized that the "stars" were actually moons orbiting the giant planet.

Galileo Galilei (1564–1642), a contemporary and correspondent of Kepler, answered all three objections.

Galileo (nearly always known by only his first name) defused the first objection with experiments that almost single-handedly overturned the Aristotelian view of physics. In particular, he demonstrated that a moving object remains in motion *unless* a force acts to stop it (an idea now codified in Newton's first law of motion [Section 4.2]). This contradicted Aristotle's claim that the natural tendency of any moving object is to come to rest. Galileo concluded that objects such as birds, falling stones, and clouds that are moving with Earth should *stay* with Earth unless some force knocks them away. This same idea explains why passengers in an airplane stay with the moving airplane even when they leave their seats.

Galileo's experiments and telescopic observations overcame remaining scientific objections to the Copernican idea, sealing the case for the Sun-centered solar system.

Tycho's supernova and comet observations already had challenged the validity of the second objection by showing that the heavens could change. Galileo shattered the idea of heavenly perfection after he built a telescope in late 1609. (Galileo did *not* invent the telescope. It was invented in 1608 by Hans Lippershey. However, Galileo took what was little more than a toy and turned it into a scientific instrument.) Through his telescope, Galileo saw sunspots on the Sun, which were considered "imperfections" at the time. He also used his telescope to prove that the Moon has mountains and valleys like the "imperfect" Earth by noticing the shadows cast near the dividing line between the light and dark portions of the lunar face (Figure 3.19). If the heavens were in fact not perfect, then the idea of elliptical orbits (as opposed to "perfect" circles) was not so objectionable.

The third objection—the absence of observable stellar parallax—had been of particular concern to Tycho. Based on his estimates of the distances of stars, Tycho believed that his naked-eye observations were sufficiently precise to detect stellar parallax if Earth did in fact orbit the Sun. Refuting Tycho's argument required showing that the stars were more distant than Tycho had thought and therefore too distant for him to have observed stellar parallax. Although Galileo didn't actually prove this fact, he provided strong evidence in its favor. For example, he saw with his telescope that the Milky Way resolved into countless individual stars. This discovery helped him argue that the stars were far more numerous and more distant than Tycho had imagined.

In hindsight, the final nails in the coffin of the Earth-centered universe came with two of Galileo's earliest discoveries through the telescope. First, he observed four moons clearly orbiting Jupiter, *not* Earth (Figure 3.20). Soon thereafter, he observed that Venus goes through phases in a way that proved that it must orbit the Sun and not Earth (Figure 3.21).

Although we now recognize that Galileo won the day, the story was more complex in his own time, when Catholic Church doctrine still held Earth to be the center of the universe. On June 22, 1633, Galileo was brought before a Church inquisition in Rome and ordered to recant his claim that Earth orbits the Sun. Nearly 70 years old and fearing for his life, Galileo did as ordered. His life was spared. However, legend has it that as he rose from his knees he whispered under his breath, *Eppur si muove*—Italian for "And yet it moves." (Given the likely consequences if Church officials had heard him say this, most historians doubt the legend.)

Galileo was not formally vindicated by the Church until 1992, but the Church gave up the argument long before that. Galileo's book, *Dia-*

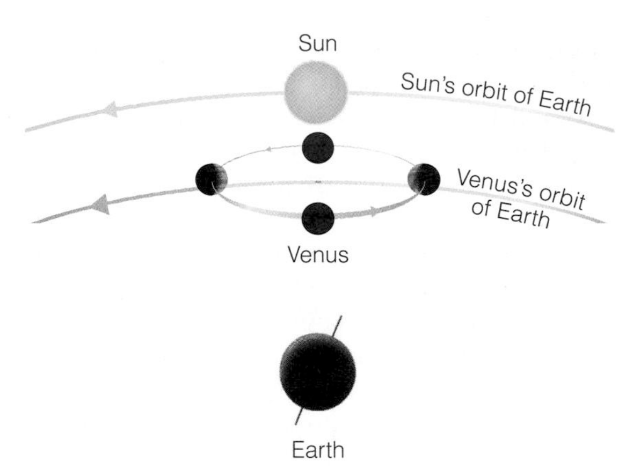

a In the Ptolemaic system, Venus orbits Earth, moving around a small circle on its larger orbital circle. But Venus always stays between Earth and the Sun in the sky, and therefore if this view were correct its phases would range only from new to crescent.

b In reality, Venus orbits the Sun, so from Earth we see it go through a complete set of phases. This is just want Galileo observed, allowing him to prove that Venus really does orbit the Sun.

Figure 3.21 Interactive Figure

Galileo's telescopic observations of Venus proved that it orbits the Sun rather than Earth.

logue Concerning the Two Chief World Systems, was removed from its index of banned books in 1824. Today, Catholic scientists are at the forefront of much astronomical research, and official Church teachings are compatible not only with Earth's planetary status but also with the theories of the Big Bang and the subsequent evolution of the cosmos and of life.

3.4 THE NATURE OF SCIENCE

The story of how our ancestors gradually figured out the basic architecture of the cosmos exhibits many features of what we now consider "good science." For example, we have seen how models were formulated and tested against observations, and modified or replaced when they failed those tests. The story also illustrates some classic mistakes, such as the failure of anyone before Kepler to question the belief that orbits must be circles. The ultimate success of the Copernican revolution led scientists, philosophers, and theologians to reassess the various modes of thinking that played a role in the 2,000-year process of discovering Earth's place in the universe. Let's examine how the principles of modern science emerged from the lessons learned in the Copernican revolution.

• How can we distinguish science from nonscience?

Perhaps surprisingly, it turns out to be quite difficult to define the term *science* precisely. The word comes from the Latin *scientia,* meaning "knowledge," but not all knowledge is science. For example, you may know what music you like best, but your musical taste is not a result of scientific study. So how can we tell what is science and what is not?

Approaches to Science One reason science is difficult to define is that not all science works in the same way. Sometimes, science means

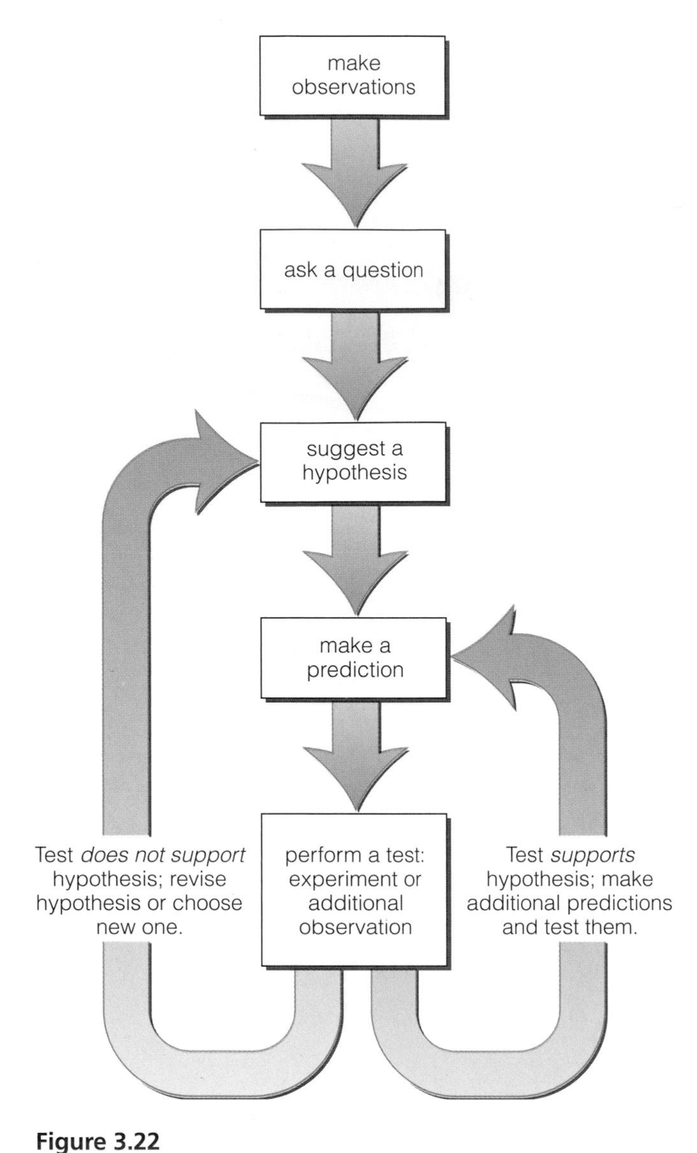

Figure 3.22

This diagram illustrates what we often call the *scientific method.*

proposing an idea and then performing experiments or observations to test it. As an example, consider what you would do if your flashlight suddenly stopped working. You might question why it has stopped working, and you might *hypothesize* that the reason is that the batteries have died. In other words, you've created a tentative explanation, or **hypothesis**, for the flashlight's failure. A hypothesis is sometimes called an *educated guess*—in this case it is "educated" because you already know that flashlights need batteries. Your hypothesis then allows you to make a simple prediction: If you replace the batteries with new ones, the flashlight should work. You can test this prediction by replacing the batteries. If the flashlight now works, you've confirmed your hypothesis. If it doesn't, you must revise or discard your hypothesis, usually in favor of some other one that you can also test (such as that the bulb is dead). Figure 3.22 illustrates the basic flow of this process, often referred to as the "scientific method.

The scientific method is a useful idealization of scientific thinking, but science rarely progresses in such an orderly way.

The scientific method is a useful idealization of scientific thinking, but science rarely progresses in such an orderly way. Scientific progress often begins with someone going out and looking at nature in a general way in hopes of learning something new and unexpected, rather than conducting a careful set of experiments. Furthermore, scientists are human beings, and their intuition and personal beliefs inevitably influence their work. Copernicus, for example, adopted the idea that Earth orbits the Sun not because he had carefully tested it but because he believed it made more sense than the prevailing view of an Earth-centered universe. While his intuition guided him to the right general idea, he erred in the specifics because he still clung to Plato's ancient belief that heavenly motion must be in perfect circles.

Given that the idealized scientific method is an overly simplistic characterization of science, how do we decide whether a claim is scientific? To answer this question, we must look a little deeper at the distinguishing characteristics of scientific thinking.

Hallmarks of Science One way to define scientific thinking is to list the criteria that scientists use when they judge competing models of nature. Historians and philosophers of science have examined (and continue to examine) this issue in great depth, and different experts express somewhat different viewpoints on the details. Nevertheless, everything we now consider to be science shares the following three basic characteristics, which we will refer to as the "hallmarks" of science (Figure 3.23):

- Modern science seeks explanations for observed phenomena that rely solely on natural causes.
- Science progresses through the creation and testing of models of nature that explain the observations as simply as possible.
- A scientific model must make testable predictions about natural phenomena that would force us to revise or abandon the model if the predictions do not agree with observations.

Each of these hallmarks is evident in the story of the Copernican revolution. The first shows up in the way Tycho's exceptionally careful measurements of planetary motion motivated Kepler to come up with a better explanation for those motions. The second is evident in the way several competing models were compared and tested, most notably those of

Ptolemy, Copernicus, and Kepler. We see the third in the fact that each model could make precise predictions about the future motions of the Sun, Moon, planets, and stars in our sky. When a model's predictions failed, the model was modified or ultimately discarded. Kepler's model gained acceptance in large part because its predictions matched Tycho's observations much better than Ptolemy's model.

Science seeks to explain observed phenomena using testable models of nature that explain the observations as simply as possible.

The criterion of simplicity in the second hallmark deserves further explanation. Remember that the original model of Copernicus did *not* match the data noticeably better than Ptolemy's model. Thus, if scientists had judged Copernicus's model solely on the accuracy of its predictions, they might have rejected it immediately. However, many scientists found elements of the Copernican model appealing, such as the simplicity of its explanation for apparent retrograde motion. They therefore kept the model alive until Kepler found a way to make it work.

In fact, if agreement with data were the sole criterion for judgment, we could imagine a modern-day Ptolemy adding millions or billions of additional circles to the geocentric model in an effort to improve its agreement with observations. A sufficiently complex geocentric model could in principle reproduce the observations with almost perfect accuracy—but it still would not convince us that Earth is the center of the universe. We would still choose the Copernican view over the geocentric view because its predictions would be just as accurate yet would follow from a much simpler model of nature. The idea that scientists should prefer the simpler of two models that agree equally well with observations is called *Occam's razor,* after the medieval scholar William of Occam (1285–1349).

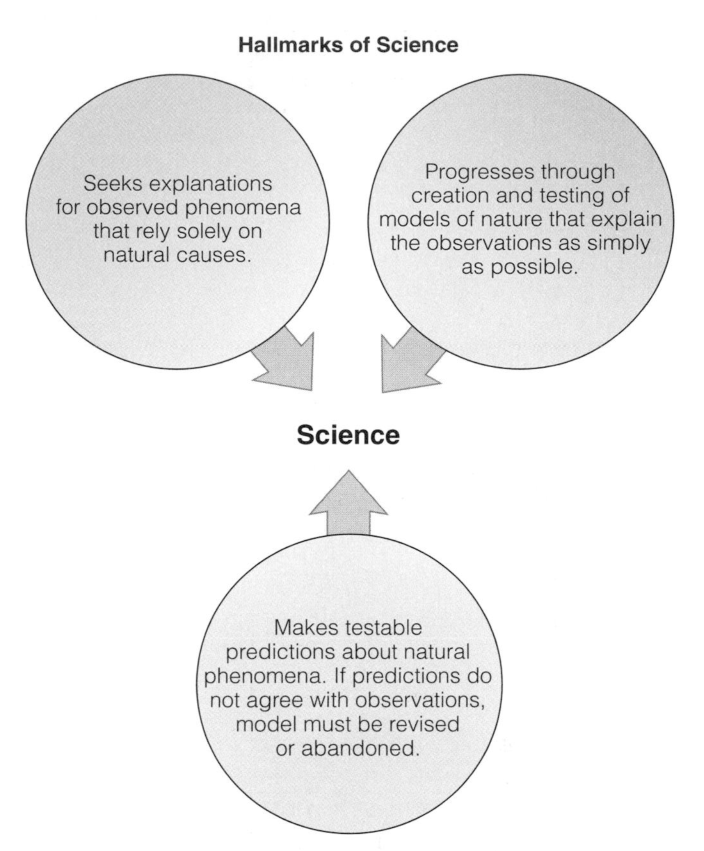

Figure 3.23

Hallmarks of science.

Objectivity in Science It's important to realize that science is not the only valid way of seeking knowledge. For example, suppose you are shopping for a car, learning to play drums, or pondering the meaning of life. In each case, you might make observations, exercise logic, and test hypotheses. Yet these pursuits clearly are not science, because they are not directed at developing testable explanations for observed natural phenomena. As long as nonscientific searches for knowledge make no claims about how the natural world works, they do not conflict with science.

The boundaries between science and nonscience are sometimes blurry. In particular, because science is practiced by human beings, individual scientists bring their personal biases and beliefs to their scientific work. These biases can influence the way a scientist proposes or tests a model. In some cases, scientists have been known to cheat—either deliberately or subconsciously—to obtain the result they desire. For example, in the late nineteenth and early twentieth centuries, some astronomers (including, most notably, Percival Lowell) claimed to see artificial canals in their blurry telescopic images of Mars. They hypothesized that Mars was home to a dying civilization that used the canals to transport water to thirsty cities (leading H. G. Wells to write his novel *The War of the Worlds*). No such canals actually exist. These astronomers apparently allowed their beliefs about extraterrestrial life to influence the way they interpreted blurry images—in essence, a form of cheating, though certainly not intentional.

Bias can sometimes show up even in the thinking of the scientific community as a whole. Some valid ideas may not be considered by any scientist because the ideas fall too far outside the general patterns of thought, or **paradigm**, of the time. Einstein's theory of relativity provides an ex-

EGGS ON THE EQUINOX

One of the hallmarks of science holds that you needn't take scientific claims on faith. In principle, at least, you can always test them for yourself. Consider the claim, repeated in news reports every year, that the spring equinox is the only day on which you can balance an egg on its end. Many people believe this claim, but you'll be immediately skeptical if you think about the nature of the spring equinox. The equinox is merely a point in time at which sunlight strikes both hemispheres equally (see Figure 2.13). It's difficult to see how sunlight could affect an attempt to balance eggs (especially if the eggs are indoors).

More important, you can test this claim directly. It's not easy to balance an egg on its end, but with practice you'll find that you can do it on any day of the year, not just on the spring equinox. Not all scientific claims are so easy to test for yourself, but the basic lesson should be clear: Before you accept any scientific claim, you should demand at least a reasonable explanation of the evidence that backs it up.

ample. Many scientists in the decades before Einstein had gleaned hints of the theory but did not investigate them, at least in part because they seemed too outlandish.

Individual scientists inevitably carry personal biases into their work, but the collective action of many scientists should ultimately make science objective.

The beauty of science is that it encourages continued testing by many people. Even if personal biases affect some results, tests by others will eventually uncover the mistakes. Similarly, if a new idea is correct but falls outside the accepted paradigm, sufficient testing and verification of the idea will eventually force a change in the paradigm. Thus, although individual scientists rarely follow the idealized scientific method, the collective action of many scientists over many years ensures that science as a whole remains objective.

• What is a scientific theory?

The most successful scientific models explain a wide variety of observations in terms of just a few general principles. When a powerful yet simple model makes predictions that survive repeated and varied testing, scientists elevate its status and call it a **theory**. Some famous examples are Isaac Newton's theory of gravity, Charles Darwin's theory of evolution, and Albert Einstein's theory of relativity.

Note that the use of the word *theory* in science contrasts with our everyday usage, which equates theories more closely with speculations or hypotheses. In everyday life, someone might get a new idea and say, for example, "I have a new theory about why people enjoy the beach." Without the support of a broad range of evidence that has been tested and confirmed by others, this idea is really only a hypothesis. Newton's theory of gravity qualifies as a scientific theory because it can be stated in

SPECIAL TOPIC: ASTROLOGY

Although the terms *astrology* and *astronomy* sound very similar, today they describe very different practices. In ancient times, however, astrology and astronomy often went hand in hand, and astrology played an important role in the historical development of astronomy. Indeed, astronomers and astrologers were usually one and the same.

In brief, the basic tenet of astrology is that human events are influenced by the apparent positions of the Sun, Moon, and planets among the stars in our sky. The origins of this idea are easy to understand. After all, there is no doubt that the position of the Sun in the sky influences our lives—it determines the seasons and hence the times of planting and harvesting, of warmth and cold, and of daylight and darkness. Similarly, the Moon determines the tides, and the cycle of lunar phases coincides with many biological cycles. Because the planets also appear to move among the stars, it seemed reasonable to imagine that planets also influence our lives, even if these influences were much more difficult to discover.

Ancient astrologers hoped that they might learn *how* the positions of the Sun, Moon, and planets influence our lives. They charted the skies, seeking correlations with events on Earth. For example, if an earthquake occurred when Saturn was entering the constellation of Leo, might Saturn's position have been the cause of the earthquake? If the king became ill when Mars appeared in the constellation Gemini and the first-quarter moon appeared in Scorpio, might it mean another tragedy for the king when this particular alignment of the Moon and Mars next recurred? Surely, the ancient astrologers thought, the patterns of influence eventually would become clear. Thus, the astrologers hoped that they might someday learn to forecast human events with the same reliability with which astronomical observations of the Sun could forecast the coming of spring.

This hope was never realized. Although many astrologers still attempt to predict future events, scientific tests have shown that their predictions come true no more often than would be expected by pure chance, Moreover, in light of our current understanding of the universe, the original ideas behind astrology no longer make sense. For example, today we use ideas of gravity and energy to explain the influences of the Sun and the Moon, and these same ideas tell us that the planets are too far from Earth to have a similar influence on our lives.

Of course, many people continue to practice astrology, perhaps because of its ancient and rich traditions. Scientifically, we cannot say anything about such traditions, because traditions are not testable predictions. But if you want to understand the latest discoveries about the cosmos, you'll need a science that can be tested and refined—and astrology can't meet these requirements.

simple mathematical terms and it explains a great many observations and experiments.

A scientific theory is a simple yet powerful model whose predictions have been borne out by repeated and varied testing.

Despite its success in explaining observed phenomena, a scientific theory can never be proved true beyond all doubt, because ever more sophisticated observations may eventually disagree with its predictions. However, anything that qualifies as a scientific theory must be supported by a large, compelling body of evidence.

In this sense, a scientific theory is not at all like a hypothesis or any other type of guess. We are free to change a hypothesis at any time, because it has not yet been carefully tested. In contrast, we can discard or replace a scientific theory only if we have an alternate way of explaining the evidence that supports it.

Again, the theories of Newton and Einstein offer a good example. Newton's theory of gravity is supported by a vast body of evidence, but by the late 1800s scientists had begun to discover cases where its predictions did not perfectly match observations. Einstein was motivated to develop his general theory of relativity in part to try to explain these failures of Newton's theory, and his new theory did indeed match the observations in these cases. Still, the many successes of Newton's theory could not be ignored, and Einstein's theory would not have gained acceptance if it had not been able to explain these successes equally well. It did, and that is why we now view Einstein's theory as a broader theory of gravity than Newton's theory. Some scientists today are seeking a theory of gravity that will go beyond Einstein's. If any new theory ever gains acceptance, it will have to match all the successes of Einstein's theory as well as working in new realms where Einstein's theory does not work.

THINK ABOUT IT When someone claims that something is "only a theory," what do you think they mean? Does this meaning of "theory" agree with the definition of a theory in science? Do scientists always use the word *theory* in its "scientific" sense? Explain.

THE BIG PICTURE

Putting Chapter 3 into Context

In this chapter, we focused on the scientific principles through which we have learned so much about the universe. Key "big picture" concepts from this chapter include the following:

- The basic ingredients of scientific thinking—careful observation and trial-and-error testing—are a part of everyone's experience. Modern science simply provides a way of organizing this everyday thinking to facilitate the learning and sharing of new knowledge.
- Although our understanding of the universe is growing rapidly today, each new piece of knowledge rests on ideas that came before. The foundations of astronomy reach far back into history and are intertwined with the general development of human culture and civilization. The ancient Greeks played a particularly important role,

including developing the idea that models can be used to explain and represent the architecture of the cosmos.

- The Copernican revolution, which overthrew the ancient Greek belief in an Earth-centered universe, did not occur instantaneously. It unfolded over a period of more than a century, during which many of the characteristics of modern science first appeared. Key figures in this revolution include Copernicus, Tycho, Kepler, and Galileo.
- Several key hallmarks distinguish science from other ways of gathering knowledge: Science seeks natural explanations for observed phenomena, it progresses through the creation and testing of models that explain the observations as simply as possible, and it revises or abandons models whose predictions disagree with observations.

3.1 THE ANCIENT ROOTS OF SCIENCE

- **In what ways do all humans employ scientific thinking?**

Scientific thinking relies on the same type of trial and error thinking that we use in our everyday lives, but in a carefully organized way.

- **How did astronomical observations benefit ancient societies?**

Ancient cultures used astronomical observations to help them keep track of time and the seasons, crucial skills for people who depended on agriculture for survival.

- **What did ancient civilizations achieve in astronomy?**

Ancient astronomers were accomplished observers who learned to tell the time of day and the time of year, to track cycles of the Moon, and to observe planets and stars. Many ancient structures aided in astronomical observations.

3.2 ANCIENT GREEK SCIENCE

- **Why does modern science trace its roots to the Greeks?**

The Greeks developed **models** of nature and emphasized the importance of having the predictions of those models agree with observations of nature.

- **How did the Greeks explain planetary motion?**

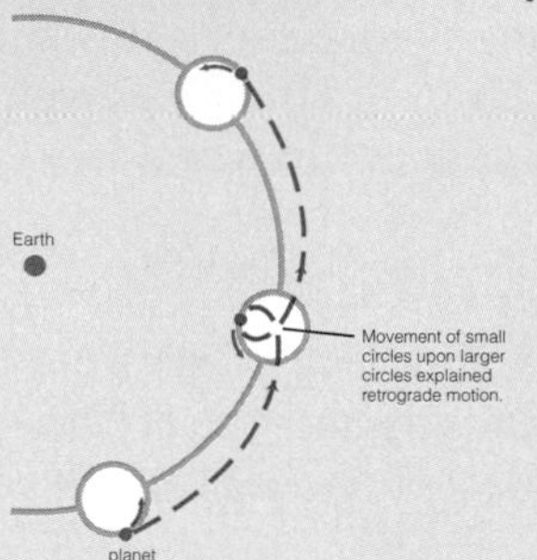

The Greek **geocentric model** reached its culmination with the model of Ptolemy. The **Ptolemaic model** explained apparent retrograde motion by having each planet move on a small circle whose center moves around Earth on a larger circle.

- **How did Islamic scientists preserve and extend Greek science?**

While Europe was in its Dark Ages, Islamic scholars in the Middle East preserved and extended ancient Greek knowledge. After the fall of Constantinople, scholars moved west to Europe, where their knowledge helped ignite the European Renaissance.

3.3 THE COPERNICAN REVOLUTION

- **How did Copernicus, Tycho, and Kepler challenge the Earth-centered idea?**

Copernicus created a Sun-centered model of the solar system designed to replace the Ptolemaic model, but it was no more accurate than Ptolemy's because Copernicus still used perfect circles. Tycho's accurate, naked-eye observations provided the data needed to improve on Copernicus's model. Kepler found a model that fit Tycho's data. Kepler's model is summarized by his three laws of planetary motion.

• **What are Kepler's three laws of planetary motion?**

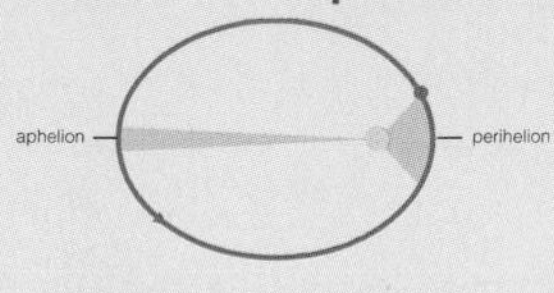

The areas swept out in 30-day periods are all equal.

(1) The orbit of each planet is an ellipse with the Sun at one focus. (2) As a planet moves around its orbit, it sweeps out equal areas in equal times. (3) More distant planets orbit the Sun at slower average speeds, obeying the precise mathematical relationship $p^2 = a^3$.

• **How did Galileo solidify the Copernican revolution?**

Galileo's experiments and telescopic observations overcame remaining objections to the Copernican idea of Earth as a planet orbiting the Sun. Although not everyone accepted his results immediately, in hindsight we see that Galileo sealed the case for the Sun-centered solar system.

3.4 THE NATURE OF SCIENCE

• **How can we distinguish science from nonscience?**

Science generally exhibits these three hallmarks: (1) Modern science seeks explanations for observed phenomena that rely solely on natural causes. (2) Science progresses through the creation and testing of models of nature that explain the observations as simply as possible. (3) A scientific model must make testable predictions about natural phenomena that would force us to revise or abandon the model if the predictions do not agree with observations.

• **What is a scientific theory?**

A scientific **theory** is a simple yet powerful model that explains a wide variety of observations in terms of just a few general principles, and has attained the status of a theory by surviving repeated and varied testing.

EXERCISES AND PROBLEMS

REVIEW QUESTIONS

1. In what way is scientific thinking natural to all of us? How does modern science differ from this everyday type of thinking?
2. Why did ancient peoples study astronomy? Describe at least four astronomical achievements of ancient cultures.
3. How are the names of the seven days of the week related to astronomical objects?
4. What is a lunar calendar? Are lunar calendars still used today? Explain.
5. What do we mean by a *model* in science? Briefly summarize how the Greek *geocentric model* explained planetary motion. What do we mean by the *Ptolemaic model*?
6. What was the *Copernican revolution*, and how did it change the human view of the universe? Briefly describe major players and events in the Copernican revolution.
7. What is an *ellipse*? What is the *focus* of an ellipse? What do we mean by the *eccentricity* of an ellipse? Explain why ellipses are important in astronomy.
8. Clearly state each of *Kepler's laws of planetary motion*. For each law, describe in your own words what it means in a way that could be understood by almost anyone.
9. What is the difference between a *hypothesis* and a *theory* in science?
10. Briefly describe each of the three hallmarks of science and how they are useful.

DOES IT MAKE SENSE?

Decide whether the statement makes sense and explain why it does or does not. (For an example, see Chapter 1, "Does It Make Sense?")

11. Ancient astronomers failed to realize that Earth goes around the Sun because they just weren't as smart as people today.
12. The date of Christmas (December 25) is set each year according to a lunar calendar.
13. When navigating in the South Pacific, the Polynesians found their latitude with the aid of the pointer stars of the Big Dipper.
14. The Ptolemaic model reproduced apparent retrograde motion by having planets move sometimes counterclockwise and sometimes clockwise in their circles.
15. In science, saying that something is a theory means that it is really just a guess.
16. A scientific theory never gains acceptance until it has been proven true beyond all doubt.
17. If the planet Uranus had been identified as a planet in ancient times, we'd probably have eight days in a week.
18. Upon its publication in 1543, the Copernican model was immediately accepted by most scientists because the idea of a Sun-centered solar system made so much more sense than the ancient, Earth-centered idea.

PROBLEMS

19. *Ancient Accomplishments.* In as much depth as possible, describe one notable astronomical achievement of an ancient culture and explain why it was significant.
20. *What Makes It Science?* Choose a single idea in the modern view of the cosmos as discussed in Chapter 1, such as "The universe is expanding," "The universe began with a Big Bang," "We are made from elements manufactured by stars," or "The Sun orbits the center of the Milky Way Galaxy once every 230 million years."
    a. Briefly describe how the idea you have chosen is rooted in each of the three hallmarks of science discussed in this chapter. (That is, explain how it is based on observations, how our understanding of it depends on a model, and how the model is testable.)

b. No matter how strongly the evidence may support a scientific idea, we can never be certain beyond all doubt that the idea is true. For the idea you have chosen, describe an observation that might cause us to call the idea into question. Then briefly discuss whether you think that, overall, the idea is likely or unlikely to hold up to future observations. Defend your opinion.

21. *The Copernican Revolution.* Based on what you have learned about the Copernican revolution, write a one- to two-page essay about how you believe it altered the course of human history.
22. *Cultural Astronomy.* Choose a particular culture of interest to you, and research the astronomical knowledge and accomplishments of that culture. Write a two- to three-page summary of your findings.
23. *Astronomical Structures.* Choose an ancient astronomical structure of interest to you (e.g., Stonehenge, Nazca lines, Pawnee lodges) and research its history. Write a two- to three-page summary of your findings. If possible, also build a scale model of the structure or create detailed diagrams to illustrate how the structure was used.
24. *Venus and the Mayans.* The planet Venus apparently played a particularly important role in Mayan society. Research the evidence and write a one- to two-page summary of current knowledge about the role of Venus in Mayan society.
25. *Scientific Test of Astrology.* Find out about at least one scientific test that has been conducted to test the validity of astrology. Write a short summary of how the test was conducted and what conclusions were reached.
26. *Your Own Astrological Test.* Devise your own scientific test of astrology. Clearly define the methods you will use in your test and how you will evaluate the results. Then carry out the test and write a report on your methods and results.

DISCUSSION QUESTIONS

27. *The Impact of Science.* The modern world is filled with ideas, knowledge, and technology that developed through science and application of the scientific method. Discuss some of these things and how they affect our lives. Which of these impacts do you think are positive? Which are negative? Overall, do you think science has benefited the human race? Defend your opinion.
28. *The Importance of Ancient Astronomy.* Why was astronomy important to people in ancient times? Discuss both the practical importance of astronomy and the importance it may have had for religious or other traditions. Which do you think was more important in the development of ancient astronomy, its practical or its philosophical role? Defend your opinion.
29. *Astronomy and Astrology.* Why do you think astrology remains so popular around the world even though it has failed all scientific tests of its validity? Do you think the popularity of astrology has any positive or negative social consequences? Defend your opinions.

MATH HELP AND EXERCISES

For additional help and practice with mathematical concepts applicable to this chapter, the Astronomy Place web site has the following resources with detailed explanations, worked examples, and practice problems:

- The Metonic Cycle

For a complete list of media resources available, go to **www.astronomyplace.com** and choose Chapter 3 from the pull-down menu.

ASTRONOMY PLACE WEB TUTORIALS

Tutorial Review of Key Concepts

Use the interactive **Tutorial** at **www.astronomyplace.com** to review key concepts from this chapter.

Orbits and Kepler's Laws Tutorial

Lesson 2 Kepler's First Law
Lesson 3 Kepler's Second Law
Lesson 4 Kepler's Third Law

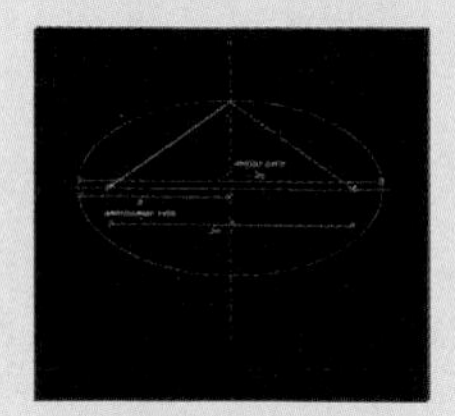
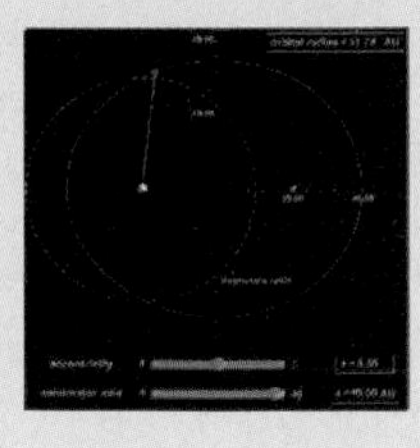
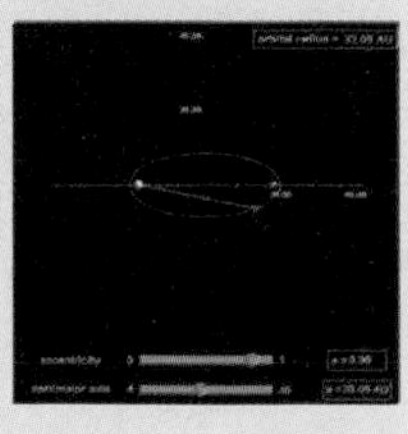

Supplementary Tutorial Exercises

Use the interactive **Tutorial Lesson** to explore the following questions.

Orbits and Kepler's Laws Tutorial, Lesson 2

1. When is an ellipse a circle?
2. Use the ellipse tool in Lesson 2 to show that the semimajor axis of an ellipse is a planet's average orbital radius:
   a. Orbital radius when the planet is closest to the Sun: perihelion = ____ AU
   b. Orbital radius when the planet is farthest from the Sun: aphelion = ____ AU
   c. Average of perihelion and aphelion = ____ AU
   d. Length of the major axis of the ellipse = ____ AU
   e. Length of the semimajor axis of the ellipse = ____ AU

   Are the average orbital radius (C) and the semimajor axis (E) equal?

EXPLORING THE SKY AND SOLAR SYSTEM

Of the many activities available on the ***Voyager: SkyGazer*** **CD-ROM** accompanying your book, use the following files to observe key phenomena covered in this chapter.

Go to the **File: Basics** folder for the following demonstrations.

1. Ptolemy on Venus
2. Phase of Mercury
3. Pluto's Orbit

Go to the **File: Demo** folder for the following demonstrations.

1. Hale–Bopp Path
2. Hyakutake Nears Earth
3. Venus–Earth–Moon

Go to the **Explore** menu for the following demonstrations.

1. Solar System
2. Paths of the Planets

WEB PROJECTS

Take advantage of the useful Web links on **www.astronomyplace.com** to assist you with the following projects.

1. *Easter.* Find out when Easter is celebrated by different sects of Christianity. Then research how and why different sects set different dates for Easter. Summarize your findings in a one- to two-page report.
2. *Greek Astronomers.* Many ancient Greek scientists had ideas that, in retrospect, seem well ahead of their time. Choose one or more of the following ancient Greek scientists, and learn enough about their work in science and astronomy to write a one- to two-page "scientific biography."

Thales	Anaximander	Pythagoras
Anaxagoras	Empedocles	Democritus
Meton	Plato	Eudoxus
Aristotle	Callipus	Aristarchus
Archimedes	Eratosthenes	Apollonius
Hipparchus	Seleucus	Ptolemy
Hypatia		

3. *The Ptolemaic Model.* This chapter gives only a very brief description of Ptolemy's model of the universe. Investigate the model in greater depth. Using diagrams and text as needed, give a two- to three-page description of the model.
4. *The Galileo Affair.* In recent years, the Roman Catholic Church has devoted a lot of resources to learning more about the trial of Galileo and to understanding past actions of the Church in the Galileo case. Learn more about such studies, and write a short report about the current Vatican view of the case.
5. *Science or Pseudoscience.* Choose some pseudoscientific claim that has been in the news recently, and learn more about it and how scientists have "debunked" it. Write a short summary of your findings.

4
Making Sense of the Universe: Understanding Motion, Energy, and Gravity

LEARNING GOALS

4.1 DESCRIBING MOTION: EXAMPLES FROM DAILY LIFE
- How do we describe motion?
- How is mass different from weight?

4.2 NEWTON'S LAWS OF MOTION
- How did Newton change our view of the universe?
- What are Newton's three laws of motion?

4.3 CONSERVATION LAWS IN ASTRONOMY
- What keeps a planet rotating and orbiting the Sun?
- Where do objects get their energy?

4.4 THE FORCE OF GRAVITY
- What determines the strength of gravity?
- How does Newton's law of gravity extend Kepler's laws?
- How do gravity and energy together allow us to understand orbits?
- How does gravity cause tides?

The history of the universe is essentially a story about the interplay between matter and energy since the beginning of time. Interactions between matter and energy began in the Big Bang and continue today in everything from the microscopic jiggling of atoms to gargantuan collisions of galaxies. Understanding the universe therefore depends on becoming familiar with how matter responds to the ebb and flow of energy.

You might guess that it would be difficult to understand the many interactions that shape the universe, because they occur on so many different size scales. However, we now know that just a few physical laws govern the movements of everything from atoms to galaxies. The Copernican revolution spurred the discovery of these laws, and Galileo deduced some of them from his experiments. But it was Sir Isaac Newton, one of the most influential human beings of all time, who put all of the pieces together into a simple system of laws describing both motion and gravity.

In this chapter, we'll discuss the laws that govern motion and energy in the universe, including Newton's laws of motion, the laws of conservation of angular momentum and of energy, and the universal law of gravitation. By understanding these laws, you will be able to make sense of the wide-ranging phenomena you will encounter as you continue your study of astronomy.

4.1 DESCRIBING MOTION: EXAMPLES FROM DAILY LIFE

Our primary goal in this chapter is to learn how we can make sense of motion in the universe. We all have a great deal of experience with motion and a natural intuition as to how motion works, but in science we need to define our ideas and terms precisely. In this section, we'll use examples from everyday life to help us define a few of the most important terms that arise in astronomy, including the terms that describe motion. We'll also explore the difference between mass and weight, which we often view as the same on Earth but which actually differ from one another in an important way.

• How do we describe motion?

You are probably familiar with the terms used to describe motion in science—terms such as *velocity, acceleration,* and *momentum.* However, their scientific definitions may differ subtly from those you use in casual conversation. Let's investigate the precise meanings of these terms.

Speed, Velocity, and Acceleration We use three basic terms to describe how an object is moving, all of which are familiar to you from driving a car:

- The **speed** of the car tells us how far it will go in a certain amount of time. For example, 100 km/hr (about 60 mi/hr) is a speed, and it tells us that you will cover a distance of 100 kilometers if you drive at this speed for an hour.
- The **velocity** of the car tells us both its speed and direction. For example, "100 km/hr going due north" describes a velocity.
- The car has an **acceleration** if its velocity is changing in any way, whether in speed or direction or both.

Figure 4.1

Speeding up, turning, and slowing down are all examples of acceleration.

30 km/hr 60 km/hr

We say that this car is accelerating because its velocity is increasing.

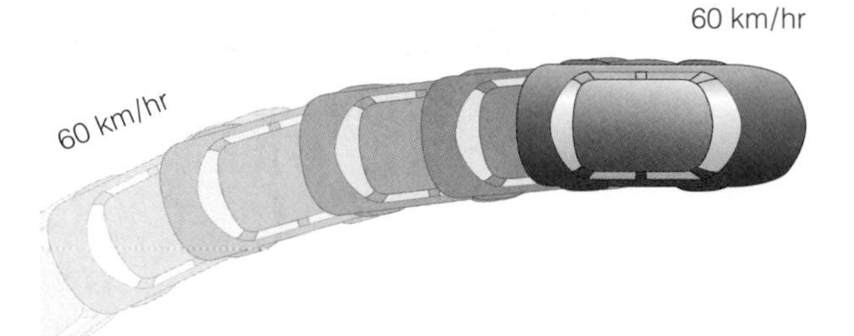

We say that this car is accelerating because its direction is changing as it turns, which means its velocity is changing even though its speed stays constant.

60 km/hr 30 km/hr 0 km/hr

We say that this car is accelerating because its velocity is decreasing. Decreasing velocity is still acceleration, although it is a negative acceleration.

An object is accelerating if either its speed or direction is changing.

You are undoubtedly familiar with the term *acceleration* as it applies to increasing speed. In science, we also say that you are accelerating when you slow down or turn (Figure 4.1). Slowing occurs when acceleration is in a direction opposite to the motion. In this case, we say that your acceleration is negative, causing your velocity to decrease. Turning changes your velocity because it changes the direction in which you are moving. Thus, turning is a form of acceleration even if your speed remains constant.

You can often feel the effects of acceleration. For example, as you speed up in a car you feel yourself being pushed back into your seat. As you slow down you feel yourself being pulled forward from the seat. As you drive around a curve you feel yourself being pushed away from the direction of your turn. In contrast, you don't feel such effects when moving at *constant velocity*. That is why you don't feel any sensation of motion when you're traveling in an airplane on a smooth flight.

The Acceleration of Gravity One of the most important types of acceleration is that caused by gravity, which makes objects accelerate as they fall. In a legendary experiment in which he supposedly dropped weights from the Leaning Tower of Pisa, Galileo demonstrated that gravity accelerates all objects by the same amount, regardless of their mass. This fact may be surprising because it seems to contradict everyday experience: A feather floats gently to the ground, while a rock plummets. However, air resistance causes this difference in acceleration. If you dropped a feather and a rock on the Moon, where there is no air, both would fall at exactly the same rate.

THINK ABOUT IT Find a piece of paper and a small rock. Hold both at the same height, one in each hand, and let them go at the same instant. The rock, of course, hits the ground first. Next, crumple the paper into a small ball and repeat the experiment. What happens? Explain how this experiment suggests that gravity accelerates all objects by the same amount.

The acceleration of a falling object is called the **acceleration of gravity**, abbreviated g. On Earth, the acceleration of gravity causes falling objects to fall faster by 9.8 meters per second (m/s), or about 10 m/s, with each passing second. For example, suppose you drop a rock from a tall building. At the moment you let it go, its speed is 0 m/s. After 1 second, the rock will be falling downward at about 10 m/s. After 2 seconds, it will be falling at about 20 m/s. In the absence of air resistance, its speed will continue to increase by about 10 m/s each second until it hits the ground (Figure 4.2). We therefore say that the acceleration of gravity is about 10 *meters per second per second*, or 10 *meters per second squared*, which we write as 10 m/s^2. (More precisely, g = 9.8 m/s^2.)

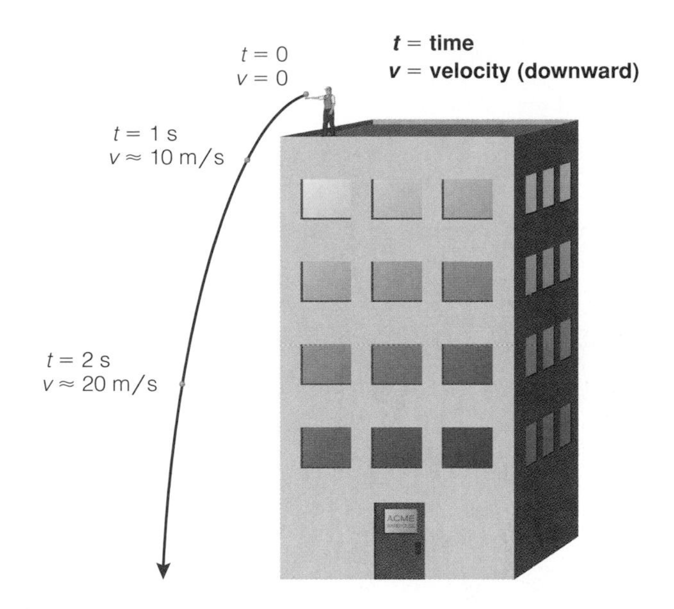

Figure 4.2

On Earth, gravity causes an unsupported object to accelerate downward at about 10 m/s^2, which means its downward velocity increases by about 10 m/s with each passing second. (Gravity does not affect horizontal velocity.)

Momentum and Force The concepts of speed, velocity, and acceleration describe how an individual object moves, but most of the interesting phenomena we see in the universe are the result of interactions between objects. We need two additional concepts to describe these interactions:

- An object's **momentum** is its combination of mass and velocity. (Mathematically, momentum = mass × velocity.)
- The only way to change an object's momentum is to apply a **force** to it.

We can understand these concepts by considering the effects of collisions. Imagine that you're innocently stopped in your car at a red light when a bug flying at a velocity of 30 km/hr due south slams into your windshield. What will happen to your car? Not much, except perhaps a bit of a mess on your windshield. Next imagine that a 2-ton truck runs the red light and hits you head-on with the same velocity as the bug. Clearly, the truck will cause far more damage. We can understand why by considering the momentum and force in each collision.

Before the collisions, the truck's much greater mass means it has far more momentum than the bug, even though both the truck and the bug are moving with the same velocity. During the collisions, the bug and the truck each transfer some of their momentum to your car. The bug has very little momentum to give to your car, so it does not exert much of a force. In contrast, the truck imparts enough of its momentum to cause a dramatic and sudden change in your car's momentum. You feel this sudden change in momentum as a force, and it can do great damage to you and your car.

The mere presence of a force does not always cause a change in momentum. For example, a moving car is always affected by forces of air resistance and friction with the road—forces that will slow your car if you take your foot off the gas pedal. However, you can maintain a constant velocity, and hence constant momentum, if you step on the gas pedal hard enough to overcome the slowing effects of these forces.

In fact, forces of some kind are always present, such as the force of gravity or the electromagnetic forces acting between atoms. The **net force** (or *overall force*) acting on an object represents the combined effect of all the individual forces put together. There is no net force on your car when you are driving at constant velocity, because the force generated by the engine to turn the wheels precisely offsets the forces of air resistance and road friction. A change in momentum occurs only when the net force is not zero.

An object must accelerate whenever a net force acts on it.

Changing an object's momentum means changing its velocity, as long as its mass remains constant. Thus, the presence of a net force causes an object to accelerate. Moreover, anytime an object accelerates a net force must be causing the acceleration. That is why you feel forces (pushing you forward, backward, or to the side) when you accelerate in your car. We can use the same ideas to understand many astronomical processes. For example, planets are always accelerating as they orbit the Sun, because their direction of travel constantly changes as they go around their orbits. We can therefore conclude that some force must be causing this acceleration. As we'll discuss shortly, Isaac Newton identified this force as gravity.

• How is mass different from weight?

In discussing momentum and force, we've referred several times to the idea of *mass*. In daily life, we usually think of mass as something you can measure with a bathroom scale, but technically the scale measures your weight, not your mass. The distinction between mass and weight rarely matters when we are talking about objects on Earth, but it is very important in astronomy:

- Your **mass** is the amount of matter in your body.
- Your **weight** describes a *force* that is acting on your mass.* In other words, it is what a scale measures when you stand on it.

To understand the difference between mass and weight, imagine standing on a scale in an elevator (Figure 4.3). Your mass will be the same no matter how the elevator is moving, but your weight can vary. When the elevator is stationary or moving at constant velocity, the scale reads your "normal" weight. When the elevator accelerates upward, the floor exerts a greater force than it does when you are at rest. You feel heavier, and the scale verifies your greater weight. When the elevator accelerates

*Many physics texts distinguish between *true weight,* which is due only to the effects of gravity on mass, and the *apparent weight* that a scale reads when other forces (such as in an accelerating elevator) also act. In this book, "weight" refers to apparent weight, except when stated otherwise.

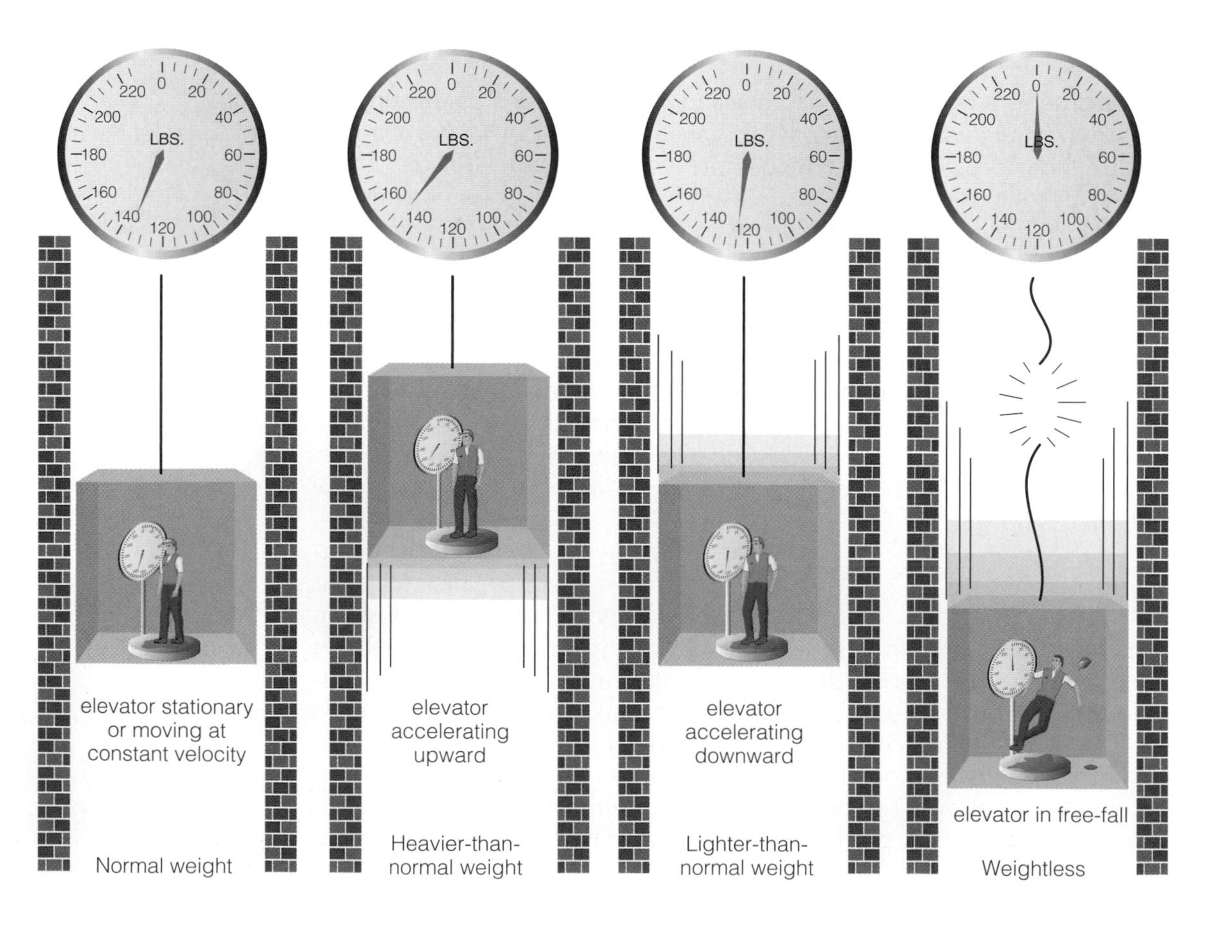

Figure 4.3 Interactive Figure

Mass is not the same as weight. The man's mass never changes, but his weight is different when the elevator accelerates.

downward, the floor and the scale exert a weaker force on you, so the scale registers less weight. Be sure to note that the scale shows a weight different from your "normal" weight only when the elevator is *accelerating,* not when it is going up or down at constant speed.

THINK ABOUT IT Find a small bathroom scale and take it with you on an elevator ride. Does your weight change when the elevator accelerates upward or downward? Does it change when the elevator is moving at constant speed? Explain your observations.

Your mass is the same no matter where you are, but your weight can vary.

Thus, your mass depends only on the amount of matter in your body and is the same anywhere, but your weight can vary because the forces acting on you can vary. For example, your mass would be the same on the Moon as on Earth, but you would weigh less on the Moon because of its weaker gravity.

Free-Fall and Weightlessness Now consider what happens if the elevator cable breaks (see the last frame in Figure 4.3). The elevator and you are suddenly both in **free-fall**—falling without any resistance to slow you down. The floor drops away at the same rate that you fall, allowing you to "float" freely above it, and the scale reads zero because you are no longer held to it. In other words, your free-fall has made you **weightless**.

In fact, you are in free-fall whenever there's nothing to *prevent* you from falling. For example, you are in free-fall when you jump off a chair or spring from a diving board or trampoline. Surprising as it may seem, you have therefore experienced weightlessness many times in your life. You can experience it right now simply by jumping off your chair. Of course, your weightlessness lasts for only the very short time until you hit the ground.

Weightlessness in Space You've probably seen video of astronauts floating weightlessly in the Space Shuttle or the Space Station. But why are they weightless? Many people guess that there's no gravity in space, but that's not true. After all, it is gravity that makes the Space Shuttle and Space Station orbit Earth. Astronauts are weightless for the same reason you are weightless when you jump off a chair: they are in free-fall.

People or objects are weightless whenever they are falling freely, and astronauts in orbit are weightless because they are in a constant state of free-fall.

Astronauts are weightless the entire time they are orbiting Earth because they are in a constant state of free-fall. The idea that an orbiting object is in a "constant state of free-fall" may sound a bit strange, but it's easy to understand. Imagine building a tower that went all the way to the Space Station's orbit, about 350 kilometers above Earth. Figure 4.4 shows the idea. If you stepped off the tower, you would fall downward, remaining weightless until you hit the ground (or until air resistance had a noticeable effect on you). Now, imagine that instead of stepping off the tower, you ran and jumped out of the tower. You'd still fall to the ground, but because of your forward motion you wouldn't fall straight down. You'd land a short distance away from the base of the tower.

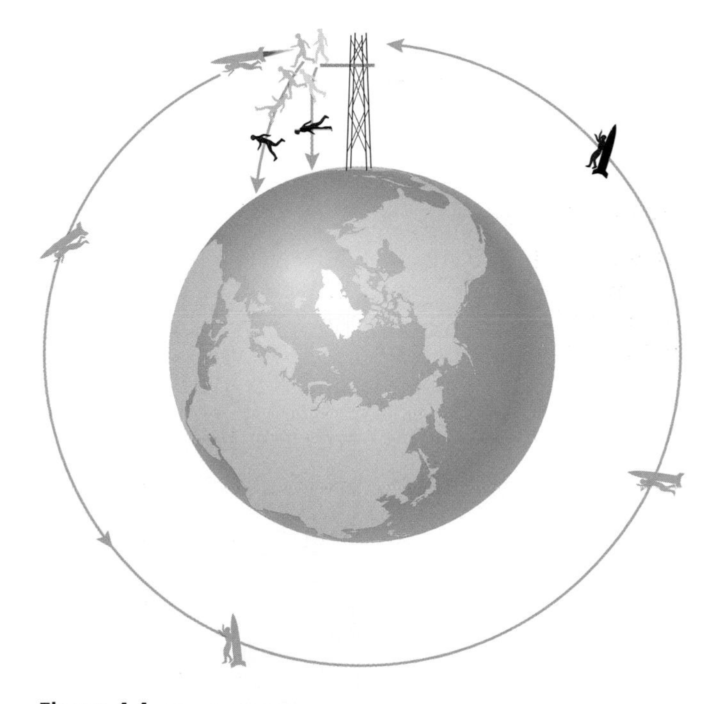

Figure 4.4 Interactive Figure

If you stepped off a tall tower, you'd fall to Earth. If you ran off, you'd still fall, but you'd land farther from the tower. If you could somehow run off fast enough (perhaps with the aid of a rocket), you'd continually "fall around" the Earth, meaning you'd be in orbit. In all cases, you are in free-fall after your initial leap, and thus would feel weightless and would float freely. *Note:* On the scale shown here, the tower extends far higher than the Space Station's orbit. (Adapted from *Space Station Science* by Marianne Dyson.)

COMMON MISCONCEPTIONS

NO GRAVITY IN SPACE?

If you ask people why astronauts are weightless in space, one of the most common answers is "There is no gravity in space." But you can usually convince people that this answer must be wrong by following with another simple question: Why does the Moon orbit Earth? Most people know that the Moon orbits Earth because of gravity, proving that there is gravity in space. In fact, at the altitude of the Space Station's orbit, the acceleration of gravity is scarcely less than it is on Earth's surface.

The real reason astronauts are weightless is that they are in a constant state of free-fall. Imagine being an astronaut. You'd have the sensation of free-fall—just as when you are falling from a diving board—the entire time you were in orbit. This constantly falling sensation makes most astronauts sick to their stomach when they first experience weightlessness. Fortunately, they quickly get used to the sensation, which allows them to work hard and enjoy the view.

The faster you ran out of the tower, the farther you'd go before landing. If you could somehow run fast enough—about 28,000 km/hr (17,000 mi/hr) at the orbital altitude of the Space Station—a very interesting thing would happen: By the time gravity had pulled you downward as far as the length of the tower, you'd already have moved far enough around Earth that you'd no longer being going down at all. Instead, you'd be just as high above Earth as you'd been all along, but a good portion of the way around the world. In other words, you'd be orbiting Earth.

Thus, the Space Shuttle, the Space Station, and all other orbiting objects stay in orbit because they are constantly "falling around" the Earth. Their constant state of free-fall makes these spacecraft and everything in them weightless.

Motion and Gravity Tutorial, Lesson 1

4.2 NEWTON'S LAWS OF MOTION

We've now covered the basic concepts of motion that we'll need in astronomy, so we are ready to explore the laws that govern motion. These laws will help us understand exactly how and why objects move through the universe.

The complexity of motion in daily life might lead you to guess that the laws governing motion would also be complex. For example, if you watch a falling piece of paper waft lazily to the ground, you'll see it rock irregularly back and forth in a seemingly unpredictable pattern. However, the complexity of this motion arises because the paper is affected by a variety of forces, including gravity and the changing forces caused by air currents. If you could analyze the forces individually, you'd find that each force affects the paper's motion in a simple, predictable way. Sir Isaac Newton (1642–1727) discovered the remarkably simple system of laws that govern all this motion.

• How did Newton change our view of the universe?

Sir Isaac Newton (1642–1727)

Newton was born prematurely in Lincolnshire, England, on Christmas Day in 1642. He had a difficult childhood and showed few signs of unusual talent. He attended Trinity College at Cambridge, where he earned his keep by performing menial labor, such as cleaning the boots and bathrooms of wealthier students and waiting on their tables.

The plague hit Cambridge shortly after Newton graduated, and he returned home. By his own account, he experienced a moment of inspiration in 1666 when he saw an apple fall to the ground. He suddenly realized that the gravity making the apple fall was the same force that held the Moon in orbit around Earth. In that moment, Newton shattered the remaining vestiges of the Aristotelian view of the world, which for centuries in Europe had been taken as near-gospel truth.

Aristotle had made many claims about the physics of motion, using his ideas to support his belief in an Earth-centered cosmos. He had also maintained that the heavens were totally distinct from Earth, so that physical laws on Earth did not apply to heavenly motion. By the time Newton saw the apple fall, the Copernican revolution had displaced Earth from a central position, and Galileo's experiments had shown that the laws of physics were not what Aristotle had believed.

Newton showed that the same physical laws that operate on Earth also operate in the heavens.

Newton's sudden insight delivered the final blow to Aristotle's physics. When Newton realized that gravity operated in the heavens as well as on Earth, he eliminated the distinction Aristotle had claimed between the two realms. For the first time in history, we saw the heavens and Earth brought together as one *universe.* Newton's insight also heralded the birth of the modern science of *astrophysics* (although the term wasn't coined until much later). Astrophysics applies physical laws discovered on Earth to phenomena throughout the cosmos.

Over the next 20 years, Newton's work completely revolutionized mathematics and science. He quantified the laws of motion and gravity, conducted crucial experiments regarding the nature of light, built the first reflecting telescopes, and invented the mathematics of calculus. We'll discuss his laws of motion in the rest of this section, and later in the chapter we'll turn our attention to Newton's discoveries about gravity.

• What are Newton's three laws of motion?

Newton published the laws of motion and gravity in 1687, in a book usually called *Principia,* short for *Philosophiae Naturalis Principia Mathematica* ("Mathematical Principles of Natural Philosophy"). He enumerated three laws that apply to all motion, so we now call them **Newton's laws of motion**.These laws govern the motion of everything from our daily movements here on Earth to the movements of planets, stars, and galaxies throughout the universe. Figure 4.5 summarizes the three laws, which we now discuss in a little more detail.

Newton's First Law Newton's first law of motion states that in the absence of a net force, an object will move with constant velocity. Thus, objects at rest (velocity = 0) tend to remain at rest, and objects in motion tend to remain in motion with no change in either their speed or their direction.

The idea that an object at rest should remain at rest is rather obvious: A car parked on a flat street won't suddenly start moving for no reason. But what if the car is traveling along a flat, straight road? Newton's first law says that the car should keep going at the same speed forever *unless* a

Figure 4.5

Newton's three laws of motion.

A spaceship needs no fuel to keep moving in space.

Newton's first law of motion: An object moves at constant velocity unless a net force acts to change its speed or direction.

A baseball accelerates as the pitcher applies a force by moving his arm. (Once released, this force and acceleration cease, so the ball's path changes only due to gravity and effects of air resistance.)

Newton's second law of motion: Force = mass × acceleration

A rocket is propelled upward by a force equal and opposite to the force with which gas is expelled out its back.

Newton's third law of motion: For any force, there is always an equal and opposite reaction force.

force acts to slow it down. You know that the car eventually will come to a stop if you take your foot off the gas pedal, so we must conclude that one or more forces are stopping the car—in this case forces arising from friction and air resistance. If the car were in space, and therefore unaffected by friction or air, it would keep moving forever (though gravity would eventually alter its speed and direction). That is why interplanetary spacecraft need no fuel to keep going after they are launched into space, and why astronomical objects don't need fuel to travel through the universe.

Newton's first law: An object moves at constant velocity if there is no net force acting upon it.

Newton's first law also explains why you don't feel any sensation of motion when you're traveling in an airplane on a smooth flight. As long as the plane is traveling at constant velocity, no net force is acting on it or on you. Therefore, you feel no different from the way you would feel at rest. You can walk around the cabin, play catch with a person a few rows forward, or relax and go to sleep just as though you were "at rest" on the ground.

Newton's Second Law Newton's second law of motion tells us what happens to an object when a net force *is* present. We have already seen that a net force will change an object's momentum, accelerating it in the direction of the force. Newton's second law quantifies this relationship, telling us that the amount of the acceleration depends on the object's mass and the strength of the net force. We usually write this law as an equation: force = mass × acceleration, or $F = ma$ for short.

Newton's second law: Force = mass × acceleration ($f = ma$).

This law explains why you can throw a baseball farther than you can throw a shot-put. The force your arm delivers to both the baseball and the shot-put equals the product of mass and acceleration. Because the mass of the shot-put is greater than that of the baseball, the same force from your arm gives the shot-put a smaller acceleration. Because of its smaller acceleration, the shot-put leaves your hand with less speed than the baseball and thus travels a shorter distance before hitting the ground.

Newton's second law also explains why large planets such as Jupiter have a greater effect on asteroids and comets than small planets such as Earth [Section 9.4]. Jupiter exerts a stronger gravitational force on passing asteroids than does Earth and therefore sends them scattering with a greater acceleration.

Newton's Third Law Think for a moment about standing still on the ground. Your weight exerts a downward force, so if this force were acting alone, Newton's second law would demand that you be accelerating downward. The fact that you are not falling means that the ground must be pushing back up on you with exactly the right amount of force to offset your downward weight. This fact is embodied in Newton's third law of motion, which tells us that any force is always opposed by an equal and opposite reaction force.

Newton's third law: For any force, there is always an equal and opposite reaction force.

This law is very important in astronomy, because it tells us that objects always attract *each other* through gravity. For example, your body always exerts a gravitational force on the Earth identical to the force that Earth exerts on you, except that it acts in the opposite direction. Of course, the same force means a much greater acceleration for you than for Earth (because your mass is so much smaller than Earth's), which is why we see you falling toward Earth when you jump off a chair, rather than Earth falling toward you.

Newton's third law also explains how a rocket works: A rocket engine generates a force that drives hot gas out the back, which creates an equal and opposite force that propels the rocket forward.

COMMON MISCONCEPTIONS

WHAT MAKES A ROCKET LAUNCH?

If you've ever watched a rocket launch, it's easy to see why many people believe that the rocket "pushes off" the ground. In fact, the ground has nothing to do with the rocket launch. The rocket's launch is explained by Newton's third law of motion. To balance the force driving gas out the back of the rocket, an equal and opposite force must propel the rocket forward. Thus, rockets can be launched horizontally as well as vertically, and a rocket can be "launched" in space (for example, from a space station) with no need for any nearby solid ground.

4.3 CONSERVATION LAWS IN ASTRONOMY

Newton's laws of motion are easy to state, but they may seem a bit arbitrary. Why, for example, should every force be opposed by an equal and opposite reaction force? In the centuries since Newton first stated his laws, we have learned that they are not arbitrary at all, but instead reflect deeper aspects of nature known as *conservation laws.*

Consider what happens when two objects collide. Newton's second law tells us that object 1 exerts a force that will change the momentum of object 2. At the same time, Newton's third law tells us that object 2 exerts an equal and opposite force on object 1—which means that object 1's momentum changes by precisely the same amount as object 2's momentum, but in the opposite direction. Thus, the total combined momentum of objects 1 and 2 remains the same both before and after the collision. We say that the total momentum of the colliding objects is conserved, reflecting a principle that we call *conservation of momentum.*

In essence, the law of conservation of momentum tells us that the total momentum of all interacting objects always stays the same. An individual object can gain or lose momentum only when a force causes it to exchange momentum with another object.

Conservation of momentum is one of several important conservation laws that underlie Newton's laws of motion and other physical laws in the universe. Two other conservation laws are especially important in astronomy. They go by the names *conservation of angular momentum* and *conservation of energy.* Let's see how these important laws work.

• What keeps a planet rotating and orbiting the Sun?

Perhaps you've wondered how Earth manages to keep rotating and going around the Sun day after day and year after year. The answer relies on a special type of momentum that we use to describe objects turning in circles or going around curves. This special type of "circling momentum" is called **angular momentum**. (The term *angular* arises because a circle turns through an *angle* of 360°.) The **law of conservation of angular momentum** tells us that total angular momentum can never change. An individual object can change its angular momentum only by transferring some angular momentum to or from another object.

Conservation of angular momentum: An object's angular momentum cannot change unless it transfers angular momentum to or from another object.

Consider Earth's orbit around the Sun. A simple formula tells us Earth's angular momentum at any point in its orbit:

$$\text{angular momentum} = m \times v \times r$$

where m is Earth's mass, v is its speed (or velocity) around the orbit, and r is the "radius" of the orbit, by which we mean its distance from the Sun (Figure 4.6). Because there are no objects around to give or take angular momentum from Earth as it orbits the Sun, Earth's orbital angular

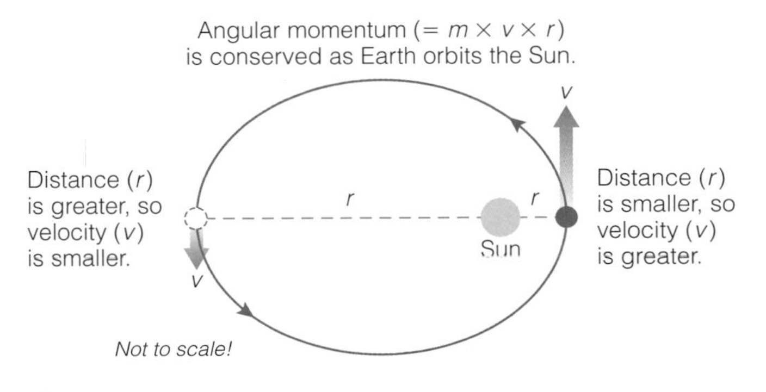

Figure 4.6

Earth's angular momentum always stays the same as it orbits the Sun, so it moves faster when it is closer to the Sun and slower when it is farther from the Sun. It needs no fuel to keep orbiting because no forces are acting in a way that could change its angular momentum.

momentum must always stay the same. This explains two key facts about Earth's orbit:

1. Earth needs no fuel or push of any kind to keep orbiting the Sun—it will keep orbiting as long as nothing comes along to take angular momentum away.
2. Because Earth's angular momentum at any point in its orbit depends on the product of its speed and radius, Earth's orbital speed must be faster when it is nearer to the Sun (and the radius is shorter) and slower when it is farther from the Sun (and the radius is longer).

The second fact is just what Kepler's second law of planetary motion tells us [Section 3.3]. Thus, the law of conservation of angular momentum tells us *why* Kepler's law is true.

The same idea explains why Earth keeps rotating. As long as Earth isn't transferring any of the angular momentum of its rotation to another object, it keeps rotating at the same rate. (In fact, Earth is very gradually transferring some of its rotational angular momentum to the Moon, and as a result Earth's rotation is gradually slowing down; see box, p. 96.)

Earth is not exchanging substantial angular momentum with any other object, so its rotation rate and orbit must stay about the same.

Conservation of angular momentum also explains why we see so many spinning disks in the universe, such as the disks of galaxies like the Milky Way and disks of material orbiting young stars. The idea is easy to illustrate with an ice skater spinning in place (Figure 4.7). Because there is so little friction on ice, the angular momentum of the ice skater remains essentially constant. When she pulls in her extended arms, she decreases her radius—which means her velocity of rotation must increase. Stars and galaxies are both born from clouds of gas that start out much larger in size. These clouds almost inevitably have some small net rotation, though it may be imperceptible. However, like the spinning skater as she pulls in her arms, they must spin faster as gravity makes them shrink in size. (We'll discuss why this also makes them flatten into disks in Chapter 6.)

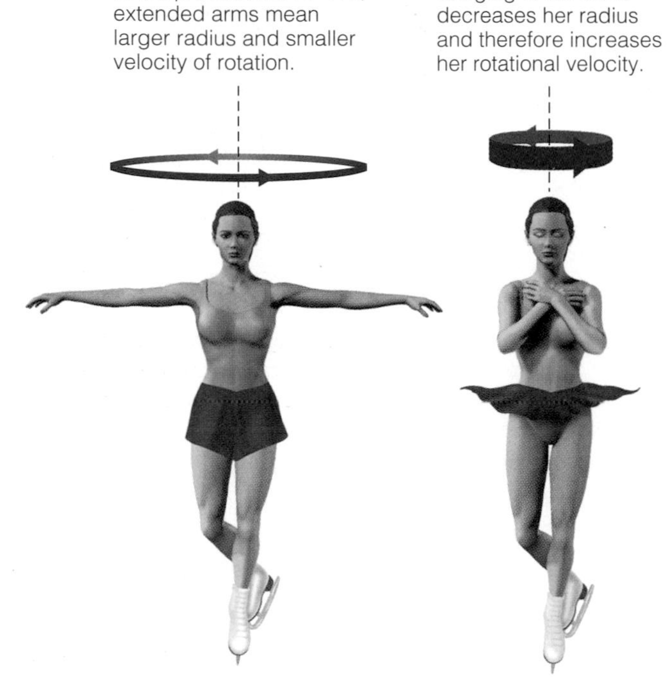

Figure 4.7

A spinning skater conserves angular momentum.

How does conservation of angular momentum explain the spiraling of water going down a drain?

• Where do objects get their energy?

The second crucial conservation law in astronomy is the **law of conservation of energy**. This law tells us that, like momentum and angular momentum, energy cannot appear out of nowhere or disappear into nothingness. Thus, objects can gain or lose energy only by exchanging energy with other objects. Because of this law, the story of the universe is a story of the interplay of energy and matter: All actions in the universe involve exchanges of energy or the conversion of energy from one form to another.

Energy can be transferred from one object to another or transformed from one type to another, but the total amount of energy is always conserved.

Throughout the rest of this book, we'll see numerous cases in which we can understand astronomical processes simply by tracing the way in which energy is transformed and exchanged. For example, we'll see that planetary interiors cool with time only because they radiate energy into space, and that the Sun became hot because of energy released by the gas that formed it. It's no exaggeration to say that, by using the laws of conservation of angular momentum and conservation of energy, we can understand almost every major process that occurs in the universe.

Basic Types of Energy Before we can fully understand the law of conservation of energy, we need to know exactly what energy is. In essence, energy is what makes matter move. Because this statement is so broad, we often distinguish between many different types of energy. For example, we talk about the energy we get from the food we eat, the energy that makes our cars go, and the energy put out by a light bulb. Fortunately, scientists have found a way to classify all these various types of energy into just three major categories (Figure 4.8):

- Energy of motion, or kinetic energy (*kinetic* comes from a Greek word meaning "motion"). Falling rocks, orbiting planets, and the molecules moving in the air around us are all examples of objects with kinetic energy.
- Energy carried by light, or **radiative energy** (the word *radiation* is often used as a synonym for *light*). All light carries energy, which is why light can cause changes in matter. For example, light can alter molecules in our eyes—thereby allowing us to see—or warm the surface of a planet.
- Stored energy, or **potential energy**, which might later be converted into kinetic or radiative energy. For example, a rock perched on a ledge has *gravitational* potential energy because it will fall if it slips off the edge, and gasoline contains *chemical* potential energy that can be converted into the kinetic energy of the moving car.

There are three basic categories of energy: energy of motion (kinetic), energy of light (radiative), and stored energy (potential).

Regardless of which type of energy we are dealing with, we can measure the amount of energy with the same standard units. For Americans, the most familiar units of energy are Calories, which are shown on food labels to tell us how much energy our bodies can draw from the food. A typical adult needs about 2,500 Calories of energy from food each day. In science, the standard unit of energy is the **joule**. One food Calorie is equivalent to about 4,184 joules, so the 2,500 Calories used daily by a typical adult is equivalent to about 10 million joules. Table 4.1 compares various energies in joules.

Thermal Energy—The Kinetic Energy of Many Particles Although there are only three major categories of energy, sometimes it is convenient to work with a subcategory of one of those three. In astronomy, the most important subcategory of kinetic energy is **thermal energy**, which represents the collective kinetic energy of the many individual particles (atoms and molecules) in a substance like a rock or the air or the gas within a distant star. In these cases, it is much easier to talk about the thermal

Figure 4.8

The three basic categories of energy. Energy can be converted from one form to another, but it can never be created or destroyed, an idea embodied in the law of conservation of energy.

TABLE 4.1 *Energy Comparisons*

Item	*Energy (joules)*
Energy of sunlight at Earth (per m^2 per second)	1.3×10^3
Energy from metabolism of a candy bar	1×10^6
Energy needed to walk for 1 hour	1×10^6
Kinetic energy of average car traveling at 60 mi/hr	1×10^6
Daily energy needs of average adult	1×10^7
Energy released by burning 1 liter of oil	1.2×10^7
Energy released by fission of 1 kg of uranium-235	5.6×10^{13}
Energy released by fusion of hydrogen in 1 liter of water	7×10^{13}
Energy released by 1-megaton H-bomb	5×10^{15}
Energy released by major earthquake (magnitude 8.0)	2.5×10^{16}
Annual U.S. energy consumption	10^{20}
Annual energy generation of Sun	10^{34}
Energy released by a supernova	$10^{44} - 10^{46}$

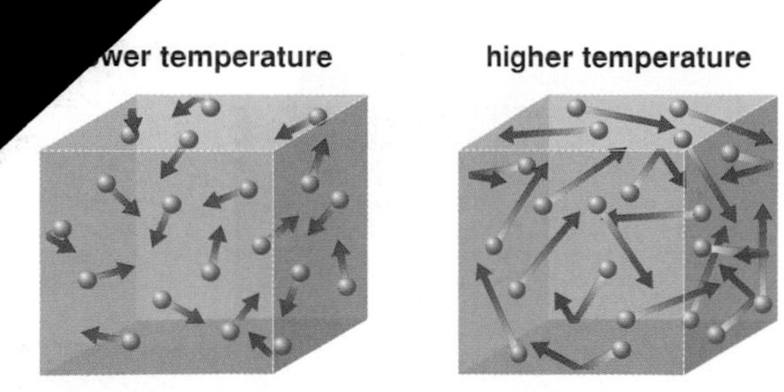

Figure 4.9

Temperature is a measure of the average kinetic energy of the particles (atoms and molecules) in a substance. The particles in the box on the right have a higher temperature because their average speeds are higher (assuming that both boxes contain particles of the same mass).

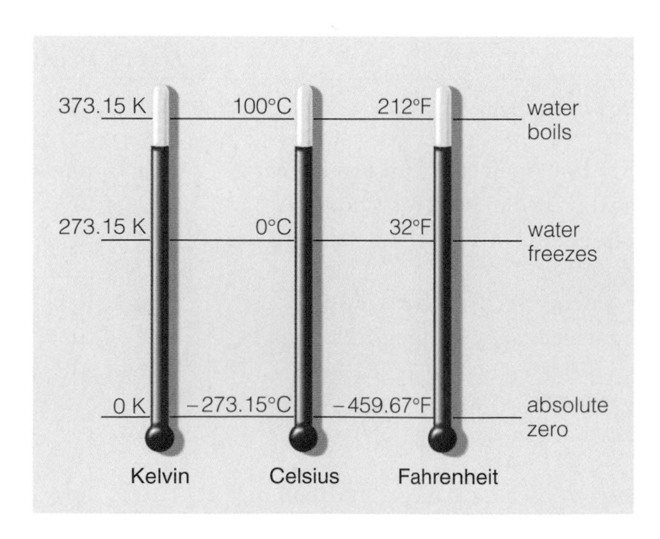

Figure 4.10

Three common temperature scales: Kelvin, Celsius, and Fahrenheit. Scientists prefer the Kelvin scale because it does not have negative temperatures. (The degree symbol ° is not usually used with the Kelvin scale.)

energy of the object rather than about the kinetic energies of its billions upon billions of individual particles.

Thermal energy gets its name because it is related to temperature, but temperature and thermal energy are not quite the same thing. Thermal energy essentially measures the *total* kinetic energy of all the particles in a substance, while **temperature** measures the *average* kinetic energy of the particles. Thus, for a particular object, a higher temperature simply means that the particles on average have more kinetic energy and hence are moving faster (Figure 4.9). You're probably familiar with temperatures measured in Fahrenheit or Celsius, but in science we often use the Kelvin temperature scale (Figure 4.10). The Kelvin scale never has negative temperatures, because it starts from the coldest possible temperature, known as absolute zero (0 K).

Thermal energy is the total kinetic energy of many individual particles.

Thermal energy depends on temperature, because a higher average kinetic energy for the particles in a substance must also lead to a higher total energy. However, thermal energy also depends on the number or density of the particles. We can see why by considering what would happen if you quickly thrust your arm in and out of a hot oven and a pot of boiling water.

The air in a hot oven is much hotter in temperature than the water boiling in a pot (typically 400°F for the oven versus 212°F for boiling water). However, the boiling water would scald your arm almost instantly, while you can safely put your arm into the oven air for a few seconds. The reason for this difference is density (Figure 4.11). If air or water is hotter than your body, molecules striking your skin transfer thermal energy to molecules in your arm. The higher temperature in the oven means that the air molecules strike your skin harder, on average, than the molecules in the boiling water. However, because the *density* of water is so much higher than the density of air (meaning water has far more molecules in the same amount of space), many more molecules strike your skin each second in the water. Thus, while each individual molecule that strikes your skin transfers a little less energy in the boiling water than in the oven, the sheer number of molecules hitting you in the water means that more thermal energy is transferred to your arm. That is why the water causes a rapid burn.

THINK ABOUT IT In air or water that is colder than your body temperature, thermal energy is transferred from you to the surrounding cold air or water. Use this fact to explain why falling into a 32°F (0°C) lake is much more dangerous than standing naked outside on a 32°F day.

Potential Energy in Astronomy Just as kinetic energy has subcategories such as thermal energy, potential energy also has numerous subcategories. Many types of potential energy are important in astronomy, but two are particularly important: gravitational potential energy and the potential energy of mass itself, or mass-energy. Let's examine each of these two types of potential energy in a little more depth.

An object's **gravitational potential energy** depends on how far it can fall as a result of gravity.An object has more gravitational potential energy when it is higher up and less when it is lower down. For example, if you throw a ball up into the air, it has more potential energy when it is high up than it does near the ground. Because energy must be conserved during the ball's flight, the ball's kinetic energy must increase when its gravitational potential energy decreases, and vice versa (Figure 4.12a). That is why the ball travels fastest (has the most kinetic energy) when it is closest to the ground, where it has the least gravitational potential energy. The higher up it is, the more gravitational potential energy it has and thus the slower the ball travels (less kinetic energy).

An object's gravitational potential energy increases when it moves higher and decreases when it moves lower.

The same general idea explains how stars become hot (Figure 4.12b). Before a star forms, its matter is spread out in a large, cold cloud of gas. Most of the individual gas particles are far from the center of this large cloud and therefore have a lot of gravitational potential energy. The particles lose gravitational potential energy as the cloud contracts under its own gravity, and this "lost" potential energy ultimately gets converted into thermal energy, making the center of the cloud hot.

The second particularly important form of potential energy in astronomy is the energy contained in mass itself, often called **mass-energy**. Einstein discovered that mass is a form of potential energy, and he expressed the idea that mass can be converted to other forms of energy with his famous formula

$$E = mc^2$$

where E is the amount of potential energy, m is the mass of the object, and c is the speed of light.

Mass itself is a form of potential energy, as described by Einstein's equation $E = mc^2$.

A small amount of mass contains a huge amount of energy. For example, the energy released by a 1-megaton H-bomb—enough to destroy a major city—comes from converting only about 0.1 kilogram of mass (about 3 ounces) into energy (Figure 4.13). The energy generated by the Sun also comes from converting a small amount of the Sun's mass into energy through the process of nuclear fusion.

Just as Einstein's formula tells us that mass can be converted into other forms of energy, it also tells us that energy can be transformed into mass. This process is especially important to understanding what we think happened during the early moments in the history of the universe,

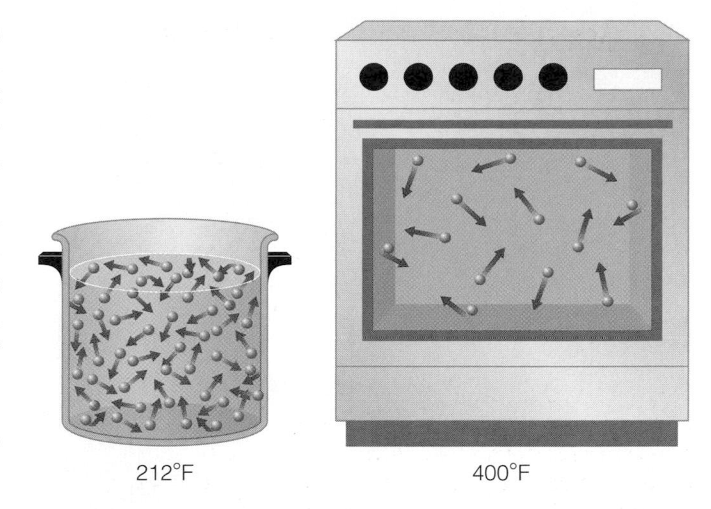

Figure 4.11

Thermal energy depends on both the temperature and density of particles in a substance. The air in the oven is hotter than the boiling water in the pot, but the water contains more thermal energy because of its much higher density (more particles in the same amount of space).

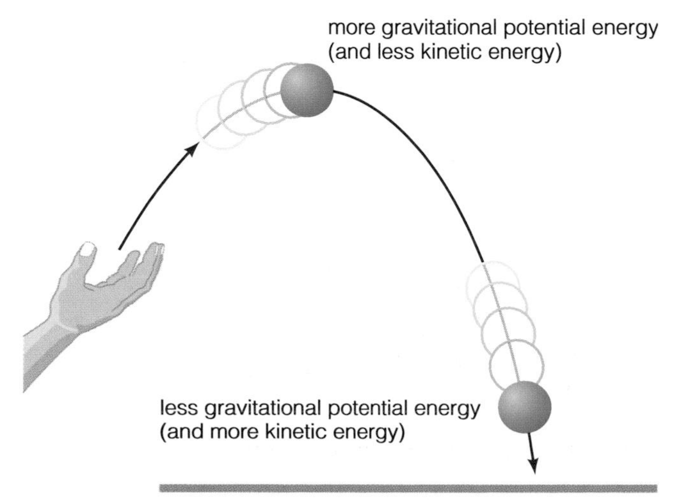

a The ball has more gravitational potential energy when it is high up than when it is near the ground.

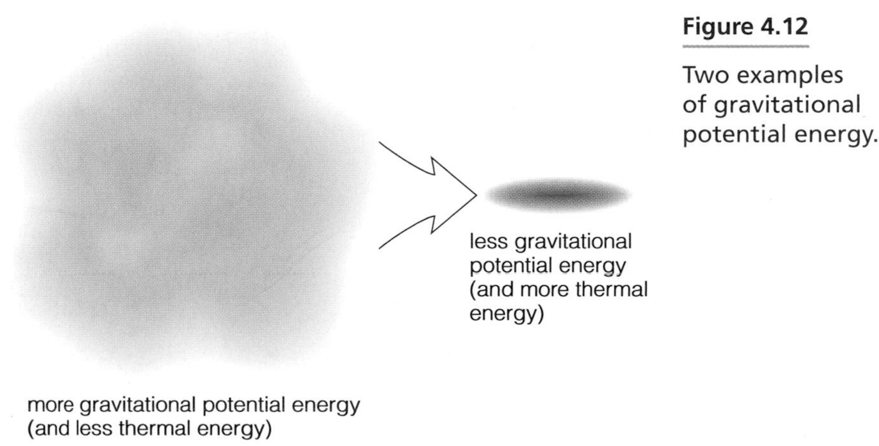

b A cloud of interstellar gas has more gravitational potential energy when it is spread out than when it shrinks in size due to gravity.

Figure 4.12

Two examples of gravitational potential energy.

Figure 4.13

The energy released by this H-bomb comes from converting only about 0.1 kg of mass into energy in accordance with the formula $E = mc^2$.

when the energy of the Big Bang turned into the mass from which the universe is now made [Section 17.1]. Scientists also use this idea to search for undiscovered particles of matter, by creating subatomic particles from energy in large machines called *particle accelerators.*

Conservation of Energy We have seen that energy comes in three basic categories—kinetic, radiative, and potential—and explored several subcategories that are especially important in astronomy: thermal energy, gravitational potential energy, and mass-energy. Now we are ready to return the question of "Where do objects get their energy?" Because energy cannot be created or destroyed, objects always get their energy from other objects. Ultimately, we can always trace an object's energy back to the Big Bang [Section 1.1], the beginning of the universe in which all matter and energy is thought to have come into existence.

The energy of any object can be traced back to the origin of the universe in the Big Bang.

For example, imagine that you've thrown a baseball. It is moving, so it has kinetic energy. Where did this kinetic energy come from? The baseball got its kinetic energy from the motion of your arm as you threw it. Your arm, in turn, got its kinetic energy from the release of chemical potential energy stored in your muscle tissues. Your muscles got this energy from the chemical potential energy stored in the foods you ate. The energy stored in the foods came from sunlight, which plants convert into chemical potential energy through photosynthesis. The radiative energy of the Sun was generated through the process of nuclear fusion, which releases some of the mass-energy stored in the Sun's supply of hydrogen. The mass-energy stored in the hydrogen came from the birth of the universe in the Big Bang. After you throw the ball, its kinetic energy will ultimately be transferred to molecules in the air or ground. According to present understanding, the total energy content of the universe was determined in the Big Bang. It remains the same today and will stay the same forever into the future.

4.4 THE FORCE OF GRAVITY

Newton's three laws of motion describe how objects in the universe move in response to forces. The laws of conservation of momentum, angular momentum, and energy offer an alternative and often simpler way of thinking about what happens when a force causes some change in the motion of one or more objects. Either way, however, we cannot fully understand motion unless we also understand the forces that lead to changes in motion. In astronomy, the most important force is gravity, which governs virtually all large-scale motion in the universe. Isaac Newton discovered the basic law that describes how gravity works.

Motion and Gravity Tutorial, Lesson 2

• What determines the strength of gravity?

Newton expressed the force of gravity mathematically with his **universal law of gravitation**. Three simple statements summarize this law:

- Every mass attracts every other mass through the force called *gravity.*
- The strength of the gravitational force attracting any two objects is *directly proportional* to the product of their masses. For example,

doubling the mass of *one* object doubles the force of gravity between the two objects.

- The strength of gravity between two objects decreases with the *square* of the distance between their centers. That is, the gravitational force follows an **inverse square law** with distance. For example, doubling the distance between two objects weakens the force of gravity by a factor of 2^2, or 4.

Doubling the distance between two objects weakens the force of gravity by a factor of 2^2, or 4.

These three statements tell us everything we need to know about Newton's universal law of gravitation. Mathematically, all three statements can be combined into a single equation, usually written like this:

$$F_g = G\frac{M_1 M_2}{d^2}$$

where F_g is the force of gravitational attraction, M_1 and M_2 are the masses of the two objects, and d is the distance between their centers (Figure 4.14). The symbol G is a constant called the **gravitational constant**, and its numerical value has been measured to be $G = 6.67 \times 10^{-11}\ \text{m}^3/(\text{kg} \times \text{s}^2)$.

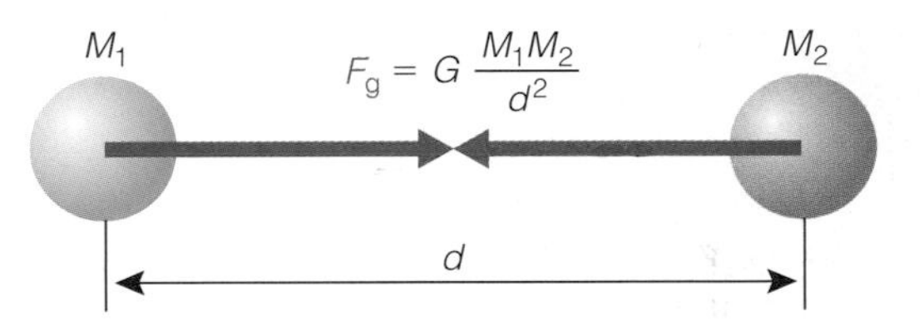

Figure 4.14

The universal law of gravitation. It is an *inverse square law* because the force of gravity declines with the square of the distance d between two objects. For example, doubling the distance d weakens the force of gravity by a factor of $2^2 = 4$.

How does the gravitational force between two objects change if the distance between them triples? If the distance between them drops in half?

Orbits and Kepler's Law Tutorial, Lessons 1–4

• How does Newton's law of gravity extend Kepler's laws?

By the time Newton published *Principia* in 1687, Kepler's three laws of planetary motion [Section 3.3] had already been known and tested for some 70 years. There was little doubt among scientists that Kepler's laws were correct, but there was great debate about *why* these laws hold true. Newton solved the mystery by explaining planetary motion through the laws of motion and the universal law of gravitation. For example, we've already seen how we can understand Kepler's second law—that a planet moves faster when it is closer to the Sun—by thinking about conservation of angular momentum. Kepler's third law—that average orbital speed is higher for planets with smaller average orbital distance—comes directly from the fact that closer-in planets have a stronger gravitational attraction to the Sun.

In essence, Newton used the mathematical expressions of his universal law of gravitation and the laws of motion to show that Kepler's laws are consequences of these more fundamental laws. In doing so, he also found that Kepler's laws were only part of the story of how objects move in response to gravity. In particular, he was able to extend Kepler's laws in three crucial ways:

- *Newton generalized Kepler's first two laws of planetary motion to apply to all orbiting objects.* For example, the orbits of a satellite around Earth, of a moon around a planet, and of an asteroid around the Sun are all ellipses in which the orbiting object moves faster at the nearer points in its orbit and slower at the farther points.

- *Newton found that ellipses are not the only possible orbital paths* (Figure 4.15). Kepler was right when he found that ellipses (which include circles) are the only possible shapes for **bound orbits**—orbits in which an object goes around another object over and over again. (The term *bound orbit* comes from the idea that gravity creates a *bond* that holds the objects together.) However, Newton discovered that objects can also follow **unbound orbits**—paths that bring an object close to another object just once. For example, some comets that enter the inner solar system follow unbound orbits.
- *Newton found that Kepler's third law could be generalized in a way that enables us to calculate the masses of distant objects.* This generalization, which we call **Newton's version of Kepler's third law**, allows us to calculate the mass of a distant object if we can observe another object orbiting it. All we need to know is the amount of time the orbiting object takes for each orbit (its orbital period) and its average distance from the object it orbits.

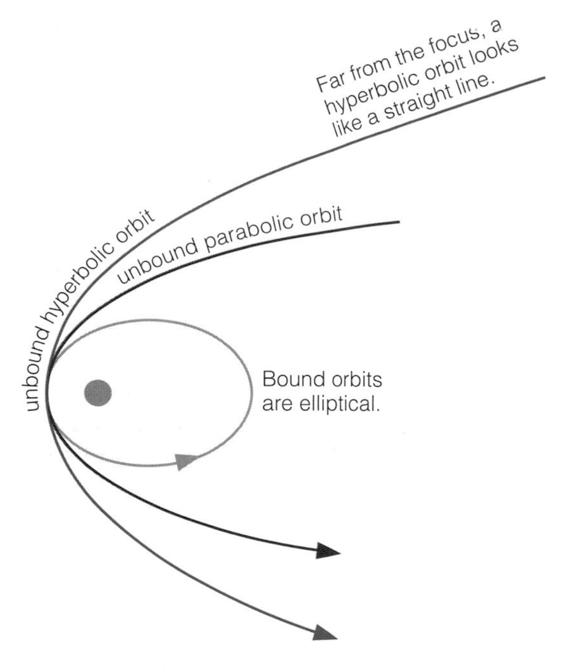

Figure 4.15

Newton showed that ellipses are not the only possible orbital paths. Orbits can also be unbound, taking the mathematical shapes of either parabolas or hyperbolas.

Newton's version of Kepler's third law allows us to calculate the masses of distant objects.

Newton's version of Kepler's third law is the primary means by which we determine masses throughout the universe. For example, it allows us to calculate the mass of the Sun from Earth's orbital period (1 year) and its average distance from the Sun (1 AU). Similarly, measuring the orbital period and average distance of one of Jupiter's moons allows us to calculate Jupiter's mass, and measuring the orbital periods and average distances of stars in binary star systems allows us to determine their masses.

Newton's version of Kepler's third law also explains another important characteristic of orbital motion. It shows that the orbital period of a *small* object orbiting a much more massive object depends only on its orbital distance, not on its mass. That is why an astronaut does not need a tether to stay close to the Space Shuttle or the Space Station during a space walk (Figure 4.16). The spacecraft and the astronaut are both much smaller than Earth and thus stay together because they have the same orbital distance and hence the same orbital period.

Figure 4.16

Newton's version of Kepler's third law enables us to measure masses throughout the universe. In addition, it shows that when one object orbits a much more massive object, the orbital period depends only on its average orbital distance. Thus, the astronaut and the Space Shuttle share the same orbit and stay together—even as both orbit Earth at a speed of some 25,000 km/hr.

• How do gravity and energy together allow us to understand orbits?

We've seen that Newton's law of universal gravitation explains Kepler's laws of planetary motion, which describe the simple and stable orbits of the planets. By extending Kepler's laws, Newton also explained many other stable orbits, such as the orbit of a satellite around Earth or of a moon around a planet. But orbits do not always stay the same. For example, you've probably heard of satellites crashing to Earth from orbit, proving that orbits can sometimes change dramatically. To understand how and why orbits sometimes change, we need to consider the role of energy in orbits.

Orbital Energy Consider the orbit of a planet around the Sun. An orbiting planet has both kinetic energy (because it is moving around the Sun) and gravitational potential energy (because it would fall toward the

Sun if it stopped orbiting). The planet's kinetic energy depends on its orbital speed, and its gravitational potential energy depends on its distance from the Sun. Because the planet's distance and speed both vary as it orbits the Sun, the amounts of its gravitational potential energy and kinetic energy also vary (Figure 4.17). However, the planet's total **orbital energy**—the sum of its kinetic and gravitational potential energies—always stays the same. This fact is a consequence of the law of conservation of energy. As long as no other object causes the planet to gain or lose orbital energy, its orbital energy cannot change and its orbit must remain the same.

Orbits cannot change spontaneously—an object's orbit can change only if it somehow gains or loses orbital energy.

Generalizing from planets to other objects leads us to a very important idea about motion throughout the cosmos: *Orbits cannot change spontaneously.* Left undisturbed, planets would forever keep the same orbits around the Sun, moons would keep the same orbits around planets, and stars would keep the same orbits around their galaxies.

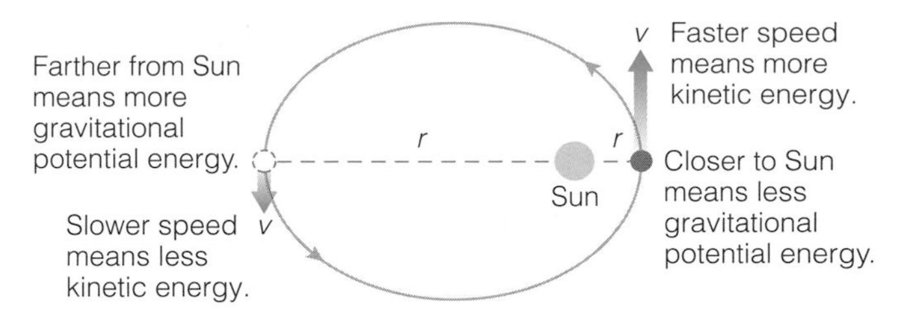

Figure 4.17

The total orbital energy of a planet stays the same throughout its orbit, because its gravitational potential increases when its kinetic energy decreases, and vice versa.

Gravitational Encounters Although orbits cannot change spontaneously, they can change through exchanges of energy. One way that two objects can exchange orbital energy is through a **gravitational encounter**, in which they pass near enough so that each can feel the effects of the other's gravity. For example, in the rare cases in which a comet happens to pass near a planet, the comet's orbit can change dramatically. Figure 4.18 shows a comet headed toward the Sun on an unbound orbit. The comet's close passage by Jupiter allows the comet and Jupiter to exchange energy. In this case, the comet loses so much orbital energy that its orbit changes from unbound to a bound, elliptical orbit. Jupiter gains exactly as much energy as the comet loses, but the effect on Jupiter is unnoticeable because of its much greater mass.

Spacecraft engineers can use the same basic idea in reverse. Each of the two Voyager spacecraft that visited the outer planets in the 1980s was deliberately sent past Jupiter on a path that caused it to gain orbital energy at Jupiter's expense. This extra orbital energy helped the spacecraft reach more distant planets. Of course, the effect of the tiny spacecraft on Jupiter was unnoticeable.

A similar dynamic sometimes occurs naturally, and may explain why most comets orbit so far from the Sun. Astronomers think that most comets once orbited in the same region of the solar system as the large outer planets [Section 9.2]. Gravitational encounters with Jupiter or the other large planets then caused some of these comets to be "kicked" into much more distant orbits around the Sun. In some cases, the comets may have been ejected from the solar system completely.

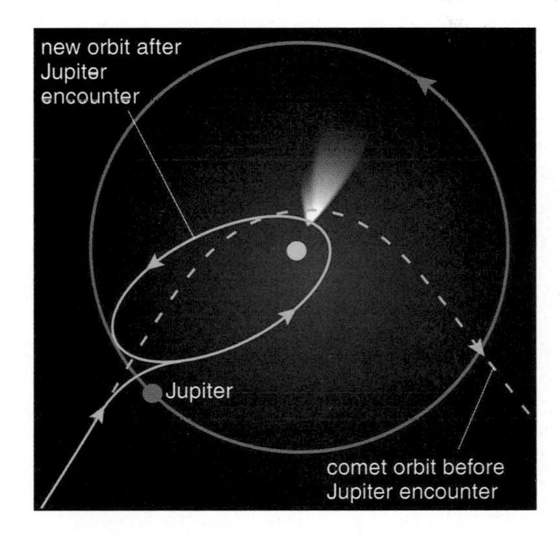

Figure 4.18

Depiction of a comet in an unbound orbit of the Sun that happens to pass near Jupiter. The comet loses orbital energy to Jupiter, changing its unbound orbit to a bound orbit around the Sun.

Atmospheric Drag Friction can cause objects to lose orbital energy. For example, consider a satellite orbiting Earth. If the orbit is fairly low—say, just a few hundred kilometers above Earth's surface—the satellite experiences a bit of drag from Earth's thin upper atmosphere. This drag gradually causes the satellite to lose orbital energy until it finally plummets to Earth. The satellite's lost orbital energy is converted to thermal energy in the atmosphere, which is why a falling satellite usually burns up.

Friction may also have played a role in shaping the current orbits of some moons and planets. For example, some of the small moons of Jupiter and the other outer planets may once have orbited the Sun independently. Their orbits could not have changed spontaneously. However, the outer planets probably once were surrounded by clouds of gas [Section 6.4], and friction would have slowed objects passing through this gas. Some of these small objects probably lost just enough energy to friction to allow them to be "captured" as moons.

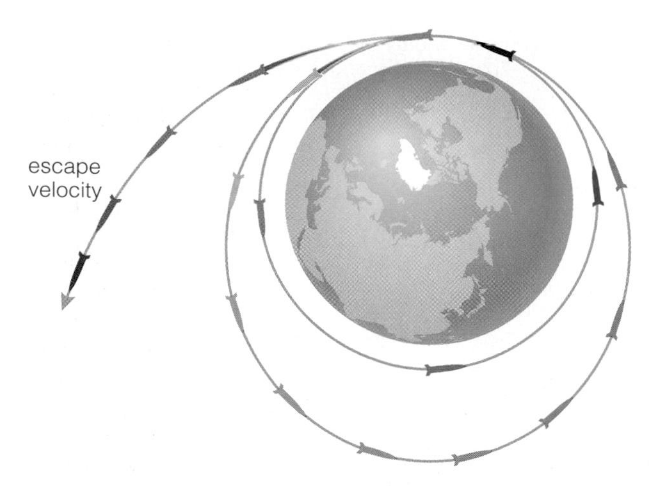

Figure 4.19 Interactive Figure

If an object orbiting Earth gains orbital energy, it moves to a higher or more elliptical orbit. With enough extra orbital energy, it may achieve escape velocity and escape Earth completely. Escape velocity depends on how high the object is when it starts. From Earth's surface, escape velocity is about 11 km/s.

Escape Velocity An object that gains orbital energy moves into a higher orbit. For example, if we want to boost the orbital altitude of a spacecraft, we can give it more orbital energy by firing a rocket. The chemical potential energy released by the rocket fuel is converted to orbital energy for the spacecraft. The Hubble Space Telescope gets periodic boosts when it is visited by the Space Shuttle. Without these boosts, atmospheric drag would by now have caused the telescope to fall to Earth.

A spacecraft that achieves escape velocity can escape Earth's orbit completely.

A spacecraft that gains orbital energy moves to a higher or more elliptical orbit. If we give a spacecraft enough orbital energy, it may end up in an unbound orbit that allows it to *escape* Earth completely (Figure 4.19). For example, if we want to send a space probe to Mars, we must use a large rocket that gives the probe enough energy to leave Earth orbit. Although it would probably make more sense to say that the probe achieves "escape energy," we instead say that it achieves **escape velocity**. The escape velocity from Earth's surface is about 40,000 km/hr, or 11 km/s, meaning that this is the minimum velocity required to escape Earth's gravity if you start near the surface.

Notice that the escape velocity does not depend on the mass of the escaping object—*any* object must travel at a velocity of 11 km/s to escape from Earth, whether it is an individual atom or molecule escaping from the atmosphere, a spacecraft being launched into deep space, or a rock blasted into the sky by a large impact. Escape velocity *does* depend on whether you start from the surface or from someplace high above the surface. Because gravity weakens with distance, it takes less energy—and hence a lower escape velocity—to escape from a point high above Earth than from Earth's surface.

• How does gravity cause tides?

Newton's universal law of gravitation has applications that go far beyond explaining Kepler's laws and orbits. For our purposes in this book, however, there is just one more topic we need to cover: how gravity leads to tides.

If you've spent time near an ocean, you're probably aware of the rising and falling of the tides. In most places, tides rise and fall twice each day. We can understand the basic cause of tides by examining the gravitational attraction between Earth and the Moon.

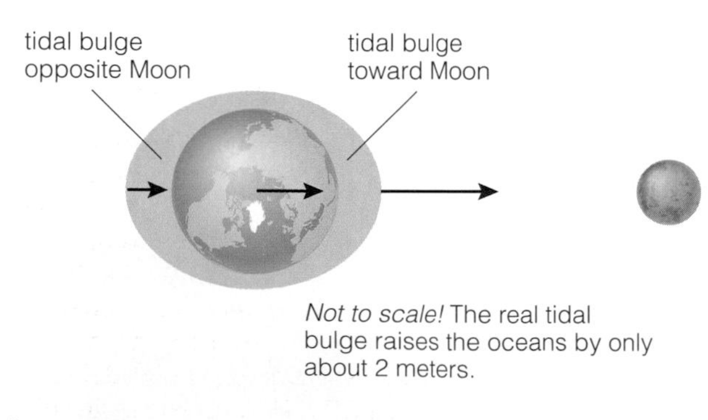

Figure 4.20

Tidal bulges face toward and away from the Moon because of the difference in the strength of the Moon's gravitational attraction on opposite sides of Earth. (Arrows show that the gravitational attraction toward the Moon is stronger on the side of Earth closer to the Moon.) There are two daily high tides as any location on Earth rotates through the two tidal bulges.

Gravity attracts Earth and the Moon toward each other (with the Moon staying in orbit as it "falls around" Earth), but it affects different parts of Earth slightly differently: Because the strength of gravity declines with distance, the side of Earth facing the Moon feels a slightly stronger gravitational attraction than the side facing away from the Moon. This varying attraction creates a "stretching force," or **tidal force**, that stretches the entire Earth to create two tidal bulges—one facing the Moon and one opposite the Moon (Figure 4.20). If you are still unclear about why there are *two* tidal bulges, think about a rubber band: If you pull on a rubber band it will stretch in both directions relative to its center, even if you pull on only one side. In the same way, Earth stretches on both sides even though the Moon is tugging harder on only one side. Earth's rotation carries us

Figure 4.21

Photographs of high and low tide at the abbey at Mont-Saint-Michel, France, one of the world's most popular tourist destinations. Here the tide rushes in much faster than a person can swim. Before a causeway was built (visible to the left, with cars on it), the Mont was accessible by land only at low tide. At high tide, it became an island.

through each of the two bulges each day, so there are two daily high tides and two daily low tides.

Tidal forces cause the entire Earth to stretch along the Earth–Moon line, creating two tidal bulges.

Tides affect both land and ocean, but we generally notice only the ocean tides because water flows much more readily than land. In addition, the height and timing of ocean tides vary considerably from place to place around the Earth. For example, while the tide rises gradually in most locations, the incoming tide near the famous abbey on Mont-Saint-Michel, France, moves much faster than a person can swim (Figure 4.21). In centuries past, the Mont was an island twice a day at high tide but was connected to the mainland at low tide. Many pilgrims drowned when they were caught unprepared by the tide rushing in. (Today, a human-made land bridge keeps the island connected to the mainland.) Another unusual tidal pattern occurs in coastal states along the northern shore of the Gulf of Mexico, where topography and other factors combine to make only one noticeable high tide and low tide each day.

The Sun also affects the tides (Figure 4.22). Although the Sun is much more massive than the Moon, its tidal effect on Earth is smaller because its much greater distance means that the *difference* in the Sun's pull on

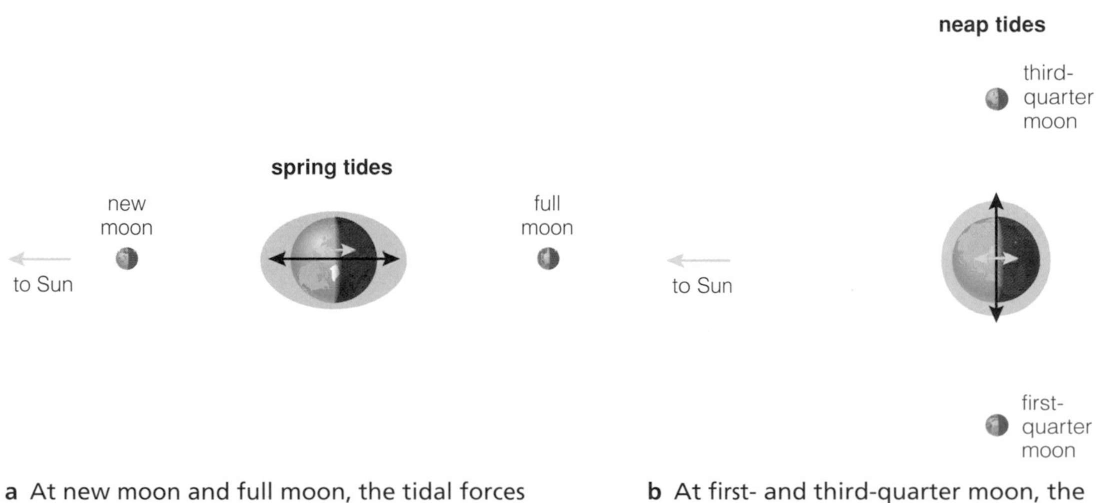

a At new moon and full moon, the tidal forces from the Sun and Moon work together, leading to enhanced *spring tides.*

b At first- and third-quarter moon, the tidal forces from the Sun and Moon work at odds, leading to smaller *neap tides.*

Figure 4.22 Interactive Figure

The Sun exerts a tidal force on Earth about one-third as strong as that from the Moon. In these diagrams, yellow arrows represent tidal force due to the Sun, which causes Earth to stretch along the Sun–Earth line, and black arrows represent tidal force due to the Moon, which causes Earth to stretch along the Earth–Moon line.

SPECIAL TOPIC: WHY DOES THE MOON ALWAYS SHOW THE SAME FACE TO THE EARTH?

You are probably aware that we always see (nearly) the same face of the Moon. This happens because the Moon rotates on its axis in exactly the same time period that it takes to orbit Earth, a trait called **synchronous rotation**. A simple demonstration shows the idea (Figure 1). Place a ball on a table to represent Earth while you represent the Moon. The only way you can face the ball at all times is by completing exactly one rotation while you complete one orbit. But why does the Moon have this synchronous rotation? We can trace the answer directly to tides.

It's easiest to start by considering the effects of tides on Earth. We might expect Earth's two tidal bulges to lie directly on the Earth–Moon line, but they don't quite: Friction between the tidal bulges and the rotating Earth pulls the bulges slightly ahead of the Earth–Moon line (Figure 2). Because the Moon's gravity tries to pull the bulges back into the Earth–Moon line, Earth's rotation slows very gradually. At the same time, the net effect of gravity from the bulges pulls the Moon slightly ahead in its orbit, causing the Moon to move gradually farther from Earth. These effects are barely noticeable on human time scales, but they add up over billions of years. Early in Earth's history, a day may have been only 5 or 6 hours long and the Moon may have been one-tenth or less of its current distance from Earth. These changes also provide a great example of conservation of angular momentum: The Moon's growing orbit gains the angular momentum that Earth loses as its rotation slows.

The gravity of the bulges pulls the Moon ahead, increasing its orbital distance.
If Earth didn't rotate, tidal bulges would be oriented along the Earth-Moon line.
Friction with the rotating Earth pulls the tidal bulges slightly ahead of the Earth-Moon line.
The Moon's gravity tries to pull the bulges back into line, slowing Earth's rotation.
Moon
Not to scale!

Figure 2

Earth's rotation pulls its tidal bulges slightly ahead of the Earth–Moon line, leading to gravitational effects that gradually slow Earth's rotation and increase the Moon's orbital distance.

Now, let's turn the situation around to see how tides affect the Moon. Because Earth is more massive than the Moon, Earth exerts a greater tidal force on the Moon than the Moon does on Earth. This tidal force gives the Moon two tidal bulges along the Earth–Moon line, much like the two tidal bulges that the Moon creates on Earth. (The Moon does not have visible tidal bulges, but it does indeed have excess mass along the Earth–Moon line.) Thus, if the Moon were rotating through its tidal bulges in the same way as Earth, the resulting friction would cause the Moon's rotation to slow down. This is exactly what we think happened long ago.

The Moon probably once rotated much faster than it does today. As a result, it *did* rotate through its tidal bulges, and its rotation gradually slowed. Once the Moon's rotation slowed to the point at which the Moon and its bulges rotated at the same rate—that is, synchronously with the orbital period—there was no further source for tidal friction. Thus, the Moon's synchronous rotation was a natural outcome of Earth's tidal effects on the Moon.

Similar tidal friction has led to synchronous rotation in many other cases. For example, Jupiter's four large moons (Io, Europa, Ganymede, and Callisto) keep nearly the same face toward Jupiter at all times, as do many other moons. Pluto and its moon Charon *both* rotate synchronously: Like two dancers, they always keep the same face toward each other. Many binary star systems also rotate in this way. Tidal forces may be most familiar because of their effects on our oceans, but they are important throughout the universe.

Figure 1

The fact that we always see the same face of the Moon means that the Moon must rotate once in the same amount of time that it takes to orbit Earth once. You can see why by walking around a model of Earth while imagining that you are the Moon.

a If you do not rotate while walking around the model, you will not always face it.

b You will face the model at all times only if you rotate exactly once during each orbit.

the near and far sides of Earth is relatively small. Overall, the Sun's tidal effect is about one-third as strong as the Moon's. When the tidal forces of the Sun and the Moon work together, as is the case at both new moon and full moon, we get the especially pronounced *spring tides* (so named because the water tends to "spring up" from the Earth). When the tidal forces of the Sun and the Moon oppose each other, as is the case at first- and third-quarter moon, we get the relatively small tides known as *neap tides.*

Tidal forces affect not only Earth, but also many other objects. For example, Earth exerts tidal forces on the Moon that explain why the Moon always shows the same face to Earth (see box on page 96), and in Chapter 9 we'll see how tidal forces have led to the astonishing volcanic activity of Jupiter's moon Io and the possiblity of a subsurface ocean on its moon Europa.

COMMON MISCONCEPTIONS

THE ORIGIN OF TIDES

Many people believe that tides arise because the Moon pulls Earth's oceans toward it. But if that were the whole story, there would be a bulge only on the side of Earth facing the Moon, and hence only one high tide each day. The correct explanation for tides must account for why Earth has two tidal bulges.

Only one explanation works: Earth must be stretching from its center in both directions (toward and away from the Moon). The stretching force, or tidal force, arises from the *difference* between the force of gravity on one side of Earth and that on the other. In fact, stretching due to tides affects many objects, not just Earth. Many moons are stretched into slightly oblong shapes by tidal forces caused by their parent planets, and mutual tidal forces stretch close binary stars into teardrop shapes. In regions where gravity is extremely strong, such as near a black hole, tides could even stretch spaceships or people.

THINK ABOUT IT

Explain why any tidal effects on Earth caused by the other planets would be extremely small (in fact, so small as to be unnoticeable).

THE BIG PICTURE

Putting Chapter 4 into Context

We've covered a lot of ground in this chapter, from the scientific terminology of motion to the story of how universal motion was understood by Newton. Be sure you understand the following "big picture" ideas:

- Understanding the universe requires understanding motion. Motion may seem complex, but it can be understood simply through Newton's three laws of motion.
- Today, we understand Newton's laws of motion through deeper principles, including the laws of conservation of angular momentum and of energy. We can understand many processes in the universe by following how energy is exchanged between different objects.
- Newton also discovered the universal law of gravitation, which explains how gravity holds planets in their orbits and much more—including how satellites can reach and stay in orbit, the nature of tides, and why the Moon rotates synchronously with Earth.
- Perhaps most important, Newton's discoveries showed that the same physical laws we observe on Earth apply throughout the universe. The universality of physics opens up the entire cosmos as a possible realm of human study.

4.1 DESCRIBING MOTION: EXAMPLES FROM DAILY LIFE

- **How do we describe motion?**

Several key terms describe motion. **Speed** is the rate at which an object is moving. **Velocity** is speed in a certain direction. **Acceleration** is a change in velocity, meaning a change in either speed or direction. **Momentum** is mass × velocity. A **force** can change an object's momentum, causing it to accelerate.

- **How is mass different from weight?**

An object's mass is the same no matter where it is located, but its weight varies with the strength of gravity or other forces acting on the object. An object becomes **weightless** when it is in **free-fall**, even though its mass is unchanged.

4.2 NEWTON'S LAWS OF MOTION

- **How did Newton change our view of the universe?**

Newton discovered the laws of motion and the law of universal gravitation. He showed that the same physical laws that operate on Earth also operate in the heavens, making it possible for us to learn about the universe by studying physical laws here on Earth.

- **What are Newton's three laws of motion?**

Newton's first law of motion

Newton's second law of motion

Newton's third law of motion

(1) An object moves at constant velocity if there is no net force acting upon it.

(2) Force = mass × acceleration ($F = ma$).
(3) For any force, there is always an equal and opposite reaction force.

4.3 CONSERVATION LAWS IN ASTRONOMY

- **What keeps a planet rotating and orbiting the Sun?**

Conservation of angular momentum means that a planet's rotation and orbit cannot change unless it transfers angular momentum to another object. The planets in our solar system do not exchange substantial angular momentum with each other or anything else, so their orbits and rotation rates remain quite steady.

- **Where do objects get their energy?**

Energy is always conserved—it can be neither created nor destroyed. Thus, objects must have gotten whatever energy they now have from exchanges of energy with other objects. Energy comes in three basic categories — **kinetic**, **radiative**, and **potential** — though sometimes we use subcategories for convenience.

4.4 THE FORCE OF GRAVITY

- **What determines the strength of gravity?**

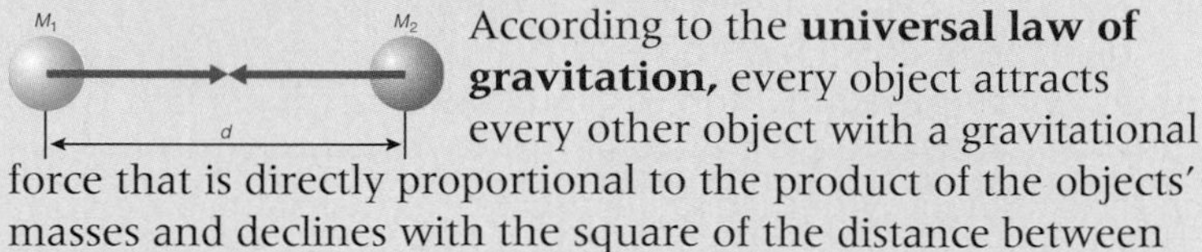

According to the **universal law of gravitation,** every object attracts every other object with a gravitational force that is directly proportional to the product of the objects' masses and declines with the square of the distance between their centers:

$$F_g = G\frac{M_1 M_2}{d^2}$$

- **How does Newton's law of gravity extend Kepler's laws?**

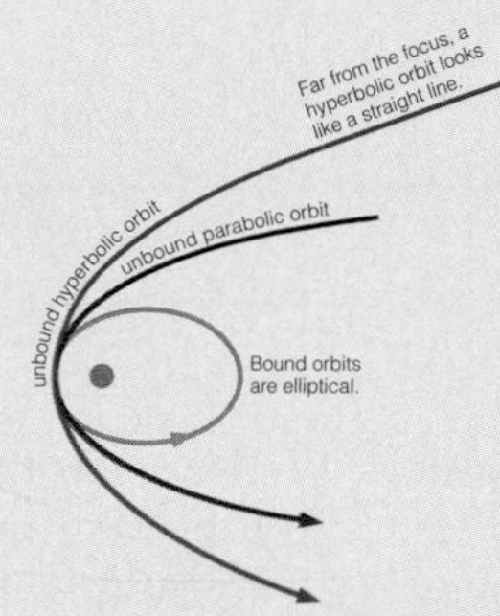

Newton extended Kepler's laws in three main ways: (1) He showed that Kepler's first two laws apply to all orbiting objects, not just planets. (2) He showed that **bound orbits** in ellipses (which include circles) are not the only possible orbital shape—orbits can also be **unbound** (in the shape of parabolas or hyperbolas). (3) **Newton's version of Kepler's third law** allows us to calculate the masses of orbiting objects from their orbital periods and distances.

- **How do gravity and energy together allow us to understand orbits?**

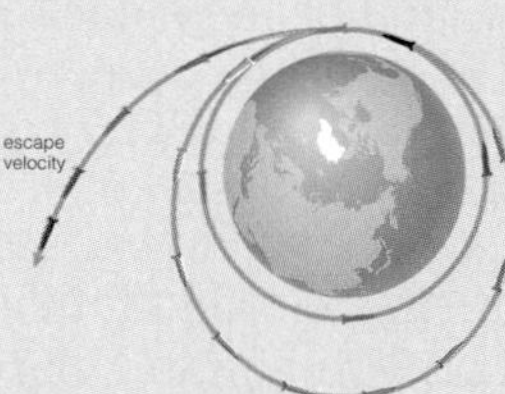

Gravity determines orbits, and an object cannot change its orbit unless it gains or loses **orbital energy** — the sum of its kinetic and gravitational potential energy — through energy transfer with other objects. If an object gains enough orbital energy, it may achieve **escape velocity** and leave the gravitational influence of the object it was orbiting.

- **How does gravity cause tides?**

The Moon's gravity stretches Earth along the Earth–Moon line, causing Earth to bulge both toward and away from the Moon. Earth's rotation carries us through each of the two bulges each day, so there are two daily high tides and two daily low tides.

EXERCISES AND PROBLEMS

REVIEW QUESTIONS

1. How does *speed* differ from *velocity?* Give an example in which you can be traveling at constant speed but not at constant velocity.
2. What do we mean by *acceleration?* Explain what we mean when we state an acceleration in units of m/s^2.
3. What is the *acceleration of gravity?* If you drop a rock from very high up, approximately how fast will it be falling after 4 seconds?
4. What is *momentum?* How can momentum be affected by a *force?* What do we mean when we say that momentum will be changed only by a *net force?*
5. Suppose you visit another planet. Will your mass be the same as it is on Earth? Will your weight be the same? Explain.
6. What is *free-fall*, and why does it make you *weightless?* Briefly describe why astronauts are weightless in the Space Station.
7. State each of *Newton's three laws of motion*. For each law, give an example of its application.
8. Briefly describe and differentiate between *kinetic energy, radiative energy,* and *potential energy*. For each type of energy, give at least two examples of objects that either have it or use it. Explain clearly.
9. Briefly describe the laws of *conservation of energy* and *conservation of angular momentum*. For each, give an example of how it is important in astronomy.
10. Define *temperature* and *thermal energy*. How are they related? How are they different? Explain, using examples as needed.
11. Which has more gravitational potential energy: a rock on the ground or a rock that you hold out the window of a 10-story building? Explain.
12. What do we mean by *mass-energy?* Is it a form of kinetic, radiative, or potential energy? How is the idea of mass-energy related to the formula $E = mc^2$?
13. What is the *universal law of gravitation?* Summarize the law in words, and then state the formula mathematically. Be sure to explain what each symbol means in the formula.
14. Does an orbiting object always return over and over again to the same place? Explain. (*Hint:* Study the difference between bound and unbound orbits.)
15. What do we need to know if we want to measure an object's mass with *Newton's version of Kepler's third law?* Explain.
16. Briefly explain why orbits cannot change spontaneously. How can atmospheric drag affect an orbit? How can a *gravitational encounter* cause an orbit to change?
17. Explain how the Moon creates tides on Earth. How do the tides vary with the phase of the Moon? Why?

DOES IT MAKE SENSE?

Decide whether each statement makes sense and explain why it does or does not.

18. If you could go shopping on the Moon to buy a pound of chocolate, you'd get a lot more chocolate than if you bought a pound on Earth. (Hint: pounds are a unit of weight, not mass.)
19. Suppose you could enter a vacuum chamber (on Earth), that is, a chamber with no air in it. Inside this chamber, if you dropped a hammer and a feather from the same height at the same time, both would hit the bottom at the same time.
20. When an astronaut goes on a space walk outside the Space Station, she will quickly float away from the station unless she has a tether holding her to the station or constantly fires thrusters on her space suit.
21. Newton's version of Kepler's third law allows us to calculate the mass of Saturn from orbital characteristics of its moon Titan.
22. If we could magically replace the Sun with a giant rock that has precisely the same mass, Earth's orbit would not change.
23. The fact that the Moon rotates once in precisely the time it takes to orbit Earth once is such an astonishing coincidence that scientists probably never will be able to explain it.
24. Venus has no oceans, so it could not have tides even if it had a moon (which it doesn't).
25. If an asteroid passed by Earth at just the right distance, it would be captured by Earth's gravity and become our second moon.
26. When I drive my car at 30 miles per hour, it has more kinetic energy than it does at 10 miles per hour.
27. Someday soon, scientists are likely to build an engine that produces more energy than it consumes.

PROBLEMS

(Quantitative problems are marked with an asterisk.)

28. *Einstein's Famous Formula.*
    a. What is the meaning of the formula $E = mc^2$? Be sure to define each variable.
    b. How does this formula explain the generation of energy by the Sun?
    c. How does this formula explain the destructive power of nuclear bombs?
29. *Head-to-Foot Tides.* You and Earth attract each other gravitationally, so you should also be subject to a tidal force resulting from the difference between the gravitational attraction felt by your feet and that felt by your head (at least when you are standing). Explain why you can't feel this tidal force.

*30. *The Gravitational Law.* Use the universal law of gravitation to answer each of the following questions.
    a. How does quadrupling the distance between two objects affect the gravitational force between them?
    b. Compare the gravitational force between Earth and the Sun to that between Jupiter and the Sun. Jupiter's mass is 318 times Earth's mass, and its distance from the Sun is 5.2 times Earth's distance.
    c. Suppose the Sun were magically replaced by a star with twice as much mass. What would happen to the gravitational force between Earth and the Sun?

*31. *Energy Comparisons.* Use the data in Table 4.1 to answer each of the following questions.
    a. Compare the energy of a 1-megaton hydrogen bomb to the energy released by a major earthquake.

b. If the United States obtained all its energy from oil, how much oil would be needed each year?
c. Compare the Sun's annual energy output to the energy released by a supernova.

*32. *Fusion Power.* No one has yet succeeded in creating a commercially viable way to produce energy through nuclear fusion. However, suppose we could build fusion power plants using the hydrogen in water as a fuel. Based on the data in Table 4.1, how much water would we need each minute in order to meet U.S. energy needs? Could such a reactor power the entire United States with the water flowing from your kitchen sink? Explain. (*Hint:* Use the annual U.S. energy consumption to find the energy consumption per minute, and then divide by the energy yield from fusing 1 liter of water to figure out how many liters would be needed each minute.)

*33. *Measuring Masses.* Mathematically, Newton's version of Kepler's third law reads:

$$p^2 = \frac{4\pi^2}{G(M_1 + M_2)} a^3$$

where p is the orbital period in *seconds*, a is the average distance in *meters*, M_1 and M_2 are the masses of the two objects in kilograms, and the gravitational constant is

$$G = 6.67 \times 10^{-11} \frac{\text{m}^3}{\text{kg} \times \text{s}^2}$$

Use this law to answer each of the following questions.

a. The Moon orbits Earth in an average time of 27.3 days at an average distance of 384,000 kilometers. Use these facts to determine the mass of Earth. You may neglect the mass of the Moon and assume $M_{\text{Earth}} + M_{\text{Moon}} - M_{\text{Earth}}$.
b. Jupiter's moon Io orbits Jupiter every 42.5 hours at an average distance of 422,000 kilometers from the center of Jupiter. Calculate the mass of Jupiter. (*Hint:* $M_{\text{Jupiter}} + M_{\text{Io}} - M_{\text{Jupiter}}$.)
c. Use Earth's orbital distance and orbital period to calculate the mass of the Sun. (*Hint:* $M_{\text{Sun}} + M_{\text{Earth}} - M_{\text{Sun}}$.)
d. Pluto's moon Charon orbits Pluto every 6.4 days with a semimajor axis of 19,700 kilometers. Calculate the *combined* mass of Pluto and Charon. Compare this combined mass to the mass of Earth, which is about 6×10^{24} kg.

DISCUSSION QUESTIONS

34. *Knowledge of Mass-Energy.* Einstein's discovery that energy and mass are equivalent has led to technological developments that are both beneficial and dangerous. Discuss some of these developments. Overall, do you think the human race would be better or worse off if we had never discovered that mass is a form of energy? Defend your opinion.

35. *Perpetual Motion Machines.* Every so often, someone claims to have built a machine that can generate energy perpetually from nothing. Why isn't this possible according to the known laws of nature? Why do you think claims of perpetual motion machines sometimes receive substantial media attention?

36. *Tidal Complications.* The ocean tides on Earth are much more complicated than they might at first seem from the simple physics that underlies tides. Discuss some of the factors that make the real tides so complicated and how these factors affect the tides. Consider the following factors: the distribution of land and oceans; the Moon's varying distance from Earth in its orbit; and the fact that the Moon's orbital plane is not perfectly aligned with the ecliptic and neither the Moon's orbit nor the ecliptic is aligned with Earth's equator.

MATH HELP AND EXERCISES

For additional help and practice with mathematical concepts applicable to this chapter, the Astronomy Place web site has the following resources with detailed explanations, worked examples, and practice problems:

- Density
- Mass-Energy
- Using Newton's Version of Kepler's Third Law
- Calculating the Escape Velocity
- The Acceleration of Gravity

For a complete list of media resources available, go to **www.astronomyplace.com** and choose Chapter 4 from the pull-down menu.

ASTRONOMY PLACE WEB TUTORIALS

Tutorial Review of Key Concepts

Use the interactive **Tutorial** at **www.astronomyplace.com** to review key concepts from this chapter.

Motion and Gravity Tutorial

Lesson 1 Newton's Law of Motion
Lesson 2 The Force of Gravity

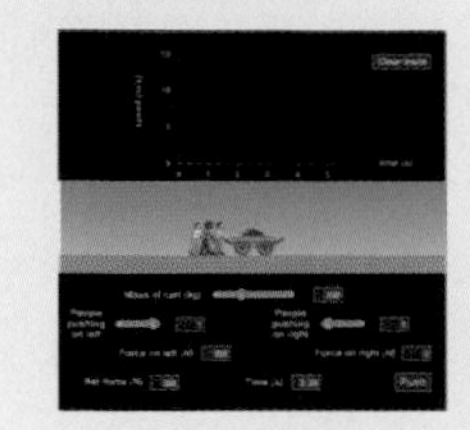
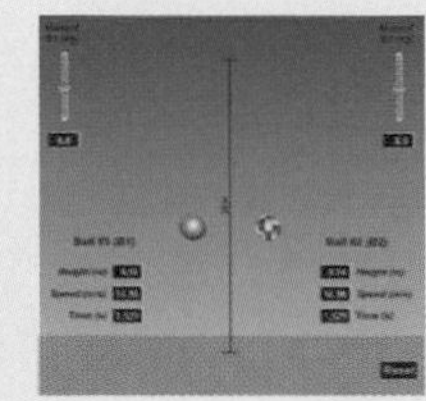
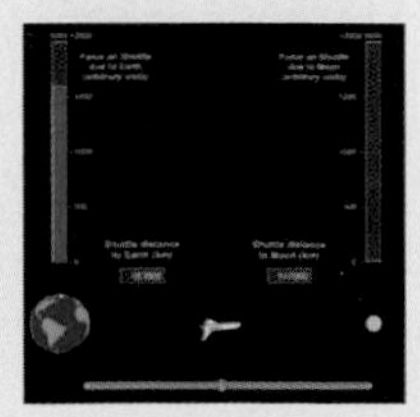

Orbits and Kepler's Laws Tutorial

Lesson 1 Gravity and Orbits
Lesson 2 Kepler's First Law
Lesson 3 Kepler's Second Law
Lesson 4 Kepler's Third Law

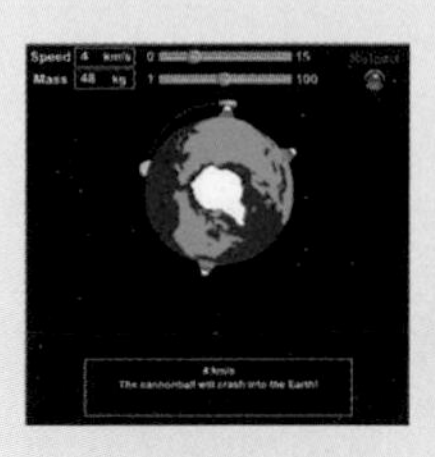

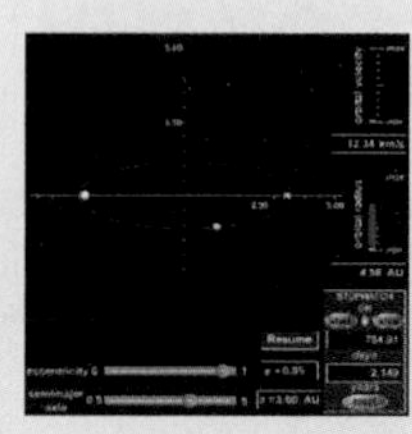

Supplementary Tutorial Exercises

Use the interactive **Tutorial Lessons** to explore the following questions.

Motion and Gravity, Lesson 2

1. Suppose Earth were twice as large in radius but still had the same mass. How much would you weigh? Explain.
2. As a spaceship moves away from Earth, does the force of gravity between Earth and the spaceship ever become zero? Explain.
3. Imagine a spaceship that somehow comes to a stop at a point precisely $\frac{3}{4}$ of the way from Earth to the Moon. Would the spaceship fall toward Earth, toward the Moon, or stay where it is? Explain.

Orbits and Kepler's Laws Tutorial, Lesson 1

1. How fast should you fire the cannonball to get it to orbit Earth?
2. How fast should you fire the cannonball to get it to escape into space?
3. How did the Apollo astronauts demonstrate that mass does not affect the rate at which objects fall?

Orbits and Kepler's Laws Tutorial, Lesson 4

1. Use the tool in Lesson 4 to predict the orbital period of an asteroid that has an orbital radius of 4 AU ($p =$ ____).
2. Use Kepler's third law, $p^2 = a^3$, to check your prediction ($a^3 =$ ____). Now take the square root of the result ($p =$ ____). Is your orbital period the same as the answer you calculated for question 1?
3. Use the tool to predict the orbital radius of an asteroid that has a period of 2 years. Does your answer agree with what you would expect from Kepler's third law?
4. Does the eccentricity of the orbit affect your answers to the last three questions?

EXPLORING THE SKY AND SOLAR SYSTEM

Of the many activities available on the ***Voyager: SkyGazer*** **CD-ROM** accompanying your book, use the following files to observe key phenomena covered in this chapter.

Go to the **File: Basics** folder for the following demonstrations.

1. Planet Paths
2. Planet Orrery
3. Follow a Planet

Go to the **File: Demo** folder for the following demonstrations.

1. Earth and Venus
2. Hyakutake at Perihelion
3. Pluto's Orbit

Go to the **Explore** menu for the following demonstrations.

1. Solar System
2. Paths of the Planets

MOVIES

Check out the following narrated and animated short documentary available on **www.astronomyplace.com** for a helpful review of key ideas covered in this chapter.

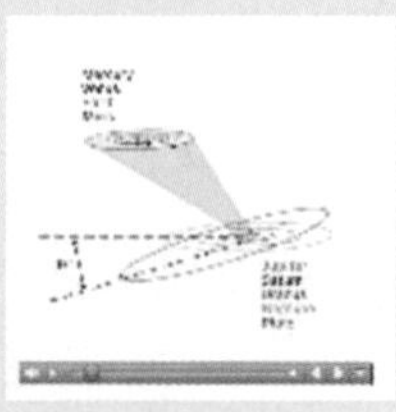

Orbits in the Solar System

WEB PROJECTS

Take advantage of the useful Web links on **www.astronomyplace.com** to assist you with the following projects.

1. *Space Station.* Visit a NASA site with pictures from the Space Station. Choose two photos that illustrate some facet of Newton's laws of motion or gravity. Explain how what is going on is related to Newton's laws.
2. *Energy Comparisons.* Using information from the Energy Information Administration Web site, choose some aspect of U.S. or world energy use that interests you. Write a short report on this issue.
3. *Nuclear Power.* There are two basic ways to generate energy from atomic nuclei: through nuclear fission (splitting nuclei) and through nuclear fusion (combining nuclei). All current nuclear reactors are based on fission, but fusion would have many advantages if we could develop the technology. Research some of the advantages of fusion and some of the obstacles to developing fusion power. Do you think fusion power will be a reality in your lifetime? Explain.

5
Light: The Cosmic Messenger

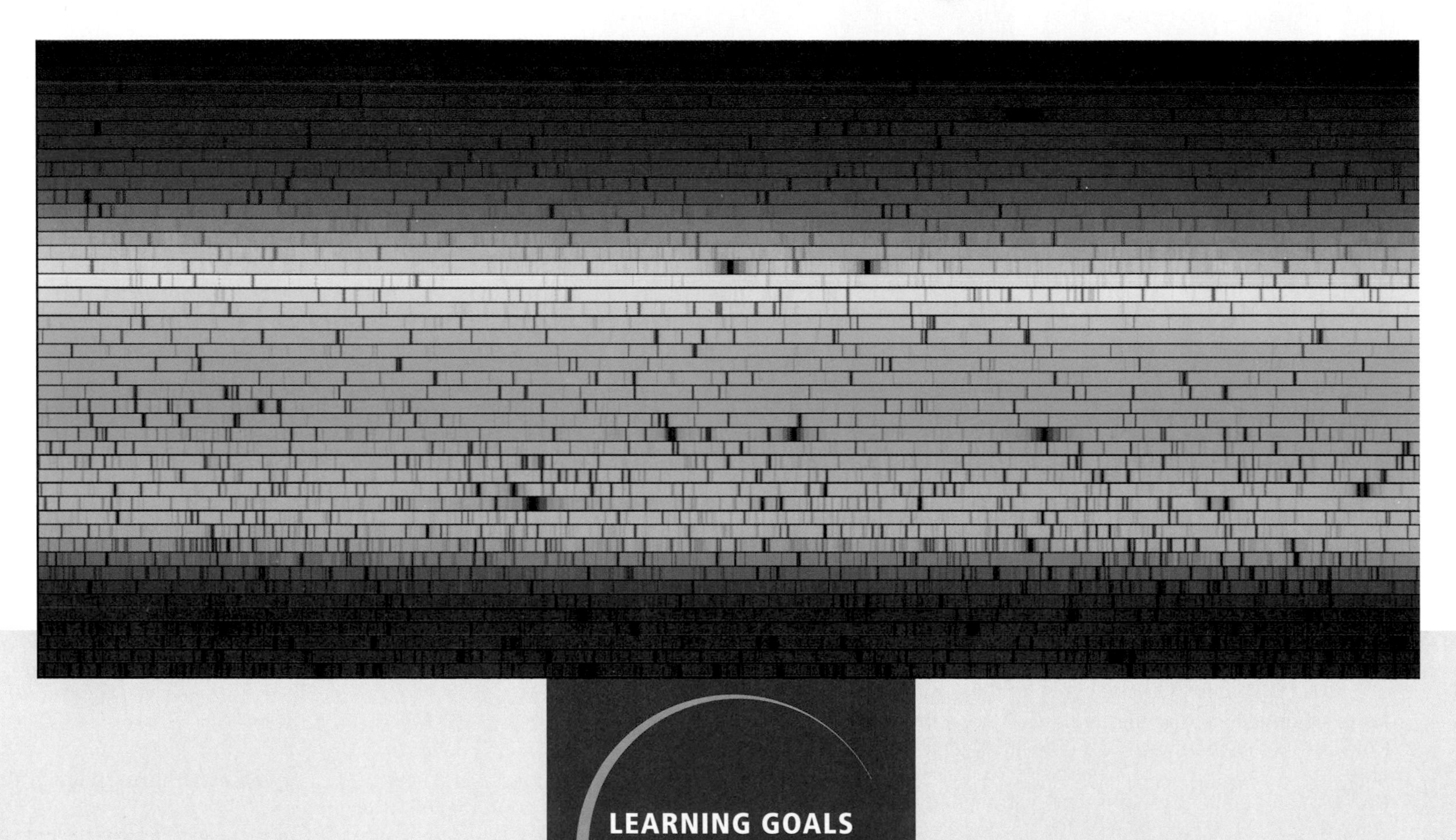

LEARNING GOALS

5.1 BASIC PROPERTIES OF LIGHT AND MATTER
- What is light?
- What is matter?
- How do light and matter interact?

5.2 LEARNING FROM LIGHT
- What types of light spectra can we observe?
- How does light tell us what things are made of?
- How does light tell us the temperatures of planets and stars?
- How does light tell us the speed of a distant object?

5.3 COLLECTING LIGHT WITH TELESCOPES
- How do telescopes help us learn about the universe?
- Why do we put telescopes in space?
- How is technology revolutionizing astronomy?

Ancient observers could discern only the most basic features of the light that they saw—such as color and brightness. Over the past several hundred years, we have discovered that light carries far more information. Today, we can analyze the light of distant objects to learn what they are made of, how hot they are, how fast they are moving, and much more. Light is truly the cosmic messenger, bringing the stories of distant objects to our home here on Earth.

Understanding the messages carried by light requires familiarity with the way light and matter interact. In this chapter, we'll explore the basic properties of light and matter that allow us to learn so much about the universe by studying light from distant objects. We'll also discuss how telescopes are used to collect that light and the technologies that make telescopes so much more powerful than our eyes.

Light and Spectroscopy Tutorial, Lesson 1

5.1 BASIC PROPERTIES OF LIGHT AND MATTER

Take a look at the photograph that opens this chapter. It is a very detailed photograph of the Sun's **spectrum**—the light from the Sun as it appears when we pass it through a prism or similar device. The rainbow of color, which stretches in horizontal rows from the upper left to the lower right of the photograph, probably reminds you of what we see whenever we pass white light through a prism (Figure 5.1). However, notice that the Sun's spectrum is not a pure rainbow. Instead, the spectrum shows hundreds of dark lines, representing places where a small piece of the rainbow is missing from the sunlight. All the features of the spectrum, including the rainbow and the dark lines, are created by interactions between light and matter in the Sun. Careful study of these features can tell us the Sun's chemical composition, its temperature, the motions of its atmosphere, and more.

We see similar dark or bright lines when we look at almost any spectrum in great detail, whether it is the spectrum of the flame from the gas grill in someone's backyard or the spectrum of a distant galaxy whose light we collect with a gigantic telescope. As long as we collect enough light to see details in the spectrum, we can learn many fundamental properties of the object we are viewing, no matter how far away the object is located.

Our primary goal in this chapter is to understand just how we can learn so much about distant objects from their spectra. But before we discuss the information encoded in spectra, we must first clarify the nature of light and matter.

• What is light?

Light is familiar to all of us, but its nature remained a mystery until quite recently in human history. Experiments performed by Isaac Newton in the 1660s provided the first real insights into the nature of light. It was already well known that passing white light through a prism produced a

Figure 5.1

When we pass white light through a prism, it disperses into a rainbow of color that we call a spectrum.

rainbow of color, but many people thought the color came from the prism rather than from the light itself. Newton proved that the colors came from the light by placing a second prism in front of the light of just one color, such as red, from the first prism. If the rainbow of color came from the prism itself, the second prism would have produced a rainbow just like the first. But it did not: When only red light entered the second prism, only red light emerged, proving that the color was a property of the light and not of the prism.

Light is also known as electromagnetic radiation.

Newton's experiment proved that white light is actually a mix of all the colors in the rainbow. Later scientists found that there is light "beyond the rainbow" as well. Just as there are sounds that our ears cannot hear (such as the sound of a dog whistle), there is light that our eyes cannot see. In fact, the **visible light** that splits into the rainbow of color is only a tiny part of the complete spectrum of light. Figure 5.2 shows the complete spectrum of light, usually called the **electromagnetic spectrum**. Light itself is often called **electromagnetic radiation**. Let's investigate why.

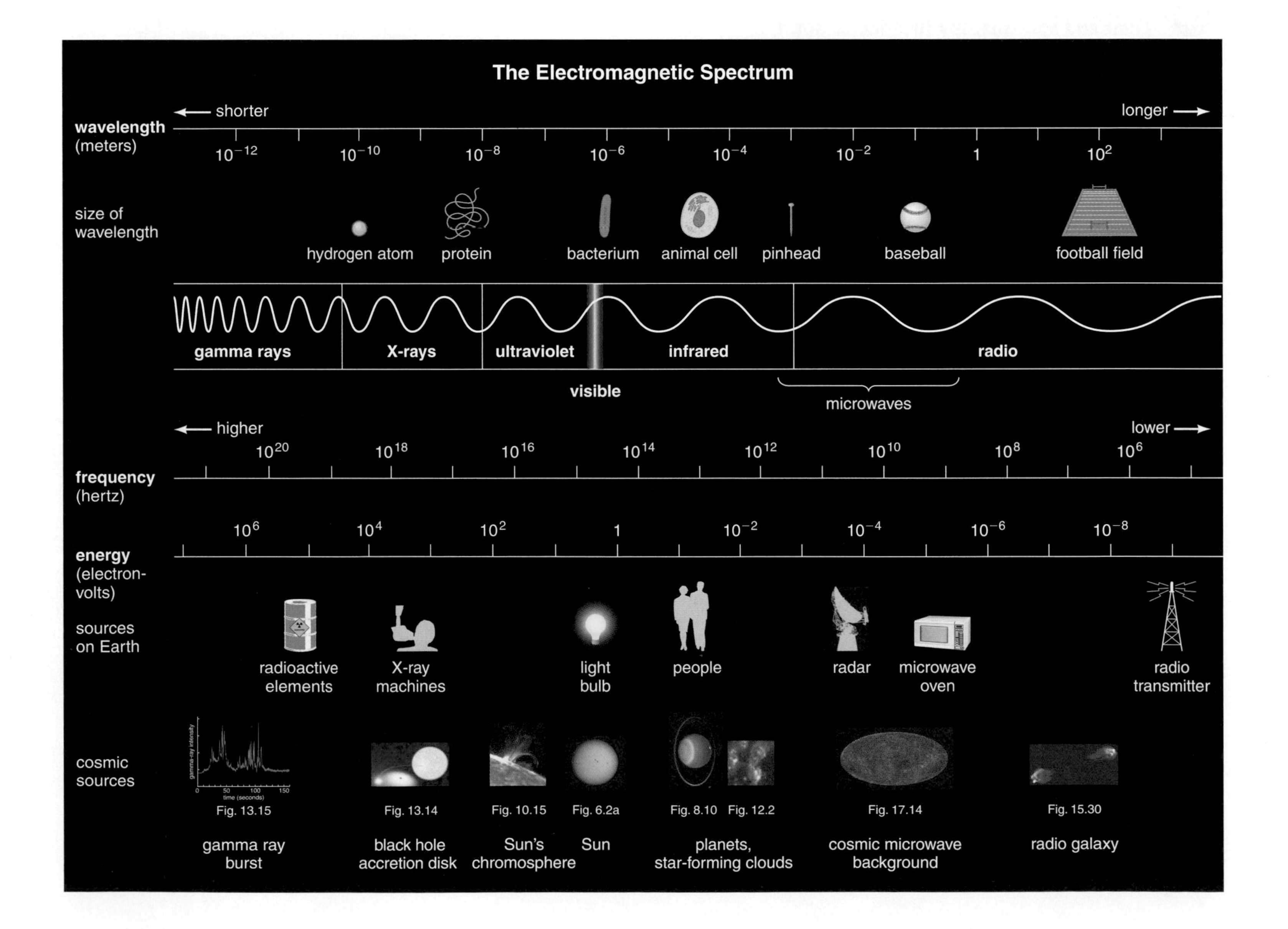

Figure 5.2

The electromagnetic spectrum. Notice that wavelength increases as we go from gamma rays to radio waves, while frequency and energy increase in the opposite direction.

Wave Properties of Light You've probably heard that light is a wave, but may not know what that means. In general, a wave is something that can transmit energy without carrying material along with it. For example, you can make waves move outward along a rope by shaking one end of it up and down (Figure 5.3a). The shaking creates a series of peaks and troughs that move along the rope, making every piece of the rope bob up and down as the peaks and troughs go by. We define the **wavelength** as the distance between adjacent peaks and the **frequency** as the number of times that any piece of the rope moves up and down each second. For example, if a piece of the rope moves up and down three times each second, we say the wave has a frequency of three cycles per second ("cycles per second" are sometimes called *hertz* for short). Notice that the rope itself stays intact as the wave moves along it, showing that it is energy and not material that is moving outward.

Light is different from waves on a rope because we cannot see anything moving up and down as it travels. However, we can tell that light is a wave from its effect on matter. If you could set up a row of electrically charged particles such as electrons, it would wriggle like a snake as a wave of light passed by (Figure 5.3b). The distance between adjacent peaks in this row of electrons would tell us the wavelength of the light wave, while the number of times each electron bobs up and down would tell us the frequency (Figure 5.3c). Because light can affect both electrically charged particles and magnets, we say that light is an *electromagnetic wave*—which is why light is called *electromagnetic radiation* and the spectrum of light is called the *electromagnetic spectrum*.

The longer the wavelength of light, the lower its frequency and energy.

All light travels through space at the same speed, which we call the **speed of light**. It is about 300,000 kilometers per second. This fact allows us to recognize an important relationship between wavelength and frequency for light: *the longer the wavelength, the lower the frequency, and vice versa.* You can readily see why: wavelength × frequency must always equal speed, because wavelength tells us the distance between peaks and frequency tells us how rapidly peaks are passing by. Because all light travels at the same speed, wavelength must go up when frequency goes down so that their product still equals the speed of light. Figure 5.2 shows these relationships across the electromagnetic spectrum. Notice, for example, that the portion of the spectrum on the far left (gamma rays) has the shortest wavelengths but the highest frequencies.

Particle Properties of Light In everyday life, waves seem to be quite different from particles. A wave exists only as a pattern of motion with a wavelength and a frequency, while a particle is an individual thing such as a marble, a baseball, or an individual atom. However, experiments show that light can behave *both* as a wave and as a particle. We've already discussed the wave nature of light. The particle nature of light arises because light comes in individual "pieces," called **photons**. Like baseballs, photons of light can be counted individually and can hit a wall one at a time. The idea that light can be both a wave and a particle may seem quite strange, but it is fundamental to modern understanding of physics.

Light comes in "pieces" called photons, each with a precise wavelength, frequency, and energy.

We therefore think of light as consisting of many individual photons, each traveling at the speed of light and characterized by a wavelength and frequency. In addition, each photon carries a particular amount

Figure 5.3 Interactive Figure

These diagrams explain the wave properties of light.

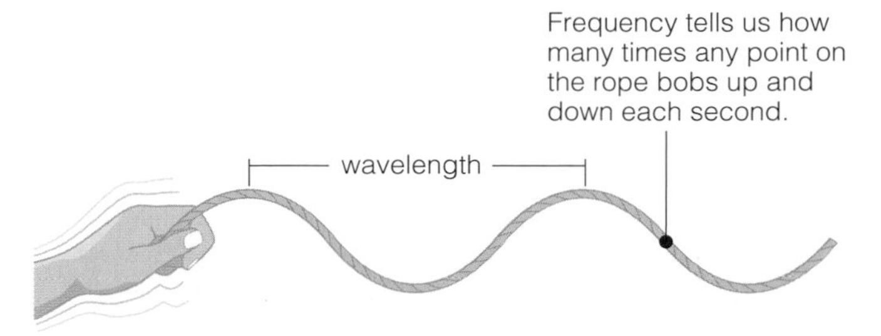

a Shaking one end of a rope up and down generates waves moving along it. All waves are characterized by a wavelength and a frequency.

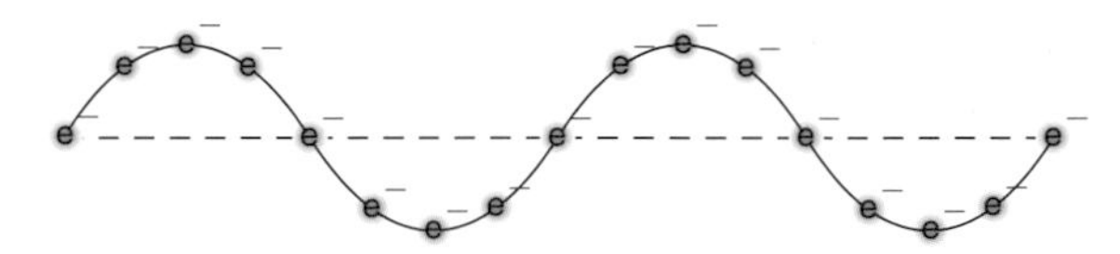

b If you could line up electrons, they would wriggle up and down as light passes by, demonstrating that light is a wave.

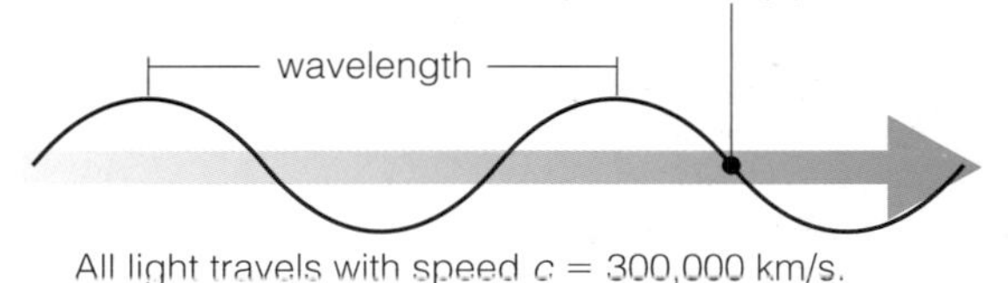

c Light can affect both electrically charged particles and magnets, so we say that light is an *electromagnetic wave.*

of energy that depends on its frequency: the higher the frequency of the photon, the more energy it carries. That is why the energy arrow in Figure 5.2 goes in the same direction as the frequency arrow.

THINK ABOUT IT How does the energy of a photon depend on its wavelength? Briefly explain why.

The Many Forms of Light Figure 5.2 also shows that we give special names to different portions of the electromagnetic spectrum. Visible light has wavelengths ranging from about 400 nm at the blue or violet end of the rainbow to about 700 nm at the red end. (A nanometer [nm] is a billionth of a meter.) Light with wavelengths somewhat longer than red light is called **infrared**, because it lies beyond the red end of the rainbow. **Radio waves** are the longest-wavelength light. Thus, radio waves are a form of light, *not* a form of sound. The region near the border between infrared and radio waves, where wavelengths range from micrometers to millimeters, is sometimes given the name **microwaves**.

On the other side of the spectrum, light with wavelengths somewhat shorter than blue light is called **ultraviolet**, because it lies beyond the blue (or violet) end of the rainbow. Light with even shorter wavelengths is called **X rays**, and the shortest-wavelength light is called **gamma rays**. Notice that visible light is an extremely small part of the entire electromagnetic spectrum: The reddest red that our eyes can see has only about twice the wavelength of the bluest blue, but the radio waves from your favorite radio station are a billion times longer than the X rays used in a doctor's office.

Radio waves, microwaves, infrared, visible light, ultraviolet, X rays, and gamma rays are all forms of light.

The different energies of different forms of light explain many familiar effects in everyday life. Radio waves carry so little energy that they have no noticeable effect on our bodies. However, radio waves can make electrons move up and down in an antenna, which is how the antenna of your car radio receives the radio waves coming from a radio station. Molecules moving around in a warm object emit infrared light, which is why we sometimes associate infrared light with heat. Visible-light photons happen to have enough energy to activate receptors in our eyes, making vision possible. Ultraviolet photons carry enough energy to harm cells in our skin, causing sunburn or skin cancer. X-ray photons have enough energy to penetrate through skin and muscle but can be blocked by bones or teeth. That is why doctors and dentists can see our underlying bone structures on photographs taken with X-ray light.

• What is matter?

Light carries information about matter across the universe, but we are usually more interested in the matter the light is coming from than we are in the light itself. Planets, stars, and galaxies are made of matter, and we must understand the nature of matter if we are to decode the messages we receive in light.

Like the nature of light, the nature of matter remained mysterious for most of human history. The ancient Greeks imagined that all material was made of four elements: fire, water, earth, and air. Some Greeks, beginning with the philosopher Democritus (c. 470–380 B.C.), further imagined that these four elements ultimately came in the form of tiny particles

COMMON MISCONCEPTIONS

IS RADIATION DANGEROUS?

Many people associate the word *radiation* with danger. However, the word *radiate* simply means "to spread out from a center" (note the similarity between *radiation* and *radius* [of a circle]). Radiation is simply energy being carried through space. If energy is being carried by particles of matter, such as protons or neutrons, we call it *particle radiation.* If energy is being carried by light, we call it *electromagnetic radiation.*

High-energy forms of radiation, such as particles from radioactive materials or X rays, are dangerous because they can penetrate body tissues and cause cell damage. Low-energy forms of radiation, such as radio waves, are usually harmless. The visible light radiation from the Sun is necessary to life on Earth. Thus, while some forms of radiation are dangerous, others are harmless or beneficial.

COMMON MISCONCEPTIONS

CAN YOU HEAR RADIO OR SEE AN X RAY?

Most people associate the term *radio* with sound, but radio waves are a form of *light* with long wavelengths—too long for our eyes to see. Radio stations encode sounds (such as voices and music) as electrical signals, which they broadcast as radio waves. What we call "a radio" in daily life is an electronic device that receives these radio waves and decodes them to re-create the sounds played at the radio station. Television is also broadcast by encoding information (both sound and pictures) in the form of light called radio waves.

X rays are also a form of light, with wavelengths far too short for our eyes to see. In a doctor's or dentist's office, a special machine works somewhat like the flash on an ordinary camera but emits X rays instead of visible light. This machine flashes the X rays at you, and a piece of photographic film (or an electronic detector) records the X rays that are transmitted through your body. You never see the X rays—you see only an image left on film by the transmitted X rays.

they called *atoms,* a Greek term meaning "indivisible." Although these beliefs bear some similarity to our modern understanding of matter, they were incorrect in many details.

Today, we have identified more than 100 chemical elements, and fire, water, earth, and air are *not* among them. Some of the most familiar chemical elements are hydrogen, helium, carbon, oxygen, silicon, iron, gold, silver, lead, and uranium. Appendix D gives the periodic table of all the elements.

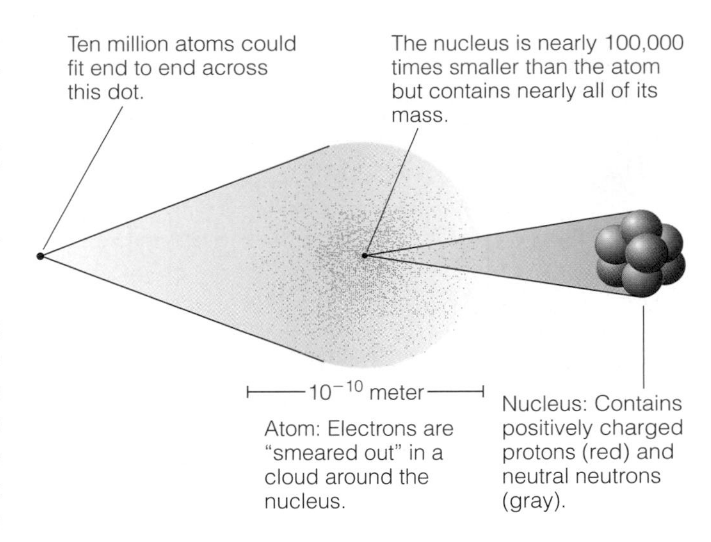

Figure 5.4

The structure of a typical atom.

Atomic Structure Each chemical element is made from a different type of **atom**, and atoms are in turn made of particles that we call **protons**, **neutrons**, and **electrons** (Figure 5.4). Protons and neutrons are found in the tiny **nucleus** at the center of the atom. The rest of the atom's volume contains the electrons that surround the nucleus. Although the nucleus is very small compared to the atom as a whole, it contains most of the atom's mass, because protons and neutrons are each about 2,000 times more massive than an electron. Note that atoms are incredibly small: Millions could fit end to end across the period at the end of this sentence. The number of atoms in a single drop of water (typically, 10^{22} to 10^{23} atoms) may exceed the number of stars in the observable universe.

The chemical elements are made of atoms, which in turn are made of protons, neutrons, and electrons.

The properties of an atom depend mainly on the amount of **electrical charge** in its nucleus. Electrical charge is a fundamental physical property that is always conserved, just as energy is always conserved. We define the electrical charge of a proton as the basic unit of positive charge, which we write as +1. The electron has an electrical charge that is precisely opposite that of a proton, so we say it has negative charge (−1). Neutrons are electrically neutral, meaning that they have no charge.

Oppositely charged particles attract one another, and similarly charged particles repel one another. The attraction between the positively charged protons in the nucleus and the negatively charged electrons that surround it is what holds an atom together. Ordinary atoms have identical numbers of electrons and protons, making them electrically neutral overall. (You may wonder why electrical repulsion doesn't cause the positively charged protons in a nucleus to fly apart from one another. The answer is that an even stronger force, called the *strong force,* overcomes electrical repulsion and holds the nucleus together [Section 10.2].)

Although we can think of electrons as tiny particles, they are not quite like tiny grains of sand, and they don't really orbit the nucleus the way planets orbit the Sun. Instead, the electrons in an atom form a kind of "smeared out" cloud that surrounds the nucleus and gives the atom its apparent size. The electrons aren't really cloudy, but it is impossible to pinpoint their positions.

In Figure 5.4, you can see that the electrons give the atom a size far larger than its nucleus even though they represent only a tiny portion of the atom's mass. If we imagine an atom on a scale that makes its nucleus the size of your fist, its electron cloud would be many miles wide.

Atomic Terminology Astronomical objects contain many different types of atoms, and it's important to have a way to describe the atoms. You've probably learned the basic terminology of atoms in past science classes, but let's review it just to be sure.

Each different chemical element contains a different number of protons in its nucleus. This number is its **atomic number**. For example, a

COMMON MISCONCEPTIONS

THE ILLUSION OF SOLIDITY

Bang your hand on a table. Although the table feels solid, it is made almost entirely of empty space! Nearly all the mass of the table is contained in the nuclei of its atoms. But the volume of an atom is more than a trillion times the volume of its nucleus, so relatively speaking the nuclei of adjacent atoms are nowhere near to touching one another. The solidity of the table comes about from a combination of electrical interactions between the charged particles in its atoms and the strange quantum laws governing the behavior of electrons. If we could somehow pack all the table's nuclei together, the table's mass would fit into a microscopic speck. Although we cannot pack matter together in this way, nature can and does—in *neutron stars,* which we will study in Chapter 13.

hydrogen nucleus contains just one proton, so its atomic number is 1. A helium nucleus contains two protons, so its atomic number is 2.

Atoms of different chemical elements have different numbers of protons.

The *combined* number of protons and neutrons in an atom is called its **atomic mass number**. The atomic mass number of ordinary hydrogen is 1 because its nucleus is just a single proton. Helium usually has two neutrons in addition to its two protons, giving it an atomic mass number of 4. Carbon usually has six protons and six neutrons, giving it an atomic mass number of 12.

Isotopes of a particular chemical element all have the same number of protons but different numbers of neutrons.

Every atom of a given element contains exactly the same number of protons, but the number of neutrons can vary. For example, all carbon atoms have six protons, but they may have six, seven, or eight neutrons. Versions of an element with different numbers of neutrons are called **isotopes** of the element. Isotopes are named by listing their element name and atomic mass number. For example, the most common isotope of carbon has 6 protons and 6 neutrons, giving it atomic mass number 6 + 6 = 12, so we call it carbon-12. The other isotopes of carbon are carbon-13 (six protons and seven neutrons give it atomic mass number 13) and carbon-14 (six protons and eight neutrons give it atomic mass number 14). We can also write isotopes by writing the atomic mass number as a superscript to the left of the element symbol: ^{12}C, ^{13}C, ^{14}C. We read ^{12}C as "carbon-12." Figure 5.5 summarizes some of this basic atomic terminology.

The symbol ^{4}He represents helium with an atomic mass number of 4. ^{4}He is the most common form of helium, containing two protons and two neutrons. What does the symbol ^{3}He represent?

The number of different material substances is far greater than the number of chemical elements because atoms can combine to form **molecules**. Some molecules consist of two or more atoms of the same element. For example, we breathe O_2, oxygen molecules made of two oxygen atoms. Other molecules, such as water, are made up of atoms of two or more different elements. The symbol H_2O tells us that a water molecule contains two hydrogen atoms and one oxygen atom. The chemical properties of a molecule are different from those of its individual atoms. For example, water behaves very differently than pure hydrogen or pure oxygen.

Figure 5.5

Terminology of atoms.

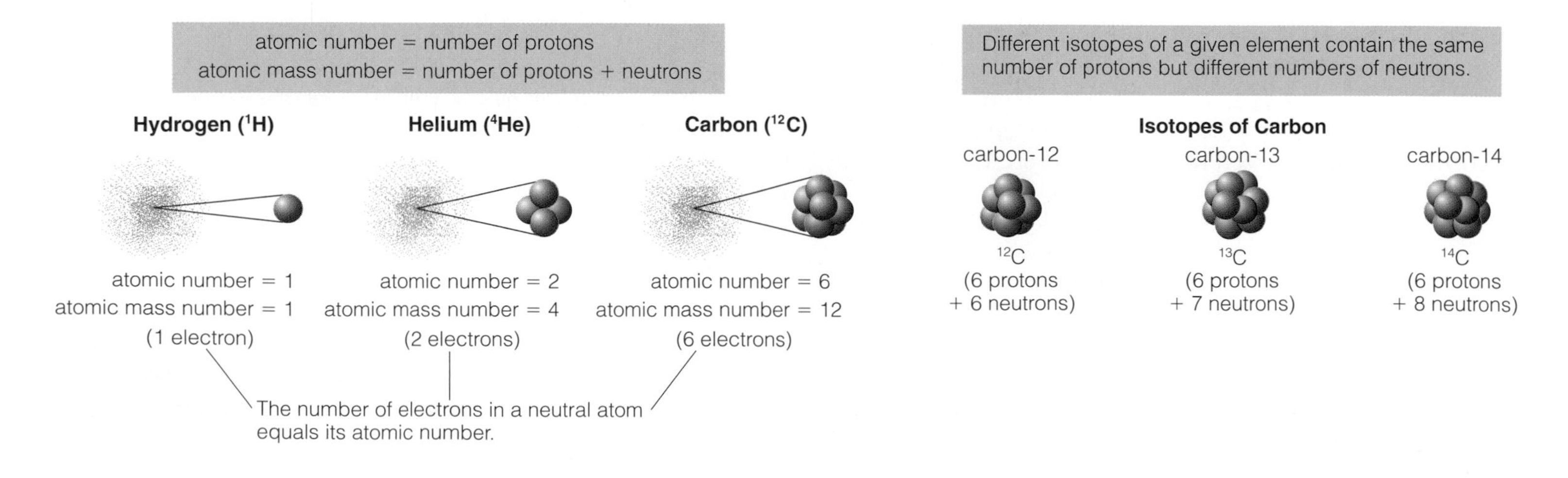

• How do light and matter interact?

Now that we have discussed the nature of light and of matter individually, we are ready to explore how light and matter can interact. These interactions provide information about the world around us and the distant universe. Recall, for example, that light and matter interact to form the spectrum that opens this chapter and that gives us so much information about the Sun. Before we study spectra in detail, let's focus on the interactions of light and matter that we see in everyday life.

Energy carried by light can interact with matter in four general ways:

- **Emission**: When you turn on a lamp, electric current flowing through the filament of the light bulb heats it to a point at which it *emits* visible light.
- **Absorption**: If you place your hand near a lit light bulb, your hand *absorbs* some of the light, and this absorbed energy warms your hand.
- **Transmission**: Some forms of matter, such as glass or air, *transmit* light. That is, they allow light to pass through.
- **Reflection/scattering**: Photons of light can bounce off matter, leading to what we call *reflection* (when the bouncing is all in the same general direction) or *scattering* (when the bouncing is more random).

Matter can emit, absorb, transmit, or reflect light.

Materials that transmit light are said to be *transparent,* and materials that absorb light are called *opaque*. Many materials are neither perfectly transparent nor perfectly opaque. For example, dark sunglasses and clear eyeglasses are both at least partially transparent, but the dark glasses absorb more light and transmit less. In addition, many materials affect different colors of light differently. For example, red glass transmits red light but absorbs other colors, while a green lawn reflects (scatters) green light but absorbs all other colors.

We can put these ideas together to understand what happens when you walk into a room and turn on the light switch (Figure 5.6). The light bulb begins to emit white light, which is a mix of all the colors in the

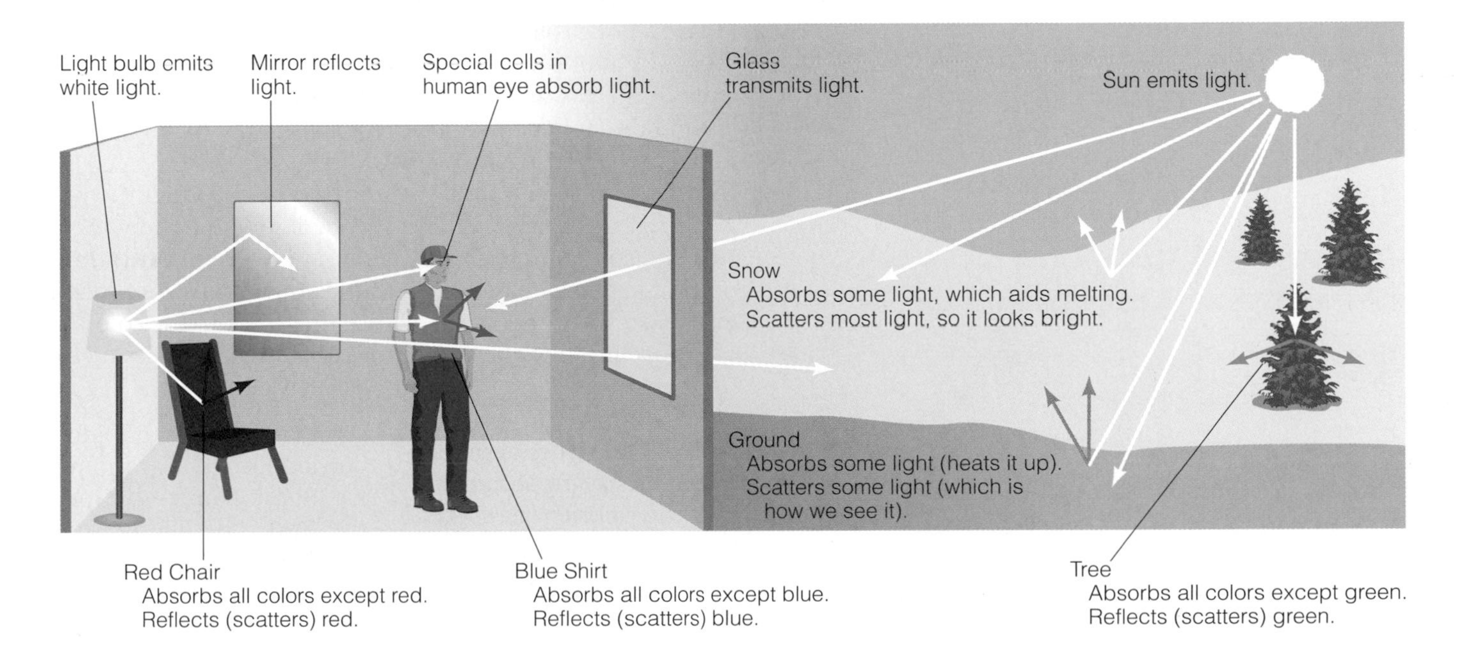

Figure 5.6

When light strikes any piece of matter in the universe, that matter reacts in one or a combination of four ways: emission, absorption, transmission, and reflection.

spectrum. Some of this light exits the room, transmitted through the windows. The rest of the light strikes the surfaces of objects inside the room, and each object's material properties determine the colors absorbed or reflected. The light coming from each object therefore carries an enormous amount of information about the object's location, shape and structure, and material makeup. You acquire this information when light enters your eyes, where it is absorbed by special cells that send signals to your brain. Your brain interprets the messages carried by the light, recognizing materials and objects in the process we call vision.

5.2 LEARNING FROM LIGHT

Light carries much more information than our naked eyes can recognize. Modern instruments can reveal otherwise hidden details in the spectrum of light, and specially equipped telescopes can record forms of light that are invisible to our eyes. In this section, we'll learn how detailed studies of light help us unlock the secrets of the universe. The key to unlocking those secrets lies in learning how to read astronomical spectra. Let's start by learning the basic types of spectra and how they are produced, and then we will be ready to see what we can learn from them.

Light and Spectroscopy Tutorial, Lessons 2–4

• What types of light spectra can we observe?

Scientists learn about spectra by studying interactions of light and matter in the laboratory. These studies show that spectra come in three basic types; real astronomical spectra are usually a combination of these types. Figure 5.7 shows an example of each type, along with the conditions under which we can see it. (The rules that specify these conditions are often called *Kirchhoff's laws.*) Study the figure carefully to notice the following key ideas:

1. The spectrum of a common (incandescent) light bulb is a continuous rainbow of color. For reasons we will discuss shortly, we call such a spectrum a **thermal radiation spectrum**.
2. If a cloud of gas lies between us and a light bulb, the cloud can absorb light of specific colors from the rainbow. Thus, the spectrum of the gas will show dark **absorption lines** over the background rainbow coming from the light bulb, making what we call an **absorption line spectrum**.
3. If we look at the cloud of gas from the side, so that the light bulb is not right behind it, we can see the light emitted by the gas itself. The spectrum of this light consists of bright **emission lines** of color against a black background, and is therefore called an **emission line spectrum**. Notice that the specific colors of the emission lines are the very same colors that are missing from the rainbow in the absorption line spectrum.

There are three basic types of spectra: thermal radiation spectra, absorption line spectra, and emission line spectra.

We can use these ideas to understand many astronomical spectra. For example, the spectrum of the Sun that opens this chapter shows an absorption line spectrum over a background rainbow of color. This tells us that we are essentially looking at a hot light source (like the

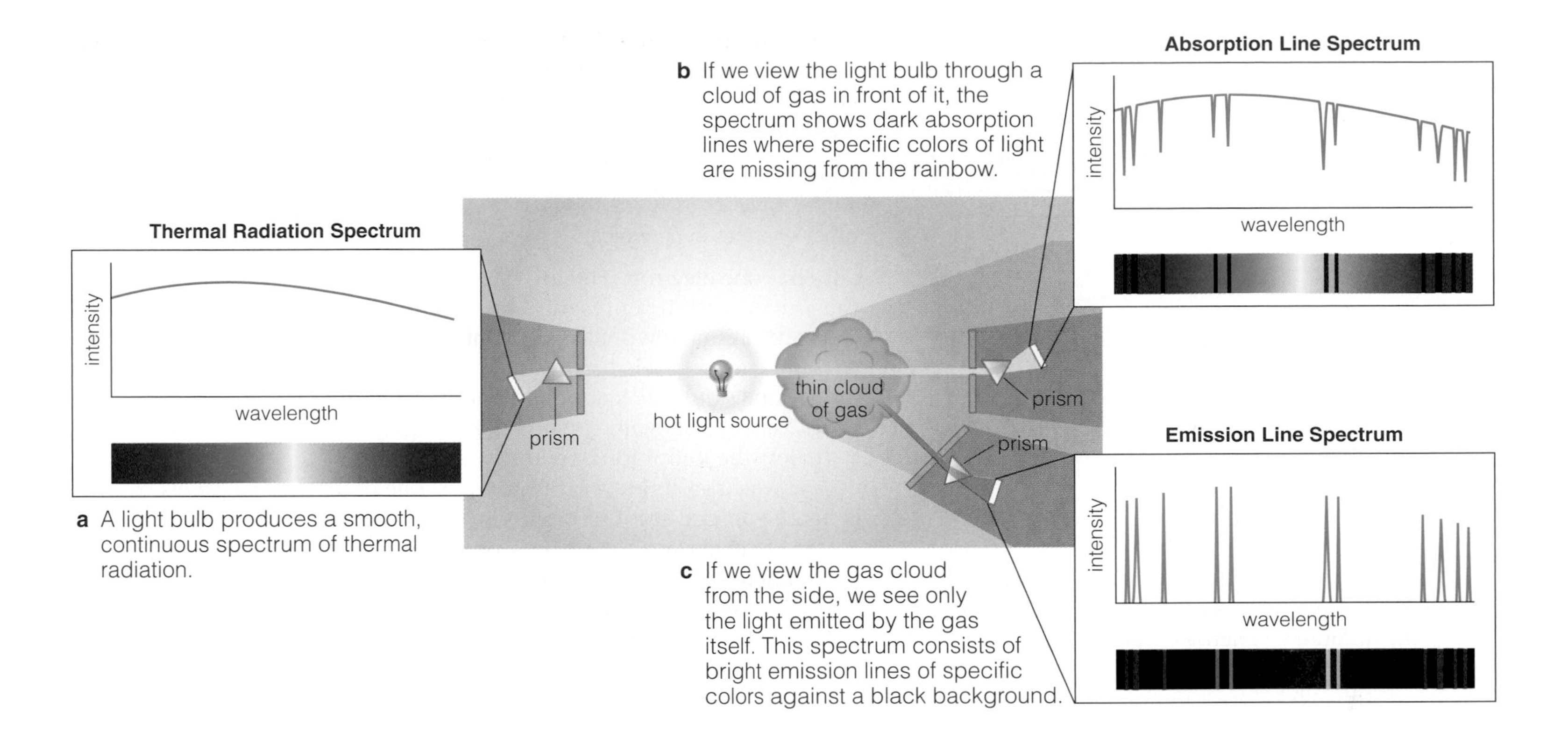

Figure 5.7 Interactive Figure

This diagram shows examples of the conditions under which we see the three basic types of spectra.

light bulb in Figure 5.7) through gas that is absorbing some of the colors (like the cloud in Figure 5.7). The hot light source is the hot interior of the Sun, and the gas is the cooler and lower-density layer of gas just above the Sun's surface.

Figure 5.7 also shows that spectra can be displayed both as bands of light and as graphs. If you simply project the light that passes through a prism onto a wall, you'll see the spectrum as a band of color. If you then carefully measure the amount or **intensity** of the light at each wavelength in the spectrum, you can display your results as a graph. The intensity is high at wavelengths where there is a lot of light and low where there is little light. For example, notice how the graph of the absorption line spectrum shows dips in intensity at the wavelengths where the band of light shows dark lines. Astronomers usually display spectra as graphs because they make it easier to tell how the precise intensity of the light varies across the spectrum.

• How does light tell us what things are made of?

We have seen *how* different viewing conditions lead to different types of spectra, so we are now ready to discuss *why*. Let's start with absorption and emission line spectra. As we'll see, the positions of the lines in these spectra can tell us what distant objects are made of.

Energy Levels in Atoms To understand why we sometimes see emission and absorption lines, we must first discuss a strange fact about electrons in atoms: The electrons can have only particular amounts of energy, and not other energies in between. As an analogy, suppose you're washing windows on a building. If you use an adjustable platform to reach high windows, you can stop the platform at any height above the ground. But if you use a ladder, you can stand only at *particular* heights—the

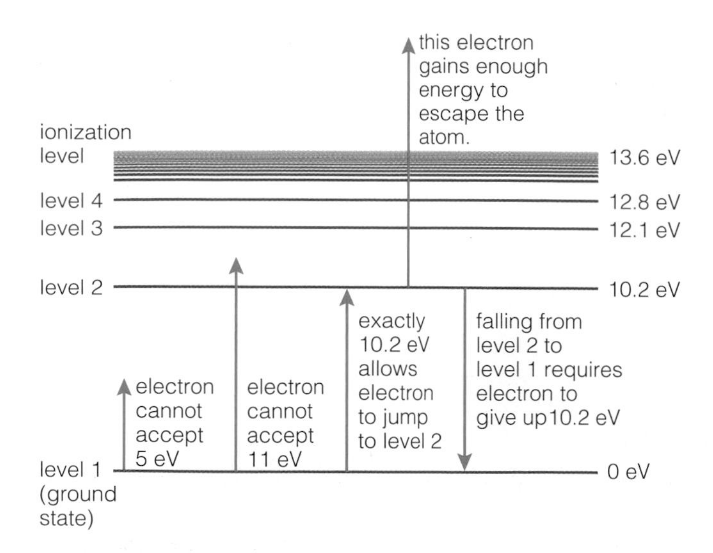

Figure 5.8

Energy levels for the electron in a hydrogen atom. The electron can jump between energy levels only if it gains or loses the amount of energy separating the levels. If the electron gains enough energy to reach the ionization level, it can leave the atom, leaving behind a positively charged ion. (The many levels between level 4 and the ionization level are not labeled.)

heights of the rungs of the ladder—and not at any height in between. The possible energies of electrons in atoms are like the possible heights on a ladder. Only a few particular energies are possible, and energies between these special few are not possible.

Electrons in atoms can have only particular amounts of energy, and not other energies in between.

Let's consider energy levels in hydrogen, shown in Figure 5.8. Hydrogen has the simplest set of energy levels of any atom, because it only has one electron. The lowest possible energy level is called level 1 or the *ground state*. It represents the minimum amount of energy that the electron can have when it is part of a hydrogen atom. Each of the other possible energy levels has a fairly specific amount of energy compared to the ground state, labeled to the right in an energy unit called the eV. (1 eV, short for electron-volt, is equivalent to 1.6×10^{-19} joule; see Section 4.3 to review the definition of a joule.)

An electron in level 1 can jump to one of the higher levels only if it gains the precise amount of energy separating the levels. For example, if you try to give the electron 5 eV of energy, it won't accept it because that is not enough energy to reach level 2. Similarly, if you try to give it 11 eV, it won't accept it because it is too much for level 2 but not enough to reach level 3. However, because level 2 lies 10.2 eV above level 1, you can make an electron jump from level 1 to level 2 by giving it exactly 10.2 eV of energy. If it then returns to level 1, it must somehow give away the same 10.2 eV of energy.

Notice that the amount of energy separating the various levels gets smaller near the top. For example, it takes more energy for the electron to jump from level 1 to level 2 than from level 2 to level 3. If the electron gains enough energy to reach the *ionization level*, it escapes the atom completely. Because the escaping electron carries away negative electrical charge, the atom is left with positive electrical charge. Electrically charged atoms are called **ions**, so we say that the escape of the electron *ionizes* the atom.

Are there any circumstances under which an electron in a hydrogen atom can gain 2.6 eV of energy? Explain.

Other atoms also have distinct energy levels, but the levels correspond to different amounts of energy than those of hydrogen. Furthermore, not only does every type of atom have its own distinct set of energy levels, but so does every type of ion and every type of molecule.

Emission and Absorption Lines The fact that each type of atom, ion, or molecule possesses a unique set of energy levels is what causes emission and absorption lines to appear at specific wavelengths in spectra. It is also what allows us to learn the compositions of distant objects in the universe. To see how, let's consider what happens in a cloud of gas consisting solely of hydrogen atoms.

If you heat this hydrogen gas, the atoms will constantly collide with one another, exchanging energy in each collision. Most of the collisions simply send the atoms careening off in new directions. However, a few of the collisions transfer the right amount of energy to bump an electron from a low energy level to a higher energy level.

Electrons can't stay in higher energy levels for long. They always fall back down to level 1, usually in a tiny fraction of a second. The energy the electron loses when it falls to the lower energy level must go somewhere, and often it goes to *emitting* a photon of light. The emitted photon must have the same amount of energy that the electron loses, which means that it has a specific wavelength (and frequency). Figure 5.9a again shows the energy levels in hydrogen that we saw in Figure 5.8. This time, however, it is also labeled with the wavelengths of the photons emitted by various downward *transitions* of an electron from a higher energy level to a lower one. For example, the transition from level 2 to level 1 emits an ultraviolet photon of wavelength 121.6 nm, and the transition from level 3 to level 2 emits a red visible-light photon of wavelength 656.3 nm.

The photons in emission lines are created when electrons jump to lower energy levels.

Although electrons bumped to higher energy levels in warm hydrogen gas quickly fall back to level 1, new collisions continually bump other electrons into the higher levels. Thus, there are always some electrons falling into lower energy levels and emitting photons with the wavelengths shown in Figure 5.9a. The gas therefore emits light only with these specific wavelengths. That is why we see an emission line spectrum, as shown in Figure 5.9b. The bright emission lines appear at the wavelengths that correspond to downward jumps of electrons, and the rest of the spectrum is dark (black).

THINK ABOUT IT If nothing continues to heat the hydrogen gas, all the electrons eventually will end up in the lowest energy level (the ground state, or level 1). Use this fact to explain why we should *not* expect to see an emission line spectrum from a very cold cloud of hydrogen gas.

Absorption lines occur when photons cause electrons to jump to higher energy levels.

Now, suppose a light bulb illuminates the hydrogen gas from behind (as in Figure 5.7b). The light bulb emits light of all wavelengths, producing a spectrum that looks like a rainbow of color. However, the hydrogen atoms will absorb those photons that have the right amount of energy needed to bump an electron from a low energy level to a higher one. Figure 5.9c shows the result. It is an absorption line spectrum, because the light bulb produces a rainbow of color while the hydrogen atoms remove (absorb) light at specific wavelengths. (You might wonder what happens to the electrons after they absorb photons and jump to a higher energy level: The electrons quickly fall back down, emitting photons of the same energy in random directions. Thus, we see absorption lines because most of the emitted photons are not sent along our line-of-sight.)

You can also now see why the dark absorption lines in Figure 5.9c occur at the same wavelengths as the emission lines in Figure 5.9b: Both types of lines represent the same energy level transitions, except in opposite directions. For example, electrons moving downward from level 3 to level 2 in hydrogen can emit photons of wavelength 656.3 nm (producing an emission line at this wavelength), while electrons absorbing photons with this wavelength can jump up from level 2 to level 3 (producing an absorption line at this wavelength).

Chemical Fingerprints The fact that hydrogen emits and absorbs at specific wavelengths makes it possible to detect its presence in distant objects. For example, imagine that you look through a telescope at an interstellar gas cloud, and its spectrum looks like that shown in Figure 5.9b.

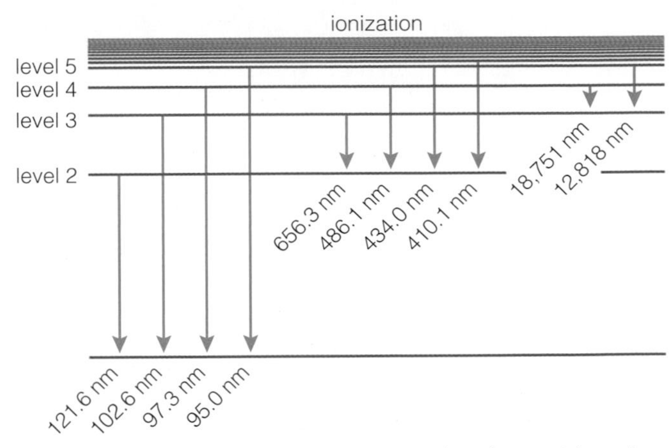

a Photons emitted by various energy-level transitions in hydrogen.

b The visible emission line spectrum from heated hydrogen gas. These lines come from transitions in which electrons fall from higher energy levels to level 2.

c If we pass white light through a cloud of cool hydrogen gas, we get this absorption line spectrum. These lines come from transitions in which electrons jump from energy level 2 to higher levels.

Figure 5.9 Interactive Figure

An atom emits or absorbs light only at specific wavelengths that correspond to changes in the atom's energy as an electron jumps between its allowed energy levels.

Because this particular set of lines is produced only by hydrogen, you can conclude that the cloud is made of hydrogen. In essence, the spectrum contains a "fingerprint" made only by hydrogen atoms.

Every kind of atom, ion, and molecule produces a unique spectral "fingerprint."

Real interstellar clouds, like all astronomical objects, are not made solely of hydrogen. However, the other chemical constituents in the cloud leave fingerprints on the spectrum in much the same way. Every type of atom, ion, and molecule has its own unique spectral fingerprint, because it has its own unique set of energy levels. Over the past century, scientists have done laboratory experiments to identify the spectral lines of every chemical element and many ions and molecules. Thus, when we see lines in the spectrum of a distant object, we can usually determine what chemicals produced them. For example, if we see spectral lines of hydrogen, helium, and carbon in the spectrum of a distant star, we know that all three elements are present in the star. Moreover, with detailed analysis, we can determine the relative proportions of the various elements. That is how we have learned the chemical compositions of objects throughout the universe.

• How does light tell us the temperatures of planets and stars?

We have seen how emission and absorption line spectra form, and how we can use them to determine the composition of a cloud of gas. Now we are ready to turn our attention to thermal radiation spectra. As we'll see, thermal radiation spectra get their name because they can help us determine the temperature of the objects that produce them.

Thermal Radiation: Every Body Does It In a cloud of gas that produces a simple emission or absorption line spectrum, the individual atoms or molecules are essentially independent of one another. Most photons pass easily through such a gas, except those that cause energy-level transitions in the atoms or molecules of the gas. However, the atoms and molecules within most of the objects we encounter in everyday life, such as rocks, light bulb filaments, and people, cannot be considered independent. Thus, their energy levels are far more complex. They tend to absorb light across a broad range of wavelengths, meaning that any light striking them cannot easily pass through and any light emitted inside them cannot easily escape. In fact, the same is true of almost any large or dense object, including planets and stars.

In order to understand the spectra of such objects, let's consider an idealized case, in which an object absorbs all the photons that strike it and does not allow photons inside it to escape easily. Photons tend to bounce randomly around inside such an object, constantly exchanging energy with its atoms or molecules. By the time the photons finally escape the object and fly off into space, their radiative energies have become randomized so that they are spread over a wide range of wavelengths. The wide wavelength range of the photons explains why the spectrum of light from such an object is smooth or *continuous,* like a pure rainbow without any absorption or emission lines.

Most important, the spectrum from such an object depends on only one thing: the object's *temperature.* To understand why, recall that temperature represents the average kinetic energy of the atoms or molecules in an object [Section 4.3]. The many random bounces of the photons inside the object mean that the photons end up with energies that match the kinetic energies of the object's atoms or molecules—which means the photon energies depend on the object's temperature. The temperature dependence of this light explains why we call it **thermal radiation** (sometimes known as *blackbody* radiation), and why its spectrum is called a thermal radiation spectrum.

Planets, stars, rocks, and people emit thermal radiation that depends only on temperature.

No real object emits a perfect thermal radiation spectrum, but almost all familiar objects—including the Sun, the planets, rocks, and even you—emit light that approximates thermal radiation. Figure 5.10 shows graphs of the idealized thermal radiation spectra of three stars and a human, each with its temperature given on the Kelvin scale (see Figure 4.10). Be sure to notice that these spectra show the intensity of light *per unit surface area,* not the total amount of light emitted by the object. For example, while the 3,000 K star emits much less light per unit surface area than the 15,000 K star, its total light output could be greater if it is much larger in size.

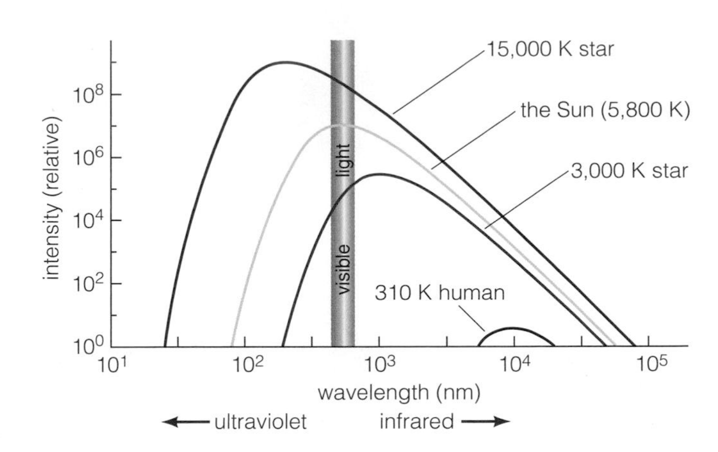

Figure 5.10 Interactive Figure

Graphs of idealized thermal radiation spectra. Note that hotter objects emit more radiation per unit surface area at every wavelength and that the peaks of their spectra occur at shorter wavelengths (higher energies). The graph uses power-of-10 scales on both axes, so that we can see all the curves even though the differences between them are quite large.

The Two Properties of Thermal Radiation

If you compare the spectra in Figure 5.10, you'll see that temperature has two effects on the spectrum:

- *Hotter objects show greater intensity at all wavelengths, indicating that they emit more total radiation per unit area.* For example, the 15,000 K star emits a lot more light at every wavelength than the 3,000 K star, and the hotter star emits light at some ultraviolet wavelengths that the cooler star does not emit at all.
- *Hotter objects emit photons with a higher average energy,* which means a shorter average wavelength. That is why the "humps" of the spectra are at shorter wavelengths for hotter objects. For example, the hump for the 15,000 K star is in ultraviolet light, the hump for the 5,800 K Sun is in visible light, and the hump for the 3,000 K star is in the infrared.

You can see these properties in action by playing with a light that has a dimmer switch. The first property is apparent as you turn the dimmer switch up: The filament in the light bulb gets hotter and the light brightens. When you turn the switch back down, the filament gets cooler and the light dims. (You can verify the changing temperature by placing your hand near the bulb.) The color of the light illustrates the second property. Before you turn on the switch, the filament is too cool to emit visible light at all, instead emitting only infrared radiation that our eyes cannot see. When you first turn the switch on low, the filament begins to glow red or orange—the longest wavelength visible light. As you turn the switch higher and the filament gets hotter, the average wavelength of the emitted photons moves toward the blue end of the visible spectrum. The mix of colors emitted at this higher temperature makes the filament and its light look white to our eyes.

An object's thermal radiation spectrum tells us its temperature.

Because thermal radiation spectra depend only on temperature, we can use them to measure the temperatures of distant objects. In many cases we can estimate temperatures simply from the object's colors. Notice that while hotter objects emit more light at *all* wavelengths, the biggest difference appears at the shortest wavelengths. An object with a temperature of 310 K, which is about human body temperature, emits mostly in the infrared and emits no visible light at all—which explains why we don't glow in the dark! A relatively cool star, with a 3,000 K surface temperature, emits mostly red light, which is why some bright stars in our sky appear reddish, such as Betelgeuse (in Orion) and Antares (in Scorpius). The Sun's 5,800 K surface emits most strongly in green light (around 500 nm), but the Sun looks yellow or white to our eyes because it also emits other colors throughout the visible spectrum. Hotter stars emit mostly in the ultraviolet, but because our eyes cannot see ultraviolet they appear blue-white in color. If an object were heated to a temperature of millions of degrees, it would radiate mostly X rays. Some astronomical objects are indeed hot enough to emit X rays, such as disks of gas encircling exotic objects like neutron stars and black holes (see Chapter 13).

The Doppler Effect Tutorial, Lessons 1–2

• How does light tell us the speed of a distant object?

We've already discussed how to analyze light to learn an object's composition and its temperature, but light can tell us still more. We can also learn about the object's speed. Changes in the object's spectrum caused by the **Doppler effect** tell us how quickly it is moving toward or away from us.

You've probably noticed the Doppler effect on the *sound* of a train whistle near train tracks. If the train is stationary, the pitch of its whistle sounds the same no matter where you stand (Figure 5.11a). But if the train is moving, the pitch will sound higher when the train is coming toward you and lower when it's moving away. Just as the train passes by, you can hear the dramatic change from high to low pitch—a sort of "weeeeeeee–oooooooooh" sound. To understand why, we have to think about what happens to the sound waves coming from the train (Figure 5.11b). When the train is moving toward you, each pulse of a sound wave is emitted a little closer to you. The result is that waves are bunched up between you and the train, giving them a shorter wavelength and higher frequency (pitch). After the train passes you by, each pulse comes

Figure 5.11

The Doppler effect.

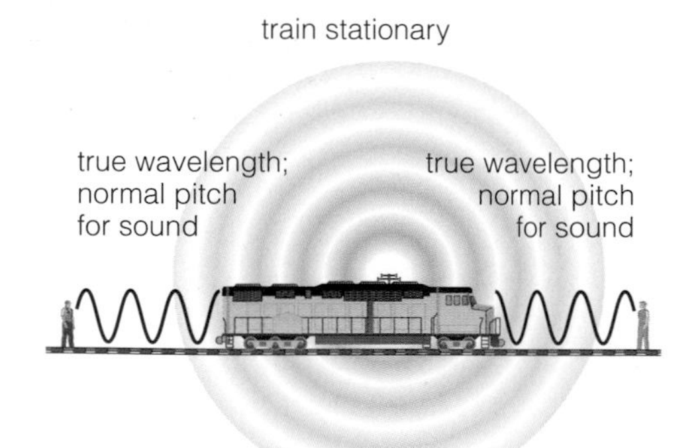

a Each circle represents the crests of sound waves going in all directions from the train whistle and emitted by the train at a different time. For example, the circles might represent waves emitted $\frac{1}{10}$ second apart.

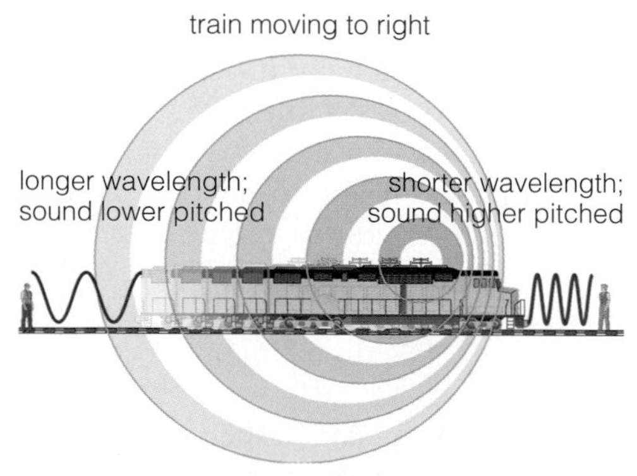

b If the train is moving, each pulse of the sound wave comes from a different location. Thus, the waves appear bunched up to a shorter wavelength (higher frequency) in front of the train and stretched out to a longer wavelength (lower frequency) behind the train.

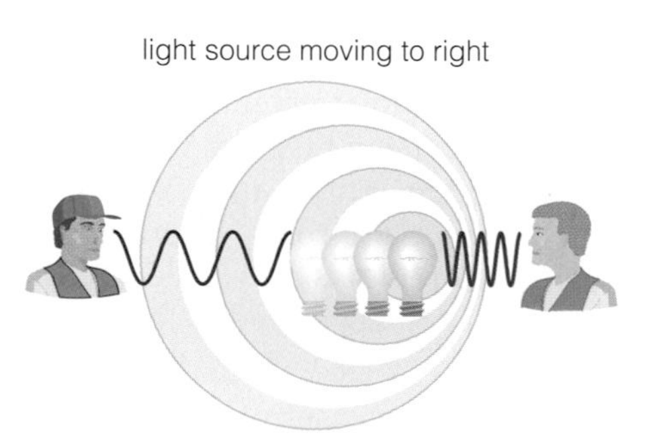

c We get the same basic effect from a moving light source.

from farther away, stretching out the wavelengths and giving the sound a lower frequency.

Spectral lines shift to shorter wavelengths when an object is moving toward us, and to longer wavelengths when an object is moving away from us.

The Doppler effect causes similar shifts in the wavelengths of light (Figure 5.11c). If an object is moving toward us, the light waves bunch up between us and the object, so that its entire spectrum is shifted to shorter wavelengths. Because shorter wavelengths of visible light are bluer, the Doppler shift of an object coming toward us is called a **blueshift**. If an object is moving away from us, its light is shifted to longer wavelengths. We call this a **redshift** because longer wavelengths of visible light are redder. For convenience, astronomers use the terms *blueshift* and *redshift* even when they aren't talking about visible light.

Spectral lines provide the reference points we use to identify and measure Doppler shifts (Figure 5.12). For example, suppose we recognize the pattern of hydrogen lines in the spectrum of a distant object. We know the **rest wavelengths** of the hydrogen lines—that is, their wavelengths in stationary clouds of hydrogen gas—from laboratory experiments in which a tube of hydrogen gas is heated so the wavelengths of the spectral lines can be measured. If the hydrogen lines from the object appear at longer wavelengths, then we know they are redshifted and the object is moving away from us. The larger the shift, the faster the object is moving. If the lines appear at shorter wavelengths, then we know they are blueshifted and the object is moving toward us.

Laboratory spectrum Lines at rest wavelengths.

Object 1 Lines redshifted: Object moving away from us.

Object 2 Greater redshift: Object moving away faster than Object 1.

Object 3 Lines blueshifted: Object moving toward us.

Object 4 Greater blueshift: Object moving toward us faster than Object 3.

Figure 5.12 Interactive Figure

Spectral lines provide the crucial reference points for measuring Doppler shifts.

THINK ABOUT IT Suppose the hydrogen emission line with a rest wavelength of 121.6 nm (the transition from level 2 to level 1) appears at a wavelength of 120.5 nm in the spectrum of a particular star. Given that these wavelengths are in the ultraviolet, is the shifted wavelength closer to or farther from blue visible light? Why, then, do we say that this spectral line is *blueshifted?*

Notice that the Doppler shift tells us only about the part of an object's full motion that is directed toward or away from us (the object's *radial* component of motion). Doppler shifts do not give us any information about how fast an object is moving across our line of sight (the object's *tangential* component of motion). For example, consider three stars all moving at the same speed, with one moving directly away from us, one moving across our line of sight, and one moving diagonally away from us (Figure 5.13). The Doppler shift will tell us the full speed only of the first star. The Doppler shift will not measure any speed for the second star, because none of its motion is directed toward or away from us. For the third star, the Doppler shift will tell us only the part of the star's speed that is directed away from us. (To measure how fast an object is moving across our line of sight, we must observe it long enough to notice how its position gradually shifts across our sky.)

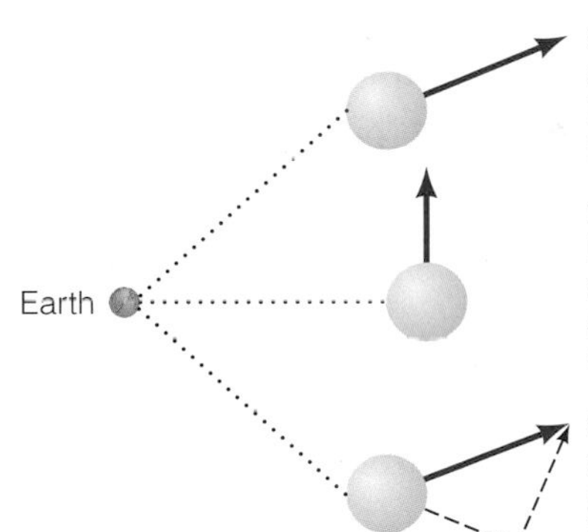

Star 1 is moving directly away from us, so the Doppler shift tells us its full speed.

Star 2 is moving across our line of sight, but not toward or away from us. Doppler shift measures no speed at all.

Star 3 is moving diagonally away from us. Doppler shift measures only the part of its speed directed away from us.

Figure 5.13

The Doppler shift measures only the portion of an object's speed that is directed toward or away from us. It does not give us any information about how fast an object is moving across our line of sight.

Telescopes Tutorial, Lessons 1–2

5.3 COLLECTING LIGHT WITH TELESCOPES

We've seen that light carries a great deal of information, but only a little of that information is obvious when we look at the sky with our naked eyes. Most of the great advances that have taken place in astronomy in

MAGNIFICATION AND TELESCOPES

You are probably familiar with how binoculars and telephoto camera lenses make objects appear larger in size—the phenomenon we call magnification. It's useful to know the magnification of binoculars and telephoto lenses, because it tells you what you can expect to see when you look through them. However, magnification is not a very useful way to describe a telescope's power, because magnification won't do you any good if the telescope collects too little light or has poor angular resolution. No matter how much you magnify the image, you cannot see details if the telescope does not collect enough light to show them, and you can never see details smaller than the angular resolution of the telescope. That is why magnifying an image too much just makes it look grainy, without revealing any new details, and why a telescope's light-collecting area and angular resolution are much more important than its magnification.

the past three centuries have been made possible through the use of telescopes. In this section, we'll briefly explore how telescopes work and how they help us learn about the universe.

• How do telescopes help us learn about the universe?

Telescopes are essentially giant eyes that can collect far more light than our naked eyes. Telescopes therefore allow us to see objects that are too faint for our eyes to see. In addition, we can connect scientific instruments to telescopes, allowing us to analyze the collected light in great detail. For example, sophisticated cameras can make high-quality images of light collected by a telescope, and spectrographs can disperse the light into spectra that can reveal an object's chemical composition, temperature, speed, and more.

The Two Key Properties of a Telescope Telescopes can differ in a number of ways, but the two most fundamental properties of any telescope are its *light collecting area* and its *angular resolution.* Let's investigate what these mean.

A telescope's **light collecting area** tells us how much total light the telescope can collect at one time. Telescopes are generally round, so we usually characterize a telescope's "size" by stating the *diameter* of its light-

TABLE 5.1 *Largest Optical (Visible-Light) Telescopes*

Size	*Name*	*Sponsor*	*Location*	*Opened*	*Special Features*
10.4 m	Gran Telescopio Canarias	Spain, Mexico, U. Florida	Canary Islands	2005	Segmented primary mirror based on mirrors for Keck telescopes
10 m	Keck I and Keck II	Cal Tech, U. California, NASA	Mauna Kea, HI	1993/1996	Two identical 10-m telescopes, each with a primary mirror consisting of 36 1.8-m hexagonal segments
9.2 m	Hobby-Eberly	U. Texas, Penn State, Stanford, Germany	Mt. Locke, TX	1997	Consists of 91 1-m segments, for a total diameter of 11 m, but only 9.2 m can be used at a time; designed primarily for spectroscopy
9.2 m	South African Large Telescope	South Africa, Rutgers, UW–Madison, UNC–Chapel Hill, Dartmouth, Carnegie-Mellon, 5 others	South Africa	2004	Based on design of Hobby-Eberly telescope
2×8.4 m	Large Binocular Telescope	U. Arizona, Ohio State U., Italy, Germany	Mt. Graham, AZ	2004	Two 8.4-m mirrors on a common mount, giving light-collecting area of 11.8-m telescope
4×8.2 m	Very Large Telescope	European Southern Observatory	Cerro Paranal, Chile	2000	Four separate 8-m telescopes designed to work individually or together as the equivalent of a 16-m telescope
8.3 m	Subaru	Japan	Mauna Kea, HI	1999	Japan's first large telescope project
8 m	Gemini North and South	U.S., U.K., Canada, Chile, Brazil, Argentina	Mauna Kea, HI (North); Cerro Pachon, Chile (South)	1999	Twin telescopes, one in each hemisphere
6.5 m	Magellan I and II	Carnegie Institute, U. Arizona, Harvard, U. Michigan, MIT	Las Campanas, Chile	2000/2002	Twin 6.5-m telescopes, known respectively as the Walter Baade and Landon Clay telescopes
6.5 m	MMT	Smithsonian Institution, U. Arizona	Mt. Hopkins, AZ	2000	Replaced an older telescope in the same observatory

collecting area. For example, a "10-meter telescope" has a light-collecting area that is 10 meters in diameter. Such a telescope has a light-collecting area more than a million times that of the human eye. Because area is proportional to the *square* of a telescope's diameter, a relatively small increase in diameter can mean a big increase in light collecting area. A 10-meter telescope has five times the diameter of a 2-meter telescope, so its light collecting area is $5^2 = 25$ times as great.

Telescopes collect far more light and allow us to see far more detail than does the naked eye.

Angular resolution gets its name because it tells us the smallest angle over which we can tell that two dots—or two stars—are distinct. For example, the human eye has an angular resolution of about 1 arcminute $(\frac{1}{60}^{\circ})$, which means that two stars will appear distinct only if they have at least this much angular separation in the sky. If the stars are separated by less than 1 arcminute, your eye will not be able to distinguish them individually and you will think you are looking at a single star.

Large telescopes can have amazing angular resolution. For example, the 2.4-meter Hubble Space Telescope has an angular resolution of about 0.05 arcsecond (for visible light), which would allow you to read this book from a distance of about 800 meters (a half mile). In principle, larger telescopes should have even better (smaller) angular resolution. However, all larger telescopes today are located on the ground, where effects of Earth's atmosphere can limit their angular resolution.

We are in the midst of a revolution in the building of large telescopes. Before the 1990s, the 5-meter Hale telescope on Mount Palomar (outside San Diego) reigned for more than 40 years as the most powerful telescope in the world. Today, it does not even make the top-10 list for telescope size (Table 5.1). Astronomers are currently working on designs for even larger telescopes, and it's quite likely that we'll see telescope diameters of 30 meters or more within the next couple of decades.

Basic Telescope Design Telescopes come in two basic designs: *refracting* and *reflecting*. A **refracting telescope** operates much like an eye, using transparent glass lenses to focus the light from distant objects (Figure 5.14). The earliest telescopes, including those built by Galileo, were refracting telescopes. The world's largest refracting telescope, completed in 1897, has a lens that is 1 meter (40 inches) in diameter and a telescope tube that is 19.5 meters (64 feet) long.

A **reflecting telescope** uses a precisely curved *primary mirror* to gather light (Figure 5.15). This mirror reflects the gathered light to a *secondary mirror* that lies in front of it. The secondary mirror then reflects the light to a focus at a place where it can be observed by the eye or instruments—sometimes through a hole in the primary mirror, and sometimes out the side of the telescope (perhaps with the aid of additional small mirrors). The fact that the secondary mirror prevents some light from reaching the primary mirror might seem like a drawback to reflecting telescopes, but in practice it is not a problem because only a small fraction of the incoming light is blocked.

Nearly all telescopes used in current astronomical research are reflectors. For a long time, the main factor limiting the size of reflecting telescopes was the sheer weight of the glass needed for their primary mirrors. Recent technological innovations have made it possible to build lighter-weight mirrors, such as the one in the Gemini telescope shown in Figure 5.15, or to make many small mirrors work together as one large one. For example, Figure 5.16 shows the primary mirror of one of the

Figure 5.14

Refracting telescopes.

lens
starlight
eyepiece
focus

a A refracting telescope collects light with a large transparent lens. The lens and eyepiece bend the light to provide a clear, focused image.

b The 1-meter refractor at the University of Chicago's Yerkes Observatory is the world's largest refracting telescope.

a The Gemini North telescope, located on the summit of Mauna Kea, Hawaii. The primary mirror, visible at the bottom of the larger lattice tube, is 8 meters in diameter. The secondary mirror, located in the smaller central lattice, reflects light back down through the hole visible in the center of the primary mirror.

b A reflecting telescope collects light with a precisely curved primary mirror. The primary mirror reflects light to the secondary mirror, which reflects the light to an eyepiece or instruments. This diagram shows a design in which the secondary reflects the light through a hole in the primary mirror, so that the light can be observed with cameras or instruments beneath the telescope.

Figure 5.15

Reflecting telescopes.

10-meter Keck telescopes, which consists of 36 smaller mirrors that function together as one.

Telescopes Across the Spectrum If we studied only visible light, we'd be missing much of the picture. Planets are relatively cool and emit primarily infrared light. The hot upper layers of stars such as the Sun emit ultraviolet and X-ray light. Some violent events even produce gamma rays that travel through space to Earth. Indeed, most objects emit light over a broad range of wavelengths. Today, astronomers study light across the entire spectrum.

Telescopes specialized to observe different wavelengths of light allow us to learn far more than we could learn from visible light alone.

The basic idea behind all telescopes is the same: to collect as much light as possible with as much resolution as possible. However, telescopes for most nonvisible wavelengths require different designs than visible-light telescopes. For example, the long wavelengths of radio waves mean that very large telescopes are necessary to achieve reasonable angular resolution. The largest single telescope in the world, the Arecibo radio dish, stretches 305 meters (1,000 feet) across a natural valley in Puerto Rico (Figure 5.17). Despite its large size, Arecibo's angular resolution is only about 1 arcminute at commonly observed radio wavelengths (for example, 21 cm [Section 15.2])—a few hundred times worse than the visible-light resolution of the Hubble Space Telescope.

Near the other end of the spectrum, X rays present a different type of problem for telescopes. Trying to focus X rays is somewhat like trying to focus a stream of bullets. If the bullets are fired directly at a metal sheet, they will puncture or damage the sheet. However, if the metal sheet is angled so that the bullets barely graze its surface, then it will slightly deflect the bullets. The mirrors of X-ray telescopes, such as NASA's Chandra X-Ray Observatory, are designed to deflect X rays in much the same way (Figure 5.18).

Every wavelength range poses its own unique challenges in building telescopes, and many new technologies have been invented to meet these challenges. Today we have the technology to observe nearly every wavelength of light coming from the cosmos. The greater problem for most observations is not the technology for the telescope itself, but dealing with problems created by Earth's atmosphere. We turn to this topic next.

Figure 5.16

This photo shows the primary mirror of one of the Keck telescopes, with a man in the center for scale. The primary mirror is made up of 36 smaller, hexagonal mirrors, arranged in a honeycomb pattern.

• Why do we put telescopes in space?

In our discussion so far, we have already encountered two telescopes that operate in space rather than on the ground: the Hubble Space Telescope and the Chandra X-Ray Observatory. Many other telescopes have also been launched into space during the past few decades. Given that it costs much more to launch a telescope into space than to build one on the ground, you might wonder why we bother to put telescopes in space. The answer is that Earth's atmosphere creates several significant problems that hinder our ability to observe from the ground.

Atmospheric Effects on Visible Light Some of the problems created by Earth's atmosphere are obvious. The brightness of the daytime sky limits visible-light observations to the night (ground-based radio telescopes can observe both day and night), and we cannot see the stars on cloudy nights. Another problem is that our atmosphere scatters the bright lights of cities, creating **light pollution** that can obscure the view even for the best telescopes (Figure 5.19). For example, the 2.5-meter telescope at Mount Wilson, the world's largest when it was built in 1917, would be much more useful today if it weren't located so close to the lights of what was once the small town of Los Angeles.

Telescopes in space are above the distorting effects of Earth's atmosphere.

A somewhat less obvious problem is the distortion of light by the atmosphere. The ever-changing motion, or *turbulence,* of air in the atmosphere bends light in constantly shifting patterns. This turbulence causes the familiar twinkling of stars. Twinkling may be beautiful to the naked eye, but it causes problems for astronomers because it blurs astronomical images.

THINK ABOUT IT Put a coin at the bottom of a cup of water. If you stir the water, the coin will appear to move around, even if it is still stationary on the bottom. What makes the coin appear to move? How is this similar to the way that our atmosphere causes stars to twinkle?

Figure 5.17

The Arecibo radio telescope stretches across a natural valley in Puerto Rico. At 305 meters across, it is the world's largest single telescope.

COMMON MISCONCEPTIONS

CLOSER TO THE STARS?

Many people mistakenly believe that space telescopes are advantageous because their locations above Earth make them closer to the stars. You can quickly realize the error of this belief by thinking about scale. On the scale of the Voyage model solar system that we discussed in Section 1.2, the Hubble Space Telescope is so close to the surface of the millimeter-diameter Earth that you would need a microscope to resolve its altitude, while the nearest stars are thousands of kilometers away. Thus, the distances to the stars are effectively the same whether a telescope is on the ground or in space. The real advantages of space telescopes all arise from their being above Earth's atmosphere and thus not subject to the many observational problems it presents.

a Artist illustration of the Chandra X-Ray Observatory, which orbits Earth.

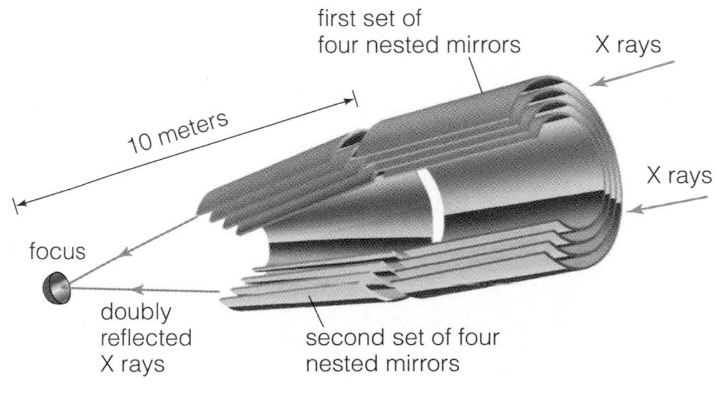

Mirror elements are 0.8 m long and from 0.6 m to 1.2 m in diameter.

b This diagram shows the arrangement of Chandra's X-ray mirrors. There are two sets of cylindrical mirrors, one near the front of the telescope and one farther back. An X ray entering straight into the telescope first hits a mirror in the first set, which deflects it to a mirror in the second set, which deflects it to the focus.

Figure 5.18

The Chandra X-Ray Observatory focuses X rays that enter the front of the telescope by deflecting them twice so that they end up focused at the back of the telescope.

Figure 5.19

VIS

Earth at night: It's pretty, but to astronomers it's light pollution. This image, a composite made from hundreds of satellite photos, shows the bright lights of cities around the world as they appear from Earth orbit at night.

Astronomers can partially mitigate effects of weather, light pollution, and atmospheric distortion by choosing appropriate sites for observatories. The key criteria are that the sites be dark (limiting light pollution), dry (limiting rain and clouds), calm (limiting turbulence), and high (placing them above at least part of the atmosphere). Islands are often ideal, and the 4,300-meter (14,000-foot) summit of Mauna Kea on the Big Island of Hawaii is home to many of the world's best observatories (Figure 5.20).

Of course, the ultimate solution to atmospheric distortion is to put telescopes in space, above the atmosphere. That is one reason why the Hubble Space Telescope (Figure 5.21) was built and why it has been so successful despite the relatively small size of its primary mirror.

Figure 5.20

Observatories on the summit of Mauna Kea in Hawaii. Mauna Kea meets the all the key criteria for an observing site: It is far from big city lights, high in altitude, and in an area where the air tends to be calm and dry.

Atmospheric Absorption of Light As we'll discuss shortly, some new technologies make it possible for ground-based observatories to equal or better the visible-light observations of the Hubble Space Telescope, at least in some cases. However, Earth's atmosphere poses one major problem that no Earth-bound technology can overcome: It prevents most forms of light from reaching the ground at all.

Figure 5.22 shows the depth to which different forms of light penetrate Earth's atmosphere. Notice that only radio waves, visible light, and small parts of the infrared spectrum can be observed from the ground. Thus, the most important reason for putting telescopes in space is to allow us to observe light that does not penetrate Earth's atmosphere. That is why the Chandra X-Ray Observatory is in space—an X-ray telescope would be completely useless on the ground. The same is true for other observatories in space. Indeed, the Hubble Space Telescope often observes in ultraviolet or infrared wavelengths that do not reach the ground,

Much of the electromagnetic spectrum can be observed only from space and not from the ground.

which is why it would remain a valuable observatory even if ground-based telescopes were some day to match all its visible-light capabilities.

• How is technology revolutionizing astronomy?

Astronomers today are making new discoveries at an astonishing rate, driven largely by the availability of more and larger telescopes, including space telescopes that can observe previously inaccessible portions of the electromagnetic spectrum. However, larger telescopes are not the only fuel for the current astronomical revolution.

Some new technologies make it possible to obtain better images or spectra with existing telescopes. For example, modern electronic detectors are much more sensitive to light than photographic film. As a result, a relatively small telescope equipped with the latest camera technology can record images as good as those that could be captured only by much larger telescopes in the past. Other technologies make it possible to record and analyze data more efficiently. For example, obtaining spectra of distant galaxies used to be a very time-consuming and labor-intensive task. Today, astronomers can sometimes obtain hundreds of spectra simultaneously in a single telescopic observation. Once astronomers have these spectra, they use computers to analyze the vast amount of data they contain.

Adaptive optics allows ground-based telescopes to overcome atmospheric distortion.

One of the most amazing new technologies is **adaptive optics**, which can allow ground-based telescopes to obtain visible-light images comparable to those from the Hubble Space Telescope. Remember that atmospheric blurring occurs because air motions cause the light of a star to dance around as it enters a telescope. Adaptive optics essentially make the telescope's mirrors do an opposite dance, canceling out the atmospheric distortions (Figure 5.23). The mirror shape (often the secondary mirror) is changed slightly many times each second to compensate for the rapidly changing atmospheric distortions. A computer calculates the necessary changes by monitoring distortions in the image of a bright star near the object under study. In some cases, if there is no bright star near the

Figure 5.21

The Hubble Space Telescope orbits Earth. Its position above the atmosphere allows it to get undistorted views of the heavens. Hubble can observe infrared and ultraviolet light as well as visible light.

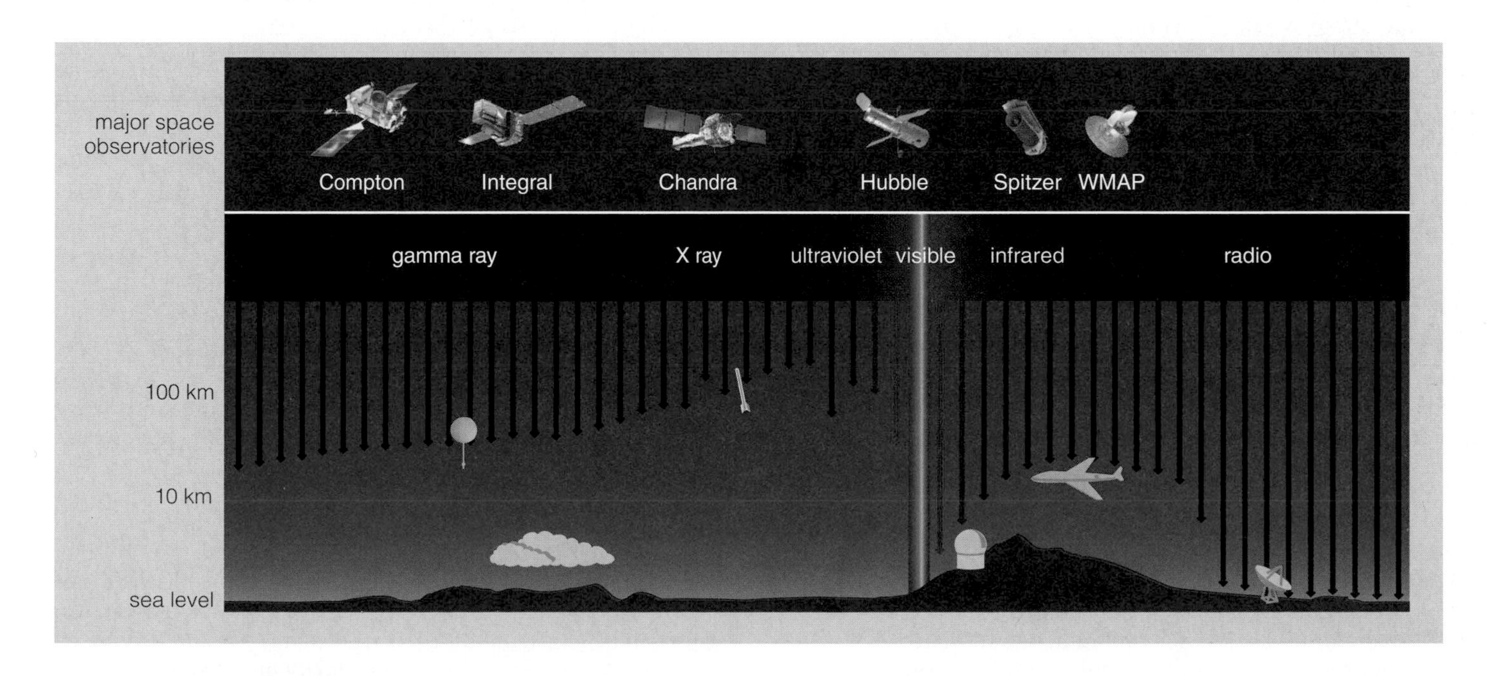

Figure 5.22

This diagram shows the approximate depths to which different wavelengths of light penetrate Earth's atmosphere. Note that most of the electromagnetic spectrum—except for visible light, a small portion of the infrared, and radio—can be observed only from very high altitudes or from space.

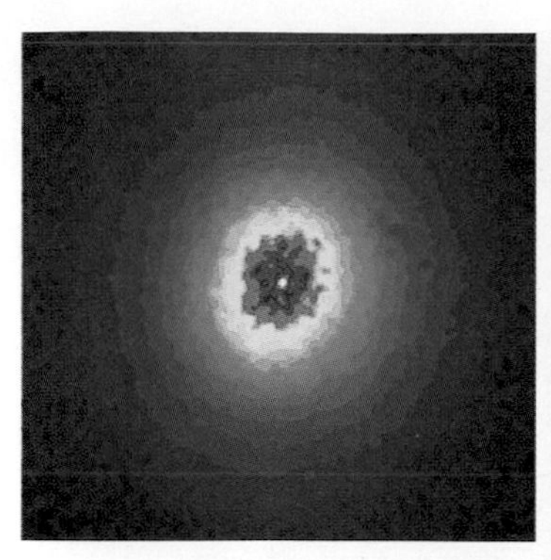

a Atmospheric distortion makes this ground-based image of a double star look like a single star.

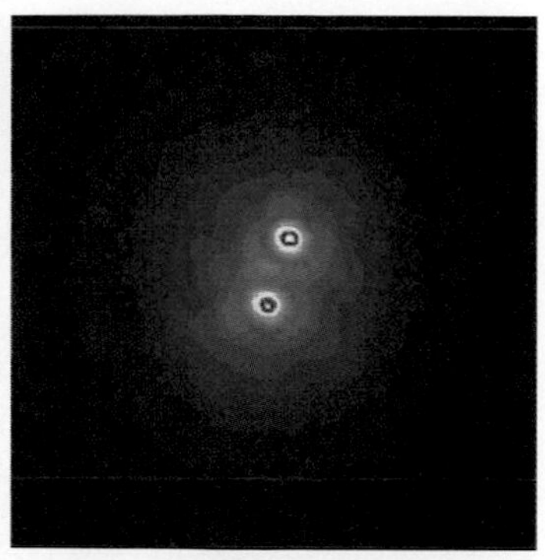

b When the same telescope is used with adaptive optics, the two stars are clearly distinguishable. The angular separation between the two stars is 0.38 arcsecond.

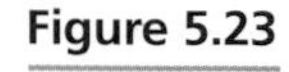

Figure 5.23

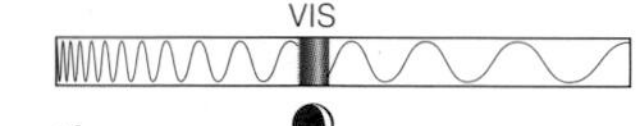

The technology of adaptive optics can enable a ground-based telescope to overcome most of the blurring caused by Earth's atmosphere. (Both of these images were taken in near-infrared light with the Canada-France-Hawaii telescope and are shown in false color.)

object of interest, the observatory shines a laser into the sky to create an *artificial star* (a point of light in Earth's atmosphere) that it can monitor for distortions.

Another technique for improving angular resolution is not particularly new, but it is becoming increasingly powerful. Since the 1950s, radio astronomers have used a technique called **interferometry** to allow two or more individual telescopes to achieve the angular resolution of a much larger telescope. For example, the Very Large Array (VLA) in New Mexico links 27 individual radio dishes laid out in the shape of a Y (Figure 5.24). When the 27 dishes are spaced as widely as possible, the VLA can achieve an angular resolution that otherwise would require a single radio telescope with a diameter of almost 40 kilometers. Astronomers sometimes link radio telescopes around the world for interferometry, achieving an angular resolution equivalent to that of a telescope the size of Earth.

Interferometry allows small telescopes to work together to obtain the angular resolution of a much larger telescope.

Interferometry is more difficult for shorter-wavelength light, but astronomers are rapidly learning to use the technique at infrared and visible wavelengths, and are testing technologies to extend it to X rays. New telescopes are now often built in pairs (such as the Keck and Magellan telescope pairs) or with more than one telescope on a common mount (such as the Large Binocular Telescope) so that they can be used for infrared and optical interferometry. In the future, astronomers hope to have telescopes in space or on the Moon

SPECIAL TOPIC: WOULD YOU LIKE YOUR OWN TELESCOPE?

Just a few years ago, a decent telescope for personal use would have set you back a few thousand dollars and taken weeks of practice to learn to use effectively. Today, you can get a good-quality telescope for as little as about $300, and built-in computer drives can make it very easy to use.

Before you start thinking about what telescope to buy, it's important to understand what a personal telescope can and cannot do. A telescope will allow you to look for yourself at light that has traveled vast distances through space to reach your eyes. It can be a rewarding experience, but the images in your telescope will *not* look like the beautiful photographs in this book—those are obtained with much larger telescopes and sophisticated cameras. In addition, while your telescope can in principle let you see many distant objects, including star clusters, nebulae, and galaxies, it won't allow you to find anything unless you first set it up properly. Even computer-driven telescopes, usually called "goto" telescopes, typically take 15 minutes to a half-hour to set up for each use (and longer when you are first learning).

If your goal is just to see the Moon and a few other objects with relatively little effort, you should probably go for a good pair of binoculars rather than a telescope. Indeed, it's generally a good idea for everyone to buy binoculars before buying a telescope, since they will help you learn about viewing the sky and are a lot less expensive. Binoculars are generally described by two numbers, such as 7x35 or 12x50. The first number is the magnification; for example "7x" means that objects will look 7 times closer through the binoculars than to your eye. The second number is the diameter of each lens in millimeters. As with telescopes, larger lenses mean more light and better views. However, larger lenses also tend to be heavier and more difficult to hold steady. If you buy a large pair of binoculars, you should also get a tripod to help hold it steady.

If you decide to go ahead with a telescope, the first rule to remember is that magnification is *not* the key factor in telescope selection. Telescopes that are advertised by their magnification, such as "650 power," are usually junk. Forget them, and instead focus on three factors when choosing your telescope:

1. *The light-collecting area* (also called *aperture*). Most personal telescopes are reflectors, so a "6-inch" telescope means a primary mirror 6 inches in diameter.
2. *Optical quality.* A poorly made telescope is worthless. Your best bet if you cannot do side-by-side comparisons is to stick with a major telescope manufacturer (such as Meade, Celestron, or Orion).
3. *Portability.* A large, bulky telescope can be great if you plan to keep it on your roof, but it won't be much fun on camping trips. Depending on how you plan to use your telescope, you'll need to make trade-offs between size and portability.

Most important, remember that a telescope is an investment: It will cost real money to buy, and you will probably keep it for many years. As you should with any investment, learn all you can before you settle on a particular model. Read reviews of telescopes in magazines such as *Astronomy*, *Mercury*, and *Sky and Telescope*. Talk to knowledgeable salespeople at stores that specialize in telescopes. And find a nearby astronomy club that holds observing sessions at which you can try out some telescopes and learn from experienced telescope users. Astronomy clubs can be found almost everywhere, and you can usually find them by typing "astronomy club" and your city into a Web search engine.

Figure 5.24

The Very Large Array (VLA) in New Mexico consists of 27 telescopes that can be moved along train tracks. The telescopes work together through interferometry and can achieve an angular resolution equivalent to that of a single radio telescope almost 40 kilometers across.

TWINKLE, TWINKLE, LITTLE STAR

Twinkling, or apparent variation in the brightness and color of stars, is not intrinsic to the stars. Instead, just as light is bent by water in a swimming pool, starlight is bent by Earth's atmosphere. Air turbulence causes twinkling because it constantly changes how the starlight is bent. Hence, stars tend to twinkle more on windy nights and at times when they are near the horizon (and therefore are viewed through a thicker layer of atmosphere).

Planets also twinkle, but not nearly as much as stars. Because planets have a measurable angular size, the effects of turbulence on any one ray of light are compensated for by the effects of turbulence on others, reducing the twinkling seen by the naked eye (but making planets shimmer noticeably in telescopes).

The blurring of images caused by twinkling is one of the primary reasons for putting telescopes in space. There, above the atmosphere, stars do not twinkle and telescopes can record sharp astronomical images.

working together as giant interferometers, offering views of distant objects that may be as detailed in comparison to Hubble Space Telescope images as Hubble's images are in comparison to the naked eye.

THE BIG PICTURE

Putting Chapter 5 into Context

This chapter was devoted to one essential purpose: understanding how we learn about the universe by observing the light of distant objects. "Big picture" ideas that will help you keep your understanding in perspective include the following:

- Light and matter interact in ways that allow matter to leave "fingerprints" on light. Thus, we can learn a great deal about the objects we observe by carefully analyzing their light. Most of what we know about the universe comes from information that we receive from light.
- The visible light that our eyes can see is only a small portion of the complete electromagnetic spectrum. Different portions of the spectrum may contain different pieces of the story of a distant object, so it is important to study all forms of light.
- There is far more to light than meets the eye. By dispersing the light of a distant object into a spectrum, we can determine the object's composition, surface temperature, motion toward or away from us, and more.
- Technology drives astronomical discovery. Every time we build a bigger telescope, develop a more sensitive detector, or open up a new wavelength region to study, we learn more about the universe than was previously possible.

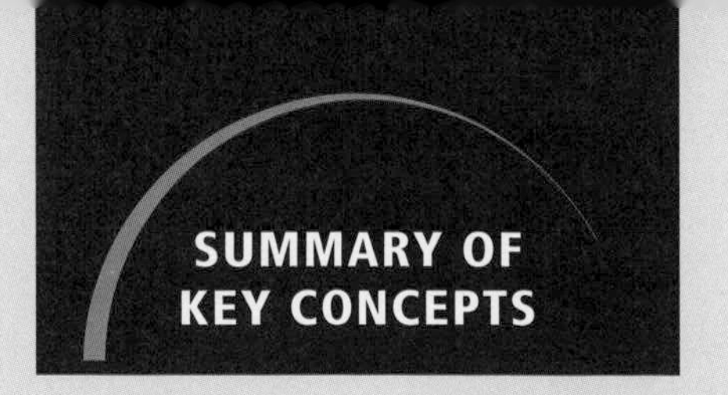

5.1 BASIC PROPERTIES OF LIGHT AND MATTER

• What is light?

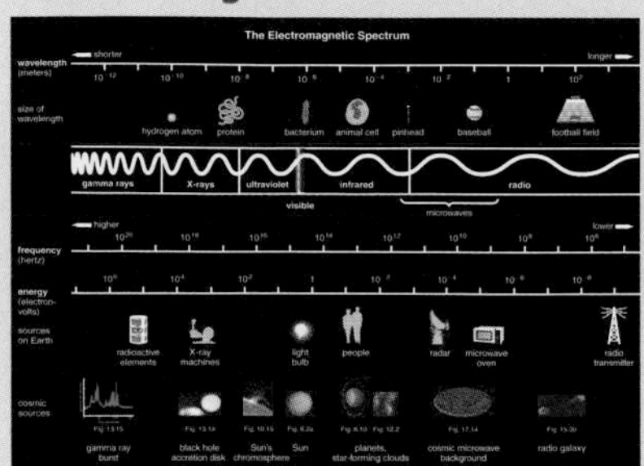

Light is an **electromagnetic wave**, but also comes in individual "pieces" called photons. Each photon has a precise wavelength, frequency, and energy: the shorter the wavelength, the higher the frequency and energy. In order of decreasing wavelength, the forms of light are **radio waves**, **microwaves**, **infrared**, **visible light**, **ultraviolet**, **X rays**, and **gamma rays**.

• What is matter?

Ordinary matter is made of atoms, which are made of protons, neutrons, and electrons. Atoms of different **chemical elements** have different numbers of protons. **Isotopes** of a particular chemical element all have the same number of protons but different numbers of neutrons.

• How do light and matter interact?

Matter can emit light, absorb light, transmit light, or reflect (or scatter) light.

5.2 LEARNING FROM LIGHT

• What types of light spectra can we observe?

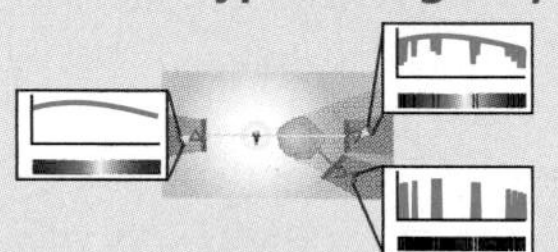

There are three basic types of spectra: a **thermal radiation** spectrum, which looks like a rainbow of light; an **absorption line** spectrum, in which specific colors are missing from the rainbow; and an **emission line** spectrum, in which we see light only with specific colors against a black background. Spectra of astronomical objects are usually combinations of these three types.

• How does light tell us what things are made of?

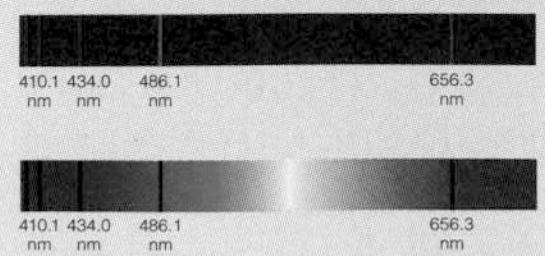

Emission or absorption lines occur only at specific wavelengths corresponding to particular energy-level transitions in atoms or molecules. Every kind of atom, ion, and molecule produces a unique set of spectral lines, so we can determine composition by identifying these lines.

• How does light tell us the temperatures of planets and stars?

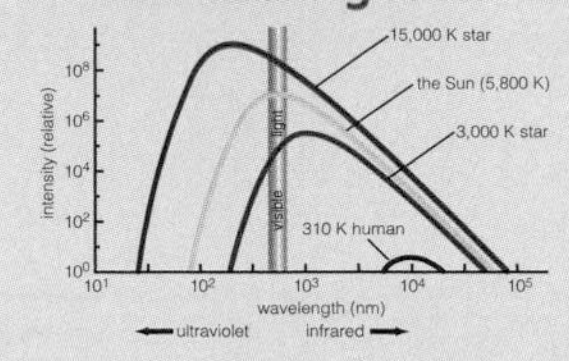

Objects such as planets and stars emit thermal radiation. We can determine temperature from the spectrum of this thermal radiation, because hotter objects emit more total radiation per unit area and emit photons with a higher average energy.

• How does light tell us the speed of a distant object?

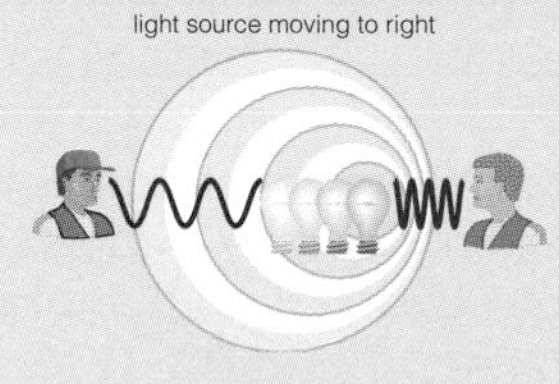

The **Doppler effect** tells us how fast an object is moving toward or away from us. Spectral lines are shifted to shorter wavelengths (a **blueshift**) in objects moving toward us and to longer wavelengths (a **redshift**) in objects moving away from us.

5.3 COLLECTING LIGHT WITH TELESCOPES

• How do telescopes help us learn about the universe?

Telescopes allow us to see fainter objects and to see more detail than we can see by eye. In addition, telescopes specialized to different wavelengths of light allow us to learn far more than we could from visible light alone.

• Why do we put telescopes in space?

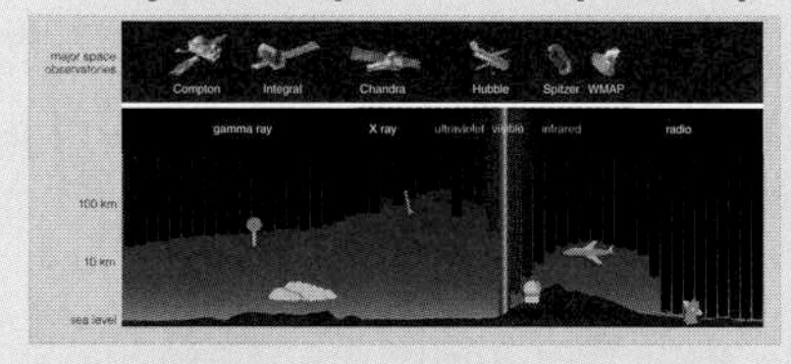

Telescopes in space are above Earth's atmosphere, and therefore not subject to problems caused by light pollution, atmospheric distortion of light, or the fact that most forms of light do not penetrate through the atmosphere to the ground.

• How is technology revolutionizing astronomy?

New technologies make it possible to build more powerful telescopes and to enhance the capabilities of existing telescopes. Two key technologies are **adaptive optics**, which can overcome the distorting effects of Earth's atmosphere, and **interferometry**, in which individual telescopes are linked in a way that allows them to obtain the angular resolution of a much larger telescope.

EXERCISES AND PROBLEMS

REVIEW QUESTIONS

1. Define *wavelength, frequency,* and *speed* for light waves. If light has a long wavelength, what can you say about its frequency? Explain.
2. What is a *photon?* In what way is a photon like a particle? In what way is it like a wave?
3. List the different forms of light in order from lowest to highest energy. Would the list be different if you went in order from lowest to highest frequency? From shortest to longest wavelength? Explain.
4. Briefly describe the structure of an atom. How big is an atom? How big is the *nucleus* in comparison to the entire atom?
5. What determines the atom's *atomic number?* What determines its *atomic mass number?* Under what conditions are two atoms different *isotopes* of the same element?
6. What is *electrical charge?* Will an electron and a proton attract or repel one another? Will two electrons attract or repel one another? Explain.
7. What are the four major ways in which light and matter can interact? Give an example from everyday life of each type of interaction.
8. What are the three basic types of spectra? Give an example of conditions that would cause us to see each of the types.
9. Why do atoms emit or absorb light of specific wavelengths? Briefly explain how we can use emission or absorption lines to determine the chemical composition of a distant object.
10. Briefly describe two ways in which the thermal radiation spectrum of an 8,000 K star would differ from that of a 4,000 K star.
11. Describe the *Doppler effect* for light and what we can learn from it. What does it mean to say that radio waves are *blueshifted?*
12. What are the two key properties of a telescope, and why is each important?
13. Suppose that two stars are separated in the sky by 0.1 arcsecond. What will you see if you look at them with a telescope that has an angular resolution of 0.01 arcsecond? What will you see if you look at them with a telescope that has an angular resolution of 0.5 arcsecond?
14. Briefly describe the differences between a refracting telescope and a reflecting telescope. Which type is more commonly used by professional astronomers?
15. List at least three ways in which Earth's atmosphere can hinder astronomical observations, and explain why putting a telescope in space helps in each case.
16. Briefly describe how adaptive optics and interferometry can improve astronomical observations.

DOES IT MAKE SENSE?

Decide whether each statement makes sense and explain why it does or does not.

17. If you could view a spectrum of light reflecting off a blue sweatshirt, you'd find the entire rainbow of color (looking the same as a spectrum of white light).
18. Because of their higher frequency, X rays must travel through space faster than radio waves.
19. Two isotopes of the element rubidium differ not only in their number of neutrons, but also in their number of protons.
20. If the Sun's surface became much hotter (while the Sun's size remained the same), the Sun would emit more ultraviolet light but less visible light than it currently emits.
21. If you could see infrared light, you would see the backs of your eyelids when you closed your eyes.
22. If you had X-ray vision, then you could read this entire book without turning any pages.
23. If a distant galaxy has a substantial redshift (as viewed from our galaxy), then anyone living in that galaxy would see a substantial redshift in a spectrum of the Milky Way Galaxy.
24. Thanks to adaptive optics, the telescope on Mount Wilson can now make ultraviolet images of the cosmos.
25. New technologies will soon allow astronomers to use X-ray telescopes on Earth's surface.
26. Thanks to interferometry, a properly spaced set of 10-meter radio telescopes can achieve the angular resolution of a single, 100-kilometer radio telescope.
27. I have a reflecting telescope in which the secondary mirror is bigger than the primary mirror.
28. An observatory on the Moon's surface could have telescopes monitoring light from all regions of the electromagnetic spectrum.

PROBLEMS

29. *Atomic Terminology Practice.*
    a. The most common form of iron has 26 protons and 30 neutrons in its nucleus. State its atomic number, atomic mass number, and number of electrons if it is electrically neutral.
    b. Consider the following three atoms: Atom 1 has 7 protons and 8 neutrons; atom 2 has 8 protons and 7 neutrons; atom 3 has 8 protons and 8 neutrons. Which two are *isotopes* of the same element?
    c. Oxygen has atomic number 8. How many times must an oxygen atom be ionized to create an O^{15} ion? How many electrons are in an O^{15} ion?
    d. Consider fluorine atoms with 9 protons and 10 neutrons. What are the atomic number and atomic mass number of this fluorine? Suppose we could add a proton to this fluorine nucleus. Would the result still be fluorine? Explain. What if we added a neutron to the fluorine nucleus?
    e. The most common isotope of gold has atomic number 79 and atomic mass number 197. How many protons and neutrons does the gold nucleus contain? If it is electrically neutral, how many electrons does it have? If it is triply ionized, how many electrons does it have?
    f. The most common isotope of uranium is ^{238}U, but the form used in nuclear bombs and nuclear power plants is ^{235}U. Given that uranium has atomic number 92, how many neutrons are in each of these two isotopes of uranium?

30. *Energy Level Transitions.* The following labeled transitions represent an electron moving between energy levels in hydrogen. Answer each of the following questions and explain your answers.

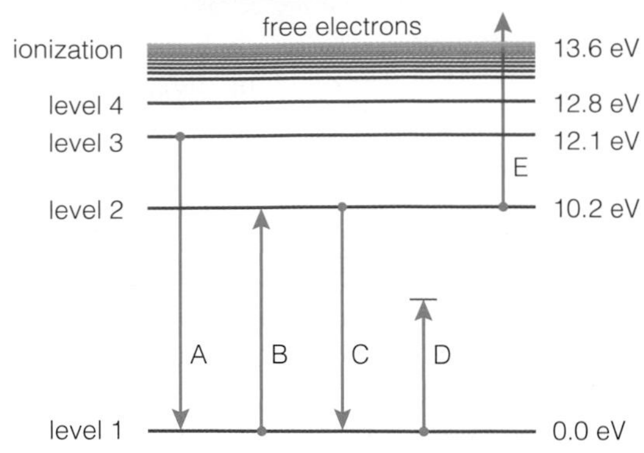

a. Which transition could represent an atom that *absorbs* a photon with 10.2 eV of energy?
b. Which transition could represent an atom that *emits* a photon with 10.2 eV of energy?
c. Which transition represents an electron that is breaking free of the atom?
d. Which transition, as shown, is *not* possible?
e. Would transition A represent emission or absorption of light? How would the wavelength of the emitted or absorbed photon compare to that of the photon involved in transition C? Explain.

31. *Spectral Summary.* Clearly explain how studying an object's spectrum can allow us to determine each of the following properties of the object.
a. The object's surface chemical composition.
b. The object's surface temperature.
c. Whether the object is a low-density cloud of gas or something more substantial.
d. The speed at which the object is moving toward or away from us.

32. *The Doppler Effect.* In hydrogen, the transition from level 2 to level 1 has a rest wavelength of 121.6 nm. Suppose you see this line at a wavelength of 120.5 nm in Star A, at 121.2 nm in Star B, at 121.9 nm in Star C, and at 122.9 nm in Star D. Which stars are coming toward us? Which are moving away? Which star is moving fastest relative to us (either toward or away from)? Explain your answers without doing any calculations.

33. *Telescope Technology.* Suppose you were building a space-based observatory consisting of five individual telescopes. Which would be the best way to use these telescopes: as five individual telescopes with adaptive optics, or as five telescopes without adaptive optics linked together for interferometry? Explain your reasoning clearly.

34. *Project: Twinkling Stars.* Using a star chart, identify 5–10 bright stars that should be visible in the early evening. On a clear night, observe each of these stars for a few minutes. Note the date and time, and for each star record the following information: approximate altitude and direction in your sky, brightness compared to other stars, color, and how much the star twinkles compared to other stars. Study your record. Can you draw any conclusions about how brightness and position in your sky affect twinkling? Explain.

DISCUSSION QUESTIONS

35. *The Changing Limitations of Science.* In 1835, French philosopher Auguste Comte stated that the composition of stars could never be known by science. Although spectral lines had been seen in the Sun's spectrum by that time, not until the mid-1800s did scientists recognize that spectral lines give clear information about chemical composition (primarily through the work of Foucault and Kirchhoff). Why might our present knowledge have seemed unattainable in 1835? Discuss how new discoveries can change the apparent limitations of science. Today, other questions seem beyond the reach of science, such as the question of how life began on Earth. Do you think such questions will ever be answerable by science? Defend your opinion.

36. *Science and Technology Funding.* Technological innovation clearly drives scientific discovery in astronomy, but the reverse is also true. For example, Newton's discoveries were made in part to explain the motions of the planets, but they have had far-reaching effects on our civilization. Congress often must make decisions between funding programs with purely scientific purposes ("basic research") and programs designed to develop new technologies. If you were a member of Congress, how would you try to allocate spending between basic research and technology? Why?

37. *Your Microwave Oven.* A *microwave oven* emits microwaves that have just the right wavelength needed to cause energy level jumps in water molecules. Use this fact to explain how a microwave oven cooks your food. Why doesn't a microwave oven make a plastic dish get hot? Why do some clay dishes get hot in the microwave? Why do dishes that aren't themselves heated by the microwave oven sometimes still get hot when you heat food on them?

MATH HELP AND EXERCISES

For additional help and practice with mathematical concepts applicable to this chapter, the Astronomy Place web site has the following resources with detailed explanations, worked examples, and practice problems:

- Wavelength, Frequency, and Energy
- Laws of Thermal Radiation
- The Doppler Shift
- Angular Separation
- The Diffraction Limit

For a complete list of media resources available, go to **www.astronomyplace.com** and choose Chapter 5 from the pull-down menu.

ASTRONOMY PLACE WEB TUTORIALS

Tutorial Review of Key Concepts

Use the interactive **Tutorials** at **www.astronomyplace.com** to review key concepts from this chapter.

Light and Spectroscopy Tutorial

Lesson 1 Radiation, Light, and Waves
Lesson 2 Spectroscopy
Lesson 3 Atomic Spectra—Emission and Absorption Lines
Lesson 4 Thermal Radiation

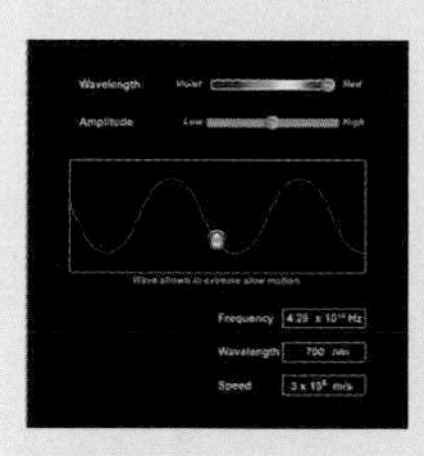

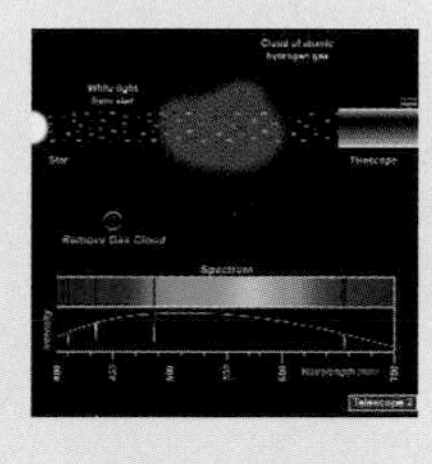

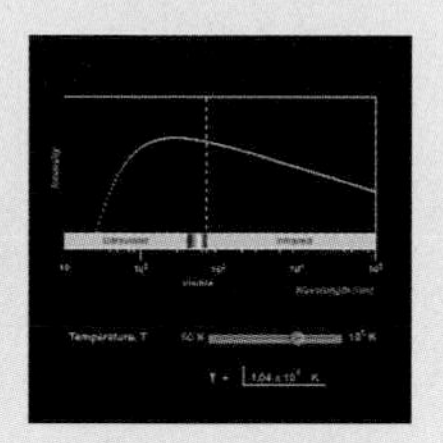

Doppler Effect Tutorial

Lesson 1 Understanding the Doppler Shift
Lesson 2 Using Emission and Absorption Lines to Measure the Doppler Shift

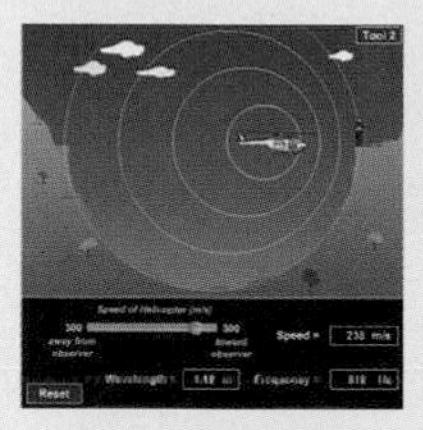

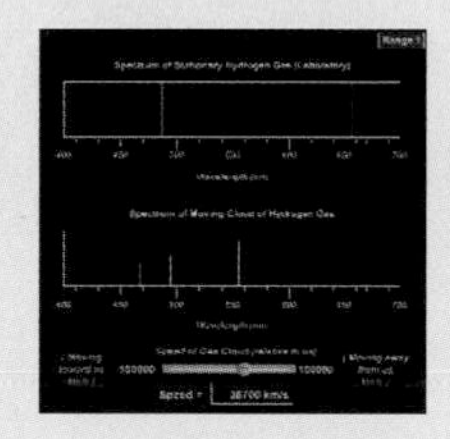

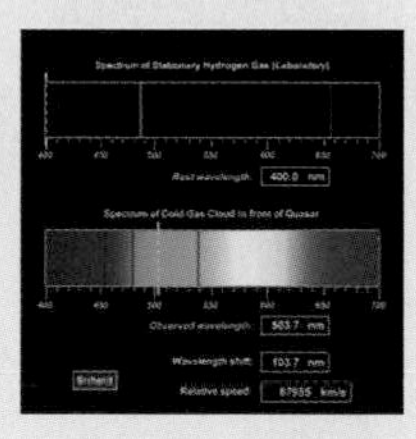

Telescopes Tutorial

Lesson 1 Optics and Light Gathering Power
Lesson 2 Angular Resolution

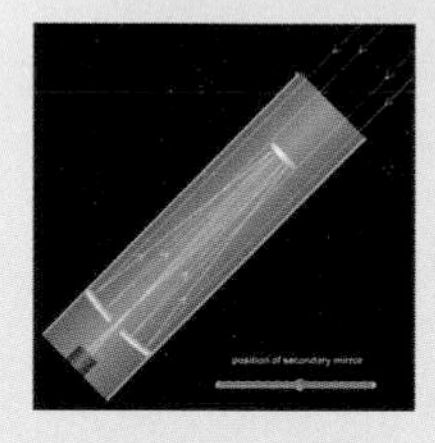

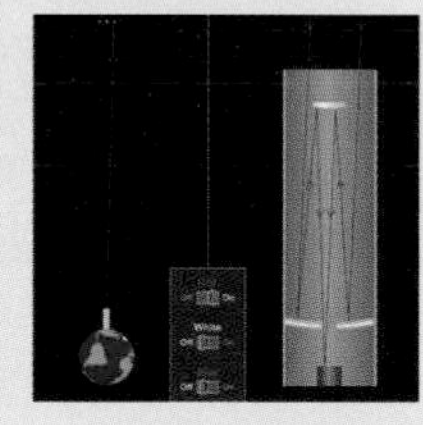

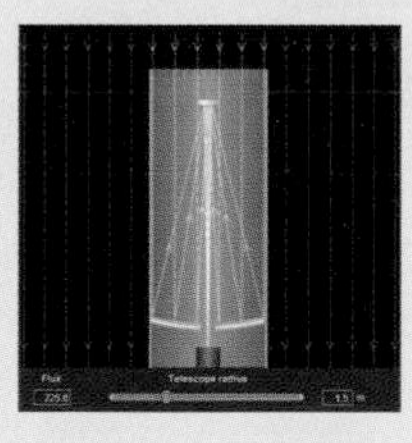

Supplementary Tutorial Exercises

Use the interactive **Tutorial Lessons** to explore the following questions.

Light and Spectroscopy Tutorial, Lesson 3

1. Why do neon lights come in so many different colors?
2. How can we tell the difference between lithium and carbon in the spectrum of a distant object?
3. How will an interstellar cloud affect the spectrum of light that we see from a star that lies behind it?

Light and Spectroscopy Tutorial, Lesson 4

1. How would an object's spectrum influence the choice of telescope used to observe it?
2. Consider the spectra of hot stars as compared to those of cooler stars. Which stars would be best observed with an ultraviolet telescope? Explain.

Doppler Effect Tutorial, Lesson 1

1. As the source of a sound passes you, which two properties will you hear change? Which change is due to the Doppler effect?
2. Which property of a sound wave is responsible for the pitch that your ears hear? How does the Doppler effect change the sound you hear?
3. How can the Doppler effect for light be used to measure the speed of a car?

Doppler Effect Tutorial, Lesson 2

1. How can we use precise wavelengths of spectral lines to measure the speed of a distant object?
2. Does it matter whether lines are in emission or absorption for the Doppler effect? Explain.

WEB PROJECTS

Take advantage of the useful Web links on **www.astronomyplace.com** to assist you with the following projects.

1. *Kids and Light.* Visit one of the many Web sites designed to teach middle and high school students about light. Read the content and try the activities. If you were a teacher, would you find the site useful for your students? Why or why not? Write a one-page summary of your conclusions.
2. *Major Ground-Based Observatories.* Take a virtual tour of one of the world's major astronomical observatories. Write a short report on why the observatory is useful to astronomy.
3. *Space Observatory.* Visit the Web site of a major space observatory, either existing or under development. Write a short report about the observatory, including its purpose, its orbit, and how it operates.
4. *Really Big Telescopes.* Several studies are under way in hopes of building telescopes far larger than any now in operation. Learn about one or more of these projects (such as OWL, the Swedish 50-m Optical Telescope, or the Thirty Meter Telescope (TMT)), and write a short report about the telescope's prospects and potential capabilities.

6
Our Solar System and Its Origin

LEARNING GOALS

6.1 A BRIEF TOUR OF THE SOLAR SYSTEM
- What does the solar system look like?

6.2 CLUES TO THE FORMATION OF OUR SOLAR SYSTEM
- What features of our solar system provide clues to how it formed?
- What theory best explains the features of our solar system?

6.3 THE BIRTH OF THE SOLAR SYSTEM
- Where did the solar system come from?
- What caused the orderly patterns of motion in our solar system?

6.4 THE FORMATION OF PLANETS
- Why are there two types of planets?
- Where did asteroids and comets come from?
- How do we explain the existence of our Moon and other "exceptions to the rules"?
- When did the planets form?

6.5 OTHER PLANETARY SYSTEMS
- How do we detect planets around other stars?
- What have other planetary systems taught us about our own?

Now that we have discussed some of the key laws of nature, we can apply these laws to the study of objects throughout our universe. We will begin by studying our solar system in this and the next three chapters, and later turn our attention to stars, galaxies, and the universe itself.

In this chapter, we'll explore the nature of our solar system and current scientific ideas about its birth. After a brief overview of the solar system and its individual worlds, we'll focus on characteristics of the solar system that offer key clues about how it formed. These clues will help us understand the current theory of solar system formation. Finally, we'll learn how astronomers have begun to discover planets around other stars, and how these other planetary systems may help us understand our own.

Scale of the Universe Tutorial, Lesson 1

6.1 A BRIEF TOUR OF THE SOLAR SYSTEM

Our ancestors long ago recognized the motions of the planets through the sky, but it has been only a few hundred years since we learned that Earth is also a planet that orbits the Sun. Even then, we knew little about the other planets until the advent of large telescopes. More recently, the dawn of space exploration has brought us far greater understanding of other worlds. We've lived in this solar system all along, but only now are we getting to know it. Let's begin with a quick tour of our planetary system, which will provide context for the more detailed study that will follow.

• What does the solar system look like?

The first step in getting to know our solar system is to visualize what it looks like as a whole. Imagine viewing the solar system from beyond the orbit of Pluto. What would we see?

Without a telescope, the answer would be "not much." Remember that the Sun and planets are all quite small compared to the distances between them [Section 1.2]—so small that if we viewed them from beyond Pluto, the planets would be only pinpoints of light and even the Sun would be just a small bright dot in the sky. But if we magnify the sizes of the planets by about a million times compared to their distances from the Sun, and show their orbital paths, we'd get the picture shown in Figure 6.1. Note that, even with this magnification, we do not see the many asteroids and comets that orbit the Sun (although the figure shows the location of the asteroid belt). The figure does not show moons, either, although the few largest of them would be barely visible on this scale.

The planets are tiny compared to the distances between them, but they exhibit clear patterns of composition and motion.

If you study Figure 6.1, you'll quickly see that our solar system is *not* a random collection of worlds. For example, all the planets orbit the Sun in the same direction and in nearly the same plane, and the four inner planets are much smaller and closer together than the next four planets.

In science, the existence of patterns like these demands an explanation. We will therefore devote most of this chapter to understanding how our modern theory of solar system formation explains these and other features of the solar system. Before we do so, however, let's look a little more closely at the Sun and its major family members.

The ten pages that follow Figure 6.1 offer a brief tour through our solar system, beginning at the Sun and continuing to each of the nine planets. The tour highlights just a few of the most interesting features of each world we visit, so that you will be familiar with each world before we study it in more depth later. The side of each page shows the planets on the 1-to-10-billion scale introduced in Chapter 1. Table 6.1, which follows the tour, summarizes key planetary data.

Orbits and Kepler's Laws Tutorial, Lessons 2–4

6.2 CLUES TO THE FORMATION OF OUR SOLAR SYSTEM

Our primary goal in this chapter is to understand the modern scientific theory of our solar system's formation. Let's begin by taking a more in-depth look at the general features of our solar system that must be explained by any successful theory of its origin. In essence, these features will be the clues that point us to a solar system formation theory.

• What features of our solar system provide clues to how it formed?

We have already seen that our solar system is not a random collection of worlds but rather is a family of worlds exhibiting many traits that would be difficult to attribute to coincidence. Thus, when we seek a theory of our solar system's formation, we are looking for a way to explain all these family traits. We could make a long list of these traits, but it is easier to develop a scientific theory by focusing on the more general structure of our solar system. For our purposes, four features stand out:

1. **Patterns of motion among large bodies**. The Sun, planets, and large moons generally orbit and rotate in a very organized way.
2. **Two major types of planets:** With the exception of Pluto, the planets divide clearly into two groups: the small, rocky planets that are close together and close to the Sun, and the large, gas-rich planets that are farther apart and farther out.
3. **Asteroids and comets**. A third major feature of our solar system is the existence of huge numbers of asteroids and comets, whose locations, orbits, and compositions follow distinct patterns.
4. **Exceptions to the rules**. Finally, there are some notable exceptions to the general "rules" we otherwise observe in our solar system. For example, Earth is unique among the inner planets in having a large moon, and Uranus has an odd, sideways tilt. We will consider these exceptions as a feature in themselves, because a successful theory must make allowances for such exceptions even as it explains the general rules.

Let's investigate each of these four features of our solar system in a little more detail.

(*continued on p. 145*)

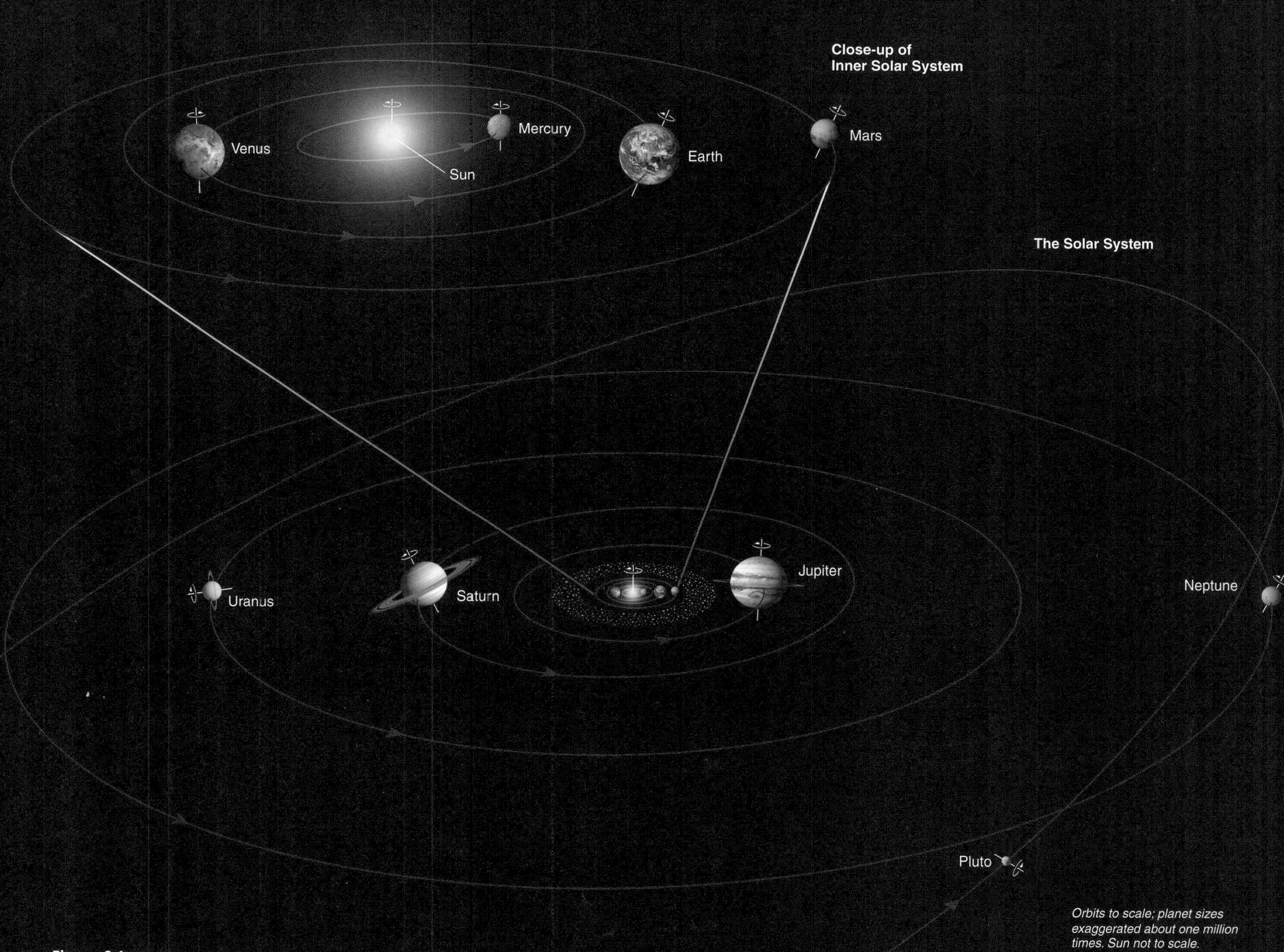

Figure 6.1 Interactive Figure

The layout of our solar system as it would appear from beyond Pluto, if we could magnify the sizes of the planets by about a million times. Notice that all the planets orbit the Sun in the same direction. The tilt of each planet's rotation axis is also shown, with a circling arrow to indicate the direction of rotation. The dots between Mars and Jupiter represent the asteroid belt.

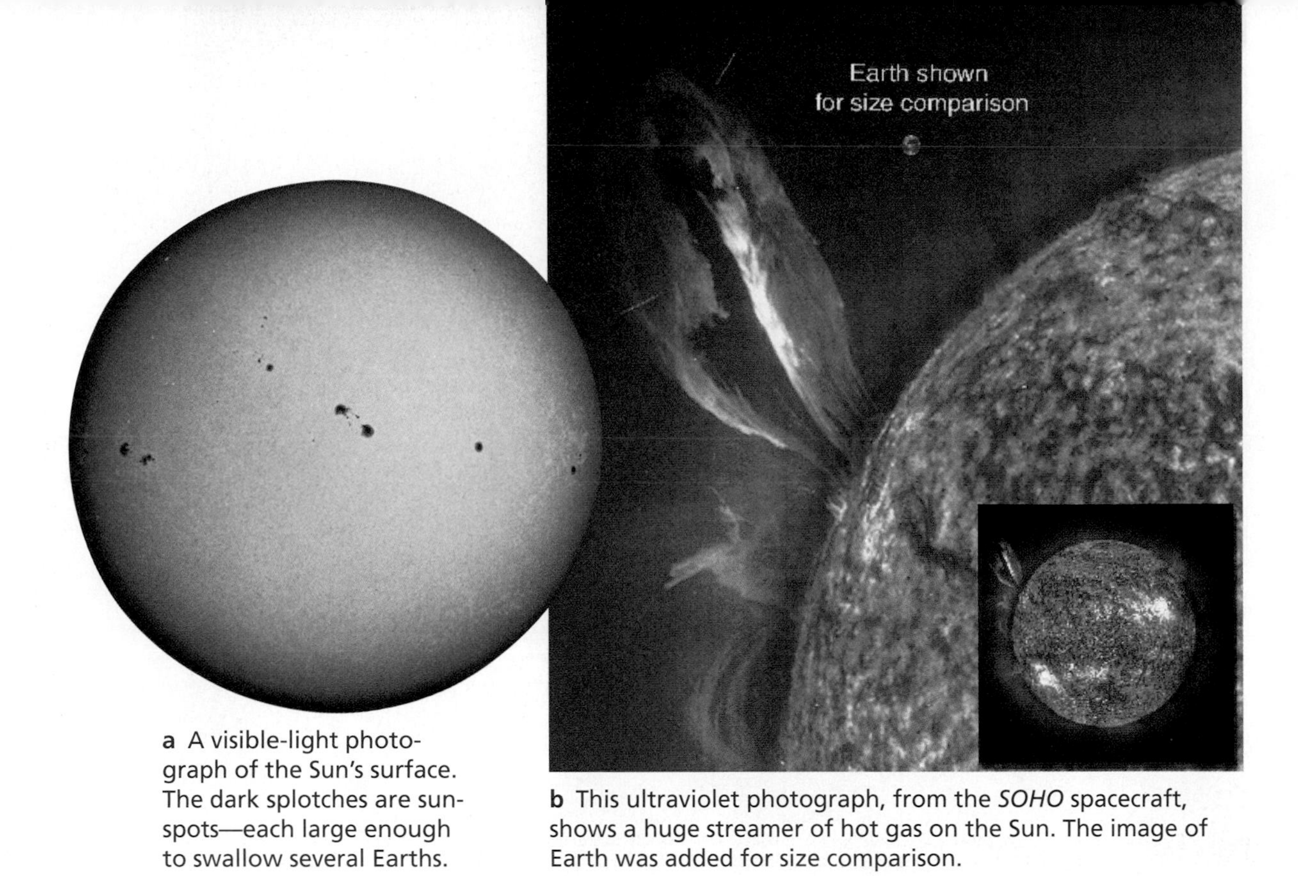

a A visible-light photograph of the Sun's surface. The dark splotches are sunspots—each large enough to swallow several Earths.

b This ultraviolet photograph, from the *SOHO* spacecraft, shows a huge streamer of hot gas on the Sun. The image of Earth was added for size comparison.

Figure 6.2

The Sun contains more than 99.9% of the total mass in our solar system.

• The Sun

- Radius: 695,000 km
- Mass (Earth = 1): 333,000
- Composition (by mass): 98% hydrogen and helium, 2% other elements

The Sun is by far the largest and brightest object in our solar system. It contains more than 99.9% of the solar system's total mass, making it more than a thousand times more massive than everything else in the solar system combined.

The Sun's surface looks solid in photographs (Figure 6.2), but it is actually a roiling sea of hot (about 5,800 K, or 6,100°C or 11,000°F) hydrogen and helium gas. The surface is speckled with sunspots that appear dark in photographs only because they are slightly cooler than their surroundings. Solar storms sometimes send streamers of hot gas soaring far above the surface.

The Sun is gaseous throughout. If you could plunge into the Sun, you'd find ever-higher temperatures as you went deeper. The source of the Sun's energy lies deep in its core, where the temperatures and pressures are so high that the Sun becomes a nuclear fusion power plant. Each second, fusion transforms about 600 million tons of the Sun's hydrogen into 596 million tons of helium. The "missing" 4 million tons becomes energy in accord with Einstein's famous formula $E = mc^2$ [Section 4.3]. Despite losing 4 million tons of mass each second, the Sun contains so much hydrogen that it has already shone steadily for almost 5 billion years and will continue to shine for some 5 billion years more.

The Sun is certainly the most influential object in our solar system. Its gravity governs the orbits of the planets. Its heat is the primary influence on planetary temperatures, and it is the source of virtually all the visible light in our solar system—the Moon and planets shine only by virtue of the sunlight they reflect. In addition, charged particles flowing outward from the Sun (the solar wind) help shape the magnetic fields of the planets and can influence planetary atmospheres. Nevertheless, we can understand almost all the present characteristics of the planets without knowing much more about the Sun than what we have just discussed. In Chapter 10, we'll study the Sun as our prototype for understanding other stars.

Figure 6.3

This image shows what it would look like to be orbiting a few hundred kilometers above Mercury's surface with your back toward the Sun. Among the stars, you can see Earth and the Moon as the blue speck and its tiny companion. The view was created using imagery from NASA's *Mariner 10* spacecraft but with computer manipulation to provide color and the orbital viewpoint. The inset (right) is a composite photograph of the full disk of Mercury by *Mariner 10;* the blank strip at the upper right was not photographed. (Image above from the Voyage scale model solar system, developed by the Challenger Center for Space Science Education, the Smithsonian Institution, and NASA. Image created by ARC Science Simulations © 2001.)

Mercury

- Average distance from the Sun: 0.39 AU
- Radius: 2,440 km
- Mass (Earth = 1): 0.055
- Average density: 5.43 g/cm^3
- Composition: rocks, metals
- Average surface temperature: 700 K (day), 100 K (night)
- Moons: 0

Mercury, the innermost planet of our solar system, is also the smallest planet except for Pluto. It is a desolate, cratered world with no active volcanoes, no earthquakes, no wind, no rain, and no life. Because there is virtually no air to scatter sunlight or color the sky, you could see stars even in the daytime if you stood on Mercury with your back toward the Sun. You wouldn't want to stay long, however, because the ground on Mercury's day side is nearly as hot as hot coals (about 425°C). Nighttime would not be much more comfortable. With no atmosphere to retain heat during the long nights (which last about 3 months), temperatures plummet to some −170°C (about −270°F)—far colder than Antarctica in winter.

Mercury is the least studied of the inner planets. Its proximity to the Sun makes it difficult to observe through telescopes, and it has been visited by just one spacecraft. *Mariner 10* collected data during three rapid flybys of Mercury in 1974–1975, obtaining images of only one hemisphere (shown in Figure 6.3). Mercury has craters almost everywhere, ancient lava flows, and tall, steep cliffs that run hundreds of kilometers in length. As we'll discuss in Chapter 7, the cliffs may in fact be wrinkles from an episode of "planetary shrinking" early in Mercury's history.

The influence of Mercury's gravity on *Mariner 10*'s orbit allowed scientists to determine Mercury's mass and to make inferences about its interior composition and structure. Mercury appears to be made mostly of iron, making it the most metal-rich of the planets. Scientists hope to learn much more about Mercury soon. A NASA spacecraft called *Messenger* is scheduled to begin Mercury observations in 2011, and a European mission to Mercury called *BepiColombo* is also being planned.

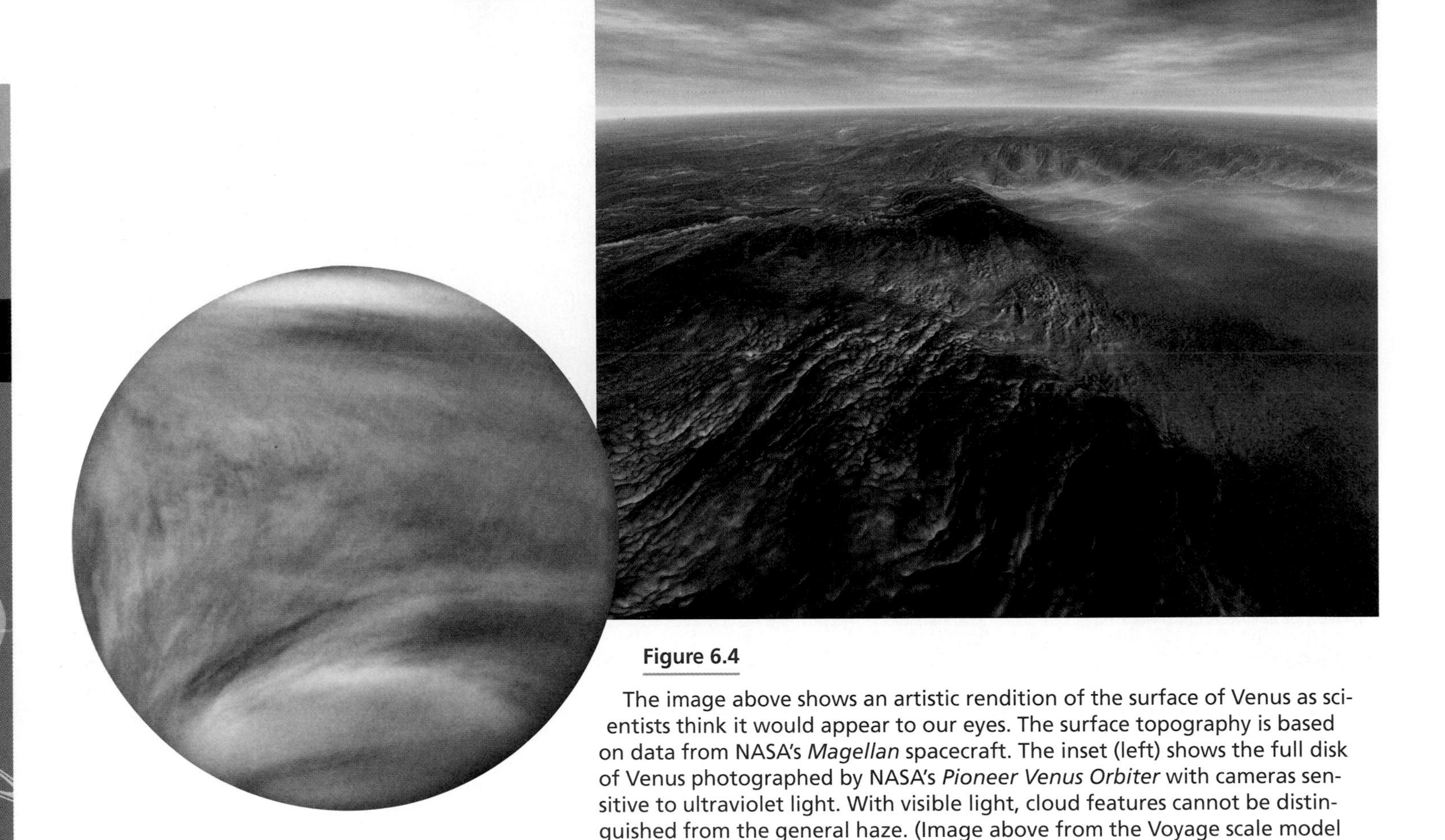

Figure 6.4

The image above shows an artistic rendition of the surface of Venus as scientists think it would appear to our eyes. The surface topography is based on data from NASA's *Magellan* spacecraft. The inset (left) shows the full disk of Venus photographed by NASA's *Pioneer Venus Orbiter* with cameras sensitive to ultraviolet light. With visible light, cloud features cannot be distinguished from the general haze. (Image above from the Voyage scale model solar system, developed by the Challenger Center for Space Science Education, the Smithsonian Institution, and NASA. Image by David P. Anderson, Southern Methodist University © 2001.)

• Venus

- Average distance from the Sun: 0.72 AU
- Radius: 6,051 km
- Mass (Earth = 1): 0.82
- Average density: 5.24 g/cm^3
- Composition: rocks, metals
- Average surface temperature: 740 K
- Moons: 0

Venus, the second planet from the Sun, is nearly identical in size to Earth. Its surface is completely hidden from view by dense clouds, so we knew little about it until a few decades ago, when cloud-penetrating radar finally allowed us to study Venus in detail (Figure 6.4). Because we knew so little about it, science fiction writers used its Earth-like size, thick atmosphere, and closer distance to the Sun to speculate that it might be a lush, tropical paradise—a "sister planet" to Earth.

The reality is far different. We now know that an extreme *greenhouse effect* bakes Venus's surface to an incredible 470°C (about 880°F), trapping heat so effectively that nighttime offers no relief. Day and night, Venus is hotter than a pizza oven. All the while, the thick atmosphere bears down on the surface with a pressure equivalent to what exists nearly a kilometer (0.6 mile) beneath the ocean's surface on Earth. Even if you could somehow survive these hazards, you'd find no oxygen to breathe and no water to drink. Far from being a beautiful sister planet to Earth, Venus resembles a traditional view of hell.

The surface of Venus has mountains, valleys, and craters and shows many signs of past (and possibly present) volcanic activity. Aside from these superficial similarities, however, the geology of Venus's surface appears to be quite different from that of Earth. We will devote much of Chapter 7 to understanding how and why Venus became so unlike Earth, both in its surface geology and in its atmosphere.

a This image (left), computer generated from satellite data, shows the striking contrast between the daylight and night-time hemispheres of Earth. The day side reveals little evidence of human presence, but at night our presence is revealed by the lights of human activity (mostly from cities as well as from agricultural, oil, and gas fires). (From the Voyage scale model solar system, developed by the Challenger Center for Space Science Education, the Smithsonian Institution, and NASA. Image created by ARC Science Simulations © 2001.)

b Earth and the Moon, shown to scale. The Moon is about one-fourth as large as Earth in diameter, while its mass is about 1/80 Earth's mass. If you wanted to show the distance between Earth and the Moon on the same scale, you'd need to hold these two photographs about 6.1 meters (20 feet) apart.

Figure 6.5

Earth, our home planet.

• Earth

- Average distance from the Sun: 1.00 AU
- Radius: 6,378 km
- Mass (Earth = 1): 1.00
- Average density: 5.52 g/cm^3
- Composition: rocks, metals
- Average surface temperature: 290 K
- Moons: 1

Beyond Venus, we next encounter our home planet. Although Earth is a barely visible speck when viewed on the scale of our entire solar system, it is the only known oasis of life. Even if we someday discover microscopic life on other worlds in our solar system, Earth is the only world on which humans could survive without a protective enclosure. It is the only planet with oxygen for us to breathe and ozone to shield us from deadly solar radiation. It is the only planet with abundant surface water to nurture life. Temperatures are pleasant for us because Earth's atmosphere contains just enough carbon dioxide and water vapor to maintain a moderate greenhouse effect.

Despite Earth's small size, its beauty is striking (Figure 6.5a). Blue oceans cover nearly three-fourths of the surface, broken by the continental land masses and scattered islands. The polar caps are white with snow and ice, and white clouds are scattered above the surface. At night, the glow of artificial lights clearly reveals the presence of an intelligent civilization.

Earth is the first planet on our tour with a moon. Moreover, our Moon is surprisingly large in comparison to Earth (Figure 6.5b)—most moons are much smaller in comparison to the planets they orbit. How Earth acquired such a large moon has long been one of the major mysteries of our solar system.

• Mars

- Average distance from the Sun: 1.52 AU
- Radius: 3,397 km
- Mass (Earth = 1): 0.11
- Average density: 3.93 g/cm^3
- Composition: rocks, metals
- Average surface temperature: 240 K
- Moons: 2 (very small)

The next planet on our tour is Mars, the last of the four inner planets of our solar system (Figure 6.6). Mars is larger than Mercury and the Moon but smaller than Venus and Earth. About half Earth's size in diameter, Mars has a mass about 10% that of Earth. Mars has two tiny moons, Phobos and Deimos, that look much like typical asteroids and may once have roamed freely in the asteroid belt. These moons are too small to influence Mars in any of the important ways that our Moon influences Earth (such as tides).

Mars is a world of wonders, with extinct volcanoes that dwarf the largest mountains on Earth, a great canyon that runs nearly one-fifth of the way around the planet, and polar caps made of frozen carbon dioxide ("dry ice") and water ice. Although Mars is frozen today, the presence of dried-up riverbeds and rock-strewn floodplains offers clear evidence that Mars had at least some warm and wet periods in the distant past. Thus, Mars may once have been hospitable for life, though major flows of liquid water probably ceased at least 3 billion years ago.

Mars looks almost Earth-like in photographs taken by spacecraft on its surface, but you wouldn't want to visit without a space suit. The air pressure is far less than that on top of Mount Everest, the temperature is usually well below freezing, the trace amounts of oxygen would not be nearly enough to breathe, and the lack of atmospheric ozone would leave you exposed to deadly ultraviolet radiation from the Sun.

Mars is the most studied planet besides Earth. More than a dozen spacecraft have flown past, orbited, or landed on Mars, and plans are in the works for many more missions to Mars. We may even send humans to Mars within our lifetime. Overturning rocks in ancient riverbeds or chipping away at ice in the polar caps, explorers will help us learn whether Mars has ever been home to life.

Figure 6.6

The globe (left) shows the full disk of Mars; the horizontal "gash" across the center is the giant canyon Valles Marineris. Below we see the surface of Mars photographed by NASA's *Pathfinder* lander in 1997. Part of the lander is visible in the foreground, while the rover named *Sojourner* studies a rock near the upper right. The photo that opens this chapter (p. 130), taken by the *Opportunity* rover in 2004, also shows Mars's surface; notice the tracks leading away from the casing that held the lander as it descended to the surface.

Figure 6.7

This image shows what it would look like to be orbiting near Jupiter's moon Io as Jupiter comes into view. Notice the Great Red Spot to the left of Jupiter's center. The extraordinarily dark rings discovered in the Voyager missions are exaggerated to make them visible. This computer visualization was created using data from both NASA's Voyager and Galileo missions. (From the Voyage scale model solar system, developed by the Challenger Center for Space Science Education, the Smithsonian Institution, and NASA. Image created by ARC Science Simulations © 2001.)

• Jupiter

- Average distance from the Sun: 5.20 AU
- Radius: 71,492 km
- Mass (Earth = 1): 318
- Average density: 1.33 g/cm^3
- Composition: mostly hydrogen and helium
- Cloud-top temperature: 125 K
- Moons: at least 63

To reach the orbit of Jupiter from Mars, we must traverse a distance that is more than double the total distance from the Sun to Mars, passing through the asteroid belt along the way. Upon our arrival, we find a planet much larger than any we have seen so far (Figure 6.7).

Jupiter is so different from the planets of the inner solar system that we must adopt an entirely new mental image of the term *planet*. Its mass is more than 300 times that of Earth, and its volume is more than 1,000 times that of Earth. Its most famous feature—a long-lived storm called the Great Red Spot—is itself large enough to swallow two or three Earths. Like the Sun, Jupiter is made primarily of hydrogen and helium and has no solid surface. If you plunged deep into Jupiter, you would be crushed by the increasing gas pressure long before you ever reached its core.

Jupiter reigns over at least 60 moons and a thin set of rings (too faint to be seen in most photographs). The four largest moons—Io, Europa, Ganymede, and Callisto—are often called the *Galilean moons* because they were discovered by Galileo shortly after he first turned his telescope toward the heavens [Section 3.3]. Each of these four moons is similar to or larger in size than our Moon, and Ganymede is larger than the planet Mercury. Io is the most volcanically active place in the solar system. Europa has an icy crust that may hide a subsurface ocean of liquid water, making it a promising place to search for life. Ganymede and Callisto may also have subsurface oceans, and their surfaces have many features that remain mysterious.

Figure 6.8

This computer simulation, built on imagery from the Voyager 1 mission, re-creates the striking view as the spacecraft passed by Saturn. We see the shadow of the rings on Saturn's sunlit face, and the rings become lost in Saturn's shadow on the night side. (From the Voyage scale model solar system, developed by the Challenger Center for Space Science Education, the Smithsonian Institution, and NASA. Image created by ARC Science Simulations © 2001.)

• Saturn

- Average distance from the Sun: 9.54 AU
- Radius: 60,268 km
- Mass (Earth = 1): 95.2
- Average density: 0.70 g/cm^3
- Composition: mostly hydrogen and helium
- Cloud-top temperature: 95 K
- Moons: at least 31

Saturn is the second-largest planet in our solar system after Jupiter, and nearly twice as far from the Sun. Saturn is only slightly smaller than Jupiter in diameter, but it is considerably less massive (about one-third Jupiter's mass) because it is less dense. Like Jupiter, Saturn is made mostly of hydrogen and helium and has no solid surface.

Saturn is famous for its spectacular rings (Figure 6.8). Although all four of the giant outer planets have rings, only Saturn's rings can be seen easily through a small telescope. The rings may look solid from a distance, but this appearance is deceiving. If you could wander into the rings, you'd find yourself surrounded by countless individual particles of rock and ice, ranging in size from dust grains to city blocks. Each ring particle orbits Saturn like a tiny moon. The rings do not touch Saturn's surface—any ring particle that wandered into the atmosphere would quickly burn up.

Like Jupiter, Saturn is orbited by many moons. Most are the size of small asteroids or comets, but a few are much larger. Saturn's largest moon, Titan, is bigger than the planet Mercury. Titan is blanketed by a thick atmosphere. On Titan's surface, you'd find an atmospheric pressure even greater than on Earth, and you could inhale air with roughly the same nitrogen content as air on Earth. However, you'd need to bring your own oxygen, and you'd certainly want a warm space suit for protection against the frigid outside temperatures. NASA's *Cassini* spacecraft, designed to explore Saturn and its rings and moons, carries a probe scheduled to be dropped to Titan's surface in January, 2005.

THINK ABOUT IT By the time you read this book, *Cassini* should have arrived at Saturn. Find out the current status of the mission (from news reports or the mission Web site). Has the mission been successful so far? If so, what have we learned?

Uranus

- Average distance from the Sun: 19.19 AU
- Radius: 25,559 km
- Mass (Earth = 1): 14.5
- Average density: 1.32 g/cm^3
- Composition: hydrogen, helium, hydrogen compounds
- Cloud-top temperature: 60 K
- Moons: at least 27

Uranus (normally pronounced YUR-uh-nus) is much smaller than either Jupiter or Saturn but still much larger than Earth (Figure 6.9). It is made largely of hydrogen, helium, and hydrogen compounds such as water (H_2O), ammonia (NH_3), and methane (CH_4). Methane gas gives Uranus its pale blue-green color. Like the other giants of the outer solar system, Uranus lacks a solid surface. At least 21 moons orbit Uranus, along with a set of rings somewhat similar to those of the other jovian planets—though they are dark and difficult to see.

The entire Uranus system—planet, rings, and moon orbits—is tipped on its side compared to the rest of the planets. This unusual orientation may be the result of a cataclysmic collision suffered by Uranus as it was forming some 4.6 billion years ago. It also makes for the most extreme pattern of seasons on any planet. If you lived on a platform floating in Uranus's atmosphere near the north pole, you'd have continuous daylight for half of each orbit, or 42 years. Then, after a very gradual sunset, you'd enter into a 42-year-long night.

Only one spacecraft has visited Uranus. *Voyager 2* flew past all four of the jovian planets before heading out of the solar system. Much of our current understanding of Uranus comes from that mission, though powerful new telescopes are also capable of studying this planet. Scientists would love an opportunity to study Uranus and its interesting rings and moons in much greater detail, but no missions to Uranus are currently under development.

Figure 6.9

This image shows a view from a vantage point high above Uranus's moon Ariel. The ring system is shown, although it would actually be too dark to see from this vantage point. This computer simulation is based on data from NASA's Voyager 2 mission. (From the Voyage scale model solar system, developed by the Challenger Center for Space Science Education, the Smithsonian Institution, and NASA. Image created by ARC Science Simulations © 2001.)

Figure 6.10

This image shows what it would look like to be orbiting Neptune's moon Triton as Neptune itself comes into view. The dark rings are exaggerated to make them visible in this computer simulation using data from NASA's Voyager 2 mission. (From the Voyage scale model solar system, developed by the Challenger Center for Space Science Education, the Smithsonian Institution, and NASA. Image created by ARC Science Simulations © 2001.)

• Neptune

- Average distance from the Sun: 30.06 AU
- Radius: 24,764 km
- Mass (Earth = 1): 17.1
- Average density: 1.64 g/cm^3
- Composition: hydrogen, helium, hydrogen compounds
- Cloud-top temperature: 60 K
- Moons: at least 13

Neptune looks nearly like a twin of Uranus, with very similar size and composition, although it is more strikingly blue (Figure 6.10). Like Uranus, Neptune has been visited only by the *Voyager 2* spacecraft. No further missions are currently planned. Neptune has rings and at least 11 moons. Its largest moon, Triton, is larger than the planet Pluto and is one of the most fascinating moons in the solar system. Its icy surface has features that appear to be somewhat like geysers but spew nitrogen gas into the sky. Even more surprisingly, Triton is the only large moon in the solar system that orbits its planet "backward"—that is, in a direction opposite to the direction in which Neptune rotates. Understanding how such a large moon ended up orbiting Neptune backward has given us new insights into the history of the outer solar system [Section 9.3].

• Pluto

- Average distance from the Sun: 39.54 AU
- Radius: 1,160 km
- Mass (Earth = 1): 0.0022
- Average density: 2.0 g/cm^3
- Composition: ices, rock
- Average surface temperature: 40 K
- Moons: 1

At its average distance from the Sun, Pluto lies far beyond the orbit of Neptune—as far as the distance between the orbits of Uranus and Neptune. It is easy to imagine that Pluto must be cold and dark. From Pluto, the Sun would appear as little more than a bright light among the stars, offering little comfort or warmth.

Pluto appears to be somewhat of a misfit among the planets. It is neither large and gaseous like the other outer planets nor rocky like the inner planets. Instead, it is very small—the smallest planet by far—and is made of ices and rock. Pluto's orbit is also unusual, being both quite elliptical and substantially inclined to the plane of the other planets. Pluto actually comes closer to the Sun than Neptune for 20 years out of each 248-year orbit. The last such period ended in 1999, and the next won't begin until 2263. Its oddball nature gives it more in common with many comets than with the other eight planets, leading to debate over whether it should be called a "planet" at all [Section 9.3].

Pluto has a moon, Charon, that is about half Pluto's size in diameter and one-eighth Pluto's mass. Our best telescopic views of Pluto and Charon reveal little detail (Figure 6.11), and no spacecraft has yet visited these distant worlds. However, a Pluto mission called New Horizons is well into development. If all goes well, it will be launched in 2006 and will arrive at Pluto in 2015.

Figure 6.11

Pluto and Charon, as photographed by the Hubble Space Telescope. Because no spacecraft has been to Pluto, we do not yet have clear pictures of this tiny planet or its moon.

TABLE 6.1 *Planetary Data*[*]

Photo	Planet	Relative Size	Average Distance from Sun (AU)	Average Equatorial Radius (km)	Mass (Earth = 1)	Average Density (g/cm^3)	Orbital Period	Rotation Period	Axis Tilt	Average Surface (or Cloud Tops) Temperature[†]	Composition	Known Moons (2004)	Rings?
	Mercury		0.387	2,440	0.055	5.43	87.9 days	58.6 days	0.0°	700 K (day) 100 K (night)	Rocks, metals	0	No
	Venus		0.723	6,051	0.82	5.24	225 days	243 days	177.3°	740 K	Rocks, metals	0	No
	Earth		1.00	6,378	1.00	5.52	1.00 year	23.93 hours	23.5°	290 K	Rocks, metals	1	No
	Mars		1.52	3,397	0.11	3.93	1.88 years	24.6 hours	25.2°	240 K	Rocks, metals	2	No
	Jupiter		5.20	71,492	318	1.33	11.9 years	9.93 hours	3.1°	125 K	H, He, hydrogen compounds[§]	63	Yes
	Saturn		9.54	60,268	95.2	0.70	29.4 years	10.6 hours	26.7°	95 K	H, He, hydrogen compounds[§]	31	Yes
	Uranus		19.2	25,559	14.5	1.32	83.8 years	17.2 hours	97.9°	60 K	H, He, hydrogen compounds[§]	24	Yes
	Neptune		30.1	24,764	17.1	1.64	165 years	16.1 hours	29.6°	60 K	H, He, hydrogen compounds[§]	13	Yes
	Pluto		39.5	1,160	0.0022	2.0	248 years	6.39 days	112.5°	40 K	Ices, rock	1	No

[*] Appendix E gives a more complete list of planetary properties.

[†] Surface temperatures for all objects except Jupiter, Saturn, Uranus, and Neptune, for which cloud-top temperatures are listed.

[§] Includes water (H_2O), methane (CH_4), and ammonia (NH_3).

(*continued from p. 132*)

Patterns of Motion Among Large Bodies Figure 6.1 showed a "side view" of a solar system, and Figure 6.12 adds a view of our solar system from above. Together, these figures reveal several clear patterns of motion. For example:

- All planetary orbits are nearly circular and lie nearly in the same plane.
- All planets orbit the Sun in the same direction—counterclockwise as viewed from high above Earth's North Pole.
- Most planets rotate in the same direction in which they orbit (counterclockwise as viewed from above the North Pole), with fairly small axis tilts. The Sun also rotates in this same direction.
- Most of the solar system's large moons exhibit similar properties in their orbits around their planets—for example, orbiting in their planet's equatorial plane in the same direction that the planet rotates.

The Sun, planets, and large moons orbit and rotate in an organized way.

We will consider these orderly patterns together as the first major feature of our solar system. As we'll see shortly, they are all consequences of processes that occurred early in the birth of our solar system.

The Existence of Two Types of Planets Our brief planetary tour showed that the four inner planets are quite different from the four large outer planets. We say that these two groups represent two distinct planetary classes: *terrestrial* and *jovian*.

Terrestrial planets are small, rocky, and close to the Sun. Jovian planets are large, gas-rich, and far from the Sun.

The **terrestrial planets** are the four planets of the inner solar system—Mercury, Venus, Earth, and Mars. (*Terrestrial* means "Earth-like.") The four terrestrial planets are relatively small and dense, with rocky surfaces and an abundance of metals deep in their interiors. They have few moons, if any, and none have rings.

The **jovian planets** are the four large planets of the outer solar system—Jupiter, Saturn, Uranus, and Neptune. (*Jovian* means "Jupiter-like.") The jovian planets are much larger in size and lower in average density than the terrestrial planets. They have rings and numerous moons. Their compositions are also quite different from those of the terrestrial worlds. They are made mostly of hydrogen, helium, and **hydrogen compounds**—compounds containing hydrogen, such as water (H_2O), ammonia (NH_3), and methane (CH_4).

Because the jovian planets are made mostly of substances that are gases under earthly conditions, they are sometimes called "gas giants." They do not have solid surfaces and look like balls of gas from the outside. However, the intense pressures and temperatures of the jovian planet interiors transform these gases into forms unlike anything we ordinarily see on Earth. Table 6.2 contrasts the general traits of the terrestrial and jovian planets.

Notice that Pluto is left out in the cold, both literally and figuratively. It is much smaller and farther from the Sun (on average) than any other planet, and it is the only planet with an ice-rich composition. Pluto's orbit is also more elliptical and more inclined to the ecliptic plane (the plane of Earth's orbit) than the orbit of any other planet. For a long time, scientists considered Pluto to be a lone misfit. However, recent discoveries suggest that Pluto is just one of many similar objects that roam the solar system beyond the orbit of Neptune [Section 9.3].

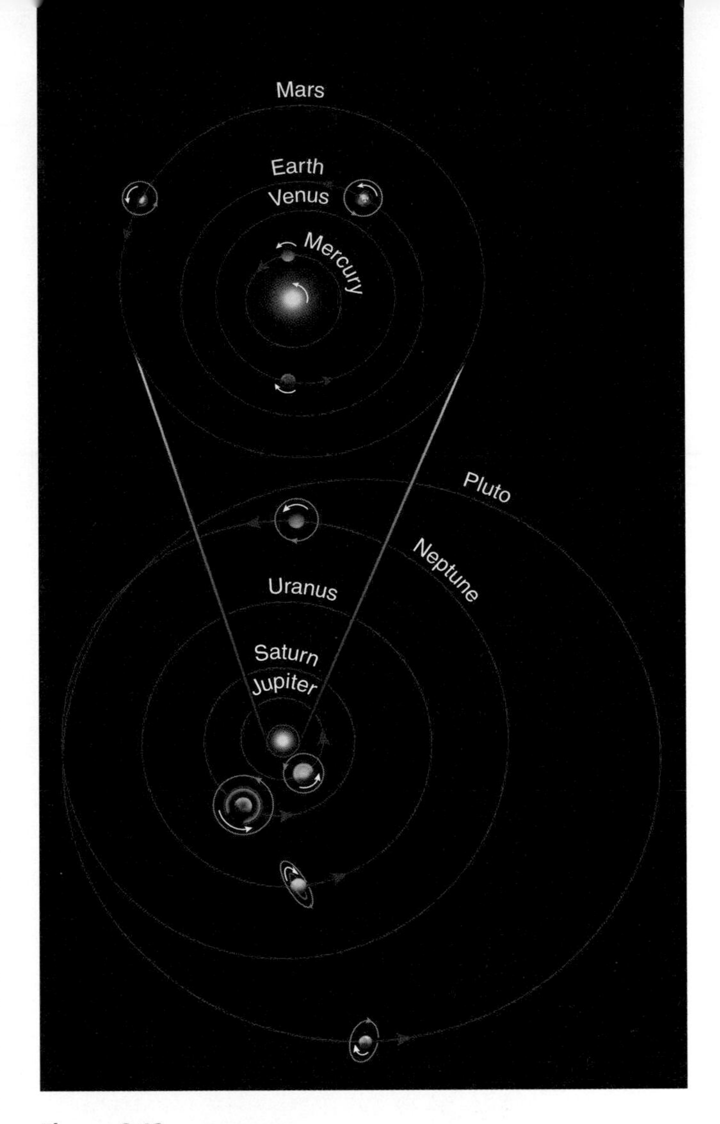

Figure 6.12 Interactive Figure

View of the solar system from high above Earth's North Pole. Notice that all orbits except Mercury and Pluto are nearly circular. The short, curved arrow for each planet indicates the planet's direction of rotation. The circles around each planet except Mercury and Venus indicate the orbital direction of the planet's major moons. (Mercury and Venus have no moons.) Most major moons orbit in the same direction that their planet rotates.

TABLE 6.2 *Comparison of Terrestrial and Jovian Planets*

Terrestrial Planets	*Jovian Planets*
Smaller size and mass	Larger size and mass
Higher density	Lower density
Made mostly of rock and metal	Made mostly of hydrogen, helium, and hydrogen compounds
Solid surface	No solid surface
Few (if any) moons and no rings	Rings and many moons
Closer to the Sun (and closer together), with warmer surfaces	Farther from the Sun (and farther apart), with cool temperatures at cloud tops

Asteroids and Comets Our third major feature of the solar system is the existence of vast numbers of small objects orbiting the Sun. These objects fall into two major groups: asteroids and comets.

Rocky asteroids and icy comets far outnumber the planets and their moons.

Asteroids are small, rocky bodies that orbit the Sun much like planets, but they are much smaller than planets (Figure 6.13). Even the largest of the asteroids have radii of only a few hundred kilometers, which means they are dwarfed by our Moon. Most asteroids are found within the **asteroid belt** between the orbits of Mars and Jupiter (see Figure 6.1).

Comets are also small objects that orbit the Sun, but they are made largely of ices (such as water ice, ammonia ice, and methane ice) rather than rock. You are probably familiar with the occasional appearance of comets in the inner solar system, where they may become visible to the naked eye with long, beautiful tails (Figure 6.14). These visitors, which may delight sky watchers for a few weeks or months, are actually quite rare among comets. The vast majority of comets never visit the inner solar system, instead orbiting the Sun in the extreme outer reaches of the solar system. As we'll discuss further in Chapter 9, there are two major groupings of comets in our solar system: One group of comets orbits the Sun in the same direction and nearly the same plane as the planets, occupying a broad region beyond the orbit of Neptune that we call the *Kuiper belt* (pronounced *koy-per*). The other group of comets does not show any simple pattern to its orbits, and resides much farther from the Sun in a region called the *Oort cloud* (*Oort* rhymes with *court*).

Figure 6.13 VIS

The asteroid Eros (photographed from the NEAR spacecraft). Its appearance is probably typical of most asteroids. Eros is about 40 kilometers in length. Like other small objects in the solar system, it is not spherical.

Exceptions to the Rules The fourth key feature of our solar system is that the general rules have a few notable exceptions. For example, while most of the planets rotate in the same direction as they orbit, Uranus and Pluto rotate nearly on their sides, and Venus rotates "backward"—clockwise, rather than counterclockwise, as viewed from high above Earth's North Pole. Similarly, while most large moons orbit their planets in the same direction as their planets rotate, many small moons have much more unusual orbits.

A successful theory of solar system formation must allow for exceptions to general rules, such as the existence of Earth's surprisingly large Moon.

One of the most interesting exceptions concerns our own Moon. While the other terrestrial planets have either no moons (Mercury and Venus) or very tiny moons (Mars, with two moons), Earth has one of the largest moons in the solar system.

Figure 6.14

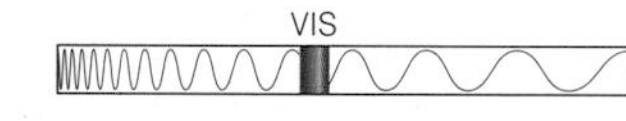

Comet Hale–Bopp, photographed over Boulder, Colorado, during its appearance in 1997.

Summary of the Four Features Figure 6.15 summarizes the four general features of the solar system that we have discussed. We are now ready to turn our attention to the task of finding a theory that can explain all these features.

• What theory best explains the features of our solar system?

Ancient Greek ideas about Earth's origins probably seemed quite reasonable when people assumed that Earth was the center of the universe, but they no longer made sense after Kepler and Galileo proved that Earth is a planet going around the Sun. By the end of the seventeenth century,

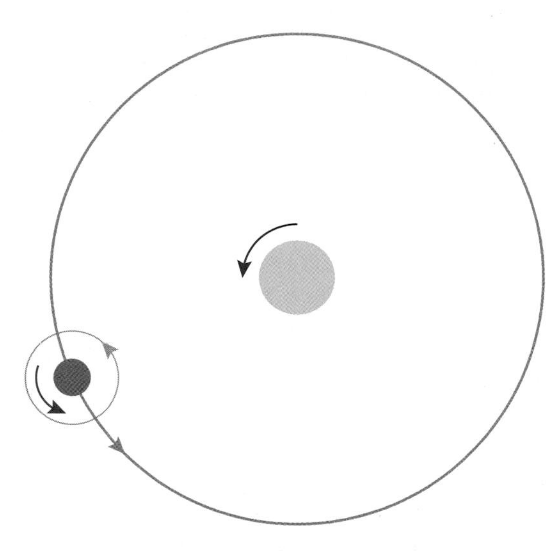

Large bodies in the solar system have orderly motions. All planets and most satellites have nearly circular orbits going in the same direction in nearly the same plane. The Sun and most of the planets rotate in this same direction as well.

Planets fall into two main categories: small, rocky terrestrial planets near the Sun and large, hydrogen-rich jovian planets farther out. The jovian planets have many moons and rings made of rock and ice. Pluto does not fit in either category.

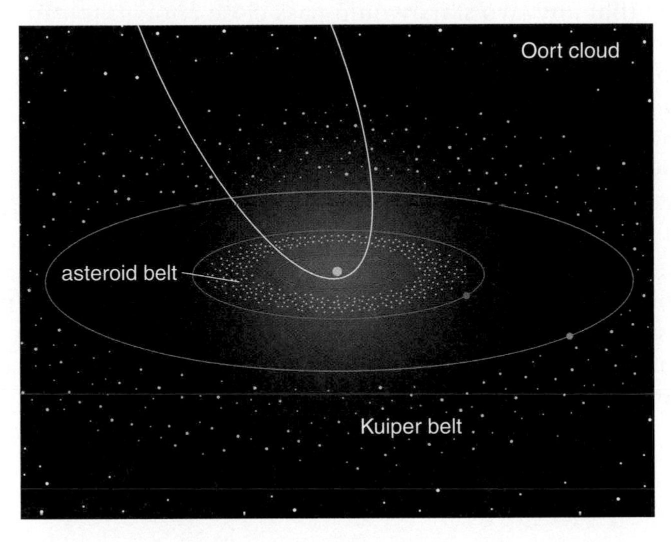

Swarms of asteroids and comets populate the solar system. Asteroids are concentrated in the asteroid belt, and comets populate the regions known as the Kuiper belt and the Oort cloud.

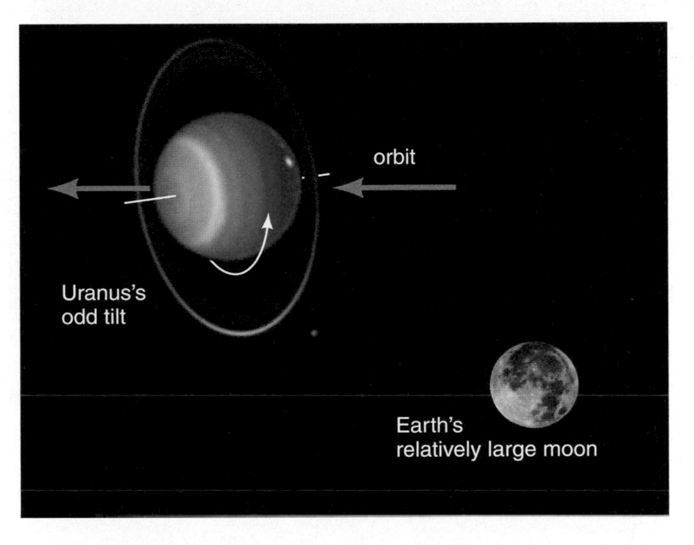

Several notable exceptions to these general trends stand out, such as planets with unusual axis tilts or surprisingly large moons, and moons with unusual orbits.

Figure 6.15

Four major features of the solar system.

the Copernican revolution [Section 3.3] and Newton's discovery of the universal law of gravitation [Section 4.4] had given us a basic understanding of the layout and motion of the planets and moons in our solar system. Clearly, new ideas were needed if we wanted to understand how our solar system had come to be.

We generally credit two eighteenth-century scientists with proposing the hypothesis that blossomed into our modern scientific theory of the origin of the solar system. Around 1755, German philosopher Immanuel Kant proposed that our solar system formed from the gravitational collapse of an interstellar cloud of gas. The same idea was put forth independently about 40 years later by French mathematician Pierre-Simon Laplace (who apparently was unaware of Kant's proposal). Because an interstellar cloud is usually called a *nebula* (Latin for "cloud"), their idea became known as the *nebular hypothesis*.

Our modern theory holds that our solar system formed from the gravitational collapse of a great cloud of gas, and it explains all the general features of our solar system.

The nebular hypothesis remained popular throughout the nineteenth century. By the early twentieth century, however, scientists had found a few aspects of our solar system that the nebular hypothesis did not seem to explain very well—at least in its original form as described by Kant and Laplace. While some scientists sought to modify the nebular hypothesis, others looked for entirely different ideas about how the solar system might have formed.

During much of the first half of the twentieth century, the nebular hypothesis faced stiff competition from a hypothesis proposing that the planets represent debris from a near-collision between the Sun and another star. According to this *close encounter hypothesis,* the planets formed from blobs of gas that had been gravitationally pulled out of the Sun during the near-collision.

Today, the close encounter hypothesis has been discarded. Calculations showed that the close encounter hypothesis could not account for either the observed orbital motions of the planets or the neat division of the planets into two major categories (terrestrial and jovian). In addition, the chance that any two stars would pass close enough to cause a substantial gravitational disruption is extremely remote. It's inconceivable that it could have happened often enough to account for the many other stars now known to have planets.

Meanwhile, so much evidence has accumulated in favor of the nebular hypothesis that it has achieved the status of a scientific theory—the **nebular theory** of our solar system's birth. As scientists investigated the implications of the idea that our solar system formed from a large cloud of gas, they found that the idea led to natural explanations for all four general features of our solar system. Moreover, the nebular theory also predicts what we should find when we study other star systems. While recent discoveries of planets around other stars have revealed a few surprises, they remain consistent with what we expect if the nebular theory is correct.

Formation of the Solar System Tutorial, Lessons 1–2

6.3 THE BIRTH OF THE SOLAR SYSTEM

We are now ready to examine the nebular theory of solar system formation in more depth. In this section, we'll begin at the beginning, examining the origins of the cloud of gas that gave birth to our solar system and explaining how the gravitational collapse of this cloud led to orderly patterns of motion. Remember that the same ideas are likely to apply to the births of other star systems as well.

• Where did the solar system come from?

The nebular theory begins with the idea that our solar system was born from a cloud of gas that collapsed under its own gravity. We usually refer to this cloud as the **solar nebula**. But where did this gas come from?

Galactic Recycling According to the evidence as we understand it today, the universe was born in the Big Bang [Section 1.1]. Hydrogen and helium were the only chemical elements present when the universe was young. All the heavier elements have been produced since that time by the stars. When stars die, they release much of their content back to space, including newly produced heavy elements. This material can then be recycled into new generations of stars (Figure 6.16).

Although this process of creating heavy elements in stars and recycling them within the galaxy has probably gone on for most of the 14-billion-year history of our universe, only a small fraction of the original hydrogen and helium has been converted into heavier elements. By the time our solar system formed about 4.6 billion years ago, about 2% (by mass) of the original hydrogen and helium in the galaxy had been converted to heavier elements. The Sun, which represents nearly all the mass in our solar system, still has this basic composition today.

The cloud of gas that gave birth to our solar system resulted from the recycling of gas through many generations of stars within our galaxy.

Thus, the gas that made up the solar nebula was the result of billions of years of galactic recycling that occurred before our solar system was born. The Sun and planets were born from this gas, and the terrestrial worlds were made from the 2% proportion of heavier elements mixed within it. As we saw in Chapter 1, we are therefore "star stuff," because we and our planet are made of elements forged in stars that lived and died long ago.

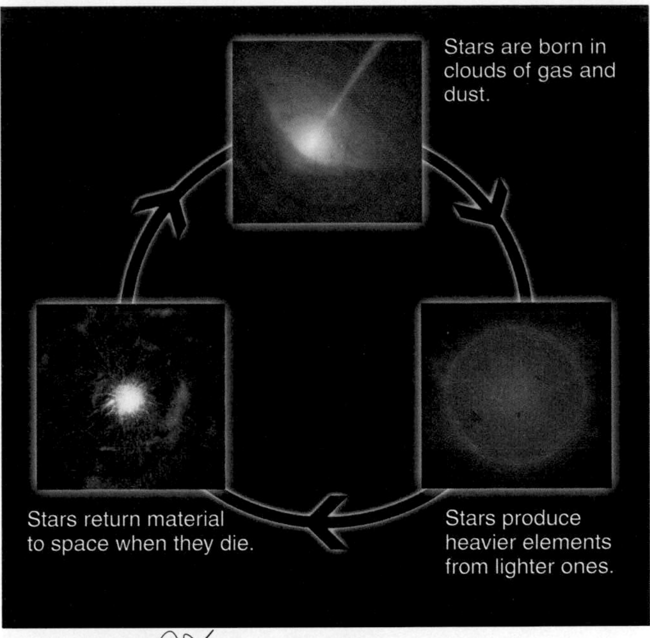

Figure 6.16

This figure, which is a portion of Figure 1.3 earlier in the book, summarizes the galactic recycling process.

Could a solar system like ours have formed with the first generation of stars after the Big Bang? Explain.

Evidence from Other Gas Clouds Strong observational evidence supports the idea that our solar system was born from an interstellar cloud of gas, because stars that appear to be in the process of formation today are always found within interstellar clouds.

The existence of many star-forming clouds today strongly supports the idea that our solar system was born in such a cloud long ago.

Figure 6.17 shows the Orion Nebula, one of the most famous star-forming clouds. Some stars have already been born within this vast cloud of gas, and thousands more will be born there over the next few million years. Moreover, each star we observe in the process of formation is surrounded by relatively dense gas that appears to be collapsing under its own gravity, suggesting that it is forming within its own "solar nebula," just as we think our own solar system formed some 4.6 billion years ago.

Figure 6.17 VIS

This photograph shows the Orion Nebula, an interstellar cloud in which new star systems are forming. Over the next few million years, thousands of stars will be born in this gas cloud. Some, perhaps even most, of these stars may end up with their own planetary systems.

• What caused the orderly patterns of motion in our solar system?

The solar nebula probably began as a large, roughly spherical cloud of very cold and very low-density gas. Initially, the cloud was likely so spread out that gravity alone could not make it shrink in size. Debris from the explosion of a nearby star (a supernova) may have rammed into the cloud with enough force to start its collapse.

Once the collapse started, the law of gravity ensured that it would continue. Remember that the strength of gravity follows an inverse square law with distance [Section 4.4]. Because the mass of the cloud remained the same as it shrank, the strength of gravity increased as the

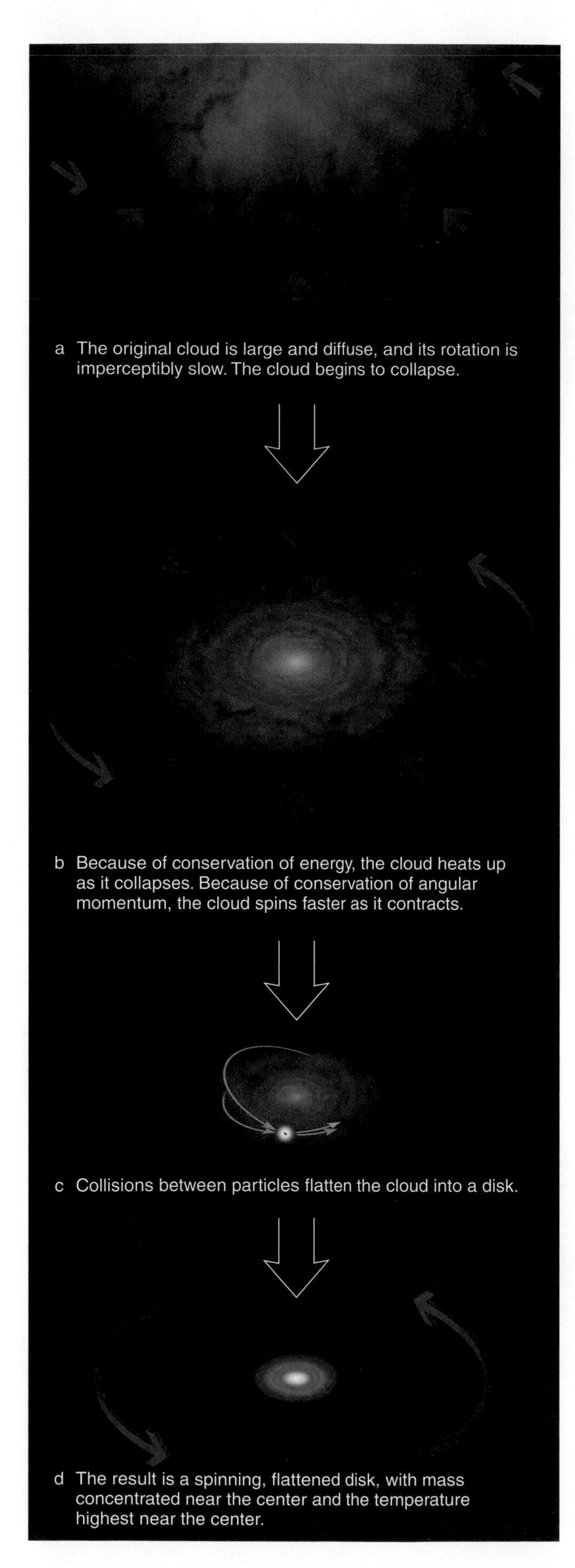

diameter of the cloud decreased. For example, when the diameter decreased by half, the force of gravity increased by a factor of four.

Because gravity pulls in all directions, you might at first guess that the solar nebula remained spherical as it shrank. Indeed, the idea that gravity pulls in all directions explains why the Sun and the planets are spherical. However, gravity is not the only physical law that affects the collapse of a cloud of gas. When we consider everything that should have occurred in the collapsing cloud, we find that the solar nebula should have become a spinning, flattened disk surrounding a central star.

Heating, Spinning, and Flattening As the solar nebula shrank in size, three important processes altered its density, temperature, and shape, changing it from a large, diffuse (spread-out) cloud to a much smaller spinning disk (Figure 6.18):

- *Heating.* The temperature of the solar nebula increased as it collapsed. Such heating represents energy conservation in action [Section 4.3]. As the cloud shrank, its gravitational potential energy was converted into thermal energy (see Figure 4.12b). The Sun formed in the center, where temperatures and densities were highest.
- *Spinning.* Like an ice skater pulling in her arms as she spins, the solar nebula rotated faster and faster as it shrank in radius. This increase in rotation rate represents conservation of angular momentum in action [Section 4.3]. The rotation of the cloud may have been imperceptibly slow before its collapse began, but the cloud's shrinkage made fast rotation inevitable. The rapid rotation helped ensure that not all the material in the solar nebula collapsed into the center: The greater the angular momentum of a rotating cloud, the more spread out it will be.
- *Flattening.* The solar nebula flattened into a disk. This flattening is a natural consequence of collisions between particles in a spinning cloud. A cloud may start with any size or shape, and different clumps of gas within the cloud may be moving in random directions at random speeds. These clumps collide and merge as the cloud collapses, and each new clump has the average velocity of the clumps that formed it. Thus, the random motions of the original cloud become more orderly as the cloud collapses, changing the cloud's original lumpy shape into a rotating, flattened disk. Similarly, collisions between clumps of material in highly elliptical orbits reduce their eccentricities, making their orbits more circular.

The orderly motions of our solar system today are a direct result of the solar system's birth in a spinning, flattened cloud of gas.

The formation of the spinning disk explains the orderly motions of our solar system today. The planets all orbit the Sun in nearly the same plane because they formed in the flat disk. The direction in which the disk was spinning became the direction of the Sun's rotation and the orbits of the planets. It was also the preferred direction of rotation for planets—which is why most plan-

Figure 6.18 Interactive Figure

This sequence of paintings shows how the gravitational collapse of a large cloud of gas causes it to become a spinning disk of matter. The hot, dense central bulge becomes a star, while planets can form in the surrounding disk.

ets rotate the same way today—though the small sizes of planets compared to the entire disk allowed some exceptions to arise. The fact that collisions in the disk tended to make orbits more circular explains why most planets in our solar system have nearly circular orbits.

Testing the Model If we are correct about the processes that occurred in the solar nebula, the same processes ought to occur in other clouds of gas that collapse to form stars. Thus, the formation of a disk in which planets can form should be a natural part of the star formation process, and we can test whether our model makes sense by searching for such disks elsewhere.

Observations of disks around other stars support the idea that our own solar system was once a spinning disk of gas.

Observational evidence does indeed support the idea that our solar nebula collapsed in the way envisioned by the nebular model. The heating that occurs in a collapsing cloud of gas means the gas should emit thermal radiation [Section 5.2], primarily in the infrared. We've detected infrared radiation from many nebulae where star systems appear to be forming today. We've even seen structures around other stars that appear to be flattened, spinning disks (Figure 6.19).

Other support for the model comes from computer simulations of the formation process. A simulation begins with a set of data representing the conditions we observe in interstellar clouds. Then, with the aid of a computer, we apply the laws of physics to predict the changes that should occur over time. These computer simulations successfully reproduce most of the general characteristics of motion in our solar system, suggesting that the nebular theory is on the right track.

Finally, evidence that our ideas about the formation of flattened disks are correct comes from many other structures in the universe. We expect flattening to occur anywhere that orbiting particles can collide, which explains why we find so many cases of flat disks, including the disks of spiral galaxies like the Milky Way, the disks of planetary rings, and the *accretion disks* surrounding many neutron stars and black holes [Section 13.3].

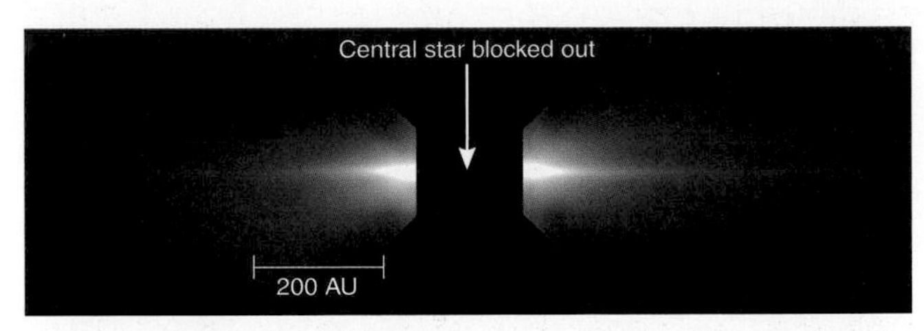

a This Hubble Space Telescope photo shows an edge-on view of a disk of dust surrounding the star Beta Pictoris. The light of the star itself is blacked out, allowing the disk to be seen.

b Each disk-shaped "blob" in this photograph is a disk of material orbiting a star, perhaps much like the disk in which the planets of our own solar system formed.

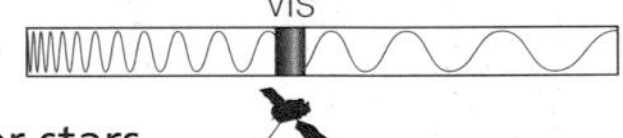

Figure 6.19

Evidence of disks around other stars.

Formation of the Solar System Tutorial, Lesson 3

6.4 THE FORMATION OF PLANETS

The planets began to form after the solar nebula had collapsed into a flattened disk of perhaps 200 AU in diameter (about twice the present-day diameter of Pluto's orbit). In this section, we'll discuss how the planets formed within the spinning disk of gas. Along the way, we'll see how the process of planet formation led to the three major features of our solar system that we have not yet explained: the existence of two types of planets, the existence of asteroids and comets, and the exceptions to the rules.

• Why are there two types of planets?

The swirling disk of gas in which the Sun and planets formed should have had the same composition throughout: 98% hydrogen and helium plus 2% heavier elements. Yet the solar system ended up with two very different types of planets: the small, rocky terrestrial planets and the large, gaseous jovian planets. How did the compositions of the planets end up being so different, when they were all born in the same cloud of gas? The locations of these two classes of planets provide the key to understanding how they formed. The terrestrial planets formed in the warm, inner regions of the swirling disk and the jovian planets formed in the colder, outer regions. Let's explore how temperature differences led directly to different types of planets.

Condensation: Sowing the Seeds of Planets In the center of the collapsing solar nebula, gravity drew together enough material to form the Sun. In the surrounding disk, however, the gaseous material was too spread out for gravity alone to clump it up. Instead, material had to begin clumping in some other way and to grow in size until gravity could start pulling it together into planets. In essence, planet formation required the presence of "seeds"—solid bits of matter around which gravity could ultimately build planets.

Planet formation began around tiny "seeds" of solid metal, rock, or ice.

The basic process of seed formation was probably much like the formation of snowflakes in clouds on Earth: When the temperature is low enough, some atoms or molecules in a gas may bond and solidify. The general process in which solid (or liquid) particles form in a gas is called **condensation**—we say that the particles *condense* out of the gas. The particles that initially condense from a gas are microscopic in size, but they can grow larger with time.

Different materials condense at different temperatures. On the basis of their condensation properties, the ingredients of the solar nebula fell into four major categories, summarized in Table 6.3:

- **Hydrogen and helium gas (98% of the solar nebula).** These gases never condense under the conditions present in a nebula.
- **Hydrogen compounds (1.4% of the solar nebula).** Materials such as water (H_2O), methane (CH_4), and ammonia (NH_3) can solidify into **ices** at low temperatures (below about 150 K under the low pressure of the solar nebula).
- **Rock (0.4% of the solar nebula).** Rocky material is gaseous at high temperatures, but condenses into solid bits of mineral at temperatures between about 500 K and 1,300 K, depending on the type of rock. (A *mineral* is a piece of rock with a particular chemical composition and structure.)
- **Metal (0.2% of the solar nebula).** Metals such as iron, nickel, and aluminum are also gaseous at very high temperatures, but condense into solid form at higher temperatures than rock—typically in the range of 1,000 K to 1,600 K.

Because hydrogen and helium gas made up 98% of the solar nebula's mass and did not condense, the vast majority of the nebula remained gaseous at all times. However, hydrogen compounds, rock, and metal could condense into solid form wherever the temperature allowed (Figure 6.20). In the innermost regions of the nebula near the forming Sun, where the temperature was above 1,600 K, it was too hot for any material to condense. Near what is now Mercury's orbit, the temperature was low enough for metals and some types of rock to condense into tiny solid particles, but other types of rock and all the hydrogen compounds remained gaseous. More types of rock could condense, along with the metals, at the distances from the Sun where Venus, Earth, and Mars

TABLE 6.3 *Materials in the Solar Nebula*

A summary of the four types of materials present in the solar nebula. The squares represent the relative proportions of each type (by mass).

	Examples	*Typical Condensation Temperature*	*Relative Abundance (by mass)*
Hydrogen and Helium Gas	hydrogen, helium	do not condense in nebula	■ 98%
Hydrogen Compounds	water (H_2O) methane (CH_4) ammonia (NH_3)	<150 K	■ 1.4%
Rock	various minerals	500–1,300 K	▪ 0.4%
Metals	iron, nickel, aluminum	1,000–1,600 K	▪ 0.2%

would form. Hydrogen compounds could condense into ices only beyond the **frost line,** which lay between the present-day orbits of Mars and Jupiter.

THINK ABOUT IT Consider a region of the solar nebula in which the temperature was about 1,300 K. What fraction of the material in this region was gaseous? What were the solid particles in this region made of? After you have answered these questions, do the same for a region with a temperature of 100 K. Would the 100 K region be closer to or farther from the Sun? Explain.

The solid seeds in the inner solar system were made only of metal and rock, but in the outer solar system they included far more abundant ices.

Thus, the frost line marked the key transition between the warm inner regions of the solar system where terrestrial planets would form and the cool outer regions where jovian planets would form. Inside the frost line, only metal and rock could condense into solid "seeds," which is why the terrestrial planets ended up being made of metal and rock. Beyond the frost line, where it was cold enough for hydrogen compounds to condense into ices, the solid seeds were built of ice along with metal and rock. Moreover, because hydrogen compounds were nearly three times as abundant in the nebula as metal and rock combined (see Table 6.3), the total amount of solid material was far greater beyond the frost line than within it. As we'll see, this greater amount of solid material probably led to the formation of the gaseous, jovian planets.

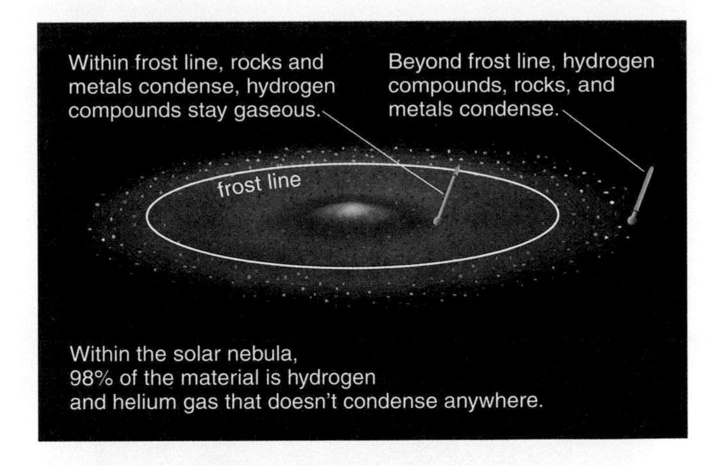

Figure 6.20 Interactive Figure

Temperature differences in the solar nebula led to different kinds of condensed materials, sowing the seeds for two different kinds of planets.

Building the Terrestrial Planets

From this point, the story of the terrestrial planets seems fairly clear: The solid seeds of metal and rock in the inner solar system ultimately grew into the full-fledged planets we see today. Because rock and metal made up such a small amount of the material in the solar nebula, the planets they built ended up relatively small in size.

The process by which small "seeds" grew into planets is called **accretion**. Accretion began with the microscopic solid particles that condensed from the gas of the solar nebula. These particles orbited the forming Sun with the same orderly, circular paths as the gas from which they condensed. Individual particles therefore moved at nearly the same speed as neighboring particles, so "collisions" were more like gentle touches. Although the particles were far too small to attract each other gravitationally at this point, they were able to stick together through electrostatic forces—the same "static electricity" that makes hair stick to a comb. Small particles thereby began to combine into larger ones. As the particles grew in mass, gravity began to aid the process of their sticking together, accelerating their growth into boulders large enough to count as **planetesimals**, which means "pieces of planets."

The terrestrial planets were made from the solid bits of metal and rock that condensed in the inner solar system.

The planetesimals grew rapidly at first. As a planetesimal grew larger, it would have both more surface area to make contact with other planetesimals and more gravity to attract them. Some planetesimals probably grew to hundreds of kilometers in size in only a few million years—a long time in human terms, but only about 1/1,000 the present

COMMON MISCONCEPTIONS

SOLAR GRAVITY AND THE DENSITY OF PLANETS

You might think that the dense rocky and metallic materials were pulled to the inner part of the solar nebula by the Sun's gravity or that gases escaped from the inner nebula because gravity couldn't hold them. But this is not the case—all the ingredients were orbiting the Sun together under the influence of the Sun's gravity. The orbit of a particle or a planet does *not* depend on its size or density, so the Sun's gravity cannot be the cause of the different kinds of planets. Rather, the different temperatures in the solar nebula are the cause.

age of the solar system. However, once the planetesimals reached these relatively large sizes, further growth became more difficult.

Gravitational encounters [Section 4.4] between planetesimals tended to alter their orbits, particularly those of the smaller planetesimals. With different orbits crossing each other, collisions between planetesimals tended to occur at higher speeds and hence became more destructive. Such collisions tended to produce fragmentation more often than accretion. Only the largest planetesimals avoided being shattered and thus were able to grow into full-fledged, terrestrial planets. Figure 6.21 summarizes the growth of the terrestrial planets through accretion.

Theoretical evidence in support of this model comes from computer simulations of the accretion process, which support the scenario described above. Observational evidence comes from meteorites, many of which appear to be surviving fragments from the early period of condensation in our solar system [Section 9.1]. Meteorites often contain metallic grains embedded in a variety of rocky minerals (Figure 6.22), just as we expect for the planetesimals of the inner solar system.

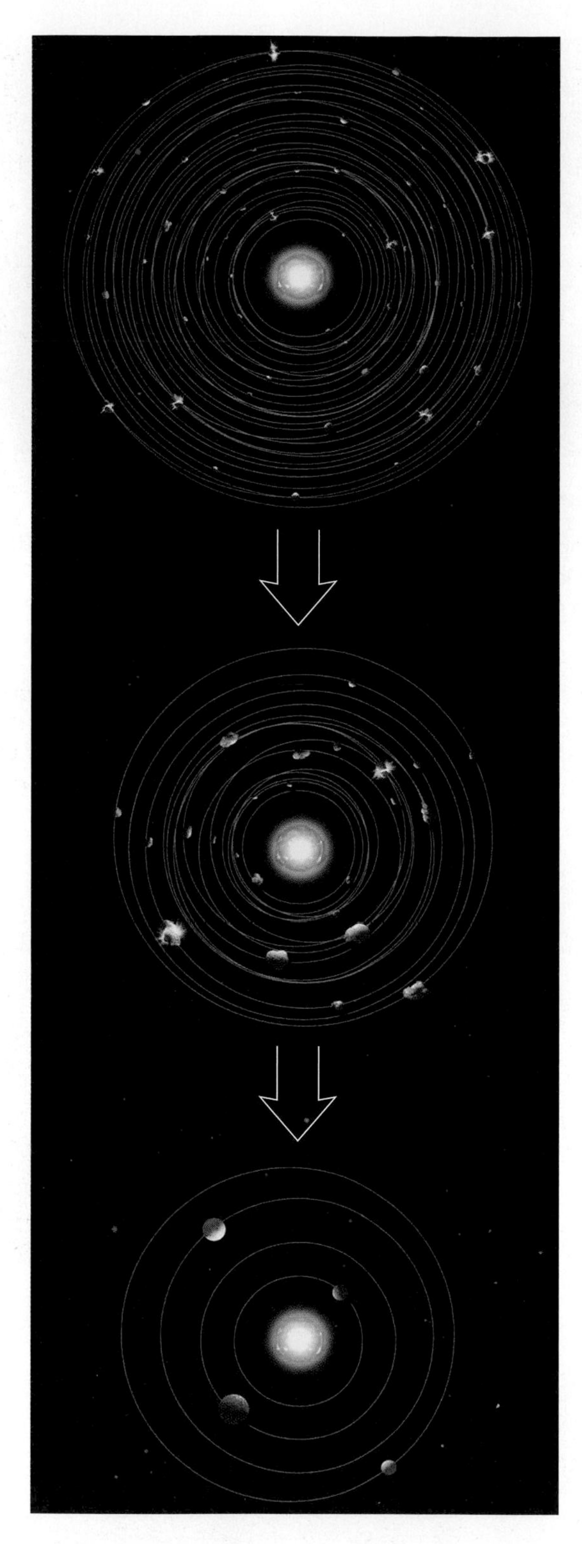

Figure 6.21 Interactive Figure

These diagrams show how planetesimals gradually accrete into terrestrial planets. Early in the accretion process, there are many relatively large planetesimals on crisscrossing orbits (top). As time passes, a few planetesimals grow larger by accreting smaller ones, while others shatter in collisions (center). Ultimately, only the largest planetesimals avoid shattering and grow into full-fledged planets (bottom). (Diagram not to scale.)

Making the Jovian Planets Accretion should have occurred similarly in the outer solar system, but condensation of ices meant that there was more solid material. The solid objects that reside in the outer solar system today, such as comets and the moons of the jovian planets, still show this icy composition. However, the growth of icy planetesimals cannot be the whole story of jovian planet formation, because the jovian planets themselves are *not* made mostly of ice. Instead, they contain large amounts of hydrogen and helium gas.

The jovian planets began as large, icy planetesimals, which then captured hydrogen and helium gas from the solar nebula.

The precise way in which the jovian planets formed remains a topic of scientific debate, but in the leading model they acquired their gas after beginning existence as large, icy planetesimals. According to this model, some of the abundant, icy planetesimals of the outer solar system grew to masses many times that of Earth. With these large masses, their gravity became strong enough to capture and hold some of the hydrogen and helium gas that made up the vast majority of the surrounding solar nebula. As the growing planets accumulated gas, their gravity grew stronger still, allowing them to capture even more gas. Ultimately, the jovian planets grew so much that they bore little resemblance to the icy seeds from which they started, instead ending up with large abundances of hydrogen and helium gas.

This model also explains the many large moons of the jovian planets. The same processes of heating, spinning, and flattening that made the disk of the solar nebula should also have affected the gas drawn by gravity to the young jovian planets. Thus, each jovian planet probably was surrounded by its own disk of gas, spinning in the same direction that the planet rotated (Figure 6.23). Moons accreted from icy planetesimals within these disks, which is why these moons tend to orbit in the same direction as their planet rotates, with nearly circular orbits lying close to the equatorial plane of their parent planet.

This model of accretion followed by gas capture explains the observed features of the jovian planets quite well. Nevertheless, some scientists have recently advanced a competing model in which disturbances (technically, instabilities) within the disk of the solar nebula could have led clumps of gas to collapse and form jovian planets (without first forming an icy planetesimal). Because this model is relatively new, scientists have not yet worked out enough details to know whether it can explain all the characteristics of the jovian planets. Most planetary scientists therefore still favor the accretion and gas capture model, but good practice of science demands that we remain open to considering new ideas.

Clearing the Nebula The vast majority of the hydrogen and helium gas in the solar nebula never became part of any planet. What became of it? Apparently, it was swept into interstellar space by the *solar wind,* which consists of charged particles (such as protons and electrons) that are continually blown off the surface of the Sun in all directions. Although the solar wind is fairly weak today, observations of other stars show that winds tend to be much stronger in young stars. Thus, the young Sun should have had a strong solar wind—strong enough to have swept huge quantities of gas out of the solar system.

Remaining gas in the solar nebula was swept into space, ending the era of planet formation.

The clearing of the gas sealed the compositional fate of the planets. If the gas had not been cleared soon after the planets formed, it might have continued to cool until hydrogen compounds could have condensed into ices even in the inner solar system. In that case, the terrestrial planets might have accreted abundant ice, and perhaps hydrogen and helium gas as well, changing their basic nature. At the other extreme, if the gas had been blown out too early, the raw materials of the planets might have been swept away before the planets could fully form. Although these extreme scenarios did not occur in our solar system, they may sometimes happen around other stars.

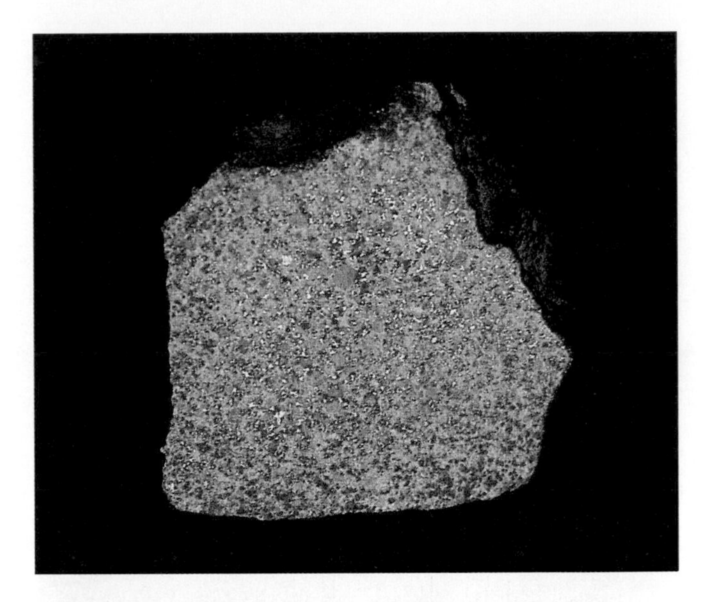

Figure 6.22

Shiny flakes of metal are clearly visible in this slice through a meteorite (a few centimeters across), mixed in among the rocky material. Such metallic flakes are just what we would expect to find if condensation really occurred in the solar nebula as described by the nebular theory.

• Where did asteroids and comets come from?

The process of planet formation also explains the origin of the many asteroids and comets that populate our solar system: They are "leftovers" from the era of planet formation. Asteroids are the rocky leftover planetesimals of the inner solar system, while comets are the icy leftover planetesimals of the outer solar system. We'll see in Chapter 9 why most asteroids ended up grouped in the asteroid belt while most comets ended up split between two regions (the Kuiper belt and the Oort cloud).

Rocky asteroids and icy comets are leftover planetesimals from the era of planet formation.

Evidence that asteroids and comets really are leftover planetesimals comes from analysis of meteorites, spacecraft visits to comets and asteroids, and computer simulations of solar system formation. In fact, the nebular theory allowed scientists to make predictions about the locations of comets that weren't verified until decades later, when we discovered large comets orbiting in the vicinity of Neptune and Pluto.

• How do we explain the existence of our Moon and other "exceptions to the rules"?

We have now explained all the major features of our solar system except for the "exceptions to the rules," such as our surprisingly large Moon and the odd rotation of Venus and Uranus. Today, we think that most of these exceptions arose from collisions or close encounters with leftover planetesimals.

The Heavy Bombardment The asteroids and comets that exist today probably represent only a small fraction of the vast numbers of leftover

Figure 6.23

The young jovian planets were surrounded by disks of gas, much like the disk of the entire solar nebula but smaller in size. According to the leading model, the planets grew as large icy planetesimals captured hydrogen and helium gas from the solar nebula. This painting shows the gas and planetesimals surrounding one jovian planet in the larger solar nebula.

planetesimals that roamed the young solar system. The rest are now gone. Some of these "lost" planetesimals may have been flung into deep space, but many others must have collided with the planets. The vast majority of these collisions occurred in the first few hundred million years of our solar system's history, during the period we call the **heavy bombardment.**

Leftover planetesimals battered the planets during the solar system's first few hundred million years.

Every world in our solar system must have been pelted by impacts during the heavy bombardment (Figure 6.24). Some of these impacts left scars that we can still see today as impact craters. These impacts did more than just batter the planets. They also brought materials from other regions of the solar system—a fact that is critical to our existence on Earth today.

Figure 6.24

Around 4 billion years ago, Earth, its Moon, and the other planets were heavily bombarded by leftover planetesimals. This painting shows the young Earth and Moon, with an impact in progress on Earth.

Remember that the terrestrial planets were built from planetesimals made of metal and rock. These planetesimals probably contained no water or other hydrogen compounds at all, because it was too hot for these compounds to condense in our region of the solar nebula. How, then, did Earth come to have the water that makes up our oceans? The answer is that water, along with other hydrogen compounds, must have been brought to Earth and the other terrestrial planets by the impact of planetesimals that formed farther from the Sun. We don't yet know whether these planetesimals came primarily from the outer asteroid belt, where rocky planetesimals contained small amounts of water and other hydrogen compounds, or whether they were comets containing huge amounts of ice. In either case, the water we drink and the air we breathe probably once were part of planetesimals floating beyond the orbit of Mars.

Captured Moons We can easily explain the orbits of most large jovian planet moons by their formation in a disk that swirled around the forming planet. However, some moons have unusual orbits—orbits in the "wrong" direction (opposite the rotation of their planet) or with large inclinations to the planet's equator. These unusual moons are probably leftover planetesimals that were captured into orbit around a planet.

It's not easy for a planet to capture a moon. An object cannot switch from an unbound orbit (for example, an asteroid whizzing by Jupiter) to a bound orbit (for example, a moon orbiting Jupiter) unless it somehow loses orbital energy [Section 4.4]. For the jovian planets, captures could have occurred when passing planetesimals were slowed by friction with the dense gas that surrounded these planets as they formed. Computer models suggest that this capture process would have worked only on objects of a few kilometers in size. Most of the small moons of the jovian planets are a few kilometers across, supporting the idea that they were captured in this way. Mars may have captured its two small moons, Phobos and Deimos, in a similar way at a time when the planet had a much more extended atmosphere than it does today (Figure 6.25).

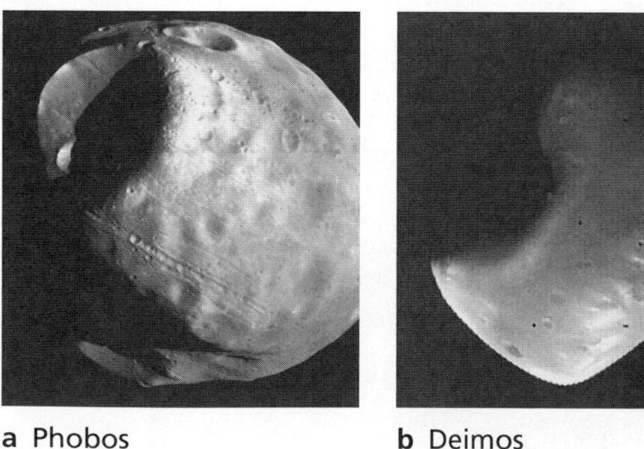

a Phobos   b Deimos

Figure 6.25

The two moons of Mars, shown here in photos taken by the Viking spacecraft, are probably captured asteroids. Phobos is only about 13 km across and Deimos is only about 8 km across—making each of these two moons small enough to fit within the boundaries of a typical large city.

The only exceptional moon that cannot be explained by a capture process is our own. Our Moon is much too large to have been captured by a small planet like Earth. We can also rule out the possibility that our Moon formed simultaneously with Earth. If the Moon and Earth had formed together, both should have accreted from planetesimals of the same type and should therefore have approximately the same composition and density. That is not the case. The Moon's density is considerably lower than Earth's, indicating that it has a very different average composition and could not have formed in the same way or at the same time as our planet. We must look for another way to explain the origin of our surprisingly large Moon.

Giant Impacts and the Formation of Our Moon The largest leftover planetesimals may have been huge, perhaps the size of Mars. When one of these planet-size planetesimals collided with a planet, the spectacle would have been awesome. Such a **giant impact** could have significantly altered a planet's fate.

What would have happened if a Mars-size object had collided with the young Earth? Depending on exactly where and how fast the object struck Earth, the blow might have tilted Earth's axis, changed its rotation rate, or completely shattered our planet. The most interesting case arises when we consider what would have happened if the impact had blasted away rock from Earth's outer layers (mantle and crust) and sent this material into orbit around our planet. According to computer simulations, this orbiting material could have reaccreted to form our Moon (Figure 6.26).

Our Moon is probably the result of a giant impact that blasted Earth's outer layers into orbit, where the material reaccreted to form the Moon.

Today, such a giant impact is the leading hypothesis for explaining the origin of our Moon. Strong support for this hypothesis comes from two features of the Moon's composition. First, the Moon's overall composition is quite similar to that of Earth's outer layers—just as we should expect if it were made from material blasted away from those layers. Second, the Moon has a much smaller proportion of easily vaporized ingredients (such as water) than Earth. This fact supports the hypothesis because these ingredients would have been vaporized by the heat of the impact. As gases, they would not have participated in the process of reaccretion that formed the Moon.

Giant impacts may also explain many of the other exceptions to the general trends. For example, a giant impact may have contributed to the slow, backward rotation of Venus, which may have had a "normal" rotation before the impact occurred. Giant impacts may also have been responsible for tilting the axes of many planets (including Earth) and perhaps for tipping Uranus on its side. Pluto's moon Charon may have formed in a giant impact similar to the one that formed our Moon. Mercury's surprisingly high density may be the result of a giant impact that blasted away its rocky outer layers, leaving behind a planet made almost entirely of the metal that was once in its core.

We've now seen that the nebular theory accounts for all the major features of our solar system. Figure 6.27 (see p. 158) summarizes the way in which the nebular theory says that our solar system formed.

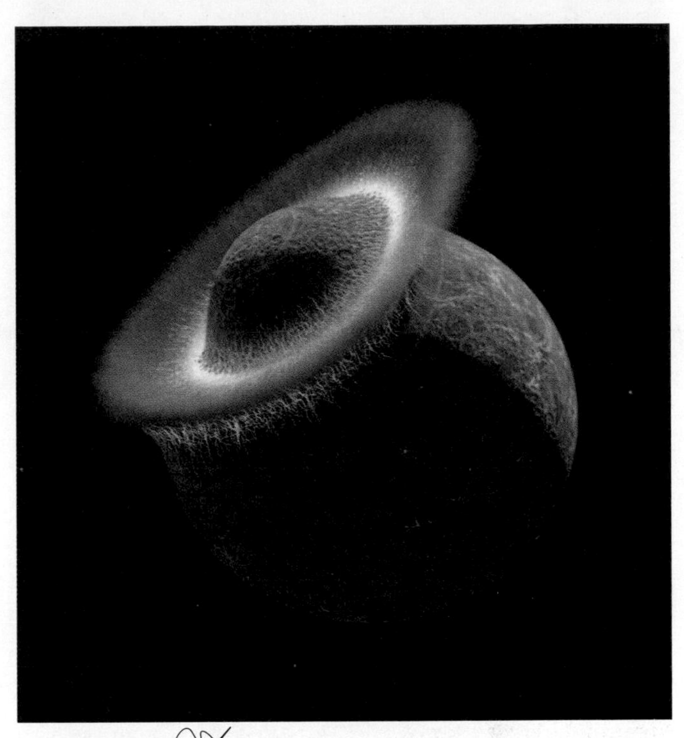

Figure 6.26

Artist's conception of the impact of a Mars-size object with Earth, as may have occurred soon after Earth's formation. The ejected material comes mostly from the outer rocky layers and accretes to form the Moon, explaining why the Moon is poor in metal.

• When did the planets form?

Computer models of planetary formation suggest that the entire process took no more than a few tens of millions of years, and perhaps significantly less. But when did it all occur, and how do we know? The answer is that the planets began to form through accretion about 4.6 billion years ago, a fact we learn by determining the age of the oldest rocks in the solar system.

Dating Rocks The method by which we measure the age of a rock is known as **radiometric dating**. This method relies on careful measurement of the proportions of various atoms and isotopes in the rock.

Remember that each chemical element is uniquely characterized by the number of protons in its nucleus. Different *isotopes* of the same element differ only in their number of neutrons [Section 5.1]. A **radioactive** isotope has a nucleus prone to spontaneous change, or *decay*, such as breaking apart or having one of its protons turn into a neutron. Decay can

Figure 6.27

A summary of the process by which our solar system formed, according to the nebular theory.

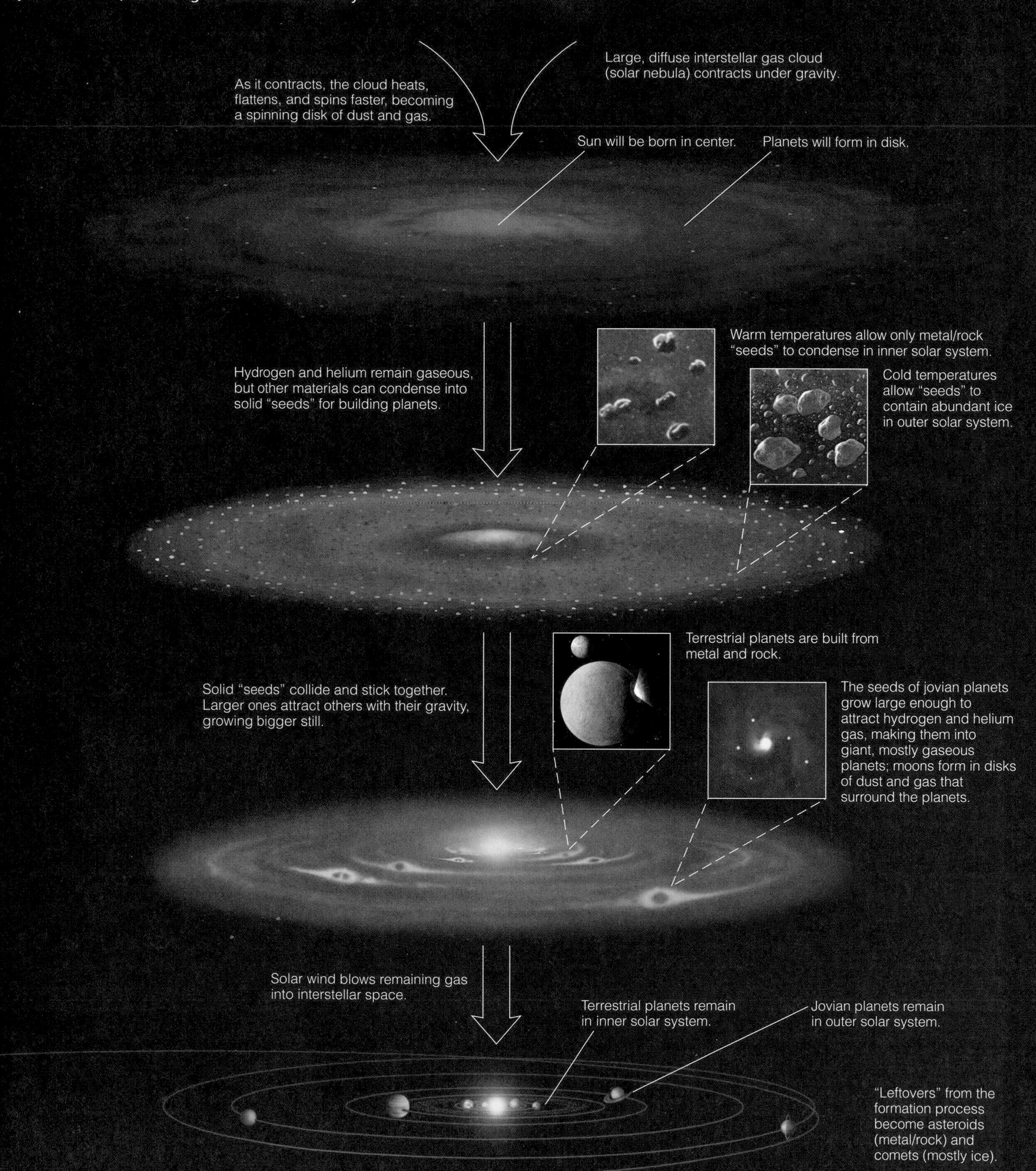

change one element into an entirely different one, with different chemical properties. For example, potassium-40 is a radioactive isotope with nuclei that decay when a proton turns into a neutron, changing the potassium-40 into argon-40. The rate at which a large collection of potassium-40 atoms decays into argon is always the same. We characterize this rate by a **half-life**—the length of time it would take for half the potassium-40 nuclei in the collection to decay. Laboratory measurements show the half-life of potassium-40 to be 1.25 billion years.

Because we can measure decay rates so precisely, in principle it is easy to determine the age of a rock. Consider a small piece of rock that contained 1 microgram of potassium-40 and no argon-40 when it formed (solidified) long ago. Because potassium-40 has a half-life of 1.25 billion years, half the original potassium-40 would have decayed into argon-40 by the time the rock was 1.25 billion years old. Thus, after 1.25 billion years, the rock contained 1/2 microgram of potassium-40 and 1/2 microgram of argon-40. Half again would have decayed by the end of the next 1.25 billion years, so after 2.5 billion years the rock contained 1/4 microgram of potassium-40 and 3/4 microgram of argon-40. After three half-lives, or 3.75 billion years, only 1/8 microgram of potassium-40 remained, while 7/8 microgram had become argon-40. Figure 6.28 summarizes the gradual decrease in the amount of potassium-40 and the corresponding rise in the amount of argon-40.

We can determine the age of a rock through careful analysis of the proportions of various atoms and isotopes within it.

We can now see the essence of radiometric dating. Suppose you find a rock that contains equal numbers of atoms of potassium-40 and argon-40. If you *assume* that all the argon came from potassium decay, then it must have taken precisely one half-life for the rock to end up with equal amounts of the two isotopes. You could therefore conclude that the rock is 1.25 billion years old (see Figure 6.28). The only question is whether you are right in assuming that the rock lacked argon-40 when it formed. In this case, knowing a bit of "rock chemistry" helps. Potassium-40 is a natural ingredient of many minerals in rocks, but argon-40 is a gas that never combines with other elements and did not condense in the solar nebula. Therefore, if you find argon-40 gas trapped inside minerals, you can be sure that it came from radioactive decay of potassium-40.

Radiometric dating is possible with many other radioactive isotopes as well, because each isotope has its own unique half-life. For example, we have been able to date many old rocks from Earth and the Moon by analyzing their proportions of uranium and lead, because the most common radioactive isotope of uranium decays (in several steps) into lead with a half-life of about 4.5 billion years. In many cases, we can date a rock that contains more than one radioactive isotope, so agreement between the ages calculated from the different isotopes gives us confidence that we have dated the rock correctly.

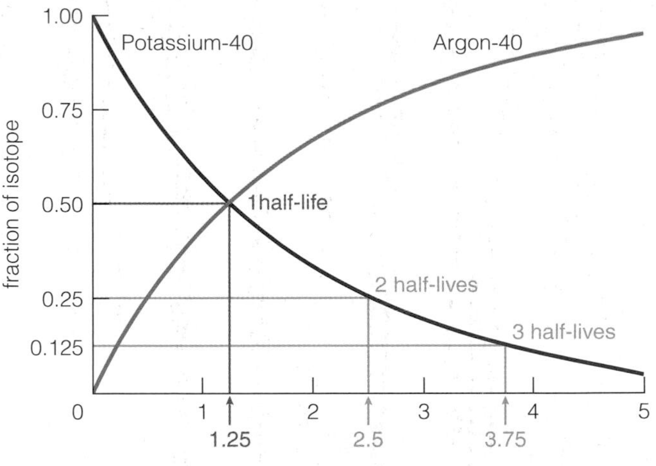

Figure 6.28

Potassium-40 is radioactive, decaying into argon-40 with a half-life of 1.25 billion years. The red line shows the decreasing amount of potassium-40, and the blue line shows the increasing amount of argon-40. The remaining amount of potassium-40 drops in half with each successive half-life.

Earth Rocks, Moon Rocks, and Meteorites Radiometric dating tells us only how long it has been since a rock solidified. Thus, we find rocks of many different ages on Earth. Some rocks are quite young because they formed recently from molten lava; others are much older. The oldest Earth rocks are about 4 billion years old, but even these are not as old as Earth itself, because Earth's entire surface has been reshaped through time.

Moon rocks brought back by the Apollo astronauts date to as far back as 4.4 billion years ago. Although they are older than Earth rocks, these Moon rocks must still be younger than the Moon itself. The ages of these rocks also tell us that the giant impact thought to have created the Moon must have occurred more than 4.4 billion years ago.

Age dating of meteorites that are unchanged since they condensed and accreted tell us that the solar system is about 4.6 billion years old.

To go all the way back to the origin of the solar system, we must find rocks that have not melted or vaporized since they first condensed in the solar nebula. Meteorites that have fallen to Earth are our source of such rocks. Many meteorites appear to have remained unchanged since they condensed and accreted in the early solar system. Careful analysis of radioactive isotopes in meteorites shows that the oldest ones formed about 4.55 billion years ago, so this time must mark the beginning of accretion in the solar nebula. Rounding upward, we say that our solar system is about 4.6 billion years old.

Detecting Extrasolar Planets Tutorial, Lessons 1–3

6.5 OTHER PLANETARY SYSTEMS

Barely a decade ago, we had no conclusive proof that planets existed around any star besides our Sun. Today, we know of more than 100 planets orbiting other stars. The discovery of these planets represents a triumph of modern technology. It also allows us to test our theory of the solar system's birth in new settings. If the nebular theory is correct, it should be able to explain the observed properties of planets that orbit other stars.

• How do we detect planets around other stars?

You might think that the easiest way to discover **extrasolar planets,** or planets around other stars, would be to photograph them through powerful telescopes. Unfortunately, direct detection of extrasolar planets remains beyond our capabilities. To understand the difficulty, imagine trying to see planets like Earth or Jupiter orbiting a nearby star. As we discussed in Chapter 1, seeing an Earth-like planet orbiting the nearest star besides the Sun would be like looking from San Francisco to see a pinhead orbiting just 15 meters from a grapefruit in Washington, D.C. Seeing a Jupiter-like planet would be like trying to see a marble about 80 meters from the grapefruit-size star.

The scale alone would make the task quite challenging, but it is further complicated by the fact that a Sun-like star would be a *billion times* as bright as the light reflected from any planets. Because even the best telescopes blur the light from stars at least a little, the small blips of planetary light would be overwhelmed by the glare of scattered starlight. Astronomers are working on technologies that may overcome this problem, such as special interferometers [Section 5.3] that can block out the bright starlight and allow us to see the dim light of planets. Nevertheless, all discoveries of extrasolar planets to date have been made on the basis of indirect evidence—evidence acquired not by seeing the planets themselves but by studying the light of the stars they orbit.

Every extrasolar planet detected to date has been found indirectly, by observing the planet's effects on the star it orbits.

Most of the extrasolar planets discovered to date have been found by observing small Doppler shifts [Section 5.2] in the spectrum of the star they orbit (Figure 6.29). An orbiting planet exerts a small gravitational pull on its star. This pull causes

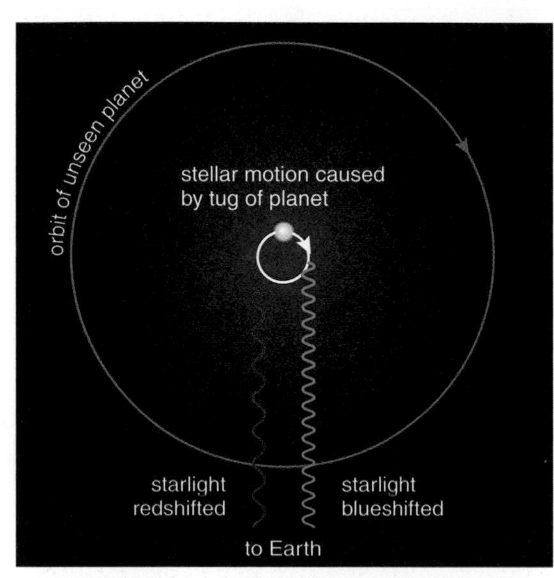

a Doppler shifts allow us to detect the slight motion of a star caused by an orbiting planet.

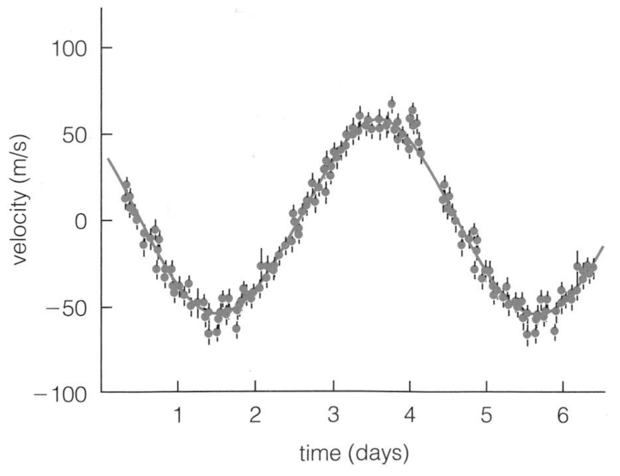

b A periodic Doppler shift in the spectrum of the star 51 Pegasi shows the presence of a large planet with an orbital period of about 4 days. Dots are actual data points; bars through dots represent measurement uncertainty.

Figure 6.29 Interactive Figure

The Doppler technique for discovering extrasolar planets.

the star to alternately move slightly toward and away from us, which makes the star's spectral lines shift alternately toward the blue and toward the red. Careful analysis of the star's motion can allow us to identify an orbiting planet, determine its orbit, and estimate its mass. In some cases, the Doppler measurements can even be used to identify more than one planet around the same star.

The Doppler technique has yielded many planet discoveries, but it has limitations. For example, it is best suited to identifying massive planets that orbit relatively close to their star, because these planets have the strongest gravitational effect on their star. With current technology it cannot be used to find Earth-size planets at all, because small planets have such a weak gravitational effect on their stars. Thus, the current lack of evidence for Earth-like planets may not necessarily mean that such planets are rare—it could be simply that they're hard to find.

Fortunately, other indirect techniques may soon allow us to discover Earth-size planets around other stars, assuming that they exist. One promising technique searches for tiny, repeated changes in a star's brightness. If an extrasolar planet happens to have an edge-on orbit as viewed from Earth, the planet will pass directly in front of its star with each orbit, creating a sort of mini-eclipse that astronomers call a **transit** (Figure 6.30). The amount of star light blocked during the transit allows us to calculate the diameter of the transiting planet. The transit shown in Figure 6.30 provided the first direct proof that one of the planets discovered to date is indeed jovian in size as well as mass.

Only a tiny fraction of planets are expected to have orbits that would cause transits, but if we observe enough stars we should still find numerous planets in this way. NASA is currently developing a mission called Kepler, scheduled for launch in 2007, which will look for transits caused by Earth-size (or even smaller) planets. Kepler will carefully monitor the brightness of 100,000 stars over a period of 4 years. If small planets are common, Kepler should find hundreds of them.

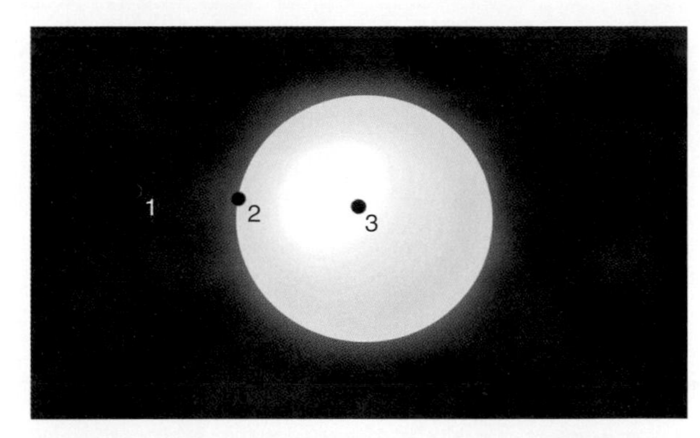

a Artist's conception of the planet as it passes directly in front of its star as seen from Earth. The numbered positions correspond to the same numbers in the data in part (b).

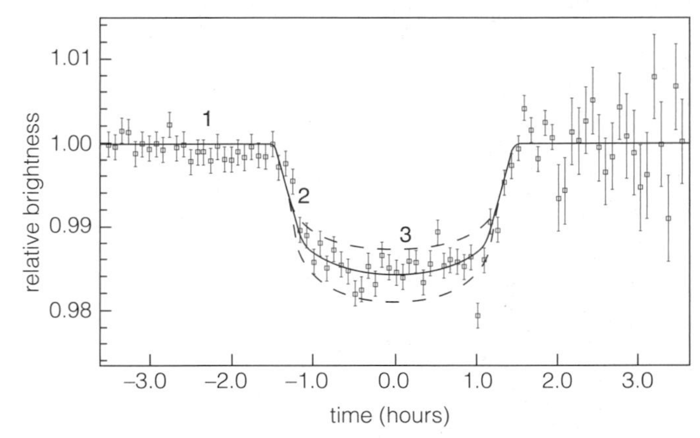

b These data show a small but measurable drop in the star's brightness during the transit. The transit repeats with every orbit of the planet, allowing us be sure it really is due to a planet and not something else.

Figure 6.30 Interactive Figure

A planet that orbits edge-on as viewed from Earth will pass directly in front of its star, creating a *transit* that causes the star's brightness to dip slightly as the planet passes in front of it. The data shown here are for a star called HD209458, which transits its star every $3\frac{1}{2}$ days. For about 2 hours, it causes a 1.7% drop in the star's brightness—which also tells us how the planet's diameter compares to the diameter of its star.

• What have other planetary systems taught us about our own?

The discovery of extrasolar planets presents us with an opportunity to test our theory of solar system formation. Can our existing theory explain other planetary systems, or will we have to go back to the drawing board? The discoveries have already presented at least one significant challenge.

In our solar system, the terrestrial planets orbit close to the Sun while the large jovian planets orbit much farther out. As we've seen, the nebular theory explains this fact by telling us that rocky planets formed in the warm inner regions of the solar nebula while gaseous planets formed in the cooler outer regions. But many of the recently discovered extrasolar planets don't seem to fit this picture: They are massive like the jovian planets, yet they orbit quite close to their star.

At first, this surprise caused scientists to revisit the entire basis of the nebular theory. However, their reexamination did not turn up any fundamental flaws in our basic understanding of the way planets form. As a result, most scientists now suspect that these close-in but massive planets originally formed far from their stars, just as the nebular theory predicts. They then *migrated* to their current orbits by somehow losing orbital energy during the time since they first formed.

How might such planetary migration occur? Calculations show that the abundant gas and dust in the disks of newly formed solar systems can exert a drag on young planets (through friction and/or gravitational attraction), causing them to spiral slowly toward their sun. In our own solar system, this drag is not thought to have played a significant role because the solar wind cleared out the gas before it could have much effect. But the wind may kick in later in at least some other solar systems, allowing time for jovian planets to migrate substantially inward. In some cases, the wind might kick in so late that the planets end up crashing into their stars. Other possibilities include the idea that waves propagating through a gaseous disk could lead to inward migration, or that a jovian planet could migrate inward as a result of multiple close encounters with much smaller planetesimals.

Studies of other planetary systems show that planets can migrate inward from their birthplaces.

The bottom line is that discoveries of extrasolar planets have shown us that our nebular theory of solar system formation was incomplete. It explained the simple layout of our solar system but it requires new features—such as planetary migration—to explain the differing layouts of other solar systems. A much wider range of solar system arrangements now seems possible than we had guessed before the discovery of extrasolar planets.

THINK ABOUT IT Look back at the discussion of the nature of science in Chapter 3, especially the definition of a scientific theory. Does our theory of solar system formation qualify as a scientific theory, even though we have recently learned that it needs modification? Does this mean the theory was "wrong" before the modifications were made? Explain.

THE BIG PICTURE

Putting Chapter 6 into Perspective

In this chapter, we've introduced the major features of our solar system and described the current scientific theory of its formation. We've seen how this theory explains the major features we observe and how it can be extended to other planetary systems. As you continue your study of the solar system, keep in mind the following "big picture" ideas:

- Our solar system is not a random collection of objects moving in random directions. Rather, it is highly organized, with clear patterns of motion and with most objects falling into simply defined categories.
- We can explain the major features of our solar system with a theory that holds that the solar system formed from the gravitational collapse of an interstellar cloud.
- Most of the general features of the solar system were determined by processes that occurred very early in the solar system's history, which began some 4.6 billion years ago.
- Planet-forming processes are apparently universal. Discoveries of planets around other stars have begun an exciting new era in planetary science.

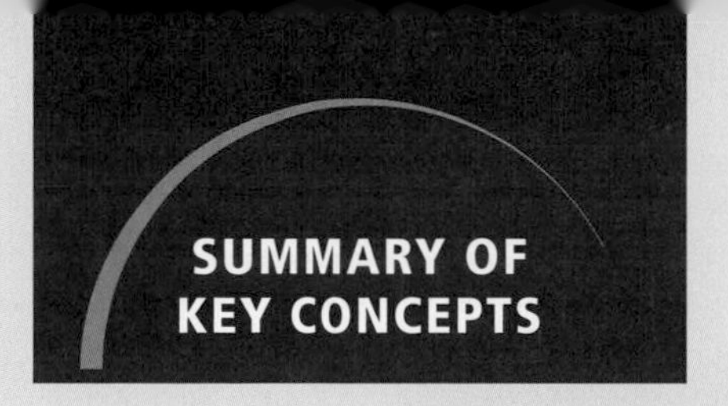

SUMMARY OF KEY CONCEPTS

6.1 A BRIEF TOUR OF THE SOLAR SYSTEM

- **What does the solar system look like?**

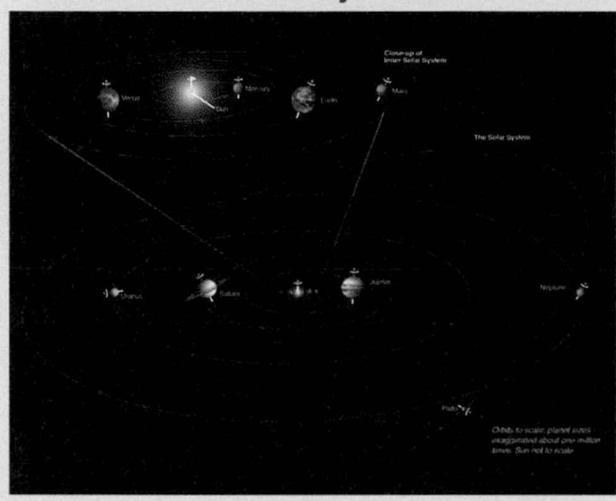

Our solar system consists of the Sun, nine planets and their moons, and vast numbers of asteroids and comets. Each world has its own unique character, but there are many clear patterns among the worlds.

6.2 CLUES TO THE FORMATION OF OUR SOLAR SYSTEM

- **What features of our solar system provide clues to how it formed?**

Four major features provide clues: (1) The Sun, planets, and large moons generally rotate and orbit in a very organized way. (2) With the exception of Pluto, the planets divide clearly into two groups: terrestrial and jovian. (3) The solar system contains huge numbers of asteroids and comets. (4) There are some notable exceptions to these general patterns.

- **What theory best explains the features of our solar system?**

The nebular theory, which holds that the solar system formed from the gravitational collapse of a great cloud of gas.

6.3 THE BIRTH OF THE SOLAR SYSTEM

- **Where did the solar system come from?**

The cloud of gas that gave birth to our solar system was the product of recycling of gas through many generation of stars within our galaxy. This gas consisted of 98% hydrogen and helium and 2% everything else combined.

- **What caused the orderly patterns of motion in our solar system?**

A collapsing gas cloud naturally tends to heat up, spin faster, and flatten out as it shrinks in size. Thus, our solar system began as a spinning disk of gas. The orderly motions we observe today all came from the orderly motion of this spinning disk of gas.

6.4 THE FORMATION OF PLANETS

- **Why are there two types of planets?**

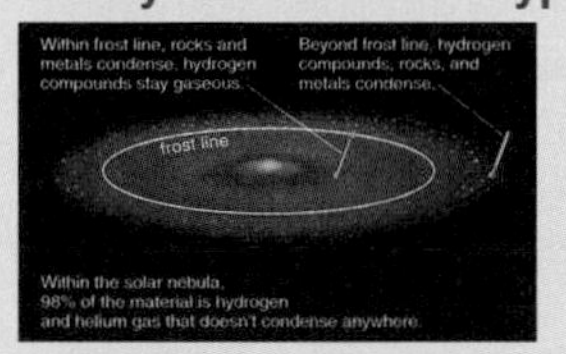

Planets formed around solid "seeds" that condensed from gas and then grew through accretion. In the inner solar system, temperatures were so high that only metal and rock could condense, which explains why terrestrial worlds are made of metal and rock. In the outer solar system, cold temperatures allowed more abundant ices to condense along with metal and rock. Icy planetesimals grew large enough for their gravity to draw in hydrogen and helium gas, building massive jovian planets.

- **Where did asteroids and comets come from?**

Asteroids are the rocky leftover planetesimals of the inner solar system, and comets are the icy leftover planetesimals of the outer solar system.

- **How do we explain the existence of our Moon and other "exceptions to the rules"?**

Most of the exceptions probably arose from collisions or close encounters with leftover planetesimals, especially during the heavy bombardment that occurred early in the solar system's history. Our Moon is probably the result of a giant impact between a Mars-size planetesimal and the young Earth.

- **When did the planets form?**

The planets began to accrete in the solar nebula about 4.6 billion years ago, a fact we determine from radiometric dating of the oldest meteorites.

6.5 OTHER PLANETARY SYSTEMS

- **How do we detect planets around other stars?**

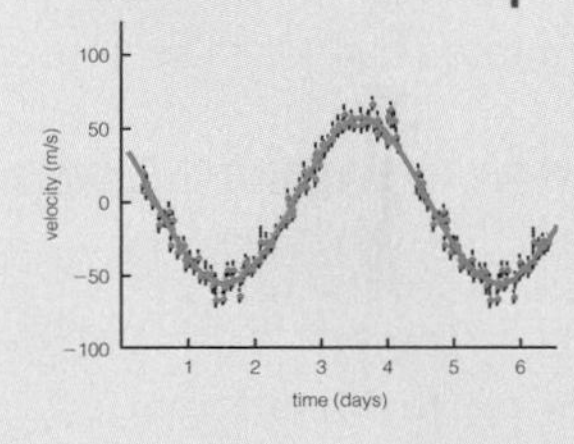

So far, we are only able to detect extrasolar planets indirectly by observing the planet's effects on the star it orbits. Most discoveries to date have been made with the Doppler technique, in which Doppler shifts reveal the gravitational tug of a planet (or more than one planet) on a star.

- **What have other planetary systems taught us about our own?**

Planetary systems exhibit a surprising range of layouts, suggesting that jovian planets sometimes migrate inward from where they are born. This lesson has taught us that despite the successes of the nebular theory, it remains incomplete.

EXERCISES AND PROBLEMS

REVIEW QUESTIONS

1. Briefly describe the layout of the solar system as it would appear from beyond the orbit of Pluto.
2. For the Sun and each of the nine planets in our solar system, briefly describe at least two features that you find interesting.
3. What are the four major features of our solar system that provide clues to how it formed? Describe each one briefly.
4. What is the *nebular theory,* and why is it widely accepted by scientists today?
5. What do we mean by the *solar nebula?* What was it made of, and where did it come from?
6. Describe each of the three key processes that led the solar nebula to take the form of a spinning disk. What observational evidence supports this scenario?
7. List the four categories of materials in the solar nebula by their condensation properties and abundance. Which ingredients are present in terrestrial planets? In jovian planets? Explain why.
8. What was the *frost line* in the solar nebula? Explain how temperature differences led to the formation of two distinct types of planets.
9. Briefly describe the process by which terrestrial planets are thought to have formed. How was the formation of jovian planets similar? How was it different? Why did the jovian planets end up with so many moons?
10. What is the *solar wind,* and what roles did it play in the early solar system?
11. What was the *heavy bombardment?* How were the objects that bombarded the young planets related to the asteroids and comets that we find in the solar system today?
12. How do we think the Moon formed, and why?
13. Briefly explain the technique of radiometric dating, and describe how we use it to determine the age of the solar system.
14. Have we ever actually seen a planet orbiting a star besides our Sun? Briefly summarize current techniques for detecting extrasolar planets.
15. How have discoveries of extrasolar planets forced scientists to modify the nebular theory of solar system formation?
16. *True or False.* Decide whether each statement is true or false, and explain why.
    a. On average, Venus has the hottest surface temperature of any planet in the solar system—even hotter than Mercury.
    b. Our Moon is about the same size as moons of the other terrestrial planets.
    c. The weather conditions on Mars today are much different than they were at some times in the distant past.
    d. Moons cannot have atmospheres, active volcanoes, or liquid water.
    e. Saturn is the only planet in the solar system with rings.
    f. If Pluto were as large as the planet Mercury, we would classify it as a terrestrial planet.
    g. Asteroids are made of essentially the same materials as the terrestrial planets.
    h. When scientists say that our solar system is about 4.6 billion years old, they are making a rough estimate based on guesswork about how long it should have taken planets to form.

SURPRISING DISCOVERIES?

Suppose we found a solar system with the property described. (These are *not* real discoveries.) In light of our theory of solar system formation, decide whether the discovery should be considered reasonable or surprising. Explain.

17. A solar system has five terrestrial planets in its inner solar system and three jovian planets in its outer solar system.
18. A solar system has four large jovian planets in its inner solar system and seven small planets made of rock and metal in its outer solar system.
19. A solar system has 10 planets that all orbit the star in approximately the same plane. However, 5 planets orbit in one direction (e.g., counterclockwise), while the other 5 orbit in the opposite direction (e.g., clockwise).
20. A solar system has 12 planets that all orbit the star in the same direction and in nearly the same plane. The 15 largest moons in this solar system orbit their planets in nearly the same direction and plane as well. However, several smaller moons have highly inclined orbits around their planets.
21. A solar system has six terrestrial planets and four jovian planets. Each of the six terrestrial planets has at least five moons, while the jovian planets have no moons at all.
22. A solar system has four Earth-size terrestrial planets. Each of the four planets has a single moon that is nearly identical in size to Earth's Moon.

PROBLEMS

23. *Planetary Tour.* Based on the brief planetary tour in this chapter, which planet besides Earth do you think is the most interesting, and why? Defend your opinion clearly in two or three paragraphs.
24. *Patterns of Motion.* In one or two paragraphs, explain why the existence of orderly patterns of motion in our solar system should suggest that the Sun and the planets all formed at one time from one cloud of gas, rather than as individual objects at different times.
25. *A Cold Solar Nebula.* Suppose the entire solar nebula had cooled to 50 K before the solar wind cleared it away. How would the composition and sizes of the planets of the inner solar system be different from what we see today? Explain your answer in a few sentences.
26. *No Gas Capture.* Suppose the solar wind had cleared away the solar nebula before the seeds of the jovian planets could gravitationally draw in hydrogen and helium gas. How would the planets of the outer solar system be different? Would they still have many moons? Explain your answer in a few sentences.
27. *Angular Momentum.* Suppose our solar nebula had begun with much more angular momentum than it did. Do you think planets could still have formed? Why or why not? What if the solar

nebula had started with zero angular momentum? Explain your answers in one or two paragraphs.

28. *Solar System Trends.* Study the planetary data in Table 6.1 to answer each of the following.
    a. Notice the relationship between distance from the Sun and surface temperature. Describe the trend, explain why it exists, and explain any notable exceptions to the trend.
    b. The text says that planets can be classified as either terrestrial or jovian, with Pluto as a misfit. Describe in general how the columns for density, composition, and distance from the Sun support this classification.
    c. Which column of data would you use to find out which planet has the shortest days? Do you see any notable differences in the length of a day for the different types of planets? Explain.
    d. Describe the trend you see in orbital periods and explain the trend in terms of Kepler's third law.
    e. Which planets would you expect to have seasons? Why?

*29. *Radiometric Dating.* You are analyzing rocks that contain small amounts of potassium-40 and argon-40. The half-life of potassium-40 is 1.25 billion years.
    a. You find a rock that contains equal amounts of potassium-40 and argon-40. How old is it? Explain.
    b. You find a rock that contains three times as much argon-40 as potassium-40. How old is it? Explain.

DISCUSSION QUESTIONS

30. *Planetary Priorities.* Suppose you were in charge of developing and prioritizing future planetary missions for NASA. What would you choose as your first priority for a new mission, and why?
31. *Theory and Observation.* Discuss the interplay between theory and observation that has led to our modern theory of solar system formation. What role does technology play in allowing us to test this theory?
32. *Lucky to Be Here?* Considering the overall process of solar system formation, do you think it was likely for a planet like Earth to have formed? Could random events in the early history of the solar system have prevented our being here today? What implications do your answers have for the possibility of Earth-like planets around other stars? Defend your opinions.

MATH HELP AND EXERCISES

For additional help and practice with mathematical concepts applicable to this chapter, the Astronomy Place web site has the following resources with detailed explanations, worked examples and practice problems:

- The Mathematics of Radioactive Decay

For a complete list of media resources available, go to **www.astronomyplace.com** and choose Chapter 6 from the pull-down menu.

ASTRONOMY PLACE WEB TUTORIALS

Tutorial Review of Key Concepts

Use the following interactive **Tutorials** at **www.astronomyplace.com** to review key concepts from this chapter.

Scale of the Universe Tutorial

Lesson 1 Distances of Scale: Our Solar System

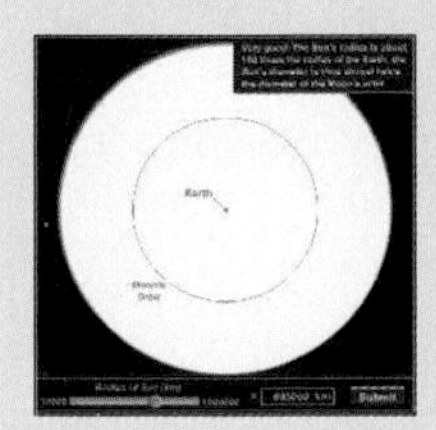
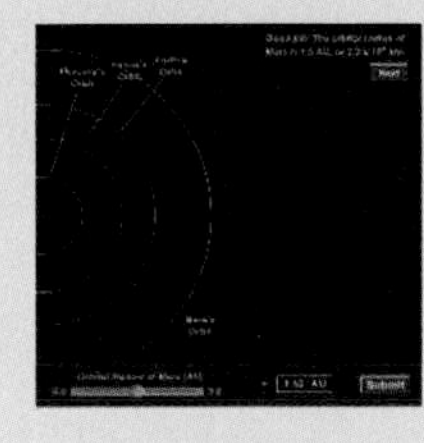

Formation of the Solar System Tutorial

Lesson 1 Comparative Planetology
Lesson 2 Formation of the Protoplanetary Disk
Lesson 3 Formation of Planets

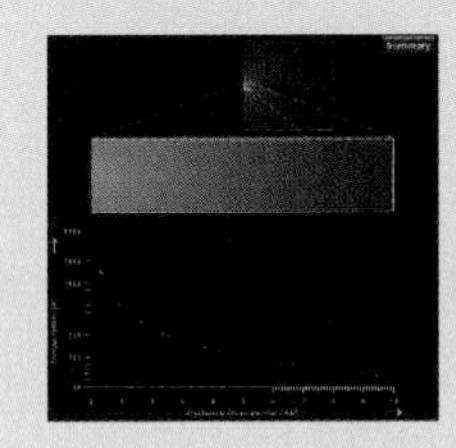
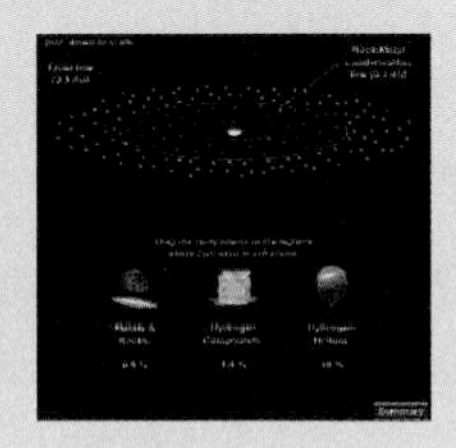
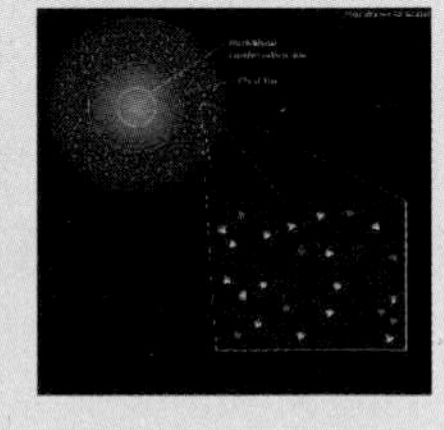

Detecting Extrasolar Planets Tutorial

Lesson 1 Taking a Picture of a Planet
Lesson 2 Stars' Wobbles and Properties of Planets
Lesson 3 Planetary Transits

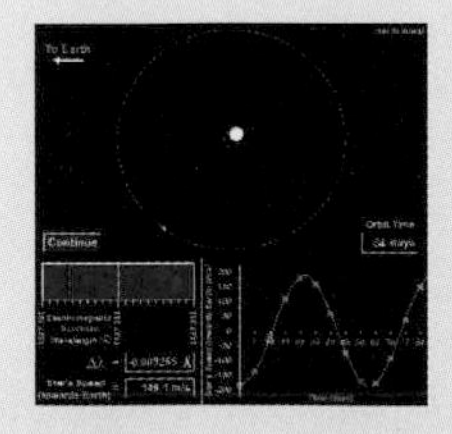
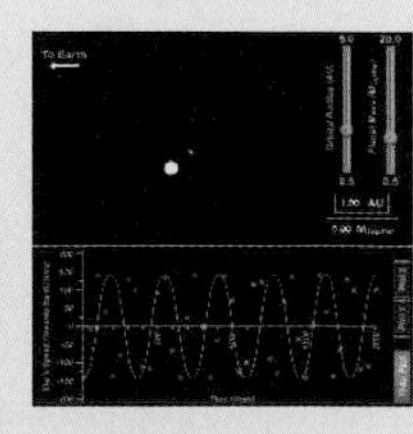
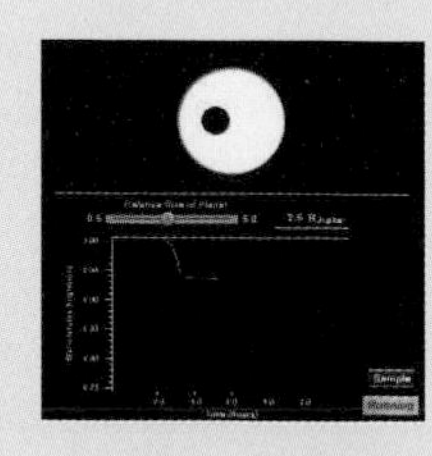

Supplementary Tutorial Exercises

Use the interactive **Tutorial Lessons** to explore the following questions.

Formation of the Solar System Tutorial, Lesson 1

1. What features distinguish the terrestrial planets from the jovian planets?
2. What is the difference between axis tilt, orbital eccentricity, and orbital inclination? What patterns do we see in these properties among the planets?
3. Which solar system bodies constitute "exceptions to the rules," and in what ways?

Formation of the Solar System Tutorial, Lesson 2

1. Study the animations of the formation of the solar nebula. Why does the nebula flatten as it collapses?
2. Use the tool in the tutorial to make a graph of temperatures in the solar nebula. How does the temperature vary with the distance from the center?
3. Why didn't any planets form within 0.3 AU of the Sun?

Detecting Extrasolar Planets Tutorial, Lesson 2

1. Using the tool provided, explain how weekly measurements allow us to determine the orbital period of the extrasolar planet.
2. Use the tool to vary the mass of the planet. How does its mass affect the Doppler shifts in its star's light?
3. Use the tool to vary the orbital radius of the planet. How does the orbital radius affect the Doppler shifts in its star's light?

Detecting Extrasolar Planets Tutorial, Lesson 3

1. Under what conditions can we view a planetary transit of another star?
2. How does the change in brightness during a transit depend on the planet's properties?

EXPLORING THE SKY AND SOLAR SYSTEM

Of the many activities available on the ***Voyager: SkyGazer* CD-ROM** accompanying your book, use the following files to observe key phenomena covered in this chapter.

Go to the **File: Basics** folder for the following demonstrations.

1. Saturn's Phases
2. Tracking Venus
3. Planet Panel

Go to the **File: Demo** folder for the following demonstrations.

1. Earth and Venus
2. Trailing Saturn
3. Triple Conjunction of 7 BC

Go to the **Explore** menu for the following demonstrations.

1. Solar System
2. Paths of Planets

MOVIES

Check out the following narrated and animated short documentaries available on **www.astronomyplace.com** for a helpful review of key ideas covered in this chapter.

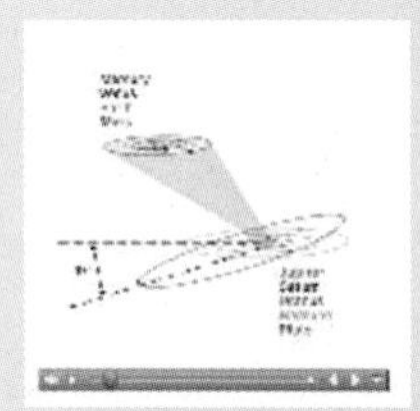

Orbits in the Solar System

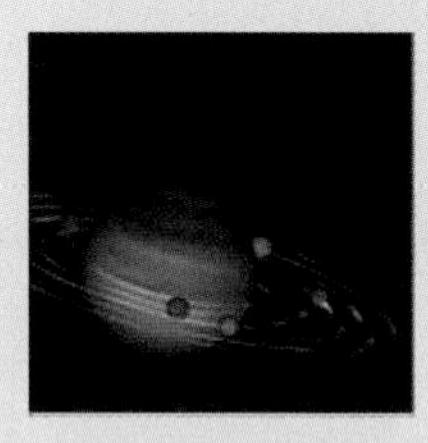

History of the Solar System

WEB PROJECTS

Take advantage of the useful Web links on **www.astronomyplace.com** to assist you with the following projects.

1. *Current Planetary Mission.* Find out what missions are currently underway to the planets of our own solar system. Visit the Web page for one of the current missions. Write a one- to two-page summary of the mission's basic design, goals, and current status.
2. *The Kepler Mission.* The Kepler mission is designed expressly to look for Earth-size planets around other stars. Go to the Kepler Web site and learn more about the mission. Write a one- to two-page summary of the mission's goals and its current status.
3. *Planet-Hunting Interferometers.* Other future missions will use interferometry to learn about extrasolar planets. Go to the Web site for one future interferometry mission under consideration, such as SIM, TPF, or Darwin. For the mission you choose, write a one- to two-page summary of the mission's goals and its current status.

7
Earth and the Terrestrial Worlds

LEARNING GOALS

7.1 EARTH AS A PLANET
- Why is Earth geologically active?
- What processes shape Earth's surface?
- How does Earth's atmosphere affect the planet?

7.2 MERCURY AND THE MOON: GEOLOGICALLY DEAD
- Was there ever geological activity on the Moon or Mercury?

7.3 MARS: A VICTIM OF PLANETARY FREEZE-DRYING
- What geological features tell us that water once flowed on Mars?
- Why did Mars change?

7.4 VENUS: A HOTHOUSE WORLD
- Is Venus geologically active?
- Why is Venus so hot?

7.5 EARTH AS A LIVING PLANET
- What unique features of Earth are important for life?
- How might human activity change our planet?
- What makes a planet habitable?

It's easy to take for granted the qualities that make Earth so suitable for human life: a temperature neither boiling nor freezing, abundant water, a protective atmosphere, and a relatively stable environment. But we need look only as far as our neighboring terrestrial worlds to see how fortunate we are. As we learned in Chapter 6, the Moon, likely made of debris from Earth, is airless and barren. Mercury is much the same. Venus is a searing hothouse, while Mars has an atmosphere so thin and cold that liquid water cannot last on its surface today.

How did the terrestrial worlds come to be so different, when all were made of essentially the same ingredients that had condensed in the solar nebula? Why did Earth alone develop the conditions that permit abundant life? We'll seek answers to these questions in this chapter. We'll begin by exploring the key processes that have shaped Earth and the other terrestrial worlds through time, and then consider the history of each world individually. Our studies will show that the histories of the worlds are not random accidents, but consequences of properties endowed at their births. Moreover, once we understand what has happened on other worlds, we'll be ready to return to Earth at the end of the chapter, seeing it in an entirely different way than we could have before the era of planetary exploration.

Mercury is heavily cratered, but also has long, steep cliffs—one is visible here as the long curve that passes through the center of the image.

The central structure is a tall, twin-peaked volcano on Venus. Both images are based on radar data from the Magellan space craft, because Venus's thick clouds prevent us from seeing the surface in visible light.

Earth has a variety of geological features visible in this photo from orbit.

Formation of the Solar System Tutorial, Lesson 1

7.1 EARTH AS A PLANET

Earth's surface seems solid and steady, but every so often it offers us a reminder that nothing about it is permanent. If you live in Alaska or California, you've probably felt the ground shift beneath you in an earthquake. In Washington State, you can see the scars of the 1980 eruptions of Mount St. Helens. In Hawaii, a visit to the still-active Kilauea volcano will remind you that you are standing on mountains of volcanic rock protruding from the ocean floor.

Volcanoes and earthquakes are not the only processes acting to reshape Earth's surface. They are not even the most dramatic: Far greater change can occur on the rare occasions when an asteroid or a comet slams into Earth. Much more gradual processes can also have spectacular effects. The Colorado River causes only small changes in the landscape from year to year, but its unrelenting flow over the past few million years carved the Grand Canyon. The Rocky Mountains were once twice as tall as they are today, having been cut down in size through tens of millions of years of erosion by wind, rain, and ice. Entire continents even move slowly about, completely rearranging the map of Earth every few hundred million years.

Earth is not alone in having undergone tremendous change since its birth. The surfaces of all five terrestrial worlds—Mercury, Venus, Earth, the Moon, and Mars—must have looked quite similar when they were young. All five were made of rocky material that condensed in the solar nebula, and all five were subjected early on to the impacts of the heavy bombardment [Section 6.4]. The great differences in their present-day appearance must therefore be the result of changes that have occurred through time. Ultimately, these changes must be traceable to fundamental properties of the planets.

Figure 7.1 shows global views of the terrestrial surfaces to scale, along with sample close-up views from orbit. Profound differences between these worlds are immediately obvious. Mercury and the Moon show the

Earth's Moon

The Moon's surface is heavily cratered in most places.

Mars has impact craters like the one near the upper right, but it also has features that look much like dried up riverbeds.

Figure 7.1

Global views to scale, along with sample close-ups viewed from orbit, of each of the five terrestrial worlds. All the photos were taken with visible light, except the Venus photos, which are based on radar data.

scars of their battering during the heavy bombardment: They are densely covered by craters except in areas that appear to be volcanic plains. Bizarre bulges and odd volcanoes dot the surface of Venus. Mars, despite its middling size, has the solar system's largest volcanoes and a huge canyon cutting across its surface, along with numerous features that appear to have been shaped by running water. Earth has surface features similar to all those on the other terrestrial worlds, and more—including a unique layer of living organisms that covers almost the entire surface of the planet.

Our primary goal in this chapter is to gain a deeper understanding of our own planet Earth by investigating how the terrestrial worlds came to be so different. We'll begin by examining the basic nature of our planet.

• Why is Earth geologically active?

All the terrestrial worlds have changed since their birth, but Earth is unique in the degree to which it continues to change today. We say that Earth is *geologically active,* meaning that its surface is continually being reshaped by volcanic eruptions, earthquakes, erosion, and other geological processes. Most of this geological activity is the result of what goes on deep inside our planet. Thus, to understand why Earth is so much more geologically active than other worlds, we must examine what the terrestrial worlds are like inside.

Interior Structure Studies of internal structure (see box, p. 173) show that all the terrestrial worlds have layered interiors. You are probably familiar with the idea of dividing these layers by density into three major categories:

- **Core**: The highest-density material, consisting primarily of metals such as nickel and iron, resides in the central core.
- **Mantle**: Rocky material of moderate density—mostly minerals that contain silicon, oxygen, and other elements—forms the thick mantle that surrounds the core.
- **Crust**: The lowest-density rock, such as granite and basalt (a common form of volcanic rock), forms the thin crust, essentially representing the world's outer skin.

Figure 7.2 shows these layers for the five terrestrial worlds. Although not shown in the figure, Earth's metallic core actually consists of two distinct regions: a solid *inner core* and a molten (liquid) *outer core.*

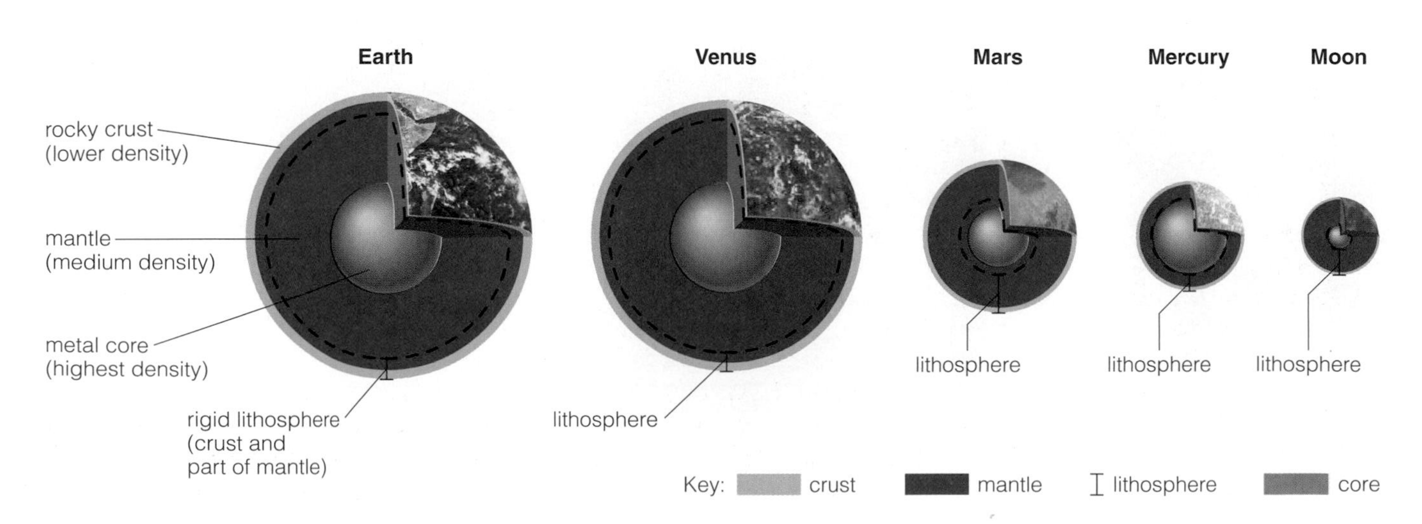

Figure 7.2

Interior structures of the terrestrial worlds, shown to scale and in order of decreasing size. The color-coded regions show the core-mantle-crust layering, which is based on density. Dashed lines indicate the lithosphere, which encompasses the crust and part of the mantle on each world and is defined by the strength of the rock rather than the density. (The thicknesses of the crust and lithosphere on Venus and Earth are exaggerated to make them visible in this figure.)

In geology, it's often more useful to categorize interior layers by rock strength rather than by density. The idea that rock can vary in strength may sound strange, since we often think of rock as the very definition of strength. However, remember that rocks feel solid only because of the electrical bonds between their molecules [Section 5.1]. Although these bonds are strong, they can still break and re-form when subjected to sustained stress. Thus, over millions and billions of years, rocky material can slowly deform and flow. The long-term behavior of rock is much like that of the popular toy Silly Putty, which breaks like a brittle solid when you pull it sharply but deforms and stretches when you pull it slowly (Figure 7.3). Also like Silly Putty, rock becomes softer and easier to deform when it is warmer.

The coolest and most rigid layer of rock near a planet's surface is called the *lithosphere.*

In terms of rock strength, Earth's outer layer consists of relatively cool and rigid rock, called the **lithosphere** (*lithos* is Greek for "stone"), that "floats" on warmer, softer rock beneath. The lithosphere encompasses the crust and part of the upper mantle on Earth, and extends deeper into the mantle on smaller worlds.

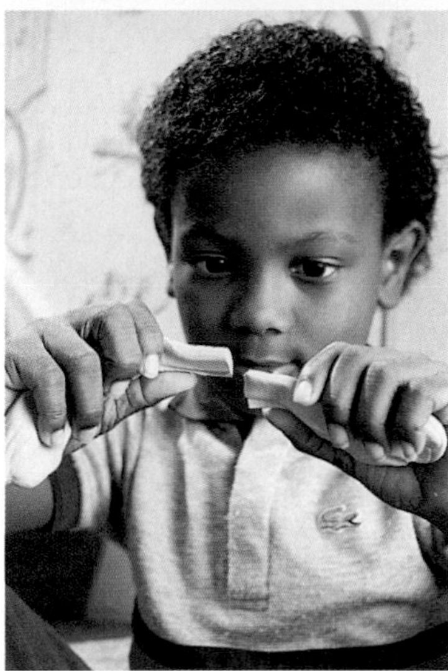

Figure 7.3

Silly Putty stretches when pulled slowly but breaks cleanly when pulled rapidly. Rock behaves just the same, but on a longer time scale.

Differentiation and Internal Heat

The distinct layering of the terrestrial worlds occurs for the same reason that liquids separate by density. For example, in a mixture of oil and water, gravity pulls the denser water to the bottom, driving the less dense oil to the top. The process by which gravity separates materials by density is called **differentiation** (because it results in layers made of *different* materials). The layered interiors of the terrestrial worlds tell us that all the worlds underwent differentiation at some time in the past, which means all these worlds must once have been hot enough inside for their interior rock and metal to melt and separate by density. Dense metals like iron sank toward the center, driving less dense rocky material toward the surface.

When they were young, Earth and the other terrestrial worlds were hot inside for two major reasons. First, the planets gained heat from the process of formation itself. During the later stages of accretion, incoming planetesimals collided at high speed with the forming planets, depositing large amounts of energy that turned into heat. The process of differentiation added further heat, released as heavy materials sank to the core. Second, the metal and rock that made up the terrestrial planets included small but important amounts of radioactive elements. As these radioactive materials decay, they release heat directly into the planetary interiors. Radioactive decay still supplies heat to the terrestrial interiors, though at a lower level than it did when the planets were young because some of the radioactive material has already decayed.

Earth and the other terrestrial worlds were once hot enough inside for their interiors to melt, allowing material to settle into layers of differing density.

None of the terrestrial worlds are still hot enough to remain liquid throughout their interiors. However, they differ considerably in the amount of heat they still retain. The most important factor that governs how long a planet takes to cool is its size: Just as a hot potato remains hot inside much longer than a hot pea, a large planet can stay hot inside much longer than a small one. You can see why size is the critical factor by picturing a large planet as a smaller planet wrapped in extra layers of rock. The extra rock acts as insulation, so it takes much longer for interior heat to reach the surface.

EARTH IS NOT FULL OF MOLTEN LAVA

Many people guess that Earth is full of molten lava (more technically known as *magma*). This misconception may arise partially because we see molten lava emerging from inside Earth when a volcano erupts. However, Earth's mantle and crust are almost entirely solid. The lava that erupts from volcanoes comes only from a narrow region of partially molten material beneath the lithosphere. Indeed, the only part of Earth's interior that is fully molten is the outer core, which is so deep that core material never erupts directly to the surface.

Internal Heat and Geological Activity

Interior heat is the primary driver of geological activity, because this heat supplies the energy needed

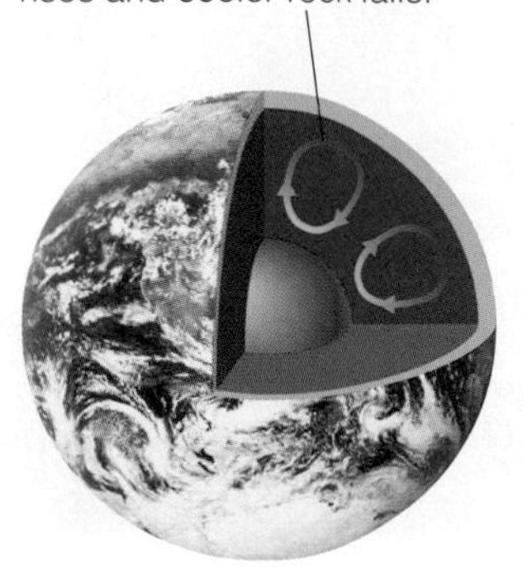

Figure 7.4

Earth's hot interior allows the mantle to undergo convection, in which hot rock gradually rises upward while cool rock gradually falls. Arrows indicate the direction of flow in a portion of the mantle.

to move rock and reshape the surface. Inside a planet, temperature increases with depth. If the deep interior is hot enough, hot rock can gradually rise within the mantle, slowly cooling as it rises. Cooler rock at the top of the mantle gradually falls (Figure 7.4). The process by which hot material expands and rises while cooler material contracts and falls is called **convection**. Keep in mind that mantle convection primarily involves solid rock, not molten rock. Because solid rock flows quite slowly, mantle convection is a very slow process. At the typical rate of mantle convection on Earth, it takes about 100 million years for a piece of rock to be carried from the base of the mantle to the top.

Larger planets retain internal heat much longer than smaller ones, and this heat drives geological activity.

Just as planetary size determines how long a planet stays hot, it is also the primary factor in the strength of mantle convection and lithospheric thickness. As a planet's interior gradually cools, the rigid lithosphere grows thicker and convection occurs only deeper inside the planet. A thick lithosphere inhibits volcanic and tectonic activity, because any molten rock is too deeply buried to erupt to the surface and the strong lithosphere resists distortion by tectonic stresses. If the interior cools enough, convection may stop entirely, leaving the planet geologically "dead."

We can thereby understand the differences in lithospheric thickness shown in Figure 7.2, which go along with differences in geological activity. Earth, the largest of the terrestrial planets, remains quite hot inside and therefore has a thin lithosphere. Venus is probably quite similar to Earth in its internal heat, though it may have a thicker lithosphere (for reasons we will discuss later). With their small sizes, Mercury and the Moon have very thick lithospheres and no geological activity. Mars, intermediate in size, has cooled significantly but probably retains some internal heat.

The Magnetic Field In addition to driving geological activity, interior heat is also responsible for Earth's global **magnetic field**. You are probably familiar with the general pattern of the magnetic field created by an iron bar (Figure 7.5a). Earth's magnetic field is generated by a process more similar to that of an *electromagnet,* in which the magnetic field arises as a battery forces charged particles (electrons) to move along a coiled wire (Figure 7.5b).

Figure 7.5

Sources of magnetic fields.

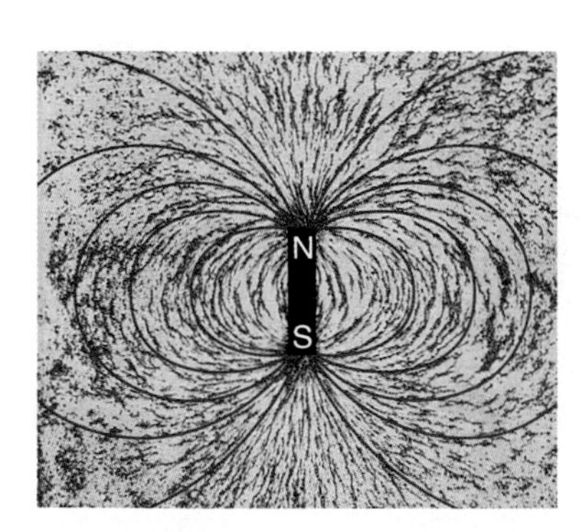

a This photo shows how a bar magnet influences iron filings (small black specks) around it. The *magnetic field lines* (red) represent this influence graphically.

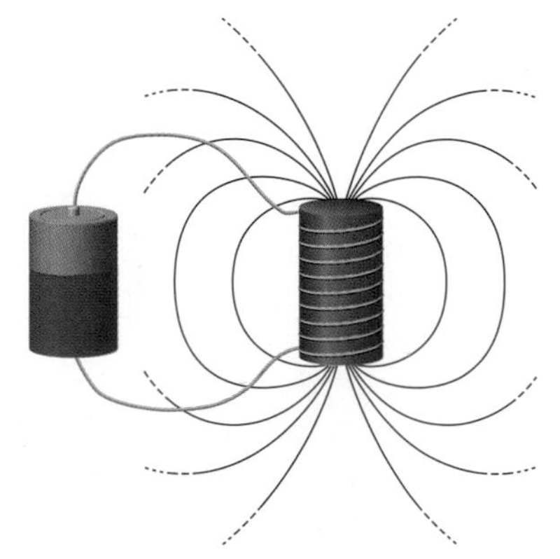

b A similar magnetic field is created by an electromagnet, which is essentially a wire wrapped around a bar and attached to a battery. The field is created by the battery-forced motion of charged particles (electrons) along the wire.

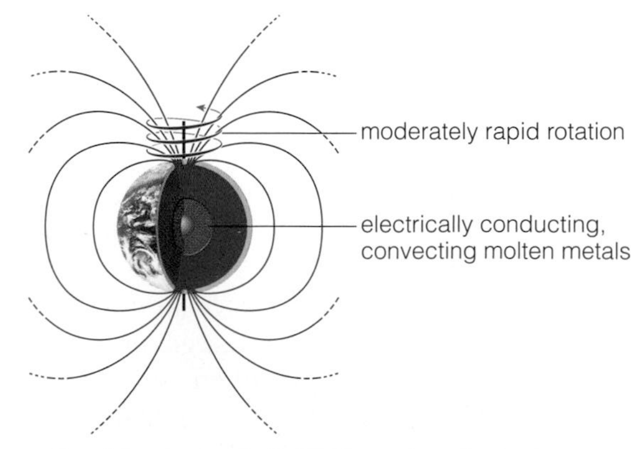

c Earth's magnetic field also arises from the motion of charged particles. The charged particles move within Earth's liquid outer core, made of electrically conducting, convecting molten metals.

Earth's magnetic field is generated by the motions of molten metals in Earth's liquid outer core.

Earth does not contain a battery, but charged particles move with the molten metals in its liquid outer core (Figure 7.5c). Internal heat causes the liquid metals to rise and fall (convection), while Earth's rotation twists and distorts the convection pattern of these molten metals. The result is that electrons in the molten metals move within Earth's outer core in much the same way they move in an electromagnet, generating Earth's magnetic field.

SPECIAL TOPIC: HOW DO WE KNOW WHAT'S INSIDE EARTH?

Our deepest drills have barely pricked Earth's surface, penetrating much less than 1% of the way into the interior. How, then, can we claim to know what our planet is like on the inside?

Our most detailed information about Earth's interior comes from *seismic waves,* vibrations that travel both through Earth's interior and along its surface after an earthquake (*seismic* comes from the Greek word for "shake"). Seismic waves come in two basic types that are analogous to two ways you can generate waves in a Slinky (Figure 1).

Pushing and pulling on one end of a Slinky (while someone holds the other end still) generates a wave in which the Slinky is bunched up in some places and stretched out in others. Waves like this in rock are called P waves. The *P* stands for *primary,* because these waves travel fastest and are the first to arrive after an earthquake, but it is easier to think of it as meaning *pressure* or *pushing.* P waves can travel through almost any material—whether solid, liquid, or gas—because molecules can always push on their neighbors no matter how weakly they are bound together. (Sound travels as a pressure wave and thus is quite similar to P waves.)

Shaking a Slinky slightly up and down generates an up-and-down motion all along its length. Such up-and-down (or side-to-side) waves in rock are called S waves. The *S* stands for *secondary* but is easier to remember as meaning *shear* or *side-to-side.* S waves generally travel only through solids, because the bonds between neighboring molecules in a liquid or gas are too weak to transmit up-and-down or sideways forces.

The precise speeds and directions in which seismic waves travel through Earth depend on the composition, density, pressure, temperature, and phase (solid or liquid) of the material they pass through. For example, P waves reach the side of the world opposite an earthquake, but S waves do not. This tells us that the S waves have been stopped by a liquid layer, which is how we know that Earth has a liquid layer in its outer core (Figure 2). More careful analysis of seismic waves has allowed geologists to build up a detailed picture of what Earth looks like on the inside.

We have also used seismic waves to study the Moon's interior, thanks to monitoring stations left behind by the Apollo astronauts. To learn about the interiors of other worlds, we use less direct clues. For example, knowing that the density of surface rock is much less than a planet's overall average density tells us that the planet must contain much more dense rock or metal inside. We can also learn about a planet's interior from precise measurements of its gravity, which tell us how mass is distributed within the planet; from studies of its magnetic field, which is generated deep inside the planet; and from observations of surface rocks that appear to have emerged from deep within the interior.

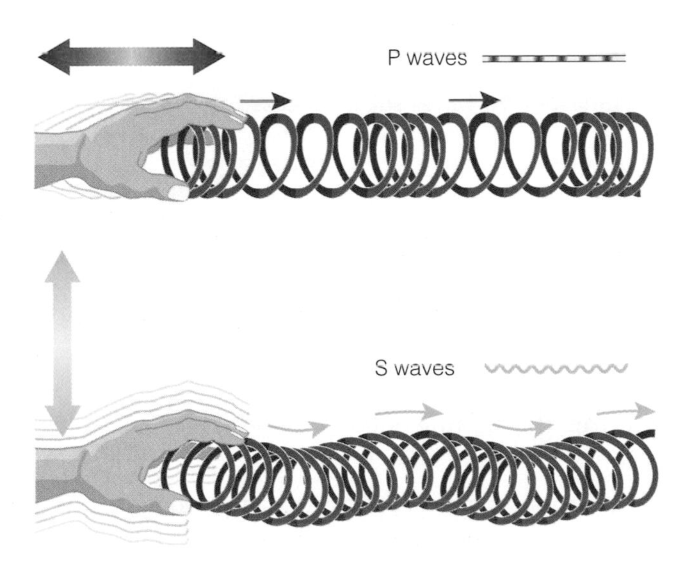

Figure 1

Slinky examples demonstrating P and S waves.

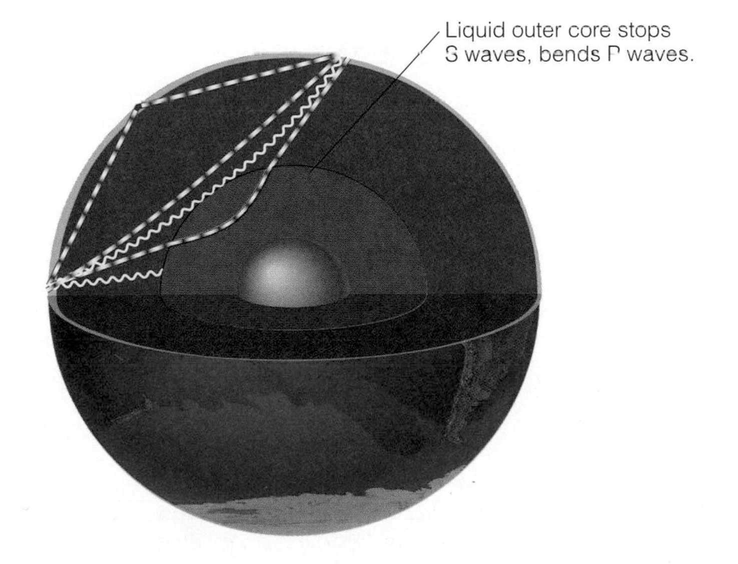

Figure 2

Because S waves do not reach the side of Earth opposite an earthquake, we infer that part of Earth's core is liquid.

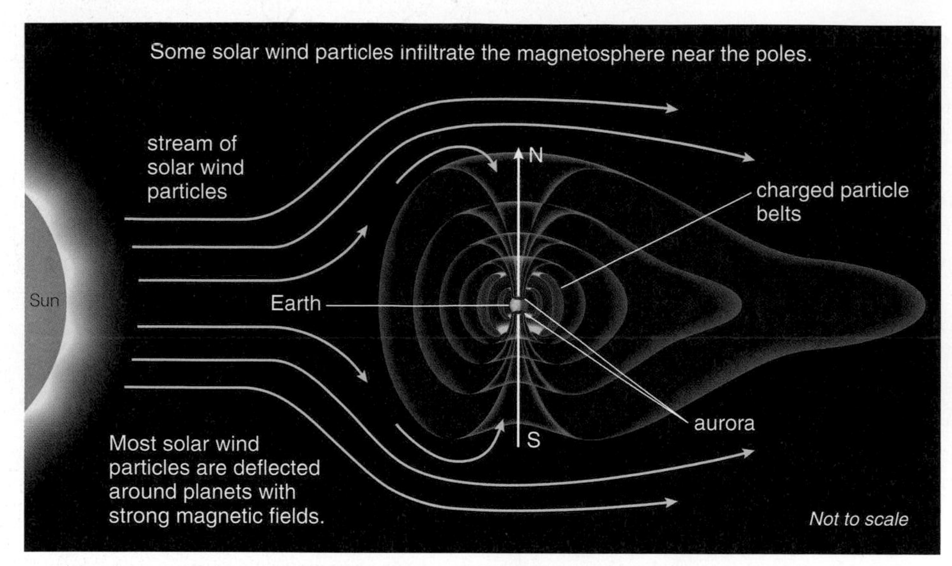

a This diagram shows how Earth's magnetosphere deflects solar wind particles. Some particles accumulate in *charged particle belts* encircling our planet.

b This photograph shows the aurora along the coast of Norway. In a video, you would see these lights dancing about in the sky.

Figure 7.6

Earth's magnetosphere acts like a protective bubble, shielding the surface from charged particles coming from the solar wind. Charged particles penetrate the magnetosphere only near the poles, where they follow magnetic field lines into the atmosphere. There, they can collide with atoms and molecules to create the aurora.

Earth's magnetic field helps protect the surface from energetic particles that continually flow outward from the Sun in what we call the *solar wind* [Sections 6.4, 10.3]. These particles could strip away atmospheric gas and cause genetic damage to living organisms. The magnetic field protects us by creating a **magnetosphere**—a kind of protective bubble that surrounds our planet (Figure 7.6a). The magnetosphere deflects most of the charged particles from the Sun around our planet. The relatively few particles that make it through the magnetosphere tend to be channeled toward the poles, where they collide with atoms and molecules in our atmosphere and produce the beautiful lights of the **aurora** (Figure 7.6b).

None of the other terrestrial worlds have a magnetic field as strong as Earth's, and thus they also lack protective magnetospheres. As we'll see when we study Venus and Mars, their lack of a magnetosphere has had a profound effect on their planetary histories.

Shaping Planetary Surfaces Tutorial, Lessons 1–3

• What processes shape Earth's surface?

Now that we have discussed how Earth and other terrestrial worlds work on the inside, we are ready to turn to their surfaces. Earth offers a huge variety of geological surface features, and the variety only increases when we survey other worlds. But on closer examination, geologists have found that almost all surface features can be explained by just four major geological processes:

- **Impact cratering**: the blasting of bowl-shaped *impact craters* by asteroids or comets striking a planet's surface.
- **Volcanism**: the eruption of molten rock, or *lava*, from a planet's interior onto its surface.
- **Tectonics**: the disruption of a planet's surface by internal stresses.
- **Erosion**: the wearing down or building up of geological features by wind, water, ice, and other phenomena of planetary weather.

Virtually all geological features originate from either impact cratering, volcanism, tectonics, or erosion.

Let's examine each of these processes in a little more detail. Before we begin, notice that impact cratering is the only one of the four processes with an external cause (impacts of objects from space). The remaining three processes are generally driven by the planet itself, and thus represent what we usually think of as *geological activity.*

Impact Cratering An impact crater forms when an asteroid or comet slams into a world with a solid surface (Figure 7.7). Impacting objects typically hit the surface at a speed between about 40,000 and 250,000 km/hr. At such a tremendous speed, the impact releases enough energy to vaporize solid rock and excavate a crater (the Greek word for "cup"). Craters generally end up circular because an impact blasts out material in all directions, no matter the direction of the incoming object. Laboratory experiments show that craters are generally about 10 times as wide as the objects that create them and about 10–20% as deep as they are wide. For example, an asteroid 1 kilometer in diameter will blast out a crater about 10 kilometers wide and 1–2 kilometers deep.

We have never witnessed a major impact on Earth (though we have witnessed one on Jupiter [Section 9.4]), but we have studied the results (Figure 7.8). We can also see numerous impact craters on other worlds, especially the Moon and Mercury (see Figure 7.1).

Comparing the number of impact craters on the Moon and Earth leads us to an important insight. Throughout its history, Earth must have been pelted by at least as many impacts as the Moon, since we occupy the same region of the solar system. Why, then, are there so many more impact craters on the Moon? The answer is that most of Earth's impact craters have been erased with time by geological activity such as volcanic eruptions and erosion.

Like all terrestrial worlds, Earth was bombarded by impacts when it was young, but most ancient craters have been erased by other geological processes.

The idea that craters can be erased with time offers a powerful technique for estimating the age of a world's surface—that is, how long it has been since craters were last erased on the surface. Remember that all the planets were battered by impacts during the heavy bombardment that occurred early in our solar system's history [Section 6.4]. Most impact craters were made during that time, and relatively few impacts have occurred since. Thus, in places where we see numerous craters, such as on much of the Moon's surface, we must be looking at a surface that has stayed virtually unchanged for billions of years. In contrast, when we see very few craters, as we do on Earth, we must be looking at a surface that has undergone recent change. Careful studies of craters on the Moon, where different parts of the surface have different ages (determined by radiometric dating of Moon rocks), have allowed planetary scientists to determine the rate at which craters were made during much of the solar system's history. As a result, we can estimate the age of any planetary surface just by photographing it from orbit and counting its craters.

Figure 7.7 Interactive Figure

Artist's conception of the impact process.

Figure 7.8

Meteor Crater in Arizona was created about 50,000 years ago by the impact of an asteroid about 50 meters across. The crater is more than a kilometer across and almost 200 meters deep.

Figure 7.9 Interactive Figure

Volcanism. This photo shows the eruption of an active volcano on the flanks of Kilauea on the Big Island in Hawaii. The inset shows the underlying process: Molten rock collects in a "magma chamber" and can erupt upward.

Figure 7.10

This photo shows the eruption of Mount St. Helens (Washington State) on May 18, 1980. Note the tremendous outgassing that accompanies the eruption.

Volcanism Volcanism occurs when underground molten rock finds a path through the lithosphere to the surface (Figure 7.9). Molten rock tends to erupt for three main reasons. First, molten rock is generally less dense than solid rock, so it has a natural tendency to rise. Second, remember that most of Earth's interior is not molten. Thus, a "chamber" of molten rock may be squeezed by the surrounding solid rock, which drives the molten rock upward under pressure. Third, molten rock often contains trapped gases that expand as it rises, which can lead to dramatic eruptions. Erupting lava can make tall, steep volcanoes if the lava is very thick, or vast flat lava plains if the lava is very runny.

Earth's atmosphere and oceans were made from gases released from the interior by volcanic outgassing.

Volcanic mountains are the most obvious result of volcanism, but volcanism has had a much more profound effect on our planet: It explains the existence of our atmosphere and oceans. Recall that Earth accreted from rocky and metallic planetesimals, while water and other ices were brought in by planetesimals from more distant reaches of the solar system [Section 6.4]. Water and gases became trapped beneath the surface in much the same way the gas in a carbonated beverage is trapped in a pressurized bottle. Volcanic eruptions later released some of this gas into the atmosphere in the process we call **outgassing** (Figure 7.10).

Measurements show that the most common gases released by outgassing are water vapor (H_2O), carbon dioxide (CO_2), nitrogen (N_2), and sulfur-bearing gases (H_2S or SO_2). The outgassed water vapor rained down to form our oceans, while the other gases helped make our atmosphere. Much of the nitrogen remains in the atmosphere to this day, where it is now the dominant ingredient (77%) of Earth's air. We'll discuss how oxygen came to make up most of the rest of our atmosphere in Section 7.5.

Earth's ongoing volcanic activity can be traced to a single property of our planet: its large size for a terrestrial world, which allows it to retain enough internal heat to drive active volcanism. If Earth had been born much smaller in size, it would have cooled off long ago and could not have active volcanoes today.

Tectonics The third major geological process, tectonics, refers to any surface reshaping that results from stretching, compression, or other forces acting on the lithosphere. Figure 7.11 shows two examples of tectonic features on Earth, one created by surface compression (the Himalayas) and one created by surface stretching (the Red Sea).

Much of the tectonic activity on any planet is a direct or indirect result of mantle convection. Tectonics is particularly important on Earth, because the underlying mantle convection fractured Earth's lithosphere into more than a dozen pieces, or *plates*. These plates move over, under, and around each other, leading to a special brand of tectonics that we call **plate tectonics**. While some type of tectonics has affected every terrestrial world, plate tectonics appears to be unique to Earth. Moreover, as we'll discuss in Section 7.5, plate tectonics may be crucial to explaining life's abundance on Earth.

Tectonics and volcanism generally occur together because both require internal heat and therefore depend on a planet's size.

Tectonic activity usually goes hand in hand with volcanism, because both require internal heat. Thus, like volcanism, Earth's ongoing tectonics is possible only because of our planet's relatively large size, which has allowed it to retain plenty of internal heat.

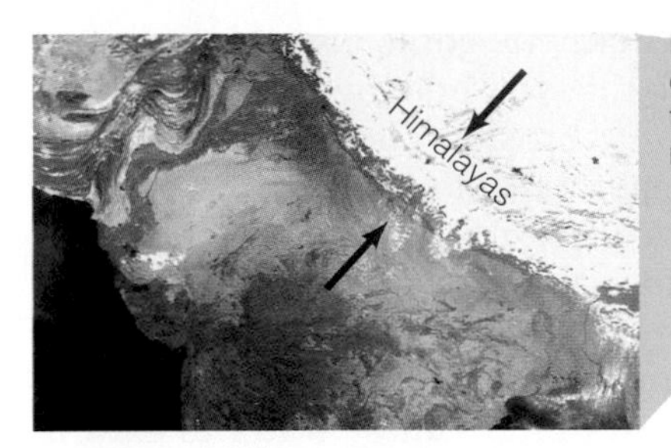

Earth's tallest mountain range, the Himalayas, created as India pushes into the rest of Asia

Internal stresses can cause compression in the crust, which can make mountains.

Internal stresses can pull the crust apart. This extension can make cracks, valleys, and seas.

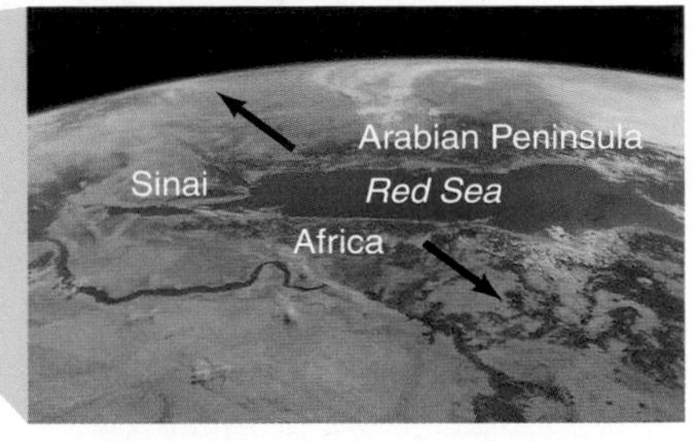

The Red Sea, created as the Arabian Peninsula was torn away from Africa

Figure 7.11 Interactive Figure

Tectonic forces can produce a wide variety of features. Mountains created by tectonic compression and valleys or seas created by tectonic stretching are among the most common. Both images are satellite photos.

Erosion The last of our four major geological processes is erosion. *Erosion* is a blanket term for a variety of processes that break down or transport rock through the action of ice, liquid, or gas. The shaping of valleys by glaciers (ice), the carving of canyons by rivers (liquid), and the shifting of sand dunes by wind (gas) are all examples of erosion (Figure 7.12).

We often associate erosion with the breakdown of existing features, but erosion also builds things. Sand dunes, river deltas, and lake-bed deposits are all examples of features built by erosion. Indeed, much of the surface rock on Earth was built by erosion. Over long periods of time, erosion has piled sediments into layers on the floors of oceans and seas,

Figure 7.12

A few examples of erosion on Earth.

a The Colorado River continues to carve the Grand Canyon after millions of years.

b Glaciers created Yosemite Valley during ice ages.

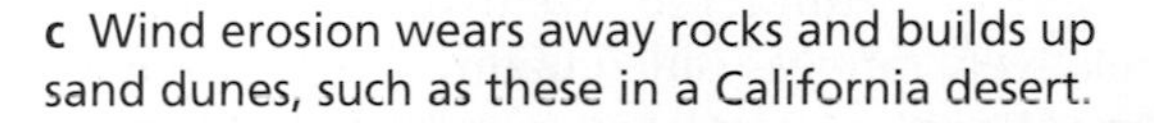

c Wind erosion wears away rocks and builds up sand dunes, such as these in a California desert.

d Erosional debris also creates geological features, as seen in this river delta.

forming what we call **sedimentary rock**. The gorgeous layered rock of the Grand Canyon is sedimentary rock that was built by erosion.

Erosion can both break down and build up geological features.

Erosion plays a far more important role on Earth than on any other terrestrial world, primarily because our planet has both strong winds and plenty of water. The strong winds are driven largely by our planet's relatively rapid rotation. The water exists as a result of outgassing by volcanism, and it causes erosion because our planet's temperature is just right to allow water to exist in both liquid and solid form on the surface.

Surface Temperatures of Terrestrial Planets Tutorial, Lessons 1–4

• How does Earth's atmosphere affect the planet?

As we have just seen, Earth's atmosphere is a major reason why erosion plays such a great role in our planet's geology. In fact, the atmosphere affects the surface in many ways. Most people realize that we need the atmosphere to breathe and to supply the pressure that allows liquid water to flow, but the importance of our atmosphere goes far deeper. Without the atmosphere, Earth's surface would be rendered lifeless by dangerous solar radiation and would be so cold that all water would be perpetually frozen.

Remarkably, our atmosphere makes life possible despite being very thin compared to the planet. About two-thirds of the air in Earth's atmosphere lies within 10 kilometers of the surface. You could represent this air on a standard globe with a layer only as thick as a dollar bill. Let's look more closely at the fundamental ways in which this thin atmosphere keeps the surface comfortable for life.

Surface Protection The Sun emits the visible light that allows us to see, but it also emits dangerous ultraviolet and X-ray radiation. In space, astronauts need thick spacesuits to protect them from the hazards of this radiation. On Earth, we are protected by our atmosphere (Figure 7.13a).

X-ray photons carry enough energy to knock electrons free from almost any atom or molecule. That is, they ionize [Section 5.2] the atoms or molecules they strike. Their high energy is what makes X rays so dangerous, but it also explains how our atmosphere protects us from them. All the X rays from the Sun are absorbed by atoms and molecules high in our atmosphere, and therefore none are left to reach the ground. That is why X-ray telescopes must be placed in space (see Figure 5.22).

Ozone absorbs dangerous solar ultraviolet radiation, while X rays are absorbed by atoms and molecules higher up in Earth's atmosphere.

Ultraviolet photons from the Sun are not so easily absorbed. Most gases are transparent to ultraviolet light, allowing it to pass through unhindered. We owe our protection from ultraviolet light to a relatively rare gas called **ozone** (O_3). Ozone resides only in the middle layer of Earth's atmosphere (more technically called the *stratosphere*), and it absorbs virtually all the dangerous ultraviolet radiation from the Sun. Without ozone, life on land would be impossible.

The Sun emits far more visible light than any other form of light, and visible light passes easily through the atmosphere. Fortunately, this visible light is not dangerous to us. In fact, it is critical to our existence not only because it allows us to see and provides energy for photosynthesis, but more importantly because some of it is absorbed by the ground and therefore is the primary source of heat for Earth's surface.

COMMON MISCONCEPTIONS

WHY IS THE SKY BLUE?

If you ask around, you'll find a wide variety of misconceptions about why the sky is blue. For example, some people think the sky is blue because of light reflecting from the oceans. Anyone who lives in Nebraska will immediately know that this must be wrong—Nebraska is too far from any ocean for its sky to be influenced by such reflection. Other people claim that "air is blue." It's not obvious what someone might mean by this statement, but it's clearly wrong. If air molecules emitted blue light, then air would glow blue even in the dark. If air molecules were blue because they reflected blue light and absorbed red light, then no red light could reach us at sunset. The real explanation for the blue sky is light scattering, as shown in Figure 7.13b. This explanation not only makes sense for our blue sky but also explains our red sunsets.

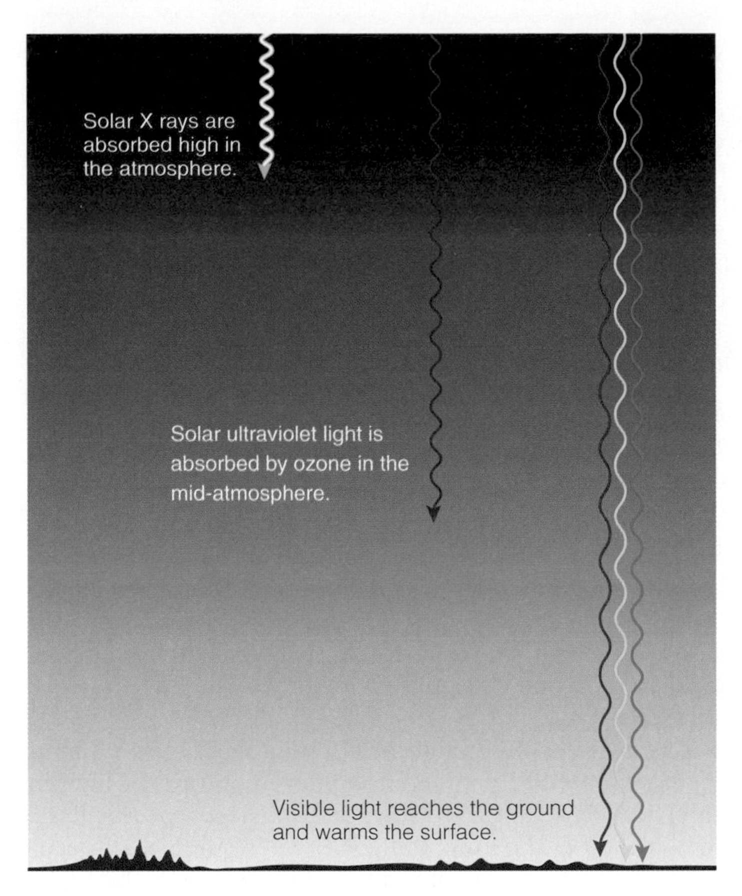

a This diagram summarizes how different forms of light from the Sun are affected by Earth's atmosphere.

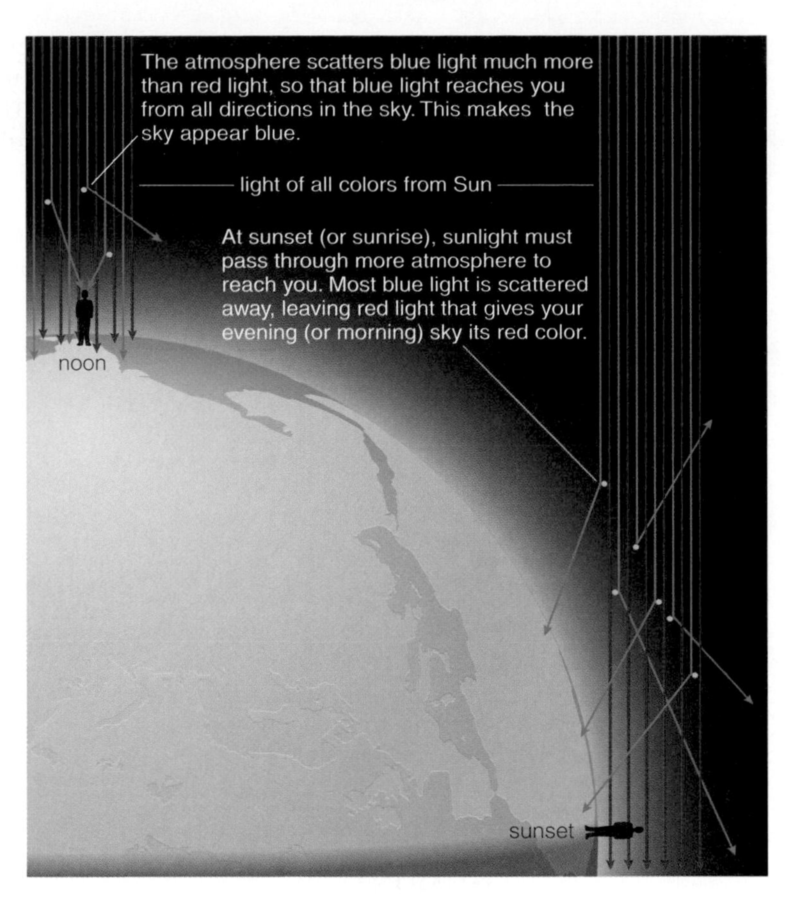

b This diagram summarizes why the sky is blue and sunsets (and sunrises) are red.

Figure 7.13

Our atmosphere protects us from dangerous radiation from the Sun, while allowing visible light to pass through and warm the surface.

Although most visible-light photons pass through Earth's atmosphere without being disturbed, a small proportion is scattered [Section 5.1] randomly around the sky. This scattering is the reason our sky is bright rather than dark (which is why we cannot see stars in the daytime). Without scattering, our sky would look like the lunar sky does to an astronaut, with the Sun just a very bright circle set against a black, star-studded background.

Scattering also explains why our sky is blue (Figure 7.13b). Visible light consists of all the colors of the rainbow, but not all the colors are scattered equally. Gas molecules scatter blue (shorter-wavelength) light much more effectively than red (longer-wavelength) light. The difference in scattering is so great that, for practical purposes, the only scattered light we see in the sky is a small amount of the blue light from the Sun. When the Sun is overhead, this scattered blue light reaches our eyes from all directions and the sky appears blue. At sunset or sunrise, the sunlight must pass through a greater amount of atmosphere on its way to us. Most of the blue light is scattered away, leaving only red light to color the sky.

The Greenhouse Effect Visible light warms Earth's surface, but not as much as you might guess. Calculations show that, by itself, visible light would warm Earth's surface to an average temperature of only about −17°C (−1°F)—well below the freezing point of water. (These calculations are based on Earth's distance from the Sun and the percentages of visible light absorbed and reflected by the ground.) In fact, Earth's global average temperature is about 15°C (59°F), plenty warm enough for liquid

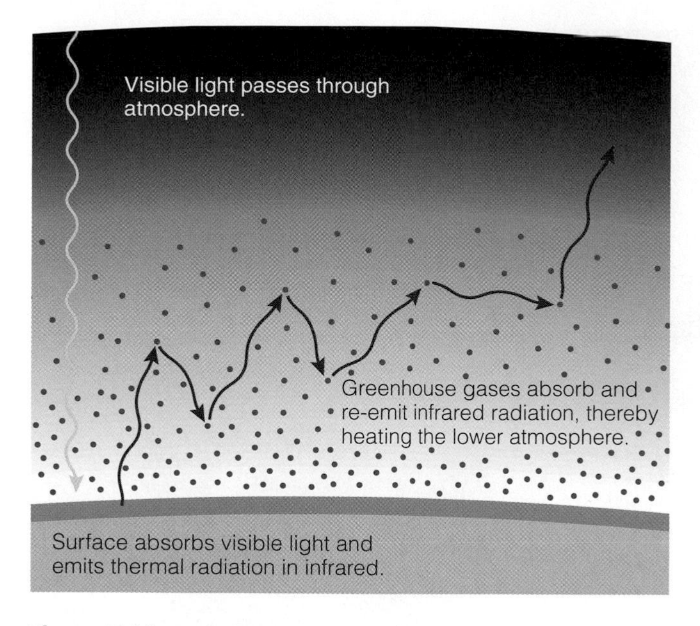

Figure 7.14 Interactive Figure

The greenhouse effect. The lower atmosphere becomes warmer than it would be if it had no greenhouse gases such as water vapor, carbon dioxide, and methane.

THE GREENHOUSE EFFECT IS BAD

The greenhouse effect is often in the news, usually in discussions about environmental problems, but in itself the greenhouse effect is not a bad thing. In fact, we could not exist without it, since it is responsible for keeping our planet warm enough for liquid water to flow in the oceans and on the surface. The "no greenhouse" temperature of Earth is well below freezing. Why, then, is the greenhouse effect discussed as an environmental problem? The reason is that human activity is adding more greenhouse gases to the atmosphere—which might change Earth's climate [Section 7.5]. While the greenhouse effect makes Earth livable, it is also responsible for the searing 470°C temperature of Venus—proving that it's possible to have too much of a good thing.

water to flow and life to thrive. How is it that Earth manages to be so much warmer than it would be from visible light warming alone?

The answer is that our atmosphere traps additional heat through what we call the **greenhouse effect**. The basic idea behind the greenhouse effect is quite simple (Figure 7.14). Some of the visible light from the Sun is absorbed by the ground (the rest is reflected). The ground returns the absorbed energy back toward space, but in the form of infrared rather than visible light. The reason planets emit infrared light is that their temperatures are too low to emit visible light. Remember that all objects emit thermal radiation by virtue of their temperatures [Section 5.2]. Planetary temperatures are in a range in which nearly all emitted energy is in the infrared.

The greenhouse effect keeps Earth's surface much warmer than it would be otherwise, allowing water to stay liquid over most of the surface.

The greenhouse effect works by "trapping" some of the infrared light emitted by the planet, slowing its return to space. This trapping occurs when atmospheric gases absorb the infrared light. Gases that are particularly good at absorbing infrared light are called **greenhouse gases**. The most important greenhouse gases include water vapor (H_2O), carbon dioxide (CO_2), and methane (CH_4). These gases absorb infrared light effectively because their molecular structures make them prone to begin rotating or vibrating when struck by a photon of infrared light.

A greenhouse gas molecule that absorbs a photon of infrared light begins to rotate and vibrate. It then reemits a photon of infrared light in some random direction. This photon is usually absorbed by another greenhouse molecule, which does the same thing. The net result is that greenhouse gases tend to slow the escape of infrared radiation from the lower atmosphere, while their molecular motions heat the surrounding air. In this way, the greenhouse effect makes the surface and the lower atmosphere warmer than they would be from sunlight alone. The more greenhouse gases present, the greater the degree of surface warming.

THINK ABOUT IT Molecules that consist of two atoms of the same type, such as N_2 and O_2, are poor infrared absorbers—which is a very good thing since these molecules make up more than 98% of Earth's atmosphere. Suppose nitrogen and oxygen were greenhouse gases that absorbed infrared light effectively. How would Earth be different?

7.2 MERCURY AND THE MOON: GEOLOGICALLY DEAD

In the rest of this chapter, we will investigate the histories of the terrestrial worlds, with the ultimate goal of learning how and why Earth became unique. We'll start in this section with the two worlds that have the simplest histories: the Moon and Mercury.

It's no accident that the Moon and Mercury have the simplest histories of the terrestrial worlds. Rather, it is a direct consequence of their small sizes. Look back at Figure 7.1, and you'll see that both these worlds are considerably smaller than Venus, Earth, or Mars. Their small size means that they long ago lost most of their internal heat, leaving them without any energy source to power ongoing geological activity. Small size also explains their lack of significant atmospheres: Their gravity is too weak to hold gas for long periods of time, and without ongoing volcanism they lack the outgassing needed to replenish gas lost in the past.

• Was there ever geological activity on the Moon or Mercury?

The most obvious surface feature on both the Moon and Mercury is their numerous impact craters, showing that impact cratering has been by far the most important geological process on both worlds. However, closer examination shows a few features that are volcanic or tectonic in origin. Apparently, the Moon and Mercury had geological activity in the past. This should make sense: Long ago, before they had a chance to cool, these worlds were hot enough inside for at least some volcanism and tectonics.

Geological Features of the Moon The familiar face of the full moon shows that not all regions of the surface look the same (see Figure 7.1). Some regions are heavily cratered. Other regions, known as the **lunar maria**, look smoother and darker. Indeed, the maria got their name because they look much like oceans when seen from afar. *Maria* (singular, *mare*) is Latin for "seas." The smooth and dark appearance of the lunar maria suggests that they were made by a flood of molten lava, and studies of moon rocks confirm this suggestion.

Figure 7.15 shows how we think the maria formed. During the heavy bombardment, the Moon's surface became covered by craters everywhere. The largest impacts of the heavy bombardment violently fractured the Moon's lithosphere beneath the huge craters they created. However, the Moon's interior had already cooled since its formation, and there was no molten rock to flood these craters immediately. Instead, the lava floods came hundreds of millions of years later, thanks to heat released by the decay of radioactive elements in the Moon's interior. This heat gradually built up during the Moon's early history, until mantle material melted about 3 to 4 billion years ago. The molten rock then welled up through the cracks in the lithosphere, flooding the largest impact craters with dark and dense lava.

The Moon's dark, smooth maria were made by floods of molten lava billions of years ago, when the Moon's interior was hot with heat released by radioactive decay.

The Moon's interior cooled quickly after that, and there was never again enough radioactive heat to cause further melting. Because the lava floods occurred after the heavy bombardment subsided,

Figure 7.15

The lunar maria formed between 3 and 4 billion years ago, when molten lava flooded large craters that had formed hundreds of millions of years earlier. This sequence of diagrams represents the formation of Mare Humorum.

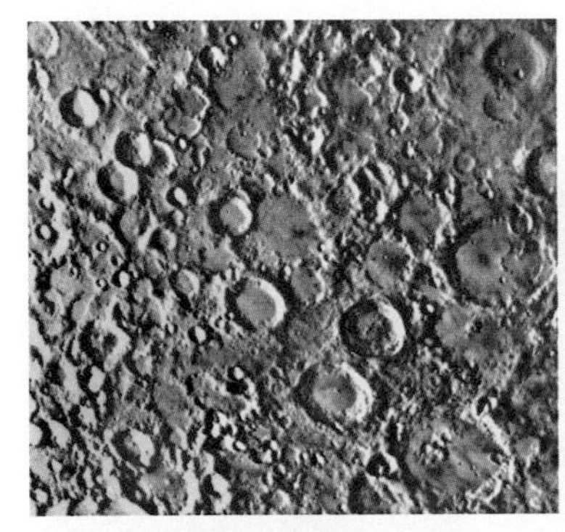

a This illustration shows the Mare Humorum region as it probably looked about 4 billion years ago, when it would have been completely covered in craters.

b Around that time, a huge impact excavated the crater that would later become Mare Humorum. The impact fractured the Moon's lithosphere and erased the many craters that had existed earlier.

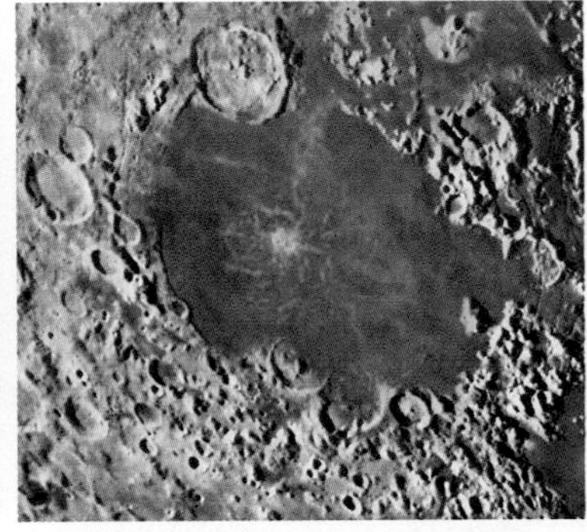

c A few hundred million years later, heat from radioactive decay built up enough to melt the Moon's upper mantle. Molten lava welled up through the lithospheric cracks, flooding the impact crater.

d This photo shows Mare Humorum as it appears today, and the inset shows its location on the Moon.

a Astronaut Gene Cernan takes the Lunar Roving Vehicle for a spin during the final Apollo mission to the Moon (Apollo 17, December 1972).

b The Apollo astronauts left clear footprints, like this one, in the Moon's powdery "soil." The powder is the result of gradual pulverization of surface rock by micrometeorites. Micrometeorites will eventually erase the astronauts' footprints, but not for millions of years.

Figure 7.16

The Moon today is geologically dead, but it is nevertheless a world from which we can learn a lot about the history of our solar system.

the maria have remained much as they were when they first formed. The relatively few impacts that we see within them today were made by impacts that occurred after the maria formed.

The Moon's era of geological activity is long gone. Today, the Moon is a desolate and nearly unchanging place. Rare impacts may occur in the future, but we are unlikely ever to witness a major one. Little happens on the Moon, aside from the occasional visit of robotic spacecraft or astronauts from Earth (Figure 7.16a).

The only ongoing geological change on the Moon is a very slow "sandblasting" of the surface by *micrometeorites*, sand-size particles from space. These tiny particles burn up as meteors in Earth's atmosphere but rain directly onto the surface of the airless Moon. The micrometeorites gradually pulverize the surface rock, which explains why the lunar surface is covered by a thin layer of powdery "soil." The Apollo astronauts left their footprints in this powdery surface (Figure 7.16b). Pulverization by micrometeorites is a very slow process, and the astronauts' footprints will last millions of years before they are finally erased.

The Moon remains a prime target of future exploration, even though it is geologically dead. Further studies of lunar geology will help us understand both the Moon's history and the history of our solar system. The Moon may also offer an excellent location for astronomical observatories, since it has no air to obstruct our view but a stable ground on which to build large telescopes. Someday, perhaps in the not too distant future, humans will build a lunar colony, with living quarters underground for protection against dangerous solar radiation.

Geological Features of Mercury Mercury looks so much like the Moon that it's often difficult at first to tell which world you are looking at in surface photos. The similarities extend to all the geological processes, though there are also a few differences. Mercury's closeness to the Sun and slow rotation make it a world of extremes. The combination of its 88-day orbit and its 59-day rotation gives Mercury days and nights that last about three Earth months each [Section 6.1]. As the Sun rises above the horizon for three months of daylight, the equatorial surface temperature soars to 425°C. At night or in shadow, the temperature falls below −150°C.

Impact craters are visible almost everywhere on Mercury, indicating an ancient surface. However, Mercury's craters are less crowded together than the craters in the most ancient regions of the Moon, suggesting that some of the craters that formed on Mercury during the heavy bombard-

Figure 7.17

Features of impact cratering and volcanism on Mercury. (Photos from *Mariner 10.*)

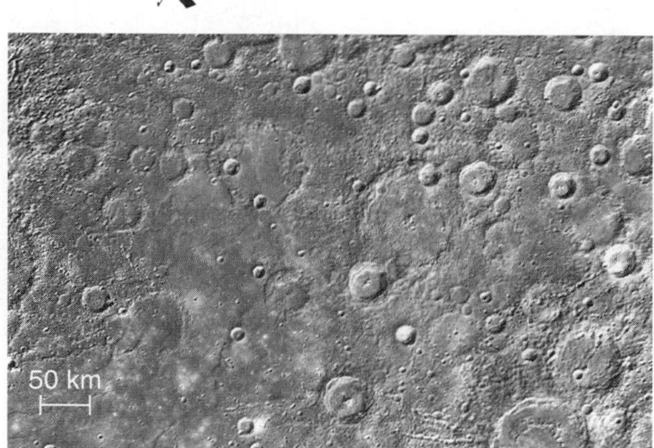

a Close-up view of Mercury's surface, showing impact craters and smooth regions where lava apparently covered up craters.

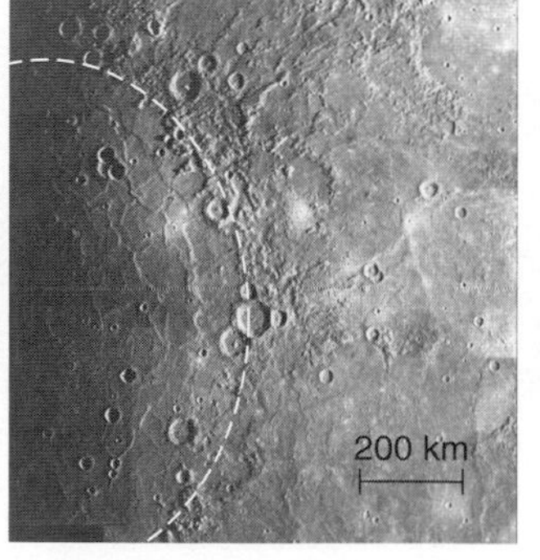

b Part of the Caloris Basin (outlined by the dashed circular ring), a large impact crater on Mercury.

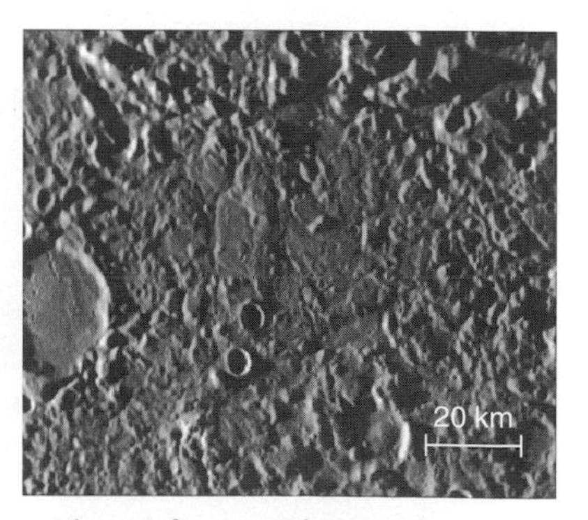

c The surface on the precise opposite side of Mercury from the Caloris Basin.

ment were later covered up by molten lava (Figure 7.17a). As on the Moon, these lava flows probably occurred when heat from radioactive decay accumulated enough to melt part of the mantle. Although we have not found evidence of lava flows as large as those that created the lunar maria, the lesser crater crowding and the many smaller lava plains suggest that Mercury had at least as much volcanism as the Moon.

The largest single surface feature on Mercury is a huge impact crater called the *Caloris Basin* (Figure 7.17b). The Caloris Basin spans more than half of Mercury's radius, and its multiple rings bear witness to the violent impact that created it. The impact must have reverberated throughout the planet—we see evidence of violent surface shaking on the precise opposite side of Mercury from the Caloris Basin (Figure 7.17c). The Caloris Basin has few craters within it, indicating that it must have formed at a time when the heavy bombardment was already subsiding.

The most surprising features of Mercury are its many tremendous cliffs—evidence of a type of past tectonics quite different from anything we have found on any other terrestrial world (Figure 7.18). Mercury's cliffs have vertical faces up to 3 or more kilometers high and typically run for hundreds of kilometers across the surface. They probably formed when tectonic forces compressed the crust, causing the surface to crumple. Because crumpling would have shrunk the portions of the surface it affected, Mercury as a whole could not have stayed the same size unless other parts of the surface expanded. However, we find no evidence of "stretch marks" on Mercury. Can it be that the whole planet simply shrank?

The planet Mercury appears to have shrunk long ago, leaving behind steep and long cliffs.

Apparently so. Early in its history, Mercury gained more internal heat than the Moon because of its larger size. This heat swelled the size of the large iron core. Later, as the core cooled, it contracted by perhaps as much as 20 kilometers in radius. The mantle and lithosphere must have contracted along with the core, generating the tectonic stresses that created the great cliffs. The contraction probably also closed off any remaining volcanic vents, ending Mercury's period of volcanism. Today, Mercury is just as geologically dead as the Moon.

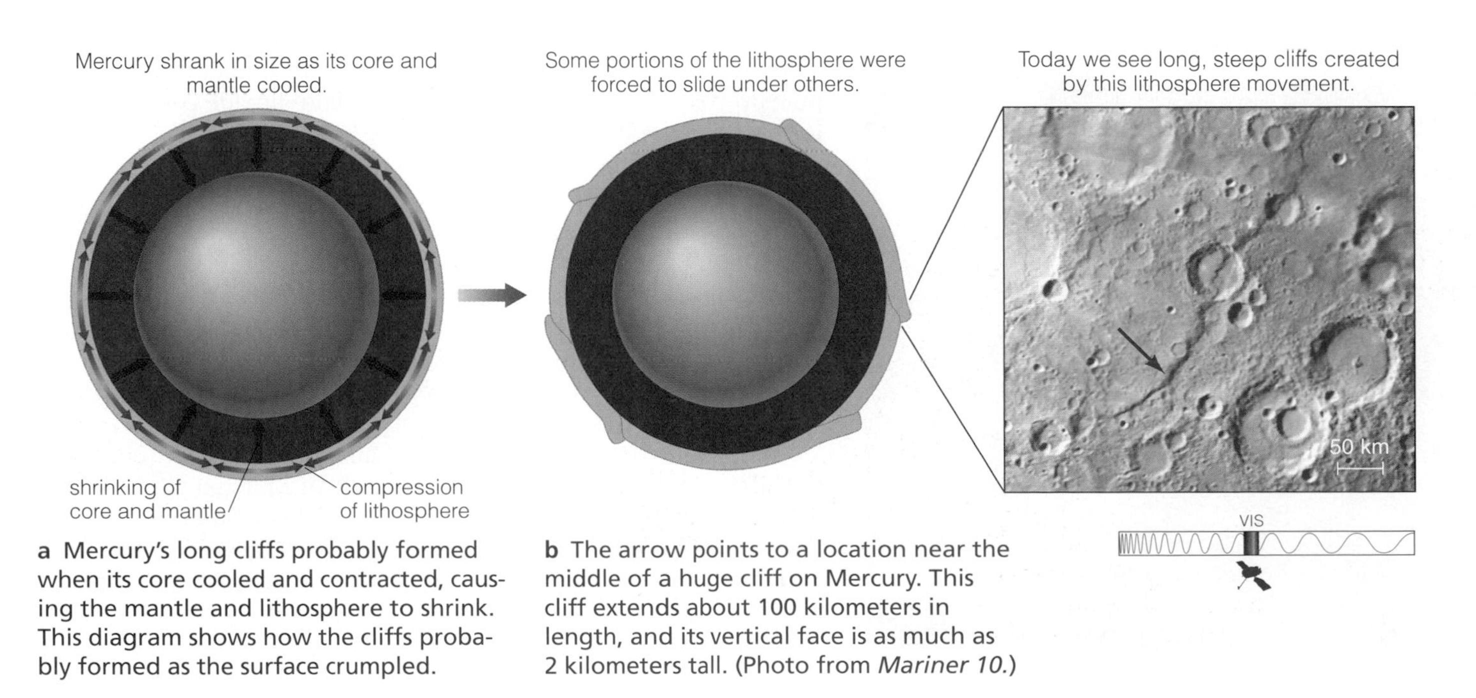

Figure 7.18

Long cliffs on Mercury offer evidence that the entire planet shrank early in its history, perhaps by as much as 20 kilometers in radius.

a Mercury's long cliffs probably formed when its core cooled and contracted, causing the mantle and lithosphere to shrink. This diagram shows how the cliffs probably formed as the surface crumpled.

b The arrow points to a location near the middle of a huge cliff on Mercury. This cliff extends about 100 kilometers in length, and its vertical face is as much as 2 kilometers tall. (Photo from *Mariner 10.*)

7.3 MARS: A VICTIM OF PLANETARY FREEZE-DRYING

Mars is much larger than the Moon and Mercury (see Figure 7.1), so we expect it to have a more interesting and varied geological history. However, it is much smaller than Earth: Its radius is about half that of Earth, and its mass is only about 10% that of Earth (see Appendix E for details). Mars is also the most distant of the terrestrial planets from the Sun, orbiting about 1.5 times as far from the Sun as Earth. Mars's size and distance from the Sun have dictated much of its geological history.

The present-day surface of Mars looks much like some deserts or volcanic plains on Earth (see Figure 6.6), and it offers several other similarities to Earth. A Martian day is less than an hour longer than an Earth day, and Mars has polar caps that resemble Earth's, although they are made primarily of frozen carbon dioxide rather than water ice. Mars's rotation axis is tilted about the same amount as Earth's, and as a result it has seasons much like those on Earth. However, the seasons last almost twice as long because a Martian year is almost twice as long as an Earth year and, unlike the case for Earth, Mars's seasons are affected by its orbit as well as its tilt. Mars's more elliptical orbit puts it significantly closer to the Sun during the southern hemisphere summer and farther from the Sun during the southern hemisphere winter. Mars therefore has more extreme seasons in its southern hemisphere—that is, shorter, hotter summers and longer, colder winters—than in its northern hemisphere.

Figure 7.19

These two Hubble Space Telescope photos contrast the appearance of Mars in the presence and absence of a global dust storm. If you look carefully at the first image, you can see localized dust storms near both polar caps (look toward the upper-right edge of the southern cap). The second image shows how, just over two months later, these storms had grown into a planetwide dust storm.

The superficial similarities between Earth and Mars have made the idea of life on Mars a staple of science fiction for more than a century. However, Mars in reality is quite different from Earth. The atmosphere is so thin that it creates only a weak greenhouse effect despite being made mostly of the greenhouse gas carbon dioxide. The temperature is usually well below freezing, with a global average of about −53°C (−63°F), and the atmospheric pressure is less than 1% that on the surface of Earth. The lack of oxygen means that Mars lacks an ozone layer, so much of the Sun's damaging ultraviolet radiation passes unhindered to the surface.

Even Martian winds are very different from those on Earth. Winds on Earth are driven primarily by effects of Earth's rotation and by heat flow from the equator to the poles. In contrast, winds on Mars are strongly affected by its extreme seasonal changes. Polar temperatures at the winter pole drop so low (about −130°C) that carbon dioxide condenses into "dry ice" at the polar cap. At the same time, frozen carbon dioxide at the summer pole sublimates into carbon dioxide gas. (*Sublimation* is the process in which an ice turns to a gas without first melting into liquid.) The atmospheric pressure therefore increases at the summer pole and decreases at the winter pole, driving strong pole-to-pole winds. Overall, as much as one-third of the total carbon dioxide of the Martian atmosphere moves seasonally between the north and south polar caps. Sometimes these winds initiate huge dust storms, particularly when the more extreme summer approaches in the southern hemisphere (Figure 7.19). The dust storms give Mars its perpetually dusty, pale pink sky.

All in all, surface conditions on Mars today make it seem utterly inhospitable to life. However, careful study of Martian geology offers evidence that Mars may once have been a much warmer and wetter place. If so, it might have had conditions under which life might have arisen and thrived. That is one of the major reasons why Mars has been visited by more spacecraft than any other planet, and why scientists are eager to study Mars for signs of past or present life.

• What geological features tell us that water once flowed on Mars?

No liquid water exists anywhere on the surface of Mars today. We know this not only because we've studied most of the surface in reasonable detail, but also because the surface conditions would not allow liquid water to be present as lakes, rivers, or even puddles. In most places and at most times, Mars is so cold that any liquid water would immediately freeze into ice. Even when the temperature rises above freezing, as it often does at midday near the equator, the air pressure is so low that liquid water would quickly evaporate. If you donned a spacesuit and took a cup of water outside your pressurized spaceship, the water would either freeze or boil away (or a combination of both) almost immediately.

Nevertheless, Mars offers ample evidence of past water flows. Because water could not have flowed for long unless the Martian atmosphere were thicker and warmer, it appears that Mars must have had at least some warm and wet periods in the distant past. Let's investigate the geological evidence for this past, more hospitable world.

The Geology of Mars In order to recognize ancient water flows on Mars and to determine when they happened, we first need to understand the general features of Martian geology. Figure 7.20 shows the full surface of Mars, based on observations from the Mars Global Surveyor mission. Aside from the polar caps, the most striking feature is the dramatic difference in terrain around different parts of the planet. Much of the

Figure 7.20

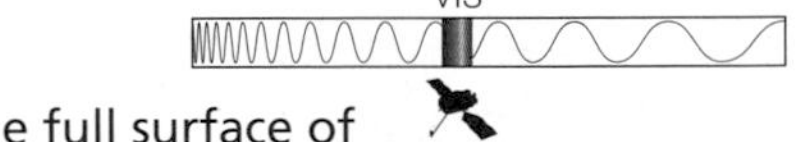

This image showing the full surface of Mars is a composite made by combining more than 1,000 images with more than 200 million altitude measurements from the Mars Global Surveyor mission. Several key geological features are labeled, and the locations of features shown in close-up photos elsewhere in this chapter are marked. The total surface area of Mars is about one-fourth that of Earth (roughly equal to the area of Earth's continents without the oceans), so a map of Earth on the same scale would fill about two pages in this book.

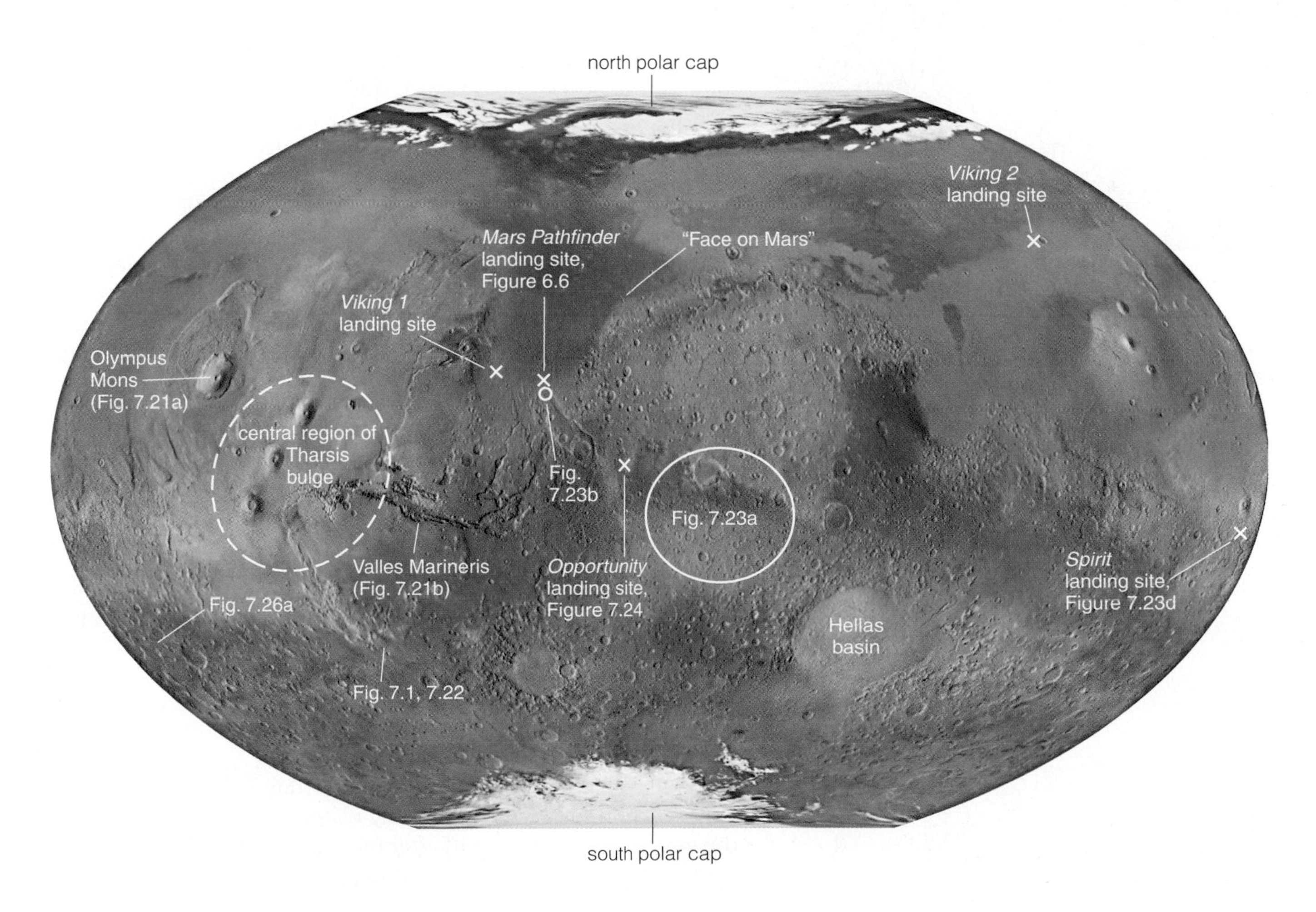

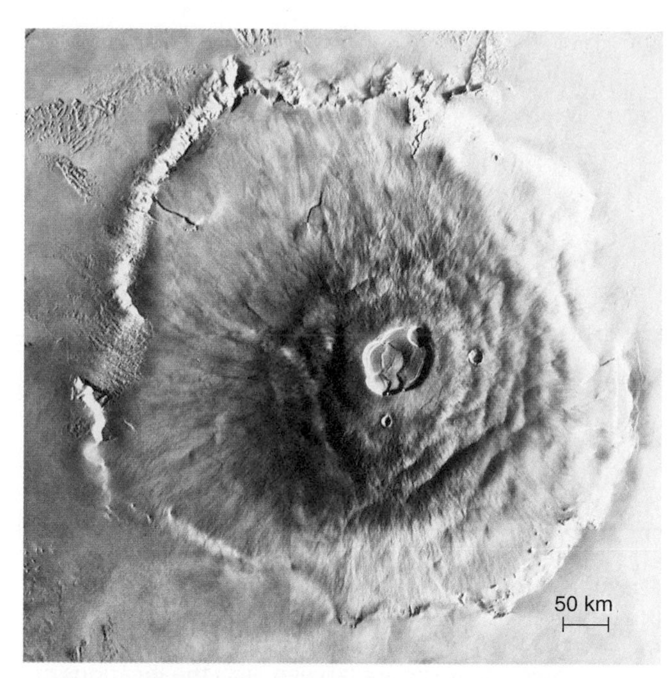

a Olympus Mons, the largest volcano in the solar system, covers an area the size of Arizona and rises higher than Mount Everest on Earth. Note the tall cliff around its rim and the central volcanic crater from which lava erupted.

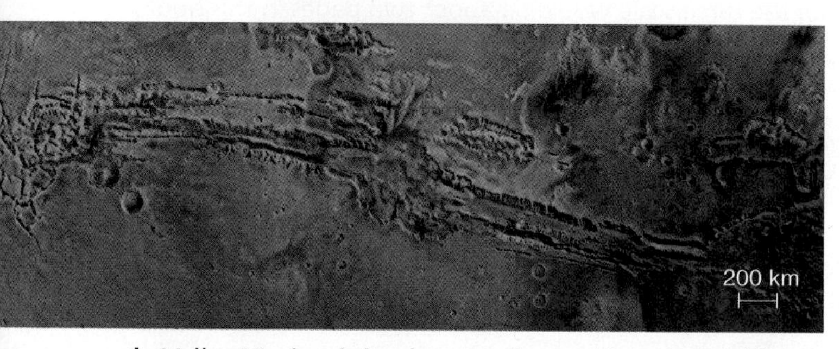

b Valles Marineris is a huge valley on Mars created in part by tectonic stresses. It extends nearly a fifth of the way around the planet (see Figures 6.6 and 7.20.)

Figure 7.21

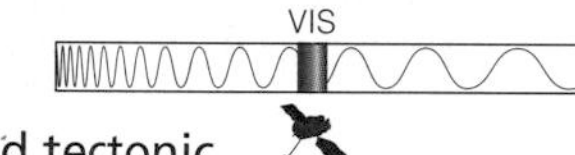

Two examples of volcanic and tectonic features on Mars.

southern hemisphere has relatively high elevation and is scarred by numerous large impact craters, including the very large crater known as the Hellas Basin. In contrast, the northern plains show few impact craters and tend to be below the average Martian surface level. The differences in crater crowding clearly show that the southern highlands are a much older surface than the northern plains, which must have had their early craters erased by other geological processes. Further study suggests that volcanism was the most important of these processes.

More dramatic evidence of volcanism on Mars comes from several towering volcanoes. One of these, Olympus Mons, is the largest known volcano in the solar system (Figure 7.21a). Its base is some 600 kilometers across, large enough to cover an area the size of Arizona. Its peak stands about 26 kilometers above the average Martian surface level, or some three times as high as Mount Everest stands above sea level on Earth. Much of Olympus Mons is rimmed by a cliff that in places is 6 kilometers high.

Mars has had very active volcanism in the past, and its surface is dotted with numerous large volcanoes.

Olympus Mons and several other large volcanoes are concentrated on or near the continent-size *Tharsis Bulge*. Tharsis, as it is usually called, is some 4,000 kilometers across, and most of it rises several kilometers above the average Martian surface level. It was probably created by a long-lived plume of rising mantle material that bulged the surface upward and provided the molten rock for the eruptions that built the giant volcanoes.

Mars also has tectonic features, though none on a global scale like the plate tectonics of Earth. The most prominent tectonic feature is the long, deep system of valleys called *Valles Marineris* (Figure 7.21b). Valles Marineris extends almost a fifth of the way along the planet's equator. It is as long as the United States is wide and almost four times as deep as Earth's Grand Canyon. No one knows exactly how Valles Marineris formed, but its location suggests a link to the Tharsis Bulge. Perhaps it formed through tectonic stresses accompanying the uplift of material that created Tharsis, cracking the surface and leaving the tall cliff walls of the valleys.

Is there any ongoing volcanic or tectonic activity on Mars? Until recently, we didn't think so. We expect Mars to be much less geologically active than Earth, because its smaller size has allowed its interior to cool much more. In addition, Martian volcanoes show enough impact craters on their slopes to suggest that they have been inactive for the past billion years. However, meteorites that appear to have come from Mars (so-called Martian meteorites [Section 18.2]) offer a different perspective. Radiometric dating [Section 6.4] of these meteorites shows some of them to be made of volcanic rock that solidified from molten lava as recently as 180 million years ago—quite recent in the 4.5-billion-year history of the solar system. Given this evidence of geologically recent volcanic eruptions, it is likely that Martian volcanoes will erupt again someday, though not necessarily in our lifetimes. Nevertheless, the Martian interior is presumably cooling and its lithosphere thickening. At best, Mars will become as geologically dead as the Moon and Mercury within a few billion years.

Ancient Water Flows Impacts, volcanism, and tectonics explain most of the major geological features of Mars, but closer examination shows ample evidence of features of erosion. For example, Figure 7.22 looks much like dry riverbeds on Earth seen from above. These channels were almost certainly carved by running water, though no one knows whether the water came from runoff after rainfall or from an underground source. However, the water apparently stopped flowing long ago. Notice that a few impact craters lie on top of the channels. From counts of the craters in and near them, it appears that these channels are at least 3 billion years old, meaning that water has not flowed through them since that time.

Other evidence also argues that Mars had rain and surface water in the distant past. In the ancient, heavily cratered terrain of the southern highlands, rainfall appears to have eroded the rims of large craters and erased smaller craters altogether (Figure 7.23a). Some craters, such as the one shown in Figure 7.23b, appear to have held lakes. Ancient rains may have filled the crater, allowing sediments to build up from material that settled to the bottom. The sculpted patterns in the crater bottom may have been created as erosion by wind exposed layer upon layer of sedimentary rock, much as erosion by the water in the Colorado River exposed the layers visible in the walls of the Grand Canyon on Earth.

Dried up riverbeds and other signs of erosion show that water flowed on Mars in the distant past.

Perhaps the most intriguing idea is that a great ocean once may have filled the low-lying regions in the north (Figure 7.23c). Careful study shows the presence of features that resemble shorelines along the boundaries of this possible ocean, although the evidence is not particularly strong. Further evidence comes from valleys that appear to have formed as lakes overflowed their shores (Figure 7.23d). Whether or not a full ocean existed, it seems that liquid water was plentiful on Mars during its first billion years.

Figure 7.22

VIS

This photo, taken by the *Viking Orbiter,* shows what appear to be dried-up riverbeds. Toward the top of the image we see many individual tributaries, which merge into the larger "river" near the lower right. Counts of craters near the channels indicate that they formed more than 3 billion years ago.

Figure 7.23

More evidence of past water on Mars.

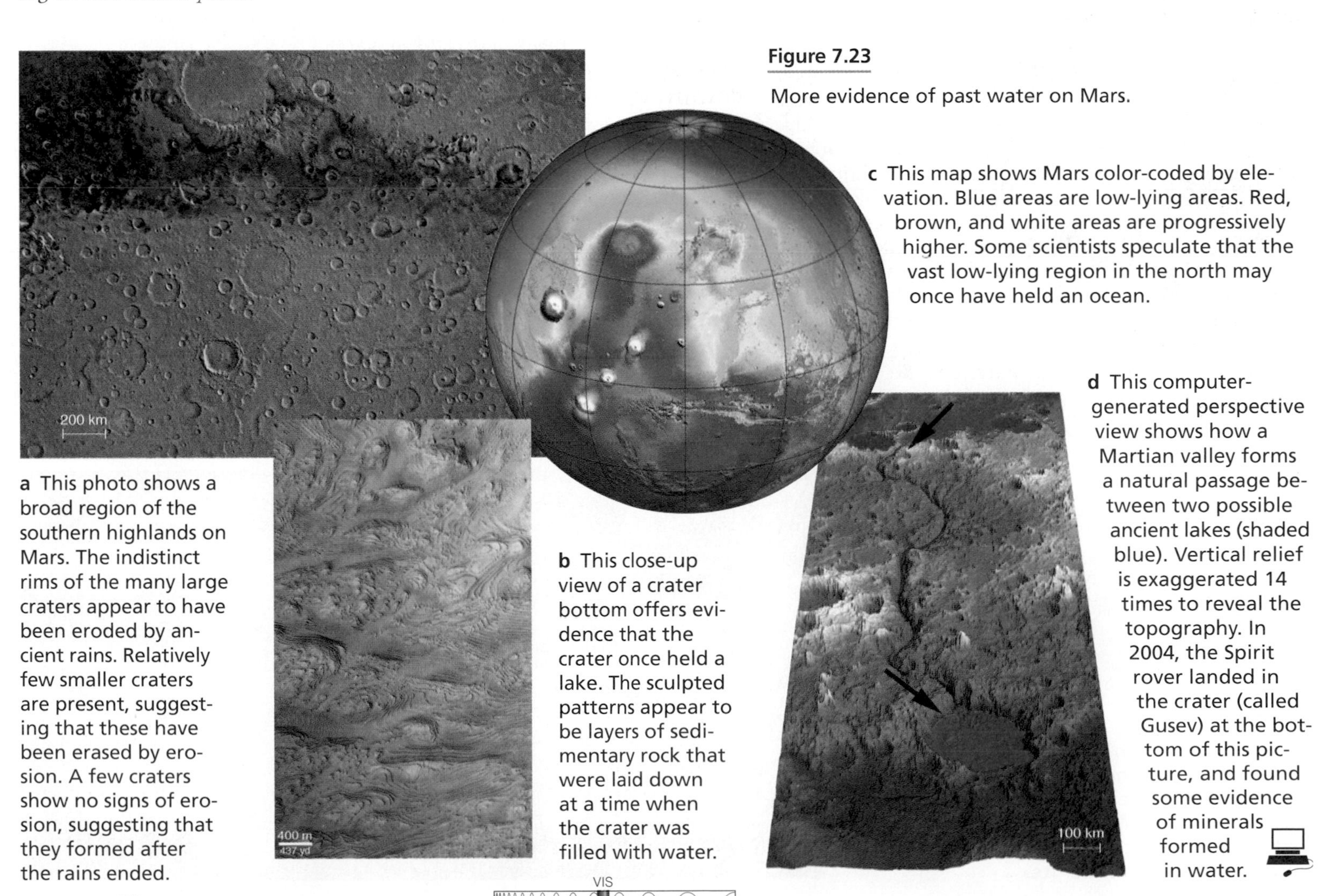

a This photo shows a broad region of the southern highlands on Mars. The indistinct rims of the many large craters appear to have been eroded by ancient rains. Relatively few smaller craters are present, suggesting that these have been erased by erosion. A few craters show no signs of erosion, suggesting that they formed after the rains ended.

b This close-up view of a crater bottom offers evidence that the crater once held a lake. The sculpted patterns appear to be layers of sedimentary rock that were laid down at a time when the crater was filled with water.

c This map shows Mars color-coded by elevation. Blue areas are low-lying areas. Red, brown, and white areas are progressively higher. Some scientists speculate that the vast low-lying region in the north may once have held an ocean.

d This computer-generated perspective view shows how a Martian valley forms a natural passage between two possible ancient lakes (shaded blue). Vertical relief is exaggerated 14 times to reveal the topography. In 2004, the Spirit rover landed in the crater (called Gusev) at the bottom of this picture, and found some evidence of minerals formed in water.

a This panorama shows an outcrop of whitish rocks that stand about knee-high. The white square near the far right identifies the rock shown in part (b).

b This rock, nicknamed Stone Mountain, was a target of a close-up study by Opportunity. The gray square is the piece of the rock shown in part (c).

c This close-up shows a piece of the rock about 3 cm across (about the width of your big toe). The layered structure, the odd indentations, and the small sphere all support the idea that the rock formed from sediments in standing water.

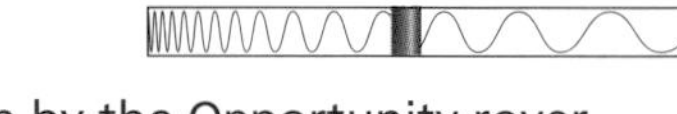

Figure 7.24

These photos were taken by the Opportunity rover, which landed on Mars in 2004.

Craters in younger, less heavily cratered regions of the Martian surface show far less erosion than the ancient craters of the southern highlands, suggesting that little or no rain has fallen in the last 3 billion years. However, a few large floodplains appear to be younger in age, with crater counts suggesting that some may have formed between 1 and 3 billion years ago. For example, the 1997 Mars Pathfinder mission landed at a site where we see rocks of many different types jumbled together and stacked against each other (see Figure 6.6), just as we find in the aftermath of floods on Earth. Because orbital photos of the region show no reservoirs from which water might have escaped, we suspect the flood occurred when volcanic heat melted underground ice, releasing water that temporarily raged across the surface.

In 2004, two robotic rovers landed on nearly opposite sides of Mars. The rovers, named *Spirit* and *Opportunity,* each traveled hundreds of meters across the surface. Both rovers made detailed studies of Martian rocks that have given us even more confidence that liquid water was once plentiful on Mars. For example, rocks at the Opportunity landing site contain tiny spheres (about the size of BB pellets) and odd indentations suggesting that they formed in standing, salty water (Figure 7.24).

Martian Water Today If water once flowed over large portions of Mars, where did it all go? As we'll discuss shortly, much of the water was probably lost to space forever. However, significant amounts of water apparently still remain, frozen at the polar caps and in the top meter or so of the surface soil around much of the rest of the planet (Figure 7.25).

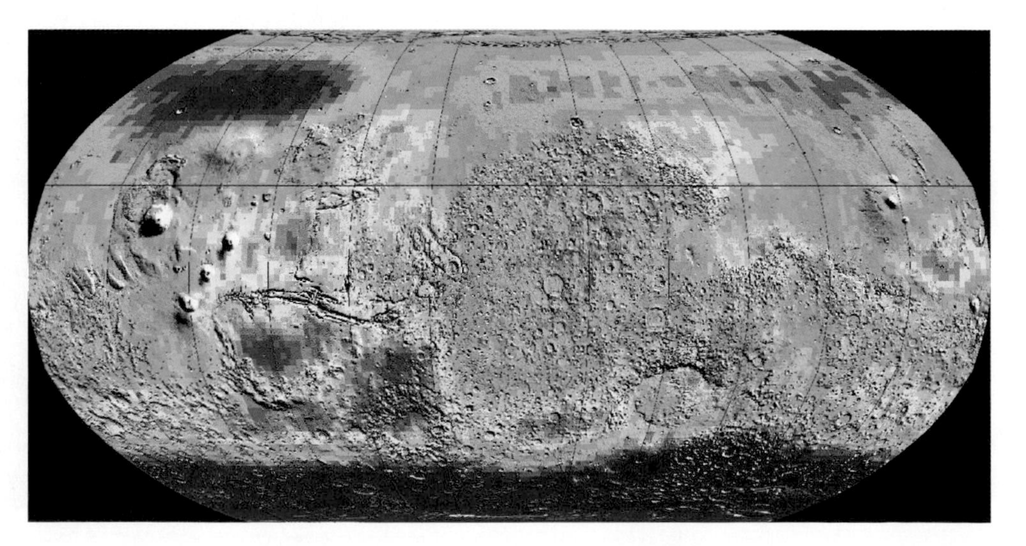

Figure 7.25

This map, made with data from *Mars Odyssey,* represents the hydrogen content of the Martian surface soil. The blue areas contain the most hydrogen, probably because they represent regions in which the top meter or so of surface soil contains water ice.

If water ice exists on Mars this close to the surface, even more water probably lies deeper underground. If there is still volcanic heat on Mars, this water may sometimes melt and flow. Although we have found no geological evidence of any large-scale water flows on Mars in the past billion years, orbital photographs offer tantalizing hints of smaller-scale water flows in much more recent times.

Gullies on crater walls suggest that water might still occasionally flow on Mars today.

The strongest evidence for liquid water in recent times comes from photos of gullies on crater and channel walls. In Figure 7.26, for example, note the striking similarity to the gullies we see on almost any eroded slope on Earth. The leading hypothesis suggests that the gullies form when snow accumulates on the crater walls in winter and melts

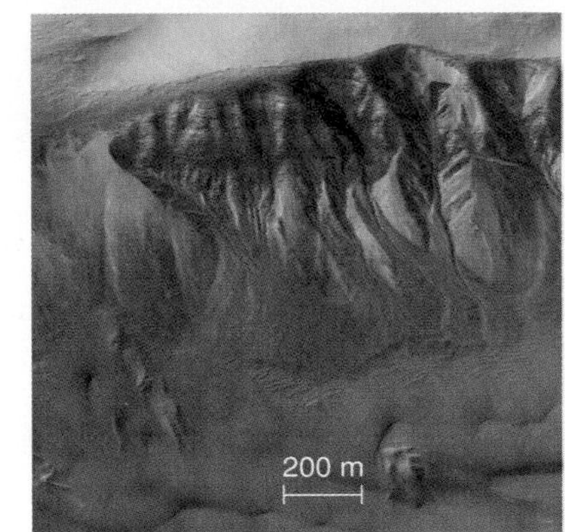

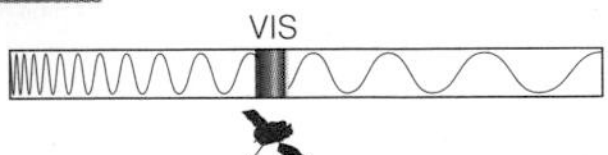

Figure 7.26

This photograph from the *Mars Global Surveyor* shows gullies on a crater wall. The gullies may have been formed by water melting under the protective cover of snowpack.

SPECIAL TOPIC: THE FACE ON MARS

Among the tens of thousands of photographs snapped as part of the Viking missions to Mars in the late 1970s, one achieved special fame. The *Viking 1* orbiter snapped a picture showing a feature on Mars that looked remarkably like a human face (Figure 1). The "face on Mars" soon spawned a cottage industry, with proponents arguing that it must be the work of a lost Martian civilization or of alien beings visiting our solar system. It still appears regularly in supermarket tabloids.

The feature in the picture certainly resembles a face. However, it seemed a near-certainty that the human likeness was a coincidence of geology and camera angle. Because of the intense public interest surrounding the face on Mars, NASA made it a target of further observations with *Mars Global Surveyor,* which entered Martian orbit in 1997.

New photographs soon showed that the scientific analysis had been correct: The feature did not look like a face when it was viewed at higher resolution. Figure 2 compares the original Viking photo with a more recent photo from *Mars Global Surveyor.* Although a few die-hard proponents of the face have not given up, the scientific conclusion is clear.

For anyone heartbroken at the thought that Mars has never harbored a civilization, Mars offers other reasons to smile. In searching through hundreds of other photos from Mars missions, NASA scientists came upon a crater that shows a "happy face" on Mars (Figure 3).

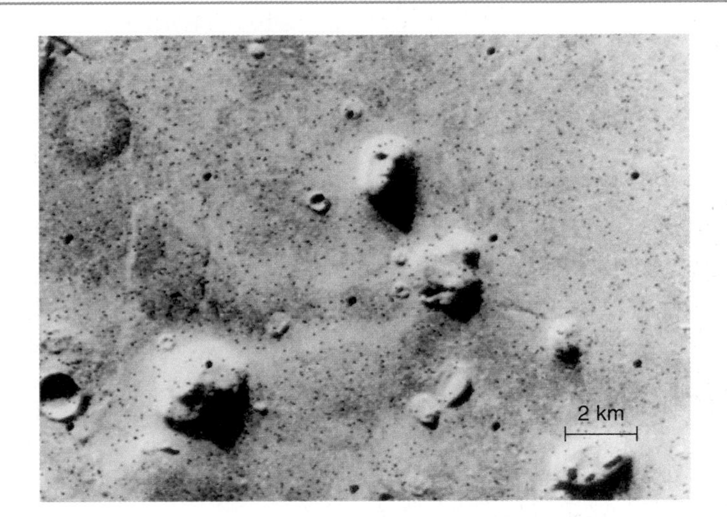

Figure 1

VIS

This *Viking 1* image shows the "face on Mars" (near top center) and the surrounding area. The face is on a tall mesa about 1 kilometer across. Some of the other features were claimed to represent a "fortress" and "pyramids." The black dots are instrument artifacts.

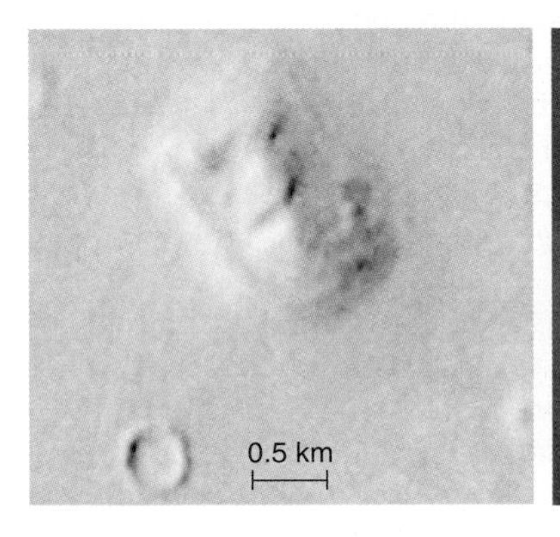

1 km

Figure 2

VIS

Two photos of the "face on Mars." The photo on the left is another image from *Viking 1* in 1976 (a blowup of the face in Figure 1). The much higher resolution photo on the right is from *Mars Global Surveyor.*

Figure 3

This Viking image shows a crater on Mars that appears to be showing a "happy face."

away in spring. Because the gullies are relatively small (note the scale bar in Figure 7.26), they should be gradually covered over by blowing sand during Martian dust storms. Thus, gullies that are still clearly visible must be no more than a few million years old. Geologically speaking, this time is short enough to make it quite likely that water flows are still forming gullies today.

• Why did Mars change?

As we've discussed, there seems little doubt that Mars had wetter and possibly warmer periods, probably with rainfall, that came to an end more than 3 billion years ago. The full extent of these periods is a topic of considerable scientific debate. Some scientists think that Mars may have been continuously warm and wet for much of its first billion years of existence. Others think that Mars may have had only intermittent periods of rainfall, perhaps triggered by the heat of large impacts, and that ancient lakes, ponds, or oceans may have been completely ice-covered. Either way, Mars apparently underwent a major and permanent climate change more than 3 billion years ago, turning a world that was at least sometimes wet and warm into a frozen wasteland.

Mars must have had a much denser and warmer atmosphere to allow rainfall of any kind. The idea that Mars once had a thicker atmosphere makes sense, because we would expect that its many volcanoes outgassed plenty of atmospheric gas. Much of this gas should have been water vapor and carbon dioxide, which are both greenhouse gases that would have helped warm the planet. If Martian volcanoes outgassed carbon dioxide and water in the same proportions as do volcanoes on Earth, Mars would have had enough water to fill oceans tens or even hundreds of meters deep. The heat of impacts may also have released water vapor into the atmosphere, enhancing the greenhouse effect until the water eventually rained out.

Early in its history, Mars probably had a dense atmosphere from volcanic outgassing, with a stronger greenhouse effect than it has today.

The bigger question is not whether Mars once had a denser atmosphere but where all the atmospheric gas went. In particular, Mars must somehow have lost most of its carbon dioxide gas. This loss would have weakened the greenhouse effect until the planet essentially froze over. Some of the carbon dioxide condensed to make the polar caps and some may be chemically bound to surface rock. However, the bulk of the gas was probably lost to space.

The precise way in which Mars lost its carbon dioxide gas is not clear, but recent data suggest a close link to a change in Mars's magnetic field (Figure 7.27). Early in its history, Mars probably had molten metals in its core, much like Earth today. Circulating with convection and Mars's rotation, these metals should have produced a magnetic field and a protective magnetosphere around Mars. However, the magnetic field weakened as the small planet cooled and the core solidified, leaving the atmosphere vulnerable to solar wind particles. These solar wind particles could have stripped gas out of the Martian atmosphere and into space.

Mars underwent permanent climate change about 3 billion years ago, when it lost much of its atmospheric carbon dioxide and water to space.

Much of the water once present on Mars is also probably gone for good. Like the carbon dioxide, some water vapor may have been stripped away by the solar wind. However, Mars also lost water in another way. Mars does not have any atmospheric gases capable of absorbing ultraviolet light in the way that

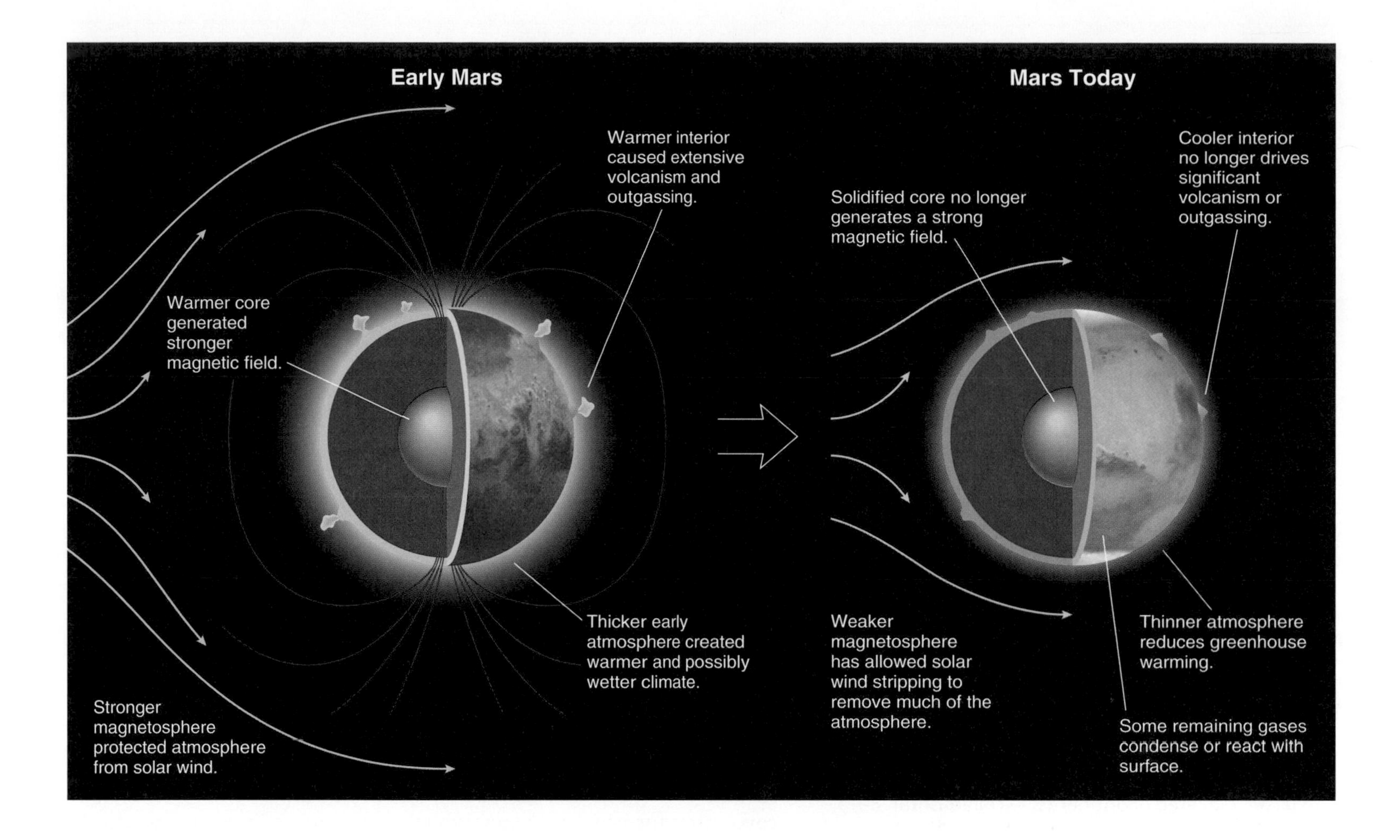

Figure 7.27

Some 3 billion years ago, Mars underwent dramatic climate change, ensuring that rain could never fall again.

ozone absorbs ultraviolet light on Earth. As a result, water molecules in the Martian atmosphere are easily broken apart by ultraviolet photons, releasing individual hydrogen atoms. Once released from water molecules, this hydrogen can escape to space.

Hydrogen escapes to space much more easily than other gases, because its light weight means that it moves faster than heavier atoms or molecules at any particular temperature. Thus, hydrogen atoms are far more likely to reach escape velocity [Section 4.4] from a planet than are heavier atoms and molecules. Mars, like all the terrestrial planets, has an escape velocity that allows atmospheric hydrogen to escape to space fairly quickly. Once the hydrogen was gone, the water molecules could never reform.

Some of the oxygen left behind by the water molecules was probably stripped away from the atmosphere by the solar wind. The rest probably was drawn out of the atmosphere through chemical reactions with surface rock. This oxygen literally rusted the Martian rocks, giving the "red planet" its distinctive tint.

In summary, Mars changed primarily because of its relatively small size. It was big enough for volcanism and outgassing to release plenty of water and atmospheric gas early in its history, but too small to maintain the internal heat needed to prevent the loss of this water and gas. As its interior cooled, its volcanoes quieted and stopped releasing gas into the atmosphere, while its relatively weak gravity and the loss of its magnetic field allowed existing gas to be stripped away to space. If Mars had been as large as Earth, so that it could still have outgassing and a global magnetic field, it might still have a moderate climate today. Mars's distance from the Sun also helped seal its fate: Even with its small size, Mars might still have some flowing water if it were significantly closer to the Sun, where the extra warmth could melt the water that remains frozen underground and at the polar caps.

THINK ABOUT IT Some people have proposed "terraforming" Mars—that is, making it more Earth-like—by finding a way to release all the carbon dioxide frozen in its polar caps into its atmosphere. Briefly explain why we would expect such a release to warm and thicken the atmosphere. If Mars once had oceans in the distant past, would terraforming allow it to have oceans again? Why or why not?

7.4 VENUS: A HOTHOUSE WORLD

We have seen that planetary size is a major factor in explaining why the histories of the Moon, Mercury, and Mars are so different from that of Earth. On the basis of size alone, we would expect Venus and Earth to be quite similar: Venus is only about 5% smaller than Earth in radius (see Figure 7.1), and its overall composition is about the same as that of Earth. However, as we saw in our planetary tour in Section 6.1, the surface of Venus is a searing hothouse, quite unlike the surface of Earth. In this section, we'll investigate how a planet so similar in size to Earth ended up so different in almost every other respect.

• Is Venus geologically active?

As we did for Mars, let's begin our study of Venus by looking at what its surface geology tells us about its planetary history. Venus's thick cloud cover prevents us from seeing through to its surface, but we can study its geological features with radar (because radio waves can pass through clouds). *Radar mapping* bounces radio waves off the surface and uses the reflections to create three-dimensional images of the surface.

Figure 7.28

This image shows the surface of Venus as revealed by radar observations from the *Magellan* spacecraft. Bright regions in this radar image represent rough areas or high altitudes, which often correspond to areas that have experienced greater geological activity. The three large, elevated "continents"—called Ishtar Terra, Lada Terra, and Aphrodite Terra—are the biggest features on the surface of Venus. The locations of features shown in close-up photos elsewhere in this chapter are marked.

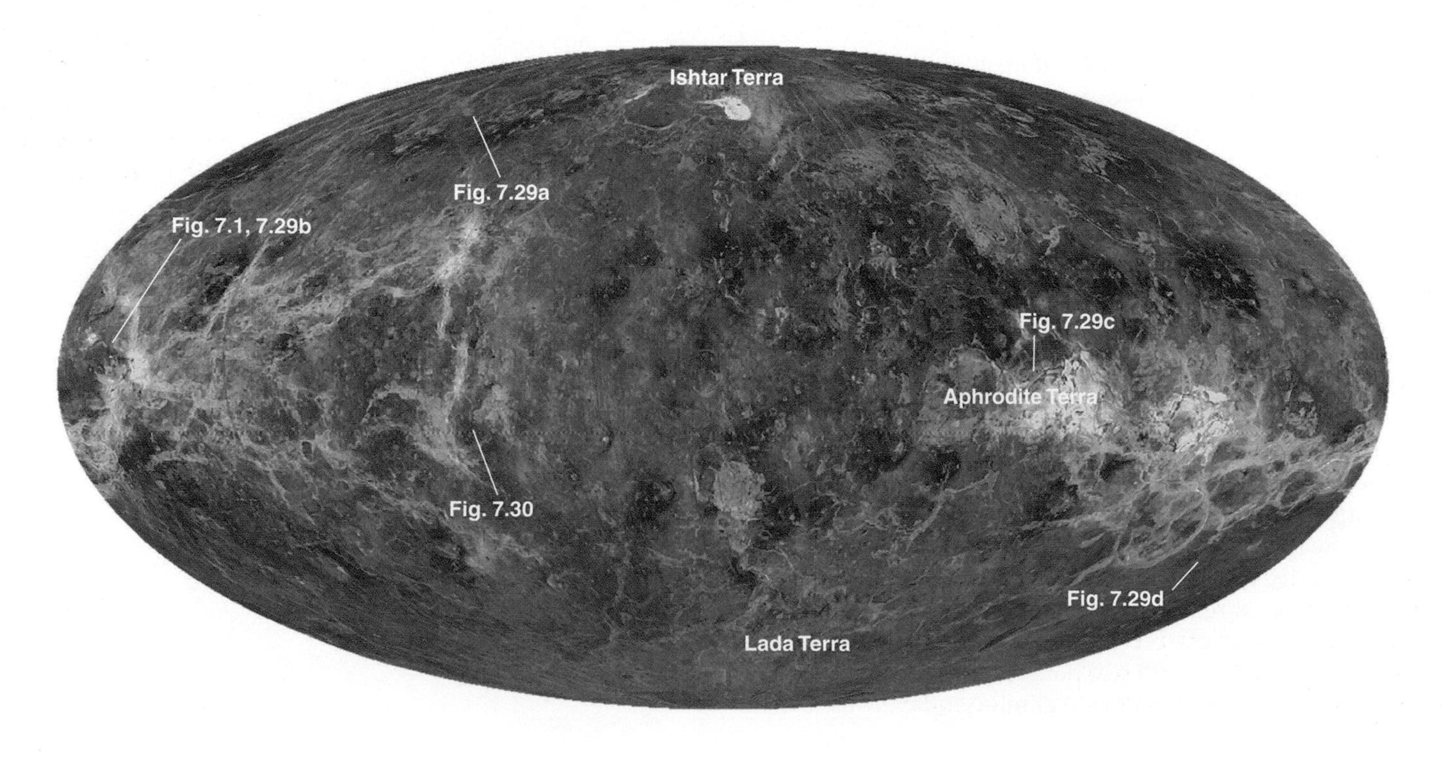

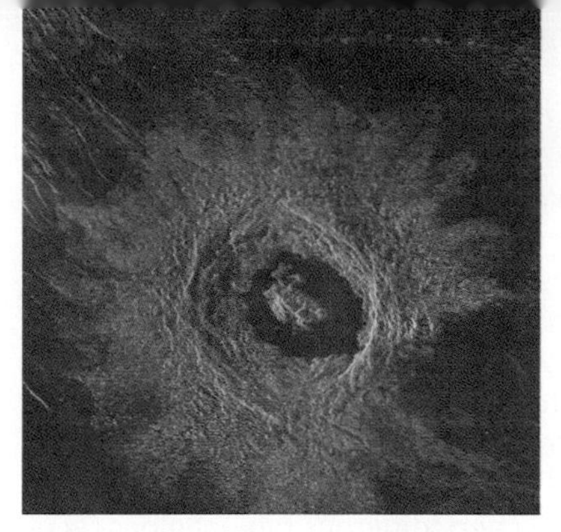

a An impact crater on Venus.

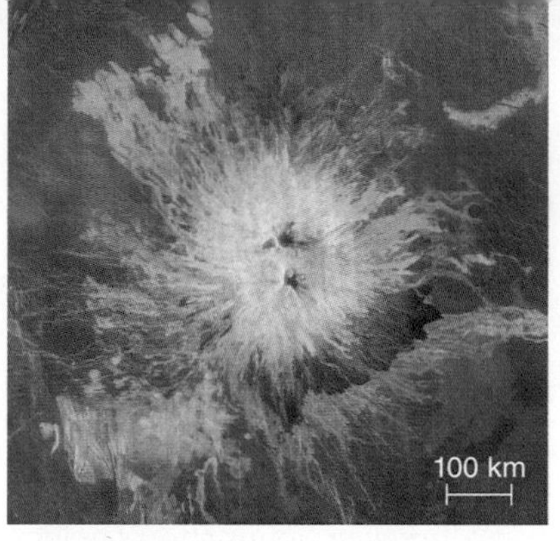

b The two peaks near the center are volcanoes, probably much like the volcanoes that created the Hawaiian islands on Earth.

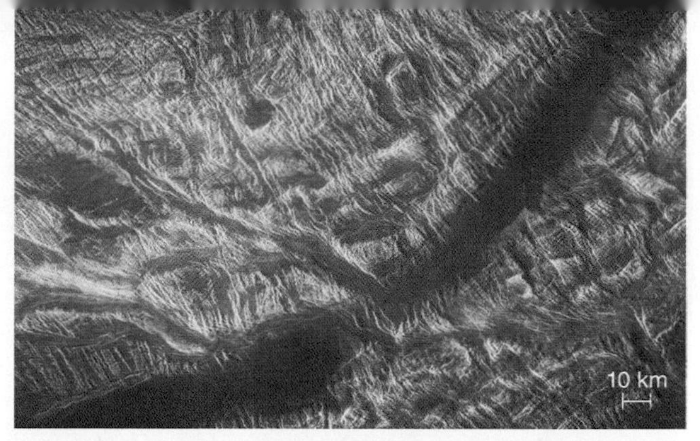

c Tectonic forces have fractured and twisted the crust in this region.

From 1990 to 1993, the *Magellan* spacecraft used radar to map the surface of Venus, discerning features as small as 100 meters across. Figure 7.28 shows a global map of Venus based on the *Magellan* radar observations. Careful study suggests that Venus is indeed geologically active, just as we would expect for a planet almost as large as Earth.

Geological Features of Venus Closer-up radar images show that Venus has many geological features similar to Earth's, including occasional impact craters, volcanoes, and a lithosphere that has been contorted by tectonic forces (Figure 7.29a–c). Venus also has some unique features, but they can still be explained by assuming that Venus has mantle convection much like Earth. For example, the circular features called *coronae* (Latin for "crowns") probably were made by a hot, rising plume in the mantle (Figure 7.29d). The plume pushed up on the crust, forming concentric tectonic stretch marks on the surface. The plume also forced lava to the surface, dotting the area with volcanoes.

Venus shows features of volcanism and tectonics, just as we expect for a planet of similar size to Earth.

Venus almost undoubtedly remains geologically active today, though we have no direct observations of active volcanoes. However, Venus should still retain nearly as much internal heat as Earth, and its relatively few impact craters tell us that its surface is geologically young. In addition, the composition of Venus's clouds suggests that volcanoes must still be active on geological time scales (erupting within the past 100 million years). The clouds contain sulfuric acid, which is made from sulfur dioxide (SO_2) and water. Sulfur dioxide enters the atmosphere through volcanic outgassing, but once in the atmosphere it is gradually removed by chemical reactions with surface rocks. Thus, the fact that sulfuric acid clouds still exist means that outgassing must continue to supply sulfur dioxide to the atmosphere.

The biggest difference between the geology of Venus and that of Earth is the lack of erosion on Venus. We might naively expect Venus's thick atmosphere to produce strong erosion, but the view both from orbit and on the surface suggests otherwise. The Soviet Union landed two probes on Venus's surface in 1975. Before they were destroyed by the intense surface heat, the probes returned images of a bleak, volcanic landscape with little evidence of erosion (Figure 7.30). We can trace the lack of erosion on Venus to two simple facts. First, Venus is far too hot for any type of rain or snow on its surface. Second, Venus has virtually no wind or weather because of its slow rotation. Venus rotates so slowly—once every 243 days—that its atmosphere barely stirs the surface. Without any glaciers, rivers, rain, or strong winds, there is very little erosion on Venus.

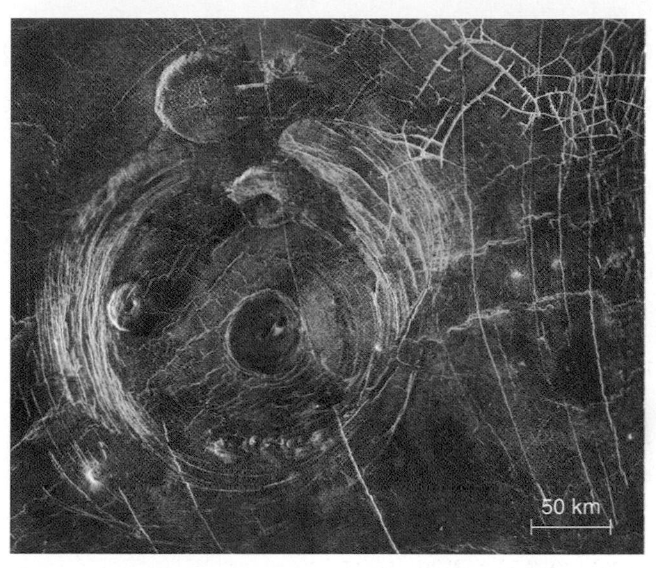

d A mantle plume probably created this round corona, which is surrounded by tectonic stress marks. The smaller round dots and blobs are volcanoes.

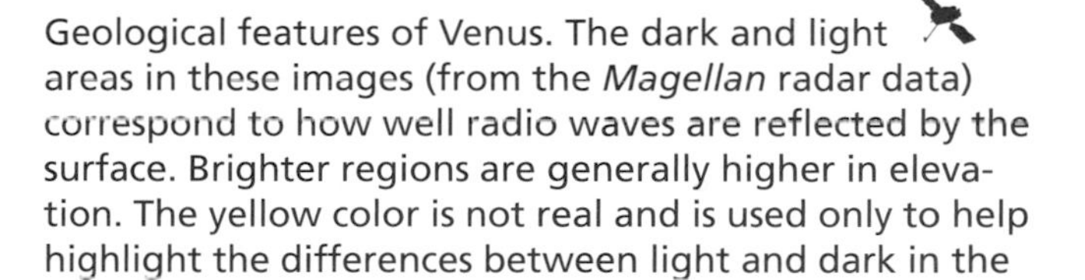

Figure 7.29

Geological features of Venus. The dark and light areas in these images (from the *Magellan* radar data) correspond to how well radio waves are reflected by the surface. Brighter regions are generally higher in elevation. The yellow color is not real and is used only to help highlight the differences between light and dark in the radar images.

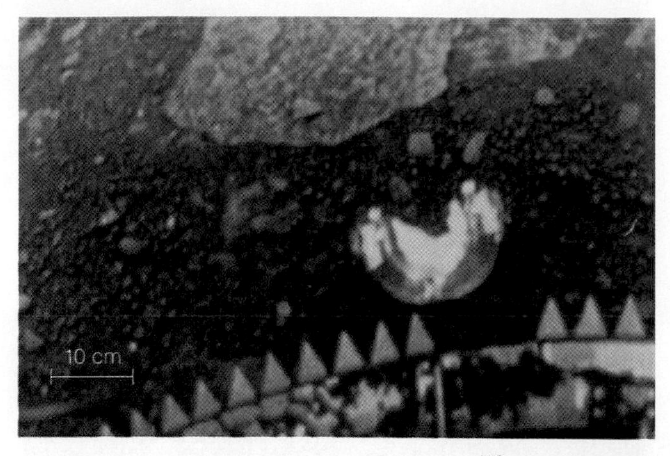

Figure 7.30

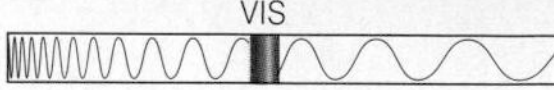

This photo from the Soviet Union's *Venera* lander shows the surface around the base of the spacecraft (visible at lower right). This region of the planet is covered in volcanic plains formed some 750 million years ago. Many volcanic rocks are visible, hardly affected by erosion despite their age.

The Absence of Plate Tectonics While we can easily explain the lack of erosion on Venus, another "missing feature" in Venus's geology is very surprising: Venus shows no evidence at all of Earth-like plate tectonics. Almost all the major features of Earth's geology are attributable to plate tectonics, including the ridges that extend thousands of kilometers along the seafloor, the deep ocean trenches found near continental boundaries, and long mountain ranges like the Rockies and Himalayas. Venus does not have any similar features.

Moreover, Venus shows evidence of a very different type of global geological change. On Earth, plate tectonics resculpts the surface gradually, so that different regions have different ages. But the relatively few impact craters on Venus are distributed fairly uniformly over the entire planet, suggesting that the surface is about the same age everywhere. Precise crater counts suggest that the surface is about 750 million years old. Apparently, the entire surface of Venus was somehow "repaved" at that time.

We do not know how much of the repaving was due to tectonic processes and how much was due to volcanism, but both probably were important. It is even possible that plate tectonics played a role before and during the repaving, only to stop for some reason after the repaving episode was over. Regardless of whether Venus may have had plate tectonics in the past, the absence of plate tectonics today poses a major mystery.

Earth's lithosphere fractured into plates because of forces arising from the underlying mantle convection. Thus, the lack of plate tectonics on Venus suggests either that it has weaker mantle convection or that its lithosphere somehow resists fracturing. The first possibility seems unlikely: Venus's similar size to Earth means it should have a similar level of mantle convection. Most scientists therefore suspect that Venus's lithosphere resists fracturing into plates because it is thicker and stronger than Earth's lithosphere, though we have no direct evidence to support this hypothesis.

Venus's lack of Earth-like plate tectonics poses a scientific mystery, but may be due to its having a thicker and stronger lithosphere than Earth.

Even if a thicker and stronger lithosphere explains the lack of plate tectonics on Venus, we are still left with the question of why the lithospheres of Venus and Earth should differ. One possible answer is Venus's high surface temperature. Venus is so hot that any water in its crust and mantle has probably been baked out over time. Water tends to soften and lubricate rock, so its loss would have tended to thicken and strengthen Venus's lithosphere. If this idea is correct, then Venus might have had plate tectonics if it had not become so hot in the first place.

• Why is Venus so hot?

It's tempting to attribute Venus's high surface temperature solely to the fact that it is closer to the Sun than Earth, but Venus would actually be quite cold without its strong greenhouse effect. Venus absorbs less sunlight than Earth, despite being closer to the Sun, because its clouds reflect so much sunlight back to space. Detailed calculations show that Venus's average surface temperature would be a frigid −43°C (−45°F) without the greenhouse effect, rather than its actual temperature of about 470°C (880°F). Thus, the real question is why Venus has such a strong greenhouse effect.

Venus's thick, carbon dioxide atmosphere creates the very strong greenhouse effect that makes Venus so hot.

On a simple level, the answer is the huge amount of carbon dioxide in Venus's atmosphere. Remember that Venus has a far thicker atmosphere than Earth—its surface pressure is about 90 times that on Earth—

and this atmosphere is about 96% carbon dioxide. The total amount of carbon dioxide in Venus's atmosphere is nearly 200,000 times as great as the amount in Earth's atmosphere. This vast quantity of carbon dioxide creates the strong greenhouse effect.

However, a deeper question still remains. Given their similar sizes and compositions, we expect Venus and Earth to have had similar levels of volcanic outgassing—and the released gas ought to have had about the same composition on both worlds. Why, then, is Venus's atmosphere so different from Earth's?

Atmospheric Composition We expect that huge amounts of water and carbon dioxide should have been outgassed into the atmospheres of both Venus and Earth. Venus's atmosphere does indeed have huge amounts of carbon dioxide, but it has virtually no water. Earth's atmosphere has very little of either gas. What happened to all the outgassed water on Venus, and to the outgassed water and carbon dioxide on Earth?

Earth has as much carbon dioxide as Venus, but it is mostly locked away in rocks rather than in our atmosphere.

We can easily account for both missing gases on Earth. The huge amounts of water vapor released into our atmosphere condensed into rain, forming our oceans. Thus, the water is still here, but in liquid rather than gaseous form. The huge amount of carbon dioxide released into our atmosphere is also still here, but in solid form: Carbon dioxide dissolves in water, where it can undergo chemical reactions to make **carbonate** rocks (rocks rich in carbon and oxygen) such as limestone. Earth has about 170,000 times as much carbon dioxide locked up in rocks as in its atmosphere—which means that Earth does indeed have about as much total carbon dioxide as Venus. Of course, the fact that Earth's carbon dioxide is mostly in rocks rather than in the atmosphere makes all the difference in the world: If this carbon dioxide were in our atmosphere, our planet would be nearly as hot as Venus and certainly uninhabitable.

We are left with the question of what happened to Venus's water. Venus certainly has no ocean and, as discussed earlier, any water in its crust and mantle was probably baked out long ago. Venus's atmosphere contains little water either. There seems to be essentially no water on Venus at all. Moreover, the absence of water explains why Venus retains so much carbon dioxide in its atmosphere: Without oceans, carbon dioxide cannot dissolve or become locked away in carbonate rocks. If it is true that a huge amount of water was outgassed on Venus, it has somehow disappeared.

Venus retains carbon dioxide in its atmosphere because it lacks oceans to dissolve the carbon dioxide and lock it away in rock.

The leading hypothesis for the disappearance of Venus's water invokes one of the same processes thought to have removed water from Mars. Water molecules in Venus's atmosphere are broken apart by ultraviolet light from the Sun. The hydrogen atoms then escape to space, ensuring that the water molecules can never reform. The oxygen from the water molecules was probably stripped away by the solar wind. Venus's atmosphere is vulnerable to the solar wind because Venus lacks a significant magnetic field, perhaps because it rotates so slowly.

Acting over billions of years, the breakdown of water molecules and the escape of hydrogen can easily explain the loss of an ocean's worth of water from Venus—as long as the water was in the atmosphere rather than in liquid oceans. Thus, our quest to understand Venus's high temperature leads to one more question: Why didn't Venus get oceans like Earth, which would have prevented its water from being lost to space?

The Runaway Greenhouse Effect To understand why Venus does not have oceans, let's consider what would happen if we could magically move Earth to the orbit of Venus.

The greater intensity of sunlight would almost immediately raise Earth's global average temperature by about 30°C, from its current 15°C to about 45°C (113°F). Although this is still well below the boiling point of water, the higher temperature would lead to increased evaporation of water from the oceans. The higher temperature would also allow the atmosphere to hold more water vapor before the vapor condensed to make rain.

The combination of more evaporation and greater atmospheric capacity for water vapor would substantially increase the total amount of water vapor in Earth's atmosphere. Because water vapor is a greenhouse gas, this added water vapor would strengthen the greenhouse effect, driving temperatures a little higher. The higher temperatures, in turn, would lead to even more ocean evaporation and more water vapor in the atmosphere—strengthening the greenhouse effect even further. In other words, we'd have a "positive feedback loop" in which each little bit of additional water vapor in the atmosphere would mean higher temperature and even more water vapor.

The process would rapidly spin out of control, resulting in a **runaway greenhouse effect** (Figure 7.31). Our planet would heat up until the oceans were completely evaporated and the carbonate rocks had released all their carbon dioxide back into the atmosphere. By the time the runaway process was complete, temperatures on our "moved Earth" would be even higher than they are on Venus today, thanks to the combined greenhouse effects of carbon dioxide and water vapor in the atmosphere. The water vapor would then gradually disappear, as ultraviolet light broke water molecules apart and the hydrogen escaped to space. In short, moving Earth to Venus's orbit would essentially turn our planet into another Venus.

We have arrived at a simple explanation of why Venus is so much hotter than Earth. Even though Venus is only about 30% closer to the

Figure 7.31

This diagram shows how, if Earth were placed at Venus's distance from the Sun, the runaway greenhouse effect would cause the oceans to evaporate completely.

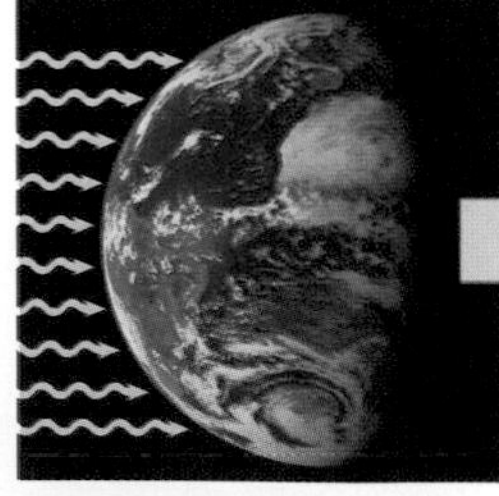

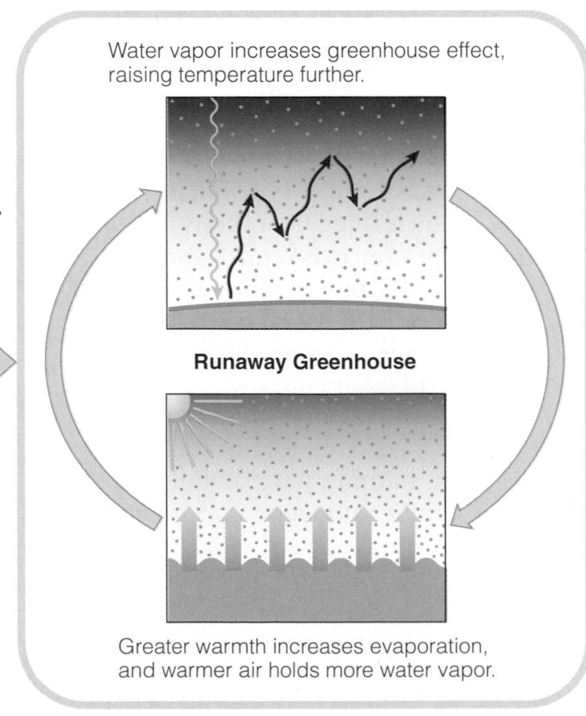

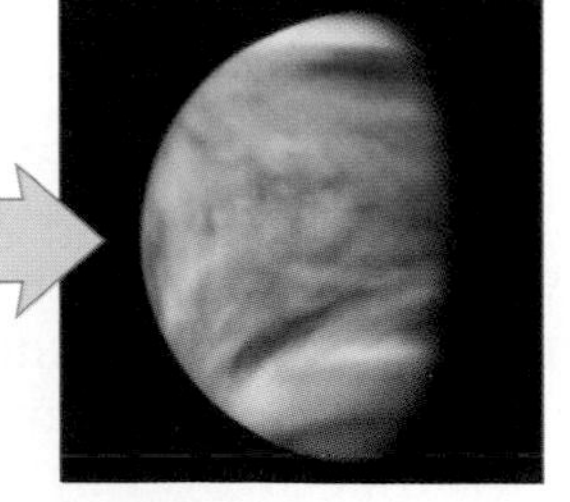

Venus is too close to the Sun to have liquid water oceans. Without water to dissolve carbon dioxide gas, Venus was doomed to its runaway greenhouse effect.

Sun than Earth, the difference in distance was apparently critical. On Earth, it was cool enough for water to rain down to make oceans. The oceans then dissolved carbon dioxide and chemical reactions locked it away in carbonate rocks, leaving our atmosphere with only enough greenhouse gases to make our planet pleasantly warm. On Venus, the greater intensity of sunlight made it just enough warmer that oceans either never formed or soon evaporated. Without oceans to dissolve carbon dioxide and make carbonate rock, carbon dioxide accumulated in the atmosphere, leading to a runaway greenhouse effect.

The next time you see Venus shining brightly as the morning or evening "star," consider the radically different path it has taken from that taken by Earth—and thank your lucky star. If Earth had formed a bit closer to the Sun or if the Sun had been slightly hotter, our planet might have suffered the same greenhouse-baked fate.

We've seen that moving Earth to Venus's orbit would cause our planet to become Venus-like. If we could somehow move Venus to Earth's orbit, would it become Earth-like? Why or why not?

7.5 EARTH AS A LIVING PLANET

We began this chapter by discussing Earth as a planet, looking at features and processes that it shares in common with some of our planetary neighbors. We then explored the histories of the other terrestrial worlds, finding that we could understand them by thinking about fundamental planetary properties (such as size and distance from the Sun) and processes familiar to us from Earth. However, we have not yet discussed the feature that makes Earth truly unique: its abundance of life, including human life. It is time for us to turn our attention back to Earth, to see how and why our planet is such a pleasant place for us to live.

• What unique features of Earth are important for life?

If you think about what we've learned about the terrestrial worlds, you can probably identify a number of features that are unique to Earth. Four unique features turn out to be particularly important to life on Earth:

- **Surface liquid water**: Earth is the only planet on which temperature and pressure conditions allow surface water to be stable as a liquid.
- **Atmospheric oxygen**: Earth is the only planet with significant oxygen in its atmosphere and an ozone layer.
- **Plate tectonics**: Earth is the only planet with a surface shaped largely by this distinctive type of tectonics.
- **Climate stability**: Earth differs from the other terrestrial worlds with significant atmospheres (Venus and Mars) in having a climate that has remained relatively stable throughout its history.

Let's examine each of these features in a little more detail.

Our Unique Atmosphere and Oceans The first and second items in our list—abundant liquid water and atmospheric oxygen—are clearly important to our existence. Water is thought to be important for life of any kind [Section 18.1], and animal life requires oxygen. We have already explained the origin of Earth's water: Water vapor outgassed from volcanoes rained down on the surface to make the oceans and neither froze nor evaporated thanks to our moderate greenhouse effect and distance from the Sun. But where did the oxygen in our atmosphere come from?

Oxygen (O_2) is not a product of volcanic outgassing. In fact, no geological process can explain how oxygen came to make up 21% of Earth's atmosphere. Moreover, oxygen is a highly reactive gas that would disappear from the atmosphere in just a few million years if it were not continuously resupplied. Many familiar chemical processes remove oxygen from the atmosphere, including fire, rust, and the discoloration of freshly cut fruits and vegetables. We must therefore explain not only how oxygen got into Earth's atmosphere in the first place, but also how the amount of oxygen remains relatively steady even while chemical reactions can remove it rapidly from the atmosphere.

Without life, there would be no oxygen in Earth's atmosphere.

The answer to the oxygen mystery is life itself. Plants and many microorganisms release oxygen through photosynthesis. Photosynthesis takes in CO_2 and releases O_2. The carbon becomes incorporated into a variety of components of living organisms. Virtually all the oxygen in Earth's atmosphere was originally produced by photosynthetic life. Today, photosynthetic organisms return oxygen to the atmosphere in approximate balance with the rate at which chemical reactions (and animals) consume it, which is why the oxygen content of the atmosphere remains steady. This oxygen is also what makes possible Earth's protective ozone layer, since ozone (O_3) is produced from ordinary oxygen (O_2).

Suppose that, somehow, all photosynthetic life (such as plants) died out. What would happen to the oxygen in our atmosphere? Could animals, including us, still survive?

Plate Tectonics The connection between tectonics and human existence is less direct than our connection with water and oxygen, but it may be equally important. Recall that Earth's lithosphere is broken into more than a dozen plates that slowly move about through the action we call plate tectonics. As we'll see shortly, plate tectonics plays a crucial role in the climate stability that has allowed life on Earth to thrive for billions of years. First, however, let's explore how plate tectonics has created the ocean basins and the continents on which we live. Figure 7.32 shows known boundaries of plates.

The motions of the plates are barely noticeable on human time scales. On average, plates move at speeds of only a few centimeters per year—about the same speed at which your fingernails grow. Nevertheless, over millions of years, the movements act like a giant conveyor belt for Earth's lithosphere, creating new crust and recycling old crust back into the mantle (Figure 7.33).

Mantle material rises upward and erupts to the surface along mid-ocean ridges, becoming new crust for the seafloor, or **seafloor crust** for short. This newly emerging material causes the seafloor to spread away

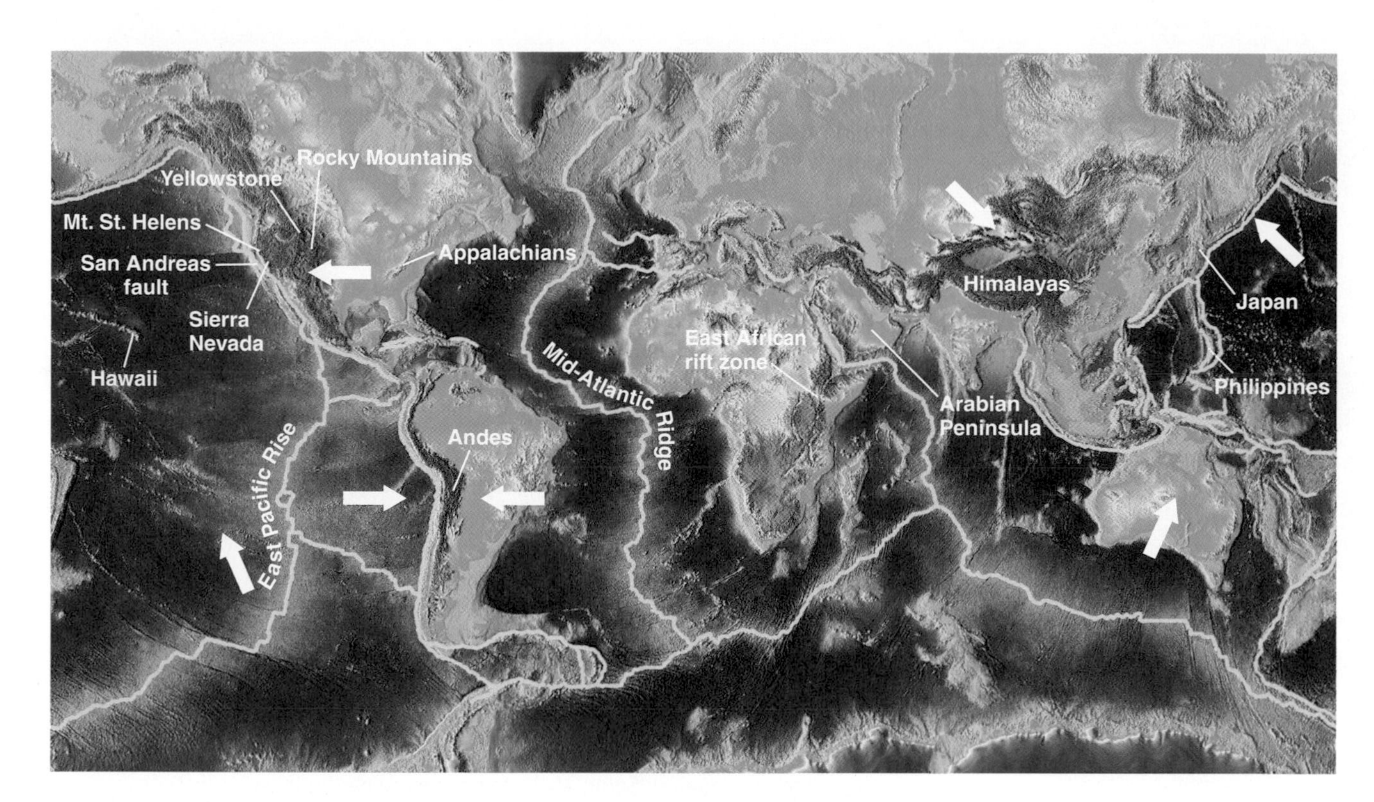

Figure 7.32

This relief map shows plate boundaries (solid yellow lines), with arrows to represent directions of plate motion. Color represents elevation, progressing from blue (lowest) to red (highest). Labels identify some of the geological features discussed in the text.

from the ridge, which is why the ridges are found in the middle of the ocean. Over tens of millions of years, any piece of seafloor crust gradually makes its way across the ocean bottom, then finally gets recycled into the mantle in the process we call **subduction**.

Plate tectonics acts like a giant conveyor belt for Earth's lithosphere, continually recycling seafloor crust and gradually building up the continents.

Subduction occurs where seafloor plates run into continental plates. As seafloor crust descends into the mantle, it is heated and may partially melt. This molten rock then erupts from volcanoes over the subduction zones, which is why

Figure 7.33 Interactive Figure

Plate tectonics acts like a giant conveyor belt that produces and recycles seafloor crust. New seafloor crust erupts from the mantle at mid-ocean ridges, where plates spread apart. At ocean trenches, the seafloor crust pushes under the less dense continental crust, returning the seafloor crust to the mantle. The subducting seafloor crust may partially melt, leading to volcanic eruptions that produce new continental crust.

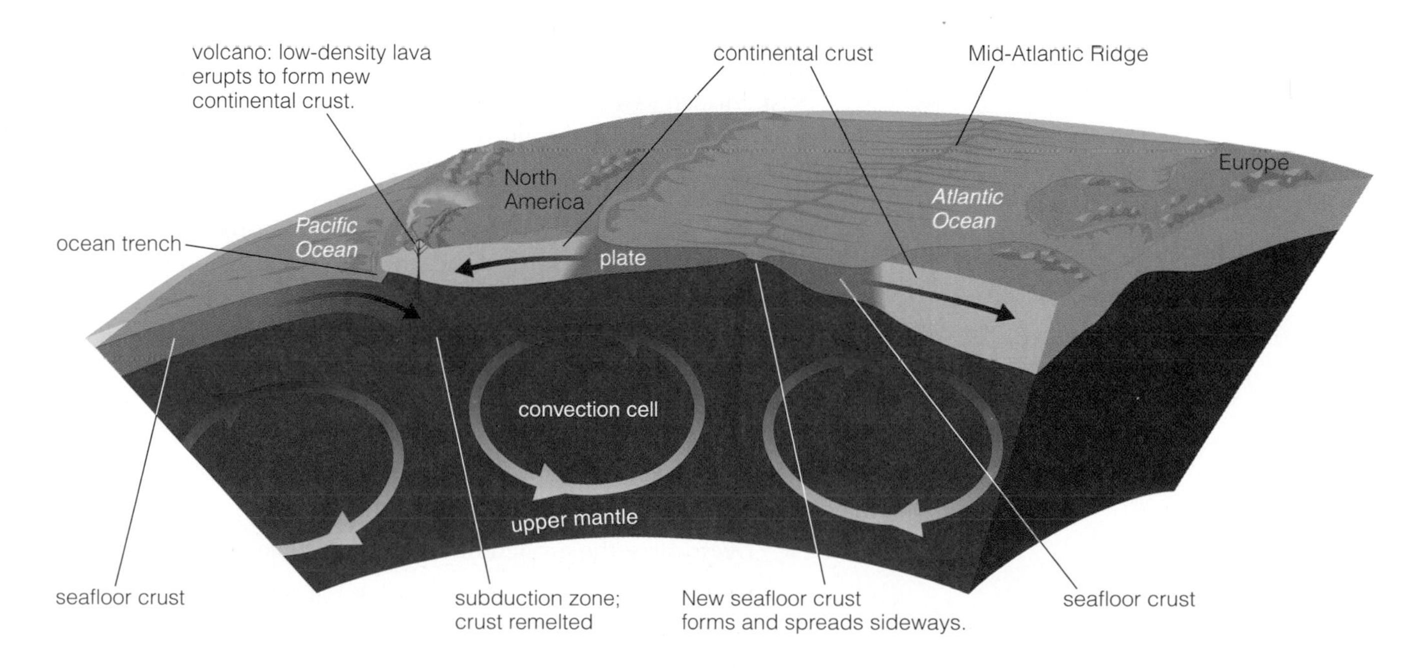

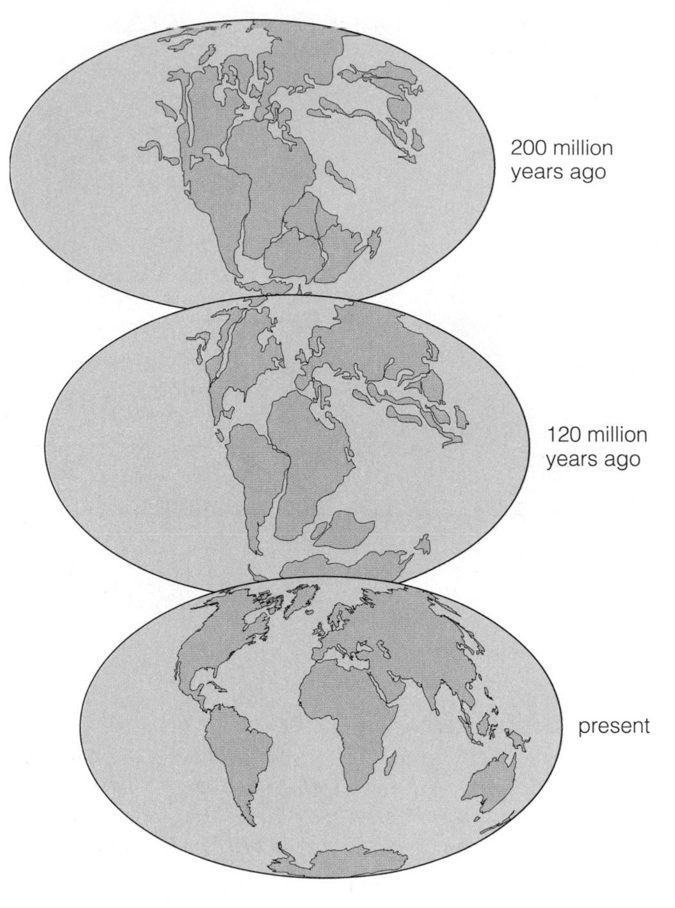

Figure 7.34 Interactive Figure

These diagrams show the changing arrangements of continents over the past 200 million years. Notice how the present shapes of South America and Africa reflect the way they fit together in the past. The continents are always in motion, so the arrangement looked different at earlier times and will look different in the future.

so many active volcanoes tend to be found along the edges of continents. Moreover, the lowest-density material tends to melt first, so the **continental crust** emerging from these landlocked volcanoes is much lower in density than seafloor crust. Note that, unlike seafloor crust, the continental crust does not get recycled back into the mantle. As a result, the continents are gradually growing with time, and their relatively low density explains why they rise above the seafloor, creating the ocean basins.

Before we examine how plate motions are tied to climate stability, it's worth noting that almost all Earth's active geology is tied to plate tectonics. For example, compression of the crust and mountain building occurs where continental plates are pushed together and valleys or seas can form where continental plates are pulling apart (see Figure 7.11). Earthquakes tend to occur when two plates get "stuck" against one another and then lurch violently when the pressure builds to the breaking point. Even the present arrangement of the continents is due to plate tectonics, since the continents are slowly pushed around as seafloors spread and subduct (Figure 7.34).

THINK ABOUT IT By studying plate boundaries in Figure 7.32, explain why California, Oregon, and Washington have more earthquakes and volcanoes than other parts of the United States. Looking at other parts of the map, would you expect Japan or South Africa to have more earthquakes? Why?

Climate Stability Earth's long-term climate stability has clearly been important to the ongoing evolution of life—and hence to our own relatively recent arrival as a species (see Figure 1.10). Had our planet undergone a runaway greenhouse effect like Venus, life would certainly have been extinguished. If Earth had suffered loss of atmosphere and a global freezing like Mars, any surviving life would have been driven to hide in underground pockets of liquid water.

Earth's climate is not perfectly stable—our planet has endured numerous ice ages and warm periods in the past. Nevertheless, even in the deepest ice ages and warmest warm periods, Earth's temperature has remained in a range in which liquid water could still exist and harbor life. On the surface, the key to this climate stability is simple: Earth has always kept just enough carbon dioxide in its atmosphere to keep the temperature in a range suitable for liquid water and life. But a deeper look reveals some complexity. In particular, we have good evidence that the Sun has brightened substantially over the past 4 billion years [Section 10.3], yet Earth's temperature has managed to stay in the same range throughout this time. Apparently, the strength of the greenhouse effect somehow self-adjusts to keep the climate stable.

The mechanism by which Earth self-regulates its temperature is called the **carbon dioxide cycle**, or the **CO_2 cycle** for short. Let's follow the cycle as illustrated in Figure 7.35, starting with the carbon dioxide in the atmosphere:

- Atmospheric carbon dioxide dissolves in the oceans.
- At the same time, rainfall erodes rocks on Earth's continents and rivers carry the eroded minerals to the oceans.
- In the oceans, the eroded minerals combine with dissolved carbon dioxide and fall to the ocean floor, making carbonate rocks such as limestone.

- Over millions of years, the conveyor belt of plate tectonics carries the carbonate rocks to subduction zones, and subduction carries them down into the mantle.
- As they are pushed deeper into the mantle, some of the subducted carbonate rock melts and releases its carbon dioxide, which then outgasses back into the atmosphere through volcanoes.

The CO_2 cycle acts as a thermostat for Earth, because the rate at which carbonate minerals form in the ocean is very sensitive to temperature. Carbonate minerals form faster at higher temperatures. As a result, a small change in Earth's temperature will be offset by changes caused through the CO_2 cycle.

Earth has remained habitable for billions of years because of a climate kept stable by the natural action of the carbon dioxide cycle.

For example, if Earth warms up a bit, carbonate minerals form in the oceans at a more rapid rate. The rate at which the oceans dissolve CO_2 gas thereby increases, pulling CO_2 out of the atmosphere. The reduced atmospheric CO_2 concentration leads to a weakened greenhouse effect that counteracts the initial warming and cools the planet back down. Similarly, if Earth cools a bit, carbonate minerals form more slowly in the oceans. The rate at which the oceans dissolve CO_2 gas thereby decreases, allowing the CO_2 released by volcanism to build back up in the atmosphere. The increased CO_2 concentration strengthens the greenhouse effect and warms the planet back up. Overall, the natural thermostat of the carbon dioxide cycle has allowed the greenhouse effect to strengthen or weaken just enough to keep Earth's climate fairly stable, regardless of what other changes have occurred on our planet.

We can now see why plate tectonics is so intimately connected to our existence. Notice that plate tectonics is a crucial part of the CO_2 cycle, since the subduction of carbonate rocks would not occur without plate movements. Thus, without plate tectonics, there could be no CO_2 cycle and Earth's climate might have undergone changes as dramatic as those that occurred on Venus and Mars.

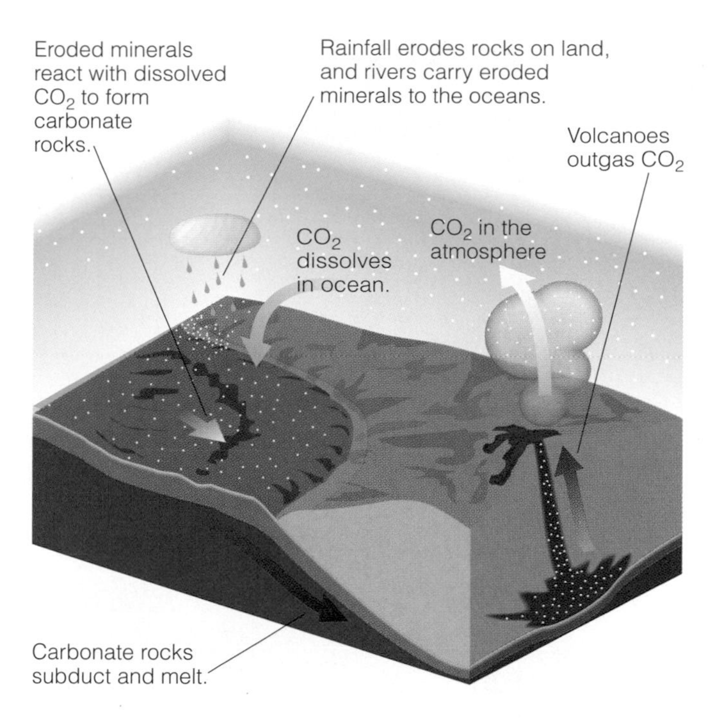

Figure 7.35

This diagram shows how the CO_2 cycle continually moves carbon dioxide from the atmosphere to the ocean to rock and back to the atmosphere. Note that plate tectonics (subduction in particular) plays a crucial role in the cycle.

• How might human activity change our planet?

We humans are well-adapted to the unique, present-day conditions on our planet. The amount of oxygen in our atmosphere, the average temperature of our planet, and the ultraviolet-absorbing ozone layer are just what we need to survive. We have seen that these "ideal" conditions are no accident, but instead consequences of our planet's unique geology and life.

Nevertheless, the stories of the dramatic and permanent climate changes that occurred on Venus and Mars should teach us to take nothing for granted. Our planet may regulate its own temperature quite effectively, and life may keep the level of oxygen and ozone in the atmosphere quite steady, but it is possible for humans to interfere with these natural mechanisms.

Some of the dangers posed by human activity are already clear. For example, we learned in the 1970s that human-made chemicals known as CFCs can rise high into the atmosphere, where they destroy some of the ozone that shields our planet from dangerous ultraviolet light. In the 1980s, we found a large and ominous "hole" forming in the ozone layer over Antarctica each spring. This clear evidence of a human role in altering a crucial aspect of our planet's habitability led to international action and treaties intended to stop ozone destruction.

Natural mechanisms maintain our planet's current conditions, but human activity can interfere with them.

Another well-recognized danger comes from extinctions. According to some estimates, human activity is driving species to extinction so rapidly that up to half of today's plant and animal species may be gone by the end of this century. This rate of extinction is comparable with that in past *mass extinctions,* such as the one that wiped out the dinosaurs some 65 million years ago [Section 9.4]. Prior mass extinctions generally led to the loss of the planet's dominant species at the time—and we are the dominant species today. Although we cannot predict the consequences of current extinctions with any certainty, it is at least conceivable that we could be unwittingly clearing the way for a new set of dominant species.

The threat that causes the most debate is **global warming**—an expected increase in Earth's global average temperature caused by human input of carbon dioxide and other greenhouse gases (such as methane and CFCs) into the atmosphere. The carbon dioxide cycle operates too slowly to absorb this excess of greenhouse gases, so it is at least possible that Earth as a whole will warm due to a strengthening greenhouse effect.

The Global Warming Debate

The potential threat of global warming is a hot political issue because alleviating the threat would require finding new energy sources and making other changes that would dramatically affect the world's economy. However, a major research effort has gradually added to our understanding of the potential threat, particularly in the past decade. Although some controversy still exists about how serious the problem of global warming might be, at least three facts are now clear:

1. Measurements show that Earth has indeed warmed up over the past 50 years, by about 0.5°C. Although this may sound small, it is a significant increase in such a short time period. The warming trend has continued and gotten stronger in recent years.
2. The burning of fossil fuels and other human activity is clearly increasing the amounts of greenhouse gases in the atmosphere. The current concentration of carbon dioxide in Earth's atmosphere is significantly higher than it has been at any time during the past 400,000 years, and the concentration is rising rapidly (Figure 7.36).
3. Because we understand the basic mechanism of the greenhouse effect (see Figure 7.14), there is no doubt that a continually rising concentration of greenhouse gases would eventually make our planet warm up.

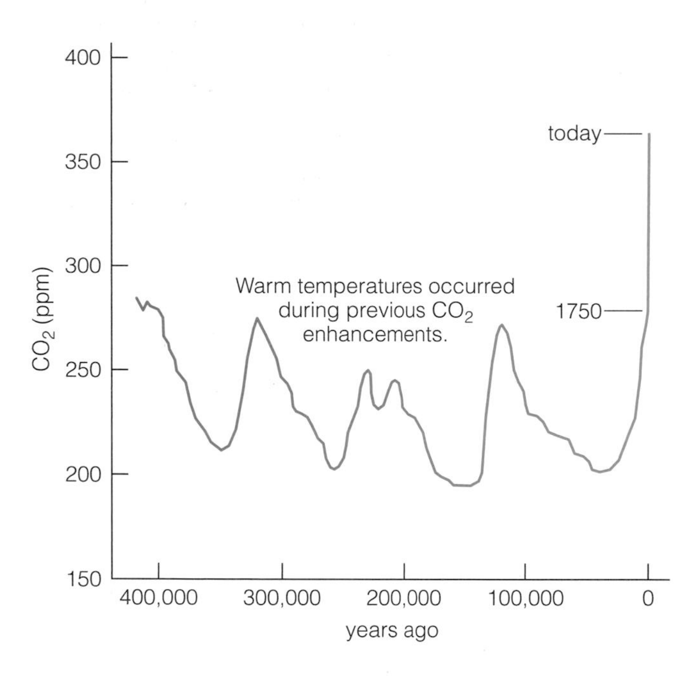

Figure 7.36

This diagram shows the atmospheric concentration of carbon dioxide over the past 400,000 years. Notice the dramatic rise in recent years. Temperature data (not shown) demonstrate that periods of higher CO_2 concentration also coincide with times of higher global average temperature. The CO_2 data for the past half-century come from direct measurements (made at Mauna Loa); most of the earlier data come from studies of air bubbles trapped in Antarctic ice (ice core samples). The concentration is measured in parts per million (ppm), which is the number of CO_2 molecules among every 1 million air molecules.

Human activity is increasing the concentration of greenhouse gases in the atmosphere, which may strengthen the greenhouse effect and lead to global warming.

However, while continued input of greenhouse gases would eventually cause global warming, we are not yet certain that human activity has caused the global warming observed to date. The problem is that numerous feedback mechanisms can counter or enhance greenhouse warming on time scales of years, decades, and even centuries, and we don't yet understand all these mechanisms. For example, increased evaporation of ocean water could enhance the greenhouse effect because water vapor is also a greenhouse gas, but it could counter global warming if it caused the formation of more clouds that prevented sunlight from reaching the ground.

Scientists seek to understand the problem better by creating sophisticated computer models of the climate and comparing the model predictions to actual observations. Current models suggest that global temperatures are increasing in a manner consistent with what we expect if human activity is the cause (Figure 7.37). Thus, while plenty of uncertainties still remain, most scientists are now convinced that human activity is indeed causing global warming.

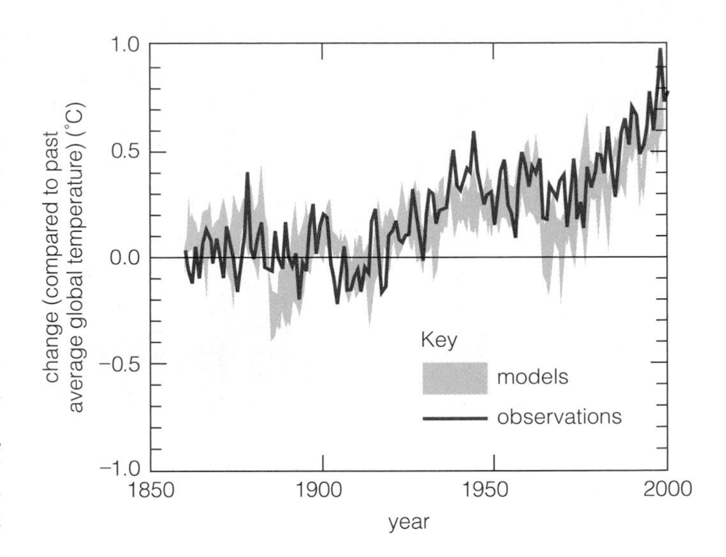

Figure 7.37

This graph compares the predictions of climate models with observed temperature changes (red line) since about 1860. The thickness of the green swath for the models represents the range of variation between several different models. These models take into account both natural factors (such as slight changes in the Sun's brightness and the effects of volcanic eruptions) and emission of greenhouse gases by humans. The agreement is not perfect—telling us we still have much to learn—but it is good enough to give us confidence that greenhouse gases are, indeed, causing global warming.

Potential Consequences of Global Warming Clearly, global warming would mean a higher global average surface temperature for Earth. Beyond that simple statement, however, the issue of consequences becomes much more complicated and the uncertainties involved in prediction even greater.

Changing weather patterns would ensure that different regions of Earth experience different degrees of warming. Some regions would even become colder. Other regions might experience more rainfall or might become deserts. The greater overall warmth of the atmosphere would tend to mean more evaporation from the oceans, leading to more numerous and more intense storms.

Another potential threat comes from rising sea level. Water expands very slightly as it warms—so slightly that we don't notice the change in a glass of water, but enough to cause changes in sea level. Sea level has already risen some 20 centimeters in the past hundred years, and could rise another meter as the oceans warm. Such increases in sea level would threaten many coastal regions with more severe flooding. A much greater increase in sea level may occur if polar ice and glaciers melt. Such melting appears to be occurring to some extent already. For example, the famous "snows" (glaciers) of Mount Kilimanjaro are already in retreat and may be gone within the next decade or so.

Secondary effects, such as those arising from changes to ocean currents or ecological changes, pose an even more intractable problem. For example, it is difficult to know how forests and other ecosystems will respond to climate changes induced by global warming. As a result, we cannot easily predict the impact of such climate changes on food production, fresh water availability, or other issues critical to the well-being of human populations.

Given the current uncertainties, no one can predict the precise impact of global warming over the next century. However, the lesson from the solar system is clear. Dramatic and deadly change can occur unexpectedly, and we do not know how our tampering might affect the finely balanced mechanisms that control Earth's climate.

If you were a political leader, how would *you* deal with the uncertain threat of global warming?

• What makes a planet habitable?

We have discussed the features of our planet that have made our world habitable for a great variety of life, including us. But why does Earth alone have these features, at least among the terrestrial worlds of our solar system?

Our comparative study of the terrestrial worlds tells us there are two primary answers. First, Earth is habitable because it is large enough to

The Role of Planetary Size

Small Terrestrial Planets

cold, solid interior
tectonic and volcanic activity end after a billion years or so
many ancient craters
little outgassing, atmosphere lost due to low gravity
no atmosphere, therefore no erosion

Large Terrestrial Planets

warm, convecting interior
ongoing tectonic and volcanic activity
most ancient craters erased
lots of outgassing
strong gravity retains atmosphere, erosion possible

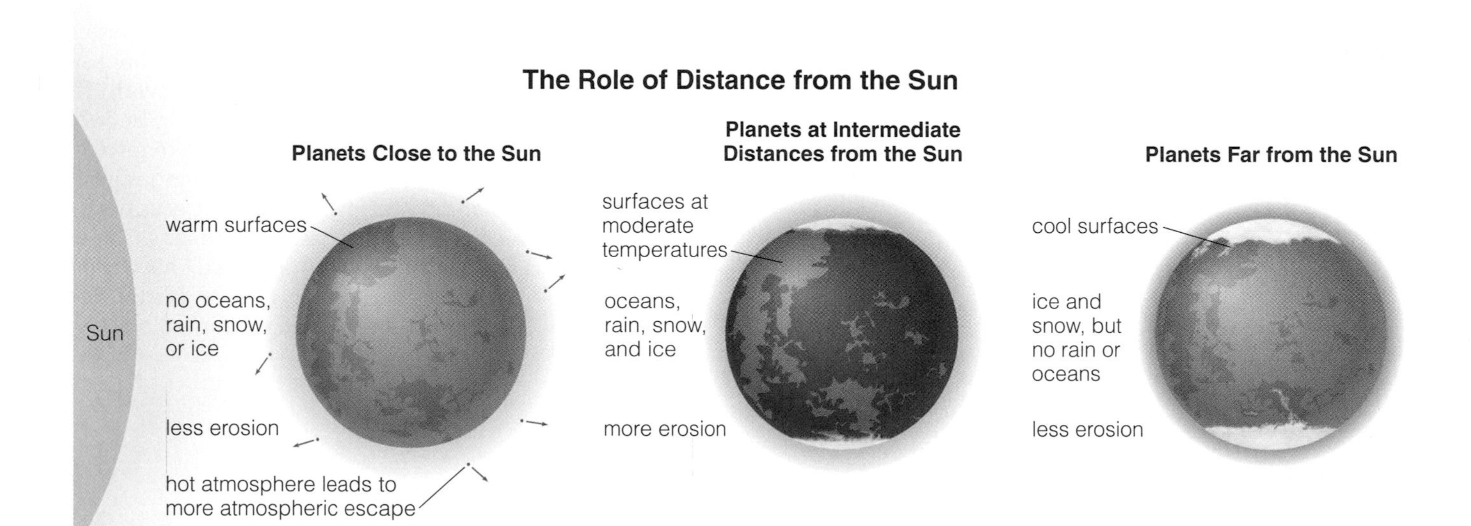

Figure 7.38

This illustration shows how a terrestrial world's size and distance from the Sun help determine its geological history and whether it has conditions suitable for life. Earth is habitable because it is large enough and at a suitably moderate distance from the Sun.

have remained geologically active since its birth, so that outgassing could release the water and gases that formed our atmosphere and oceans. Second, we are located at a distance from the Sun where outgassed water vapor was able to condense and rain down to form oceans, thus making possible the carbon dioxide cycle that regulates our climate.

Earth is habitable because it is large enough to remain geologically active and located at a distance from the Sun where oceans were able to form.

Figure 7.38 summarizes the lessons we have learned about how a planet's size and distance from the Sun will affect its fate. Figure 7.39 shows the trends we've seen for the terrestrial planets in our solar system, which should make sense as you examine the role of planetary size. In principle, these lessons mean we can now predict the geological and atmospheric properties of terrestrial worlds that we may someday find around other stars. Only a suitably large terrestrial planet located at an intermediate distance from its star is likely to have conditions under which life could thrive. Of course, we do not yet know whether simply having conditions suitable for life means that life will actually arise. We will consider current understanding of this fascinating question in Chapter 18, when we consider the prospects of finding life beyond Earth.

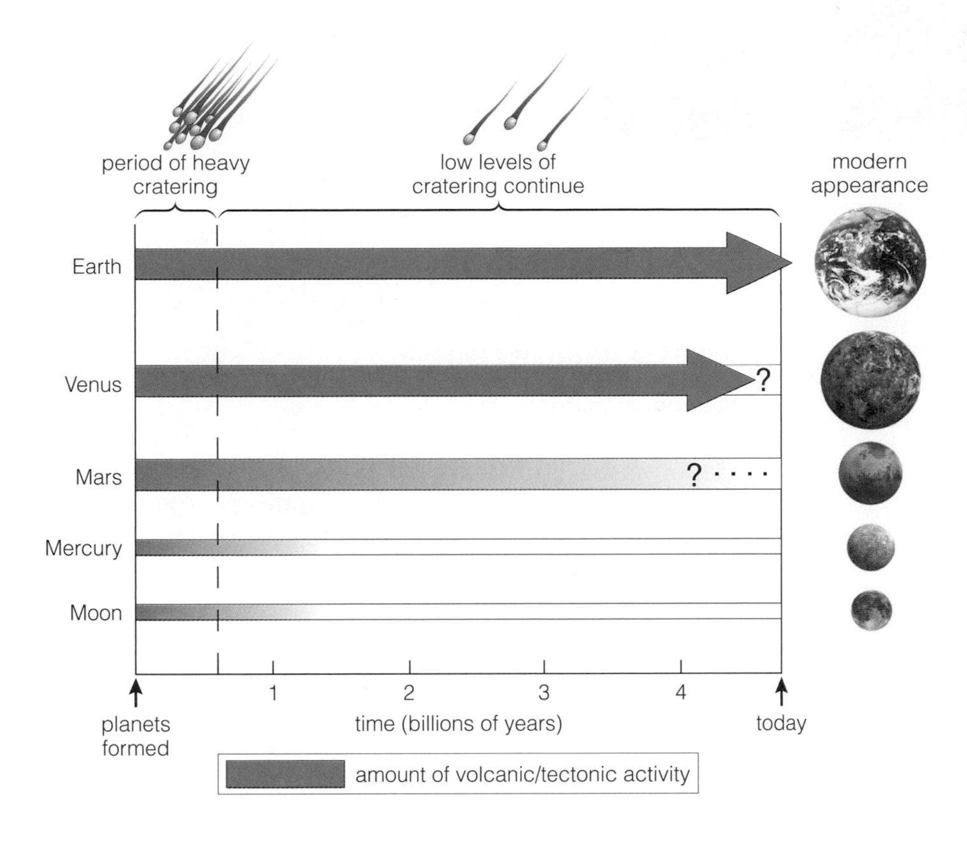

Figure 7.39

This diagram summarizes the geological histories of the terrestrial worlds. The brackets along the top indicate that impact cratering has affected all worlds similarly. The arrows represent volcanic and tectonic activity. A thicker and darker arrow means more volcanic/tectonic activity, and the arrow length tells us how long this activity persisted. Notice that the trend follows the order of planetary size: Earth remains active to this day. Venus has also been quite active, though we are uncertain whether it remains so. Mars has had an intermediate level of activity and might still have low-level volcanism. Mercury and the Moon have had very little volcanic/tectonic activity.

THE BIG PICTURE

Putting Chapter 7 into Context

In this chapter, we have explored the histories of the terrestrial worlds. As you think about the details you have learned, keep the following "big picture" ideas in mind:

- The terrestrial worlds all looked much the same when they were born, so their present-day differences are a result of geological processes that occurred in the ensuing 4.5 billion years.
- The primary factor in determining a terrestrial world's geological history is its size, because only a relatively large world can retain internal heat long enough for ongoing geological activity. However, the differences between Venus and Earth show that distance from the Sun also plays an important role.
- A planet's distance from the Sun is important to its surface temperature, but the cases of Venus and Earth show that the strength of the greenhouse effect can play an even bigger role. Humans are currently altering the balance of greenhouse gases in Earth's atmosphere, with potentially dire (if uncertain) consequences.
- The histories of Venus and Mars show that a stable climate like Earth's is more the exception than the rule. The stable climate that makes our existence possible is a direct consequence of our planet's unique geology, including its plate tectonics and carbon dioxide cycle.

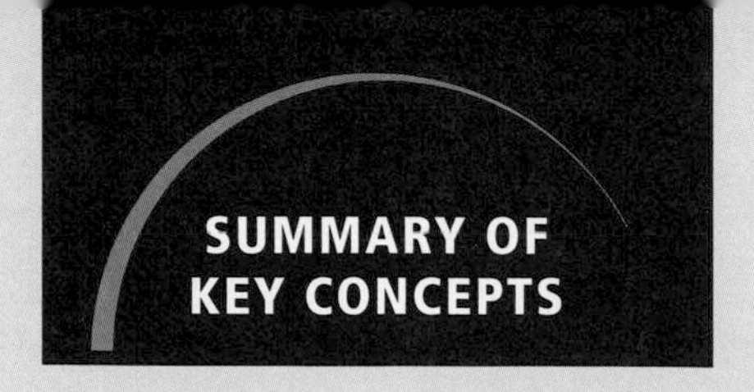

7.1 EARTH AS A PLANET

- **Why is Earth geologically active?**

Internal heat drives geological activity, and Earth retains plenty of internal heat because of its relatively large size for a terrestrial world. This heat causes mantle convection and keeps Earth's lithosphere thin, ensuring active surface geology. It also keeps part of Earth's core melted, and the circulation of this molten metal creates Earth's magnetic field.

- **What processes shape Earth's surface?**

The four major geological processes are impact cratering, volcanism, tectonics, and erosion. Earth has experienced many impacts, but most craters have been erased by other processes. We owe the existence of our atmosphere and oceans to volcanic outgassing. A special brand of tectonics—plate tectonics—shapes much of Earth's surface. Ice, water, and wind drive rampant erosion on our planet.

- **How does Earth's atmosphere affect the planet?**

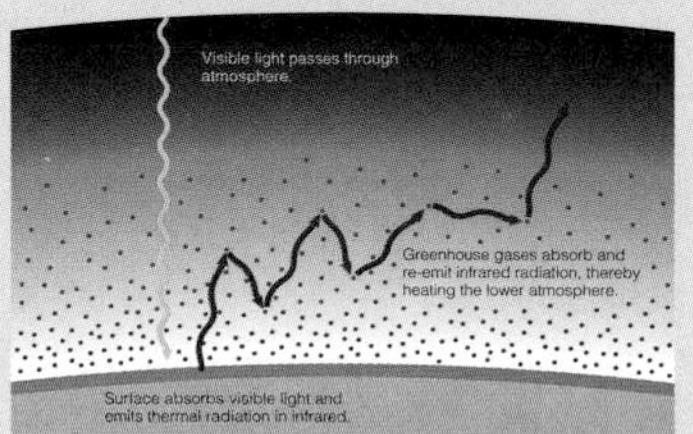

Two crucial effects are (1) protecting the surface from dangerous solar radiation—ultraviolet is absorbed by ozone and X rays are absorbed high in the atmosphere—and (2) the greenhouse effect, without which the surface temperature would be below freezing.

7.2 MERCURY AND THE MOON: GEOLOGICALLY DEAD

- **Was there ever geological activity on the Moon or Mercury?**

Both the Moon and Mercury had some volcanism and tectonics when they were young. However, because of their small sizes, their interiors long ago cooled too much for ongoing geological activity.

7.3 MARS: VICTIM OF PLANETARY FREEZE-DRYING

- **What geological features tell us that water once flowed on Mars?**

Dry river channels, rock-strewn floodplains, and eroded craters all show that water once flowed on Mars, though any periods of rainfall seem to have ended at least 3 billion years ago. Mars today still has water ice underground and in its polar caps, and could possibly have pockets of underground liquid water.

- **Why did Mars change?**

Mars's atmosphere must once have been much thicker with a much stronger greenhouse effect, so change must have occurred due to loss of atmospheric gas. Much of the lost gas probably was stripped away by the solar wind, which was able to reach the atmosphere as Mars cooled and lost its magnetic field and protective magnetosphere. Water was probably also lost because ultraviolet light could break apart water molecules in the atmosphere, and the lightweight hydrogen then escaped to space.

7.4 VENUS: A HOTHOUSE WORLD

- **Is Venus geologically active?**

Venus almost certainly remains geologically active today. Its surface shows evidence of major volcanic or tectonic activity in the past billion years, and it should retain nearly as much internal heat as Earth. However, geological activity on Venus differs from that on Earth in at least two key ways: lack of erosion and lack of plate tectonics.

- **Why is Venus so hot?**

Venus's extreme surface heat is a result of its thick, carbon dioxide atmosphere, which creates a very strong greenhouse effect. The reason Venus has such a thick atmosphere is its distance from the Sun: It was too close to develop liquid oceans like those on Earth, where most of the outgassed carbon dioxide dissolved in water and became locked away in rock. Thus, the carbon dioxide remained in the atmosphere, creating the strong greenhouse effect.

7.5 EARTH AS A LIVING PLANET

- **What unique features of Earth are important for life?**

Unique features of Earth on which we depend for survival are (1) surface liquid water, made possible by Earth's moderate temperature; (2) atmospheric oxygen, a product of photosynthetic life; (3) plate tectonics, driven by internal heat; and (4) climate stability, a result of the carbon dioxide cycle, which in turn requires plate tectonics.

- **How might human activity change our planet?**

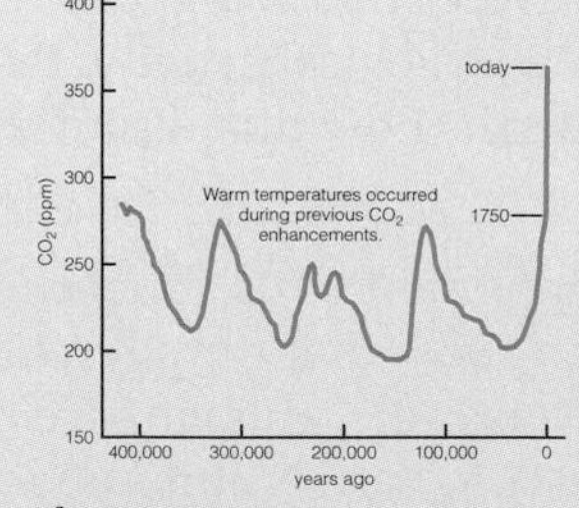

Ozone depletion can leave surface life more vulnerable to dangerous solar ultraviolet radiation, and the high rate of extinctions could have unknown consequences. The human release of greenhouse gases into the atmosphere may already be causing global warming and certainly would affect the climate if it continues.

- **What makes a planet habitable?**

We can trace Earth's habitability to its relatively large size and its distance from the Sun. Its size keeps the internal heat that allowed volcanic outgassing to lead to our oceans and atmosphere, and also drives the plate tectonics that helps regulate our climate through the carbon dioxide cycle. Its distance from the Sun is neither too close nor too far, thereby allowing liquid water to exist on Earth's surface.

EXERCISES AND PROBLEMS

REVIEW QUESTIONS

1. What are Earth's basic layers by composition? What do we mean by the *lithosphere,* and why isn't it listed as one of the three layers by composition?
2. What is differentiation, and how did it affect the internal structures of the terrestrial worlds?
3. Why do large planets retain internal heat longer than smaller planets? Briefly explain how internal heat is related to mantle convection and lithospheric thickness.
4. Why does Earth have a global *magnetic field?* What is the *magnetosphere?*
5. Define each of the four major geological processes, and give examples of features on Earth shaped by each process.
6. What is *outgassing,* and how did it lead to the existence of Earth's atmosphere and oceans?
7. Describe how Earth's atmosphere protects the surface from harmful radiation. What is the role of *ozone?*
8. What does the *greenhouse effect* do to a planet? Explain the role of greenhouse gases and describe the basic mechanism of the greenhouse effect.
9. How do crater counts tell us the age of a planetary surface? Briefly explain why the Moon is so much more heavily cratered than Earth.
10. Briefly summarize the geological history of the Moon. Be sure to explain the origin of the lunar *maria.*
11. Briefly summarize the geological history of Mercury. Be sure to explain how Mercury's great cliffs are thought to have formed.
12. Describe at least three similarities and three differences between Earth and Mars.
13. Choose five features visible on global maps of Mars (Figure 7.20) and in each case explain what we think the feature is and how it formed.
14. Explain why liquid water is not stable on Mars today, but why we nonetheless think it flowed in the distant past on Mars. Could there still be liquid water anywhere on Mars today? Explain.
15. Briefly summarize how and why Mars lost much of its atmosphere some 3 billion years ago.
16. Describe at least three major geological features of Venus. Why is it surprising that Venus lacks plate tectonics? What might explain this lack?
17. What do we mean by a *runaway greenhouse effect?* Explain why this process occurred on Venus but not on Earth.
18. List four unique features of Earth in comparison to other terrestrial worlds, and briefly explain what we mean by each one.
19. What is *plate tectonics?* How does it change the arrangement of the continents with time?
20. What is the *carbon dioxide cycle,* and why is it so crucial to life on Earth?
21. Briefly summarize the debate over global warming, and the potential consequences if it proves to be real.
22. Based on Figure 7.38, write a paragraph each on the role of planetary size and the role of distance from the Sun in explaining the current nature of the terrestrial worlds.

SURPRISING DISCOVERIES?

Suppose we were to make the following discoveries. (These are *not* real discoveries.) In light of your understanding of planetary geology, decide whether the discovery should be considered reasonable or surprising. Explain your answer, if possible tracing your logic back to the terrestrial world's formation properties. (In some cases, both views can be defended.)

23. The next mission to Mercury photographs part of the surface never seen before and detects vast fields of sand dunes.
24. New observations show that several of the volcanoes on Venus have erupted within the past few million years.
25. A Venus radar mapper discovers extensive regions of layered sedimentary rocks, similar to those found on Earth.
26. Radiometric dating of rocks brought back from one lunar crater shows that the crater was formed only a few tens of millions of years ago.
27. New orbital photographs of craters on Mars that have gullies also show pools of liquid water to be common on the crater bottoms.
28. Clear-cutting in the Amazon rain forest on Earth exposes vast regions of ancient terrain that is as heavily cratered as the lunar highlands.
29. Drilling into the Martian surface, a robotic spacecraft discovers liquid water a few meters beneath the slopes of a Martian volcano.
30. We find a planet in another solar system that has an Earth-like atmosphere with plentiful oxygen but no life of any kind.
31. We find a planet in another solar system that orbits at the same distance from its star as Earth and has Earth-like plate tectonics, but it is only the size of the Moon.
32. We find evidence that the early Earth had more carbon dioxide in its atmosphere than Earth does today.

PROBLEMS

33. *Miniature Mars.* Suppose Mars had turned out to be significantly smaller than its current size—say, the size of our Moon. How would this have affected the number of geological features due to each of the four major geological processes? Do you think Mars would still be a good candidate for harboring extraterrestrial life? Summarize your answers in two or three paragraphs.
34. *Two Paths Diverged.* By looking back to fundamental properties such as size and distance from the Sun, explain why Earth has oceans and very little atmospheric carbon dioxide, while similar-size Venus has a thick carbon dioxide atmosphere.
35. *Change in Formation Properties.* Consider Earth's size and distance from the Sun. Choose one property and suppose that it had been different (for example, smaller size or greater distance). Describe how this change might have affected Earth's subsequent history and the possibility of life on Earth.

36. *Experiment: Geological Properties of Silly Putty.* Roll room-temperature Silly Putty into a ball and measure its diameter. Place the ball on a table and gently place one end of a heavy book on it. After 5 seconds, measure the height of the squashed ball. Repeat the experiment two more times, the first time warming the Silly Putty in hot water before you start and the second time cooling it in ice water before you start. How do the different temperatures affect the rate of "squashing"? How does the experiment relate to planetary geology? Explain.

37. *Experiment: Planetary Cooling in a Freezer.* To simulate the cooling of planetary bodies of different sizes, use a freezer and two small plastic containers of similar shape but different size. Fill each container with cold water and put both in the freezer at the same time. Checking every hour or so, record the time and your estimate of the thickness of the "lithosphere" (the frozen layer) in the two tubs. How long does it take the water in each tub to freeze completely? Describe in a few sentences the relevance of your experiment to planetary geology. Extra credit: Plot your results on a graph with time on the x-axis and lithospheric thickness on the y-axis. What is the ratio of the two freezing times?

38. *Amateur Astronomy: Observing the Moon.* Any amateur telescope has resolution adequate to identify geological features on the Moon. The light highlands and dark maria should be evident, and shadowing is visible near the line between night and day. Try to observe the Moon near the first- or third-quarter phase. Sketch or photograph the Moon at low magnification, and then zoom in on a region of interest. Again sketch or photograph your field of view, label its features, and identify the geological process that created them. Look for craters, volcanic plains, and tectonic features. Estimate the size of each feature by comparing it to the size of the whole Moon (radius = 1,738 km).

DISCUSSION QUESTIONS

39. *What Is Predictable?* We've found that much of a planet's geological history is destined from its birth. Briefly explain why, and discuss the level of detail that is predictable. For example, was Mars's general level of volcanism predictable? Could we have predicted a mountain as tall as Olympus Mons or a canyon as long as Valles Marineris? Explain.

40. *Worth the Effort?* Politicians often argue over whether planetary missions are worth the expense involved. Based on what we have learned by comparing the geologies of the terrestrial worlds, do you think the missions that have given us this knowledge have been worth their expense? Defend your opinion.

41. *Lucky Earth.* The climate histories of Venus and Mars make it clear that it's not "easy" to get a pleasant climate like that of Earth. How does this affect your opinion about whether Earth-like planets might exist around other stars? Explain.

42. *Terraforming Mars.* Some people have suggested that we might be able to carry out planetwide engineering of Mars that would cause its climate to warm and its atmosphere to thicken. This type of planet engineering is called *terraforming,* because its objective is to make a planet more Earth-like and thus easier for humans to live on. Discuss possible ways to terraform Mars, at least in principle. Do any of these ideas seem practical? Does it seem like a good idea? Defend your opinions.

MATH HELP AND EXERCISES

For additional help and practice with mathematical concepts applicable to this chapter, the Astronomy Place web site has the following resources with detailed explanations, worked examples, and practice problems:

- The Surface Area-to-Volume Ratio
- "No Greenhouse" Temperatures
- Thermal Escape from an Atmosphere

For a complete list of media resources available, go to **www.astronomyplace.com** and choose Chapter 7 from the pull-down menu.

ASTRONOMY PLACE WEB TUTORIALS

Tutorial Review of Key Concepts

Use the following interactive **Tutorials** at **www.astronomyplace.com** to review key concepts from this chapter.

Shaping Planetary Surfaces Tutorial

Lesson 1 The Four Geological Processes
Lesson 2 What do Geological Processes Depend On?
Lesson 3 Planet Surface Evolution

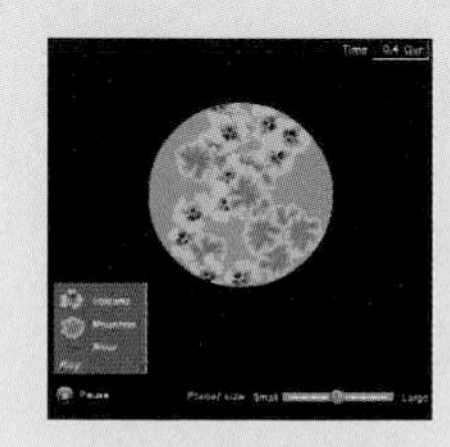

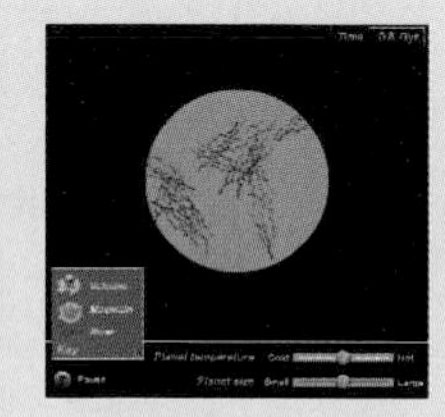

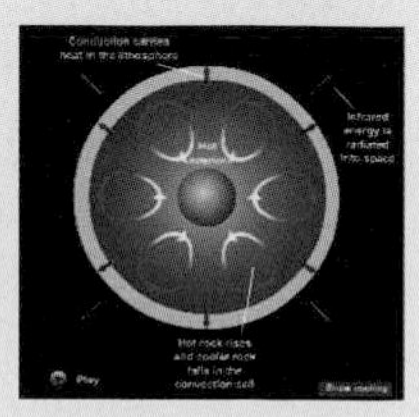

Surface Temperature of Terrestrial Planets Tutorial

Lesson 1 Energy Balance
Lesson 2 Role of Planet's Distance from the Sun
Lesson 3 Role of Planet's Albedo
Lesson 4 Role of Planet's Atmosphere

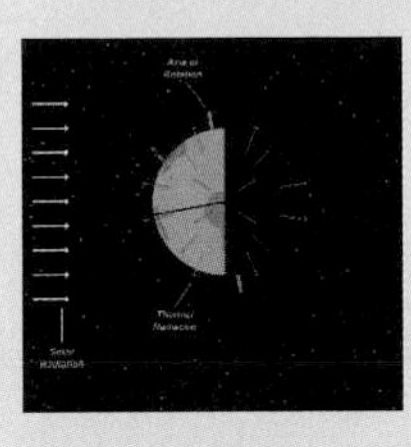 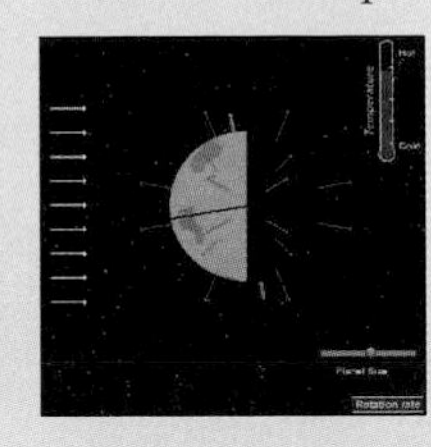 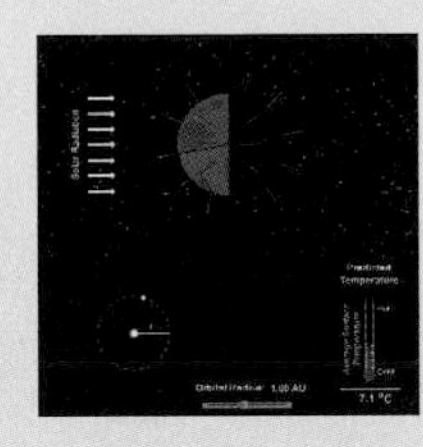

Supplementary Tutorial Exercises

Use the interactive **Tutorial Lessons** to explore the following questions.

Shaping Planetary Surfaces Tutorial, Lesson 2

1. How has the rate of crater formation in the solar system changed over time?
2. Why do we see so many more impact craters on the Moon and Mercury than on Earth?
3. Why is erosion so much more dominant on Earth than on any of the other terrestrial worlds?

Shaping Planetary Surfaces Tutorial, Lesson 3

1. How has Earth's geology differed from that of the other terrestrial planets and why?
2. Suppose Earth had been born at half its current size. How would you expect it to have been different? Would life still be possible? Explain.
3. Suppose Earth had been born 10% larger than its current size. How would you expect it to have been different? Would life still be possible? Explain.

Surface Temperature of Terrestrial Planets Tutorial, Lesson 4

1. How would Earth's temperature be different if the carbon dioxide concentration doubled but all other gas concentrations remained the same?
2. A small warming could lead to increased evaporation of water. What would happen to the temperature if, along with the doubling of carbon dioxide as in question 1, the atmospheric water vapor concentration increased by 10% (to 1.1 times its current value)?
3. How might increased cloud cover offset the warming due to the greenhouse effect?
4. Increased cloud cover also means increased atmospheric water vapor. Use the tools to explore the balance between warming due to more water vapor and cooling due to more cloud cover. If we continued to add greenhouse gases to the atmosphere for centuries, what would ultimately happen to our planet's temperature? Explain.

EXPLORING THE SKY AND SOLAR SYSTEM

Of the many activities available on the ***Voyager: SkyGazer*** **CD-ROM** accompanying your book, use the following files to observe key phenomena covered in this chapter.

Go to the **File: Demo** folder for the following demonstrations.

1. Earth and Venus
2. Venus-Earth-Moon

Go to the **Explore** menu for the following demonstrations.

1. Solar System
2. Paths of the Planets

MOVIES

Check out the following narrated and animated short documentary available on **www.astronomyplace.com** for a helpful review of key ideas covered in this chapter.

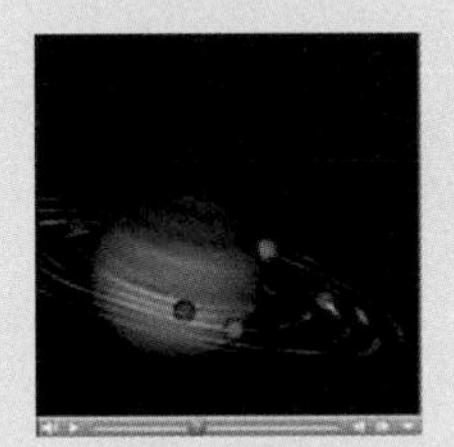

History of the Solar System

WEB PROJECTS

Take advantage of the useful Web links on **www.astronomyplace.com** to assist you with the following projects.

1. *"Coolest" Surface Photo.* Visit the Astronomy Picture of the Day Web site, and search for past images of the terrestrial worlds. Look at many of them, and choose the one you think is the "coolest." Make a printout, write a short description of what it shows, and explain what you like about it.
2. *Water on Mars.* Go to the home page for NASA's Mars Exploration Program, and look for the latest evidence concerning recent water flows on Mars. Write a few paragraphs describing the new evidence and what it tells us. How will future missions help resolve the questions?
3. *Mars Colonization.* Visit the Web site of a group that advocates human colonization of Mars, such as the Mars Society. Learn about the challenges of human survival on Mars and about prospects for terraforming Mars. Overall, do you think colonization of Mars is a good idea? Do you think terraforming is a good idea? Write a short essay describing what you've learned and defending your opinions.
4. *Human Threats to Earth.* Write an in-depth research report, three to five pages in length, about current understanding and controversy regarding global warming. Be sure to address both the latest knowledge about the issue and proposals for alleviating any dangers associated with it. End your report by making your own recommendations about what, if anything, needs to be done to prevent damage to Earth.

8
Jovian Planet Systems

LEARNING GOALS

8.1 A DIFFERENT KIND OF PLANET
- What are jovian planets made of?
- What are jovian planets like on the inside?
- What is the weather like on jovian planets?

8.2 A WEALTH OF WORLDS: SATELLITES OF ICE AND ROCK
- What kinds of moons orbit the jovian planets?
- What makes Jupiter's Galilean moons unusual?
- What makes Titan different from other moons?
- Why are small icy moons more geologically active than small rocky planets?

8.3 JOVIAN PLANET RINGS
- What are Saturn's rings like?
- Why do the jovian planets have rings?

In Roman mythology, the namesakes of the jovian planets are rulers among gods: Jupiter is the king of the gods, Saturn is Jupiter's father, Uranus is the lord of the sky, and Neptune rules the sea. However, our ancestors could not have foreseen the true majesty of the four jovian planets. The smallest, Neptune, is large enough to contain the volume of more than 50 Earths. The largest, Jupiter, has a volume some 1,400 times that of Earth. These worlds are totally unlike the terrestrial planets. They are essentially giant balls of gas, with no solid surface on which to stand.

Why should we care about a set of worlds so different from our own? Apart from satisfying natural curiosity, studies of the jovian planets and their moons help us understand the birth and evolution of our solar system—which in turn helps us understand our own planet Earth. In this chapter, we'll explore the jovian planet systems, first focusing on the planets themselves, then on their many moons, and finally on their beautifully complex rings.

Our discussion may apply to much more than the four jovian planets of our own solar system. In fact, all of the more than 100 planets so far discovered around other stars are probably jovian in nature. Although this may be the case simply because we can't yet detect smaller planets, it nevertheless means that known jovian planets far outnumber the known terrestrial worlds. Thus, what we learn about the jovian planets in our solar system may help us understand other planets and other solar systems as well.

Formation of the Solar System Tutorial, Lesson 1

8.1 A DIFFERENT KIND OF PLANET

The jovian planets are radically different from the terrestrial planets. They are far larger in size and very different in composition. They are orbited by rings and numerous moons. They even rotate much faster than the terrestrial planets.

We toured the jovian planets briefly in Section 6.1. (You might wish to review that tour before continuing.) Now we are ready to explore the jovian planets in a little more depth. As you'll see from both the discussion and the selection of photos, much of our present knowledge has come from spacecraft visits, especially from the Voyager 1 and 2 missions that flew past these planets in the late 1970s and 1980s and the *Galileo* spacecraft that orbited Jupiter from 1995 until 2003. Currently, scientists are looking forward to learning much more from the Cassini mission, which should be orbiting Saturn by the time you read this book.

• What are jovian planets made of?

The jovian planets are often called "gas giants," making it sound as if they are entirely gaseous like air on Earth. While this idea is not entirely wrong, the reality is somewhat more complex.

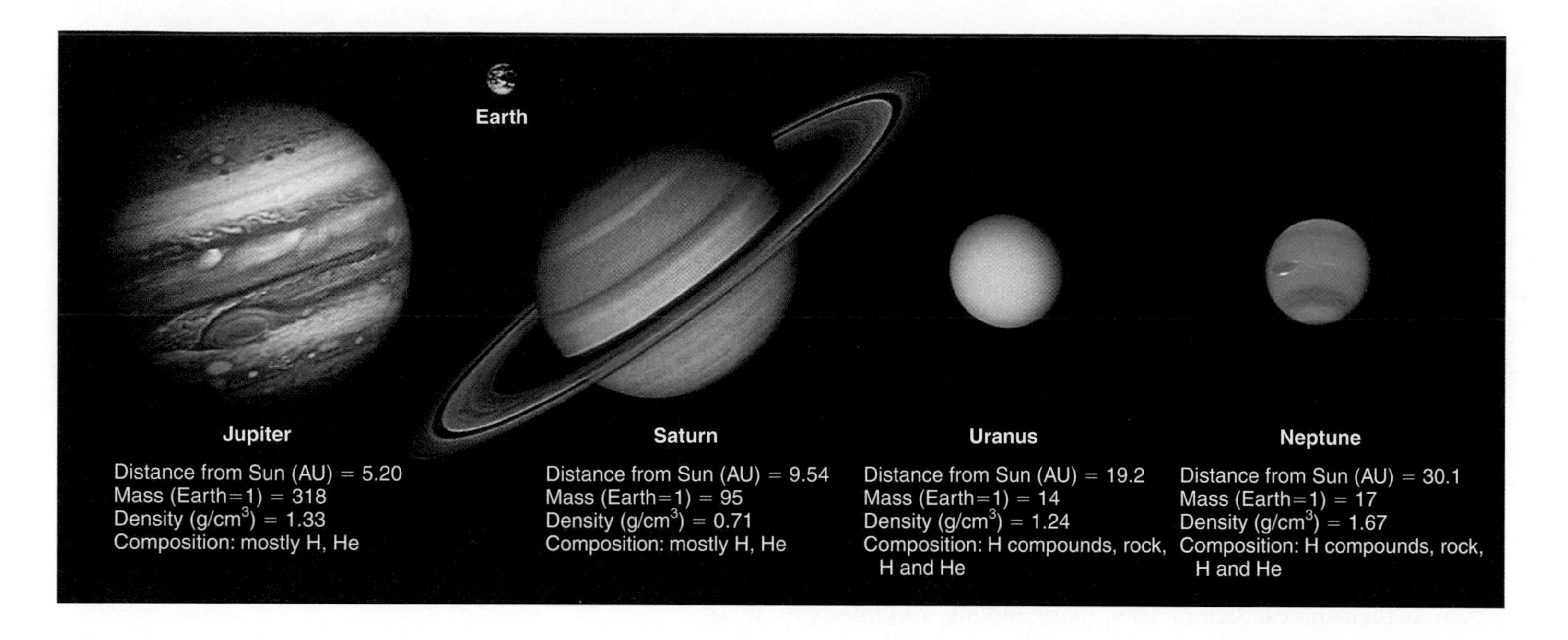

Figure 8.1

Jupiter, Saturn, Uranus, and Neptune, shown to scale with Earth for comparison.

Figure 8.1 shows a montage of the jovian planets compiled by the Voyager spacecraft, along with basic data and Earth included for scale. The immense sizes of the jovian worlds are apparent. But while all four are enormous, there are important differences between them. In particular, note that they differ substantially in mass, density, and overall composition.

General Composition of Jovian Planets Jupiter and Saturn are made almost entirely of hydrogen and helium, with just a few percent of their masses coming from hydrogen compounds and even smaller amounts of rock and metal. Thus, their overall compositions are much more similar to the composition of the Sun than to the compositions of the terrestrial planets. Some people have gone as far as to call Jupiter a "failed star" because it has a starlike composition but lacks the nuclear fusion needed to make it shine. Its lack of fusion is due to its size: Although Jupiter is large for a planet, it is much less massive than any star. As a result, its gravity is too weak to compress its interior to the extreme temperatures and densities needed for nuclear fusion. (Jupiter would have needed to grow to about 80 times its current mass to have become a star.) Of course, where some people see a failed star, others see an extremely successful planet.

The jovian planets are made mostly of hydrogen, helium, and hydrogen compounds, making them very different in composition from terrestrial worlds.

Uranus and Neptune contain proportionally much smaller amounts of hydrogen and helium. These gases make up much less than half their total masses. Instead, Uranus and Neptune are made primarily of hydrogen compounds such as water (H_2O), methane (CH_4), and ammonia (NH_3), along with smaller amounts of metal and rock.

We can understand the differences in composition among the jovian planets by looking at the way in which we think they formed.

Gas Capture in the Solar Nebula Recall that the jovian planets formed in a very different way from the terrestrial planets [Section 6.4]. The terrestrial planets grew only through accretion of solid planetesimals containing rock and metal. Because rock and metal were quite rare in the solar nebula, the terrestrial planets never grew massive enough for their gravity to hold any of the abundant hydrogen and helium gas that made up most of the nebula.

The jovian planets formed in the outer solar system (beyond the frost line), where it was cold enough for hydrogen compounds to condense into ices (see Figure 6.20). Because hydrogen compounds were so much more abundant than metal and rock, some of the ice-rich planetesimals of the outer solar system grew to great size. Once these planetesimals became sufficiently massive, their gravity allowed them to draw in the hydrogen and helium gas that surrounded them. All four jovian planets are thought to have grown from ice-rich planetesimals of about the same mass—roughly 10 times the mass of Earth. Thus, their differences in composition stem from the amount of hydrogen and helium gas that they captured.

Jupiter and Saturn captured so much hydrogen and helium gas that these gases now make up the vast majority of their masses. The ice-rich planetesimals from which they grew now represent only a small fraction of their overall masses—about 3% in Jupiter's case and about 10% in Saturn's case.

Uranus and Neptune pulled in much less hydrogen and helium gas. For example, notice in Figure 8.1 that Uranus's mass is about 14 times Earth's mass. Since it is thought to have grown around an ice-rich planetesimal that was about 10 times Earth's mass, the drawn-in hydrogen and helium gas must make up only about a third of Uranus's total mass. The bulk of its mass consists of material from the original ice-rich planetesimal: hydrogen compounds mixed with smaller amount of rock and metal. The same is true for Neptune, though its higher density suggests that it may have formed around a slightly more massive ice-rich planetesimal.

The jovian planets nearer to the Sun captured more hydrogen and helium gas, making them larger and leaving them with smaller proportions of hydrogen compounds, rock, and metal.

We can understand why the jovian planets captured different amounts of gas by considering their distances from the Sun. As the solar system formed, the solid particles that condensed far from the Sun must have been much more widely spread out than particles that condensed nearer to the Sun. Thus, at greater distances from the Sun, it took longer for small particles to accrete into large, icy planetesimals with gravity strong enough to pull in gas from the surrounding nebula. Jupiter would have been the first jovian planet to get a large enough solid planetesimal to start drawing in gas, followed by Saturn, Uranus, and Neptune. Because all the planets stopped accreting gas at the same time—when the solar wind blew all the remaining gas into interstellar space—the more distant planets had less time to capture gas and thus ended up smaller in size.

Density Differences Notice in Figure 8.1 that Saturn is considerably less dense than Uranus or Neptune. This should make sense when you compare compositions. After all, the hydrogen compounds, rock, and metal that make up Uranus and Neptune are normally much more dense than hydrogen or helium gas. By the same logic, we'd expect Jupiter to be even less dense than Saturn—but it's not. To understand Jupiter's surprisingly high density, we need to think about how massive planets are affected by their own gravity.

Building a planet of hydrogen and helium is a bit like making one out of very fluffy pillows. Imagine assembling a planet pillow by pillow. As each new pillow is added, those on the bottom are compressed more by those above. As the lower layers are forced closer together, their mutual gravitational attraction increases, compressing them even further. At

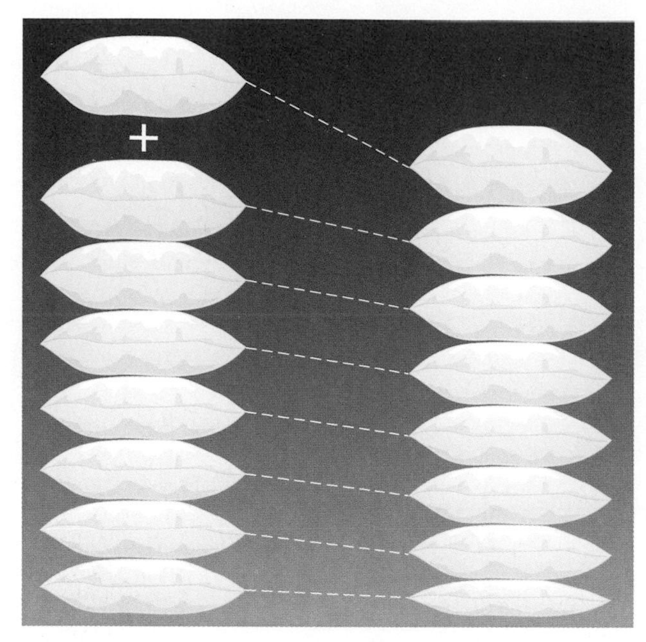

a Adding pillows to a stack may increase its height at first but eventually just compresses the stack, making its density greater. Similarly, adding mass to a jovian planet eventually just increases its density rather than increasing its radius.

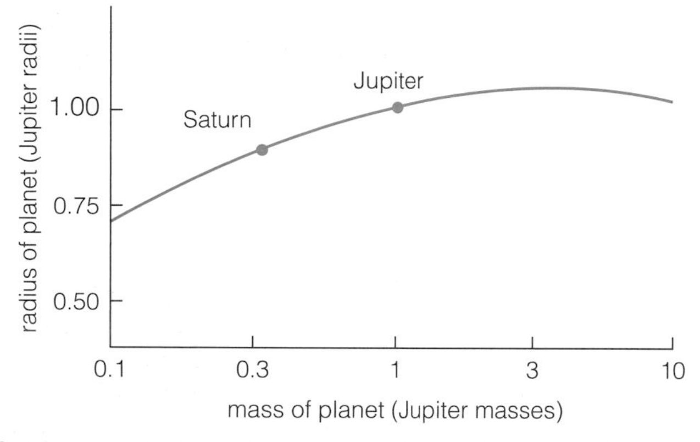

b This graph shows how radius depends on mass for a hydrogen/helium planet. Notice that Jupiter is only slightly larger in radius than Saturn, despite being three times as massive. Gravitational compression of a planet much more massive than Jupiter would actually make it smaller in size.

Figure 8.2

The relationship between mass and radius for a planet made of hydrogen and helium.

first the stack grows substantially with each pillow, but eventually the growth slows until adding pillows barely increases the height of the stack (Figure 8.2a).

This analogy explains why Jupiter is only slightly larger than Saturn in radius even though it is more than three times as massive. The extra mass of Jupiter compresses its interior to a much higher density. More precise calculations show that Jupiter's radius is almost the maximum possible radius for a jovian planet. If much more gas were added to Jupiter, its weight would actually compress the interior enough to make the planet *smaller* rather than larger (Figure 8.2b). Thus, some extrasolar planets that are larger in mass than Jupiter are probably smaller in size.

THINK ABOUT IT Saturn's average density of 0.71 g/cm^3 is less than that of water. As a result, it is sometimes said that Saturn could float on a giant ocean. Suppose there really were a gigantic planet with a gigantic ocean and we put Saturn on the ocean's surface. Would it float? If not, what would happen?

• What are jovian planets like on the inside?

Based on their compositions, Jupiter and Saturn may seem to deserve the name "gas giants." After all, they became giants primarily by capturing so much hydrogen and helium gas. The name may seem less fitting for Uranus and Neptune, since they are made mostly of other materials. However, closer inspection shows that the name is a little misleading even for Jupiter and Saturn, because their strong gravity compresses most of the "gas" into forms quite unlike anything we are familiar with in everyday life on Earth. Let's begin by considering what Jupiter is like on the inside, then extend the ideas to the other jovian planets.

Inside Jupiter Jupiter's lack of a solid surface makes it tempting to think of the planet as "all atmosphere," but you could not fly through Jupiter's interior in the way airplanes fly through air. A spacecraft plunging into Jupiter would find increasingly higher temperatures and pressures as it descended. The Galileo spacecraft dropped a scientific probe into Jupiter in 1995, and it collected measurements for about an hour before being destroyed by the ever-increasing pressures and temperatures. (Galileo survived to a depth of about 200 km, only about 0.3% of Jupiter's radius.)

If you plunged below Jupiter's clouds, you'd never encounter a solid surface—just ever-denser and hotter hydrogen/helium compressed into bizarre liquid and metallic phases.

In fact, while Jupiter has no solid surface, computer models tell us that it still has fairly distinct interior layers. The layers do not differ much in composition—all except the core are mostly hydrogen and helium—but instead they differ in the phase (such as liquid or gas) of their hydrogen. As shown in Figure 8.3, Jupiter's interior layers from cloudtops to core are:

- Gaseous hydrogen: In the outer layer, conditions are moderate enough for hydrogen to remain in its familiar, gaseous form. This layer extends about 10% of the way from the cloudtops toward the center. In the outer portions of this layer, the gas would seem much like ordinary air (except with a different composition than air on Earth), so we usually think of this region as Jupiter's atmosphere.

- Liquid hydrogen: This layer occupies about the next 10% of Jupiter's interior. The temperature exceeds 2,000 K and the pressure exceeds 500,000 times the pressure on Earth's surface. Laboratory experiments show that hydrogen acts more like a liquid than a gas under these conditions. Notice that the density in this layer (ranging from 0.5 to almost 1.0 g/cm^3) is only slightly less than the density of water.
- Metallic hydrogen: In most of the rest of Jupiter, the temperatures and pressures are so extreme that hydrogen is forced into a compact, metallic form. Just as is the case with everyday metals, electrons are free to move around in metallic hydrogen, so it conducts electricity quite effectively. As we'll see shortly, Jupiter's magnetic field is generated in this layer of metallic hydrogen.
- Core: The core is a mix of hydrogen compounds, rock, and metal. However, the high temperature and extreme pressure ensure that this mix bears little resemblance to familiar solids or liquids. The core contains about 10 times as much mass as the entire Earth, but it is only about the same size as Earth because it is compressed to such high density.

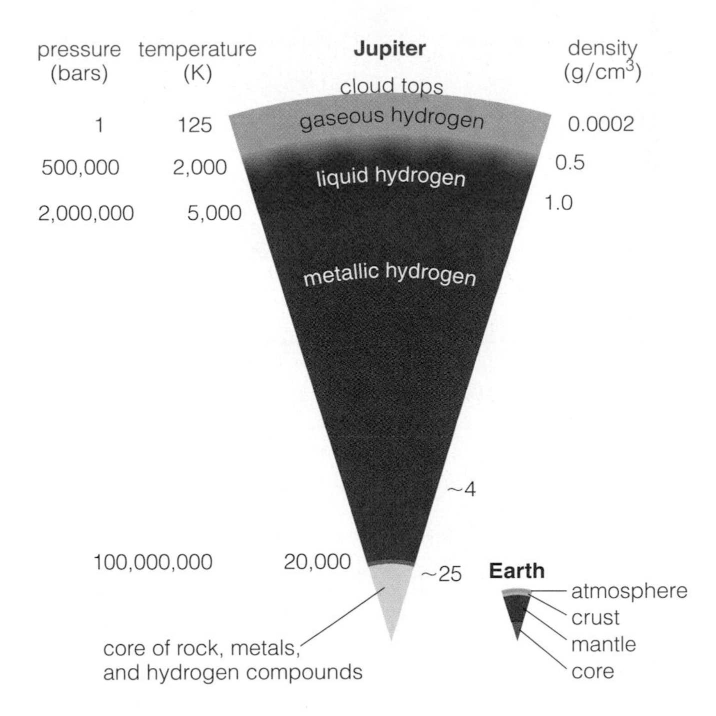

Figure 8.3

Jupiter's interior structure, labeled with the pressure, temperature, and density at various depths. Earth's interior structure is shown to scale for comparison. (Notes on units: 1 bar is approximately the atmospheric pressure at sea level on Earth; the density of liquid water is 1 gram per cubic centimeter (g/cm^3).)

Comparing Jovian Interiors Because all four jovian planets have cores of about the same mass, their interiors differ mainly in the hydrogen/helium layers that surround their cores. Figure 8.4 contrasts the four jovian interiors. (Remember that while the outer layers are named for the phase of their hydrogen, they also contain helium and hydrogen compounds.)

Saturn is the most similar to Jupiter, just as we should expect given its similar size and composition. It has the same set of four layers as Jupiter, and the layers differ from those of Jupiter only because of Saturn's lower mass and weaker gravity. The lower mass makes the weight of the overlying layers less on Saturn than on Jupiter, so you must travel deeper into Saturn to find the layer where pressure changes hydrogen from one form to another. That is why Saturn has a thicker layer of gaseous hydrogen and a much thinner and more deeply buried layer of metallic hydrogen.

Pressures within Uranus and Neptune are not high enough to form liquid or metallic hydrogen at all. Each of these two planets has only a thick layer of gaseous hydrogen surrounding its core of hydrogen compounds, rock, and metal. This core material may be liquid, making for very odd "oceans" buried deep inside Uranus and Neptune.

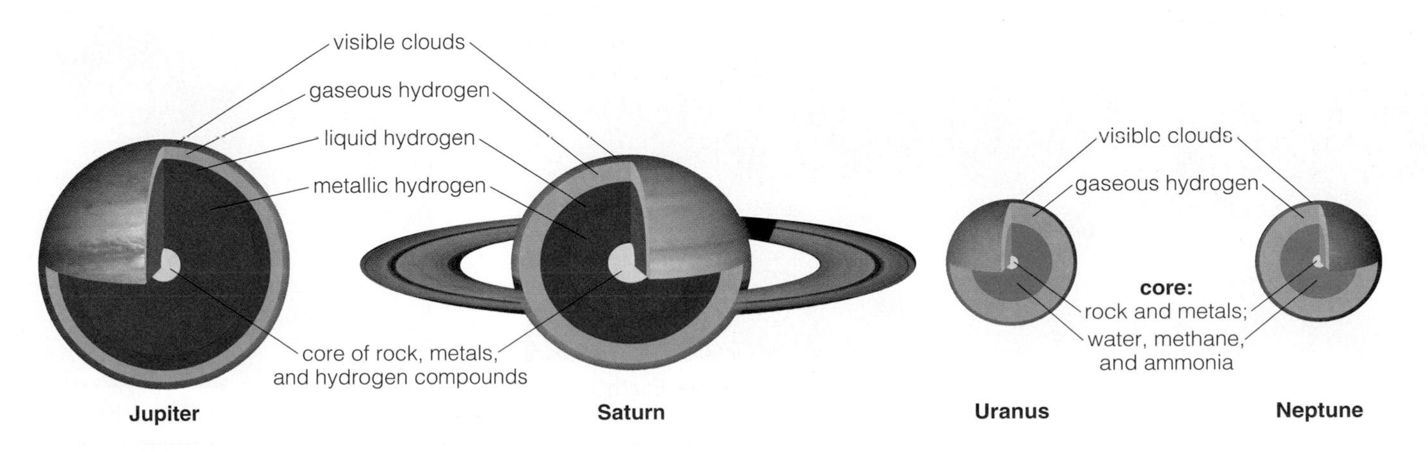

Figure 8.4

These diagrams compare the interior structures of the jovian planets, shown approximately to scale. All four planets have cores of rock, metal, and hydrogen compounds, with masses about 10 times the mass of Earth's core. They differ primarily in the depth of the hydrogen/helium layers that surround their cores. (The cores of Uranus and Neptune are differentiated into separate layers of rock/metal and hydrogen compounds.)

Magnetic Fields Recall that Earth has a global magnetic field generated by the movements of charged particles in our planet's metallic outer core (see Figure 7.5). The jovian planets also have global magnetic fields generated by motions of charged particles deep in their interiors.

Jupiter's magnetic field is by far the strongest—some 20,000 times stronger than Earth's. This strong field is generated in Jupiter's thick layer of metallic hydrogen. Just as on Earth, Jupiter's magnetic field creates a *magnetosphere* that surrounds the planet and shields it from the solar wind. But like almost everything else about Jupiter, its magnetosphere is enormous. It begins to deflect the solar wind some 3 million kilometers (about 40 Jupiter radii) before the solar wind even reaches Jupiter (Figure 8.5a). If our eyes could see this part of Jupiter's magnetosphere, it would be larger than the full moon in our sky.

Jupiter's magnetosphere traps far more charged particles than Earth's magnetosphere. These particles contribute to auroras on Jupiter (Figure 8.5b). They also create belts of very intense radiation around Jupiter, which can cause damage to orbiting spacecraft. The main source of the many charged particles is Jupiter's volcanically active moon Io. Gases escaping from Io's volcanoes become ionized and feed a donut-shaped charged particle belt (called the *Io torus*) that approximately traces Io's orbit (Figure 8.5c).

The other jovian planets also have magnetic fields and magnetospheres, but theirs are much weaker than Jupiter's (although still much

Figure 8.5

Jupiter's strong magnetic field creates an enormous magnetosphere that captures particles from the solar wind and particles released from Io's active volcanoes.

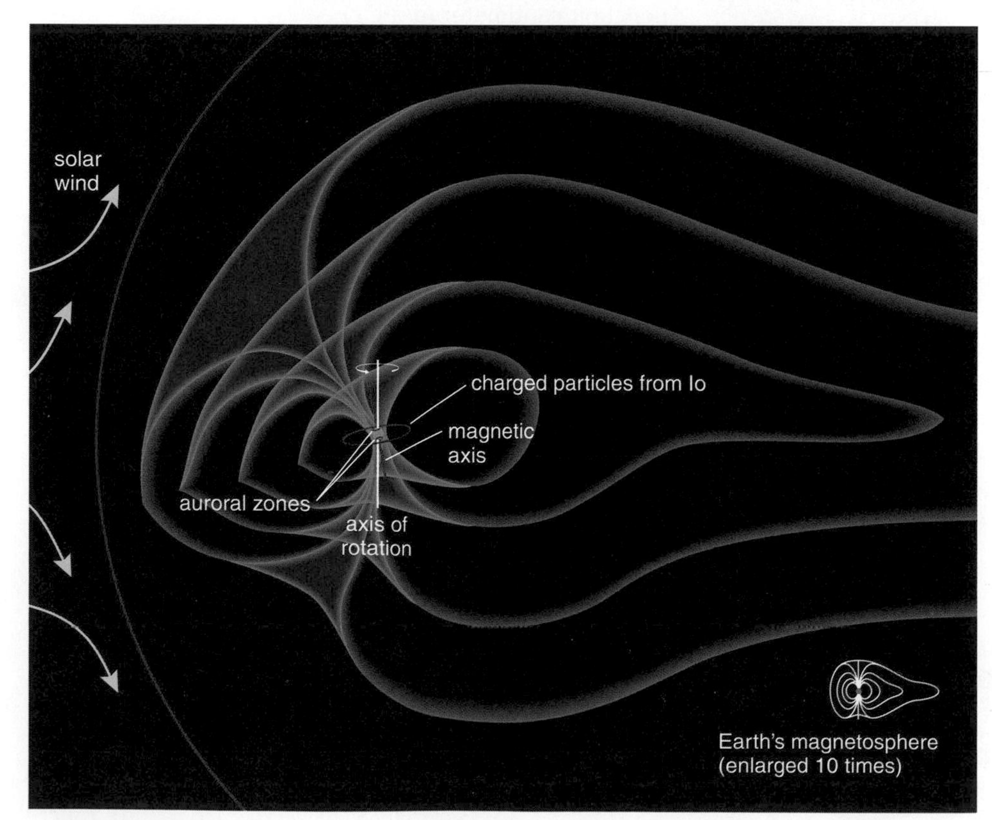

a This diagram shows key features of Jupiter's invisible magnetosphere. Earth's magnetosphere is shown for comparison.

b Jupiter's aurora can be seen in the ultraviolet images overlaid on a photograph of Jupiter (Hubble Space Telescope photos).

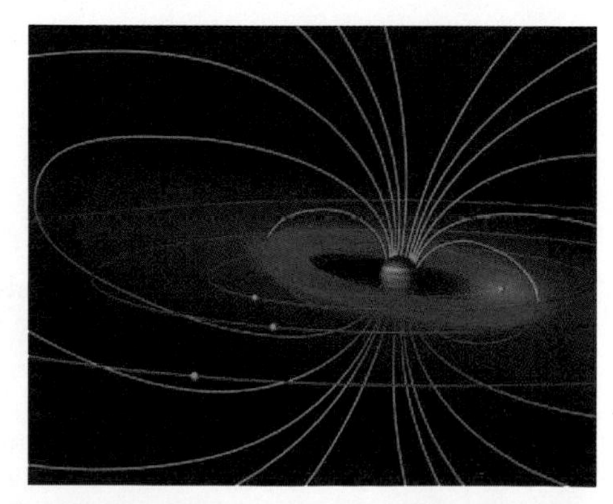

c The bright red "donut" encircling Jupiter in this artist's conception is the *Io torus,* a charged particle belt consisting of ionized gases from Io's volcanoes. The blue lines represent Jupiter's strong magnetic field.

stronger than Earth's). Saturn's magnetic field is weaker than Jupiter's because it has a thinner layer of electrically conducting metallic hydrogen. Uranus and Neptune, smaller still, have no metallic hydrogen at all. Their relatively weak magnetic fields must be generated in their core "oceans" of hydrogen compounds, rock, and metal.

• What is the weather like on jovian planets?

Jovian atmospheres have dynamic winds and weather, including colorful cloudy skies and enormous storms readily visible to telescopes and spacecraft. Weather on these planets is driven both by energy from the Sun (as on the terrestrial planets) and by heat generated within the planets themselves. All but Uranus generate a surprising amount of internal heat that contributes to the energy in their atmospheres. No one knows the precise source of the internal heat on jovian planets, but it probably comes from the conversion of gravitational potential energy to thermal energy inside them. The best guesses are that this conversion comes from a slow but imperceptible contraction in overall size, or from ongoing differentiation as heavier materials continue to sink toward the core.

As we did for the jovian planetary interiors, let's examine different aspects of their atmospheres by starting with Jupiter as the prototype for each feature. We'll then use the general differences between the jovian planets to understand differences in their weather.

Clouds and Colors The spectacular colors of the jovian planets are probably the first thing that jumps out at you when you look at the photos in Figure 8.1. Many mysteries remain about precisely why the jovian planets are so colorful, but at least some major color features are caused by clouds. Earth's clouds look white from space because they are made of water that reflects the white light of the Sun. The jovian planets have clouds of several different types, and some of these reflect light of other colors.

Clouds form when a gas condenses to make tiny liquid droplets or solid flakes. Water vapor is the only gas that can condense in Earth's atmosphere, which is why clouds on Earth are made of water droplets or flakes that can produce rain or snow. In contrast, Jupiter's atmosphere has several gases that can condense to form clouds. Each of these gases condenses at a different temperature, leading to distinctive cloud layers at different altitudes.

Jupiter has three primary cloud layers, which we can understand by considering temperatures at different altitudes (Figure 8.6). Just as the temperature tends to fall as you climb up a mountain on Earth, the temperature drops with altitude in Jupiter's atmosphere. About 100 kilometers below the highest cloudtops, the temperatures are nearly Earth-like and water can condense to form clouds. The temperature drops as we go higher, and about 50 kilometers above the water clouds it is cold enough for a gas called ammonium hydrosulfide (NH_4SH) to condense into clouds. These ammonium hydrosulfide clouds reflect brown and red light (though no one knows why), and thus produce many of the dark colors of Jupiter. Higher still, the temperature is so cold that ammonia (NH_3) condenses to make an upper layer of white clouds. The tops of the ammonia clouds are usually considered the "cloudtops" of Jupiter, which is why Figure 8.6 shows this altitude as zero altitude for Jupiter.

Saturn has the same set of three cloud layers as Jupiter, but these layers occur deeper in Saturn's atmosphere. The reason is that Saturn's outer atmosphere is colder than Jupiter's, both because Saturn is farther from the Sun and because it has weaker gravity. For example, to find the relatively warm temperatures at which water vapor can condense to form

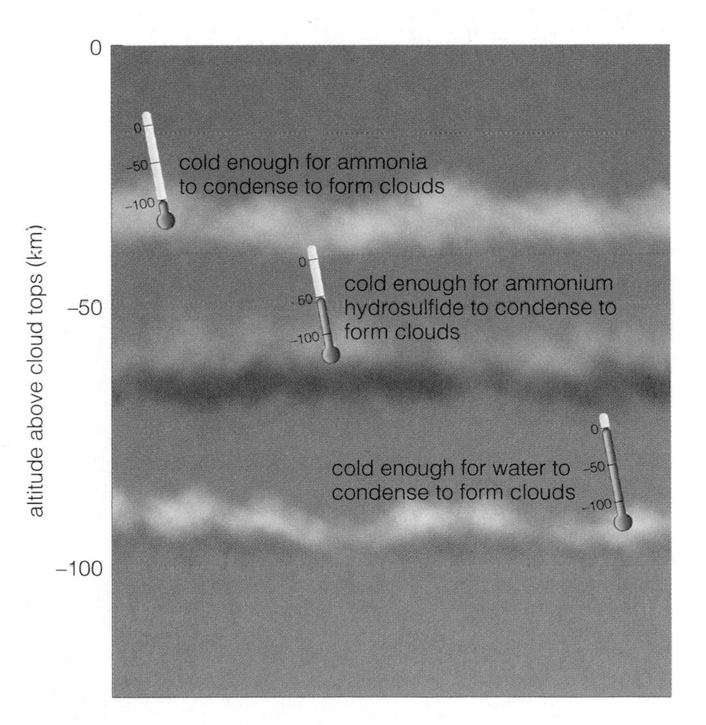

Figure 8.6

This graph shows the temperature structure of Jupiter's atmosphere. Jupiter has at least three distinct cloud layers because different atmospheric gases condense at different temperatures and hence at different altitudes.

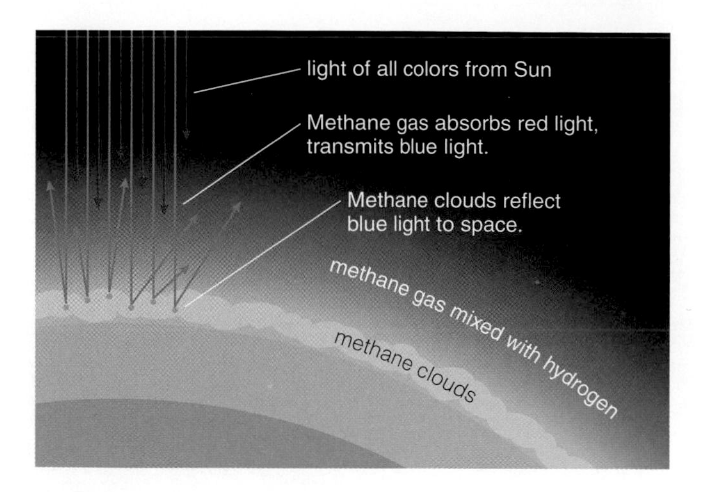

Figure 8.7

Neptune and Uranus look blue because methane gas absorbs red light but transmits blue light. Clouds of methane snowflakes reflect the transmitted blue light back to space.

water clouds, we must look about 200 kilometers deeper into Saturn than into Jupiter. The fact that Saturn's clouds lie deeper in its atmosphere than Jupiter's probably explains Saturn's more subdued colors: Less light penetrates to the depths at which Saturn's clouds are found, and the light they reflect is more obscured by the atmosphere above them.

All the jovian planets have cloudy skies, but clouds of different kinds form at different altitudes in each planet's atmosphere.

Uranus and Neptune are so cold that any cloud layers similar to those of Jupiter or Saturn would be buried too deep in their atmospheres for us to see. Instead, the colors of Uranus and Neptune come mainly from methane gas and clouds (Figure 8.7). These two planets have much more methane gas than Jupiter or Saturn, and the cold temperatures allow some of this methane to condense into clouds. Methane gas absorbs red light, allowing only blue light to penetrate to the level at which the methane clouds form. The methane clouds reflect this blue light upward, giving the planets their blue colors.

Global Winds and Storms In photographs, Jupiter shows alternating east-west stripes of white and reddish brown clouds. These stripes are the result of incredible winds that make hurricanes on Earth seem mild by comparison.

Jupiter's striped appearance comes from its cloud colors. The entire planet is blanketed with the reddish-brown clouds that make up one of the lower cloud layers (see Figure 8.6). However, in some places these lower clouds are hidden from view by the white stripes, which get their color from an upper layer of white ammonia clouds. The white clouds form only in regions where air is rising to high, cold altitudes—cold enough for white ammonia snowflakes to form. Ammonia "snow" falls from these clouds, depleting the rising air of ammonia. When the rising air reaches its highest point, it spills north or south and descends in the bands of falling air. Because this air now lacks ammonia, no ammonia clouds can form and we can see down to the reddish brown clouds (made of ammonium hydrosulfide) that lie deeper in the atmosphere. That is why the bands of falling air have their dark color.

Figure 8.8 Interactive Figure

Wind patterns on both Earth and Jupiter arise from the way planetary rotation affects rising and falling air.

a This photograph shows how storms circulate around low pressure regions (L) on Earth. Earth's rotation causes this circulation, which is in opposite directions in the two hemispheres.

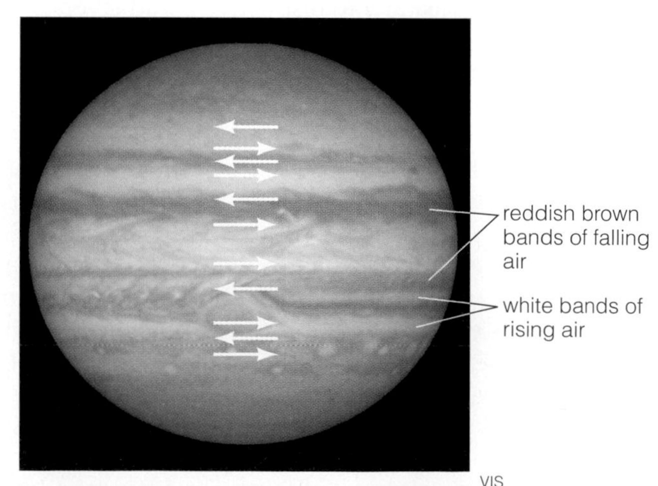

b Jupiter's faster rotation and larger size essentially stretch out the circulation patterns that occur on Earth into planet-wide bands of fast moving air.

Why do Jupiter's regions of rising and falling air make horizontal stripes all the way around the planet? The answer has to do with Jupiter's rotation. You are probably familiar with circulation of storms on Earth (Figure 8.8a). Storms around low-pressure regions tend to circulate counterclockwise in the Northern Hemisphere and clockwise in the Southern Hemisphere. Earth's rotation causes these circulation patterns by diverting north- or south-flowing air. (More technically, Earth's rotation produces something called the *Coriolis effect,* which diverts the paths of missiles or rockets as well as winds.)

The same effect occurs on Jupiter, but Jupiter's much faster rotation and larger size make the effect much stronger. In essence, the circular patterns we see in Earth's atmosphere become stretched out to the east and west on Jupiter, to such an extent that they end up going all the way around the planet (Figure 8.8b). This leads to very high east-west wind speeds—sometimes more than 400 km/hr (250 mi/hr).

The rapid rotation of the jovian planets helps drive strong winds, creating their banded appearances and sometimes giving rise to huge storms.

Just as unusually large hurricanes occasionally arise on Earth, Jupiter also has its share of powerful storms. Of course, because it is Jupiter, its storms dwarf those of Earth. White and brown ovals that frequently appear in Jupiter's atmosphere are low-pressure storms like hurricanes, but many are as large as the entire Earth. Jupiter's most famous feature—its **Great Red Spot**—is also a gigantic storm.

The Great Red Spot is more than twice as wide as Earth. It is somewhat like a hurricane on Earth, except that its winds circulate around a high-pressure region rather than a low-pressure region (Figure 8.9). It is also extremely long-lived compared to storms on Earth: Astronomers have seen it throughout the three centuries during which telescopes have been powerful enough to detect it. No one knows why the Great Red Spot has lasted so long. However, storms on Earth tend to lose their strength when they pass over land. Perhaps Jupiter's biggest storms last for centuries simply because there's no solid surface effect to sap their energy.

The other jovian planets also have dramatic weather patterns (Figure 8.10). As on Jupiter, Saturn's rapid rotation creates alternating bands

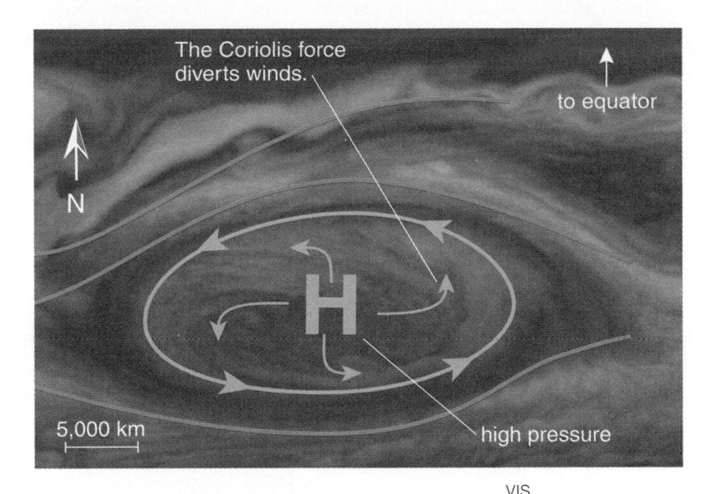

Figure 8.9 Interactive Figure

This photograph shows Jupiter's Great Red Spot, a huge high-pressure storm that is large enough to swallow two or three Earths. The overlaid diagram shows a weather map of the region.

Figure 8.10

Selected views of weather patterns on the four jovian planets.

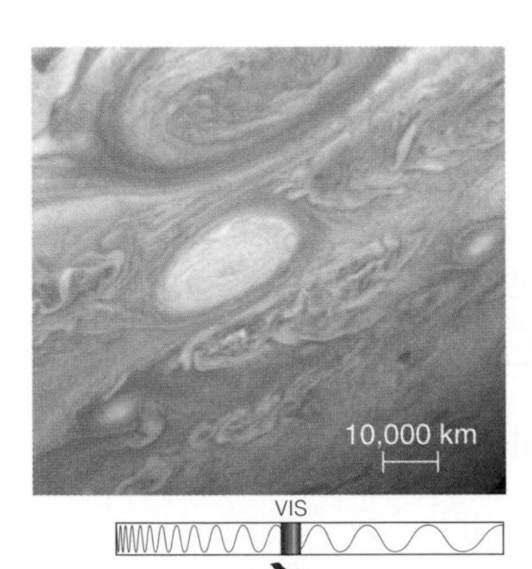

a A close-up of Jupiter's atmosphere in its southern hemisphere, with the Great Red Spot visible near the top. The white oval is a low-pressure storm.

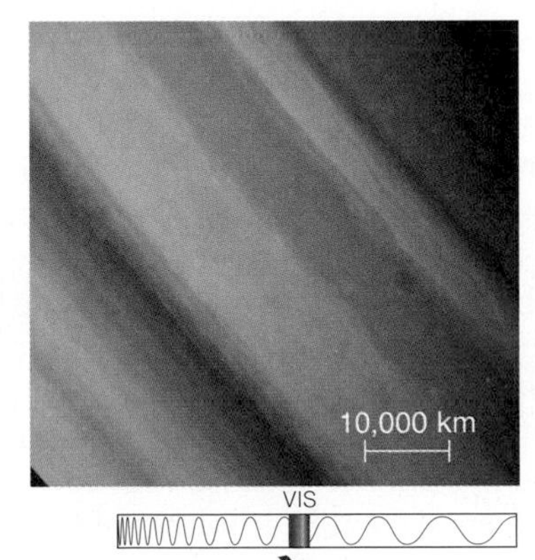

b A close-up of Saturn's atmosphere. Its banded appearance is very similar to that of Jupiter, but it has even faster winds.

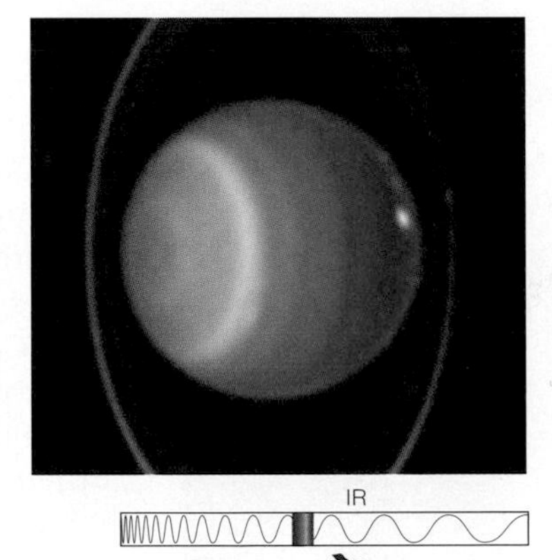

c This infrared photograph from the Hubble Space Telescope shows several storms (the bright red blotches) brewing on Uranus. (The brownish lines on either side are portions of Uranus's rings.)

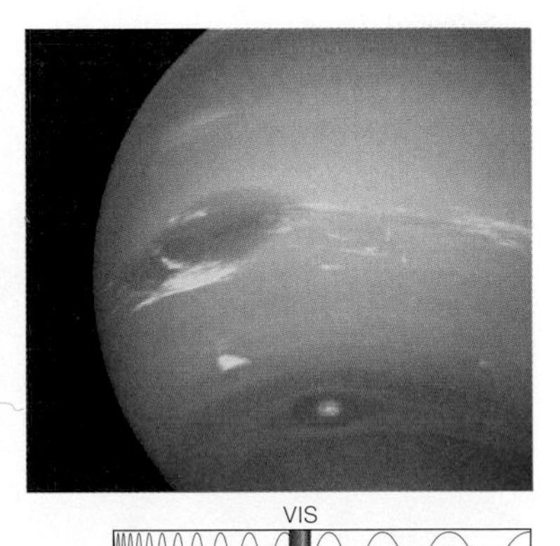

d Neptune's atmosphere has bands much like Jupiter and Saturn, and occasional strong storms. The large storm to the left of center is known as the Great Dark Spot.

of rising and falling air, along with rapid east-west winds. In fact, Saturn's winds are even faster than Jupiter's—a surprise that scientists have yet to explain. Neptune's atmosphere is also banded, and we have seen a high-pressure storm, called the Great Dark Spot, similar to Jupiter's Great Red Spot. However, the Great Dark Spot did not last so long, having disappeared from view just six years after its discovery.

The greatest surprise in jovian weather comes from Uranus. When *Voyager 2* flew past Uranus in 1986, photographs revealed virtually no clouds and no banded structure like those found on the other jovian planets. Scientists attributed the lack of weather to the lesser internal heat of Uranus. However, more recent observations from the Hubble Space Telescope show storms raging in Uranus's atmosphere. The storms may be brewing because of the changing seasons: Thanks to Uranus's extreme axis tilt and 84-year orbit of the Sun [Section 6.1], its southern hemisphere is just beginning to see sunlight for the first time in decades.

8.2 A WEALTH OF WORLDS: SATELLITES OF ICE AND ROCK

The jovian planets are majestic and fascinating, but they are only the beginning of our exploration of jovian planet *systems.* Each of the four jovian systems includes numerous moons and a set of rings. The total mass of all the moons and rings put together is minuscule compared to any one of the jovian planets, but the remarkable diversity of these satellites makes up for their lack of size. In this section, we'll explore a few of the most interesting aspects of the jovian moons. We'll then discuss rings in the following section.

• What kinds of moons orbit the jovian planets?

We now know of more than 100 moons orbiting the jovian planets. Jupiter has the most, with more than 60 moons known to date.

Since there are so many moons, it's helpful to organize them into three groups by size: small moons less than about 300 km in diameter, medium-size moons ranging from about 300 to 1,500 km in diameter, and large moons more than 1,500 km in diameter. These categories are useful because size relates to geological activity. The large moons show evidence of active and ongoing geology, and most of the medium-size moons seem to have had geological activity in the past.

Figure 8.11 shows a montage of all the medium-size and large moons. These moons resemble the terrestrial planets in many ways. Each is spherical with a solid surface and its own unique geology. Some possess atmospheres, hot interiors, and even magnetic fields. The two largest—Jupiter's moon Ganymede and Saturn's moon Titan—are larger than the planet Mercury. Four others are larger than Pluto: Jupiter's moons Io, Europa, and Callisto, and Neptune's moon Triton. However, they differ from terrestrial worlds in their compositions: Because they formed in the cold outer solar system, they contain substantial amounts of ice in addition to metal and rock.

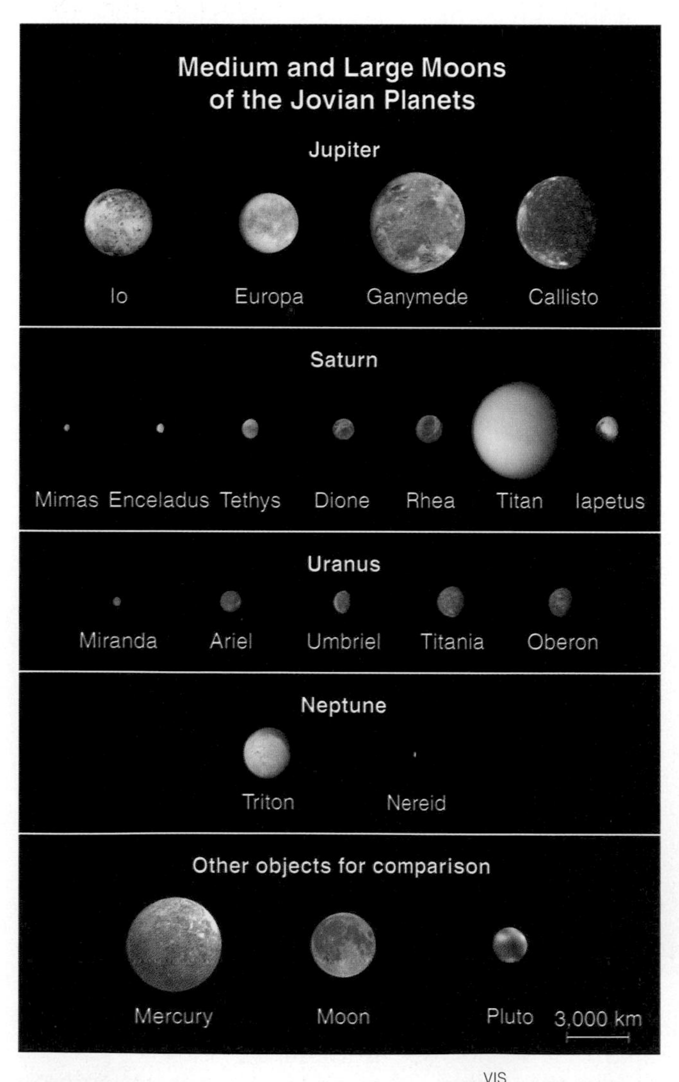

Figure 8.11

The medium-size and large moons of the jovian planets, with sizes (but not distances) shown to scale. Mercury, the Moon, and Pluto are included for comparison.

Most of the medium-size and large moons probably formed by accretion within the disks of gas surrounding individual jovian planets [Section 6.4]. That explains why their orbits are almost circular and lie close to the equatorial plane of their parent planet, and also why these moons orbit in the same direction in which their planet rotates.

The medium-size and large moons tend to be the most interesting, but they are far outnumbered by small moons. Many of the small moons are probably captured asteroids or comets, and as a result they do not follow any particular orbital patterns. Dozens of the smallest moons have been discovered only within the past few years, and many more may yet be discovered. (See Table E.3 in Appendix E for a complete list of the known moons and their general properties.)

A few jovian moons rival the smallest planets in size and geological interest, while vast numbers of smaller moons are captured asteroids and comets.

The small moons' shapes generally resemble potatoes (Figure 8.12), because their gravities are too weak to force their rigid material into spheres. We have not studied these moons in depth, but we expect their small sizes to allow for little if any geological activity. For the most part, the small moons are just chunks of ice and rock held captive by the gravity of a massive jovian planet.

Figure 8.12

These photos from the Voyager spacecraft show five of Saturn's smaller moons. All are much smaller than the smallest moons shown in Figure 8.11. Their irregular shapes are due to their small size, which makes their gravities too weak to force them into spheres.

• What makes Jupiter's Galilean moons unusual?

We are now ready to embark on a brief tour of the most interesting moons of the jovian planets. Our first stop is Jupiter. Jupiter's four largest moons, known as the *Galilean moons* because they were discovered by Galileo [Section 3.3], are all large enough that they would count as planets if they orbited the Sun (Figure 8.13).

Io: The Most Volcanically Active World in the Solar System For anyone who thinks of moons as barren, geologically dead places like our own Moon, Io shatters the stereotype. When the *Voyager* spacecraft first photographed Io up close about two decades ago, we discovered a world with a surface so young that not a single impact crater has survived from past impacts. Moreover, *Voyager* cameras recorded volcanic eruptions in progress as the spacecraft passed by. We now know that Io is by far the most volcanically active world in our solar system. Its entire surface is pockmarked by large volcanoes (Figure 8.14). It has no impact craters because its surface is continually covered by debris from volcanic eruptions.

Io's active volcanoes tell us that it must be quite hot inside. However, Io is only about the size of our geologically dead Moon, so it should have long ago lost any heat from its birth and is too small for radioactivity to provide much ongoing heat. How, then, can Io be so hot inside? The only possible answer is that some other ongoing process must be heating Io's

Figure 8.13

This set of photos, taken by the *Galileo* spacecraft orbiting Jupiter, shows global views of the four Galilean moons as we know them today (left to right): Io, Europa, Ganymede, and Callisto. Sizes are shown to scale. (Io is about the size of Earth's Moon.)

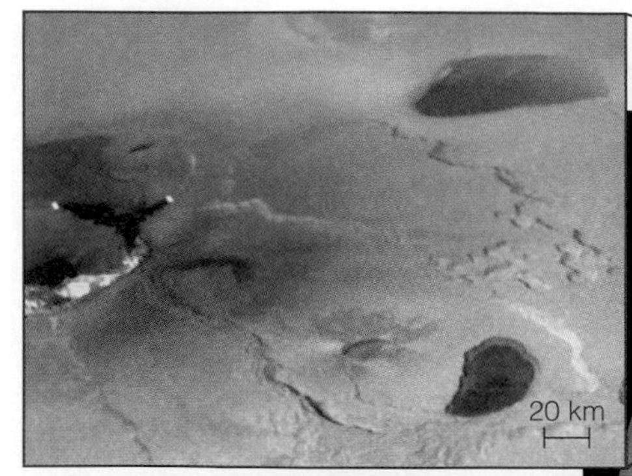

This close-up shows the glow of intensely hot lava from a volcanic eruption.

Io. Most of the black, brown, and red spots are recently active volcanic features. White and yellow areas are sulfur and sulfur dioxide deposits from volcanic gases.

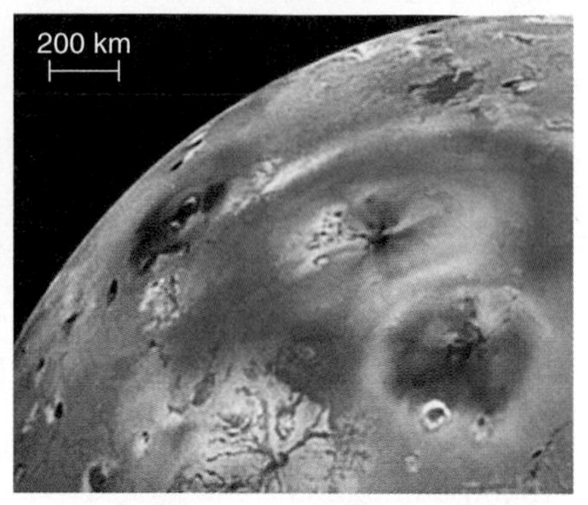

The black and red rings are fallout from volcanic eruptions; their appearance can change from month to month. (This region is located on the opposite side of Io from that shown in the central photograph.)

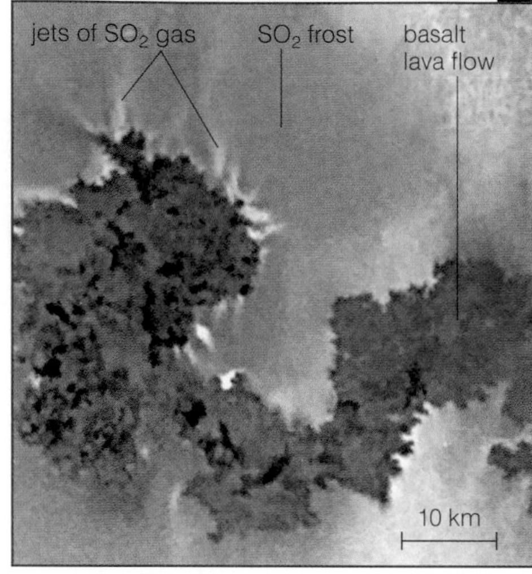

This huge plume of gas, rising some 80 km high, was created when hot lava flowed over sulfur dioxide frost, causing the frost to sublimate explosively into gas.

Figure 8.14

VIS

Io is the most volcanically active body in the solar system. (Photographs from the *Galileo* spacecraft; some colors slightly enhanced or altered.)

interior. Scientists have identified this process and call it **tidal heating**, because it arises from effects of tidal forces exerted by Jupiter.

Just as Earth exerts a tidal force that causes the Moon to keep the same face toward us at all times [Section 4.4], a tidal force from Jupiter makes Io keep the same face toward Jupiter as it orbits. But Jupiter's mass make this tidal force far larger than that which Earth exerts on the Moon. Moreover, Io's orbit is slightly elliptical, so its orbital speed and distance from Jupiter vary. This variation means that the strength and direction of the tidal force change slightly as Io moves through each orbit, which in turn changes the size and orientation of Io's tidal bulges (Figure 8.15a). The result is that Io is continuously being flexed in different directions, which generates friction inside it. The flexing heats the interior in the same way that flexing warms Silly Putty. Tidal heating generates tremendous heat on Io—detailed calculations show that it does indeed explain Io's incredible volcanic activity.

But we are still left with a deeper question: Why is Io's orbit slightly elliptical, when almost all other large satellites' orbits are virtually circular? The answer lies in an interesting dance executed by Io and its neighboring moons (Figure 8.15b).

Orbital resonances among the Galilean moons make Io's orbit slightly elliptical, leading to the tidal heating that makes Io the most volcanically active place in the solar system.

During the time Ganymede takes to complete one orbit of Jupiter, Europa completes exactly two orbits and Io completes exactly four orbits. The three moons therefore line up periodically, and the gravitational tugs they exert on one another add up over time. Because the tugs are always in the same direction with each alignment, they tend to stretch out the orbits, making them slightly elliptical. The effect is much like that

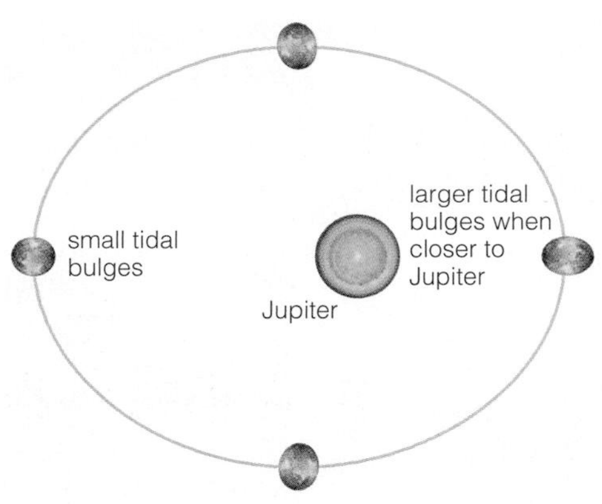

a Io has tidal bulges created by the tidal force exerted by Jupiter. Tidal heating arises because Io's elliptical orbit (exaggerated in this diagram) continually changes the size and direction of the tidal bulges. These changes generate internal friction, making Io's interior hot enough to drive its volcanic activity.

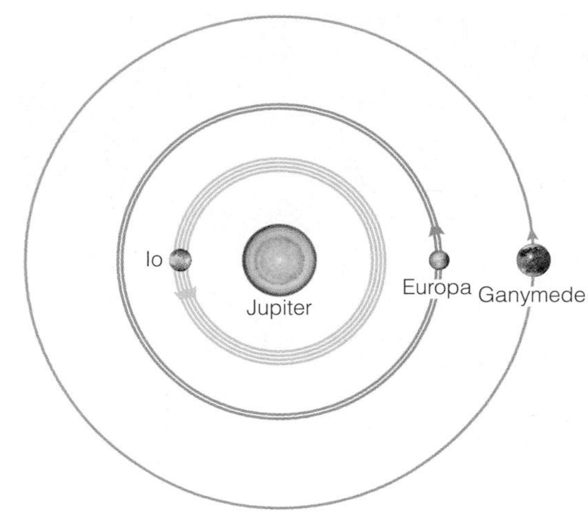

b Io's orbit is elliptical because of the way Io interacts with its neighbors Europa and Ganymede. About every seven Earth days (one Ganymede orbit, two Europa orbits, and four Io orbits), the three moons line up as shown. The small gravitational tugs that occur in this alignment add up over time, and have made all three orbits slightly elliptical.

Figure 8.15

These diagrams explain the cause of tidal heating on Io. Tidal heating also affects Europa and Ganymede, but not as much because they are farther from Jupiter, so Jupiter's tidal force is weaker.

of pushing a child on a swing. If timed properly, a series of small pushes can add up to a *resonance* that causes the child to swing quite high. For the three moons, the **orbital resonance** that makes their orbits elliptical comes from the small gravitational tugs that repeat at each alignment.

Europa: The Water World? Europa offers a stark contrast to Io. Instead of having a surface dotted by active volcanoes, Europa is covered by water ice (Figure 8.16). Nevertheless, its fractured, frozen surface must hide an interior made hot by the same type of tidal heating that powers Io's volcanoes—but tidal heating is weaker on Europa because it lies farther from Jupiter. Europa has only a handful of impact craters, which means that ongoing geological activity must have erased the evidence of nearly all past impacts. Scientists suspect that this geological activity is driven either by ice that is soft enough to undergo convection or by liquid water beneath the icy crust.

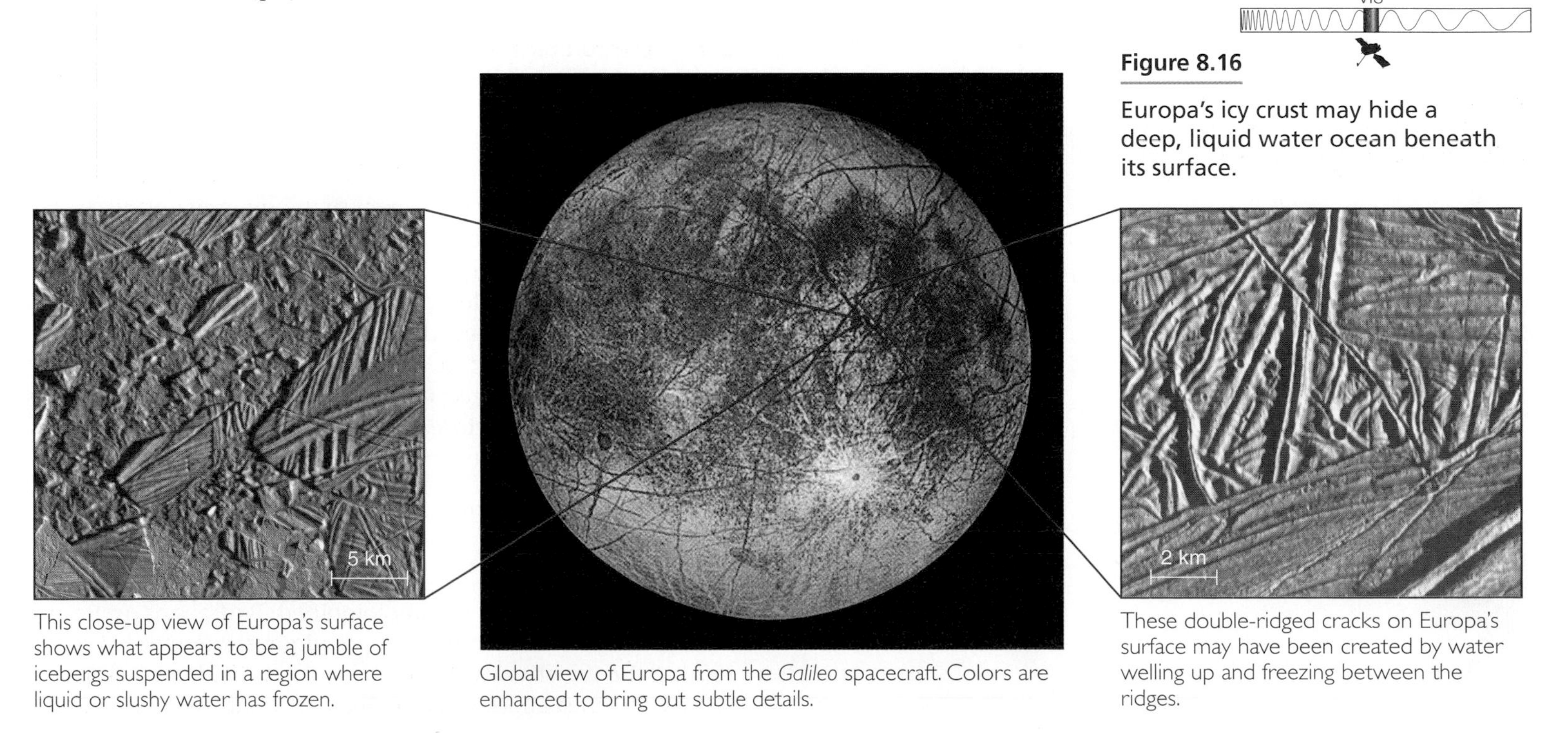

Figure 8.16

Europa's icy crust may hide a deep, liquid water ocean beneath its surface.

This close-up view of Europa's surface shows what appears to be a jumble of icebergs suspended in a region where liquid or slushy water has frozen.

Global view of Europa from the *Galileo* spacecraft. Colors are enhanced to bring out subtle details.

These double-ridged cracks on Europa's surface may have been created by water welling up and freezing between the ridges.

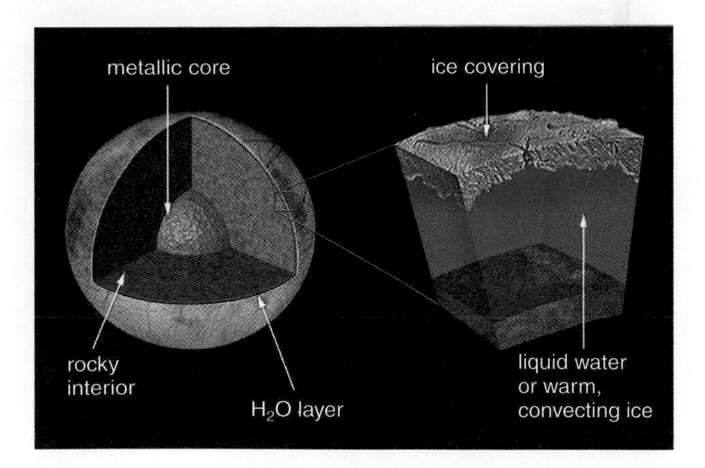

Figure 8.17

This diagram shows one model of Europa's interior structure. There is little doubt that the H_2O layer is real, but questions remain about whether the material beneath the icy crust is actually liquid water or just relatively warm, convecting ice.

Figure 8.18

Tidal heating may give Europa a subsurface ocean beneath its icy crust. This artist's conception imagines a region where the crust has been disrupted by an undersea volcano.

In fact, we have good reason to suspect that a deep water ocean lies beneath Europa's icy skin. Data collected by the *Galileo* spacecraft suggest that Europa has a metallic core and rocky mantle surrounded by a layer of water (H_2O) more than 100 kilometers thick (Figure 8.17). We do not yet have enough data to be certain that the water is liquid rather than ice, but calculations suggest that tidal heating could supply enough heat to make it liquid. Moreover, several other pieces of evidence also suggest a liquid water ocean, including close-up photos of the surface (see Figure 8.16) and careful studies of Europa's magnetic field.

Tidal heating may create a deep ocean of liquid water beneath Europa's icy crust.

If it really exists, Europa's liquid ocean may be more than 100 kilometers deep. If so, the ocean of Europa contains more than twice as much liquid water as all of Earth's oceans combined. Perhaps as on Earth's seafloor, lava erupts from vents on Europa's seafloor, sometimes violently enough to jumble the icy crust above (Figure 8.18). And, knowing that primitive life thrives near seafloor vents on Earth, we can wonder whether

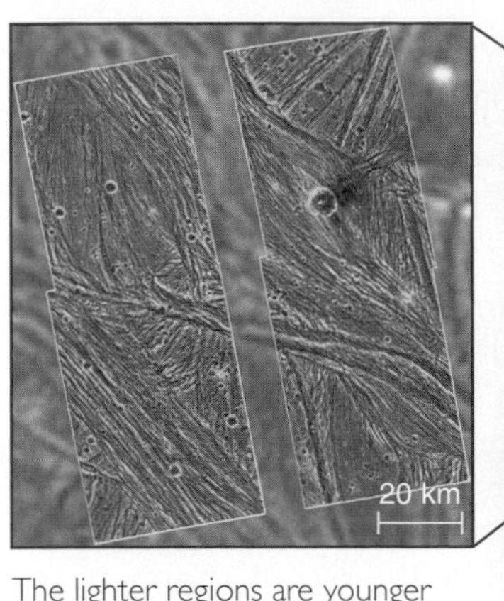

The lighter regions are younger landscapes where eruptions of melted water have covered ancient craters. These regions also show long grooves in the surface, probably formed by tectonic stresses or by water erupting along a surface crack and expanding as it refreezes.

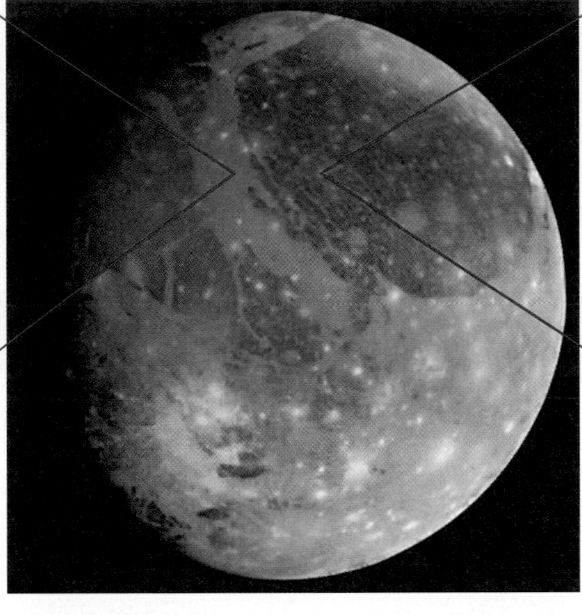

10 km

The darker regions of Ganymede's surface are heavily cratered and must be billions of years old.

Figure 8.19

Ganymede, the largest moon in the solar system, has both old and young regions on its surface of water ice.

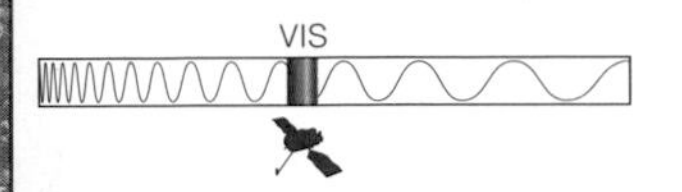

Europa might also be a home to life—a possibility we will explore further in Chapter 18.

Ganymede and Callisto Jupiter's two other large moons, Ganymede and Callisto, also show intriguing geology. Like Europa, both have surfaces of water ice.

Ganymede, the largest moon in the solar system, has a surface with what appears to be a dual personality (Figure 8.19). Some regions are dark and densely cratered, suggesting that they look much the same today as they did billions of years ago. Other regions are light-colored with very few craters, suggesting that liquid water has recently erupted and refrozen. Could Ganymede, like Europa, have a liquid water ocean beneath its surface? If so, we'd need to explain the source of the heat that melts Ganymede's subsurface ice. Ganymede has some tidal heating, but calculations suggest that it is not strong enough to account for an ocean (if one exists). Perhaps ongoing radioactive decay supplies enough additional heat to make an ocean. Or perhaps not—no one yet knows what secrets Ganymede hides.

The outermost Galilean moon, Callisto, looks most like what scientists originally expected for an outer solar system satellite: a heavily cratered iceball (Figure 8.20). The bright patches on its surface are impact craters. However, the surface still holds some surprises. Close-up images show a dark, powdery substance concentrated in low-lying areas, leaving ridges and crests bright white. The nature of this material and how it got there are unknown.

Tidal heating is weak on Ganymede and absent on Callisto, yet both moons show some evidence of subsurface oceans. No one knows why.

Even more surprising, magnetic field data suggest that Callisto, too, could hide a subsurface ocean. No one knows what might heat the interior of Callisto, since it does not participate in the orbital resonances of the other Galilean moons and therefore has no tidal heating at all. Nevertheless, the potential for an ocean raises the intriguing possibility that there could be three "water worlds" orbiting Jupiter—with far more total ocean than we find here on Earth.

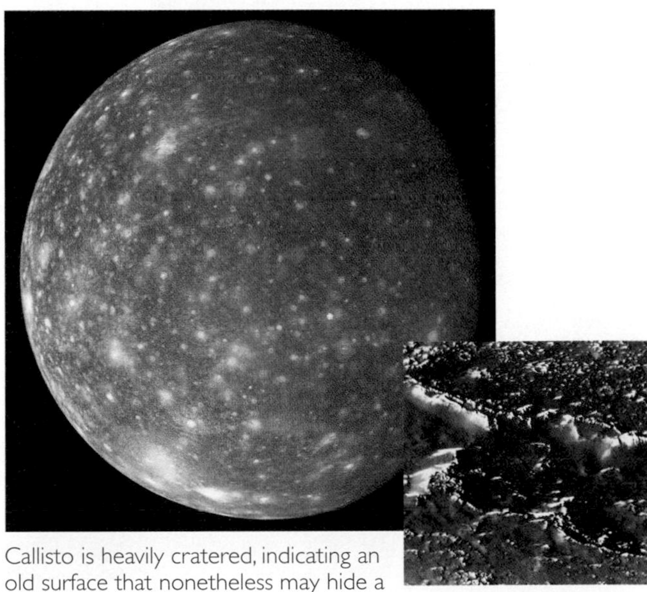

Callisto is heavily cratered, indicating an old surface that nonetheless may hide a deeply buried ocean.

Close-up photo shows a dark powder overlaying the low areas of the surface.

VIS

Figure 8.20

Callisto, the outermost of the four Galilean moons, has a heavily cratered icy surface.

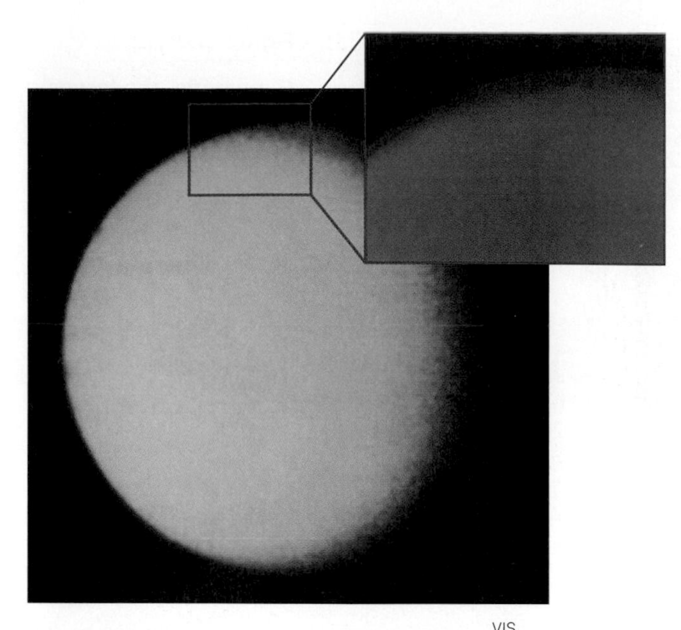

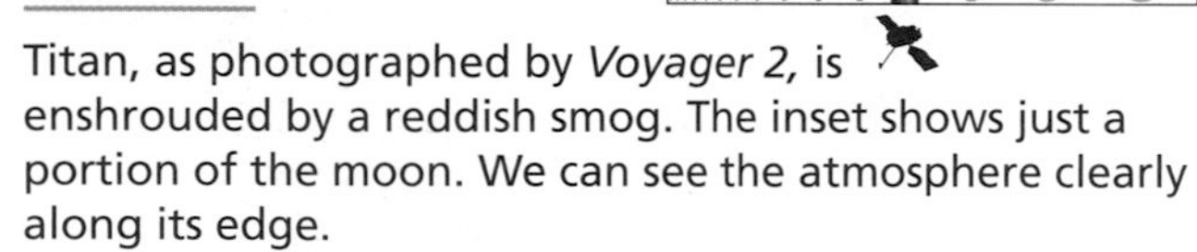

Figure 8.21

Titan, as photographed by *Voyager 2,* is enshrouded by a reddish smog. The inset shows just a portion of the moon. We can see the atmosphere clearly along its edge.

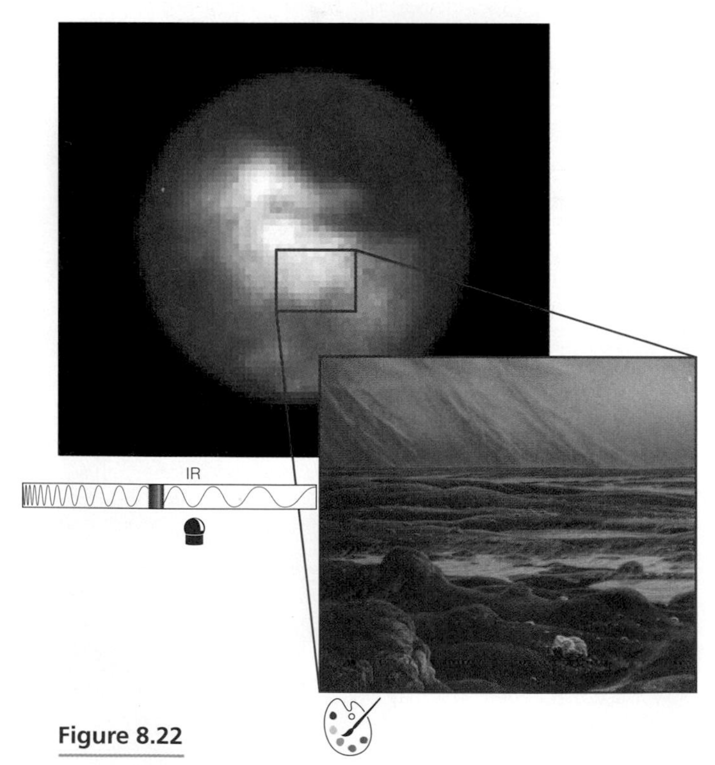

Figure 8.22

This infrared image of Titan (from the Keck telescope in Hawaii) shows brightness variations; dark areas may be oceans of liquid ethane and methane. The inset is an artist's conception of the surface of Titan, showing a possible ethane/methane sea.

• What makes Titan different from other moons?

Leaving Jupiter behind, our tour of moons takes us to Saturn. Saturn has six medium-size moons and one large moon: Titan, the second-largest moon in the solar system after Ganymede.

Smog-Covered Titan Titan is unique among the moons of our solar system in having a thick atmosphere—so thick that it completely hides the surface from view (Figure 8.21). Titan's reddish color comes from chemicals in its atmosphere much like those that make smog over cities on Earth. The atmosphere is about 90% nitrogen, not that different from the 77% nitrogen content of Earth's atmosphere. However, on Earth the rest of the atmosphere is mostly oxygen, while the rest of Titan's atmosphere consists of argon, methane (CH_4), ethane (C_2H_6), and other hydrogen compounds.

Methane and ethane are both greenhouse gases. They give Titan an appreciable greenhouse effect [Section 7.1] that makes it warmer than it would be otherwise. Still, because of its great distance from the Sun, its surface temperature is a frigid 93 K (−180°C). The surface pressure on Titan is about 1.5 times the sea level pressure on Earth, which would be fairly comfortable if not for the lack of oxygen and the cold temperatures.

A moon with a thick atmosphere would be intriguing enough, but we have at least two other reasons for special interest in Titan. First, scientists suspect that chemical reactions on Titan produce numerous organic chemicals—the chemicals that are the basis of life. Few people think we could find actual life on Titan due to the cold temperatures, but many hope we'll learn about the chemistry that may have occurred on Earth before life actually arose.

Titan is the only moon in the solar system with a thick atmosphere, and it may have lakes or oceans of liquid ethane and methane.

Second, although it is far too cold for liquid water to exist on Titan, Titan may have lakes or oceans of liquid ethane and methane. The idea that such oceans might exist originally came from detailed analysis of Titan's atmospheric composition, and gained additional support from infrared pictures of Titan. Infrared light can penetrate through the clouds, and the infrared photos show variations in surface brightness that may represent dark ethane/methane oceans and bright continents (Figure 8.22). We should learn for certain whether such frigid oceans really exist by early 2005, by which time the *Cassini* spacecraft will have begun radar mapping of Titan and will have dropped the *Huygens* probe toward Titan's surface. The probe will collect data and take photographs on the way down and has enough battery power to survive for perhaps half an hour after it reaches the surface. Just in case, the probe is designed to float in liquid ethane.

THINK ABOUT IT — Would you expect Titan's surface to have many impact craters? Explain.

Saturn's Medium-Size Moons The Cassini mission should also teach us more about Saturn's other moons, especially its six medium-size moons (Figure 8.23). These moons are all far smaller than any of the terrestrial worlds (see Figure 8.11), so if they acted like terrestrial worlds we

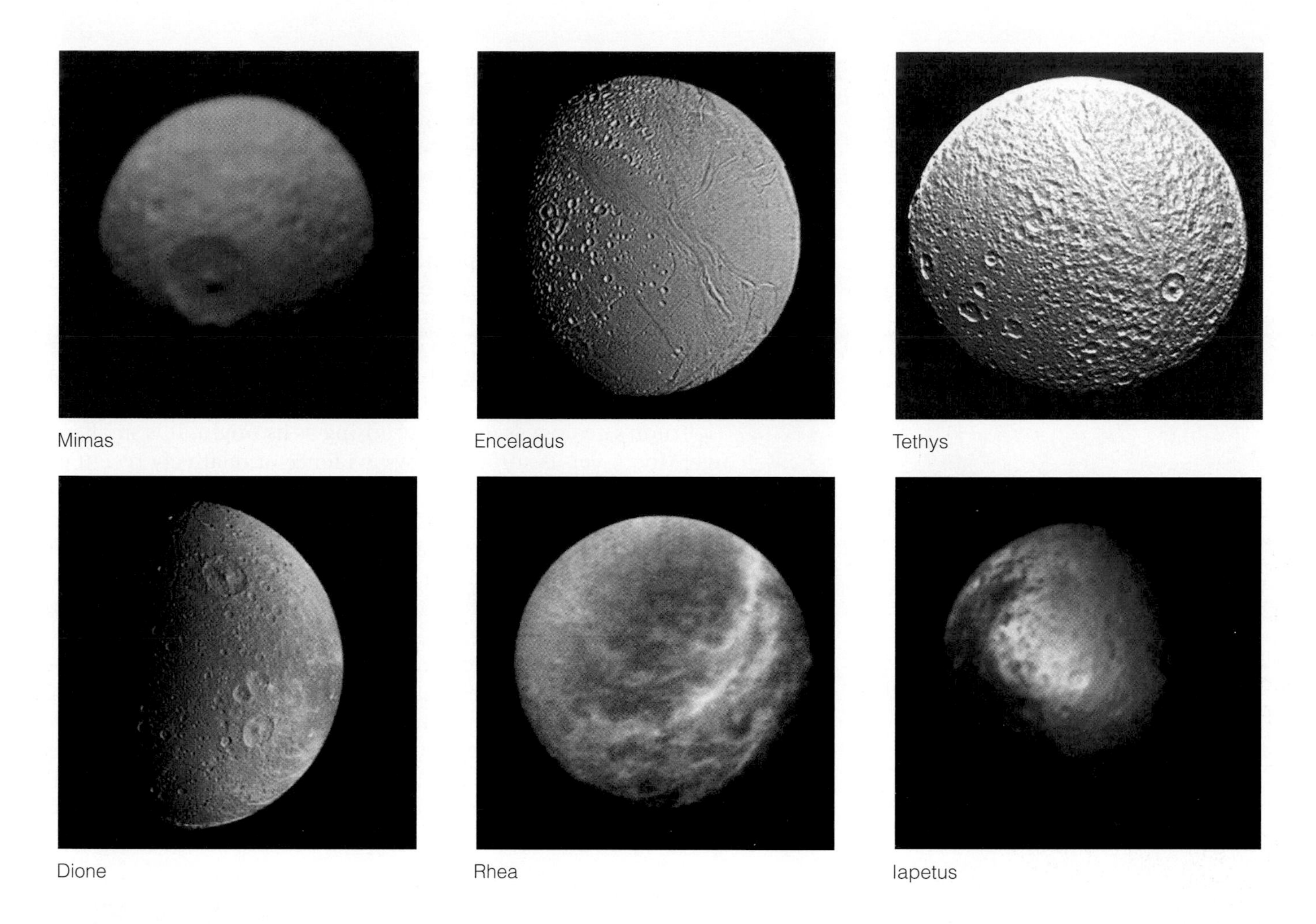

would expect their interiors to have cooled so quickly and completely that they would never have had any significant geological activity. All six have heavily cratered surfaces, confirming that they lack widespread geological activity today.

However, photos also reveal abundant evidence of past volcanism and/or tectonics on most of these icy moons. For example, the light regions visible in the photos may be places where icy lava once flowed. Enceladus even has strange grooves similar to some seen on Ganymede, while Iapetus has one side that is much darker than the other. Only Mimas, the smallest of these six moons, lacks evidence of past geological activity.

How is it possible that moons as small as these can have had such active geology? The answer lies in their icy compositions, which makes them fundamentally different from terrestrial worlds despite the outward similarities of their solid surfaces. We'll investigate this issue in more depth in a moment, but first let's look at some other examples of geological activity on small moons of the outer solar system.

Figure 8.23

VIS

Portraits taken by the *Voyager* spacecraft of Saturn's six medium-size moons (not to scale). All but Mimas show evidence of past volcanism and/or tectonics.

• Why are small icy moons more geologically active than small rocky planets?

The moons of Uranus and Neptune offer further evidence of surprising geological activity. Uranus has five medium-size moons (and no large moons), and at least three of them show evidence of past volcanism or tectonics.

Miranda, the smallest of the five, is the most surprising (Figure 8.24). Despite its small size, it shows tremendous tectonic features and relatively few craters. Apparently, it underwent geological activity well after the heavy bombardment ended [Section 6.4], erasing its early craters.

The surprises continue with Neptune's moon Triton (Figure 8.25). Triton is a strange moon to begin with: It is a large moon, but it does not follow the orbital patterns of all other large moons in the outer solar system. In fact, it orbits Neptune "backward" (opposite to Neptune's rotation) and at a high inclination to Neptune's equator. These are telltale signs of a moon that was captured rather than having formed in the disk of gas around its planet. No one knows how a moon as large as Triton could have been captured, but it still seems almost certain that Triton once orbited the Sun rather than Neptune.

Triton's geology is just as surprising as its origin. It is smaller than our own Moon, yet its surface shows evidence of relatively recent geological activity. Some regions show evidence of past volcanism, while others show wrinkly ridges (nicknamed "cantaloupe terrain") that appear tectonic in nature. Triton even has a very thin atmosphere that has left some wind streaks on its surface. It's likely that Triton was originally captured into an elliptical orbit, which may have led to enough tidal heating to explain its geological activity.

Based on what we learned when studying the geology of the terrestrial worlds, the active geology of the jovian moons seems out of character with their sizes. Numerous jovian moons remained geologically active far longer than Mercury or our Moon, yet they are no bigger and in many cases much smaller in size. However, there is a crucial difference between the jovian moons and the terrestrial worlds: composition.

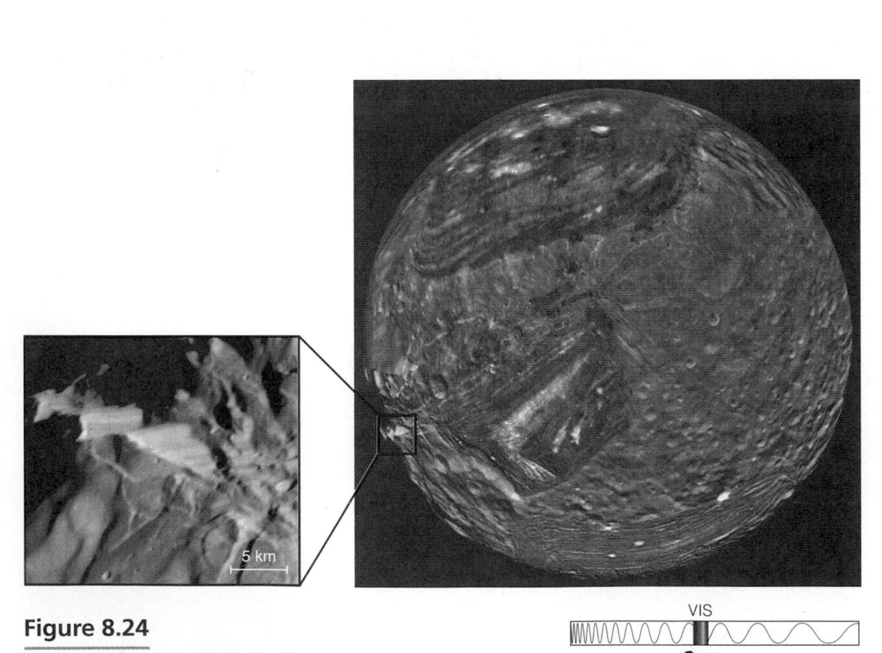

Figure 8.24

The surface of Miranda shows astonishing tectonic activity despite its small size. The cliff walls seen in the inset are higher than those of the Grand Canyon on Earth.

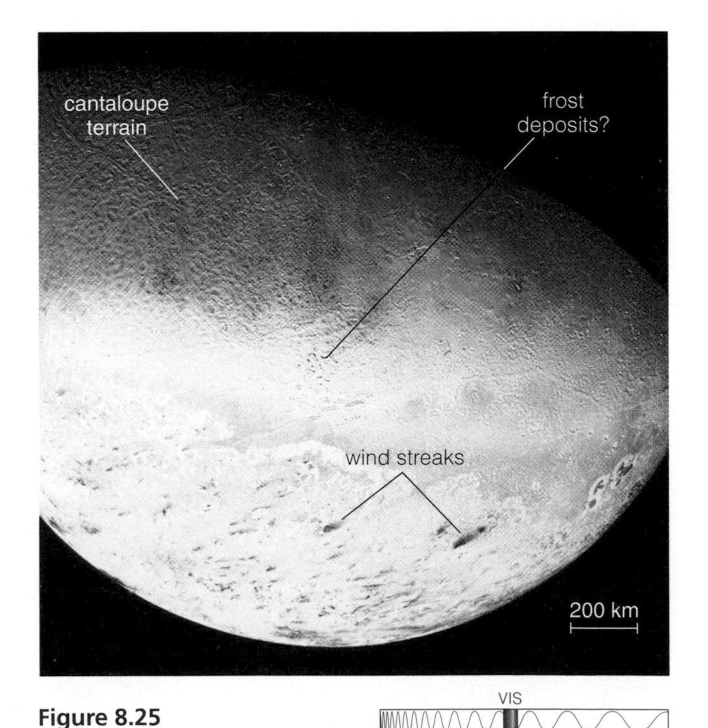

Figure 8.25

Neptune's moon Triton shows evidence of a surprising level of past geological activity. This photo of Triton's southern hemisphere was taken by *Voyager 2*.

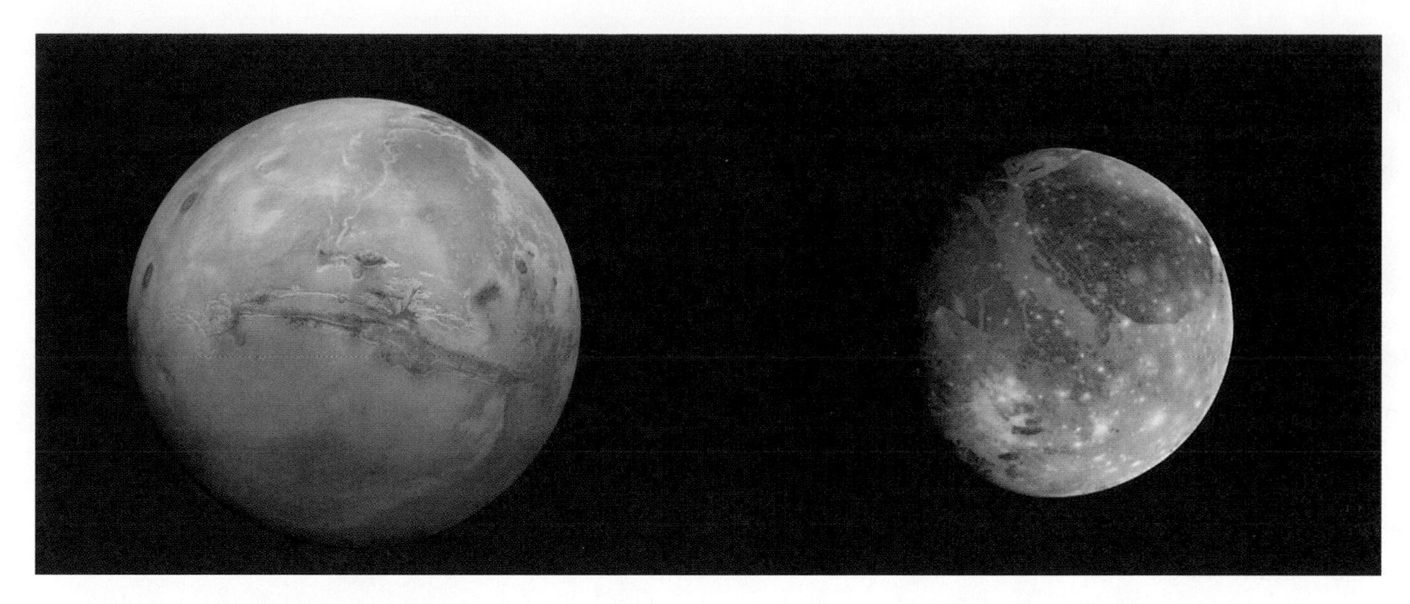

Terrestrial Planet Geology

- Internal heat, primarily from radioactive decay, can cause volcanic and tectonic activity.
- Only large planets retain enough internal heat to stay geologically active today.
- Example: Mars (photo above) probably retains some internal heat. If it had been smaller, like Mercury, it would be geologically "dead" today. If it had been larger, like Earth, it would probably have much more active and ongoing tectonics and volcanism.

Jovian Moon Geology

- Tidal heating can cause tremendous geological activity on moons on elliptical orbits around massive planets.
- Even without tidal heating, icy materials can melt and deform at lower temperatures than rock, increasing the likelihood of geological activity.
- Together, these effects explain why icy moons are much more likely to have ongoing geological activity than rocky terrestrial worlds of the same size.
- Example: Ganymede (photo above) shows evidence of recent geological activity, even though it is similar in size to the geologically dead terrestrial planet Mercury.

Figure 8.26

Jovian moons can be much more geologically active than terrestrial worlds of similar size due to their icy compositions and a heating source (tidal heating) that is not important on the terrestrial worlds.

The jovian moons contain ices that can melt or deform at far lower temperatures than rock. As a result, they can still be warm enough to experience geological activity even when their interiors have cooled to temperatures far below what they were at their births. Indeed, most of the volcanism that has occurred in the outer solar system (except on Io) probably did not produce any hot lava at all. Instead, it produced icy lava that was essentially liquid water, perhaps mixed with methane and ammonia.

Ice can deform or melt at much lower temperatures than rock, allowing even small icy moons to undergo geological activity.

The major lesson, then, is that "ice geology" is possible at far lower temperatures than "rock geology." This fact, combined with the occasional extra heating source of tidal heating, explains how the jovian moons have had such interesting geological histories despite their small sizes. Figure 8.26 summarizes the differences between the geology of jovian moons and that of the terrestrial worlds.

8.3 JOVIAN PLANET RINGS

The jovian planet systems have three major components: the planets themselves, the moons, and the rings that encircle the planets. We have already studied the planets and their moons, so we now turn our attention to their amazing rings. We'll begin by exploring the rings of Saturn, since they are by far the most spectacular.

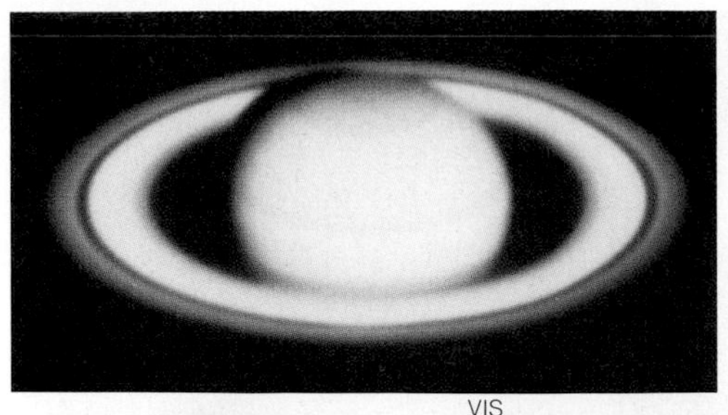

a This Earth-based telescopic view of Saturn makes the rings look like large, concentric sheets. The dark gap within the rings is called the *Cassini division.*

b This *Voyager* image of Saturn's rings reveals many individual rings separated by narrow gaps.

c Artist's conception of particles in a ring system. All the particles are moving slowly relative to one another and occasionally collide.

Figure 8.27 Interactive Figure

Zooming in on Saturn's rings.

• What are Saturn's rings like?

You can see Saturn's rings through a backyard telescope, but learning their nature requires higher resolution (Figure 8.27). Earth-based views make the rings appear to be continuous, concentric sheets of material separated by a large gap (called the *Cassini division*). Spacecraft images reveal these "sheets" to be made of many individual rings, each separated from the next by a narrow gap. But even these appearances are somewhat deceiving. If we could wander into Saturn's rings, we'd find that they are made of countless icy particles ranging in size from dust grains to large boulders. All are far too small to be photographed even from spacecraft passing nearby.

Ring Particle Characteristics Spectroscopy reveals that Saturn's ring particles are made of relatively reflective water ice. The rings look bright where they contain enough particles to intercept sunlight and scatter it back toward us. We see gaps in places where there are few particles to reflect sunlight.

Saturn's rings are made of vast numbers of icy particles ranging in size from dust to boulders, each circling Saturn according to Kepler's laws.

Each individual ring particle orbits Saturn independently in accord with Kepler's laws, so the rings are much like myriad tiny moons. The individual ring particles are so close together that they collide frequently. In the densest parts of the rings, each particle collides with another every few hours. However, the collisions are fairly gentle: Despite the high orbital speeds of the ring particles, nearby ring particles are moving at nearly the same speed and touch only gently when they collide.

Figure 8.28

This gap within rings is created by the gravity of a 20-km-wide moon, visible as little more than a dot in the inset.

THINK ABOUT IT Which ring particles travel faster: those closer to Saturn or those farther away? Explain why. (*Hint:* Review Kepler's third law.)

The frequent collisions explain why Saturn's rings are perhaps the thinnest known astronomical structure. They span more than 270,000 kilometers in diameter but are only a few tens of *meters* thick. To understand how collisions keep the rings thin, imagine what would happen to a ring particle on an orbit slightly inclined to the central ring plane. The particle would collide with other particles every time its orbit intersected the ring plane, and its orbital tilt would be reduced with every collision. Before long, these collisions would force the particle to conform to the orbital pattern of the other particles. Thus, any particle that moves away from the narrow ring plane is soon brought back within it.

Rings and Gaps Close-up photographs show an astonishing number of rings, gaps, ripples, and other features—the total number of features may be as high as 100,000. Scientists are still struggling to explain all the features, but some general ideas are now clear.

Rings and gaps are caused by particles bunching up at some orbital distances and being forced out at others. This bunching happens when gravity nudges the orbits of ring particles in some particular way. One source of nudging comes from small moons located within the gaps in the rings themselves, and hence sometimes called *gap moons* (Figure 8.28). The gravity of a gap moon can effectively keep the gap clear of smaller ring particles. In some cases, two nearby gap moons can force particles between them into a very narrow ring. (The gap moons are often called *shepherd moons* in those cases, because they act like a shepherd forcing particles into line.)

Ring particles also may be nudged by the gravity from larger, more distant moons. For example, a ring particle orbiting about 120,000 kilometers from Saturn's center will circle the planet in exactly half the time it takes the moon Mimas to orbit. Every time Mimas returns to a certain location, the ring particle will also be at its original location and therefore will experience the same gravitational nudge from Mimas. The periodic nudges reinforce one another and clear a gap in the rings—in this case, the large gap visible from Earth (the Cassini division). This type of reinforcement due to repeated gravitational tugs is another example of an *orbital resonance,* much like the orbital resonances that make Io's orbit elliptical (see Figure 8.15b). Other orbital resonances, caused by moons both within the rings and farther out from Saturn, probably explain most of the intricate structures visible in ring photos.

• Why do the jovian planets have rings?

For a long time, Saturn's rings were thought to be unique in the solar system. Most scientists therefore assumed they were formed by some kind of rare event, perhaps a moon wandering too close to Saturn long ago and being torn apart by tidal forces. However, we now know that all four jovian planets have rings (Figure 8.29). Although Saturn's rings clearly have more numerous and more reflective particles than the other ring systems, we can no longer think that rings are rare. We therefore need an explanation for rings that doesn't require rare events to have happened for all four planets.

Some scientists once guessed that the ring particles might be leftover chunks of rock and ice that condensed in the disks of gas that orbited each jovian planet when it was young. This would explain why all four jovian planets have rings, because tidal forces near each planet would have prevented these chunks from accreting into a full-fledged moon. However, we now know that the ring particles cannot be leftovers from the birth of the planets, because particles of the size we find in the rings today cannot survive for billions of years. Ring particles are continually being ground down in size, primarily by the impacts of the countless sand-size particles that orbit the Sun—the same types of particles that become meteors in Earth's atmosphere and cause micrometeorite impacts on the

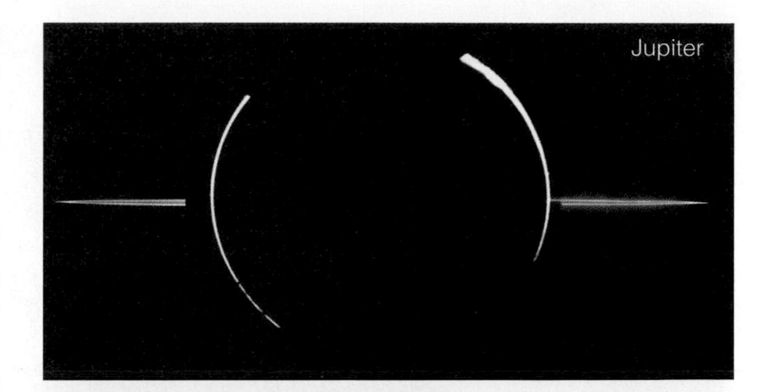

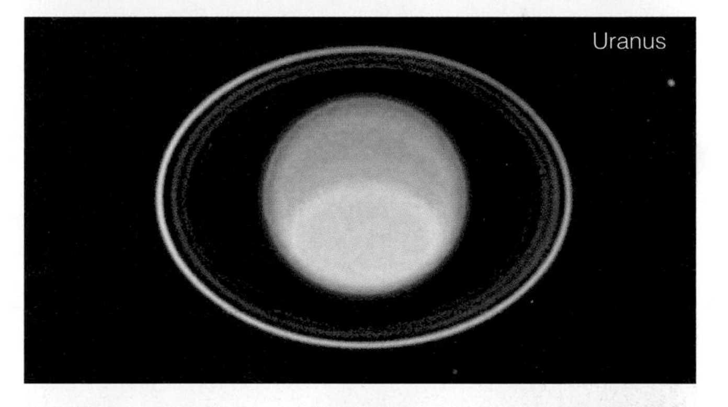

Figure 8.29

Four ring systems (not to scale). The rings differ in appearance and in the composition and sizes of the ring particles. These photos were taken with visible light except for Uranus, which was photographed in infrared light by the Hubble Space Telescope. The Saturn photo is from *Cassini* and the Jupiter and Neptune photos are from *Voyager.*

Figure 8.30

This illustration summarizes the origin of rings around the jovian planets.

Moon [Section 7.2]. Millions of years of such tiny impacts would long ago have ground the existing ring particles to dust.

Ring particles cannot last for billions of years, so the rings we see today must be made of particles created recently.

We are therefore left with only one reasonable possibility for the origin of the rings: New particles are continually supplied to the rings to replace those that are destroyed. These new particles must come from a source that lies in each planet's equatorial plane. The most likely source is numerous small moons—moons the size of gap moons (see Figure 8.28)—that formed in the disks of material orbiting the young jovian planets. Like the ring particles themselves, these small moons are gradually being ground away by tiny impacts, but they are large enough to still exist despite four and a half billion years of such sandblasting.

The small moons contribute ring particles in two ways. First, each tiny impact releases particles from a small moon's surface, and these released particles become new, dust-size ring particles. Ongoing impacts ensure that some ring particles are present at all times. Second, occasional larger impacts can shatter a small moon completely, creating a supply of boulder-size ring particles. These boulders are then slowly ground into smaller ring particles by the frequent tiny impacts. In summary, the dust- to boulder-size particles in rings all ultimately come from the gradual dismantling of small moons that formed during the birth of the solar system (Figure 8.30).

New ring particles are released by impacts on small moons within the rings.

The collisions that shatter small moons and generate a major source of ring particles must occur only occasionally and at essentially random times, which means that the number of particles in any particular ring system must change dramatically over millions and billions of years. Rings may be broad and bright when they are full of particles and almost invisible when particles are few. Thus, the brilliant spectacle of Saturn's rings may be a special treat of our epoch, one that could not have been seen a billion years ago and that may not last long on the time scale of our solar system.

THE BIG PICTURE

Putting Chapter 8 into Context

In this chapter, we saw that the jovian planets really are a different kind of planet and, indeed, a different kind of planetary system. The jovian planets dwarf the terrestrial planets. Even some of their moons are as large as terrestrial worlds. As you continue your study of the solar system, keep in mind the following "big picture" ideas:

- The jovian planets may lack solid surfaces on which geology can occur, but they are interesting and dynamic worlds with rapid winds, huge storms, strong magnetic fields, and interiors in which common materials behave in unfamiliar ways.
- Despite their relatively small sizes and frigid temperatures, many jovian moons are geologically active by virtue of their icy compositions and tidal heating. Ironically, the cold temperatures in the solar nebula led to their icy compositions and hence their geological activity.
- Ring systems probably owe their existence to small moons formed in the disks of gas that produced the jovian planets billions of years ago. The rings we see today are composed of particles liberated from those moons surprisingly recently.
- Understanding the jovian planet systems forced us to modify many of our earlier ideas about the solar system, in particular by adding the concepts of ice geology, tidal heating, and orbital resonances. Each new set of circumstances we discover offers further opportunities to learn how our universe works.

8.1 A DIFFERENT KIND OF PLANET

• What are jovian planets made of?
Jupiter and Saturn are made almost entirely of hydrogen and helium, while Uranus and Neptune are made mostly of hydrogen compounds mixed with metal and rock. These differences arose because all four planets started from ice-rich planetesimals of about the same size, but captured different amounts of hydrogen and helium gas from the solar nebula.

• What are jovian planets like on the inside?

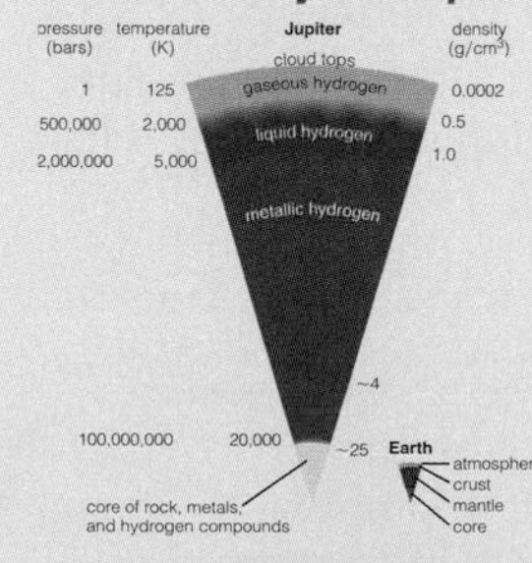

The jovian planets have layered interiors with very high internal temperatures and pressures. All have a core about 10 times as massive as Earth, consisting of hydrogen compounds, metals, and rock. They differ mainly in their surrounding layers of hydrogen and helium.

• What is the weather like on jovian planets?

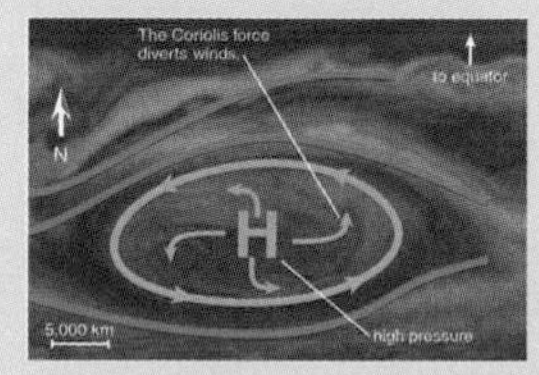

The jovian planets all have multiple cloud layers that help determine the colors of the planets, fast winds, and large storms. Some storms, such as the Great Red Spot, can apparently rage for centuries or longer.

8.2 A WEALTH OF WORLDS: SATELLITES OF ICE AND ROCK

• What kinds of moons orbit the jovian planets?
We can categorize the more than 100 known moons as small, medium-size, or large. Most of the medium-size and large moons probably formed with their planet in the disks of gas that surrounded the jovian planets when they were young. Smaller moons are often captured asteroids or comets.

• What makes Jupiter's Galilean moons unusual?

Io is the most volcanically active object in the solar system, thanks to an interior kept hot by tidal heating—which occurs because Io's close orbit is made elliptical by orbital resonances with other moons. Europa may have a deep, liquid water ocean under its icy crust, also thanks to tidal heating. Ganymede and Callisto may also have subsurface oceans.

• What makes Titan different from other moons?

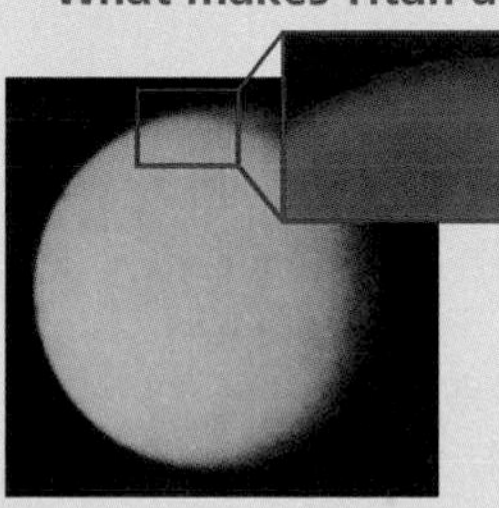

Titan is the only moon in our solar system with a thick atmosphere. It may even have lakes or oceans of liquid ethane and methane on its surface.

Continued ▶

- **Why are small icy moons more geologically active than small rocky planets?**

Ices deform and melt at much lower temperatures than rock, allowing icy volcanism and tectonics at surprisingly low temperatures.

8.3 JOVIAN PLANET RINGS

- **What are Saturn's rings like?**

Saturn's rings are made up of countless individual particles, each orbiting Saturn independently like a tiny moon. The rings lie in Saturn's equatorial plane, and they are extremely thin.

- **Why do the jovian planets have rings?**

Ring particles probably come from the dismantling of small moons formed in the disks of gas that surrounded the jovian planets billions of years ago. Small ring particles come from countless tiny impacts on the surfaces of these moons, while larger ones come from impacts that shatter the moons.

EXERCISES AND PROBLEMS

REVIEW QUESTIONS

1. Briefly describe how the differences in composition among the jovian planets can be traced to their formation in the solar nebula.
2. Why is Jupiter so much denser than Saturn? Could a planet be smaller in size than Jupiter but greater in mass? Explain.
3. Briefly describe the interior structure of Jupiter and why it is layered in this way. How do the interiors of the other jovian planets compare to Jupiter?
4. Why does Jupiter have such a strong magnetic field? Describe a few features of Jupiter's magnetosphere.
5. Briefly describe Jupiter's cloud layers. How do the cloud layers help explain Jupiter's colors? Why are Saturn's colors more subdued? Why are Uranus and Neptune blue?
6. Briefly describe Jupiter's weather patterns and contrast them with those on the other jovian planets. What is the *Great Red Spot?*
7. Briefly describe how we categorize jovian moons by size. What is the origin of most of the medium-size and large moons? What is the origin of many of the small moons?
8. What are the key features of Jupiter's four Galilean moons? Explain the role of tidal heating and orbital resonances in explaining these features.
9. Describe the atmosphere of Titan. What evidence suggests that Titan may have oceans of liquid ethane?
10. Briefly describe Triton and explain why we think it is a captured moon.
11. Briefly explain why icy moons can have active geology at much smaller sizes than rocky worlds.
12. What are rings? What are they made of, and how do they differ among the four jovian planets? Briefly describe the effects of gap moons and orbital resonances on ring systems.
13. Explain why we think that ring particles must be replenished over time. Should the jovian planet rings always look the same?

SURPRISING DISCOVERIES?

Suppose someone claimed to make the discoveries described below. (These are *not* real discoveries.) Decide whether each discovery should be considered reasonable or surprising. More than one right answer may be possible, so explain your answer clearly.

14. Saturn's core is pockmarked with impact craters and dotted with volcanoes erupting basaltic lava.
15. Neptune's deep blue color is not due to methane, as previously thought, but instead is due to its surface being covered with an ocean of liquid water.
16. A jovian planet in another star system has a moon as big as Mars.
17. An extrasolar planet is discovered that is made primarily of hydrogen and helium. It has approximately the same mass as Jupiter but is the same size as Neptune.
18. A new small moon is discovered orbiting Jupiter outside the orbits of other known moons. It is smaller than Jupiter's other moons but has several large, active volcanoes.
19. A new moon is discovered orbiting Neptune. The moon orbits in Neptune's equatorial plane and in the same direction that Neptune rotates, but it is made almost entirely of metals such as iron and nickel.
20. An icy, medium-size moon is discovered orbiting a jovian planet in a star system that is only a few hundred million years old. The moon shows evidence of active tectonics.
21. A jovian planet is discovered in a star system that is much older than our solar system. The planet has no moons at all, but it has a system of rings as spectacular as the rings of Saturn.

PROBLEMS

22. *The Importance of Rotation.* Suppose the material that formed Jupiter came together without any rotation so that no "jovian nebula" formed and the planet today wasn't spinning. How else would the jovian system be different? Think of as many effects as you can, and explain each in a sentence.
23. *Minor Ingredients Matter.* Suppose the jovian planet atmospheres were composed only of hydrogen and helium, with no hydrogen compounds at all. How would the atmospheres be different in terms of clouds, color, and weather? Explain.
24. *Hot Jupiters.* Many of the newly discovered planets orbiting other stars are more massive than Jupiter but orbit much closer to their

stars. Assuming that they would be Jupiter-like if they orbited at a greater distance from their star, how would you expect these new planets to differ from the jovian planets of our solar system? How would you expect their moons to differ? Explain.

25. *Observing Project: Jupiter's Moons.* Using binoculars or a small telescope, view the moons of Jupiter. Make a sketch of what you see, or take a photograph. Repeat your observations several times (nightly, if possible) over a period of a couple of weeks. Can you determine which moon is which? Can you measure the moons' orbital periods? Can you determine their approximate distances from Jupiter? Explain.
26. *Observing Project: Saturn's Rings.* Using binoculars or a small telescope, view the rings of Saturn. Make a sketch of what you see, or take a photograph. What season is it in Saturn's northern hemisphere? How far do the rings extend above Saturn's atmosphere? Can you identify any gaps in the rings? Describe any other features you notice.

DISCUSSION QUESTIONS

27. *Jovian Planet Mission.* We can study terrestrial planets up close by landing on them, but jovian planets have no surfaces to land on. Suppose that you are in charge of planning a long-term mission to "float" in the atmosphere of a jovian planet. Describe the technology you would use and how you would ensure survival for any people assigned to this mission.
28. *Pick a Moon.* Suppose you could choose any one moon to visit in the solar system. Which one would you pick, and why? What dangers would you face in your visit to this moon? What kinds of scientific instruments would you want to bring along for studies?

For a complete list of media resources available, go to **www.astronomyplace.com** and choose Chapter 8 from the pull-down menu.

ASTRONOMY PLACE WEB TUTORIALS

Tutorial Review of Key Concepts

Use the following interactive **Tutorial** at **www.astronomyplace.com** to review key concepts from this chapter.

Formation of the Solar System Tutorial

Lesson 1 Comparative Planetology

Supplementary Tutorial Exercises

Use the interactive **Tutorial Lessons** to explore the following questions.

Formation of the Solar System Tutorial, Lesson 1

1. Contrast the general features of the jovian planets with those of the terrestrial planets.
2. Compare the properties of each of the four jovian planets. What properties do they all share? In what ways are they different?

EXPLORING THE SKY AND SOLAR SYSTEM

Of the many activities available on the ***Voyager: SkyGazer* CD-ROM** accompanying your book, use the following files to observe key phenomena covered in this chapter.

Go to the **File: Basics** folder for the following demonstrations.

1. Tracking Jupiter and Io
2. Saturn

Go to the **File: Demo** folder for the following demonstrations.

1. Backside of Jupiter
2. Locked on Dione
3. Three Moons on Jupiter

WEB PROJECTS

Take advantage of the useful Web links on **www.astronomyplace.com** to assist you with the following projects.

1. *The Galileo Mission to Jupiter.* Learn more about the Galileo mission and how it ended, and about ongoing scientific study of the Galileo data. Write a short summary of your findings.
2. *News from Cassini.* Find the latest news about the Cassini mission to Saturn. What is the current mission status? Write a short report about the mission's status and recent findings.
3. *Oceans of Europa.* The possibility of subsurface oceans on Europa holds great scientific interest. Investigate plans for further study of Europa, either from Earth or with future spacecraft. Write a short summary of the plans and how they might help us learn whether an ocean really exists and, if so, what it might contain.

9
Remnants of Rock and Ice: Asteroids, Comets, and Pluto

9.1 ASTEROIDS AND METEORITES
- Why is there an asteroid belt?
- How are meteorites related to asteroids?

9.2 COMETS
- How do comets get their tails?
- Where do comets come from?

9.3 PLUTO: LONE DOG, OR PART OF A PACK?
- What is Pluto like?
- Is Pluto a planet or a Kuiper belt comet?

9.4 COSMIC COLLISIONS: SMALL BODIES VERSUS THE PLANETS
- Have we ever witnessed a major impact?
- Did an impact kill the dinosaurs?
- Is the impact threat a real danger or just media hype?
- How do other planets affect impact rates and life on Earth?

Asteroids and comets might at first seem insignificant in comparison to the planets and moons we've discussed so far, but there is strength in numbers. The trillions of small bodies orbiting our Sun are far more important than their small size might suggest.

The appearance of comets has more than once altered the course of human history when our ancestors acted upon superstitions related to comet sightings. More profoundly, asteroids or comets falling to Earth have scarred our planet with impact craters and have altered the course of biological evolution. Asteroids and comets are also important scientifically: As remnants from the birth of our solar system, they have taught us a lot about how our solar system formed.

In this chapter, we will explore asteroids, comets, and the rocks that fall to Earth as meteorites. We will also examine the smallest planet, Pluto, which may be a misfit among planets but is right at home among the smaller objects of our solar system. Finally, we will explore the dramatic effects of the occasional collisions between small bodies and large planets.

9.1 ASTEROIDS AND METEORITES

We begin our study of small bodies by focusing on asteroids and meteorites. Recall that asteroids are rocky leftover planetesimals—chunks of rock that still orbit the Sun because they never managed to become part of a planet [Section 6.4]. Meteorites are pieces of rock that have fallen to the ground from space. Thus, asteroids and meteorites are closely related: Most meteorites are pieces of asteroids that orbited the Sun for billions of years before falling to Earth.

Asteroids are virtually undetectable by the naked eye and went unnoticed for almost two centuries after the invention of the telescope. The first asteroids were discovered only about 200 years ago. It took 50 years to discover the first 10 asteroids. Today, advanced telescopes can discover far more than that in a single night, and more than 150,000 asteroids have been cataloged. Asteroids can be recognized in telescopic images because they move noticeably relative to the stars over a short time period (Figure 9.1).

Asteroids come in a wide range of sizes. The largest, Ceres, is just under 1,000 kilometers in diameter, or about half the diameter of Pluto. About a dozen others are large enough that we would call them medium-size moons if they orbited a planet. Smaller asteroids are much more common. There are probably more than a million asteroids with diameters greater than 1 kilometer, and many more even smaller in size.

Despite their large numbers, asteroids don't add up to much in total mass. If we could put all the asteroids together and allow gravity to compress them into a sphere, they'd make an object less than 2,000 kilometers in diameter—far smaller than any terrestrial planet. (However, as we'll see shortly, in the distant past there may have been enough asteroid mass to make a planet.)

A handful of asteroids have been photographed up close by spacecraft (Figure 9.2). As we would expect for small objects, they are not spherical because their gravity is not strong enough to reshape rocky material. The

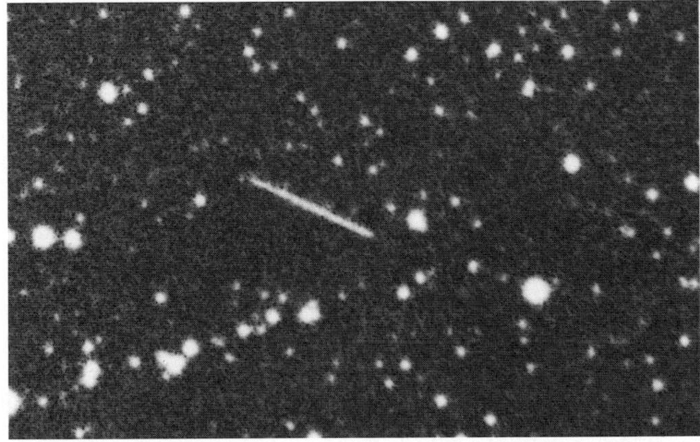

Figure 9.1 VIS

Because asteroids orbit the Sun, they move relative to the stars just as planets do. In this long-exposure photograph, stars show up as white dots, and the motion of an asteroid makes it show up as a short streak.

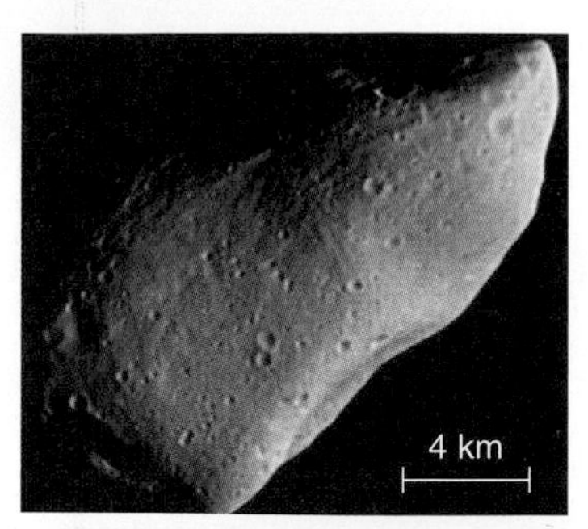

a Gaspra, photographed by the *Galileo* spacecraft.

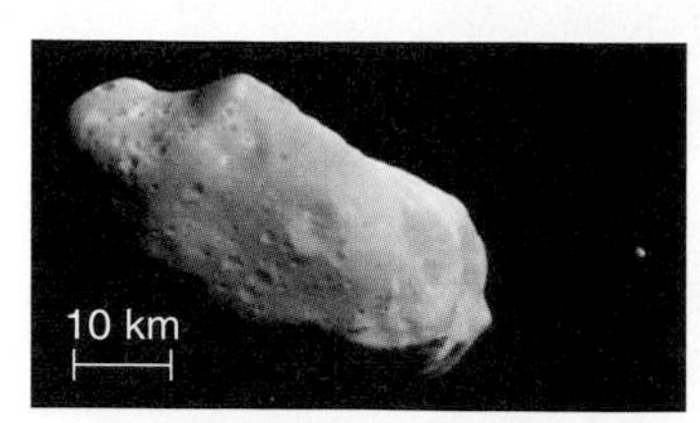

b Ida, photographed by the *Galileo* spacecraft. The small dot to the right is Dactyl, a tiny moon orbiting Ida.

c Mathilde, photographed by the Near-Earth Asteroid Rendezvous (NEAR) spacecraft.

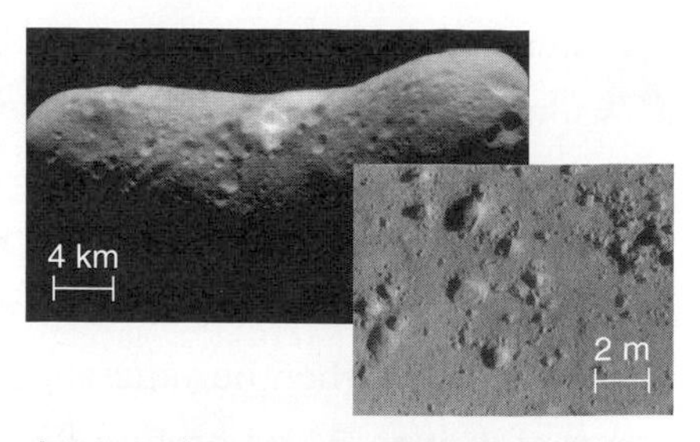

d Eros, photographed by the NEAR spacecraft, which orbited Eros for a year before ending its mission with a soft landing on the asteroid's surface. The inset photo was taken by NEAR just before it landed.

Figure 9.2

VIS

Close-up views of asteroids studied by spacecraft.

images also reveal numerous impact craters, telling us that asteroids, like planets and moons, have been battered by impacts. Indeed, many asteroids have odd shapes because they are fragments of larger asteroids that were shattered in collisions.

• Why is there an asteroid belt?

The asteroid belt between Mars and Jupiter gets its name because it is where we find the majority of asteroids (Figure 9.3). But why are asteroids located mainly in this region, rather than being spread throughout the inner solar system?

The answer is that the asteroid belt was the only place where rocky planetesimals could survive for billions of years. During the birth of the solar system, planetesimals formed throughout the inner solar system. However, most of those within Mars's orbit ultimately accreted into one of the four inner planets. The relatively few asteroids that orbit in the inner solar system today are almost certainly "impacts waiting to happen." Some of these asteroids pass near Earth's orbit and may pose a potential threat to our planet—a threat we will discuss in Section 9.4.

Rocky planetesimals survived in the asteroid belt between Mars and Jupiter because they did not accrete into a planet.

In contrast, the asteroids in the asteroid belt stay clear of any planet and can therefore survive on their current orbits for billions of years. But this leaves us with a deeper question: Why didn't another planet form in this region and sweep up asteroids just as the terrestrial worlds did in the inner solar system?

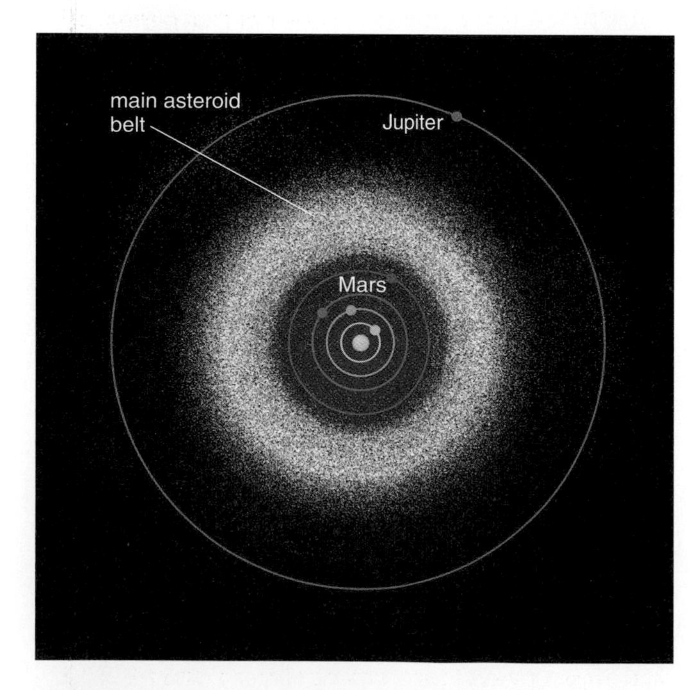

Figure 9.3

Calculated positions of 152,942 asteroids for midnight, 1 January 2004. The asteroids themselves are much smaller than the dots on this scale. The asteroids that share Jupiter's orbit, found 60° ahead of and behind Jupiter, are called *Trojan asteroids.*

THINK ABOUT IT Recall that the frost line in the solar nebula lay between the present-day orbits of Mars and Jupiter (see Figure 6.20), just outside the region where we find the asteroid belt today. Use this fact to explain why we generally do not find asteroids in the outer solar system. (*Hint:* Remember that asteroids are rocky, not icy, in composition.)

Resonances with Jupiter A key clue to understanding the asteroid belt comes from closer examination of asteroid orbits. The graph in Figure 9.4 shows the number of asteroids with various orbital periods (listed along the lower axis) that correspond to particular average orbital distances (listed along the top of the graph). Notice that most asteroids tend to share a few particular orbital periods and distances, leaving gaps between them in which there are very few asteroids. (The gaps are often called *Kirkwood gaps*, after their discoverer.)

The gaps in the asteroid belt are not random. They occur at orbital periods that bear special and simple relationships to Jupiter's nearly 12-year orbital period. For example, the arrow labeled $\frac{1}{4}$ in Figure 9.4 points to a gap at an orbital period that is exactly one-quarter the length of Jupiter's orbital period. Similarly, the arrow labeled $\frac{1}{2}$ points to a gap at an orbital period that is exactly half as long as Jupiter's.

The gaps occur at these special places because of *orbital resonances* with Jupiter. We've already encountered orbital resonances twice before: first, in explaining the elliptical orbits of Jupiter's moons Io, Europa, and Ganymede [Section 8.2]; second, in explaining the gaps in Saturn's rings [Section 8.3]. Recall that resonances arise whenever objects periodically line up with each other so that gravity affects them over and over again in the same way. Asteroids with orbital periods that are simple fractions of Jupiter's orbital period get these repeated tugs. For example, any asteroid with an orbital period that is half of Jupiter's orbital period would line up with Jupiter on every other orbit. It would receive the same gravitational nudge with each alignment and thus would soon be pushed out of this orbit. That is why there are no asteroids with orbital periods exactly half of Jupiter's, or with most of the other special orbital periods shown in Figure 9.4.

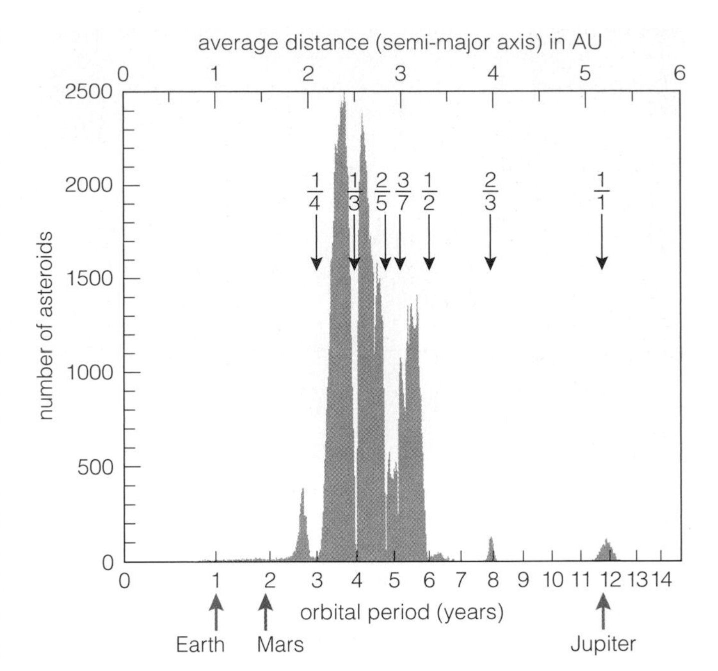

Figure 9.4

This graph shows the number of asteroids with different orbital periods, corresponding to different average distance from the Sun (labeled along the top). Several gaps—places where few if any asteroids orbit the Sun—are labeled by the ratio of orbital period at that distance to Jupiter's orbital period. For example, the label $\frac{1}{3}$ means that the orbital period of objects with this average distance is one-third of Jupiter's orbital period, or about 4 years. Most orbital resonances result in gaps, but some (such as $\frac{1}{1}$) happen to gather up asteroids. (The asteroids with the $\frac{1}{1}$ resonance share Jupiter's orbit and are called *Trojan asteroids*.)

Jupiter's Role in the Asteroid Belt Jupiter's role in shaping the asteroid belt goes much deeper than just explaining the locations of the gaps. Jupiter's gravity also explains why no planet ever formed in this region.

Jupiter's gravity, through the influence of orbital resonances, stirred up asteroid orbits and thereby prevented their accretion into a planet.

When the solar system was forming, the region between Mars and Jupiter must have contained far more rocky material than it does today—probably enough to form at least one more planet as large as Earth or Mars. However, orbital resonances with the young planet Jupiter disrupted the orbits of this region's planetesimals, preventing them from accreting into a full-fledged planet. Over the next 4.5 billion years, ongoing orbital disruptions gradually kicked pieces of this "unformed planet" out of the asteroid belt altogether. Once booted from the asteroid belt, these objects either crashed into a planet or moon or were flung out of the solar system. Thus, the asteroid belt lost most of its original mass, which explains why the total mass of all its asteroids is now less than that of any terrestrial planet.

The asteroid belt is still undergoing slow change today. Jupiter's gravity continues to nudge asteroid orbits, sending asteroids on collision courses with each other and occasionally the planets. As a result, a major collision occurs somewhere in the asteroid belt every 100,000 years or so. Thus, over long periods of time, larger asteroids continue to be broken into smaller ones, with each collision also creating numerous dust-size particles. The asteroid belt has been grinding itself down in this way for more than 4 billion years and will continue to do so for as long as the solar system exists.

• How are meteorites related to asteroids?

Because asteroids are leftovers from the birth of our solar system, studying them should teach us a lot about how Earth and the other planets formed.

DODGE THOSE ASTEROIDS!

Science fiction movies often show brave spacecraft pilots navigating through crowded fields of asteroids, dodging this way and that as they heroically pass through with only a few bumps and bruises. It's great drama, but not very realistic. The asteroid belt looks crowded when we draw it on paper as in Figure 9.3, but in reality it is an enormous region of space. Despite the large number of asteroids in the asteroid belt, they are thousands to millions of kilometers apart on average—so far apart that it would take incredibly bad luck to crash into one by accident. Indeed, spacecraft must be carefully guided to fly close enough to an asteroid to take a decent photograph. Future space travelers will have plenty of dangers to worry about, but dodging asteroids is not likely to be one of them.

Figure 9.5

This large meteorite, called the Ahnighito Meteorite, is located at the American Museum of Natural History in New York. Its dark, pitted surface comes from its fiery passage through Earth's atmosphere. Meteorites enter the atmosphere at speeds of up to 250,000 km/hr (150,000 mi/hr).

We can study asteroids with telescopes and spacecraft, but it would be far better if we could study an asteroid sample in a laboratory. No spacecraft has yet returned an asteroid sample, but we have pieces of asteroids nonetheless—the rocks called meteorites that fall from the sky.

The Difference Between Meteors and Meteorites Before we discuss what we learn from meteorites, let's be clear about what they are. In everyday language, people often use the terms *meteors* and *meteorites* interchangeably. Technically, however, a **meteor** is only a flash of light caused by a particle entering our atmosphere at high speed, not the particle itself. The vast majority of the particles that make meteors are no larger than peas and burn up completely before ever reaching the ground.

Only in rare cases are meteors caused by something large enough to survive the plunge through our atmosphere and leave a **meteorite** on the ground. Those cases make unusually bright meteors, called *fireballs*. Observers find a few meteorites each year by following the trajectories of fireballs.

Unless you actually see a meteorite fall, it can be difficult to distinguish meteorites from terrestrial rocks. Fortunately, a few clues can help. Meteorites are usually covered with a dark, pitted crust resulting from their fiery passage through the atmosphere (Figure 9.5). Some have an unusually high metal content, enough to attract a magnet hanging on a string. The ultimate judge of extraterrestrial origin is laboratory analysis. Meteorites often contain elements such as iridium that are very rare in Earth rocks, and even common elements in meteorites tend to have different ratios among their isotopes [Section 5.1] than rocks from Earth. Thus, once we've identified a suspected meteorite, chemical analysis can tell us whether it really came from space.

Types of Meteorites The precise origin of meteorites was long a mystery, but in recent decades we've been able to determine where in our solar system they come from. The most direct evidence comes from the relatively few meteorite falls that have been observed or filmed. Such observations allow scientists to calculate the orbits that led the meteorites to crash to Earth, telling us where the rocks originated. In every case so far, the results clearly show that the meteorites originated in the asteroid belt.

Detailed analysis of thousands of meteorites shows that they come in two basic types:

- **Primitive meteorites** (Figure 9.6a) are simple mixtures of rock and metal, sometimes also containing carbon compounds and small amounts of water. Radiometric dating [Section 6.4] shows them to be about 4.6 billion years old. Thus, they appear to be remnants from the birth of our solar system, essentially unchanged since they first accreted in the solar nebula.
- **Processed meteorites** (Figure 9.6b) appear to be pieces of large asteroids that, like the terrestrial worlds, underwent differentiation into a core/mantle/crust structure [Section 7.1]. Some processed meteorites are made mostly of iron, suggesting that they came from the core of a shattered asteroid. Others are rocky and either came from the mantle or crust of a shattered asteroid, or were blasted off the surface of a large asteroid by an impact. Radiometric dating shows that processed meteorites are generally younger than the primitive meteorites.

Most meteorites are pieces of asteroids, and they teach us much about the early history of our solar system.

Both types of meteorite teach us important lessons about our solar system. Primitive meteorites essentially represent samples of the first material to condense and accrete in the solar nebula. Thus, they teach us about the composition of the solidified material from which the planets originally formed. In addition, their ages tell us when the process of accretion first began in the solar nebula, which is why we use them to determine the age of the solar system [Section 6.4]. The differences in composition among primitive meteorites reflect where they condensed. Those made only of metal and rock must have condensed in the inner regions of the asteroid belt. Those with carbon compounds must have condensed farther out, where it was cool enough for such compounds and even some water to condense. Indeed, asteroids in the outer regions of the asteroid belt are much darker, as expected for their carbon-rich composition.

Processed meteorites represent direct samples of shattered worlds. Those that come from surfaces are often so close in composition to volcanic rocks on Earth that we conclude that some asteroids had active volcanoes when they were young. Processed meteorites that come from the cores or mantles of shattered asteroids tell us what those asteroids were like on the inside. They also represent a form of direct proof that large worlds really do undergo differentiation (confirming what we infer from seismic studies of Earth [Section 7.1]). Indeed, some large asteroids are apparently so similar to small terrestrial worlds that they could arguably be considered small planets, especially since they had active geology in the past. A forthcoming NASA mission called Dawn should tell us more about the connection between meteorites and asteroids. If all goes well, Dawn will visit several asteroids for close-up study, including the largest (Ceres), beginning in about 2010.

Meteorites from the Moon and Mars In a few cases, the compositions of processed meteorites do not appear to match any known asteroids. Instead they appear to match either the Moon or Mars.

Careful analysis of these meteorite compositions makes us very confident that we really do have a few meteorites that were once part of the Moon or Mars. Moderately large impacts can blast surface material from terrestrial worlds into interplanetary space. Once they are blasted into space, the rocks orbit the Sun until they come crashing down on another world. Calculations show that it is not surprising that we should have found a few meteorites chipped off the Moon and Mars in this way.

Study of these *lunar meteorites* and *Martian meteorites* is providing new insights into conditions on the Moon and Mars. In at least one case, a Martian meteorite may offer clues about whether life ever existed on Mars [Section 18.2].

Primitive Meteorites

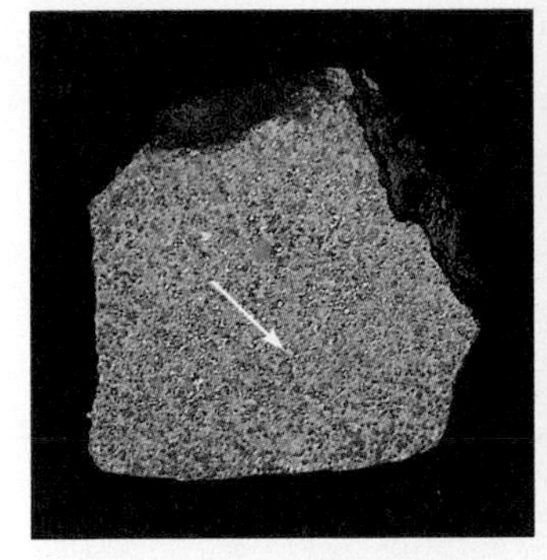

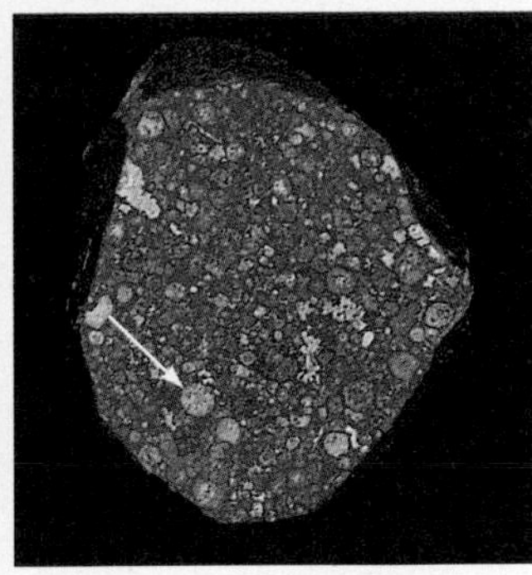

a *Left:* A stony primitive meteorite. It is made of rocky material embedded with shiny metal flakes (arrow), which were presumably among the first particles to condense in the solar nebula. *Right:* A carbon-rich primitive meteorite. The small whitish spheres (arrow) may be solidified droplets splashed out as meteorites accreted in the solar nebula.

Processed Meteorites

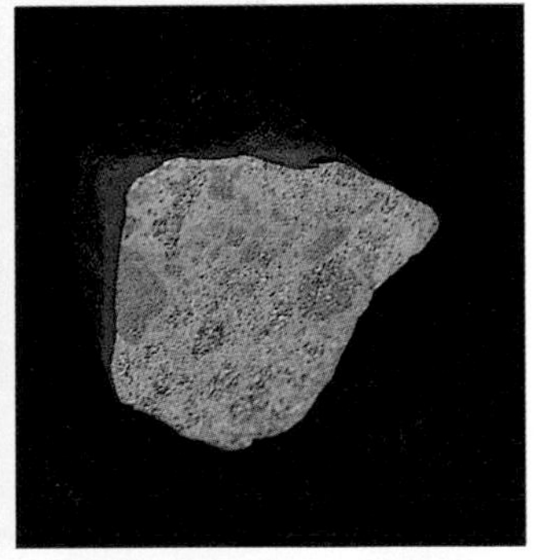

b *Left:* This meteorite is made mostly of iron, and must have come from a shattered object's core. *Right:* This meteorite is rocky and resembles earthly volcanic rocks in composition; it must have come from the mantle or crust of a larger asteroid.

Figure 9.6

There are two basic types of meteorites: primitive and processed. They are shown slightly larger than actual size. (Each photo shows a flat face because the meteorite has been sliced open with a rock saw.)

9.2 COMETS

Asteroids are one of the two major categories of small bodies in the solar system. The other is comets, to which we now turn our attention.

Asteroids and comets have much in common. They are both leftover planetesimals from the birth of our solar system. They both come in a wide range of sizes. The primary difference between them is their composition, which reflects where they formed. Asteroids are rocky because they formed in the inner solar system where metal and rock condensed.

Comets are ice-rich because they formed beyond the frost line, where abundant hydrogen compounds condensed into ice [Section 6.4].

• How do comets get their tails?

For most of human history, comets were familiar only from their occasional presence in the night sky. Every few years, a comet becomes visible to the naked eye, appearing as a fuzzy ball with a long tail. In 1996 and 1997, we were treated to back-to-back brilliant comets: comet Hyakutake and comet Hale–Bopp (Figure 9.7). In photographs, the tails make it look as if comets are racing across the sky, but they are not. If you watch a comet for minutes or hours, you'll see it staying nearly stationary relative to the stars around it in the sky. Over many days it will rise and set just like the stars, while gradually moving relative to the constellations. You'll be able to see it night after night for several weeks or more, until it finally fades from view.

Today, we know that the vast majority of comets do not have tails and are never visible in our skies. Most comets never venture anywhere close to Earth, instead remaining perpetually in the far outer reaches of our solar system. There, they slowly orbit the Sun forever unless their orbits are changed by the gravitational influence of a planet, another comet, or even a star passing by in the distance.

Most comets remain perpetually frozen in the outer solar system. Only a few enter the inner solar system, where they can grow tails.

The comets that appear with long tails in the night sky are the rare ones that have had their orbits changed, causing them to venture into the inner solar system. Most of these comets will not return to the inner solar system for thousands of years, if ever. A few happen to pass near enough to a planet to have their orbits changed further, and some end up on elliptical orbits that periodically bring them close to the Sun. The most famous example is Halley's comet, which orbits the Sun every 76 years. It last passed through the inner solar system in 1986, and it will return in 2061.

Figure 9.7 VIS

Brilliant comets can appear at almost any time, as demonstrated by the back-to-back appearances of (a) comet Hyakutake in 1996 and (b) comet Hale–Bopp in 1997, photographed at Mono Lake in California.

Comet Structure

Comets grow tails only as they enter the solar system, where they are heated by the warmth of the Sun. To see what happens, let's follow the comet path shown in Figure 9.8. Far from the Sun, the comet is completely frozen. If you could see it, the comet would look like a large "dirty snowball"—a chunk of ice mixed with rocky dust and some more complex chemicals. We call this chunk of ice the **nucleus** of the comet. Comet nuclei are a typically a few kilometers across.

As the comet accelerates toward the Sun, its surface temperature increases, and ices begin to sublimate into gas. Some of the escaping gas drags away dust particles from the nucleus. Thus, the comet gets a huge, dusty atmosphere called a **coma**. The coma is far larger than the nucleus it surrounds.

The coma grows as the comet continues into the inner solar system, and some of the gas and dust is pushed away from the Sun, forming the comet's tails. Comets have two visible tails, one made of ionized gas, or *plasma,* and the other made of dust. These tails can be hundreds of millions of kilometers in length.

When a comet nears the Sun, its ices can sublimate into gas and carry off dust, creating a coma and long tails.

The **plasma tail** consists of gas escaping from the coma. Ultraviolet light from the Sun ionizes the gas, and the solar wind then carries this gas straight outward from the Sun. Thus, the plasma tail extends almost directly away from the Sun at all times.

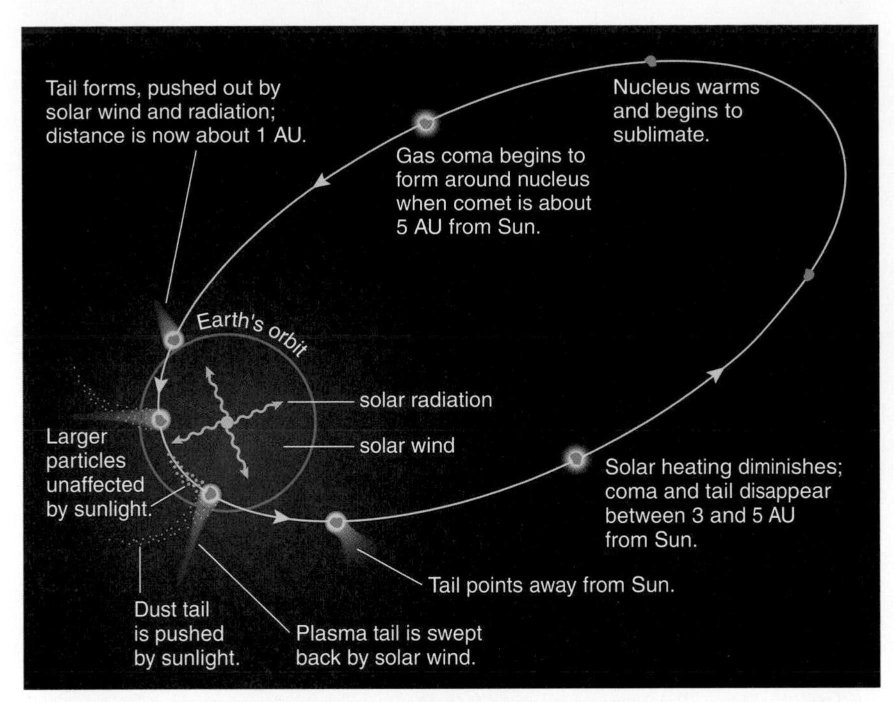

a This diagram shows the changes that occur when a comet's orbit takes it on a passage into the inner solar system. (Not to scale.)

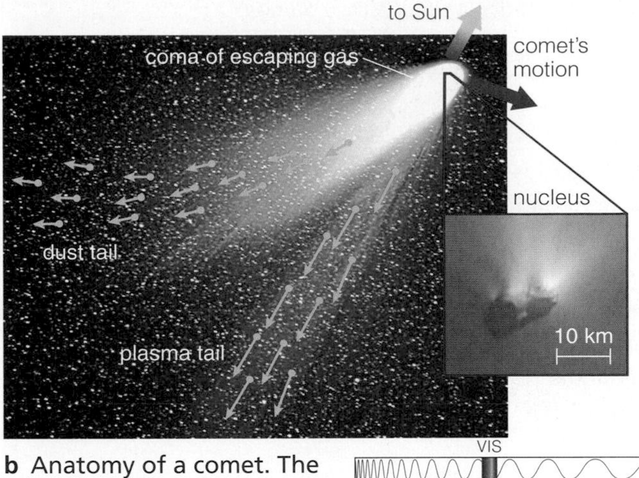

b Anatomy of a comet. The inset photo is the nucleus of Halley's comet photographed by the *Giotto* spacecraft; the coma and tails shown are those of comet Hale–Bopp from a ground-based photo.

Figure 9.8

A comet grows a coma and tail around its nucleus only if it happens to come close to the Sun. Most comets never do this, instead remaining perpetually frozen in the far outer solar system.

The **dust tail** is made of dust-size particles escaping from the coma. These particles are not affected by the solar wind and instead are pushed away from the Sun by the much weaker pressure of sunlight itself (*radiation pressure*). Thus, while the dust tail points generally away from the Sun, it has a slight curve back in the direction the comet came from.

After the comet loops around the Sun and begins to head back outward, sublimation declines, the coma dissipates, and the tails disappear. Nothing happens until the comet again comes sunward—in a century, a millennium, a million years, or perhaps never.

Comet Tails and Meteor Showers Comets also eject sand-to-pebble-size particles that are too big to be affected by either the solar wind or sunlight. These particles essentially form a third, invisible tail that follows the comet around its orbit. They are also the particles responsible for most meteors and meteor showers.

We see a meteor lighting up the sky when one of these small particles (or a similar-size particle from an asteroid) burns up in our atmosphere. The sand-to-pebble-size particles are much too small to be seen themselves, but they enter the atmosphere at such high speed that they make the surrounding air glow with heat. It is this glow that we see as the brief but brilliant flash of a meteor. The small particles are vaporized by the heat and thus never reach the ground. You can typically see a few meteors each hour on any clear night. Remember that you are watching the effects of comet dust entering our planet's atmosphere from space.

Comets eject small particles that follow the comet around in its orbit and cause meteor showers when Earth crosses the comet's orbit.

Some comet dust is sprinkled throughout the inner solar system, but the "third tails" of ejected particles make the dust much more concentrated along the orbits of comets. This dust rains down on our planet whenever we cross a comet's orbit, producing a **meteor shower**. Meteor showers recur at about the

TABLE 9.1 *Major Annual Meteor Showers*

Shower Name	*Approximate Date*
Quadrantids	January 3
Lyrids	April 22
Eta Aquarids	May 5
Delta Aquarids	July 28
Perseids	August 12
Orionids	October 22
Taurids	November 3
Leonids	November 17
Geminids	December 14
Ursids	December 23

same time each year because the orbiting Earth passes through a particular comet's orbit at the same time each year. For example, the meteor shower known as the *Perseids* occurs about August 12 each year—the time when Earth passes through the orbit of a comet called Swift–Tuttle. Table 9.1 lists major annual meteor showers.

If you go outside during one of the annual meteor showers, you may see dozens of meteors per hour. The meteors generally appear to radiate from a particular direction in the sky, for essentially the same reason that snow or heavy rain seems to come from a particular direction in front of a moving car (Figure 9.9). Because more meteors impact in front of Earth than behind it (just as more snow hits the front windshield of a moving car), meteor showers are best observed in the predawn sky, which is the time of day when part of your sky faces in the direction of Earth's motion.

• Where do comets come from?

Comets that repeatedly visit the inner solar system, like Halley's comet, cannot last long on the time scale of our solar system. A comet loses a small fraction of its material on every pass around the Sun, so there is nothing left after just a few hundred passages. Thus, the comets that humans have seen in the skies cannot be the same ones that were seen by dinosaurs, or by inhabitants of the even earlier Earth. So where do comets come from?

We've already stated that comets come from the outer solar system, but we can be much more specific. By analyzing the orbits of comets that pass close to the Sun, scientists learned that there are two major "reservoirs" of comets in the outer solar system.

Most comets that visit the inner solar system follow orbits that seem almost random. They do not orbit the Sun in the same direction as the planets, and their elliptical orbits can be pointed in any direction. Moreover, at their far points their orbits carry them far beyond the orbits of the planets—sometimes nearly a quarter of the distance to the nearest star.

Figure 9.9

A meteor shower appears to radiate from a particular point in the sky.

a When driving into a heavy rain storm or blizzard, the rain drops or snowflakes appear to come from a single direction in front of the moving car.

b A meteor shower occurs as Earth crosses a comet's orbit, which is lined with small particles ejected from the comet. Like the rain or snow seen from a moving car, the meteors seem to come at the moving Earth from a single direction in space.

c This digital composite of photos shows how meteors appear to radiate from a particular direction. Each streak of light is a meteor. The photos were taken in Australia during the 2001 Leonid meteor shower; the large rock is Uluru, also known as Ayers Rock.

These comets must come plunging sunward from a vast, spherical region of space that scientists call the **Oort cloud** (after astronomer Jan Oort). Be sure to note that the Oort cloud is not a cloud of gas, but rather a collection of many individual comets. Based on the number of Oort cloud comets that enter the inner solar system each year, we conclude that the Oort cloud must contain about a trillion (10^{12}) comets.

The comets we occasionally see in the inner solar system come from two major reservoirs of comets in the outer solar system: the Kuiper belt and the Oort cloud.

A smaller number of the comets that visit the inner solar system have a pattern to their orbits. They travel around the Sun in the same plane and direction as the planets on elliptical orbits that carry them no more than about twice as far from the Sun as Pluto. These comets must come from a ring of comets that orbit the Sun beyond the orbit of Neptune. This ring is usually called the **Kuiper belt** (pronounced koy-per; named for astronomer Gerald Kuiper). Figure 9.10 contrasts the general features of the Kuiper belt and the Oort cloud.

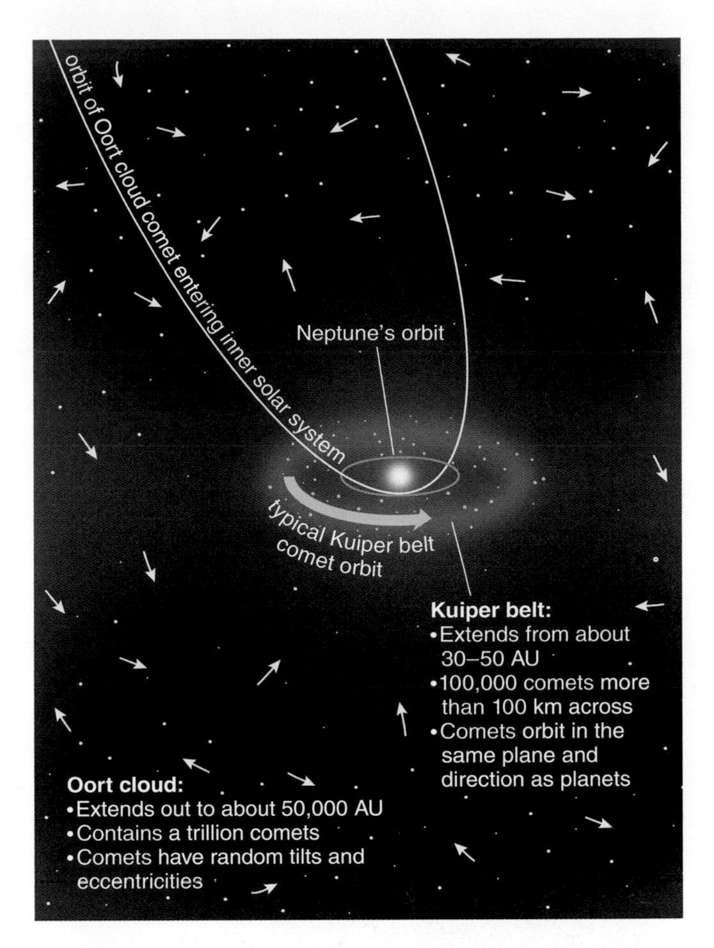

Figure 9.10

The comets we occasionally see in the inner solar system come from two major reservoirs of comets in the outer solar system: the Kuiper belt and the Oort cloud.

The Origin of Comets

How did comets end up in these far-flung regions of the solar system? The only answer that makes scientific sense comes from thinking about what happened to the icy, leftover planetesimals that roamed the region in which the jovian planets formed.

The leftover planetesimals that cruised the spaces between Jupiter, Saturn, Uranus, and Neptune were doomed to suffer either a collision or a close gravitational encounter with one of the young jovian planets. Recall that when a small object passes near a large planet, the planet is hardly affected but the small object may be flung off at high speed [Section 4.4]. Thus, the planetesimals that escaped being swallowed up by the jovian planets tended to be flung off in all directions. Some may have been cast away at such high speeds that they completely escaped the solar system and now drift through interstellar space. The rest ended up on orbits with very large average distances from the Sun. These became the comets of the Oort cloud. The random directions in which they were flung into the Oort cloud explain why this region is roughly spherical in shape. Oort cloud comets are so far away that they can be nudged by the gravity of distant stars, preventing some of them from ever returning near the planets and sending others plummeting toward the Sun.

The Kuiper belt comets orbit in the region in which they formed, just beyond Neptune's orbit. The more distant Oort cloud contains comets that once orbited among the jovian planets.

Beyond the orbit of Neptune, the icy planetesimals were much less likely to be cast off by gravitational encounters. Instead, they remained in orbits going in the same directions as planetary orbits and concentrated relatively near the ecliptic plane. These are the comets of the Kuiper belt. Thus, the comets of the Kuiper belt seem to have originated farther from the Sun than the comets of the Oort cloud, even though the Kuiper belt comets now reside much closer. Kuiper belt comets can be nudged by the gravity of the jovian planets through orbital resonances, sending some on close passes through the inner solar system.

Comet Sizes

In our study of asteroids, we found that they come in a wide range of sizes. We might therefore expect to find comets in a wide range of sizes as well, since they are just another type of leftover planetesimal. All the comets we've observed in the inner solar system are fairly small, with nuclei no larger than a few kilometers across. However, in the past couple of decades we've learned that comets can be much larger.

Most Oort cloud comets are probably fairly small. Because they formed between the jovian planets, they couldn't grow to more than a few kilometers in size before they were ejected into the Oort cloud. The Kuiper belt comets are a different story.

Because the Kuiper belt comets stayed in the region in which they formed, they should have been able to continue their accretion as long as other icy particles were nearby. In principle, one could have grown large enough to become the seed of a fifth jovian planet, beyond Neptune, but that did not occur—probably because the density of material was too low at this great distance from the Sun. Nevertheless, it's reasonable to think that many of them grew to hundreds or even thousands of kilometers in diameter.

Scientists began to find evidence for large Kuiper belt comets in the 1990s, when improvements in telescope technology made it possible to detect them directly. By 2004, more than 800 Kuiper belt comets had been discovered, allowing scientists to infer that the region contains at least 100,000 comets more than 100 kilometers across. Some are much larger. An object named Quaoar (pronounced *kwa-whar*), discovered in 2002, has a diameter of about 1,300 kilometers and orbits the Sun a billion kilometers beyond Pluto. Two more recent discoveries, known as 2004DW and Sedna, may be even larger. 2004DW, which has an orbit very similar to that of Pluto, may be two-thirds the size of Pluto in diameter. Sedna, which orbits the Sun at three times Pluto's distance and may either be a distant member of the Kuiper belt or an unusually near member of the Oort cloud, may be 1,700 kilometers across, or three-fourths the diameter of Pluto. If comets can be as big as these recently discovered objects, can they be bigger still? It's time to consider the odd planet Pluto.

Formation of the Solar System Tutorial, Lesson 3

9.3 PLUTO: LONE DOG, OR PART OF A PACK?

Pluto has seemed a misfit among the planets almost since the day it was discovered in 1930. Its 248-year orbit is more elliptical and more inclined to the ecliptic plane than that of any other planet (Figure 9.11). At its nearest, it is closer to the Sun than Neptune, as was the case between 1979 and 1999. At its farthest, it is 50 AU from the Sun, putting it far beyond the realm of the jovian planets. Its size is also out of character with the other planets. At only 2,230 kilometers in diameter, it is much smaller than any terrestrial planet and smaller than several moons of the jovian planets.

Despite the fact that Pluto sometimes comes closer to the Sun than Neptune, there is no danger of the two planets colliding. For every two Pluto orbits, Neptune circles the Sun three times. This stable orbital resonance means that Neptune is always a safe distance away whenever Pluto approaches its orbit. Neptune and Pluto will probably continue their dance of avoidance until the end of the solar system.

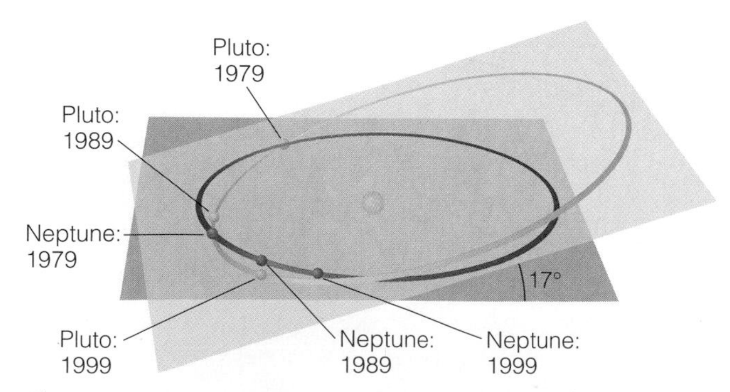

Figure 9.11 Interactive Figure

Pluto's orbit is significantly more elliptical and more tilted relative to the ecliptic than that of any other planet. It comes closer to the Sun than Neptune for 20 years in each 248-year orbit, as was the case between 1979 and 1999. There's no danger of a collision between Pluto and Neptune, because they share an orbital resonance in which Neptune completes three orbits for every two orbits by Pluto.

• What is Pluto like?

Pluto's great distance and small size make it difficult to study. We didn't even know the mass of Pluto until the 1978 discovery of its moon Charon, which orbits Pluto every 6.4 days. Once astronomers measured Charon's orbital characteristics, they were able to apply Newton's version of Kepler's third law [Section 4.4] to estimate Pluto's mass.

Pluto is smaller, icier, and more distant than any of the planets, but its surprisingly large moon, thin atmosphere, and surface markings hint at an interesting history.

Even today, our best photographs of Pluto and Charon show little more than fuzzy blobs (Figure 9.12a). However, from 1985 to 1990, Pluto and Charon happened to be aligned in a way that made them eclipse each other every few days as seen from Earth. Detailed analysis of brightness variations during these eclipses allowed the calculation of accurate sizes, masses, and densities for both Pluto and Charon. These measurements confirmed that Pluto and Charon are both made of ice mixed with rock—much like comets. The eclipse data even allowed astronomers to construct rough maps of Pluto's surface markings, which have not yet been explained (Figure 9.12b).

Curiously, Pluto is slightly higher in density than Charon. This has led astronomers to guess that Charon might have been created by a giant impact similar to the one thought to have formed our Moon [Section 6.4]. A large comet crashing into Pluto may have blasted away its low-density outer layers, which then formed a ring around Pluto and eventually reaccreted to make Charon.

Pluto currently has a thin atmosphere of nitrogen and other gases formed by sublimation of surface ices. However, the atmosphere should gradually refreeze onto the surface as Pluto's 248-year orbit carries it farther from the Sun.

a Pluto (left) and its moon Charon (right), photographed by the Hubble Space Telescope.

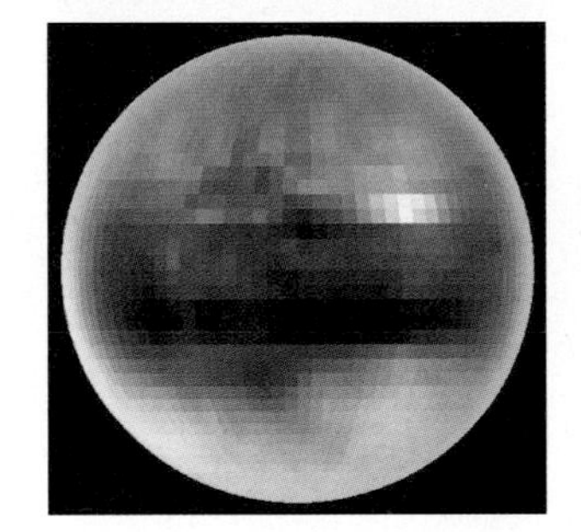

b The surface of Pluto in approximate true color, as derived with computer processing from brightness measurements made during mutual eclipses between Pluto and Charon. The many squares in the image represent the smallest regions for which we could accurately estimate colors.

Figure 9.12

Pluto's great distance and small size make it difficult to study.

• Is Pluto a planet or a Kuiper belt comet?

Pluto's small size and odd orbit make it a misfit among the planets, and many scientists now question whether Pluto should be considered a planet at all. But if it's not a planet, what is it? Let's review the basic clues:

- Pluto spends most of its time well beyond Neptune, in the same region in which we find the comets of the Kuiper belt.
- Pluto's inclined, elliptical orbit is out of character with the orbits of the other planets, but it is just what we might expect for a Kuiper belt comet.
- Pluto's composition does not match that of any of the other planets, but it appears to be a nearly perfect match to many known Kuiper belt comets.

Could Pluto simply be a large comet of the Kuiper belt? A closer look at the orbits of Kuiper belt comets adds support to this idea. Like Pluto, many Kuiper belt comets have stable orbital resonances with Neptune. In fact, more than a dozen Kuiper belt comets have the same orbital period and average distance from the Sun as Pluto itself (and are nicknamed "Plutinos"). Several Kuiper belt comets are even known to possess moons.

Pluto has much more in common with other Kuiper belt comets than with any other planet.

All things considered, Pluto stands out from other Kuiper belt comets only in its size. But, as we discussed earlier, we have already found other Kuiper belt comets that are not much smaller than Pluto, and larger ones may yet be discovered. Moreover, we are nearly certain that at least one object larger than Pluto once roamed the Kuiper belt: Neptune's moon Triton. Recall that Triton is quite similar to (and larger than) Pluto and that its orbit indicates that it must be a captured object [Section 8.2]. Thus, before Triton was captured, Pluto held no better than second place in size among the objects of the Kuiper belt.

THINK ABOUT IT The fact that Pluto seems clearly to be a member of the Kuiper belt has led to debate over whether it should still be called a planet. Do *you* think Pluto should be counted as a planet? Would your opinion change if we discovered another object in the Kuiper belt that is larger than Pluto? Would it change if we discovered 10 other objects larger than Pluto? Defend your opinions.

Planet or not, Pluto remains the largest known object in the solar system never visited by spacecraft. Close-up observations of its surface, atmosphere, and unusual moon are sure to teach us as much about icy bodies as past missions did about rocky bodies. If all goes well, the first mission to Pluto will be launched in 2006, arriving there a decade later.

9.4 COSMIC COLLISIONS: SMALL BODIES VERSUS THE PLANETS

The hordes of small bodies orbiting the solar system are slowly shrinking in number through collisions with the planets and ejection from the solar system. Many more must have roamed the solar system in the days of the heavy bombardment, when most impact craters were formed [Section 6.4]. Plenty of small bodies still remain, however, and cosmic collisions still occur on occasion. These collisions have important ramifications for Earth as well as for the other planets.

• Have we ever witnessed a major impact?

Modern scientists have never witnessed a major impact on a solid world, but in 1994 we were privileged to witness one on Jupiter. The dramatic event involved a comet named *Shoemaker–Levy 9*, or *SL9* for short.

Rather than having a single nucleus, SL9 consisted of a string of comet nuclei lined up in a row (Figure 9.13a). Apparently, a single comet nucleus had been ripped apart by tidal forces from Jupiter on a previous close pass near the planet. Crater chains on Jupiter's moons are evidence that similar breakups of comets near Jupiter have occurred in the past. For example, Figure 9.13b shows a chain of craters on Callisto, presumably formed when a string of comet nuclei crashed into its icy surface.

Comet SL9 caused a string of violent impacts on Jupiter in 1994, reminding us that catastrophic collisions still happen.

Comet SL9 was discovered more than a year before its impact on Jupiter, and orbital calculations had told astronomers precisely when the impact would occur. Thus, when the impacts finally began, they were observed with nearly every major telescope in existence, as well as by spacecraft that were in a position to get a view. The images were astonishing (Figure 9.14). Each of the individual nuclei in comet SL9 crashed into Jupiter with an energy equivalent to that of a million hydrogen bombs. Comet nuclei barely a kilometer across left scars larger than the entire Earth. The scars lasted for months before winds finally made them fade from view.

The SL9 impacts allowed us to study both the impact process and material splashed out from deep inside Jupiter. They also provided two important sociological lessons. First, "Comet Crash Week" proved to be one of the best examples of international collaboration in history. With the aid of the Internet, scientists quickly and effectively shared data from observatories around the world. Second, extensive media coverage helped the event capture the public imagination, leading to awareness that impacts are not

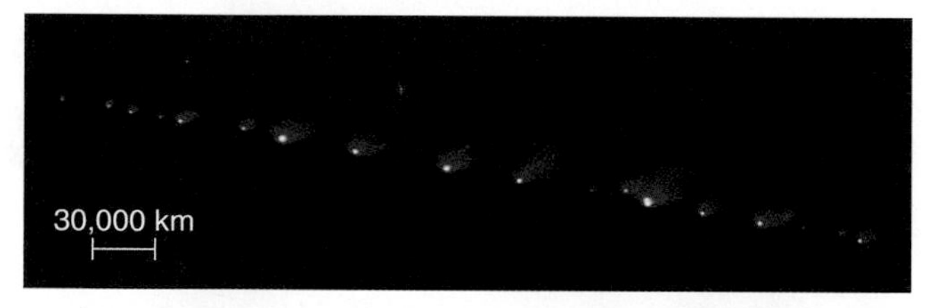

a This photo shows comet Shoemaker–Levy 9 after it was broken apart by tidal forces from Jupiter. The kilometer-sized remains were shrouded in clouds of dust and gas.

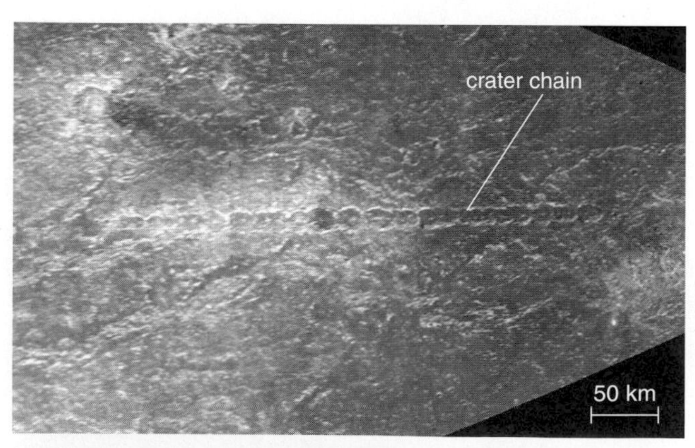

b This chain of craters on Callisto probably formed long ago from impacts by fragments of a comet that, like comet Shoemaker–Levy 9, was broken apart by tidal forces from Jupiter.

Figure 9.13

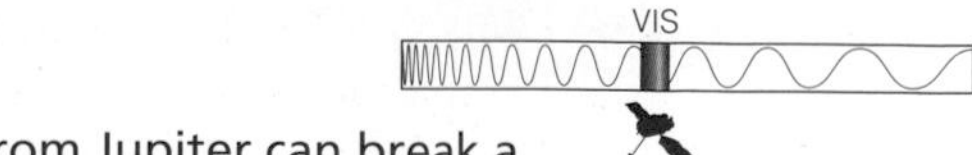

Tidal forces from Jupiter can break a single comet nucleus into a chain of smaller nuclei.

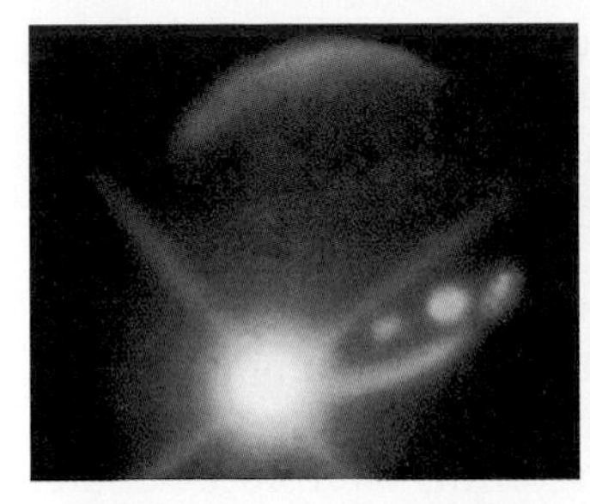

a This infrared photo shows the brilliant glow of a rising fireball from one of the impacts. Jupiter is the round disk, with the impact occurring near the lower left of the disk.

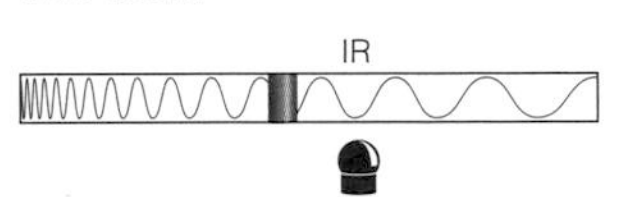

b Each of the black dots in this Hubble Space Telescope photo is a scar from the impact of one of the SL9 nuclei.

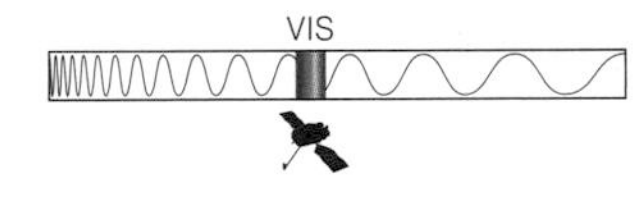

c This painting shows how the impacts might have looked from the surface of Io. The impacts occurred on Jupiter's night side.

Figure 9.14

The impacts of comet Shoemaker–Levy 9 on Jupiter allowed astronomers their first direct view of a cosmic collision.

merely relegated to ancient geological history. If such violent impacts can happen on other planets in our lifetime, could they also happen on Earth?

• Did an impact kill the dinosaurs?

There's little doubt that major impacts have occurred on Earth in the past: Geologists have identified more than 100 impact craters on our planet. So before we consider whether an impact might occur in our lifetimes, it's worth examining the potential consequences if it did. Clearly, an impact could cause widespread physical damage. But a growing body of evidence, accumulated over the past three decades, suggests that an impact can do much more—in some cases, large impacts may have altered the entire course of evolution.

In 1978, while analyzing geological samples collected in Italy, a scientific team led by Luis and Walter Alvarez (father and son) made a startling discovery. They found that a thin layer of dark sediments deposited about 65 million years ago—about the time the dinosaurs went extinct—was unusually rich in the element iridium. Iridium is a metal that is rare on Earth's surface but common in meteorites. Subsequent studies found the same iridium-rich layer in 65-million-year-old sediments around the world (Figure 9.15). The Alvarez team suggested a stunning hypothesis: The extinction of the dinosaurs was caused by the impact of an asteroid or comet.

In fact, the death of the dinosaurs was only a small part of the biological devastation that seems to have occurred 65 million years ago. The fossil record suggests that up to 99% of all living organisms died around that time and that up to 75% of all existing *species* were driven to extinction. This makes the event a clear example of a **mass extinction**—the rapid extinction of a large fraction of all living species. Could it really have been caused by an impact?

There's still some scientific controversy about whether the impact was the sole cause of the mass extinction or just one of many causes, but

Figure 9.15

The arrow points to a layer of sediment laid down on Earth by an impact 65 million years ago, which is linked to the extinction of the dinosaurs. The sediment is about 2 centimeters thick at this location.

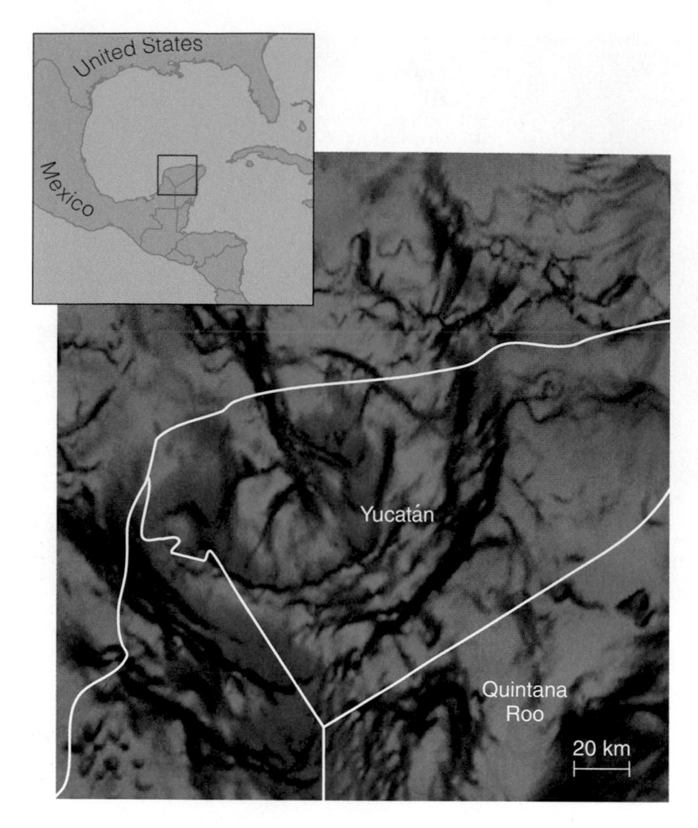

Figure 9.16

This computer-generated image, based on measurements of small local variations in the strength of gravity, shows an impact crater in the northwest corner of Mexico's Yucatán peninsula. The crater measures about 200 kilometers across, and about half of it lies underwater off the coast. The red box on the inset map shows the region covered by the image. The white lines on the image show the coastline and the borders of Mexican states.

there's little doubt that a major impact coincided with the death of the dinosaurs. In addition to the evidence within the sediments, scientists have identified a 65-million-year-old crater (Figure 9.16), apparently created by the impact of an asteroid or a comet measuring about 10 kilometers across.

An iridium-rich sediment layer and an impact crater on the Mexican coast show that a large impact occurred at the time the dinosaurs died out, 65 million years ago.

If the impact was indeed the cause of the mass extinction, here's how it probably happened. On that fateful day some 65 million years ago, the asteroid or comet slammed into Mexico with the force of a hundred million hydrogen bombs (Figure 9.17). North America may have been devastated immediately. Hot debris from the impact rained around the rest of the world, igniting fires that killed many more living organisms.

The longer-term effects were even more severe. Dust and smoke remained in the atmosphere for weeks or months, blocking sunlight and causing temperatures to fall as if Earth were experiencing a global and extremely harsh winter. The reduced sunlight would have stopped photosynthesis for up to a year, killing large numbers of species throughout the food chain. Acid rain may have been another by-product, killing vegetation and acidifying lakes around the world. Chemical reactions in the atmosphere probably produced nitrous oxides and other compounds that dissolved in the oceans and killed marine organisms.

Perhaps the most astonishing fact is not that 75% of all species died but that 25% survived. Among the survivors were a few small mammals. These mammals may have survived in part because they lived in underground burrows and managed to store enough food to outlast the global winter that followed the impact.

The evolutionary impact of the extinctions was profound. For 180 million years, dinosaurs had diversified into a great many species large and small, while mammals (which had arisen at almost the same time as the

Figure 9.17

This painting shows an asteroid or comet moments before its impact on Earth, some 65 million years ago. The impact probably caused the extinction of the dinosaurs, and without it the dinosaurs might still rule Earth today.

dinosaurs) had remained small and rodentlike. With the dinosaurs gone, mammals became the new kings of the planet. Over the next 65 million years, the small mammals rapidly evolved into an assortment of much larger mammals—ultimately including us.

• Is the impact threat a real danger or just media hype?

The dinosaur extinction is only one of several known mass extinctions in Earth's past, and some of the others also seem to coincide with times of past impacts. Thus, impacts may have altered the course of life on Earth several times in the past. If so, could a future impact endanger our own survival? Hollywood takes the threat seriously enough to have made several feature movies about it. Should we take it seriously too?

We know that space is filled with plenty of objects that could hit our planet. Astronomers have identified more than 600 asteroids larger than 1 kilometer in diameter with orbits that pass near Earth's orbit. While orbital calculations show that none of these objects will strike Earth in the foreseeable future, many other near-Earth asteroids probably remain undiscovered. NASA is currently working to identify as many of these potential asteroid threats as possible. It takes time to fully determine the orbit of an asteroid, so the discovery of a new near-Earth asteroid sometimes leads to headlines about a possible impact on a forthcoming pass of the asteroid by Earth—but you rarely hear of the reprieve once we learn that the orbit rules out the impact threat. The threat from comets is more difficult to gauge, because they reside so far from the Sun. By the time we saw a comet plunging toward us from the outer solar system, we'd have at best a few years to prepare for the impact.

The threat of a major impact is undoubtedly real, but the chance of a large impact in our lifetime is quite small. Geological data show that impacts large enough to cause mass extinctions happen many tens of millions of years apart on average. We're far more likely to do ourselves in than to be done in by a large asteroid or comet.

Smaller impacts can be expected more frequently. While such impacts would not wipe out our civilization, they could kill thousands or millions of people. We even know of one close call in modern times. In 1908, a tremendous explosion occurred over Tunguska, Siberia (Figure 9.18). The explosion, estimated to have released energy equivalent to that of several atomic bombs, is thought to have been caused by a small asteroid no more than about 40 meters across. If the asteroid had exploded over a major city instead of Siberia, it would have been the worst natural disaster in human history.

Figure 9.19 shows how often, on average, we expect Earth to be hit by objects of different sizes. Notice that objects similar in size to the one that exploded over Tunguska event strike our planet every century or so. That's a level of threat that we cannot discount. Nevertheless, these once-a-century events are unlikely to strike populated areas. After all, most of Earth's surface is ocean, and even on land humans are concentrated in relatively small urban areas.

Impacts will certainly occur in the future, and while the chance of a major impact in our lifetimes is small, the effects could be devastating.

If we were to find an asteroid or a comet on a collision course with Earth, could we do anything about it? Many people have proposed schemes to save Earth by using nuclear weapons or other means to demolish or divert an incoming asteroid, but no one knows whether current technology

Figure 9.18

This photo shows forests burned and flattened by the 1908 impact over Tunguska, Siberia. Atmospheric friction caused the small asteroid to explode completely before it hit the ground, so it left no impact crater.

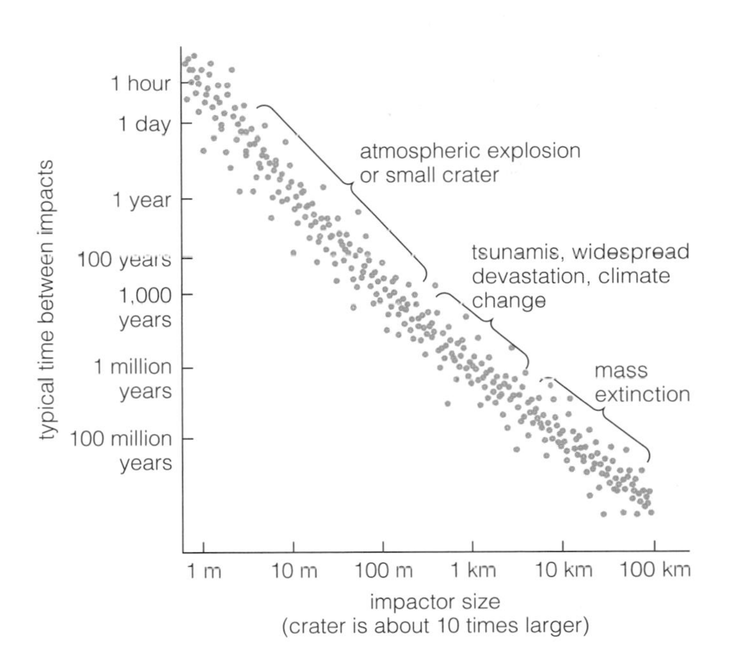

Figure 9.19

This graph shows that larger objects (asteroids or comets) hit the Earth less frequently than smaller ones. The labels describe the effects of impacts of different sizes. (The graph shows dots rather than a solid line to indicate that the relationship is only approximate.)

is really up to the task. We can only hope that the threat doesn't become a reality before we're ready.

THINK ABOUT IT Study Figure 9.19. Based on the frequency of impacts large enough to cause serious damage, do you think we should be spending time and resources to counter the impact threat? Or should we focus resources on other threats first? Defend your opinion.

• How do other planets affect impact rates and life on Earth?

Ancient people imagined that the mere movement of planets relative to the visible stars in our sky could somehow have an astrological influence on our lives. Although we no longer give credence to this ancient superstition, we now know that planets can have a real effect on life on Earth. By catapulting asteroids and comets in Earth's direction, other planets have caused cosmic collisions that have helped shape Earth's destiny.

Jupiter and the other jovian planets have had the greatest effects because of their influence on the small bodies of the solar system. As we saw earlier in this chapter, Jupiter disturbed the orbits of rocky planetesimals outside Mars's orbit, preventing a planet from forming and instead creating the asteroid belt. The jovian planets ejected icy planetesimals to create the distant Oort cloud of comets, and orbital resonances with Neptune shape the orbits of many comets in the Kuiper belt. Ultimately, every asteroid or comet that has impacted Earth since the end of the heavy bombardment was in some sense sent our way by the influence of Jupiter or one of the other jovian planets. Figure 9.20 summarizes the ways the jovian planets have controlled the motions of asteroids and comets.

Nearly every asteroid or comet that has ever struck Earth was in some sense sent our way by the influence of Jupiter or one of the other jovian planets.

Thus, because impacts of asteroids and comets have played such an important role in the history of our planet, we find a deep connection between the jovian planets and the survival of life on Earth. If Jupiter did not exist, the threat from asteroids might be much smaller, since the objects that make up the asteroid belt might instead have become part of a planet. On the other hand, the threat from comets might be much greater: Jupiter probably ejected more comets to the Oort cloud than any other jovian planet, and without Jupiter those comets might have remained dangerously close to Earth. Of course, even if Jupiter has protected us from impacts, it's not clear whether that has been good or bad for life overall. The dinosaurs appear to have suffered from an impact, but the same impact may have paved the way for our existence. Thus, while some scientists argue that more impacts would have damaged life on Earth, others argue that more impacts might have sped up evolution.

The role of Jupiter has led some scientists to wonder whether we could exist if our solar system had been laid out differently. Could it be that civilizations can arise only in solar systems that happen to have a Jupiter-like planet in a Jupiter-like orbit? No one yet knows the answer to this question. What we do know is that Jupiter has had profound effects on life on Earth and will continue to have effects in the future.

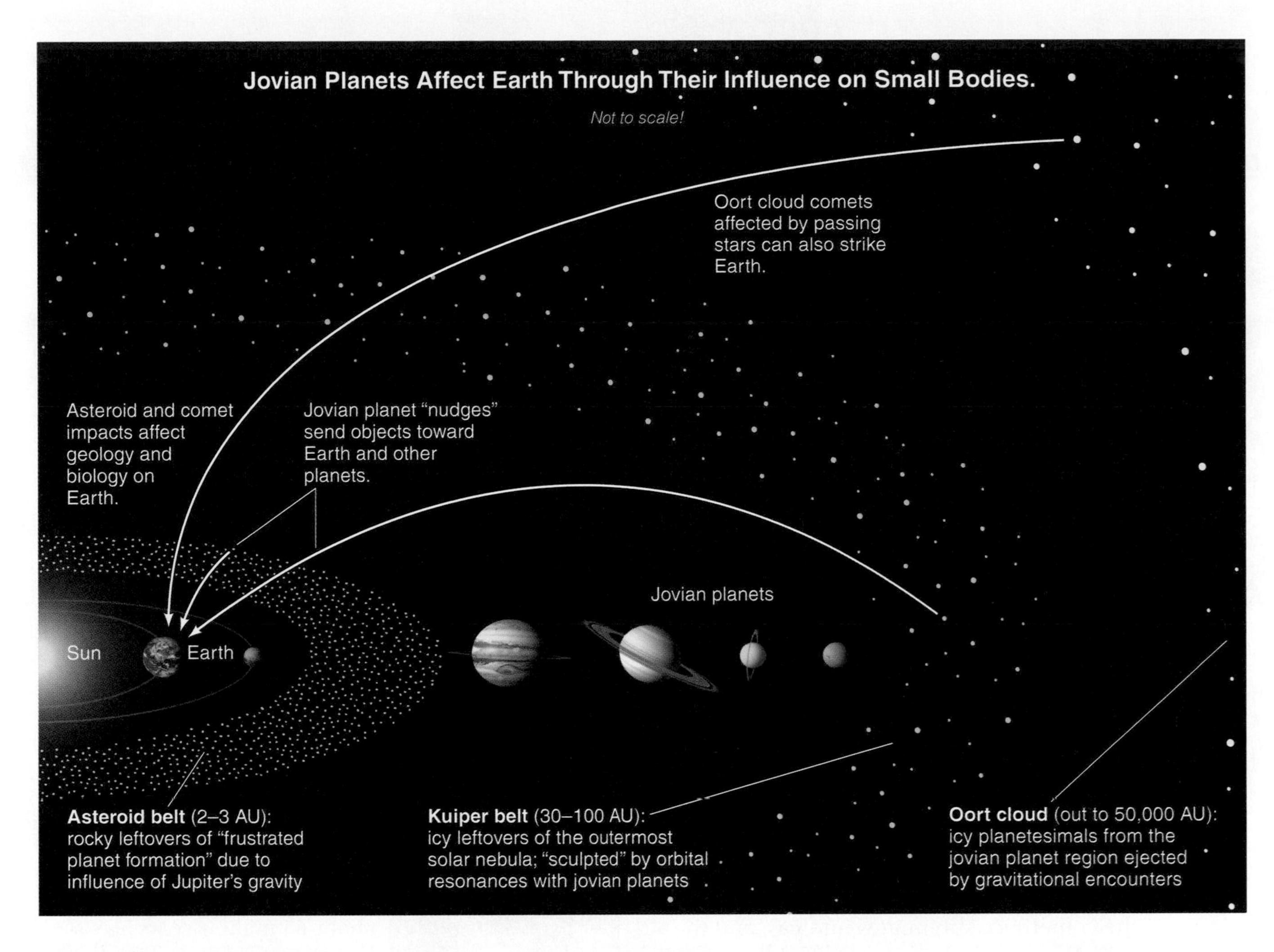

Figure 9.20

The connections between the jovian planets, small bodies, and Earth. The gravity of the jovian planets helped shape both the asteroid belt and the Kuiper belt, and the Oort cloud consists of comets ejected from the jovian planet region by gravitational encounters with these large planets. Ongoing gravitational influences sometimes send asteroids or comets heading toward Earth.

THE BIG PICTURE

Putting Chapter 9 into Context

In this chapter we concluded our study of the solar system by focusing on its smallest objects, finding that these objects can have big consequences. Keep in mind the following "big picture" ideas:

- Asteroids, comets, and meteorites may be small compared to planets, but they provide much of the evidence that has helped us understand how the solar system formed.
- Pluto is called the ninth planet, but it is much more similar to the thousands of Kuiper belt comets than to the other eight planets.
- The small bodies are subject to the gravitational whims of the largest. The jovian planets shaped the asteroid belt, the Kuiper belt, and the Oort cloud, and they continue to nudge objects onto collision courses with the planets.
- Collisions not only bring meteorites and leave impact craters but can profoundly affect life on Earth. The dinosaurs were probably wiped out by an impact, and future impacts pose a threat that we cannot ignore.

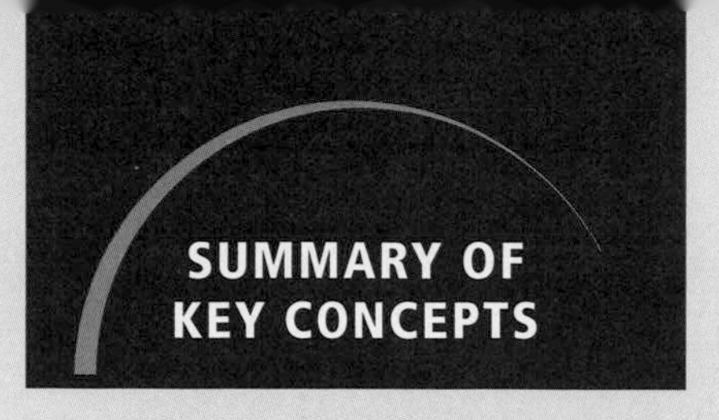

9.1 ASTEROIDS AND METEORITES

- **Why is there an asteroid belt?**

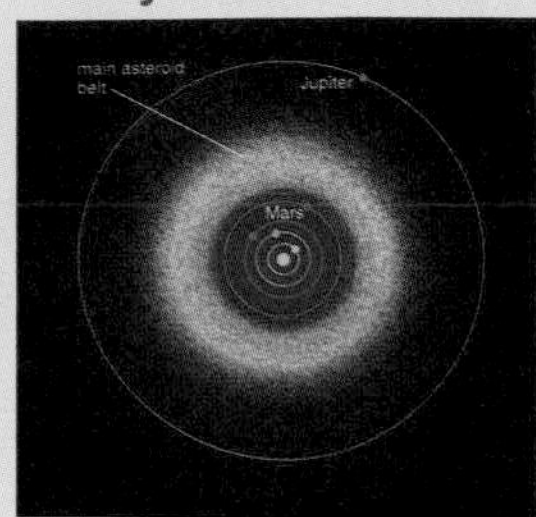

Orbital resonances with Jupiter disrupted the orbits of planetesimals, preventing them from accreting into a planet. Those that were not ejected from this region make up the asteroid belt today. Most asteroids in other regions of the inner solar system accreted into one of the planets.

- **How are meteorites related to asteroids?**

Most meteorites are pieces of asteroids. Primitive meteorites are essentially unchanged since the birth of the solar system. Processed meteorites are fragments of larger asteroids that underwent differentiation.

9.2 COMETS

- **How do comets get their tails?**

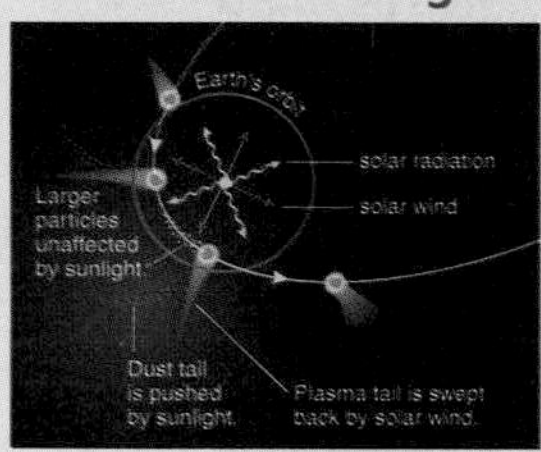

The vast majority of comets do not have tails, as they reside so far from the Sun that they are perpetually frozen. Only those few comets that enter the solar system grow tails. As the comet approaches the Sun its nucleus—all the comet consists of when it is far away and frozen—heats up. Some of the comet's ice sublimates into gas, and the escaping gases carry along some dust. The gas and dust form a coma and two tails: a plasma tail of ionized gas and a dust tail. Larger particles can also escape, becoming the particles that cause meteors and meteor showers on Earth.

- **Where do comets come from?**

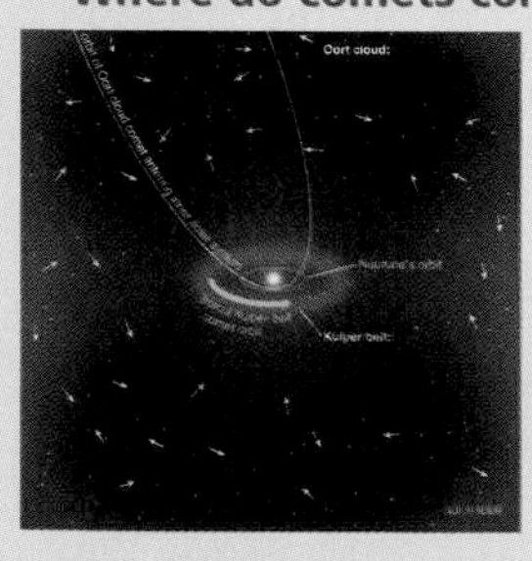

Comets that enter the solar system come from one of two reservoirs in the outer solar system: the Kuiper belt and the Oort cloud. The Kuiper belt comets still reside in the region beyond Neptune in which they formed during the birth of the solar system. The Oort cloud comets are thought to have formed in the region of the jovian planets, and were kicked out to the great distance of the Oort cloud by gravitational encounters with the planets.

9.3 PLUTO: LONE DOG, OR PART OF A PACK?

- **What is Pluto like?**

Pluto is much smaller than any other planet, with an orbit more elliptical and more inclined to the ecliptic plane than that of any other planet. It is made mostly of ices and has a very thin atmosphere of gases that are expected to freeze onto the surface as Pluto moves farther from the Sun in its 248-year orbit. It has a moon, Charon, with a slightly lower density than Pluto, suggesting that Charon may have been formed in a giant impact.

- **Is Pluto a planet or a Kuiper belt comet?**

Whether Pluto should be called a "planet" is a matter of opinion, but its properties suggest that it is a Kuiper belt comet. Its composition and orbital properties match those of other Kuiper belt comets and do not fit in with the other planets. It is the largest known Kuiper belt comet today, but there may be larger ones still awaiting discovery.

9.4 COSMIC COLLISIONS: SMALL BODIES VERSUS THE PLANETS

- **Have we ever witnessed a major impact?**

In 1994, we observed the impacts of comet Shoemaker–Levy 9 on Jupiter. The comet had fragmented into a string of individual nuclei, so there was a string of impacts that left Jupiter's atmosphere scarred for months.

- **Did an impact kill the dinosaurs?**

We are not certain whether an impact was the sole cause, but a major impact clearly coincided with the mass extinction in which the dinosaurs died out, about 65 million years ago. Sediments from the time show clear evidence of an impact, and an impact crater of the right age has been found near the coast of Mexico.

- **Is the impact threat a real danger or just media hype?**

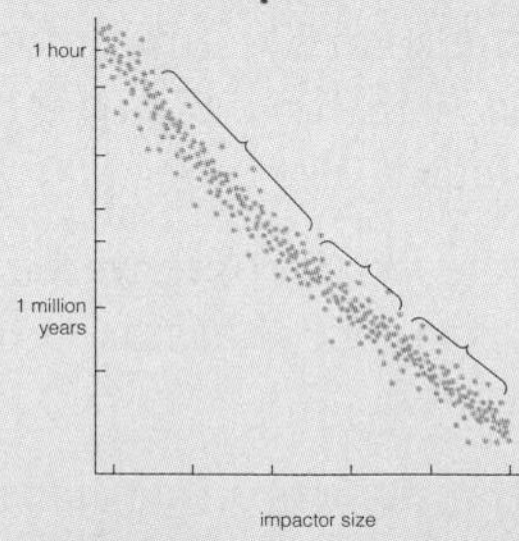

Impacts certainly pose a threat, though the probability of a major impact in our lifetimes is fairly low.

- **How do other planets affect impact rates and life on Earth?**

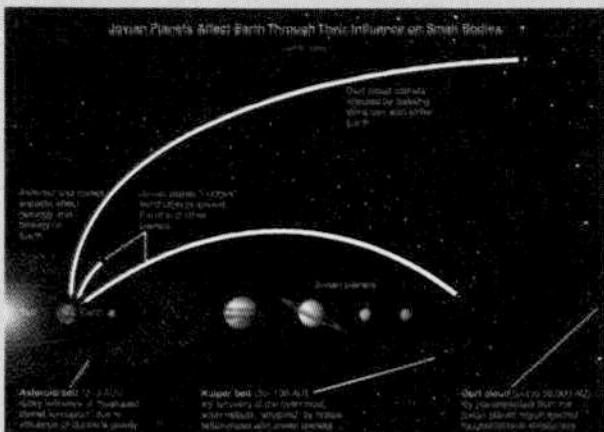

Impacts of asteroids and comets are always linked in at least some way to the gravitational influences of Jupiter and the other jovian planets. These gravitational influences have shaped the asteroid belt, the Kuiper belt, and the Oort cloud, and sometimes still help determine when an object is flung our way.

EXERCISES AND PROBLEMS

REVIEW QUESTIONS

1. Briefly explain why comets, asteroids, and meteorites are so useful in helping us understand the history of the solar system.
2. How does the largest asteroid compare in size to the planets? How does the total mass of all asteroids compare to the mass of a terrestrial world?
3. Where is the *asteroid belt* located, and why? Briefly explain how orbital resonances with Jupiter have affected the asteroid belt.
4. What is the difference between a *meteor* and a *meteorite?* How can we distinguish a meteorite from a terrestrial rock?
5. Distinguish between *primitive meteorites* and *processed meteorites* in terms of both composition and origin.
6. What does a comet look like when it is far from the Sun? How does its appearance change when it is near the Sun? What happens to comets that make many passes near the Sun?
7. What produces the *coma* and tails of a comet? What is the *nucleus?* Why do tails point away from the Sun?
8. Explain how meteor showers are linked to comets. Why do meteor showers recur at about the same time each year?
9. Describe the *Kuiper belt* and *Oort cloud* in terms of their locations and the orbits of comets within them. How did comets come to exist in these two regions?
10. Briefly describe Pluto and Charon. Why won't Pluto collide with Neptune? How do we think Charon formed?
11. Briefly summarize the evidence suggesting that Pluto is a Kuiper belt comet.
12. Briefly describe the impact of comet Shoemaker–Levy 9 on Jupiter.
13. Briefly describe the evidence suggesting that the mass extinction that killed off the dinosaurs was caused by an impact. How did the impact lead to the mass extinction?
14. How often should we expect impacts of various sizes on Earth? How serious a threat do we face from these impacts?

SURPRISING DISCOVERIES?

Suppose we were to make the following discoveries. (These are *not* real discoveries.) Would you consider the discovery to be surprising? Explain why or why not.

15. A small asteroid that orbits within the asteroid belt has an active volcano.
16. Scientists discover a meteorite that, based on radiometric dating, is 7.9 billion years old.
17. An object that resembles a comet in size and composition is discovered orbiting in the inner solar system.
18. Studies of a large object in the Kuiper belt reveal that it is made almost entirely of rocky (as opposed to icy) material.
19. Astronomers discover a previously unknown comet that will be brightly visible in our night sky about 2 years from now.
20. A mission to Pluto finds that it has lakes of liquid water on its surface.
21. Geologists discover a crater from a 5-km object that impacted Earth more than 100 million years ago.
22. Archaeologists learn that the fall of ancient Rome was caused in large part by an asteroid impact in Asia.
23. Astronomers discover three objects with the same average distance from the Sun (and the same orbital period) as Pluto.
24. Astronomers discover an asteroid with an orbit suggesting that it will impact the Earth in the year 2064.

PROBLEMS

(Quantitative problems are marked with an asterisk.)

25. *The Role of Jupiter.* Suppose that Jupiter had never existed. Describe at least three ways in which our solar system would be different, and clearly explain why.
26. *Life Story of an Iron Atom.* Imagine that you are an iron atom in a processed meteorite made mostly of iron that has recently fallen to Earth. Tell the story of how you got here, beginning from the time you were part of the gas in the solar nebula 4.6 billion years ago. Include as much detail as possible. Your story should be scientifically accurate but also creative and interesting.
27. *Project: Tracking a Meteor Shower.* Armed with an expendable star chart and a dim (preferably red) flashlight, set yourself up comfortably to watch one of the meteor showers listed in Table 9.1. Each time you see a meteor, record its path on your star chart. Record at least a few dozen meteors, and try to determine the *radiant* of the shower—that is, the point in the sky from which the meteors appear to radiate. Does the meteor shower live up to its name?

*28. *Impact Energies.* A relatively small impact crater 20 km in diameter could be made by a comet 2 km in diameter traveling at 30 km/s (30,000 m/s).
   a. Assume that the comet has a total mass of 4.2×10^{12} kg. What is its total kinetic energy? (*Hint:* The kinetic energy is equal to $\frac{1}{2} \times m \times v^2$, where m is the comet's mass and v is its speed. If you use mass in kg and velocity in m/s, the answer for kinetic energy will have units of joules.)
   b. Convert your answer from part (a) to an equivalent in megatons of TNT, the unit used for nuclear bombs. Comment on the degree of devastation the impact of such a comet could cause if it struck a populated region on Earth. (*Hint:* One megaton of TNT releases 4.2×10^{15} joules of energy.)

DISCUSSION QUESTIONS

29. *Rise of the Mammals.* Suppose the impact 65 million years ago had not occurred. How do you think our planet would be different? For example, do you think that mammals still would eventually have come to dominate Earth? Would we be here? Defend your opinions.
30. *Could an Impact Cause a Nuclear War?* Imagine that an impact like the 1908 Tunguska impact occurred over a nuclear-armed country, such as Russia, India, Pakistan, or the United States. At first, the impact would look much like the result of a nuclear attack. Some military experts worry that in the panic of the moment a leader might order a "retaliatory" nuclear attack before learning that the explosion was caused by an impact rather than a bomb. Do you think an impact could start a nuclear war? What could we do to prevent this danger?

31. *Reducing the Impact Threat.* Based on your own opinion of how the impact threat compares to other threats, how much money and resources do you think should be used to alleviate it? What types of programs would you support? (Examples include programs to search for objects that could strike Earth, to develop defenses against impacts, or to build actual defenses.) Defend your opinions.

MATH HELP AND EXERCISES

For additional help and practice with mathematical concepts applicable to this chapter, the Astronomy Place web site has the following resources with detailed explanations, worked examples, and practice problems.

- "No Greenhouse" Temperatures

For a complete list of media resources available, go to **www.astronomyplace.com** and choose Chapter 9 from the pull-down menu.

ASTRONOMY PLACE WEB TUTORIALS

Tutorial Review of Key Concepts

Use the following interactive **Tutorial** at **www.astronomyplace.com** to review key concepts from this chapter.

Formation of the Solar System Tutorial

Lesson 3 Formation of Planets

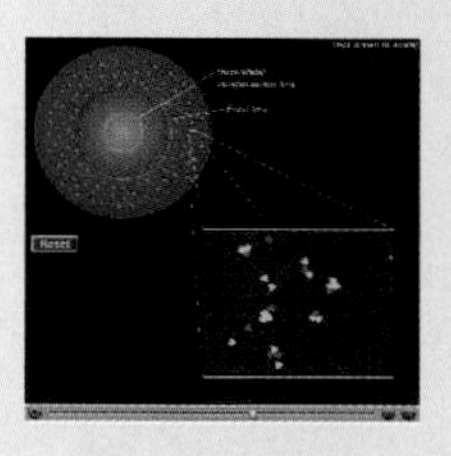

Supplementary Tutorial Exercises

Use the interactive **Tutorial Lesson** to explore the following questions.

Formation of the Solar System Tutorial, Lesson 3

1. Where were icy planetesimals able to form in the solar nebula?
2. Pluto is the smallest planet. Why isn't it made of metal and rock like the small terrestrial planets?
3. Pluto is in the outer solar system. Why didn't it capture hydrogen and helium gas like the jovian planets?
4. Suppose the solar wind had not cleared the solar nebula until much later in the history of the solar system. Do you think Pluto would have been different? Explain.

EXPLORING THE SKY AND SOLAR SYSTEM

Of the many activities available on the ***Voyager: SkyGazer* CD-ROM** accompanying your book, use the following files to observe key phenomena covered in this chapter.

Go to the **File: Basics** folder for the following demonstrations.

1. Orbit of Hale–Bopp
2. Pluto's Orbit

Go to the **File: Demo** folder for the following demonstrations:

1. Hale–Bopp Path
2. Hyakutake at Perihelion
3. Hyakutake Nears Earth
4. Pluto's Orbit

MOVIES

Check out the following narrated and animated short documentaries available on **www.astronomyplace.com** for a helpful review of key ideas covered in this chapter.

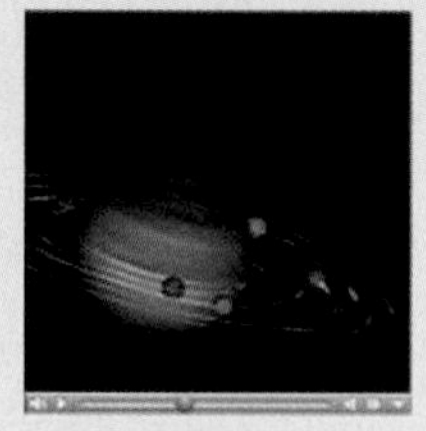

History of the Solar System Movie

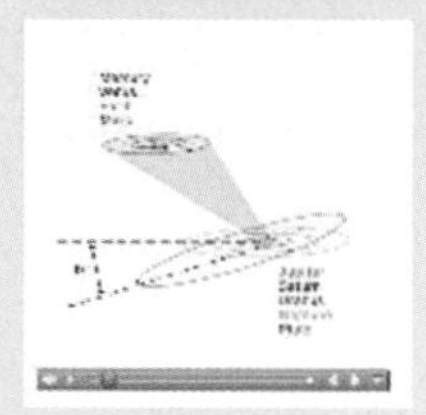

Orbits in the Solar System Movie

WEB PROJECTS

Take advantage of the useful Web links on **www.astronomyplace.com** to assist you with the following projects.

1. *The NEAR Mission.* Learn about the NEAR mission to the asteroid Eros. How did the mission reach Eros? How did the mission end? What did it accomplish? Write a one- to two-page summary of your findings.
2. *Stardust.* Learn about NASA's Stardust mission (launched in 1999) to return comet material to Earth. Write a short report about the mission status and its science.
3. *Asteroid and Comet Missions.* Learn about another proposed space mission to study asteroids or comets. For the mission you chose, write a short report about the mission plans, goals, and prospects for success.
4. *The New Horizons Mission to Pluto.* The New Horizons mission to visit Pluto is scheduled for launch in 2006. Find out the current status of the mission. What are its science goals? Summarize your findings in a few paragraphs.

10
Our Star

LEARNING GOALS

10.1 A CLOSER LOOK AT THE SUN
- Why does the Sun shine?
- What is the Sun's structure?

10.2 NUCLEAR FUSION IN THE SUN
- How does nuclear fusion occur in the Sun?
- How does the energy from fusion get out of the Sun?
- How do we know what is happening inside the Sun?

10.3 THE SUN–EARTH CONNECTION
- What causes solar activity?
- How does solar activity affect humans?
- How does solar activity vary with time?

Astronomy today encompasses the study of the entire universe, but the root of the word *astronomy* comes from the Greek word for "star." Although we have learned a lot about the universe up to this point in the book, only now do we turn our attention to the study of the stars, the namesakes of astronomy.

When we think of stars, we usually think of the beautiful points of light visible on a clear night. But the nearest and most easily studied star is visible only in the daytime—our Sun. Of course, the Sun is important to us in many more ways than as an object for astronomical study. The Sun is the source of virtually all light, heat, and energy reaching Earth, and life on Earth's surface could not survive without it.

In this chapter, we will study the Sun in some depth. We will learn how the Sun generates the energy that supports life on Earth. Equally important, we will study our Sun as a star so that in subsequent chapters we can more easily understand stars throughout the universe.

The Sun Tutorial, Lesson 1

10.1 A CLOSER LOOK AT THE SUN

We discussed the general features of the Sun in our tour of the solar system in Chapter 6, learning that it is a giant ball of very hot gas (see p. 134). Now it's time to get better acquainted with our nearest star. In doing so, we will answer questions our ancestors have asked for millennia.

Ancient peoples recognized the vital role of the Sun in their lives. Some worshipped the Sun as a god. Others created mythologies to explain its daily rise and set. But no one who lived before the 20th century ever knew how the Sun provides us with light and heat.

Most ancient thinkers imagined the Sun to be some type of fire, perhaps a lump of burning coal or wood. It was a reasonable suggestion for the times, since science had not yet advanced to the point where the idea could be put to the test. Ancient people did not know the size or distance of the Sun and thus could not imagine how incredible its energy output really is. Nor did they know how long Earth had existed and thus had no way to realize that the Sun must have provided light and heat for a very long time.

The size and distance of the Sun were reasonably well known by the mid-1800s, so scientists could seriously begin to address the question of how the Sun shines. The ancient idea of burning coal or wood was quickly ruled out: Calculations showed that such burning could not possibly account for the Sun's huge output of energy. Other ideas based on chemical processes were likewise ruled out.

In the late 1800s, astronomers came up with an idea that seemed more plausible, at least at first. They suggested that the Sun generates energy by slowly contracting in size, a process called **gravitational contraction** (or *Kelvin-Helmholtz contraction,* after the scientists who proposed the mechanism). Recall that a shrinking gas cloud heats up because some of the gravitational potential energy of gas particles far from the cloud center is converted into thermal energy as the gas moves inward (see Figure 4.12). A gradually shrinking Sun would always have some gas moving inward, converting gravitational potential energy into thermal energy. This thermal energy would keep the inside of the Sun hot. Because of its large mass, the Sun would need to contract only very slightly each year to maintain its temperature—so slightly that the contraction would have been unnoticeable to 19th century astronomers. Calculations also showed that gravitational contraction could have kept the Sun shining steadily for up to about 25 million years. For a while, some astronomers thought that this idea had solved the ancient mystery of how the Sun shines. However, geologists pointed out a fatal flaw: Studies of rocks and fossils had already shown Earth to be far older than 25 million years, which meant that gravitational contraction could not be the mechanism by which the Sun generates its energy.

Figure 10.1

An acrobat stack is in gravitational equilibrium: The lowest person supports the most weight and feels the greatest pressure, and the overlying weight and underlying pressure decrease for those higher up.

• Why does the Sun shine?

With both chemical processes and gravitational contraction ruled out as the explanation for why the Sun shines, scientists of the late 1800s were at a loss. There was no known way that an object the size of the Sun could generate so much energy for billions of years. A completely new type of explanation was needed, and it came with Einstein's publication of his special theory of relativity in 1905.

Einstein's theory included his famous discovery of $E = mc^2$ [Section 4.3]. This equation shows that mass itself contains an enormous amount of potential energy. Calculations immediately showed that the Sun's mass contained more than enough energy to account for billions of years of sunshine, if only there were some way for the Sun to convert the energy of mass into thermal energy. It took a few decades for scientists to work out the details, but by the end of the 1930s we had learned that the Sun converts mass into energy through the process of nuclear fusion [Section 1.1].

The Stable Sun Nuclear fusion requires extremely high temperatures and densities (for reasons we will discuss in the next section). In the Sun, these conditions are found deep in the core. Thus, for the Sun to shine steadily, it must have a way of keeping the core hot and dense. It maintains these internal conditions through a natural balance between two competing forces: gravity pulling inward and pressure pushing outward. This balance is called **gravitational equilibrium** (or *hydrostatic equilibrium*).

A stack of acrobats provides a simple example of gravitational equilibrium (Figure 10.1). The bottom person supports the weight of everybody above him, so his arms must push upward with enough pressure to support all this weight. At each higher level, the overlying weight is less, so it's a little easier for each additional person to hold up the rest of the stack.

Everywhere inside the Sun, the outward push of pressure balances the inward pull of gravity.

Gravitational equilibrium works much the same in the Sun, except the outward push against gravity comes from internal gas pressure rather than an acrobat's arms. The Sun's internal pressure precisely balances gravity at every point within it, thereby keeping the Sun stable in size (Figure 10.2). Because the weight of overlying layers is greater as we look deeper into the Sun, the pressure must increase with depth. Deep in the Sun's core, the pressure makes the gas hot and dense enough to sustain nuclear fusion. The energy released by fusion, in turn, heats the gas and thus generates pressure that keeps the Sun in balance against the inward pull of gravity.

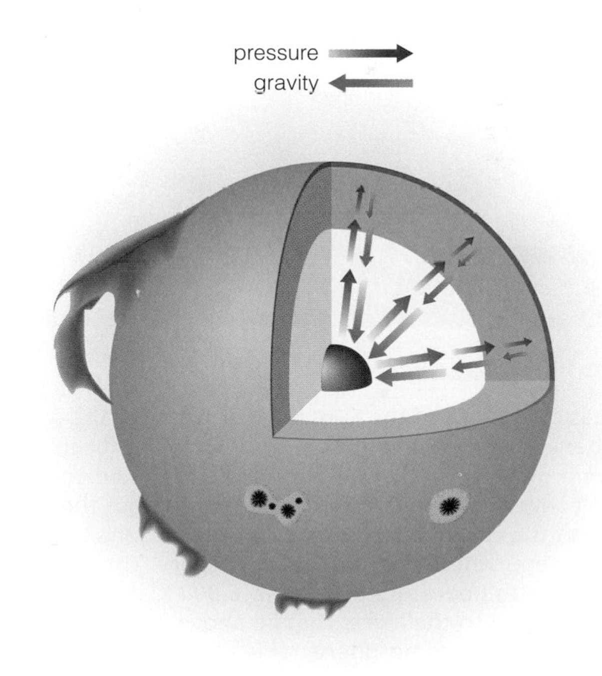

Figure 10.2

Gravitational equilibrium in the Sun: At each point inside, the pressure pushing outward balances the weight of the overlying layers.

THINK ABOUT IT

Earth's atmosphere is also in gravitational equilibrium, with the weight of upper layers supported by the pressure in lower layers. Use this idea to explain why the air gets thinner at higher altitudes.

How Fusion Started The Sun's gravitational equilibrium is fairly easy to understand if we think only about how it works at present: Gravity pushes inward while the energy released by fusion generates the pressure that pushes outward. But how did the Sun become hot enough for fusion to begin in the first place?

The answer invokes the mechanism of gravitational contraction that astronomers of the late 1800s mistakenly thought might be responsible for the Sun's heat today. Recall that our Sun was born about 4.6 billion years ago from a collapsing cloud of interstellar gas [Section 6.2]. The contraction of the cloud released gravitational potential energy, raising the interior temperature and pressure higher and higher—but not high enough to halt the contraction. The cloud continued to shrink until the central temperature and density finally grew high enough to sustain nuclear fusion.

Compression of gas in the Sun's core made it hot enough for fusion, which provides the energy that now keeps pressure and gravity in balance.

In summary, the answer to the question "Why does the Sun shine?" is that about 4.6 billion years ago *gravitational contraction* made the Sun hot enough to sustain nuclear fusion in its core. Ever since, energy liberated by fusion has maintained the Sun's *gravitational equilibrium* and kept the Sun shining steadily, supplying the light and heat that sustain life on Earth.

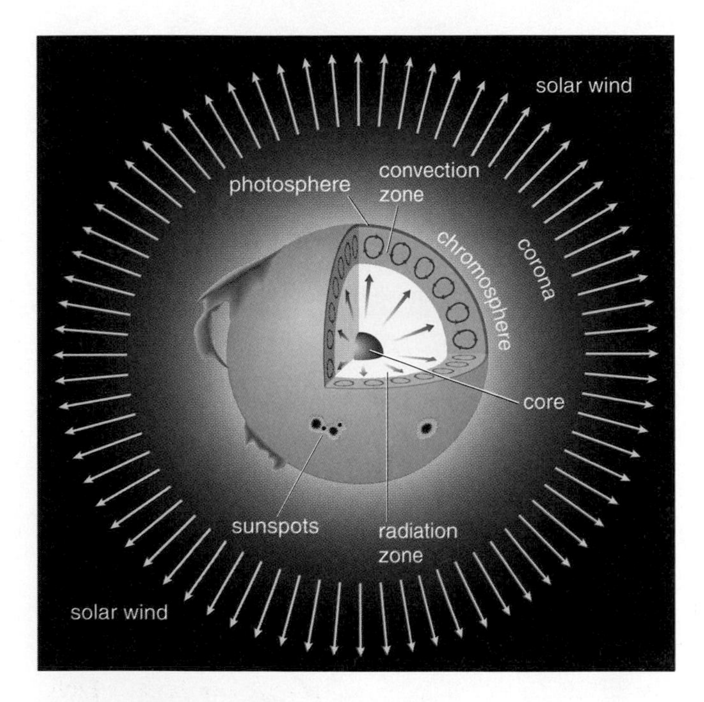

Figure 10.3 Interactive Figure

The basic structure of the Sun. Nuclear fusion in the solar core generates the Sun's energy. Photons of light carry that energy through the radiation zone to the bottom of the convection zone. Rising plumes of hot gas then transport the energy through the convection zone to the photosphere, where it is radiated into space. The photosphere, at a temperature of roughly 6,000 K, is relatively cool compared to the layers that lie above it. The temperature of the chromosphere, which is directly above the photosphere, exceeds 10,000 K. The temperature of the corona, extending outward from the chromosphere, can reach 1 million degrees. Because the coronal gas is so hot, some of it escapes the Sun's gravity, forming a solar wind that blows past Earth and out beyond Pluto.

• What is the Sun's structure?

If the Sun is not a ball of fire as ancient people imagined, what exactly is it? We've already stated that it is a giant ball of hot gas. More technically, the Sun is a ball of **plasma**—a gas in which many of the atoms are ionized [Section 5.2] because of the high temperature. A plasma behaves much like an ordinary gas, but its many positively charged ions and freely moving electrons mean it can create and respond to magnetic fields.

The differing temperatures and densities of the plasma at different depths give the Sun the layered structure shown in Figure 10.3. To make sense of what you see in the figure, let's imagine that you have a spaceship that can somehow withstand the immense heat and pressure of the Sun, and take an imaginary journey from Earth to the center of the Sun. This journey will acquaint you with the basic properties of the Sun, which we'll discuss in greater detail in the rest of this chapter.

Basic Properties of the Sun As you begin your journey from Earth, the Sun appears as a whitish ball of glowing gas. Just as astronomers have done in real life, you can use simple observations to determine basic properties of the Sun. Spectroscopy [Section 5.2] tells you that the Sun is made almost entirely of hydrogen and helium. From the Sun's angular size and distance, you can determine that its radius is just under 700,000 kilometers, or more than 100 times the radius of Earth. Even **sunspots**, which appear as dark splotches on the Sun's surface, can be larger in size than Earth.

Astronomers have measured the Sun's mass from its gravitational effects on the planets. Its mass is about 300,000 times the mass of Earth, or more than 1,000 times the mass of all the planets put together.

THINK ABOUT IT As a brief review, describe how astronomers use Newton's version of Kepler's third law to determine the mass of the Sun. What two properties of Earth do we need to know in order to apply this law? (*Hint:* see Section 4.4.)

The Sun releases an enormous amount of radiative energy into space. Recall that in science we measure energy in units of *joules* [Section 4.2]. We define **power** as the *rate* at which energy is used or released. The standard unit of power is the **watt**, defined as 1 joule of energy per second; that is 1 watt = 1 joule/s. For example, a 100-watt light bulb requires 100 joules of energy for every second it is left turned on. The Sun's total power output, or **luminosity**, is an incredible 3.8×10^{26} watts. If we could somehow capture and store just 1 second's worth of the Sun's luminosity, it would be enough to meet current human energy demands for roughly the next 500,000 years! Table 10.1 summarizes the basic properties of the Sun.

TABLE 10.1 *Basic Properties of the Sun*

Radius (R_{Sun})	696,000 km (about 109 times the radius of Earth)
Mass (M_{Sun})	2×10^{30} kg (about 300,000 times the mass of Earth)
Luminosity (L_{Sun})	3.8×10^{26} watts
Composition (by percentage of mass)	70% hydrogen, 28% helium, 2% heavier elements
Surface temperature	5,800 K (average); 4,000 K (sunspots)
Core temperature	15 million K

The Sun's Atmosphere

Even at great distances from the Sun, you and your spacecraft can feel slight effects from the **solar wind**—the stream of charged particles continually blown outward in all directions from the Sun. Recall that the solar wind helps shape the magnetospheres of planets (see Figure 7.6) and blows back the material that forms the plasma tails of comets [Section 9.2]. The solar wind also cleared away the gas of the solar nebula at the end of the era of planet formation, some four and a half billion years ago [Section 6.4].

As you approach the Sun more closely, you'll begin to encounter the low-density gas that represents what we usually think of as the Sun's atmosphere. The outermost layer of this atmosphere, called the **corona**, extends several million kilometers above the visible surface of the Sun. The temperature of the corona is astonishingly high—about 1 million K—explaining why this region emits most of the Sun's X rays. However, the corona's density is so low that your spaceship feels relatively little heat despite the million-degree temperature [Section 4.3].

The Sun's upper atmosphere is much hotter than the visible surface, or *photosphere*, but its density is much lower.

Nearer the surface, the temperature suddenly drops to about 10,000 K in the **chromosphere**, the middle layer of the solar atmosphere and the region that radiates most of the Sun's ultraviolet radiation. Then you plunge through the lowest layer of the atmosphere, or **photosphere**, which is the visible surface of the Sun. Although the photosphere looks like a well-defined surface from Earth, it consists of gas far less dense than Earth's atmosphere. The temperature of the photosphere averages just under 6,000 K, and its surface seethes and churns like a pot of boiling water. The photosphere is also where you'll find sunspots, regions of intense magnetic fields that would cause your compass needle to swing wildly about.

The Sun's Interior

Up to this point in your journey, you may have seen Earth and the stars when you looked back. But as you slip beneath the photosphere, blazing light engulfs you. You are inside the Sun, and your spacecraft is tossed about by incredible turbulence. If you can hold steady long enough to see what is going on around you, you'll notice spouts of hot gas rising upward, surrounded by cooler gas cascading down

COMMON MISCONCEPTIONS

THE SUN IS NOT ON FIRE

We are accustomed to saying that the Sun is "burning," a way of speaking that conjures up images of a giant bonfire in the sky. However, the Sun does not burn in the same sense as a fire burns on Earth. Fires generate light through chemical changes that consume oxygen and produce a flame. The glow of the Sun has more in common with the glowing embers left over after the flames have burned out. Much like hot embers, the Sun's surface shines with the visible thermal radiation produced by any object that is sufficiently hot [Section 5.2].

However, hot embers quickly stop glowing as they cool, while the Sun keeps shining because its surface is kept hot by the energy rising from the Sun's core. Because this energy is generated by nuclear fusion, we sometimes say that it is the result of "nuclear burning"—a term that suggests nuclear changes in much the same way that "chemical burning" suggests chemical changes. Nevertheless, while it is reasonable to say that the Sun undergoes nuclear burning in its core, it is not accurate to speak of any kind of burning on the Sun's surface, where light is produced primarily by thermal radiation.

from above. You are in the **convection zone**, where energy generated in the solar core travels upward, transported by the rising of hot gas and falling of cool gas called *convection* [Section 7.1]. With some quick thinking, you may realize that the photosphere above you is the top of the convection zone and that convection is the cause of the Sun's seething, churning appearance.

About a third of the way down to the center, the turbulence of the convection zone gives way to the calmer plasma of the **radiation zone**, where energy is carried outward primarily by photons of light. The temperature rises to almost 10 million K, and your spacecraft is bathed in X rays trillions of times more intense than the visible light at the solar surface.

No real spacecraft could survive, but your imaginary one keeps plunging straight down to the solar **core**. There you finally find the source of the Sun's energy: nuclear fusion transforming hydrogen into helium. At the Sun's center, the temperature is about 15 million K, the density is more than 100 times that of water, and the pressure is 200 billion times that on the surface of Earth. The energy produced in the core today will take about a million years to reach the surface.

Inside the Sun, temperature rises with depth, reaching 15 million K in the core.

With your journey complete, it's time to turn around and head back home. We'll continue this chapter by studying fusion in the solar core and then tracing the flow of the energy generated by fusion as it moves outward through the Sun.

The Sun Tutorial, Lessons 2–3

10.2 NUCLEAR FUSION IN THE SUN

We've seen that the Sun shines because of energy generated by nuclear fusion, and that this fusion occurs under the extreme temperatures and densities found deep in the Sun's core. But exactly how does fusion occur and release energy? And how can we claim to know about something taking place out of sight in the Sun's hidden interior?

Before we begin to answer these questions, it's important to realize that the nuclear reactions that generate energy in the Sun are very different from those used to generate energy in human-built nuclear reactors on Earth. Our nuclear power plants generate energy by splitting large nuclei—such as those of uranium or plutonium—into smaller ones. The process of splitting an atomic nucleus is called *nuclear fission.* In contrast, the Sun makes energy by combining, or fusing, two or more small nuclei into a larger one. That is why we call the process *nuclear fusion.* Figure 10.4 summarizes the difference between fission and fusion.

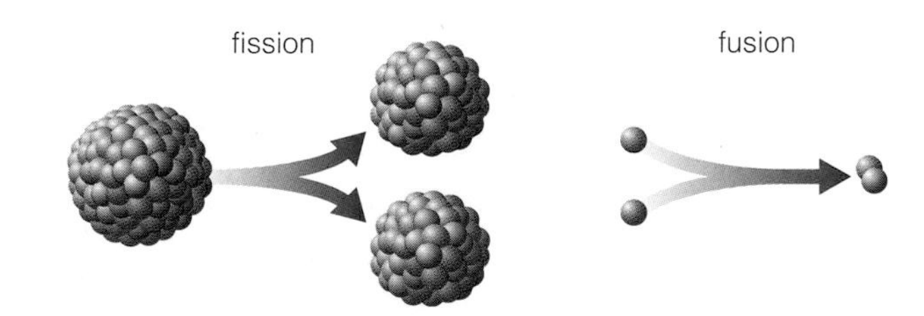

Figure 10.4

Nuclear fission splits a nucleus into smaller nuclei (not usually of equal size), while nuclear fusion combines smaller nuclei into a larger nucleus.

• How does nuclear fusion occur in the Sun?

Fusion occurs within the Sun because the 15 million K plasma in the solar core is like a "soup" of hot gas, with bare, positively charged atomic nuclei (and negatively charged electrons) whizzing about at extremely high speeds. At any one time, some of these nuclei are on high-speed collision courses with each other. In most cases, electromagnetic forces deflect the nuclei, preventing actual collisions, because positive charges repel one another. If nuclei collide with sufficient energy, however, they can stick together to form a heavier nucleus (Figure 10.5).

Sticking positively charged nuclei together is not easy. The **strong force**, which binds protons and neutrons together in atomic nuclei, is the only force in nature that can overcome the electromagnetic repulsion between two positively charged nuclei. In contrast to gravitational and electromagnetic forces, which drop off gradually as the distances between particles increase (by an inverse square law [Section 4.4]), the strong force is more like glue or Velcro: It overpowers the electromagnetic force over very small distances but is insignificant when the distances between particles exceed the typical sizes of atomic nuclei. The key to nuclear fusion, therefore, is to push the positively charged nuclei close enough together for the strong force to outmuscle electromagnetic repulsion.

Positively charged nuclei fuse together if they pass close enough for the strong force to overpower electromagnetic repulsion.

The high pressures and temperatures in the solar core are just right for fusion of hydrogen nuclei into helium nuclei. The high temperature is important because the nuclei must collide at very high speeds if they are to come close enough together to fuse. The higher the temperature, the harder the collisions, making fusion reactions more likely at higher temperatures. The high pressure of the overlying layers is necessary because without it the hot plasma of the solar core would simply explode into space, shutting off the nuclear reactions.

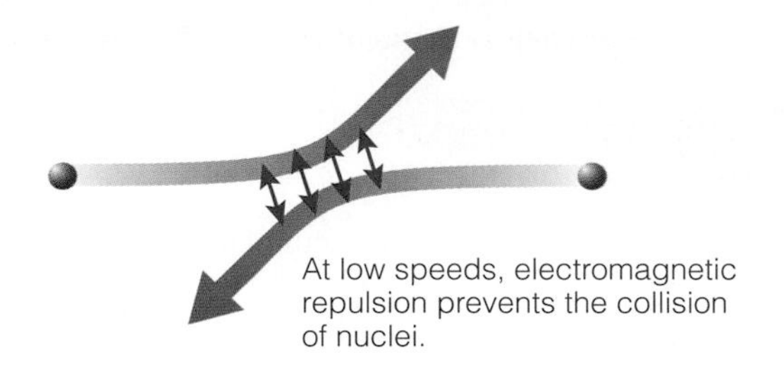

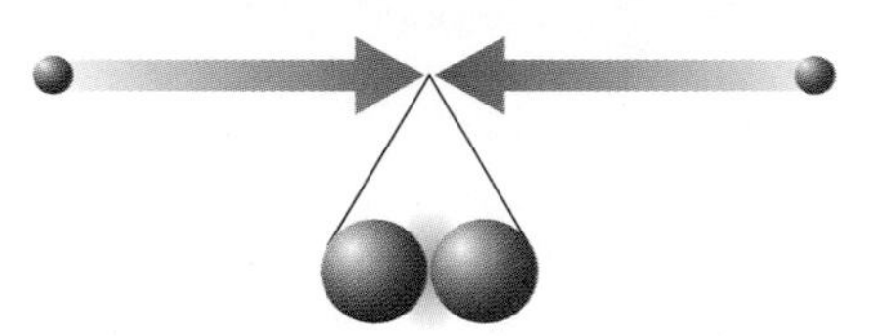

Figure 10.5

Positively charged nuclei can fuse only if a high-speed collision brings them close enough for the strong force to come into play.

THINK ABOUT IT The Sun generates energy by fusing hydrogen into helium, but as we'll see in later chapters, some stars fuse helium or even heavier elements. Do temperatures need to be higher or lower for the fusion of heavier elements than for the fusion of hydrogen? Why? (*Hint:* How does the positive charge of a nucleus affect the difficulty of fusing it to another nucleus?)

The Proton–Proton Chain

Let's investigate the fusion process in the Sun in a little more detail. Recall that hydrogen nuclei are simply individual protons, while the most common form of helium consists of two protons and two neutrons (see Figure 5.5). Thus, the overall hydrogen fusion reaction transforms four individual protons into a helium nucleus containing two protons and two neutrons:

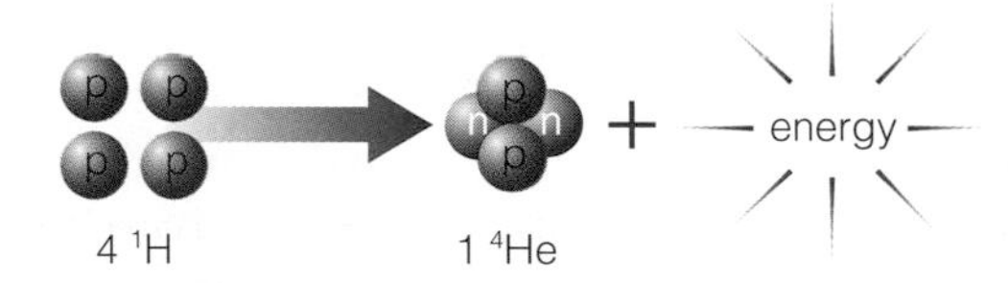

Nuclear fusion in the Sun combines four hydrogen nuclei into one helium nucleus, releasing energy in the process.

This overall reaction actually proceeds through several steps involving just two nuclei at a time. The sequence of steps that occurs in the Sun is called the **proton–proton chain**, because it begins with collisions between individual protons (hydrogen nuclei). Figure 10.6 illustrates the steps in the proton–proton chain. Notice that the overall reaction is just as described above, with four protons combining to make one helium nucleus. Energy is carried off by the gamma rays and subatomic particles (neutrinos and positrons) released in the process.

Fusion of hydrogen into helium generates energy because a helium nucleus has a mass slightly less (by about 0.7%) than the combined mass of four hydrogen nuclei. Thus, when four hydrogen nuclei fuse into a

Hydrogen Fusion by the Proton–Proton Chain

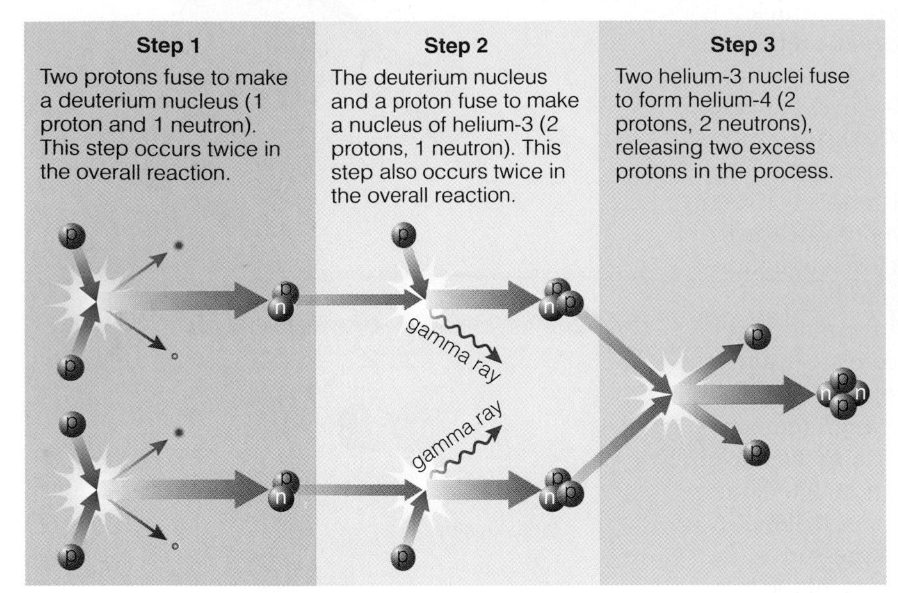

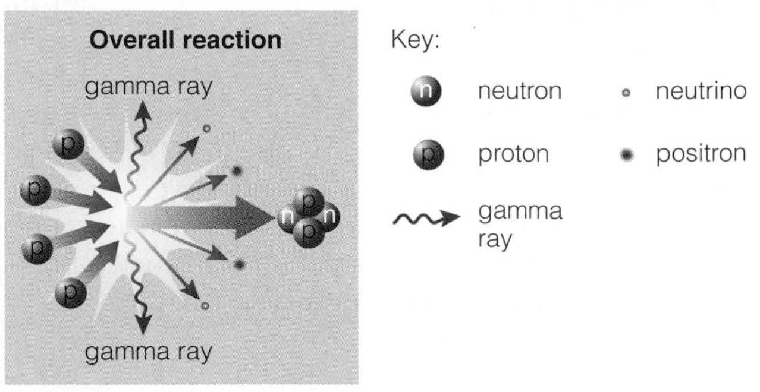

Figure 10.6 Interactive Figure

In the Sun, four hydrogen nuclei (protons) fuse into one helium-4 nucleus by way of the proton–proton chain. The energy released in the reaction is carried off by gamma rays and subatomic particles known as neutrinos and positrons.

helium nucleus, a little bit of mass disappears. The disappearing mass becomes energy in accord with Einstein's formula $E = mc^2$. Overall, fusion in the Sun converts about 600 million tons of hydrogen into 596 million tons of helium every second, which means that 4 million tons of matter is turned into energy each second. Although this sounds like a lot, it is a minuscule fraction of the Sun's total mass and thus does not affect the overall mass of the Sun in any measurable way.

The Solar Thermostat Nuclear fusion is the source of all the energy the Sun releases into space. If the fusion rate varied, so would the Sun's energy output, and large variations in the Sun's luminosity would almost surely be lethal to life on Earth. Fortunately, the Sun fuses hydrogen at a steady rate, thanks to a natural feedback process that acts as a thermostat for the Sun's interior. We have encountered a natural thermostat once before, when we discussed how Earth's carbon dioxide cycle keeps our planet's surface temperature relatively steady [Section 7.5]. The solar thermostat is even more efficient.

Solar energy production remains steady because the rate of nuclear fusion is very sensitive to temperature. A slight increase in the Sun's core temperature would mean a much higher fusion rate, and a slight decrease in temperature would mean a much lower fusion rate. Either kind of change in the fusion rate would alter the Sun's gravitational equilibrium—the balance between the pull of gravity and the push of internal pressure—in a way that would quickly restore the original temperature and fusion rate. To see how, let's examine what would happen if there were a small change in the core temperature.

Suppose the Sun's core temperature rose very slightly. The rate of nuclear fusion would soar, generating lots of extra energy. Because energy moves so slowly through the Sun, this extra energy would be bottled up in the core, causing an increase in the core pressure. The push of this pressure would temporarily exceed the pull of gravity, causing the core to expand and cool. This cooling, in turn, would cause the fusion rate to drop back down until the core was restored to its original size and temperature, thus returning the fusion rate back to its normal value.

Gravitational equilibrium acts as a thermostat that keeps the Sun's core temperature and fusion rate steady.

A slight drop in the Sun's core temperature would trigger an opposite chain of events. The reduced core temperature would lead to a decrease in the rate of nuclear fusion, causing a drop in pressure and contraction of the core. As the core shrank, its temperature would rise until the fusion rate returned to normal and restored the core to its original size and temperature. In summary, natural feedback processes automatically keep the Sun's central temperature and fusion rate steady and thus keep the Sun's overall energy output steady as well.

Figure 10.7

A photon in the solar interior bounces randomly among electrons, slowly working its way outward in a process.

• How does the energy from fusion get out of the Sun?

The solar thermostat balances the Sun's fusion rate so that the amount of nuclear energy generated in the core equals the amount of energy radiated from the surface as sunlight. However, the journey of solar energy from the core to the photosphere takes hundreds of thousands of years.

Most of the energy released by fusion starts its journey out of the solar core in the form of photons. Although photons travel at the speed of light, the path they take through the Sun's interior zigzags so much that it takes them a very long time to make any outward progress. Deep in the solar interior, the plasma is so dense that a photon can travel only a fraction of a millimeter in any one direction before it interacts with an electron. Each time a photon "collides" with an electron, the photon gets deflected into a new and random direction. Thus, the photon bounces around the dense interior in a haphazard way (sometimes called a *random walk*) and only very gradually works its way outward from the Sun's center (Figure 10.7).

Randomly bouncing photons carry energy through the deepest layers of the Sun, and convection carries energy through the upper layers to the surface.

Energy released by fusion moves outward through the Sun's radiation zone (see Figure 10.3) primarily by way of these randomly bouncing photons. At the top of the radiation zone, where the temperature has dropped to about 2 million K, the solar plasma absorbs photons more readily (rather than just bouncing them around). This absorption creates the conditions needed for convection [Section 7.1], in which rising plumes of hot gas carry energy upward to the Sun's surface, while cool gas sinks back down around the plumes. The convecting gas gives the Sun's photosphere a mottled appearance in close-up photographs (Figure 10.8). We see bright blobs where hot gas is welling up from inside the Sun and darker borders around those blobs where the cooler gas is sinking. Each bright blob lasts only a few minutes before being replaced by others.

In the photosphere, the density of the gas is low enough so that photons can escape to space. Thus, the energy produced hundreds of thousands of years earlier in the solar core finally emerges from the Sun as thermal radiation [Section 5.2] produced by the almost 6,000 K gas of the photosphere. Once in space, the photons travel straight away at the speed of light, bathing the planets in sunlight.

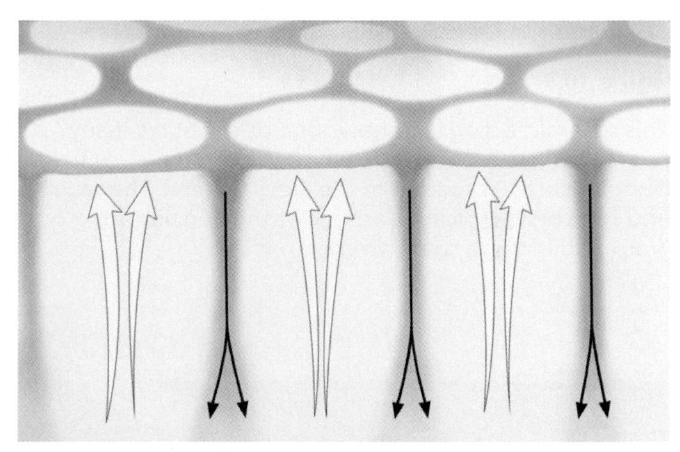

a This diagram shows convection beneath the Sun's surface: hot gas (white arrows) rises while cooler gas (orange/black arrows) descends around it.

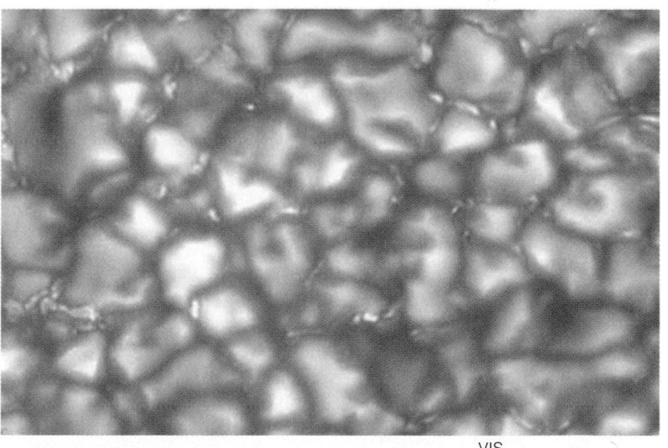

b This photograph shows the mottled appearance of the Sun's photosphere. The bright spots, each about 1,000 kilometers across, correspond to the rising plumes of hot gas in the diagram in part (a).

Figure 10.8

The Sun's photosphere churns with rising hot gas and falling cool gas as a result of underlying convection.

• How do we know what is happening inside the Sun?

We cannot see inside the Sun, so you may wonder how we can claim to know so much about what goes on underneath its surface and in its core. In fact, we can study the Sun's interior in three different ways: through

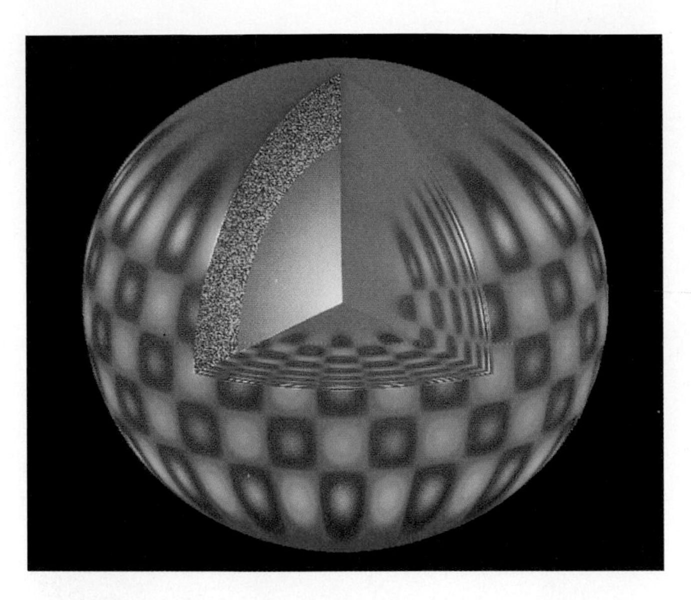

Figure 10.9

This schematic diagram shows one of the Sun's many possible vibration patterns, which we can measure from Doppler shifts in spectra of the Sun's surface. The blue and red colors indicate portions of the Sun vibrating outward and inward, respectively.

Figure 10.10

This tank of dry-cleaning fluid (visible underneath the catwalk), located deep within South Dakota's Homestake mine, was a solar neutrino detector. The chlorine nuclei in the cleaning fluid turned into argon nuclei when they captured neutrinos from the Sun.

mathematical models of the Sun, observations of "sun quakes," and observations of solar neutrinos.

Mathematical Models The primary way we learn about the interior of the Sun (and other stars) is by creating *mathematical models* that use the laws of physics to predict internal conditions. A basic model uses the Sun's observed composition and mass as inputs to equations that describe gravitational equilibrium and the rate at which solar energy moves from the core to the photosphere. With the aid of a computer, we can use the model to calculate the Sun's temperature, pressure, and density at any depth. We can then predict the rate of nuclear fusion in the solar core by combining these calculations with knowledge about nuclear fusion gathered in laboratories here on Earth.

If a model is correct, it should correctly "predict" the radius, surface temperature, luminosity, age, and many other properties of the Sun. Current models do indeed predict these properties quite accurately, giving us confidence that we really do understand what is going on inside the Sun.

Sun Quakes A second way to learn about the inside of the Sun is to observe "sun quakes"—vibrations of the Sun's surface that are somewhat similar to the vibrations that earthquakes cause on Earth. Sun quakes result from movement of gas within the Sun, which generates vibrations that travel through the Sun like sound waves moving through air. We can observe these vibrations on the Sun's surface by looking for Doppler shifts [Section 5.3]. Light from portions of the surface that are rising toward us is slightly blueshifted, while light from portions that are falling away from us is slightly redshifted. The vibrations are relatively small but measurable (Figure 10.9).

The characteristics of sun quakes confirm our mathematical models of the Sun's interior.

In principle, we can deduce a great deal about the solar interior by carefully analyzing these vibrations. (By analogy to seismology on Earth, this type of study of the Sun is called *helioseismology*—*helios* means "sun.") Results to date confirm that our mathematical models of the solar interior are on the right track. At the same time, they provide data that can be used to improve the models further.

Solar Neutrinos Another way to study the solar interior is to observe subatomic particles produced by fusion reactions in the core. If you look back at Figure 10.6, you'll see that some of the energy (about 2% of the total) released in nuclear fusion is carried off by subatomic particles called **neutrinos**. Neutrinos are a strange type of particle that rarely interacts with anything at all and which can therefore pass through almost anything. For example, while an inch of lead will stop an X ray, stopping an average neutrino would require a slab of lead more than a light-year thick! Thus, the neutrinos produced by fusion travel essentially nonstop from the solar core out into space, some of them aimed directly at Earth.

Don't panic, but as you read this sentence about a thousand trillion solar neutrinos will zip through your body. Traveling at nearly the speed of light, they reach us just minutes after being produced in fusion reactions in the Sun's core. Fortunately, they pass through our bodies and our planet as easily as they pass through anything else, doing no damage at all.

Because the neutrinos reach us so quickly, in principle they give us a way to directly monitor what is happening in the core of the Sun. However, the same properties that allow them to pass easily through the Sun and Earth make them dauntingly difficult to capture and count.

Detectors built to capture neutrinos rely on the fact that they *do* occasionally interact with matter. To distinguish neutrino captures from reactions caused by other particles, neutrino detectors are usually placed deep underground in mines. The overlying rock blocks most other particles, but the neutrinos have no difficulty passing through.

The first major solar neutrino detector, built in the 1960s, was located 1,500 meters underground in the Homestake gold mine in South Dakota (Figure 10.10). The detector for this "Homestake experiment" consisted of a 400,000-liter vat of chlorine-containing dry-cleaning fluid. It turns out that, on very rare occasions, a chlorine nucleus can capture a neutrino and change into a nucleus of radioactive argon. By looking for radioactive argon in the tank of cleaning fluid, experimenters could count the number of neutrinos captured in the detector.

From the many trillions of solar neutrinos that passed through the tank of cleaning fluid each second, experimenters expected to capture an average of just one neutrino per day. This predicted capture rate was based on measured properties of chlorine nuclei and models of nuclear fusion in the Sun. However, over a period of more than two decades, neutrinos were captured only about once every three days on average. That is, the Homestake experiment detected only about one-third of the predicted number of neutrinos. This disagreement between model predictions and actual observations came to be called the **solar neutrino problem**.

For a while, the solar neutrino problem was one of the great mysteries in astronomy. Other experiments found the same shortfall as the Homestake experiment. These results allowed only two possible conclusions: Either something was wrong with our understanding of fusion in the Sun, or some of the Sun's neutrinos were somehow hiding themselves from us.

Today, scientists are fairly convinced that the neutrinos were hiding. Here's how we think it works: Scientists have known for decades that neutrinos come in three types, called electron neutrinos, muon neutrinos, and tau neutrinos. Fusion reactions in the Sun produce only electron neutrinos, and until recently most solar neutrino detectors could detect only electron neutrinos. Thus, if some of the electron neutrinos produced by fusion change into neutrinos of the other two types on their way to Earth, then our detectors would count fewer electron neutrinos than we would have expected.

Subatomic particles called neutrinos provide a direct probe of nuclear fusion in the Sun, and recent results indicate that fusion occurs as our models predict.

Recent experiments support the idea that neutrinos can change type on their way to Earth, probably during the portion of their trip from the Sun's core to its surface. Most significantly, a new detector in Canada called the Sudbury Neutrino Observatory is capable of detecting all three neutrino types (Figure 10.11). Early results from this observatory suggest that the total number of neutrinos of all types is indeed what we expect from our models of nuclear fusion in the Sun. If these results hold up as more data are collected, it will mean not only that the solar neutrino problem is solved, but also that neutrino counts verify that nuclear fusion occurs in the Sun in just the way our models predict.

Figure 10.11

This photograph shows the main tank of the Sudbury Neutrino Observatory in Canada, located at the bottom of a mine shaft more than 2 kilometers underground. The large sphere, 12 meters in diameter, contains 1,000 tons of ultrapure "heavy water." (Heavy water is water in which the two hydrogen atoms are replaced by deuterium, making each molecule heavier than a molecule of ordinary water.) Neutrinos of all three types can cause reactions in the heavy water, and detectors surrounding the tank can record these reactions when they occur.

10.3 THE SUN–EARTH CONNECTION

Energy liberated by nuclear fusion in the Sun's core eventually reaches the solar surface. There, before it is ultimately released into space as sunlight, the energy helps create a wide variety of phenomena that we can observe from Earth. Sunspots are only the most obvious of these phenomena. Because sunspots and other features of the Sun's surface change with time, they constitute what we call *solar weather* or **solar activity**. The "storms" associated with solar weather are not just of academic interest. Sometimes they are so violent that they affect our day-to-day life on Earth. In this section, we'll explore solar activity and its far-reaching effects.

• What causes solar activity?

Most of the Sun's surface churns constantly with rising and falling gas and thus looks like the close-up photo shown in Figure 10.8. However, larger features sometimes appear, including sunspots, the huge explosions known as solar flares, and gigantic loops of hot gas extending high into the Sun's corona. All of these features are created by magnetic fields, which form and change easily in the convecting plasma in the outer layers of the Sun.

Sunspots and Magnetic Fields Sunspots are the most striking features of the solar surface (Figure 10.12a). If you could look directly at a sunspot (but you should *never* stare at the Sun), you would find it blindingly bright. However, sunspots appear dark in photographs because they are less bright than the surrounding photosphere. They are less bright because they are cooler: The temperature of the plasma in sunspots is about 4,000 K, significantly cooler than the 5,800 K plasma that surrounds them.

If you think about this for a moment, you may wonder how sunspots can be so much cooler than their surroundings. Gas can usually flow easily, so we might expect the hotter gas around a sunspot to mix with its cooler gas, quickly warming the sunspot. Thus, the fact that sunspots stay relatively cool means that something must prevent hot plasma from entering them, and that "something" turns out to be magnetic fields.

Detailed observations of the Sun's spectral lines reveal sunspots to be regions with strong magnetic fields. These magnetic fields can alter the energy levels in atoms and ions and therefore can alter the spectral lines they produce. More specifically, magnetic fields cause some spectral lines

Figure 10.12

VIS

Sunspots are regions of strong magnetic fields.

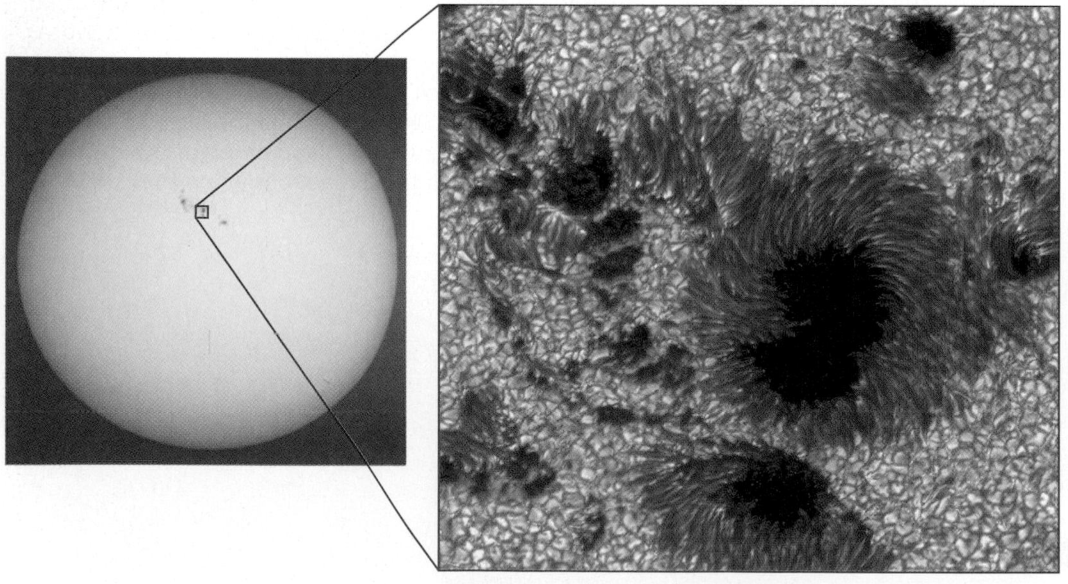

a This close-up view of the Sun's surface shows two large sunspots and several smaller ones. Each of the big sunspots is roughly as large as Earth.

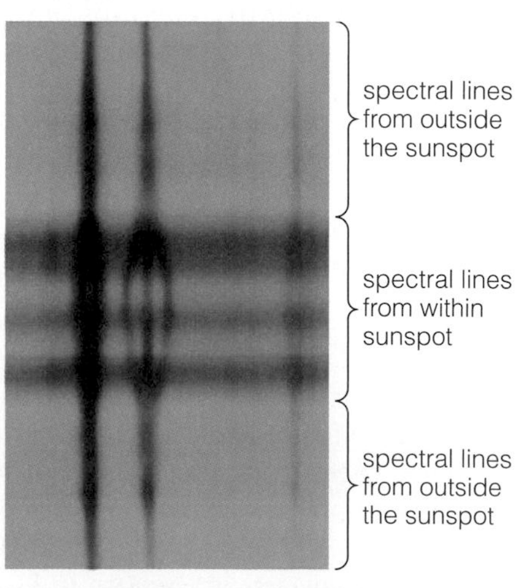

b Very strong magnetic fields split the absorption lines in spectra of sunspot regions. The dark vertical bands are absorption lines in a spectrum of the Sun. Notice that these lines split where they cross the dark horizontal bands corresponding to sunspots.

to split into two or more closely spaced lines (Figure 10.12b). Wherever we see this effect (called the *Zeeman effect*), we know that magnetic fields must be present. Thus, scientists can map magnetic fields on the Sun by looking for the splitting of spectral lines in light from different parts of the solar surface.

Sunspots are regions of strong magnetic fields, which keep the sunspots cooler than the surrounding photosphere.

To understand how sunspots stay cool, we must investigate the nature of magnetic fields in a little more depth. Magnetic fields are invisible, but we can represent them by drawing **magnetic field lines** (Figure 10.13). These lines represent the directions in which compass needles would point if we placed them within the magnetic field. The lines come closer together where the field is stronger, and move farther apart where the field is weaker. Because these imaginary field lines are so much easier to visualize than the magnetic field itself, we usually discuss magnetic fields by talking about how the field lines would appear. Charged particles, such as the ions and electrons in the solar plasma, cannot easily move perpendicular to the field lines but instead follow spiraling paths along them.

The magnetic fields explain why sunspots are cooler than their surroundings. Sunspots occur where tightly wound magnetic fields poke nearly straight out from the solar interior (Figure 10.14a). These tight magnetic field lines suppress convection within the sunspot and prevent surrounding plasma from entering the sunspot. With hot plasma unable to enter the region, the sunspot plasma becomes cooler than that of the rest of the photosphere. Individual sunspots typically last up to a few weeks, dissolving when their magnetic fields weaken and allow plasma to flow in.

Sunspots tend to occur in pairs, connected by a loop of magnetic field lines that can arc high above the Sun's surface (Figure 10.14b). Gas in the

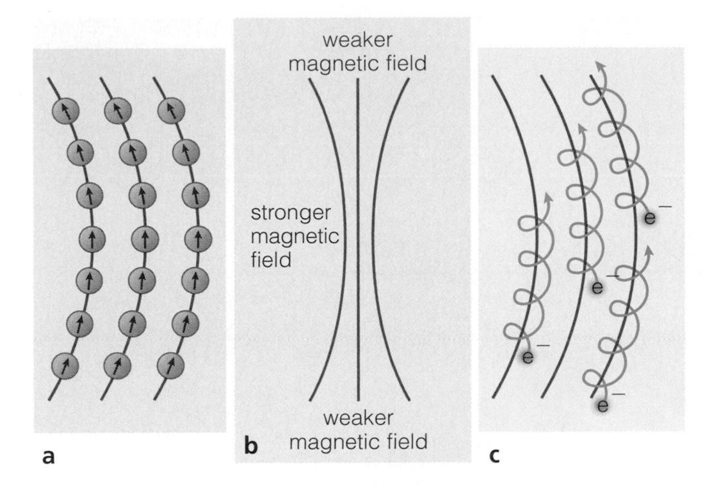

Figure 10.13

We draw magnetic field lines (red) to represent invisible magnetic fields. (**a**) Magnetic field lines represent the directions in which compass needles (black arrows in gray circles) would point. (**b**) Lines closer together indicate a stronger magnetic field. (**c**) Charged particles follow spiraling paths (blue) along magnetic field lines (red) but cannot easily move perpendicular to the lines.

Figure 10.14

Strong magnetic fields keep sunspots cooler than the surrounding photosphere, while magnetic loops can arch from the sunspots to great heights above the Sun's surface.

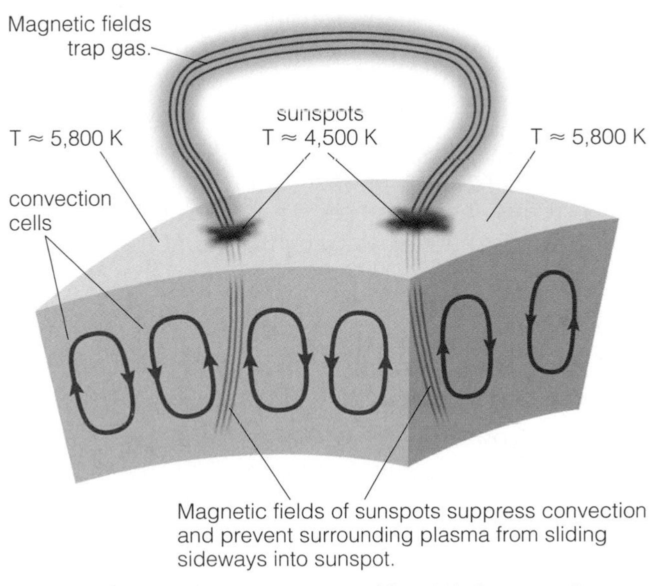

a Pairs of sunspots are connected by tightly wound magnetic field lines.

b This X-ray photo (from NASA's TRACE mission) shows hot gas trapped within looped magnetic field lines.

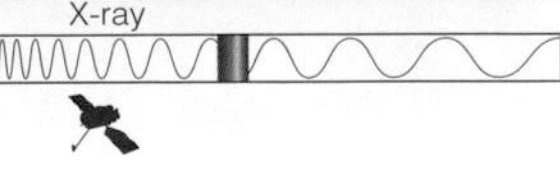

Sun's chromosphere and corona becomes trapped in these loops, making giant **solar prominences**. Some prominences, such as the one shown in the photo that opens this chapter (p. 257), rise to heights of more than 100,000 kilometers above the Sun's surface. Individual prominences can last for days or even weeks, disappearing only when the magnetic fields finally weaken and release the trapped gas.

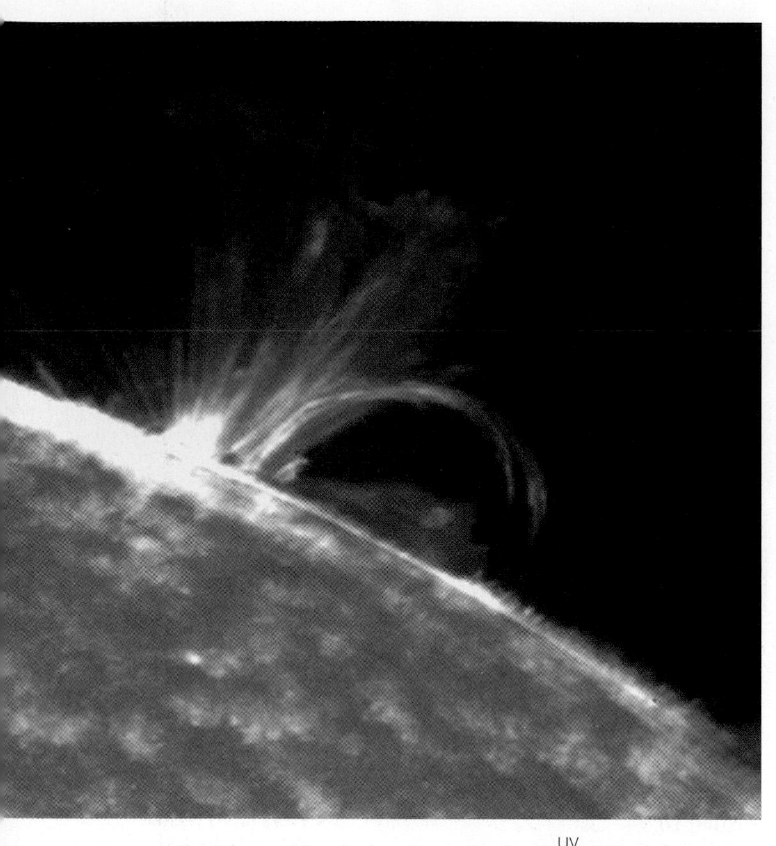

Figure 10.15

This photo (from TRACE) of ultraviolet light emitted by hydrogen atoms shows a solar flare erupting from the Sun's surface.

Solar Storms The magnetic fields winding through sunspots and prominences sometimes undergo dramatic and sudden change, producing short-lived but intense storms on the Sun. The most dramatic of these storms are **solar flares**, which send bursts of X rays and fast-moving charged particles shooting into space (Figure 10.15).

Energy released when magnetic field lines snap can lead to dramatic solar storms, which sometimes eject bursts of energetic particles into space.

Flares generally occur in the vicinity of sunspots, which is why we think they are created by changes in magnetic fields. The leading model for solar flares suggests that they occur when the magnetic field lines become so twisted and knotted that they can no longer bear the tension. They are thought to suddenly snap and reorganize themselves into a less twisted configuration. The energy released in the process heats the nearby plasma to 100 million K over the next few minutes or hours, generating X rays and accelerating some of the charged particles to nearly the speed of light.

Heating of the Chromosphere and Corona As we've seen, many of the most dramatic weather patterns and storms on the Sun involve the very hot gas of the Sun's chromosphere and corona. But why is that gas so hot in the first place? This question perplexed scientists for decades.

Remember that temperatures gradually decline as we move outward from the Sun's core to the top of its photosphere. We might expect the decline to continue in the Sun's atmosphere, but instead it reverses, making the chromosphere and corona much hotter than the Sun's surface. Some aspects of this atmospheric heating remain a mystery today, but we have at least a general explanation: The Sun's strong magnetic fields carry energy upward from the churning solar surface to the chromosphere and corona.

More specifically, the rising and falling of gas in the convection zone probably shakes tightly wound magnetic field lines beneath the solar surface. The magnetic field lines carry this energy upward to the solar atmosphere, where they deposit this energy as heat. Thus, the same magnetic fields that keep sunspots cool make the overlying plasma of the chromosphere and corona hot.

Magnetic fields deposit energy above the Sun's surface, heating the chromosphere and corona.

Observations confirm the connection between magnetic fields and the structure of the chromosphere and corona. The density of gas in the chromosphere and corona is so low that we cannot see this gas with our eyes except during a total eclipse, when we can see the faint visible light scattered by electrons in the corona (see Figure 2.24). However, we can observe the chromosphere and corona at any time with ultraviolet and X-ray telescopes in space: The roughly 10,000 K plasma of the chromosphere emits strongly in the ultraviolet, and the million K plasma of the corona is the source of virtually all X rays coming from the Sun. Figure 10.16 shows an X-ray image of the Sun. The X-ray emission is brightest in regions where hot gas is being trapped and heated in loops of magnetic field. In fact, the bright spots in the corona tend to be directly above sunspots in the photosphere, confirming that they are created by the same magnetic fields.

Notice that some regions of the corona, called **coronal holes**, barely show up in X-ray images. More detailed analyses show that the magnetic field lines in coronal holes project out into space like broken rubber bands, allowing particles spiraling along them to escape the Sun altogether. These particles streaming outward from the corona are the source of the solar wind.

• How does solar activity affect humans?

Flares and other solar storms sometimes eject large numbers of highly energetic charged particles from the Sun's corona. These particles travel outward from the Sun in huge bubbles that we call **coronal mass ejections** (Figure 10.17). The bubbles have strong magnetic fields and can reach Earth in a couple of days if they happen to be aimed in our direction. Once a coronal mass ejection reaches Earth, it can create a *geomagnetic storm* in Earth's magnetosphere. On the positive side, these storms can lead to unusually strong auroras (see Figure 7.6) that can be visible throughout much of the United States. On the negative side, they can also hamper radio communications, disrupt electrical power delivery, and damage the electronic components in orbiting satellites.

Particles ejected from the Sun during periods of high activity can hamper radio communications, disrupt power delivery, and damage orbiting satellites.

During a particularly powerful magnetic storm on the Sun in March 1989, the U.S. Air Force temporarily lost track of more than 2,000 satellites, and powerful currents induced in the ground circuits of the Quebec hydroelectric system caused it to collapse for more than 8 hours. The combined cost of the loss of power in the United States and Canada exceeded $100 million. In January 1997, AT&T lost contact with a $200 million communications satellite, probably because of damage caused by particles coming from another powerful solar storm. More recently, in the fall of 2003, a series of extremely powerful solar flares once again threatened Earth's communications and electrical systems, but they escaped major damage, due in part to our improved preparedness for solar storms.

Satellites in low-Earth orbit are particularly vulnerable during periods of strong solar activity, when the increase in solar X rays and energetic particles heats Earth's upper atmosphere, causing it to expand. The density of the gas surrounding low-flying satellites therefore rises, exerting drag that saps their energy and angular momentum. If this drag proceeds unchecked, the satellites ultimately plummet back to Earth. Satellites in low orbits, including the Hubble Space Telescope and the Space Station, require occasional boosts to prevent them from falling out of the sky.

• How does solar activity vary with time?

Solar weather is just as unpredictable as weather on Earth. Individual sunspots can appear or disappear at almost any time, and we have no way to know that a solar storm is coming until we observe it through our telescopes. However, long-term observations have revealed overall patterns in solar activity, making sunspots and solar storms more common at some times than at others.

Figure 10.16 Interactive Figure — X-ray

An X-ray image of the Sun reveals the million-degree gas of the corona. Brighter regions of this image (yellow) correspond to regions of stronger X-ray emission. The darker regions (such as near the north pole at the top of this photo) are the coronal holes from which the solar wind escapes. (From the Yohkoh space observatory.)

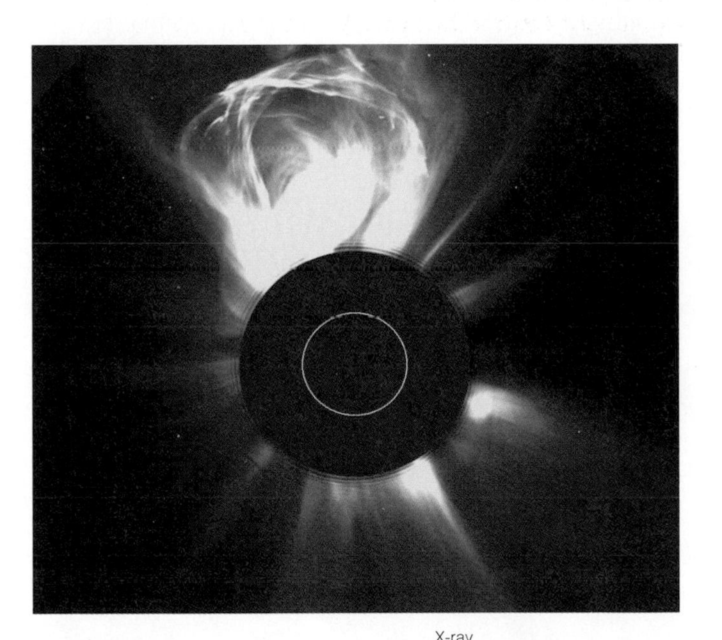

Figure 10.17 X-ray

This X-ray image from the SOHO spacecraft shows a coronal mass ejection (the bright arc of gas headed almost straight upward as oriented here) during the solar storms of 2003. The central red disk blots out the Sun itself, and the white circle represents the size of the Sun in this picture.

a This graph shows how the number of sunspots on the Sun changes with time. The vertical axis shows the percentage of the Sun's surface covered by sunspots. The cycle has a period of approximately 11 years.

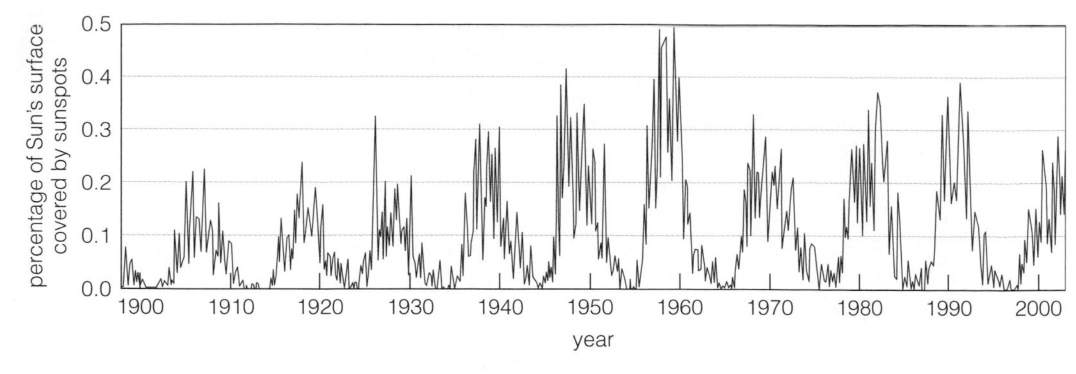

b This graph shows how the latitudes at which sunspot groups appear tend to shift during a single sunspot cycle. Each dot represents a group of sunspots and indicates the year (horizontal axis) and latitude (vertical axis) at which the group appeared.

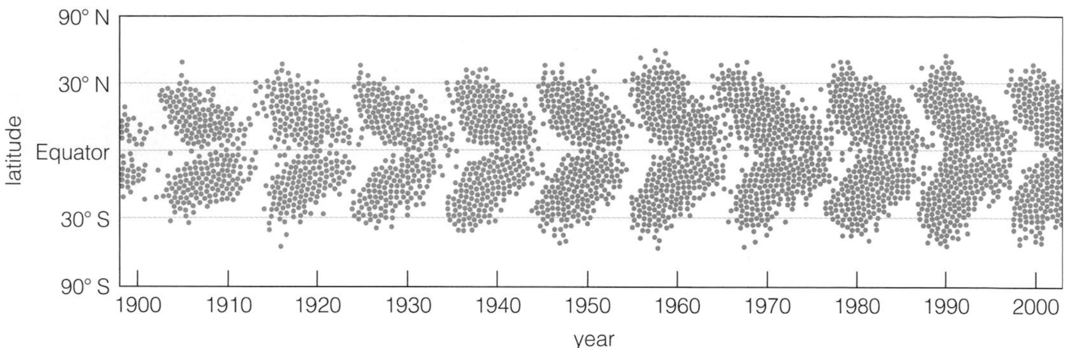

Figure 10.18

Sunspot cycle during the past century.

The Sunspot Cycle The most notable pattern in solar activity is the **sunspot cycle**—a cycle in which the average number of sunspots on the Sun gradually rises and falls (Figure 10.18). At the time of *solar maximum,* when sunspots are most numerous, we may see dozens of sunspots on the Sun at one time. In contrast, we may see few if any sunspots at the time of *solar minimum.* The frequency of prominences, flares, and coronal mass ejections also follows the sunspot cycle, with these events being most common at solar maximum and least common at solar minimum.

The average number of sunspots on the Sun rises and falls in an approximately 11-year cycle.

As you can see in Figure 10.18a, the sunspot cycle varies from one period to the next. The average length of time between maximums is 11 years, but we have observed it to be as short as 7 years and as long as 15 years. Observations going further back in time suggest that sunspot activity can sometimes cease almost entirely. For example, virtually no sunspots were observed by astronomers between the years 1645 and 1715, a period sometimes called the *Maunder minimum* (after E. W. Maunder, who identified it in historical sunspot records).

The locations of sunspots on the Sun also vary with the sunspot cycle (Figure 10.18b). As a cycle begins at solar minimum, sunspots form primarily at mid-latitudes (30° to 40°) on the Sun. The sunspots tend to form at lower latitudes as the cycle progresses, appearing very close to the solar equator as the next solar minimum approaches. Then the sunspots of the next cycle begin to form near mid-latitudes again.

The Cause of the Sunspot Cycle The reason for the sunspot cycle is not well understood, but an important clue comes from careful study of the magnetic fields in sunspots from one cycle to the next. Apparently, the Sun's entire magnetic field flip-flops with each cycle, turning magnetic north into magnetic south and vice versa. We know this because the magnetic field lines connecting pairs of sunspots (see Figure 10.14) all tend to point in the same direction throughout an 11-year cycle (within each hemisphere). For example, all compass needles might point from the easternmost sunspot to the westernmost sunspot in a pair. However, as the cycle ends at solar minimum, the magnetic field reverses: In the subsequent solar cycle, the field lines connecting pairs of sunspots point in the opposite direction.

The magnetic reversals hint that the sunspot cycle is related to the generation of magnetic fields on the Sun. They also tell us that the *complete* magnetic cycle of the Sun (sometimes called the *solar cycle)* really averages 22 years, since it takes two 11-year cycles before the magnetic field is back the way it started.

The leading model of the sunspot cycle ties it to a combination of convection and the Sun's rotation. The Sun rotates faster at its equator than at its poles. Observations of sunspots and Doppler shift measurements tell us that the Sun's equator rotates in about 25 days, and the rotation period increases with latitude to about 30 days near the poles. This difference in rotation speed stretches and shapes the Sun's magnetic fields.

THINK ABOUT IT Suppose you take a photograph of the Sun and notice two sunspots: one near the equator and one directly north of it at higher latitude. If you look again at the Sun in a few days, would you still find one sunspot directly north of the other? Why or why not? Explain how you could use this type of observation to learn how the Sun rotates.

Imagine what happens to magnetic field lines that start out running along the Sun's surface from south to north (Figure 10.19). At the equator, the lines circles the Sun every 25 days, but at higher latitudes they lag behind. As a result, the lines gradually get wound more and more tightly around the Sun. This process, operating at all times over the entire Sun, produces the contorted field lines that generate sunspots and other solar activity.

The detailed behavior of these magnetic fields is quite complex, so scientists attempt to study it with sophisticated computer models. Using these models, scientists have successfully replicated many features of the sunspot cycle, including changes in the number and latitude of sunspots and the magnetic field reversals that occur about every 11 years. However, much still remains mysterious, including why the period of the sunspot cycle varies and why solar activity is different from one cycle to the next.

Figure 10.19

The Sun rotates more quickly at its equator than it does near its poles. Because gas circles the Sun faster at the equator, it drags the Sun's north-south magnetic field lines into a more twisted configuration. The magnetic field lines linking pairs of sunspots, depicted here as green and black blobs, trace out the directions of these stretched and distorted field lines.

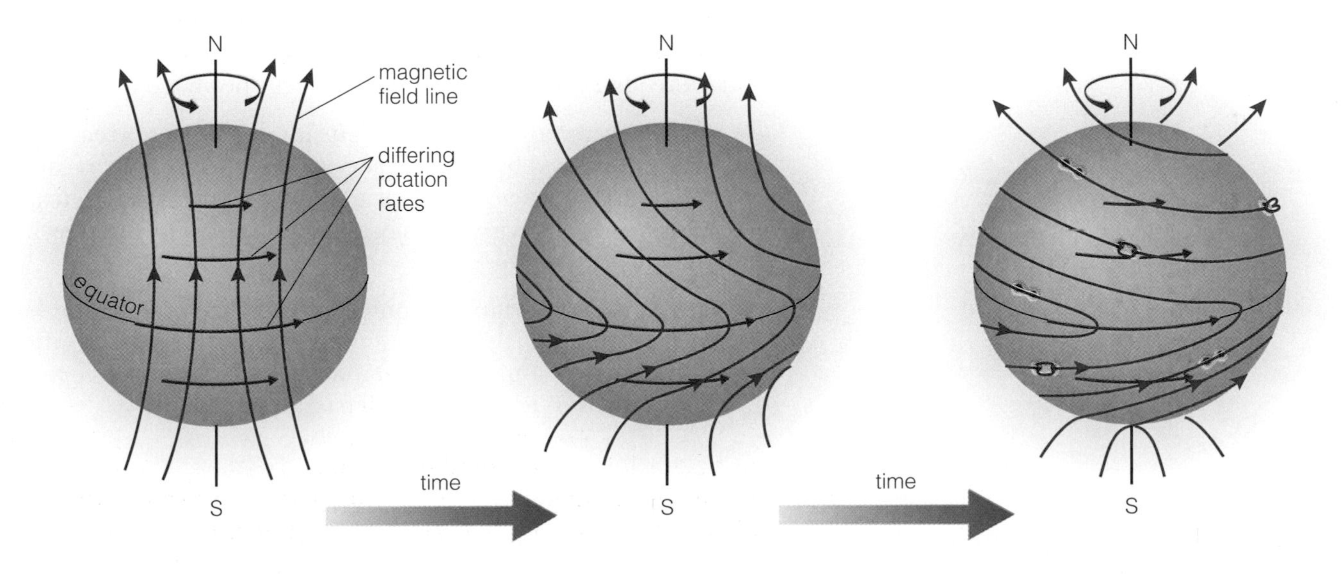

The Sunspot Cycle and Earth's Climate Despite the changes that occur during the sunspot cycle, the Sun's total output of energy barely changes at all—the largest measured changes have been less than 0.1 percent of the Sun's average luminosity. However, the ultraviolet and X-ray output of the Sun, which comes from the magnetically heated gas of the chromosphere and corona, can vary much more significantly. Could any of these changes affect the weather or climate on Earth?

Some data suggest connections between solar activity and Earth's climate. For example, the period from 1645 to 1715, when solar activity seems to have virtually ceased, was a time of exceptionally low temperatures in Europe and North America known as the *Little Ice Age*. However, no one really knows whether the low solar activity caused these low temperatures, or whether it was just a coincidence. Similarly, some researchers have claimed that certain weather phenomena, such as drought cycles or frequencies of storms, are correlated with the 11- or 22-year cycle of solar activity. A few scientists even claim that global warming [Section 7.5] has been caused more by changes in the Sun than by emissions of greenhouse gases. Again, the data supporting these claims are weak, though it is still possible that they are correct. The bottom line is that we do not yet know whether or how much solar activity affects weather on Earth. Given the importance of the question, it is certain to remain a hot topic of scientific research.

THE BIG PICTURE

Putting Chapter 10 into Context

In this chapter, we have examined our Sun, the nearest star. When you look back at this chapter, make sure you understand these "big picture" ideas:

- The ancient riddle of why the Sun shines is now solved. The Sun shines with energy generated by fusion of hydrogen into helium in the Sun's core. After a journey through the solar interior lasting several hundred thousand years and an 8-minute journey through space, a small fraction of this energy reaches Earth and supplies sunlight and heat.
- Gravitational equilibrium, the balance between pressure and gravity, determines the Sun's interior structure and helps create a natural thermostat that keeps the fusion rate steady in the Sun. If the Sun were not so steady, life on Earth might not have been possible.
- The Sun's atmosphere displays its own version of weather and climate, governed by solar magnetic fields. Some solar weather, such as coronal mass ejections, clearly affects Earth's magnetosphere. Other claimed connections between solar activity and Earth's climate may or may not be real.
- The Sun is important not only as our source of light and heat, but also because it is the only star near enough for us to study in great detail. In the coming chapters, we will use what we've learned about the Sun to help us understand other stars.

SUMMARY OF KEY CONCEPTS

10.1 A CLOSER LOOK AT THE SUN

- **Why does the Sun shine?**

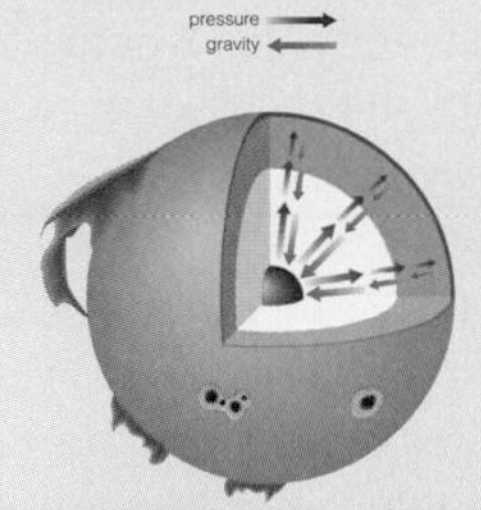

The Sun shines because **gravitational equilibrium** keeps its core hot and dense enough to release energy through nuclear fusion. The core originally became hot through the release of energy by **gravitational contraction**, as gravity made the Sun's birth cloud contract.

- **What is the Sun's structure?**

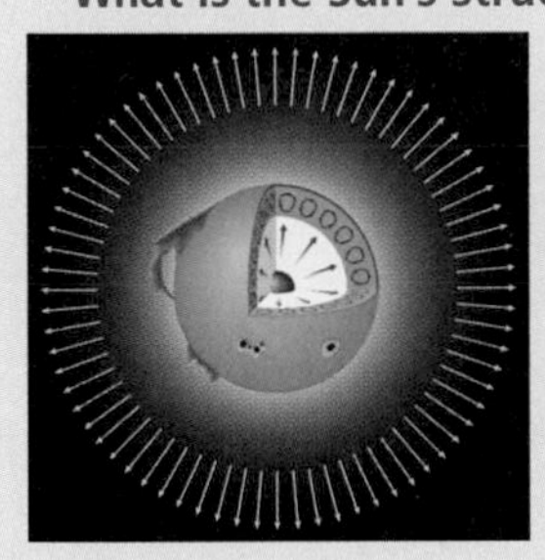

The Sun's interior layers, from the inside out, are the **core**, the **radiation zone**, and the **convection zone**. Atop the convection zone lies the **photosphere**, the layer from which photons can freely escape into space. Above the photosphere, which is essentially the surface of the Sun, are the **chromosphere** and the **corona**.

10.2 NUCLEAR FUSION IN THE SUN

- **How does nuclear fusion occur in the Sun?**

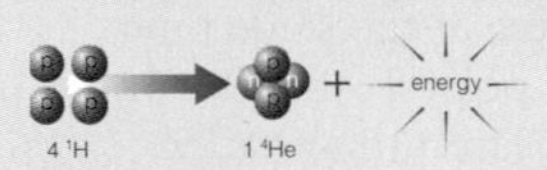

The core's extreme temperature and density are just right for fusion of hydrogen into helium, which occurs via the **proton–proton chain**. Because the fusion rate is so sensitive to temperature, gravitational equilibrium acts as a thermostat that keeps the fusion rate steady.

- **How does the energy from fusion get out of the Sun?**

Energy moves through the deepest layers of the Sun—the core and the radiation zone—in the form of randomly bouncing photons. After energy emerges from the radiation zone, convection carries it the rest of the way to the photosphere, where it is radiated into space as sunlight. Energy produced in the core takes hundreds of thousands of years to reach the photosphere.

- **How do we know what is happening inside the Sun?**

We can construct theoretical models of the solar interior using known laws of physics and then check the models against observations of the Sun's size, surface temperature, and energy output. We also use studies of "sun quakes" and solar **neutrinos**.

10.3 THE SUN–EARTH CONNECTION

- **What causes solar activity?**

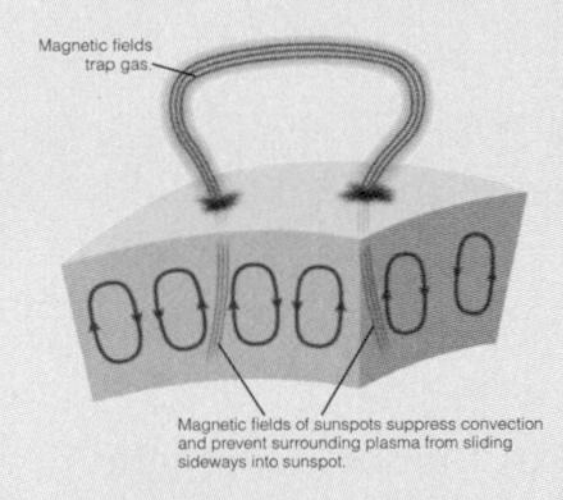

Convection combined with the rotation pattern of the Sun—faster at the equator than at the poles—causes **solar activity** because these gas motions stretch and twist the Sun's magnetic field. These contortions of the magnetic field are responsible for phenomena such as **sunspots**, **flares**, **prominences**, and **coronal mass ejections**.

- **How does solar activity affect humans?**

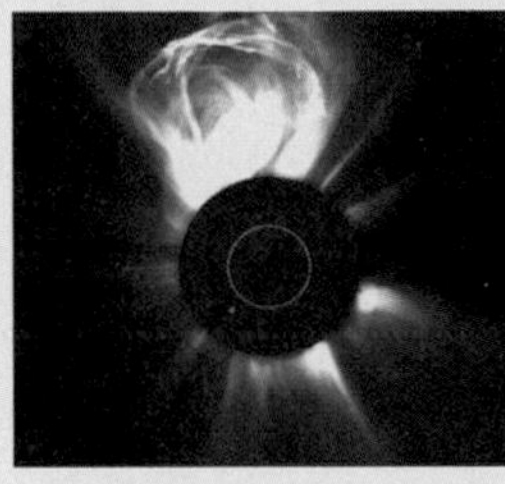

Bursts of charged particles ejected from the Sun during periods of high solar activity can hamper radio communications, disrupt electrical power generation, and damage orbiting satellites.

- **How does solar activity vary with time?**

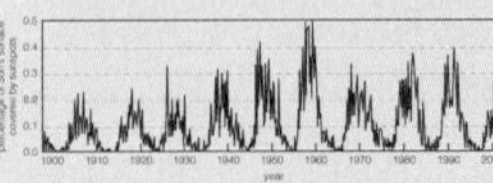

The **sunspot cycle**, or the variation in the number of sunspots on the Sun's surface, has an average period of 11 years. The magnetic field flip-flops every 11 years or so, resulting in a 22-year magnetic cycle. Sunspots first appear at mid-latitudes at solar minimum, then become increasingly more common near the Sun's equator as the next minimum approaches.

EXERCISES AND PROBLEMS

REVIEW QUESTIONS

1. Briefly describe how *gravitational contraction* generates energy. When was it important in the Sun's history? Explain.
2. What two forces are balanced in *gravitational equilibrium?* Describe how gravitational equilibrium makes the Sun hot and dense in its core.
3. State the Sun's luminosity, mass, radius, and average surface temperature, and put the numbers into perspective that makes them meaningful.
4. Briefly describe the distinguishing features of each of the layers of the Sun shown in Figure 10.3.
5. What is the difference between nuclear *fission* and nuclear *fusion?* Which one is used in nuclear power plants? Which one is used by the Sun?
6. Why does nuclear fusion require high temperatures and pressures?
7. What is the overall nuclear fusion reaction in the Sun? Briefly describe the proton–proton chain.
8. Does the Sun's fusion rate remain steady or vary wildly? Describe the feedback process that regulates the fusion rate.
9. Why does the energy produced by fusion in the solar core take so long to reach the solar surface? Describe the processes by which energy generated by fusion makes its way to the Sun's surface.
10. Explain how mathematical models allow us to predict conditions inside the Sun. How can we be confident that the models are on the right track?
11. What are *neutrinos?* What was the *solar neutrino problem,* and why do we think it has now been solved?
12. What do we mean by *solar activity?* Describe some of the features of solar activity, including *sunspots, solar prominences, solar flares,* and *coronal mass ejections.*
13. Describe the appearance and temperature of the Sun's photosphere. Why does the surface look mottled? How are sunspots different from the surrounding photosphere?
14. How do magnetic fields keep sunspots cooler than the surrounding plasma? Explain.
15. Why is the chromosphere best viewed with ultraviolet telescopes? Why is the corona best viewed with X-ray telescopes? Briefly explain how we think the chromosphere and corona are heated.
16. What is the sunspot cycle? Describe how the Sun changes during the cycle, and how the changes are thought to be related to magnetic fields and the Sun's rotation. Does the sunspot cycle influence Earth's climate? Explain.

SENSIBLE STATEMENTS?

Decide whether each of the following statements is sensible and explain why it is or is not.

17. Before Einstein, gravitational contraction appeared to be a perfectly plausible mechanism for solar energy generation.
18. A sudden temperature rise in the Sun's core is nothing to worry about, because conditions in the core will soon return to normal.
19. If fusion in the solar core ceased today, worldwide panic would break out tomorrow as the Sun began to grow dimmer.
20. Astronomers have recently photographed magnetic fields churning deep beneath the solar photosphere.
21. Neutrinos probably can't harm me, but just to be safe I think I'll wear a lead vest.
22. If you want to see lots of sunspots, just wait for solar maximum!
23. News of a major solar flare today caused concern among professionals in the fields of communications and electrical power generation.
24. By observing solar neutrinos, we can learn about nuclear fusion deep in the Sun's core.
25. If the Sun's magnetic field somehow disappeared, there would be no more sunspots on the Sun.
26. Scientists are currently building an infrared telescope designed to observe fusion reactions in the Sun's core.

PROBLEMS

(Quantitative problems are marked with an asterisk.)

27. *An Angry Sun.* A *Time* magazine cover once suggested that an "angry Sun" was becoming more active as human activity changed Earth's climate through global warming. It's certainly possible for the Sun to become more active at the same time that humans are affecting Earth, but is it possible that the Sun could be responding to human activity? Can humans affect the Sun in any significant way? Explain.

*28. *The Lifetime of the Sun.* The total mass of the Sun is about 2×10^{30} kg, of which about 75% was hydrogen when the Sun formed. However, only about 13% of this hydrogen ever becomes available for fusion in the core. The rest remains in layers of the Sun where the temperature is too low for fusion.
   a. Based on the given information, calculate the total mass of hydrogen available for fusion over the lifetime of the Sun.
   b. Combine your results from part (a) and the fact that the Sun fuses about 600 billion kilograms of hydrogen each second to calculate how long the Sun's initial supply of hydrogen can last. Give your answer in both seconds and years.
   c. Given that our solar system is now about 4.6 billion years old, when will we need to worry about the Sun running out of hydrogen for fusion?

*29. *Solar Power Collectors.* This problem leads you through the calculation and discussion of how much solar power can be collected by solar cells on Earth.
   a. Imagine a giant sphere surrounding the Sun with a radius of 1 AU. What is the surface area of this sphere, in square meters? (*Hint:* The formula for the surface area of a sphere is $4\pi r^2$.)

b. Because this imaginary giant sphere surrounds the Sun, the Sun's entire luminosity of 3.8×10^{26} watts must pass through it. Calculate the power passing through each square meter of this imaginary sphere in *watts per square meter.* Explain why this number represents the maximum power per square meter that can be collected by a solar collector in Earth orbit.
c. List several reasons why the average power per square meter collected by a solar collector on the ground will always be less than what you found in part (b).
d. Suppose you want to put a solar collector on your roof. If you want to optimize the amount of power you can collect, how should you orient the collector? (*Hint:* The optimum orientation depends on both your latitude and the time of year and day.)

DISCUSSION QUESTIONS

30. *The Role of the Sun.* Briefly discuss how the Sun affects us here on Earth. Be sure to consider not only factors such as its light and warmth, but also how the study of the Sun has led us to new understanding in science and to technological developments. Overall, how important has solar research been to our lives?
31. *The Sun and Global Warming.* One of the most pressing environmental issues on Earth concerns the extent to which human emissions of greenhouse gases are warming our planet. Some people claim that part or all of the observed warming over the past century may be due to changes on the Sun, rather than to anything humans have done. Discuss how a better understanding of the Sun might help us understand the threat posed by greenhouse gas emissions. Why is it so difficult to develop a clear understanding of how the Sun affects Earth's climate?

MATH HELP AND EXERCISES

For additional help and practice with mathematical concepts applicable to this chapter, the Astronomy Place Web site has the following resources with detailed explanations, worked examples, and practice problems.

- Mass–Energy Conversion in the Sun
- Laws of Thermal Radiation

For a complete list of media resources available, go to **www.astronomyplace.com** and choose Chapter 10 from the pull-down menu.

ASTRONOMY PLACE WEB TUTORIALS

Tutorial Review of Key Concepts

Use the following interactive **Tutorials** at **www.astronomyplace.com** to review key concepts from this chapter.

The Sun Tutorial

Lesson 1 Structure and Gravitational Equilibrium
Lesson 2 Fusion of Hydrogen into Helium
Lesson 3 Why Does Fusion Only Occur at High Temperatures?

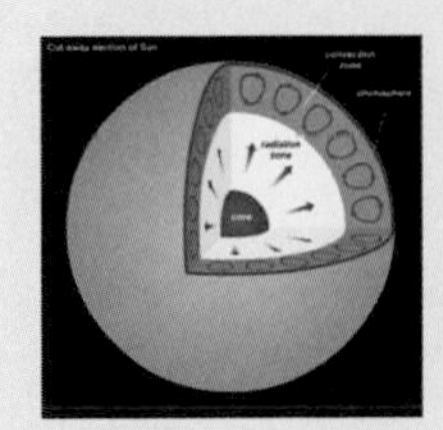
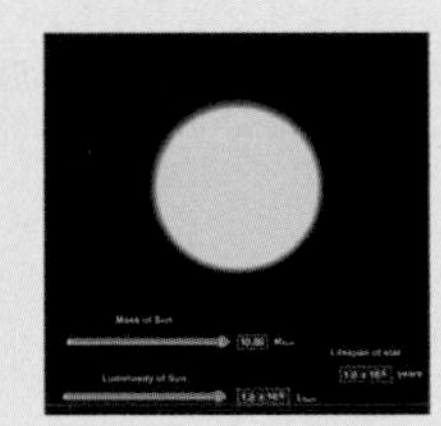
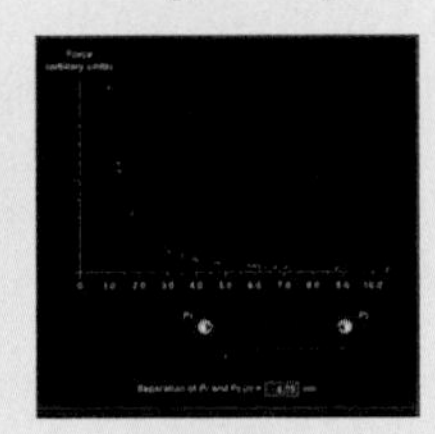

WEB PROJECTS

Take advantage of the useful Web links on **www.astronomyplace.com** to assist you with the following projects.

1. *Current Solar Weather.* Daily information about solar activity is available at NASA's Web site sunspotcycle.com. Where are we in the sunspot cycle right now? When is the next solar maximum or minimum expected? Have there been any major solar storms in the past few months? If so, did they have any significant effects on Earth? Summarize your findings in a one- to two-page report.
2. *Solar Observatories in Space.* Visit NASA's Web site for the Sun–Earth connection and explore some of the current and planned space missions designed to observe the Sun. Choose one mission to study in greater depth, and write a one- to two-page report on the mission status and goals and what it has taught or will teach us about the Sun.
3. *Sudbury Neutrino Observatory.* Visit the Web site for the Sudbury Neutrino Observatory (SNO) and learn how it has helped to solve the solar neutrino problem. Write a one- to two-page report describing the observatory, any recent results, and what we can expect from it in the future.

11
Surveying the Stars

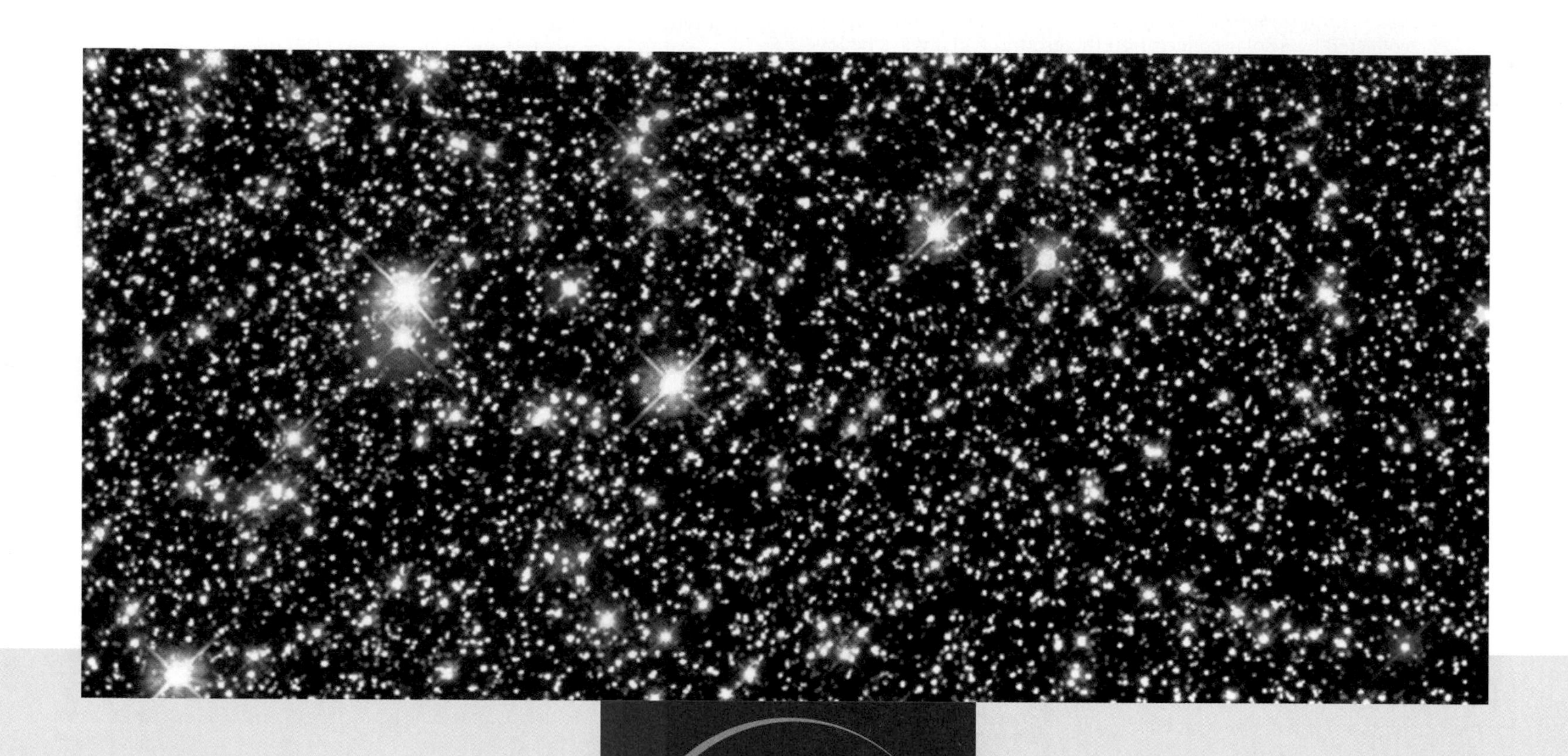

LEARNING GOALS

11.1 PROPERTIES OF STARS
- How luminous are stars?
- How hot are stars?
- How massive are stars?

11.2 CLASSIFYING STARS
- How do we classify stars?
- Why is a star's mass its most important property?
- What is a Hertzsprung–Russell diagram?

11.3 STAR CLUSTERS
- What are the two types of star clusters?
- How do we measure the age of a star cluster?

On a clear, dark night, a few thousand stars are visible to the naked eye. Many more become visible through binoculars, and a powerful telescope reveals so many stars that we could never hope to count them. Like each individual person, each individual star is unique. Like the human family, all stars share much in common.

Today, we know that stars are born from clouds of interstellar gas, shine brilliantly by nuclear fusion for millions or billions of years, and then die, sometimes in dramatic ways. This chapter outlines how we study and categorize stars and how we have come to realize that stars, like people, change over their lifetimes.

11.1 PROPERTIES OF STARS

Imagine that an alien spaceship flies by Earth on a simple but short mission: The visitors have just 1 minute to learn everything they can about the human race. In 60 seconds, they will see next to nothing of any individual person's life. Instead, they will obtain a collective "snapshot" of humanity showing people from all stages of life engaged in their daily activities. From this snapshot alone, they must piece together their entire understanding of human beings and their lives, from birth to death.

We face a similar problem when we look at the stars. Compared with stellar lifetimes of millions or billions of years, the few hundred years humans have spent studying stars with telescopes is rather like the aliens' 1-minute glimpse of humanity. We see only a brief moment in any star's life, and our collective snapshot of the heavens consists of such frozen moments for billions of stars. From this snapshot, we try to reconstruct the life cycles of stars.

Thanks to the efforts of hundreds of astronomers studying this snapshot of the heavens, stars are no longer mysterious points of light in the sky. We now know that all stars share a lot in common with the Sun. They all form in great clouds of gas and dust, and each one begins its life with roughly the same chemical composition as the Sun. Three-quarters of a star's mass at birth is hydrogen, and about one-quarter is helium, with no more than about 2% consisting of elements heavier than helium. Nevertheless, stars are not all the same; they differ in such properties as size, age, brightness, and temperature. We'll devote most of this and the next chapter to understanding how and why stars differ. First, however, let's explore how we measure three of the most fundamental properties of stars: luminosity, surface temperature, and mass.

• How luminous are stars?

If you go out on any clear night, you'll immediately see that stars vary in brightness. Some stars are so bright that we can use them to identify constellations [Section 2.1]. Others are so dim that our naked eyes cannot see them at all. However, these differences in brightness do not by themselves tell us anything about how much light these stars are really generating, because the brightness of a star depends on its distance as well as on how much light it actually emits. For example, the stars Procyon and Betelgeuse, which make up two of the three corners of the winter triangle (see Figure 2.2), appear about equally bright in our sky. However,

Betelgeuse actually emits about 5,000 times as much light as Procyon. It has about the same brightness in our sky because it is much farther away.

Because two similar-looking stars can be generating very different amounts of light, we need to distinguish clearly between a star's brightness as it appears in our sky and the actual amount of light that a star emits into space:

- When we talk about how bright stars look in our sky, we are talking about **apparent brightness**—the brightness of a star as it appears to our eyes.
- When we talk about how bright stars are in an absolute sense, regardless of their distance, we are talking about **luminosity**—the total amount of light that a star emits into space.

A star's *apparent brightness* in the sky depends on both its true light output, or *luminosity*, and its distance.

It's easy to understand the difference between the two meanings by thinking about a 100-watt light bulb. The bulb always puts out the same amount of light, so its luminosity doesn't vary. However, its apparent brightness depends on your distance from the bulb: It will look quite bright if you stand very close to it, but quite dim if you are far away.

The Inverse Square Law for Light The apparent brightness of a star or any other light source obeys an *inverse square law* with distance, much like the inverse square law that describes the force of gravity [Section 4.4]. For example, if we viewed the Sun from twice Earth's distance, it would appear dimmer by a factor of $2^2 = 4$. If we viewed it from 10 times Earth's distance, it would appear $10^2 = 100$ times dimmer.

Figure 11.1 shows why apparent brightness follows an inverse square law. The same total amount of light must pass through each imaginary sphere surrounding the star. If we focus on the light passing through the small square on the sphere located at 1 AU, we see that the same amount of light must pass through *four* squares of the same size on the sphere located at 2 AU. Thus, each square on the sphere at 2 AU receives only $\frac{1}{2^2} = \frac{1}{4}$ as much light as the square on the sphere at 1 AU. Similarly, the same amount of light passes through *nine* squares of the same size on the sphere located at 3 AU. Thus, each of these squares receives only $\frac{1}{3^2} = \frac{1}{9}$ as much light as the square on the sphere at 1 AU. Generalizing, we see that the amount of light received per unit area decreases with increasing distance by the square of the distance—an inverse square law.

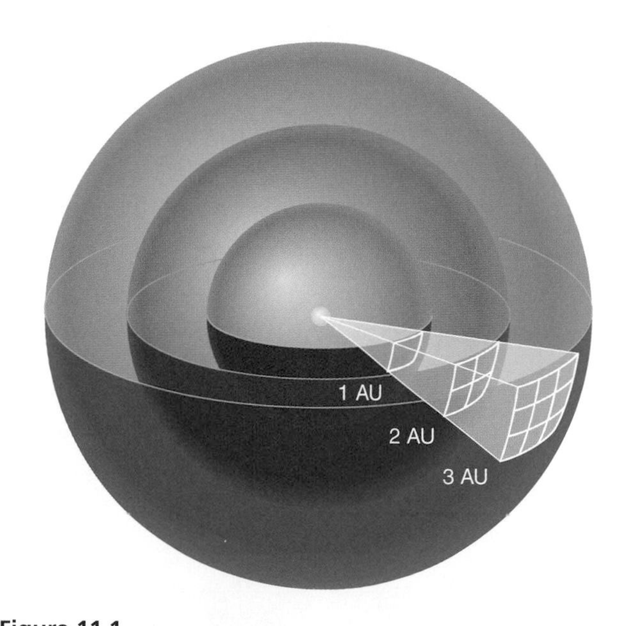

Figure 11.1

The inverse square law for light. At greater distances from a star, the same amount of light passes through an area that gets larger with the square of the distance. The amount of light per unit area therefore declines with the square of the distance.

Doubling the distance to a star would decrease its apparent brightness by a factor of 2^2, or 4.

This inverse square law leads to a very simple and important formula relating the apparent brightness, luminosity, and distance of any light source. We will call it the **inverse square law for light**:

$$\text{apparent brightness} = \frac{\text{luminosity}}{4\pi \times (\text{distance})^2}$$

Because the standard units of luminosity are watts [Section 10.1], the units of apparent brightness are *watts per square meter.* (The 4π in the formula for the inverse square law for light comes from the fact that the surface area of a sphere is given by $4\pi \times (\text{radius})^2$.)

In principle, we can always determine a star's apparent brightness by carefully measuring the amount of light we receive from the star per square meter. However, before we can use the inverse square law for light to calculate the star's luminosity, we must also know its distance.

THINK ABOUT IT Suppose Star A is four times as luminous as Star B. How will their apparent brightnesses compare if they are both the same distance from Earth? How will their apparent brightnesses compare if Star A is twice as far from Earth as Star B? Explain.

Measuring Cosmic Distances Tutorial, Lesson 2

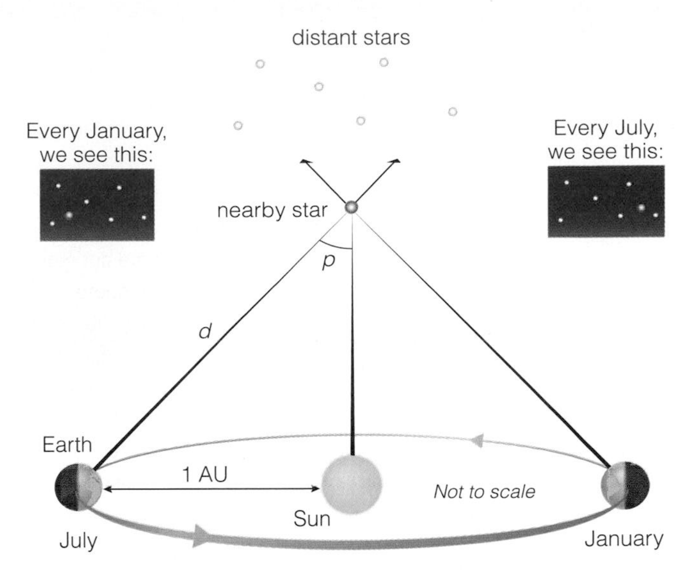

Figure 11.2 Interactive Figure

Parallax makes the apparent position of a nearby star shift back and forth with respect to distant stars over the course of each year. The angle *p*, called the *parallax angle*, represents half the total parallax shift each year. If we measure *p* in arcseconds, the distance *d* to the star in parsecs is $1/p$. The angle in this figure is greatly exaggerated: All stars have parallax angles of less than 1 arcsecond.

Measuring Distance Through Stellar Parallax The most direct way to measure the distances to stars is with *stellar parallax,* the small annual shifts in a star's apparent position caused by Earth's motion around the Sun [Section 2.4]. Recall that you can observe parallax of your finger by holding it at arm's length and looking at it alternately with first one eye closed and then the other. Astronomers measure stellar parallax by comparing observations of a nearby star made 6 months apart (Figure 11.2). The nearby star appears to shift against the background of more distant stars because we are observing it from two opposite points of Earth's orbit.

We can measure the distance to a nearby star by observing how its apparent location shifts as Earth orbits the Sun.

Astronomers can calculate a star's distance if they can measure the precise amount of the star's annual shift due to parallax. This means measuring the angle *p* in Figure 11.2, which we call the star's *parallax angle.* Notice that this angle would be smaller if the star were farther away, so we conclude that more distant stars have *smaller* parallax angles.

In fact, all stars are so far away that they have very small parallax angles, which explains why the ancient Greeks were never able to measure parallax with their naked eyes. Even the nearest stars have parallax angles smaller than 1 arcsecond—well below the approximately 1 arcminute angular resolution of the naked eye [Section 5.3]. For increasingly distant stars, the parallax angles quickly become too small to measure even with our highest-resolution telescopes. Current technology allows us to measure parallax accurately only for stars within a few hundred light-years—not much farther than what we call our *local solar neighborhood* in the vast, 100,000-light-year-diameter Milky Way Galaxy.

By definition, the distance to an object with a parallax angle of 1 arcsecond is 1 **parsec** (pc). (The word *parsec* comes from the words *parallax* and *arcsecond.*) Because all stars have parallax angles smaller than one arcsecond, they are all farther than 1 parsec away. In fact, with a bit of geometry and Figure 11.2, it's possible to show that a parsec is equivalent to 3.26 light-years. You'll often hear astronomers state distances in parsecs, kiloparsecs (1,000 parsecs), or megaparsecs (1 million parsecs). However, in this book we'll stick to giving distances in light-years.

Parallax was the first technique astronomers developed for measuring distances to stars, and it remains the only technique that tells us distances directly, without any assumptions about the nature of stars. If we know a star's distance from parallax, we can calculate its luminosity with the inverse square law for light. We have parallax measurements for thousands of stars, which is a large enough number so that astronomers have been able to draw some general conclusions about stars. As we'll

COMMON MISCONCEPTIONS

PHOTOS OF STARS

Photographs of stars, star clusters, and galaxies convey a great deal of information, but they also contain a few artifacts that are not real. For example, different stars seem to have different sizes in photographs such as that in Figure 11.3, but stars are so far away that they should all appear as mere points of light. The sizes are an artifact of how our instruments record light. Bright stars tend to be overexposed in photographs, making them appear larger in size than dimmer stars. Overexposure also explains why the centers of globular clusters and galaxies usually look like big blobs in photographs: The central regions of these objects contain many more stars than the outskirts, and the combined light of so many stars tends to get overexposed to make a big blob.

Spikes around bright stars in photographs, often making the pattern of a cross with a star at the center, are another such artifact. Again, you can see these spikes around many of the brightest stars in Figure 11.3. These spikes are not real but rather are created by the interaction of starlight with the supports holding the secondary mirror in the telescope [Section 5.3]. The spikes generally occur only with point sources of light like stars, and not with larger objects like galaxies. When you look at a photograph showing many galaxies (for example, Figure 15.1), you can tell which objects are stars by looking for the spikes.

see later, these lessons have taught astronomers how to estimate luminosities for many more stars, even without first knowing their distances. Thus, astronomers today often use the inverse square law for light to calculate distances to objects for which we can reliably estimate luminosities, as well as for calculating luminosities of objects for which we have measured distances.

The True Luminosities of Stars Now that we have discussed how we determine stellar luminosities, it's time to take a quick look at the results. We usually state stellar luminosities in comparison to the Sun's luminosity, which we write as L_{Sun} for short. For example, Proxima Centauri, the nearest of the three stars in the Alpha Centauri system and hence the nearest star besides our Sun, is only about 0.0006 times as luminous as the Sun, or $0.0006 L_{Sun}$. Betelgeuse, the bright left-shoulder star of Orion, has a luminosity of $38{,}000 L_{Sun}$, meaning that it is 38,000 times as luminous as the Sun. Overall, studies of the luminosities of many stars have taught us two particularly important lessons:

- Stars come in a wide range of luminosities, with our Sun somewhere in the middle. The dimmest stars have luminosities $^1/_{10{,}000}$ times that of the Sun ($10^{-4}\ L_{Sun}$), while the brightest stars are about 1 million times as luminous as the Sun ($10^6\ L_{Sun}$).
- Dim stars are far more common than bright stars. For example, even though our Sun is roughly in the middle of the overall range of stellar luminosities, it is brighter than the vast majority of stars in our galaxy.

Later we'll see why the luminosities of stars span such a wide range, but for now let's continue on to learn how we measure some other fundamental properties of stars.

• How hot are stars?

The question "How hot are stars?" sounds simple, but we must be careful about what we mean by a star's *temperature*. Recall that the Sun's temperature varies throughout its interior, from some 15 million K in its core to about 5,800 K at its surface (photosphere). However, only the Sun's surface temperature is directly measurable. Our knowledge of the Sun's inside

SPECIAL TOPIC: THE MAGNITUDE SYSTEM

Many amateur and professional astronomers describe stellar brightness using the ancient **magnitude system** devised by the Greek astronomer Hipparchus (c. 190–120 B.C.). The magnitude system originally classified stars according to how bright they look to human eyes—the only instruments available in ancient times. The brightest stars received the designation "first magnitude," the next brightest "second magnitude," and so on. The faintest visible stars were magnitude 6. We call these descriptions **apparent magnitudes** because they compare how bright different stars *appear* in the sky. Star charts (such as those in Appendix J) often use dots of different sizes to represent the apparent magnitudes of stars.

In modern times, the magnitude system has been extended and more precisely defined. As a result, stars can have fractional apparent magnitudes, and a few bright stars have apparent magnitudes *less than* 1—which means *brighter* than magnitude 1. For example, the brightest star in the night sky, Sirius, has an apparent magnitude of −1.46. Appendix F gives apparent magnitudes and actual luminosities for both the nearest stars and the brightest stars visible in the sky.

The modern magnitude system also defines **absolute magnitudes** as a way of describing stellar luminosities. A star's absolute magnitude is the apparent magnitude it would have *if* it were at a distance of 10 parsecs (32.6 light-years) from Earth. For example, the Sun's absolute magnitude is about 4.8, meaning that the Sun would have an apparent magnitude of 4.8 *if* it were 10 parsecs away from us—bright enough to be visible, but not conspicuous, on a dark night.

Although many articles and books still quote apparent and absolute magnitudes, comparisons between stars are much easier when we think about how the inverse square law for light makes apparent brightness depend on distance and luminosity. Thus, we'll stick to the inverse square law in this book.

temperatures comes from mathematical models of the solar interior [Section 10.2]. The same is true for other stars: We can measure only their surface temperatures, and must use models to learn about their interior temperatures. Thus, when astronomers speak of the "temperature" of a star, they usually mean the surface temperature.

Measuring a star's surface temperature is somewhat easier than measuring its luminosity, because the star's distance doesn't affect the measurement. Instead, we determine surface temperature directly from either the star's color or its spectrum. Let's briefly investigate how each technique works.

Color and Temperature Take a careful look at Figure 11.3. Notice that stars come in almost every color of the rainbow. Simply looking at the colors tells us something about the surface temperatures of the stars. For example, a red star is cooler than a yellow star, which in turn is cooler than a blue star.

Stars come in different colors because they emit thermal radiation [Section 5.2]. Recall that a thermal radiation spectrum depends only on the (surface) temperature of the object that emits it (see Figure 5.10). For example, the Sun's 5,800 K temperature causes it to emit most strongly in the middle of the visible portion of the spectrum, which is why the Sun looks yellow or white in color. A cooler star, such as Betelgeuse (surface temperature 3,400 K), looks red because it emits much more red light than blue light. A hotter star, such as Sirius (surface temperature 9,400 K), emits a little more blue light than red light and therefore has a slightly blue color to it.

Astronomers can measure surface temperature fairly precisely by comparing a star's apparent brightness in two different colors of light. For example, by comparing the amount of blue light and red light coming from Sirius, astronomers can measure how much more blue light it emits than red light. Because thermal radiation spectra have a very distinctive shape (again, see Figure 5.10), this difference in blue and red light output allows astronomers to calculate a surface temperature.

Figure 11.3 VIS

This Hubble Space Telescope picture shows a wide variety of stars that differ in color and brightness. Most of the stars in this picture are at roughly the same distance, about 2,000 light-years from the center of our galaxy. Clouds of gas and dust obscure our view of visible light from most of our galaxy's central regions, but a gap in the clouds allows us to see the stars in this photo.

Spectral Type and Temperature A star's spectral lines provide a second way to measure its surface temperature. Moreover, because stellar colors can be affected by interstellar dust, temperatures determined from spectral lines are generally more accurate than temperatures determined from colors alone. Stars displaying spectral lines of highly ionized elements must be fairly hot, because it takes a high temperature to ionize atoms. Stars displaying spectral lines of molecules must be relatively cool, because molecules break apart into individual atoms at relatively cool temperatures. Thus, the types of spectral lines present in a star's spectrum provide a direct measure of the star's surface temperature.

Astronomers classify stars according to surface temperature by assigning a **spectral type** determined from the spectral lines present in a star's spectrum. The hottest stars, with the bluest colors, are called spectral type O, followed in order of declining surface temperature by spectral types B, A, F, G, K, and M. The time-honored mnemonic for remembering this sequence, OBAFGKM, is "Oh Be A Fine Girl/Guy, Kiss Me!" Table 11.1 summarizes the characteristics of each spectral type.

Each spectral type is subdivided into numbered subcategories (e.g., B0, B1, . . . , B9). The larger the number, the cooler the star. For example, the Sun is designated spectral type G2, which means it is slightly hotter than a G3 star but cooler than a G1 star.

TABLE 11.1 *The Spectral Sequence*

Spectral Type	*Example(s)*	*Temperature Range*	*Key Absorption Line Features*	*Brightest Wavelength (color)*	*Typical Spectrum*
O	Stars of Orion's Belt	>30,000 K	Lines of ionized helium, weak hydrogen lines	<97 nm (ultraviolet)*	
B	Rigel	30,000 K–10,000 K	Lines of neutral helium, moderate hydrogen lines	97–290 nm (ultraviolet)*	
A	Sirius	10,000 K–7,500 K	Very strong hydrogen lines	290–390 nm (violet)*	
F	Polaris	7,500 K–6,000 K	Moderate hydrogen lines, moderate lines of ionized calcium	390–480 nm (blue)*	
G	Sun, Alpha Centauri A	6,000 K–5,000 K	Weak hydrogen lines, strong lines of ionized calcium	480–580 nm (yellow)	
K	Arcturus	5,000 K–3,500 K	Lines of neutral and singly ionized metals, some molecules	580–830 nm (red)	
M	Betelgeuse, Proxima Centauri	<3,500 K	Molecular lines strong	>830 nm (infrared)	

* All stars above 6,000 K look more or less white to the human eye because they emit plenty of radiation at all visible wavelengths.

Spectra of stars show that their surface temperatures range from more than 40,000 K to less than 3,000 K, corresponding to the sequence of spectral types OBAFGKM.

The range of surface temperatures for stars is much narrower than the range of luminosities. The coolest stars, of spectral type M, have surface temperatures as low as 3,000 K. The hottest stars, of spectral type O, have surface temperatures that can exceed 40,000 K. However, the cool, red stars turn out to be much more common than hot, blue stars.

THINK ABOUT IT Invent your own mnemonic for the OBAFGKM sequence. To help get you thinking, here are two examples: (1) Only Bungling Astronomers Forget Generally Known Mnemonics; and (2) Only Business Acts For Good, Karl Marx.

• How massive are stars?

The mass of a star is more difficult to measure than its temperature or luminosity, but mass is very important. As we'll discuss later, a star's mass is its single most important property. The most dependable method for "weighing" a star relies on Newton's version of Kepler's third law [Section 4.4]. Recall that this law can be applied only when we can observe one object orbiting another, and requires that we measure both the orbital period

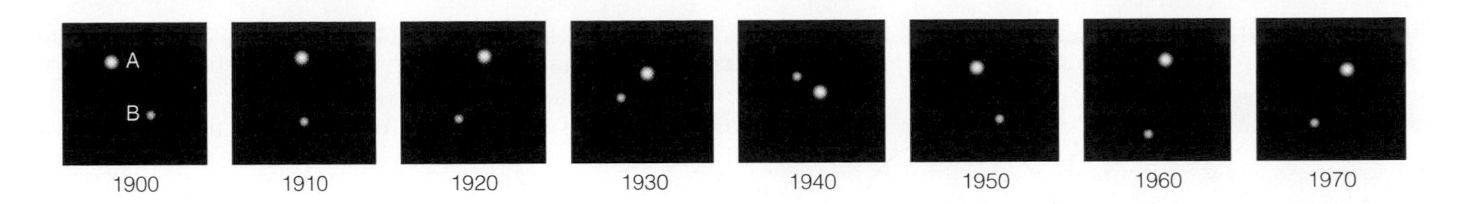

Figure 11.4

Each frame represents the relative positions of Sirius A and Sirius B at 10-year intervals from 1900 to 1970. The back-and-forth "wobble" of Sirius A allowed astronomers to infer the existence of Sirius B even before the two stars could be resolved in telescopic photos.

and the average orbital distance of the orbiting object. For stars, these requirements generally mean that we can apply the law to measure masses only in **binary star systems**—systems in which two stars continually orbit one another. Before we consider how we determine the orbital periods and distances needed to use Newton's version of Kepler's third law, let's look briefly at the different types of binary star systems that we can observe.

Types of Binary Star Systems Surveys show that about half of all stars orbit a companion star of some kind, and thus are members of binary star systems. These star systems fall into three classes:

- A *visual binary* is a pair of stars that we can see distinctly (with a telescope) as the stars orbit each other. Sometimes we observe a star slowly shifting position in the sky as if it were a member of a visual binary, but its companion is too dim to be seen. For example, slow shifts in the position of Sirius, the brightest star in the sky, revealed it to be a binary star long before its companion was discovered (Figure 11.4).
- An *eclipsing binary* is a pair of stars that orbit in the plane of our line of sight (Figure 11.5). When neither star is eclipsed, we see the combined light of both stars. When one star eclipses the other, the apparent brightness of the system drops because some of the light is blocked from our view. A *light curve,* or graph of apparent brightness against time, reveals the pattern of the eclipses. The most famous example of an eclipsing binary is Algol, the "demon star" in the constellation Perseus (*algol* is Arabic for "the ghoul"). Algol's brightness drops to only a third of its usual level for a few hours about every three days as the brighter of its two stars is eclipsed by its dimmer companion.

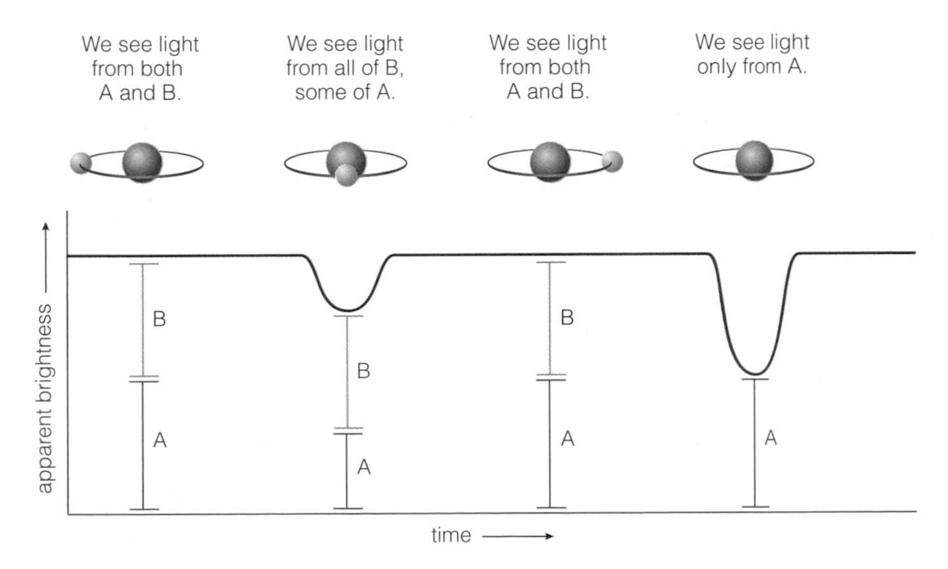

Figure 11.5

The apparent brightness of an eclipsing binary system drops when either star eclipses the other.

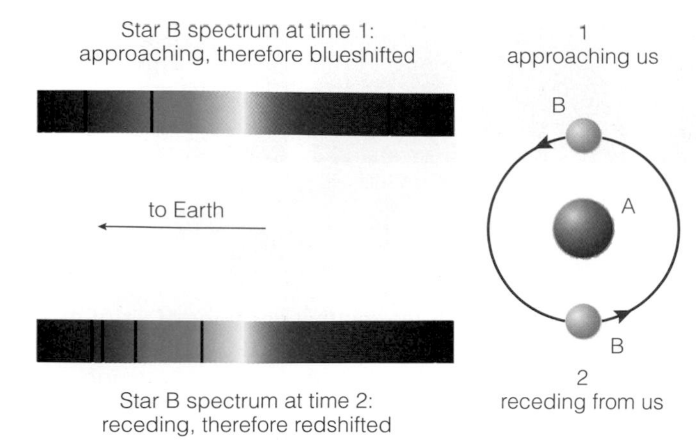

Figure 11.6 Interactive Figure

The spectral lines of a star in a binary system are alternately blueshifted as it comes toward us in its orbit and redshifted as it moves away from us.

- If a binary system is neither visual nor eclipsing, we may be able to detect its binary nature by observing Doppler shifts in its spectral lines [Section 5.2]. Such systems are called *spectroscopic binary* systems. If one star is orbiting another, it periodically moves toward us and away from us in its orbit. Its spectral lines show blueshifts and redshifts as a result of this motion (Figure 11.6). Sometimes we see two sets of lines shifting back and forth—one set from each of the two stars in the system (a *double-lined* spectroscopic binary). Other times we see a set of shifting lines from only one star because its companion is too dim to be detected (a *single-lined* spectroscopic binary).

Some star systems combine two or more of these binary types. For example, telescopic observations reveal Mizar (the second star in the handle of the Big Dipper) to be a visual binary. Spectroscopy then shows that each the two stars in the visual binary is itself a spectroscopic binary (Figure 11.7).

Measuring Masses in Binary Systems Even for a binary system, we can apply Newton's version of Kepler's third law only if we can measure both the orbital period and the separation of the two stars. Measuring orbital period is fairly easy. In a visual binary, we simply observe how long each orbit takes. In an eclipsing binary, we measure the time between eclipses. In a spectroscopic binary, we measure the time it takes the spectral lines to shift back and forth.

We can determine the masses of stars in binary systems if we can measure both their orbital period and the separation between them.

Determining the average separation of the stars in a binary system is usually much more difficult. Except in rare cases in which we can measure the separation directly, we can calculate the separation only if we know the actual orbital speeds of the stars from their Doppler shifts. Unfortunately, a Doppler shift tells us only the portion of a star's velocity that is directly toward us or away from us (see Figure 5.13). Because orbiting stars generally do not move directly along our line of sight, their actual velocities can be significantly greater than those we measure through the Doppler effect.

The exceptions are eclipsing binary stars. Because these stars orbit in the plane of our line of sight, their Doppler shifts can tell us their true orbital velocities. Eclipsing binaries are therefore particularly important to the study of stellar masses. As an added bonus, eclipsing binaries allow

Figure 11.7

Mizar looks like one star to the naked eye but is actually a system of four stars. Through a telescope Mizar appears to be a visual binary made up of two stars, Mizar A and Mizar B, that gradually change positions, indicating that they orbit every few thousand years. Moreover, each of these two "stars" is itself a spectroscopic binary, making a total of four stars. (The star Alcor appears very close to Mizar to the naked eye but does *not* orbit it.)

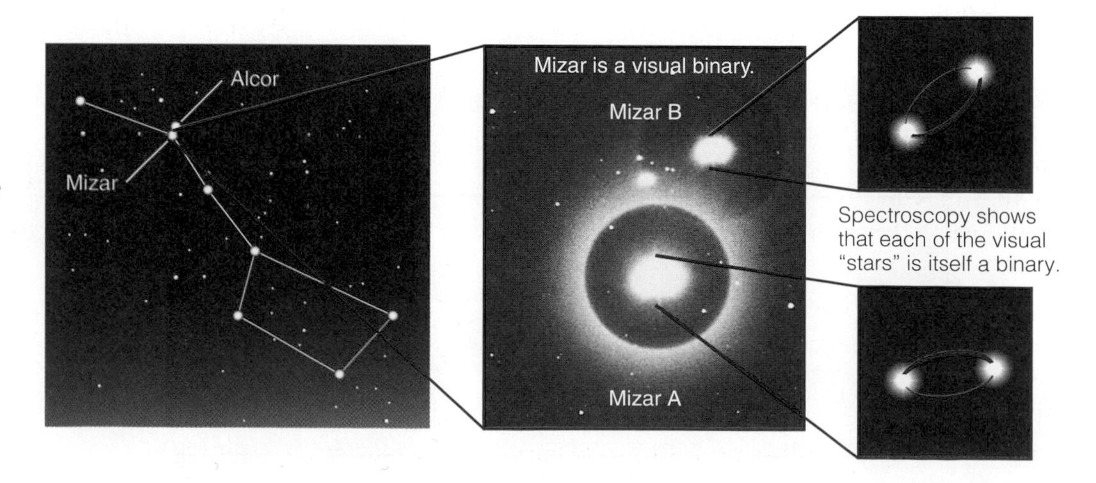

us to measure stellar radii directly. Because we know how fast the stars are moving across our line of sight as one eclipses the other, we can determine their radii by timing how long each eclipse lasts.

Through careful observations of eclipsing binaries and other binary star systems, astronomers have established the masses of many different kinds of stars. The overall range extends from as little as 0.08 times the mass of the Sun ($0.08M_{Sun}$) to about 100 times the mass of the Sun ($100M_{Sun}$). We'll discuss the reasons for that mass range in Chapter 12.

The Hertzsprung–Russell Diagram Tutorial, Lessons 1–3

11.2 CLASSIFYING STARS

We have seen that stars come in a wide range of luminosities, surface temperatures, and masses. But are these characteristics randomly distributed among stars, or can we find patterns that might tell us something about stellar lives? In this section, we'll see how the basic properties of stars are related to one another, and how astronomers have used these relationships to develop a classification system that has made it much easier for us to learn about the lives of stars.

• How do we classify stars?

Take another look at Figure 11.3 and think about how you would classify these stars. Almost all of them lie near the center of our galaxy, about 28,000 light-years from Earth. Because they are nearly all at about the same distance, we can compare their true luminosities by looking at their apparent brightnesses in the photograph. If you look closely, you might notice a couple of important patterns:

- Most of the very brightest stars are reddish in color.
- If you ignore those relatively few bright red stars, there's a general trend to the luminosities and colors among all the rest of the stars: The brighter ones are white with a little bit of blue tint, the more modest ones are similar to our Sun in color with a yellowish white tint, and the dimmest ones are barely visible specks of red.

If you remember that colors tell us about surface temperature—blue is hotter and red is cooler—you'll realize that these patterns must be telling us about relationships between surface temperature and luminosity. Let's investigate these relationships in a little more detail.

Hydrogen-Burning Stars The vast majority of the stars in Figure 11.3 follow the second of the two patterns: The brighter ones are bluer and the dimmer ones are redder. These "normal" stars are usually called **main-sequence stars** (for reasons that will become clear in the next section), and their defining characteristic is that they generate energy by fusing hydrogen into helium in their cores, just like our Sun.

All main-sequence stars generate energy by fusing hydrogen into helium in their cores. Thus, our Sun is a main-sequence star.

Because our Sun is a main-sequence star, it makes a good starting point for studying the relationship between surface temperature and luminosity among all main-sequence stars. Recall that our Sun is a star of spectral type G2. Other main-sequence stars of spectral type G are similar to our Sun in both surface temperature (around 5,800 K) and luminosity (around 1 L_{Sun}).

Main-sequence stars with cooler surface temperatures, which means those of spectral type K or M, are considerably less luminous than the Sun. For example, main-sequence stars of spectral type K have surface temperatures around 4,000 K and luminosities only about a tenth that of the Sun (0.1 L_{Sun}), while those of spectral type M typically have surface temperatures cooler than 3,500 K and luminosities as small as a thousandth that of our Sun (0.001 L_{Sun}). These dim, red stars are by far the most common kind in our galaxy, but their low luminosities make them difficult to see. In fact, the nearest star besides our Sun—Proxima Centauri—is too faint to see with the naked eye because it is one of these cool, dim, main-sequence stars of spectral type M.

The surface temperatures and luminosities of main-sequence stars are closely related: Hot main-sequence stars are much more luminous than cool ones.

At the other extreme, blue main-sequence stars produce much more light than the Sun. A main-sequence star of spectral type O has a surface temperature exceeding 30,000 K and emits more than 100,000 times as much light as our Sun (100,000 L_{Sun}). Brilliant blue stars like these are spectacular, but they are also rare. You won't find any of these in Figure 11.3, nor will you find one within a few hundred light-years of the Sun in any direction.

Giants and Supergiants The bright red stars in Figure 11.3 clearly do not follow the general pattern of the hydrogen-burning, main-sequence stars. Their red colors tell us that they have low surface temperatures, which means they would be dim if they were main-sequence stars. Instead, they are extremely bright.

The fact that these stars are cooler but much more luminous than the Sun tells us that they must be much larger in radius than the Sun. Remember that a star's surface temperature determines the amount of light it emits per unit surface area [Section 5.2]: Hotter stars emit much more light per unit surface area than cooler stars. For example, a blue star would emit far more total light than a red star of the same size. Thus, a star that is red and cool can be bright only if it has a very large surface area, which means it must be enormous in size.

Astronomers generally divide these enormous stars into two major groups: giants and supergiants. The **giants** have radii 10 to 100 times that of the Sun. Several of the brightest stars in the night sky are red giants, including Arcturus and Aldebaran (the eye of the bull in the constellation Taurus).

Supergiants are even larger. The most famous supergiant star is Betelgeuse, the left shoulder in the constellation Orion. Its red color indicates a surface temperature barely half that of the Sun (about 3,400 K), but its luminosity is 38,000 times that of the Sun. Thus, Betelgeuse must be truly enormous. In fact, the radius of Betelgeuse is roughly 500 times that of the Sun, which is equivalent to more than twice the Earth–Sun distance. If our Sun were somehow to trade places with Betelgeuse, we would suddenly find ourselves deep inside a star. Figure 11.8 compares the sizes of representative stars of different types.

Giants and supergiants clearly do not follow the relationship between surface temperature and luminosity that we observe for hydrogen-burning, main-sequence stars. As we'll discuss in the next chapter, we now know that giants and supergiants are stars nearing the ends of their lives. The supply of hydrogen fuel in the core of such a star is running out, and it responds to this fuel shortage by releasing fusion energy at a furious rate. Thus, in order to radiate away this huge amount of energy, the surface of a dying star must expand to an enormous size.

Giants and supergiants are stars that are nearing the ends of their lives.

Because giants and supergiants are so bright, we can see them even if they are not especially close to us. Many of the brightest stars visible to the naked eye are giants or supergiants. They are often identifiable by the reddish color produced by their cool surfaces. Nevertheless, giants and supergiants are considerably rarer than main-sequence stars. In our snapshot of the heavens, we catch most stars in the act of hydrogen burning and relatively few in a later stage of life.

White Dwarf Stars In addition to main-sequence stars, giants, and supergiants, there is another general class of star that is so dim as to be unnoticeable in a picture like Figure 11.3. The surfaces of these stars can be hotter than the Sun's, making them blue or white in color, but their luminosities are far lower than that of the Sun (typically $0.1\ L_{Sun}$ to $0.0001\ L_{Sun}$). These stars must be tiny in order to radiate so little light despite their high surface temperatures. Such stars are called **white dwarf** stars, and they are typically no larger in radius than our planet Earth. Nevertheless, the mass of a white dwarf is similar to that of the Sun, meaning that its matter must be compressed to an extremely high density, unlike anything found on Earth.

White dwarf stars are the cooling embers of stars that have exhausted their fuel for nuclear fusion.

What are these strange, dense objects that we call white dwarf stars? They shine because they are the hot embers of stars that have no fuel left for nuclear fusion. Because their cores no longer generate energy, they slowly cool with time, becoming dimmer until they fade from view completely. Thus, white dwarfs are essentially stars that have already died. We'll discuss how white dwarfs form in the next chapter, and discuss their bizarre nature in Chapter 13.

Luminosity Class and Stellar Classification We have found that stars generally fall into one of three basic groups: (1) main-sequence stars, such as our Sun, that generate energy by fusing hydrogen into helium in their cores; (2) stars nearing the ends of their lives that have expanded in size to become giants or supergiants as they furiously generate energy by fusion; and (3) "dead" stars, such as white dwarfs, that have finished their nuclear burning lives entirely. From the standpoint of how stars change as they live, these are the major groups. However, astronomers use a slightly more complex system of formal classification.

Stars that are still generating energy through fusion are assigned a **luminosity class,** designated with a Roman numeral. Even when we don't know a star's distance well enough to compute its luminosity, we can assign it to a general luminosity class based on details of its spectrum that depend on the star's radius. Thus, despite the name, a star's luminosity class is more closely related to its size than to its luminosity. The basic luminosity classes are I for supergiants, III for giants, and V for main-sequence stars. Stars sometimes also come in "in-between" classes. For example, some stars have radii larger than those of main-sequence stars but not quite large enough to qualify as giants. We call such stars *subgiants* and assign them luminosity class IV. Table 11.2 summarizes the luminosity classes.

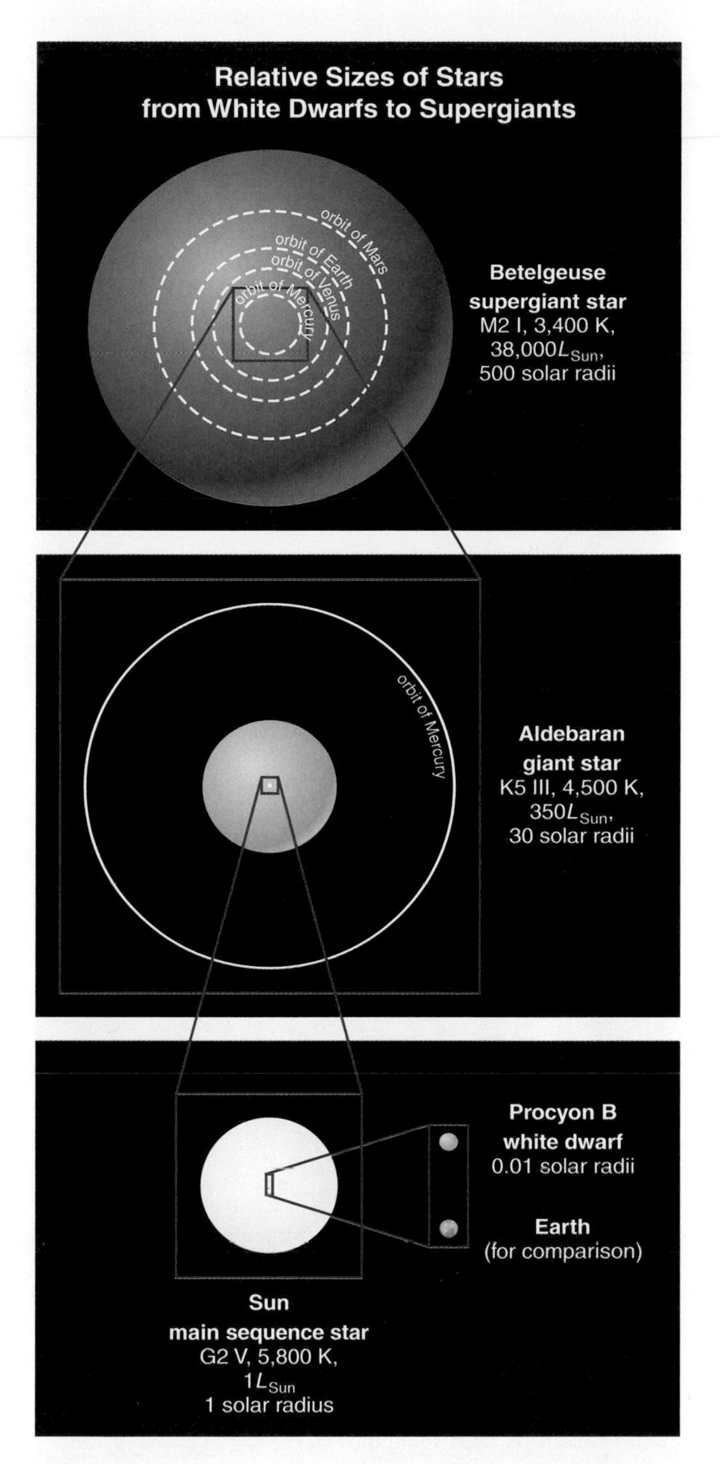

Figure 11.8

The relative sizes of stars. A supergiant like Betelgeuse would fill the inner solar system. A giant like Aldebaran would fill the inner third of Mercury's orbit. The Sun is a hundred times larger in radius than a white dwarf, which is roughly the same size as Earth.

TABLE 11.2 *Stellar Luminosity Classes*

Class	*Description*
I	Supergiants
II	Bright giants
III	Giants
IV	Subgiants
V	Main-sequence stars

Notice that we have now described two different ways of categorizing stars:

- A star's spectral type, designated by one of the letters OBAFGKM, tells us its surface temperature and color. O stars are the hottest and bluest, while M stars are the coolest and reddest.
- A star's luminosity class, designated by a Roman numeral, is based on its luminosity but also tells us about the star's radius and stage of life. Luminosity class I stars have the largest radii, with radii decreasing to luminosity class V.

The full classification of a star includes both a spectral type (OBAFGKM) and a luminosity class.

We use both spectral type and luminosity class to fully classify a star. For example, the complete classification of our Sun is G2 V. The G2 spectral type means it is yellow-white in color, and the luminosity class V means it is a hydrogen-burning, main-sequence star. Betelgeuse is M2 I, making it a red supergiant. Proxima Centauri is M5 V—similar in color and surface temperature to Betelgeuse, but far dimmer because of its much smaller size. White dwarf stars fall outside this classification system and instead are often assigned the class "wd."

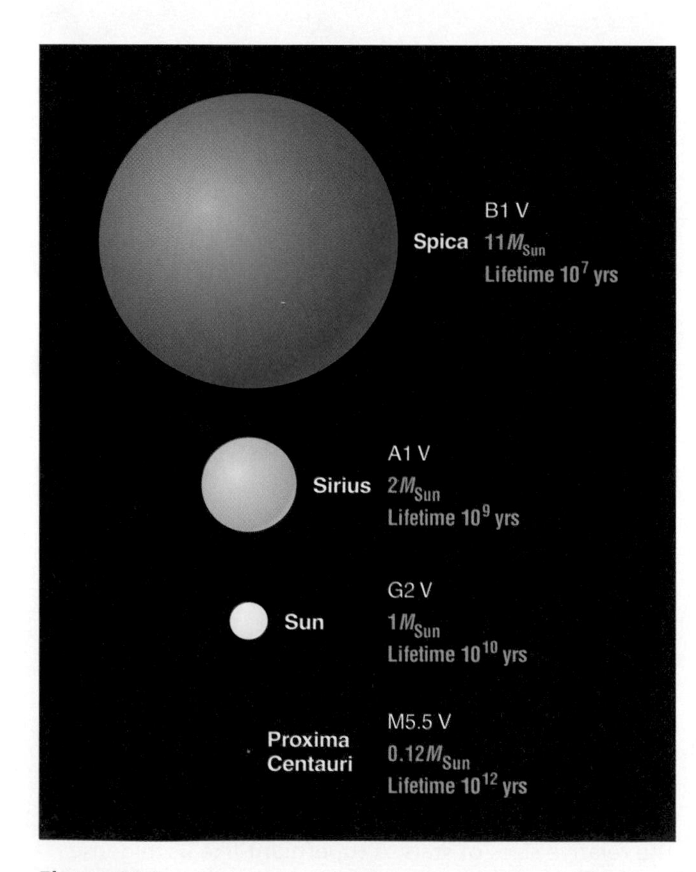

Figure 11.9

Four main-sequence stars shown to scale. The mass of a main-sequence star determines its fundamental properties of luminosity, surface temperature, radius, and lifetime. More massive main-sequence stars are hotter and brighter than less massive ones, but have shorter lifetimes.

• Why is a star's mass its most important property?

Astronomers began classifying stars by their spectral type and luminosity class before they understood *why* stars vary in these ways. Today, we know that the most fundamental property of any star is its mass. Let's investigate how mass determines other properties of stars.

Masses of Main-Sequence Stars All stars spend most of their lives as hydrogen burning, main-sequence stars. During this portion of a star's life, its mass determines both its surface temperature and luminosity. Astronomers measuring the masses of stars in binary systems have found that main-sequence stars with larger masses are always hot and bright, while main-sequence stars with small masses are always cool and dim. Figure 11.9 compares four main-sequence stars, showing how they differ because of their different masses.

A main-sequence star's mass determines both its luminosity and its surface temperature.

We can understand why mass determines luminosity by thinking about how nuclear fusion works in the Sun [Section 10.2]. Recall that nuclear fusion requires high temperatures and densities in the core, and that the Sun's internal conditions are determined by its gravitational equilibrium—a balance between the inward pull of gravity and the outward push of pressure. In a star that is more massive than the Sun, the greater weight of its overlying layers means the star must sustain a higher nuclear fusion rate to generate the additional pressure needed to maintain gravitational equilibrium. The higher nuclear fusion rate makes the star more luminous. Conversely, a star that is less massive than the Sun has less weight bearing down on its core, and hence has a lower nuclear fusion rate and lower luminosity. The great range of luminosities among main-sequence stars comes about because the nuclear fusion rate is very sensitive to mass. For example, a main-sequence star with ten times the mass of the Sun ($10\ M_{Sun}$) fuses its hydrogen at a rate about 10,000 times as fast as the Sun, making it about 10,000 times as luminous as the Sun.

The relationship between mass and surface temperature is a little subtler. In general, a very luminous star must either be very large or have a very high surface temperature, or some combination of both. The most massive main-sequence stars are thousands of times more luminous than the Sun but only about 10 times the size of the Sun in radius. Thus, their surfaces must be significantly hotter than the Sun's surface to account for their high luminosities. Main-sequence stars more massive than the Sun therefore have higher surface temperatures than the Sun, and those less massive than the Sun have lower surface temperatures.

The relationship between mass, surface temperature, and luminosity means that we can estimate a main-sequence star's mass just by knowing its spectral type. For example, any hydrogen-burning, main-sequence star that has the same spectral type as the Sun (G2) must have about the same mass and luminosity as the Sun. Similarly, any main-sequence star of spectral type B1 must have about the same mass and luminosity as Spica (see Figure 11.9). (Note that only main-sequence stars follow this simple relationship between mass, temperature, and luminosity; it does not hold for giants or supergiants.)

Stellar Lifespans A star is born with a limited supply of core hydrogen and therefore can remain as a hydrogen-fusing main-sequence star for only a limited time—the star's **hydrogen-burning lifetime**. Because stars spend the vast majority of their lives fusing hydrogen into helium, we sometimes refer to the hydrogen-burning lifetime as simply the "lifetime." Mass also determines a star's life expectancy: Massive stars have *shorter* lives than less massive stars.

Why do more massive stars have shorter lives? A star's lifetime depends on both its mass and its luminosity. Its mass determines how much hydrogen fuel the star initially contains in its core. Its luminosity determines how rapidly the star uses up its fuel. Massive stars start their lives with a larger supply of hydrogen, but they fuse this hydrogen into helium so rapidly that they end up with shorter lives. For example, a 10-solar-mass star ($10M_{\text{Sun}}$) is born with 10 times as much hydrogen as the Sun. However, its luminosity of 10,000 L_{Sun} means it burns through this hydrogen at a rate 10,000 times as fast as the rate in the Sun. Thus, because a 10-solar-mass star has only 10 times as much hydrogen and burns through it 10,000 times faster, its lifetime is only $^{1}/_{1,000}$ as long as the Sun's lifetime. Since the Sun's hydrogen-burning lifetime is about 10 billion years [Section 10.1], a 10-solar-mass star must have a lifetime of only about 10 million years. (Its actual lifetime is a little longer than this, because it can use more of its core hydrogen for fusion than can the Sun.)

The lives of more massive stars are much shorter than those of less massive stars because they quickly use up their hydrogen fuel.

On the other end of the scale, a 0.3-solar-mass main-sequence star emits a luminosity just 0.01 times that of the Sun and consequently lives roughly $0.3/0.01 = 30$ times as long as the Sun, or some 300 billion years. In a universe that is now about 14 billion years old, even the most ancient of these small, dim, red stars of spectral type M still survive and will continue to shine faintly for hundreds of billions of years to come.

A star's mass continues to control what happens to it even after all the core hydrogen has fused into helium. In the next chapter, we will look at stars' life stories in greater detail. We will see that all stars become giants once their core hydrogen is exhausted, and the amount of time a star spends in these late stages of life also depends on mass. Even near the end of life, the most massive stars always use up their fuel faster and die sooner. And at the very end, the way a star dies ultimately depends on the mass it had at birth.

• What is a Hertzsprung–Russell diagram?

All the relationships between the properties of stars that we have covered so far are summarized in Figure 11.10. This figure is an example of a Hertzsprung–Russell diagram—an **H–R diagram** for short—named after astronomers Ejnar Hertzsprung and Henry Norris Russell. Diagrams like this one are among the most important tools astronomers use to understand stars. Let's briefly examine how Figure 11.10 is constructed and why it shows so much information about stars.

Basics of the H–R Diagram All you need to know to plot a star on an H–R diagram is its luminosity and its spectral type.

- The horizontal axis represents stellar surface temperature, which, as we've discussed, corresponds to spectral type. Temperature increases *from right to left* because Hertzsprung and Russell based their diagrams on the spectral sequence OBAFGKM.
- The vertical axis represents stellar luminosity, in units of the Sun's luminosity (L_{Sun}). Stellar luminosities span a wide range, so we keep the graph compact by making each tick mark represent a luminosity 10 times as large as the prior tick mark.

An H–R diagram plots the surface temperatures of stars against their luminosities.

Each location on the diagram represents a unique combination of spectral type and luminosity. For example, the dot representing the Sun in Figure 11.10 corresponds to the Sun's spectral type, G2, and its luminosity, $1L_{Sun}$. Because luminosity increases upward on the diagram and surface temperature increases leftward, stars near the upper left are hot and luminous. Similarly, stars near the upper right are cool and luminous, stars near the lower right are cool and dim, and stars near the lower left are hot and dim.

Explain how the colors of the stars in Figure 11.10 help indicate stellar surface temperature. Do these colors tell us anything about interior temperatures? Why or why not?

The H–R diagram also provides direct information about stellar radii, because a star's luminosity depends on both its surface temperature and its surface area or radius. If two stars have the same surface temperature, one can be more luminous than the other only if it is larger in size. Stellar radii therefore must increase as we go from the high-temperature, low-luminosity corner on the lower left of the H–R diagram to the low-temperature, high-luminosity corner on the upper right.

Patterns in the H–R Diagram Stars do not fall randomly throughout an H–R diagram like Figure 11.10 but instead fall into several distinct groups:

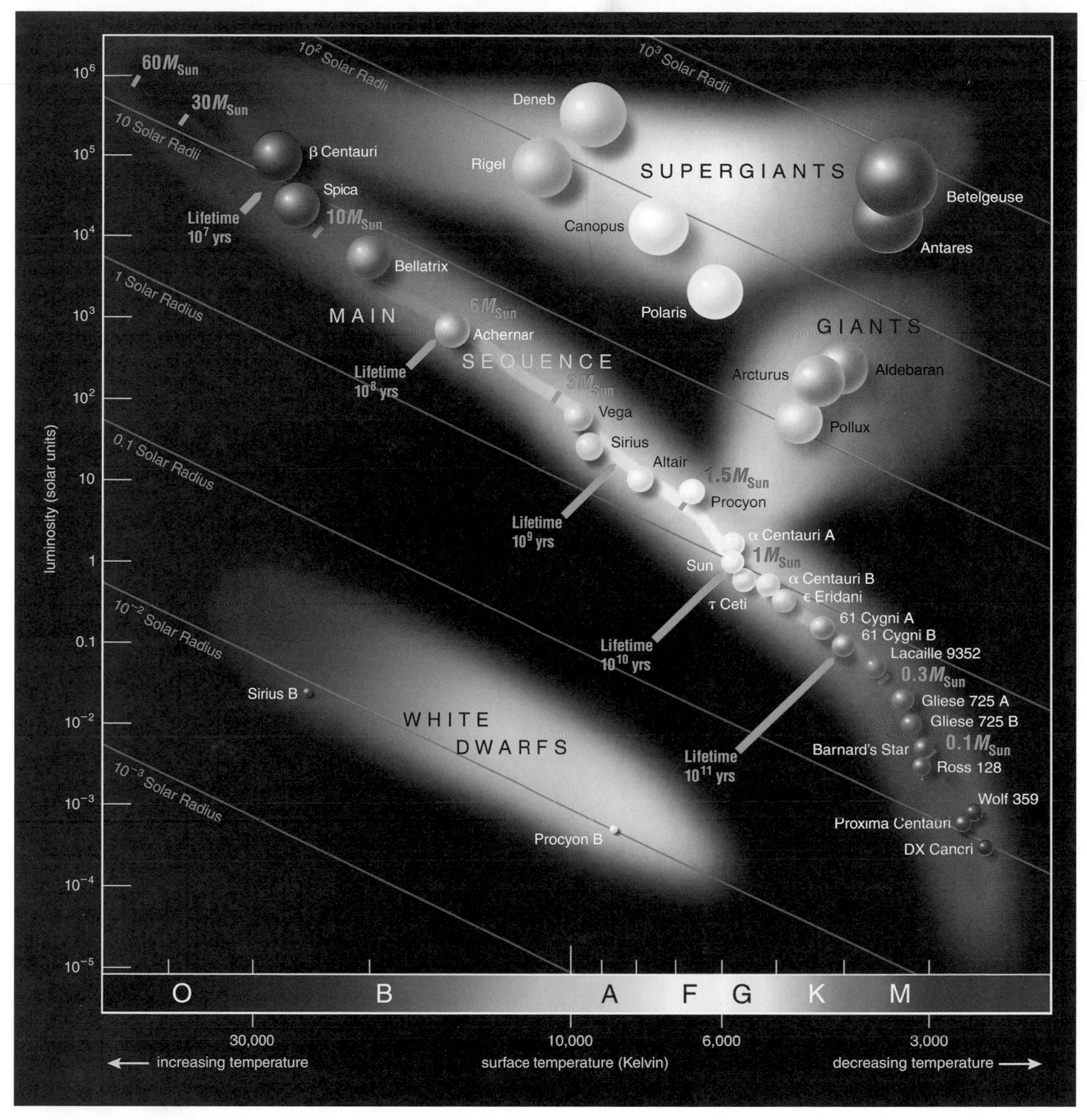

Figure 11.10 Interactive Figure

An H–R diagram, one of astronomy's most important tools, shows a star's surface temperature or spectral type along the horizontal axis and its luminosity along the vertical axis. Several of the brightest stars in the sky are plotted here, along with a few of those closest to Earth. They are not drawn to scale—the diagonal lines, labeled in solar radii, indicate how large they are compared to the Sun. The lifetime and mass labels apply only to main-sequence stars. (Star positions on this diagram are based on data from the Hipparcos satellite.)

- The vast majority of stars fall along a diagonal streak running from the upper left to the lower right of the diagram. This streak is known as the **main sequence**, which is why these hydrogen-burning stars are known as *main-sequence stars.* As we have seen, the position of a star along this sequence depends on its mass. High-mass stars are at the upper left because they are hot and luminous. Low-mass stars are at the lower right because they are cool and dim. Purple labels give the masses of stars along the main sequence, and green labels give the hydrogen-burning lifetimes.
- **Supergiants** are found near the top of the H–R diagram because they are very large in addition to being very bright.
- Just below the supergiants are the **giants**, because they are somewhat smaller in radius and lower in luminosity than the supergiants.
- **White dwarf** stars are found near the lower left because their luminosities are small and their surface temperatures are often hot. Notice that the two white dwarfs plotted in Figure 11.10 both fall close to the 10^{-2} solar radius line, implying that they are similar in radius to Earth.

In summary, we can plot any star on an H–R diagram just by knowing its spectral type and luminosity. Once we've located the star on the diagram, we can immediately tell whether it is a hydrogen-burning star on the main sequence or a star in a later stage of life such as a giant or supergiant. We can also immediately determine its radius. And, if it is a main-sequence star, its position along the main sequence tells us its mass and lifetime.

THINK ABOUT IT By studying Figure 11.10, determine the approximate spectral type, luminosity class, and radius of the following stars: Bellatrix, Vega, Antares, Pollux, and Proxima Centauri. Can you determine the masses and lifetimes of any of these stars? Explain.

The wealth of information that can be gleaned from an H–R diagram makes it a very useful tool for understanding how stars work. In particular, we have learned a lot about stellar lives by studying H–R diagrams for star clusters, the topic we turn to next.

Stellar Evolution Tutorial, Lessons 1, 4

11.3 STAR CLUSTERS

Most stars are born in groups because the giant gas clouds that form stars usually make many stars at once. A single interstellar gas cloud can contain enough material to form thousands of stars. In our snapshot of the heavens, many stars still congregate in the groups in which they formed.

These groups are known as *star clusters,* and they are extremely useful to astronomers for two key reasons:

1. All the stars in a cluster lie at about the same distance from Earth.
2. All the stars in a cluster formed at about the same time (within a few million years of one another).

Astronomers can therefore use star clusters as laboratories for comparing the properties of stars that all have similar ages, and we shall see that these features of star clusters enable us to use them as cosmic clocks.

Figure 11.11

A photo of the Pleiades, a nearby open cluster of stars. The most prominent stars in this open cluster are of spectral type B, indicating that the Pleiades are no more than 100 million years old, relatively young for a star cluster. The region shown is about 11 light-years across.

Figure 11.12

The globular cluster M 80 is more than 12 billion years old. The prominent reddish stars in this Hubble Space Telescope photo are red giant stars nearing the ends of their lives. The central region pictured here is about 15 light-years across.

• What are the two types of star clusters?

Star clusters come in two basic types: modest-size **open clusters** and densely packed **globular clusters**. Open clusters of stars are always found in the disk of the galaxy. They can contain up to several thousand stars and typically are about 30 light-years across. The most famous open cluster is the *Pleiades,* a prominent clump of stars in the constellation Taurus (Figure 11.11). The Pleiades are often called the *Seven Sisters,* although only six of the cluster's several thousand stars are easily visible to the naked eye. Other cultures have other names for this beautiful group of stars. In Japanese it is called *Subaru,* which is why the logo for Subaru automobiles is a diagram of the Pleiades.

Most globular clusters are found above or below the disk of our galaxy, in the region we call the *halo* (see Figure 1.14). A globular cluster can contain more than a million stars concentrated in a ball typically from 60 to 150 light-years across. Its central region can have 10,000 stars packed into a space just a few light-years across (Figure 11.12). The view from a planet in a globular cluster would be marvelous, with thousands of stars lying closer than Alpha Centauri is to the Sun.

• How do we measure the age of a star cluster?

We can use clusters as clocks because we can determine their ages by plotting the cluster's stars in an H–R diagram. To understand how the process works, look at Figure 11.13, which shows an H–R diagram for the Pleiades. Most of the stars in the Pleiades fall along the standard main sequence, with one important exception: The Pleiades' stars trail away to the right of the main sequence at the upper end. That is, the hot, short-lived stars of spectral type O are missing from the main sequence. Appar-

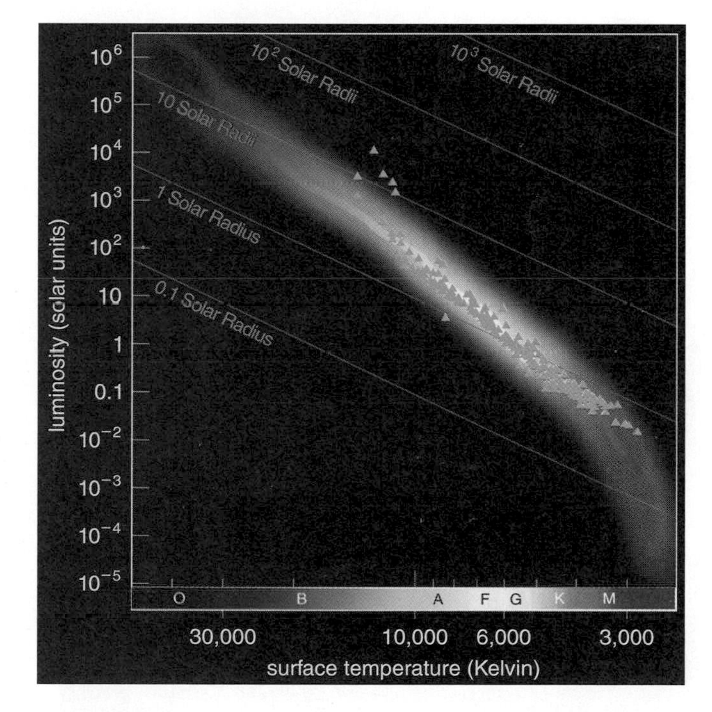

Figure 11.13

An H–R diagram for the stars of the Pleiades. Triangles represent individual stars. The Pleiades cluster is missing its upper main-sequence stars, indicating that these stars have already ended their hydrogen-burning lives. The main-sequence turnoff point at about spectral type B6 tells us that the Pleiades are about 100 million years old.

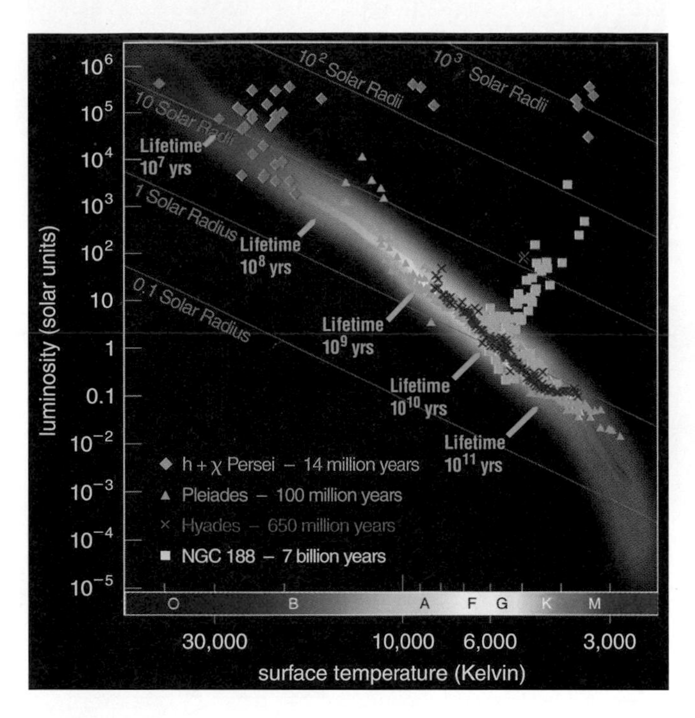

Figure 11.14

This H–R diagram shows stars from four clusters. Their differing main-sequence turnoff points indicate very different ages.

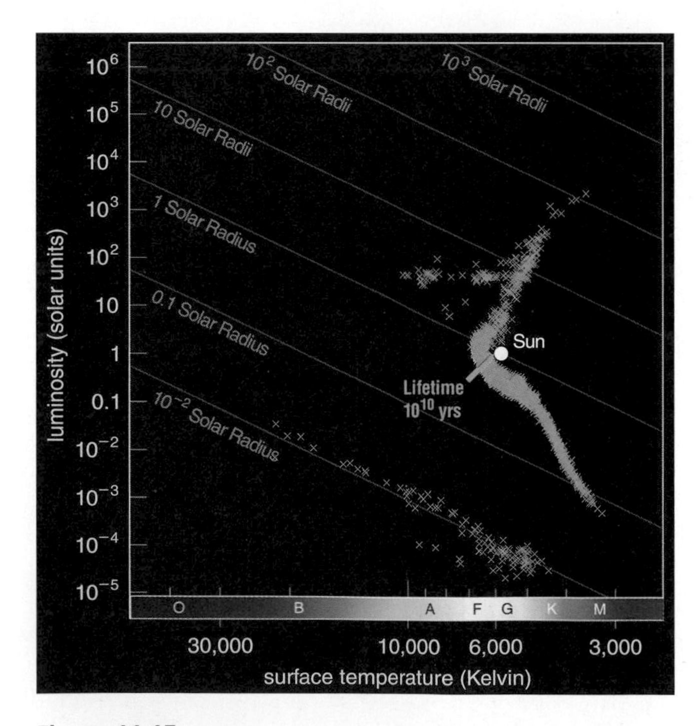

Figure 11.15

This H–R diagram shows stars from the globular cluster M4. The main-sequence turnoff point is in the vicinity of stars like our Sun, indicating an age for this cluster of around 10 billion years. A more technical analysis of this cluster places its age at around 13 billion years.

ently, the Pleiades are old enough for its main-sequence O stars to have already ended their hydrogen-burning lives. At the same time, the cluster is young enough for some of its stars of spectral type B to still survive as hydrogen-burning stars on the main sequence.

The precise point on the H–R diagram at which the Pleiades' main sequence diverges from the standard main sequence is called its **main-sequence turnoff** point. In this cluster, it occurs around spectral type B6. The main-sequence lifetime of a B6 star is roughly 100 million years, so this must be the age of the Pleiades. Any star in the Pleiades that was born with a main-sequence spectral type hotter than B6 had a lifetime shorter than 100 million years and hence is no longer found on the main sequence. Stars with lifetimes longer than 100 million years are still fusing hydrogen and hence remain as main-sequence stars. Over the next few billion years, the B stars in the Pleiades will die out, followed by the A stars and the F stars. Thus, if we could make an H–R diagram for the Pleiades every few million years, we would find that the main sequence gradually grows shorter.

The age of a star cluster approximately equals the hydrogen-burning lifetime of the most massive main-sequence stars remaining within it.

Comparing the H–R diagrams of other open clusters makes this effect more apparent (Figure 11.14). In each case, we determine the cluster's age from the lifetimes of the stars at its main-sequence turnoff point. More precisely, we can determine the age of any star cluster if we can make an H–R diagram showing the spectral types and luminosities of its stars: *The age of the cluster is equal to the lifetime of stars at its main sequence turnoff point.* Stars in a particular cluster that once resided above the turnoff point on the main sequence have already exhausted their core supply of hydrogen, while stars below the turnoff point remain on the main sequence.

THINK ABOUT IT Suppose a star cluster is precisely 10 billion years old. Where would you expect to find its main-sequence turnoff point? Would you expect this cluster to have any main-sequence stars of spectral type A? Would you expect it to have main-sequence stars of spectral type K? Explain. (*Hint:* What is the lifetime of our Sun?)

The technique of identifying main-sequence turnoff points is our most powerful tool for evaluating the ages of star clusters. We've learned that most open clusters are relatively young, with very few older than about 5 billion years. In contrast, the stars at the main-sequence turnoff points in globular clusters are usually less massive than our Sun (Figure 11.15). Because stars like our Sun have a lifetime of about 10 billion years and these stars have already died in globular clusters, we conclude that globular-cluster stars are older than 10 billion years.

More precise studies of the turnoff points in globular clusters, coupled with theoretical calculations of stellar lifetimes, place the ages of these clusters at about 13 billion years, making them the oldest known objects in the galaxy. In fact, globular clusters place a constraint on the possible age of the universe: If stars in globular clusters are 13 billion years old, then the universe must be at least this old. Recent observations suggesting that the universe is about 14 billion years old therefore fit well with the ages of these stars, and tell us that the first stars began to form by the time the universe was a billion years old.

THE BIG PICTURE
Putting Chapter 11 into Context

We have classified the diverse families of stars visible in the night sky. Much of what we know about stars, galaxies, and the universe itself is based on the fundamental properties of stars introduced in this chapter. Make sure you understand the following "big picture" ideas:

- All stars are made primarily of hydrogen and helium, at least at the time they form. The differences between stars come about primarily because of differences in mass and age.
- Stars spend most of their lives as main-sequence stars that fuse hydrogen into helium in their cores. The most massive stars, which are also the hottest and most luminous, live only a few million years. The least massive stars, which are coolest and dimmest, will survive until the universe is many times its present age.
- Much of what we know about the universe comes from studies of star clusters. For example, we can measure a star cluster's age by plotting its stars in an H–R diagram and determining the hydrogen-burning lifetime of the brightest and most massive stars still on the main sequence.

11.1 PROPERTIES OF STARS

- **How luminous are stars?**

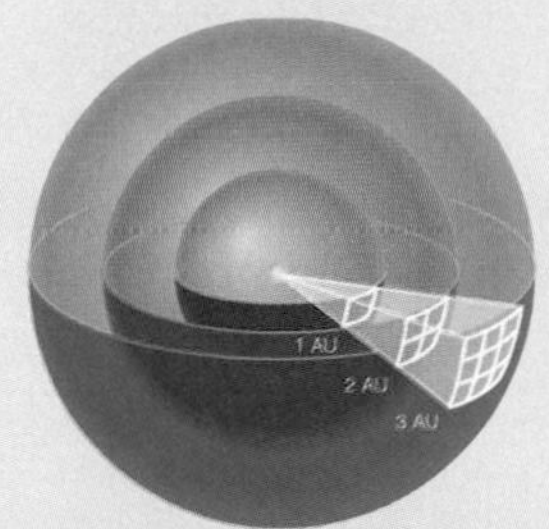

The **apparent brightness** of a star in our sky depends on both its **luminosity**—the total amount of light it emits into space—and its distance from Earth, as expressed by the **inverse square law for light**. By studying stars for which we know distances from stellar parallax or other techniques, we have learned that the most luminous stars emit more than 100,000 times as much light as the Sun, while the least luminous put out less than 0.0001 times as much light as the Sun.

- **How hot are stars?**

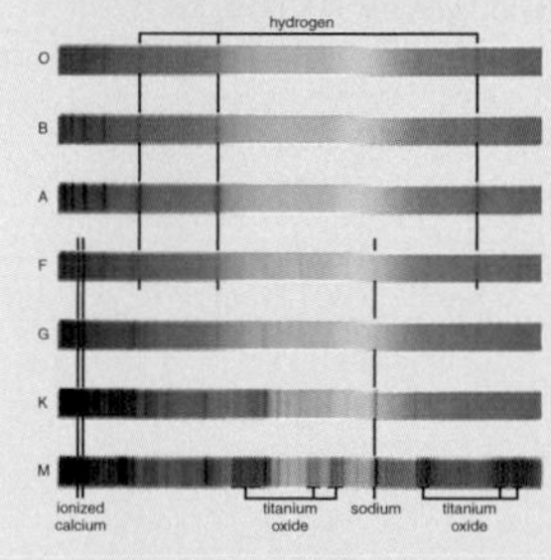

The surface temperatures of the hottest stars exceed 40,000 K and those of the coolest stars are less than 3,000 K. We measure a star's surface temperature from its color or spectrum, and we classify spectra according to the sequence of **spectral types** OBAFGKM, which runs from hottest to coolest. Cool, red stars of spectral type M are much more common than hot, blue stars of spectral type O.

- **How massive are stars?**

The overall range of stellar masses runs from 0.08 times the mass of the Sun to about 100 times the mass of the Sun. We can measure the masses of stars in **binary star systems** using Newton's version of Kepler's third law if we can measure the orbital period and separation of the two stars.

11.2 CLASSIFYING STARS

- **How do we classify stars?**

We classify stars according to their **spectral type** and **luminosity class**. The spectral type tells us the star's surface temperature and color. The luminosity class tells us how much light the star puts out. **Giant** and **supergiant** stars put out far more light than **main-sequence** stars like our Sun, even though their surface temperatures are generally lower, meaning that their radii must be much larger. All stars become giants or supergiants near the ends of their lives, and many end up as hot but dim **white dwarfs**.

Continued ▶

- **Why is a star's mass its most important property?**

A star's mass at birth determines virtually everything that happens to it throughout its life. While a star is a main-sequence star, its mass determines its luminosity, surface temperature, radius, and hydrogen-burning lifetime—which is shorter for more massive stars. Once a star exhausts its core hydrogen, its mass determines how and when it grows into a giant or supergiant, and also determines what happens to it when it finally dies.

- **What is a Hertzsprung–Russell diagram?**

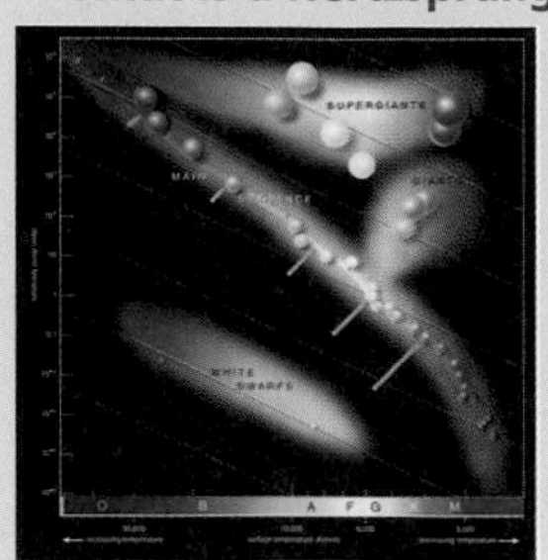

An **H–R diagram** plots stars according to their surface temperatures and luminosities. Hydrogen-burning stars occupy a narrow band in the diagram known as the **main sequence**. Giants and supergiants are to the upper right of the main sequence and white dwarfs are to the lower left.

11.3 STAR CLUSTERS

- **What are the two types of star clusters?**

Open clusters contain up to several thousand stars and are found in the disk of the galaxy. **Globular clusters** contain hundreds of thousands of stars, all closely packed together. They are found mainly in the halo of the galaxy.

- **How do we measure the age of a star cluster?**

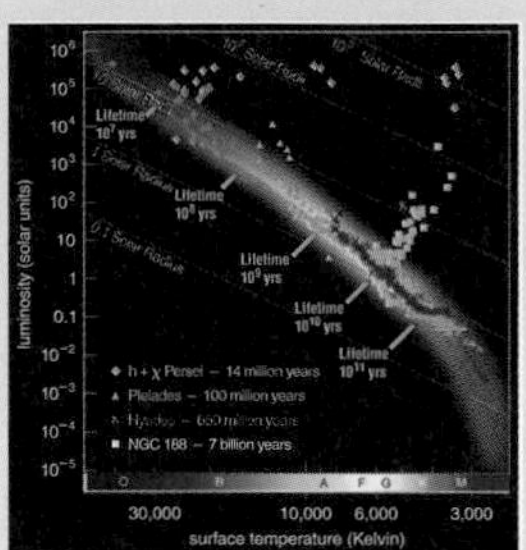

Because all of a cluster's stars were born at the same time, we can measure a cluster's age by finding the **main-sequence turnoff** point on an H–R diagram of its stars. The cluster's age is equal to the hydrogen-burning lifetime of the hottest, most luminous stars that remain on the main sequence. We've learned that open clusters are much younger than globular clusters, which can be as old as about 13 billion years.

EXERCISES AND PROBLEMS

REVIEW QUESTIONS

1. Briefly explain how we can learn about the lives of stars, even though their lives are far longer than human lives.
2. In what ways are all stars similar? In what ways can stars differ?
3. How is a star's *apparent brightness* related to its *luminosity?* Explain by describing the *inverse square law for light.*
4. Briefly explain how we use *stellar parallax* to determine a star's distance. Once we know a star's distance, how can we determine its luminosity?
5. What do we mean by a star's spectral type? How is a star's spectral type related to its surface temperature and color? Which stars are hottest and coolest in the spectral sequence OBAFGKM?
6. What are the three basic types of *binary star systems?* Why are *eclipsing binaries* so important to measuring masses of stars?
7. What is the defining characteristic of a *main-sequence* star? How is surface temperature related to luminosity for main-sequence stars?
8. How do we know that *giants* and *supergiants* are large in radius? How big are they? How do they differ from main-sequence stars?
9. What are *white dwarfs?*
10. What do we mean by a star's *luminosity class?* What does the luminosity class tell us about the star? Briefly explain how we classify stars by spectral type and luminosity class.
11. Briefly explain why massive main-sequence stars are more luminous and have hotter surfaces than less massive ones.
12. Which stars have longer lifetimes: massive stars or less massive stars? Explain why.
13. Draw a sketch of a basic Hertzsprung–Russell (H–R) diagram. Label the main sequence, giants, supergiants, and white dwarfs. Where on this diagram do we find stars that are cool and dim? Cool and luminous? Hot and dim? Hot and luminous?
14. Describe in general terms how open clusters and globular clusters differ in their numbers of stars, ages, and locations in the galaxy.
15. Explain why H–R diagrams look different for star clusters of different ages. How does the location of the main-sequence turnoff point tell us the age of the star cluster?

TRUE STATEMENTS?

Decide whether each of the following statements is true or false and clearly explain how you know.

16. Two stars that look very different must be made of different kinds of elements.
17. Two stars that have the same apparent brightness in the sky must also have the same luminosity.
18. Sirius looks brighter than Alpha Centauri, but we know that Alpha Centauri is closer because its apparent position in the sky shifts by a larger amount as Earth orbits the Sun.
19. Stars that look red-hot have hotter surfaces than stars that look blue.

20. Some of the stars on the main sequence of the H–R diagram are not converting hydrogen into helium.
21. The smallest, hottest stars are plotted in the lower left-hand portion of the H–R diagram.
22. Stars that begin their lives with the most mass live longer than less massive stars because they have so much more hydrogen fuel.
23. Star clusters with lots of bright, blue stars of spectral type O and B are generally younger than clusters that don't have any such stars.
24. All giants, supergiants, and white dwarfs were once main-sequence stars.
25. Most of the stars in the sky are more massive than the Sun.

PROBLEMS

26. *Stellar Data.* Consider the following data table for several bright stars. M_v is absolute magnitude, and m_v is apparent magnitude. (Hint: Remember that the magnitude scale runs backward, so that brighter stars have smaller (or more negative) magnitudes.)

Star	M_v	m_v	*Spectral Type*	*Luminosity Class*
Aldebaran	−0.2	+0.9	K5	III
Alpha Centauri A	+4.4	0.0	G2	V
Antares	−4.5	+0.9	M1	I
Canopus	−3.1	−0.7	F0	II
Fomalhaut	+2.0	+1.2	A3	V
Regulus	−0.6	+1.4	B7	V
Sirius	+1.4	−1.4	A1	V
Spica	−3.6	+0.9	B1	V

Answer each of the following questions, including a brief explanation with each answer.

a. Which star appears brightest in our sky?
b. Which star appears faintest in our sky?
c. Which star has the greatest luminosity?
d. Which star has the least luminosity?
e. Which star has the highest surface temperature?
f. Which star has the lowest surface temperature?
g. Which star is most similar to the Sun?
h. Which star is a red supergiant?
i. Which star has the largest radius?
j. Which stars have finished burning hydrogen in their cores?
k. Among the main-sequence stars listed, which one is the most massive?
l. Among the main-sequence stars listed, which one has the longest lifetime?

27. *Data Tables.* Study the spectral types listed in Appendix F for the 20 brightest stars and for the stars within 12 light-years. Why do you think the two lists are so different? Explain.

DISCUSSION QUESTION

28. *Classification.* About a century ago Annie Jump Cannon and other astronomers at Harvard University made many important contributions to astronomy, particularly in the area of systematic stellar classification. Why do you think rapid advances in our understanding of stars followed so quickly on the heels of their efforts? Can you think of other areas in science where huge advances in understanding followed directly from improved systems of classification?

MATH HELP AND EXERCISES

For additional help and practice with mathematical concepts applicable to this chapter, the Astronomy Place Web site has the following resources with detailed explanations, worked examples, and practice problems.

- The Inverse Square Law of Light
- The Parallax Formula
- The Modern Magnitude Scale
- Orbital Separation and Newton's Version of Kepler's Third Law
- Calculating Stellar Radii

For a complete list of media resources available, go to **www.astronomyplace.com,** and choose Chapter 11 from the pull-down menu.

ASTRONOMY PLACE WEB TUTORIALS

Tutorial Review of Key Concepts

Use the interactive **Tutorials** at **www.astronomyplace.com** to review key concepts from this chapter.

Measuring Cosmic Distances Tutorial

Lesson 2 Stellar Parallax

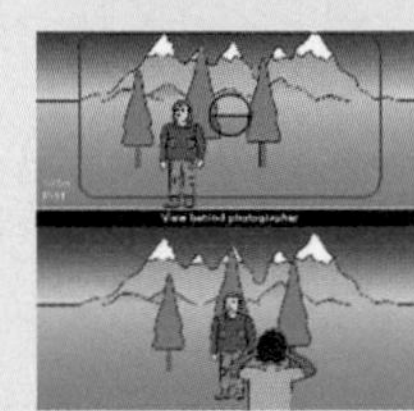

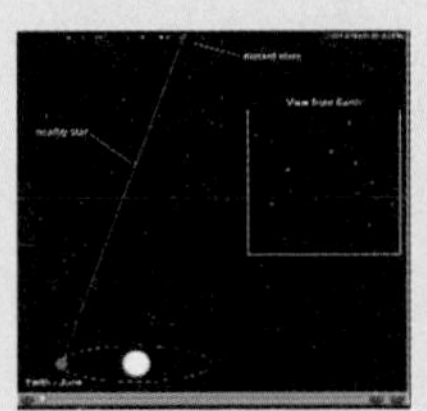

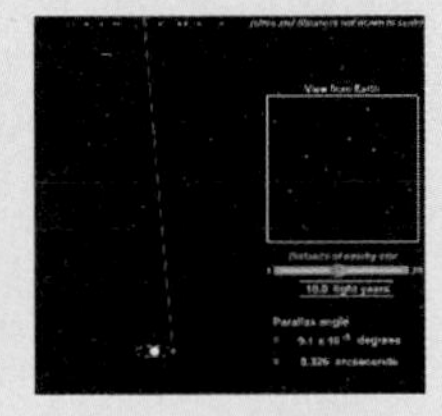

Measuring Cosmic Distances Tutorial, Lesson 2

1. Explain how we measure distances with stellar parallax. Give an example.
2. Explain why we cannot use parallax to measure the distance to *all* stars.

Hertzsprung–Russell Diagram Tutorial

Lesson 1 The Hertzsprung–Russell (H–R) Diagram
Lesson 2 Determining Stellar Radii
Lesson 3 The Main Sequence

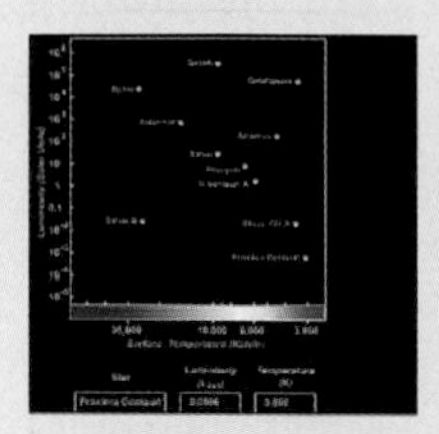 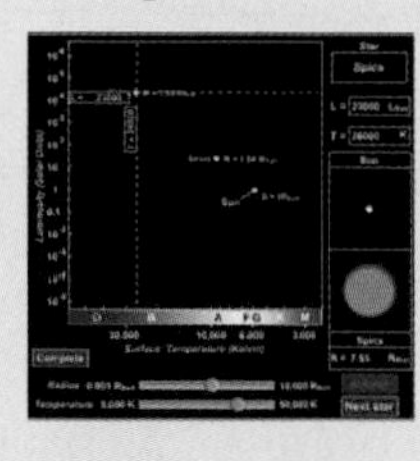 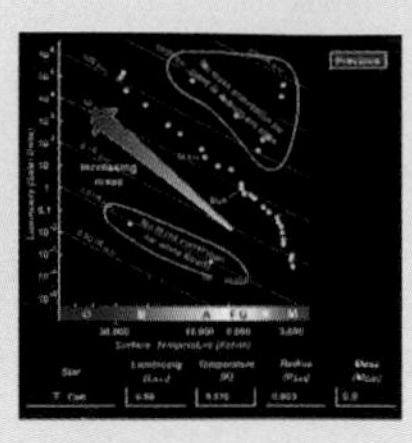

Stellar Evolution Tutorial

Lesson 1 Main-Sequence Lifetimes
Lesson 4 Cluster Dating

 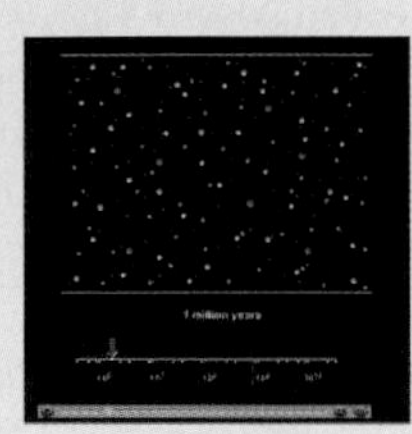 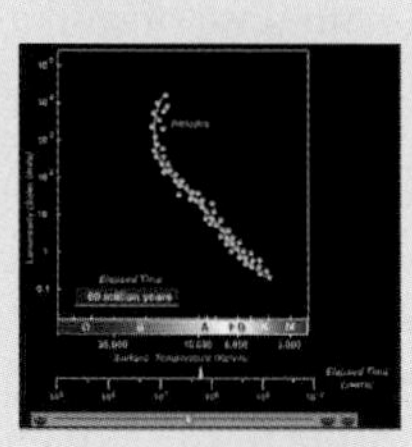

Supplementary Tutorial Exercises

Use the interactive **Tutorial Lessons** to explore the following questions.

Hertzsprung–Russell Diagram Tutorial, Lessons 1–3

1. If one star appears brighter than another, can you be sure that it is more luminous? Why or why not?
2. Answer each part of this question with either high, low, left, or right. On an H–R diagram, where will a star be if it is hot? Cool? Bright? Dim?
3. Why is there a relationship between stellar radii and locations on the H–R diagram?
4. Why is there a relationship between luminosity and mass for main-sequence stars on the H–R diagram?

Stellar Evolution Tutorial, Lesson 4

1. In the animation at the beginning of Lesson 4, list the order in which you saw the three differently colored stars in the cluster disappear, and explain why they disappeared in this order.
2. In the second animation in Lesson 4, in what order did you see stars on the main sequence disappear? Explain the reason for this.
3. How does the age of a dim star cluster of mostly small stars compare to a bright cluster with some giants in it? Explain your answer.

EXPLORING THE SKY AND SOLAR SYSTEM

Of the many activities available on the ***Voyager: SkyGazer*** **CD-ROM** accompanying your book, use the following files to observe key phenomena covered in this chapter.

Go to the **File: Basics** folder for the following demonstrations:

1. Large Stars
2. More Stars
3. Star Color and Size

Go to the **File: Demo** folder for the following demonstrations:

1. Circling the Hyades
2. Flying Around Pleiades
3. The Tail of Scorpius

WEB PROJECTS

Take advantage of the useful Web links on **www.astronomyplace.com** to assist you with the following projects.

1. *Women in Astronomy.* Until fairly recently, men greatly outnumbered women in professional astronomy. Nevertheless, many women made crucial discoveries in astronomy throughout history—including discovering the spectral sequence for stars. Do some research about the life and discoveries of a female astronomer from any time period and write a two- to three-page scientific biography.
2. *The Hipparcos Mission.* The European Space Agency's Hipparcos mission, which operated from 1989 to 1993, made precise parallax measurements for more than 40,000 stars. Learn about how Hipparcos allowed astronomers to measure smaller parallax angles than they could from the ground and how Hipparcos discoveries have affected our knowledge of the universe. Write a one- to two-page report on your findings.

12
Star Stuff

LEARNING GOALS

12.1 STAR BIRTH
- How do stars form?
- How massive are newborn stars?

12.2 LIFE AS A LOW-MASS STAR
- What are the life stages of a low-mass star?
- How does a low-mass star die?

12.3 LIFE AS A HIGH-MASS STAR
- What are the life stages of a high-mass star?
- How do high-mass stars make the elements necessary for life?
- How does a high-mass star die?

12.4 SUMMARY OF STELLAR LIVES
- How does a star's mass determine its life story?
- How are the lives of stars with close companions different?

We inhale oxygen with every breath. Iron-bearing hemoglobin in our blood carries this oxygen through our bodies. Chains of carbon and nitrogen form the backbone of the proteins, fats, and carbohydrates in our cells. Calcium strengthens our bones, while sodium and potassium ions moderate communications of the nervous system. What does all this biology have to do with astronomy? The profound answer, recognized only in the second half of the twentieth century, is that life is based on elements created by stars.

We've already discussed in general terms how the elements in our bodies came to exist. Hydrogen and helium were produced in the Big Bang, and heavier elements were created later by stars and scattered into space by stellar explosions. There, in the spaces between the stars, these elements mixed with interstellar gas and became incorporated into subsequent generations of stars.

In this chapter, we will discuss the origins of the elements in greater detail by delving into the lives of stars. As you read, keep in mind that no matter how far removed the stars may seem from our everyday lives, they actually are connected to us in the most intimate way possible: Without the births, lives, and deaths of stars, none of us would be here. We are truly made from "star stuff."

12.1 STAR BIRTH

Stars are born in interstellar clouds of gas and dust. We've already studied one example of star birth in Chapter 6, which told the story of how our Sun and solar system formed. We will now extend those ideas to the births of other stars, concentrating on the processes that occur inside a newly forming star and lead to the ignition of hydrogen fusion in its core. Later, in Chapter 14, we'll see how star birth links up with the star-gas-star cycle that recycles matter throughout out Milky Way.

VIS

Figure 12.1

The dark region in this photograph is a star-forming cloud of molecular hydrogen gas in the constellation Scorpius. It is dark because dust particles obscure the light from more distant stars. The fuzzy, blue-white blotches near the edges of the dark cloud contain newly formed stars whose light is reflecting off surrounding gas. The region pictured here is about 50 light-years across.

• How do stars form?

The clouds that give birth to stars tend to be quite cold, typically only 10–30 K. (Recall that 0 K is absolute zero, and temperatures on Earth are around 300 K; see Figure 4.10.) They also tend to be quite dense compared to the rest of the gas between the stars, although they would qualify as superb vacuums by earthly standards. Like the galaxy as a whole, star-forming clouds are made almost entirely of hydrogen and helium.

Stars are born in clouds of gas and dust that are much denser and colder than most interstellar gas.

Star-forming clouds are sometimes called **molecular clouds** (Figure 12.1), because their low temperatures allow hydrogen atoms to pair up to form hydrogen molecules (H_2). The relatively rare atoms of elements heavier than helium can form more complex molecules—such as carbon monoxide or water—or tiny, solid grains of dust. More important, the cold temperatures and high densities allow gravity to compress these clouds without resistance from the pressure that would exist in

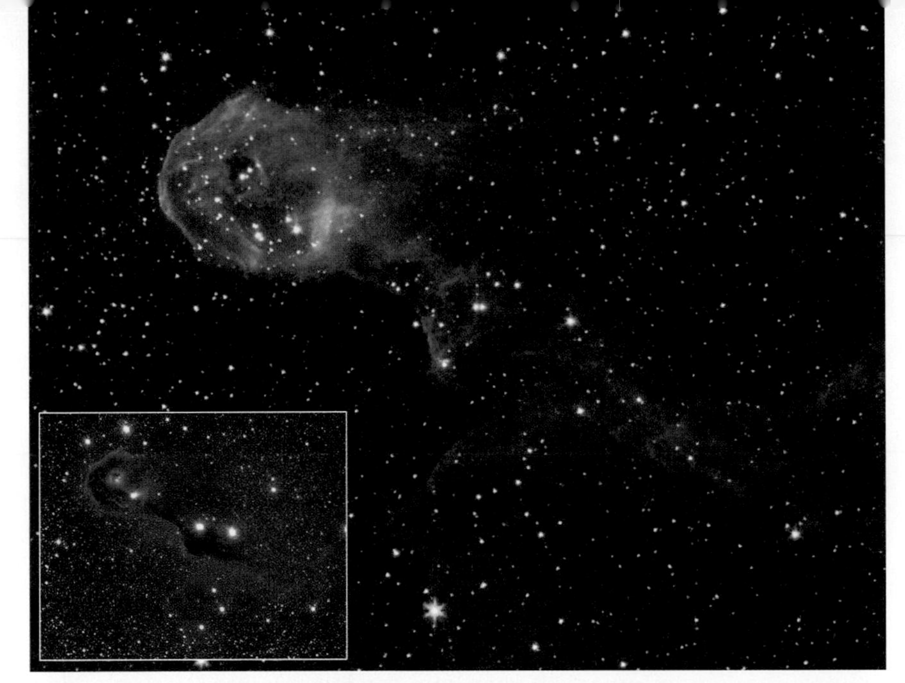

a This infrared image from the Spitzer Space Telescope shows a star-forming cloud in the constellation Cepheus. The pink color represents infrared light from the cloud's cold, dense gas. The inset photo shows the same region in visible light; notice that the cold gas that glows in the infrared is dark in the visible photo.

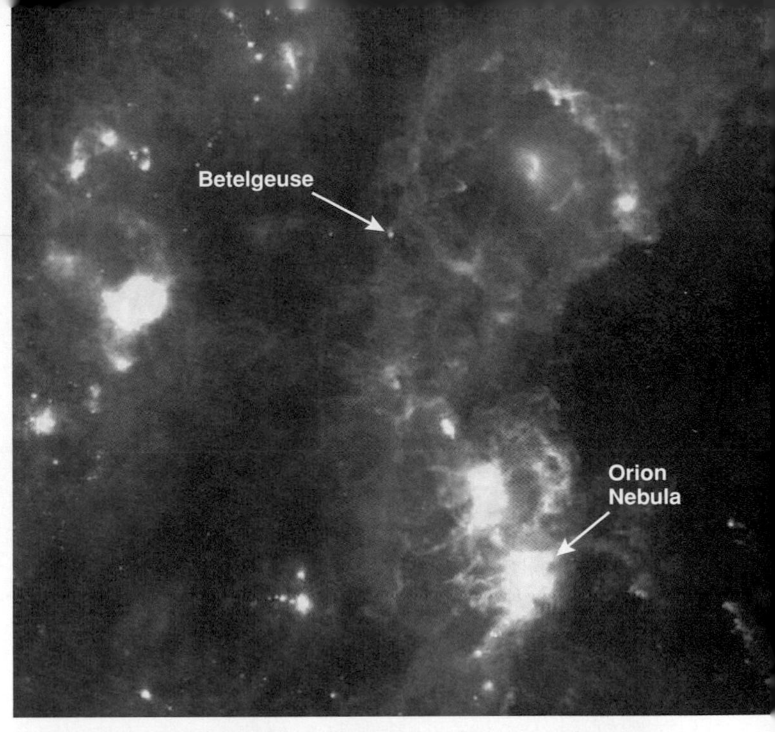

b This picture shows infrared radiation from star-forming regions in the constellation Orion. The colors correspond to the gas temperature: Red is cooler, and yellow and white are hotter. Star formation is most intense in the yellow-white regions (where temperatures are 60–100 K), including the Orion Nebula. The region pictured here is about 80 light-years across at the distance of the Orion Nebula.

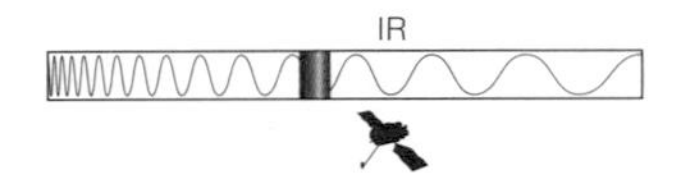

Figure 12.2

Star-forming molecular clouds are cold and therefore emit infrared rather than visible light. (Remember that the colors in the infrared photos are not real because we cannot see infrared light; the colors were chosen to make it easy for us to see where the cameras detected infrared light.)

clouds of higher temperature. Gravity pulls the molecular gas toward the densest regions of the cloud, where it fragments into numerous pieces that each form one or more new stars.

From Cloud to Protostar Each shrinking fragment of a molecular cloud heats up as it contracts. The source of this heat is the gravitational potential energy [Section 4.3] released as gravity pulls each part of the cloud fragment closer to the center of the fragment. Early in the process of star formation, the contracting gas quickly radiates away much of this energy, preventing the temperature and pressure from building enough to resist gravity. The temperature of the cloud remains below 100 K, so it glows in long-wavelength infrared light (Figure 12.2).

As the cloud continues to contract, its growing density makes the escape of radiation increasingly difficult. Eventually, the central region of the cloud fragment grows dense enough to trap infrared radiation, so that it can no longer radiate away its heat. The central temperature and pressure then begin to rise dramatically. The rising pressure pushes back against the crush of gravity, slowing the contraction. The dense center of the cloud fragment is now a **protostar**—the clump of gas that will become a new star. Gas from the surrounding cloud continues to rain down on the protostar, increasing its mass.

A molecular cloud fragment heats up and spins faster as gravity makes it contract, producing a rapidly rotating protostar at its center.

Protostars rotate rapidly when they form. Random motions of gas particles inevitably give a gas cloud some small overall rotation, although it may be imperceptibly slow. Like an ice skater pulling in her arms, the cloud rotates faster as it contracts, so that its total angular momentum remains conserved [Section 4.3]. The rapid rotation of protostars is probably responsible for the formation of some binary star systems. Protostars that spin too fast may split in two, with each fragment becoming a separate star. The resulting

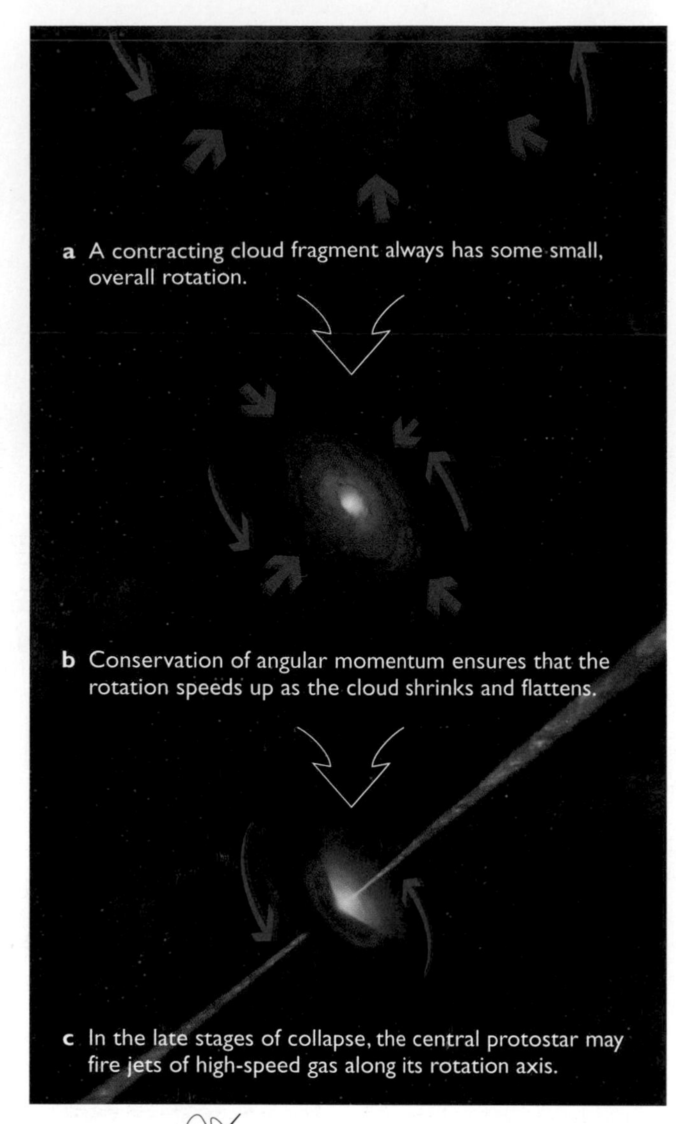

Figure 12.3

Artist's conception of star birth.

pair is called a **close binary** system, in which two stars coexist in close proximity and rapidly orbit each other.

Disks and Jets The rotation of the shrinking cloud fragment also produces a spinning disk of gas around the protostar (Figure 12.3). Because angular momentum must be conserved, the gas surrounding the protostar rotates faster as it contracts, and collisions between gas particles tend to flatten this gas into a disk [Section 6.3]. In our own solar system, the planets formed in this spinning disk. Many other star systems probably give birth to planets in the same way, though we do not yet know how commonly this occurs.

Conservation of angular momentum ensures that protostars rotate rapidly and are surrounded by spinning disks of gas.

Observations show that the late stages of a star's formation can be surprisingly violent. Many young stars fire high-speed streams of gas, or **jets**, into interstellar space (Figure 12.4). We generally see two jets, shooting in opposite directions along the protostar's rotation axis. Sometimes the jets are lined with glowing blobs of gas, which are presumably clumps of matter swept up as the jets plow into the surrounding interstellar material.

No one knows exactly how protostars generate these jets, but magnetic fields probably play an important role. A protostar's rapid rotation helps it generate a strong magnetic field, and this field may help channel the jets along the rotation axis. In addition, the strong magnetic field helps the protostar generate a strong wind—an outward flow of particles similar to the *solar wind* [Section 10.1]. Together, winds and jets probably help clear away the cocoon of gas that surrounds a forming star, revealing the protostar within. They also help the protostar shed some of its angular momentum by carrying material off into interstellar space. Thus, the protostar's rotation gradually slows, which is why most older stars rotate slowly.

From Protostar to the Main Sequence A protostar looks starlike, but its interior is not yet hot enough for fusion. The central temperature of a protostar may be only a million degrees or so when its wind and jets blow away the surrounding gas. To ignite fusion, the protostar needs to contract further to boost the central temperature.

A protostar becomes a main-sequence star when it begins to sustain hydrogen fusion in its core.

A protostar becomes a true star when its core temperature exceeds 10 million K, hot enough for hydrogen fusion to operate efficiently. The ignition of fusion halts the protostar's gravitational contraction and marks what we consider the birth of a star. The new star's interior structure stabilizes because the energy produced in the center matches the amount radiated from its surface. The star is now a hydrogen-burning, main-sequence star [Section 11.2].

The length of time from the formation of a protostar to the birth of a main-sequence star depends on the star's mass. Massive stars do everything faster. The contraction of a high-mass protostar into a main-sequence star of spectral type O or B may take only a million years or less. A star like our Sun takes about 50 million years to go from the beginning of the protostellar stage to becoming a main-sequence star. A very low-mass star of spectral type M may spend more than a hundred million years as a protostar. Thus, the most massive stars in a young star cluster may live and die before the smallest stars begin to fuse hydrogen in their cores.

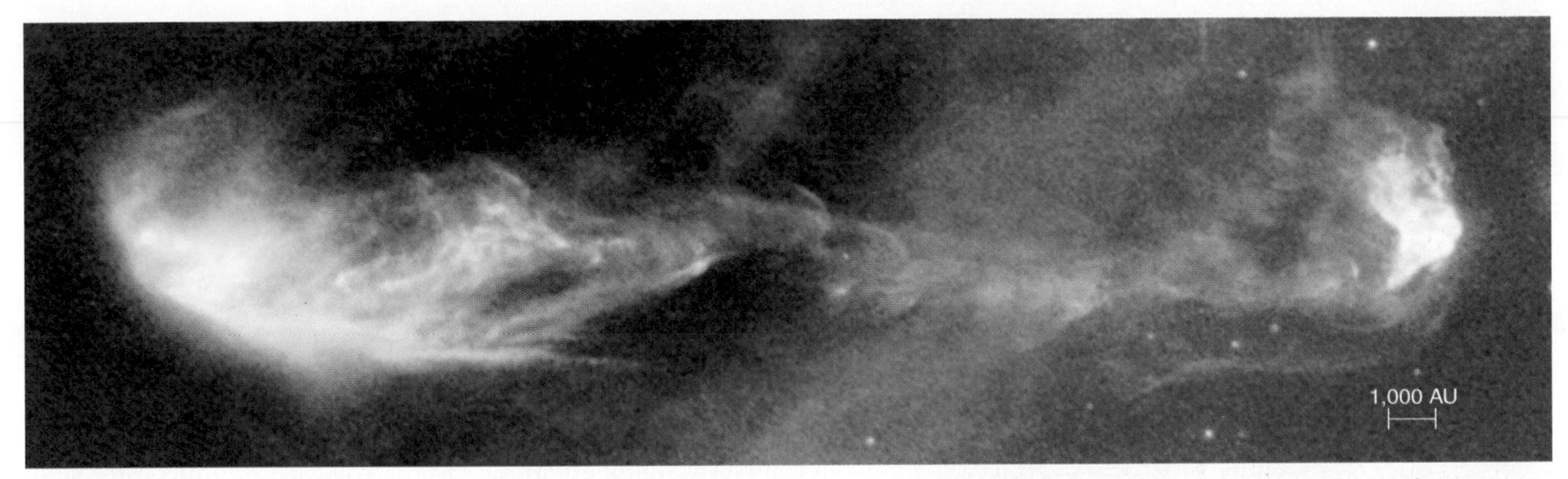

a This photograph shows a jet of material being shot far into interstellar space by a protostar. The protostar is at the left in this photo. The structure near the far right is formed as the jet material rams into surrounding interstellar gas.

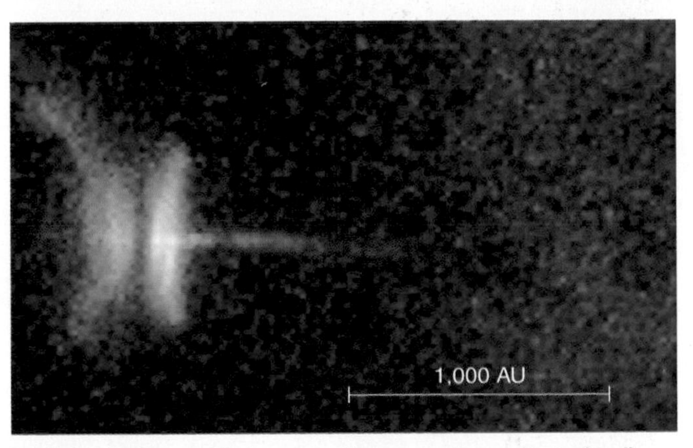

b This photograph shows a close-up view of jets (red) and a disk of gas (green) around a protostar. We are seeing the disk nearly edge-on, so that we can see its top and bottom surfaces but not the thicker and darker central region.

Figure 12.4

These photos show jets of gas shot from protostars into interstellar space.

• How massive are newborn stars?

The masses of newborn stars seem to depend on the processes that govern clumping and fragmentation in star-forming clouds. We do not yet fully understand those processes, but we can observe the results.

A single group of molecular clouds can give birth to a star cluster containing thousands of stars. In a newly formed star cluster, stars with low masses greatly outnumber stars with high masses. For every star with a mass between 10 and 100 solar masses, there are typically 10 stars with masses between 2 and 10 solar masses, 50 stars with masses between 0.5 and 2 solar masses, and a few hundred stars with masses below 0.5 solar mass. Thus, although the Sun lies toward the middle of the overall range of stellar masses, most stars in a new star cluster are less massive than the Sun. With the passage of time, the balance tilts even more in favor of the low-mass stars as the high-mass stars die away.

Limits on Stellar Masses There are limits on how massive a star can be. Theoretical models indicate that stars with masses above about $100M_{\text{Sun}}$ generate energy so furiously that gravity cannot resist their internal pressure. Such stars effectively blow themselves apart and drive their outer layers into space. Observations confirm these models, because stars much more massive than $100M_{\text{Sun}}$ simply haven't been found, and the very high luminosities of such stars ought to make them easy to find if they exist.

A protostar smaller than $0.08M_{\text{Sun}}$ never gets hot enough for efficient hydrogen fusion.

On the other end of the scale, calculations show that the central temperature of a protostar with less than $0.08M_{\text{Sun}}$ never climbs above the 10 million K threshold needed for efficient hydrogen fusion. Instead, a strange effect called *degeneracy pressure* halts the gravitational contraction of the core before hydrogen burning can begin. Degeneracy pressure can push outward in the same way as the ordinary gas pressure that we have discussed up to this point in the book. However, it has a very different origin.

Ordinary gas pressure is often called **thermal pressure** because it is closely linked to temperature: Raising the temperature increases the particle speeds and thereby raises the pressure. In contrast, **degeneracy pressure** does not depend on temperature at all. It arises when subatomic particles are packed together as closely as the laws of quantum physics allow. These are the same laws that give rise to distinct energy levels in atoms [Section 5.2].

Recall that electrons in atoms face restrictions on their energies: They can occupy only particular energy levels. In much the same way, there are restrictions on how close together electrons can be in a gas. Under most circumstances, these restrictions have little effect on the motions or locations of the electrons, and hence little effect on the pressure. However, in a protostar with a mass below $0.08M_{\text{Sun}}$, the electrons become packed closely enough for these restrictions to matter. We can see how this leads to degeneracy pressure with a simple analogy.

Unlike ordinary thermal pressure, degeneracy pressure does not depend on temperature.

Imagine an auditorium in which the spacing between chairs is dictated by the laws of quantum mechanics and people represent electrons (Figure 12.5). As though playing a game of musical chairs, the people are always moving from seat to seat, just as the electrons must remain constantly in motion. Most objects are like auditoriums with many more available chairs than people, so that the people (electrons) can easily find chairs as they move about. However, the cores of protostars with masses below $0.08M_{\text{Sun}}$ are like much smaller auditoriums with so few chairs that the people (electrons) fill nearly all of them. Because there are virtually no open seats, the people (electrons) can't all squeeze into a smaller section of the auditorium. This resistance to squeezing is the origin of degeneracy pressure. If the people were really like electrons, quantum laws would also require them to move faster and faster to find open seats as you tried to squeeze them into a smaller section. However, their speeds would have nothing to do with temperature. Thus, degeneracy pressure and the particle motion that goes with it arise *only* because of the restrictions on where the particles can go, which is why they are unaffected by temperature.

Brown Dwarfs Because degeneracy pressure halts the contraction of a protostar with less than $0.08M_{\text{Sun}}$ before fusion fully kicks in, the result is a "failed star" that slowly radiates away its internal thermal energy, gradually cooling with time. Such objects, called **brown dwarfs**, occupy a fuzzy gap between what we call a planet and what we call a star. (Note that $0.08M_{\text{Sun}}$ is about 80 times the mass of Jupiter.) Because degeneracy pressure does *not* rise and fall with temperature, the gradual cooling of a brown dwarf's interior does not weaken its degeneracy pressure. In the constant battle of any "star" to resist the crush of gravity, brown dwarfs are

Figure 12.5

The auditorium analogy to degeneracy pressure. Chairs represent available possible places (quantum states) for subatomic particles, and people moving from chair to chair represent the particles.

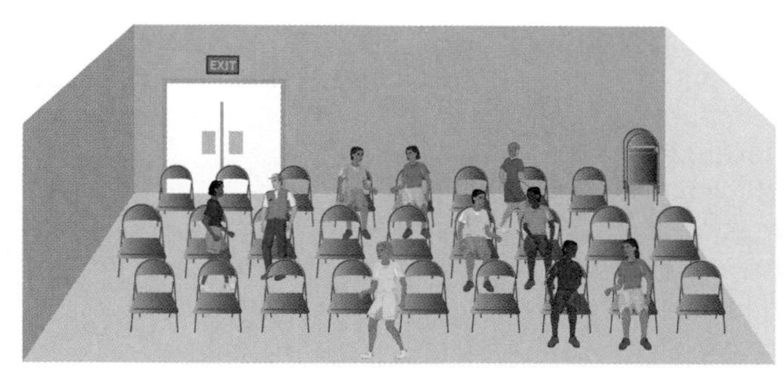

a When there are many more available places (chairs) than particles (people), a particle is unlikely to try to occupy the same place as another particle. The only pressure comes from the termperature-related motion of the particles, which is the ordinary or thermal pressure.

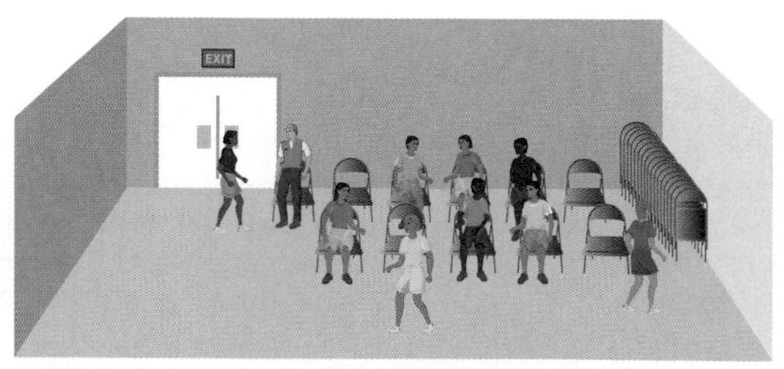

b When the number of particles (people) approaches the number of available places (chairs), finding an available place requires that the particles move faster than they would otherwise. This extra motion creates degeneracy pressure.

winners, albeit dim ones. Their degeneracy pressure will not diminish with time, so gravity will never gain the upper hand.

Brown dwarfs radiate primarily in the infrared and actually look deep red or magenta in color rather than brown. They are far dimmer than normal stars and therefore are extremely difficult to detect, even if they are quite nearby. The first brown dwarf was discovered in 1995—a $0.05M_{Sun}$ object (called Gliese 229B) in orbit around a much brighter star (Gliese 229A). Many more brown dwarfs are now known. If the trend that makes small stars far more common than massive stars continues to masses below $0.08M_{Sun}$, brown dwarfs might outnumber ordinary stars by a huge margin.

Stellar Evolution Tutorial, Lesson 2

12.2 LIFE AS A LOW-MASS STAR

As we saw in Chapter 11, a star's mass is its most important property. It determines not only a star's luminosity and surface temperature during its main-sequence (hydrogen-burning) life but also its hydrogen-burning lifetime and what happens to it after if finally exhausts its core hydrogen. Thus, all stars that start life with about the same mass lead similar lives and die in similar ways. However, the life stories of stars on the low end of the overall range of stellar masses are quite different from those on the high end of that range.

Stars of relatively low mass like our Sun shine steadily for billions of years with energy generated by hydrogen fusion. At the ends of their lives, they swell up to become red giants and then die by expelling their outer layers into space. In the end, all that remains of a low-mass star is the kind of dead core that we know as a white dwarf [Section 11.2]. Stars of much higher mass—more than about 8 to 10 times the mass of the Sun—lead much shorter and more dramatic lives. They burn rapidly through their core hydrogen, shining steadily for only a few million years. In their final stages of life, they produce many other elements by fusion and then die in the titanic explosions known as supernovae. We'll investigate the dramatic lives of high-mass stars in Section 12.3. First, in this section, we'll explore the lives and deaths of low-mass stars like our Sun.

Before we begin studying the lives of stars, you may wonder why we think we know their life stories. The answer comes from a combination of theoretical modeling and observations of stars. Astronomers use the known laws of physics to create mathematical models of stellar interiors, just as we use models to determine what is going on inside the Sun today [Section 10.2]. These models predict both what is happening in a star at any stage of its life, and how the star must change as its life continues. The models also predict what a star should look like from the outside at various stages of its life. By observing stars of different ages, we can test whether our models are successful in their predictions. The close correspondence between models and observations gives us great confidence that we really do understand the lives of stars.

• What are the life stages of a low-mass star?

Because mass is the key factor in any star's life, all low-mass stars go through life stages similar to those of our Sun. Let's therefore take our Sun as an example of a low-mass star, investigating how we think it will change in the distant future.

Our Sun is currently in the middle of its roughly 10-billion year life as a hydrogen-burning, main-sequence star. As we discussed in Chapter 10, our Sun fuses hydrogen into helium in its core slowly and steadily. The energy released by nuclear fusion gradually moves outward through the solar interior, sometimes traveling as the random bounces of photons and sometimes moving outward with convection. Overall, energy takes hundreds of thousands of years to travel from the core to the surface, where it finally escapes into space as the light that makes the Sun shine. The Sun's fusion rate is well regulated by its gravitational equilibrium—the balance between the outward push of pressure and the inward pull of gravity—so its life will remain relatively uneventful until hydrogen fuel begins to run out in its core.

Red Giant Stage Fusion supplies the thermal energy that maintains the Sun's internal pressure and balances it against gravity. But when the Sun's core hydrogen is finally depleted, nuclear fusion will cease. With no fusion to supply thermal energy and maintain the pressure, the core will no longer be able to resist the crush of gravity, and it will begin to shrink. After 10 billion years of steady hydrogen fusion, the Sun will be entering an entirely new phase of life.

The Sun's core will shrink after exhausting its hydrogen fuel, but rapid hydrogen fusion in a layer surrounding the core will cause the Sun's upper layers to expand outward.

Surprisingly, the Sun's outer layers will expand outward while its core is shrinking. The reason for this strange behavior is a new source of fusion on the outside of the shrinking core. After the core exhausts its hydrogen, it will be made almost entirely of helium, because helium is the "ash" left behind by hydrogen fusion. However, the gas surrounding the core will still contain plenty of fresh hydrogen that has never previously undergone fusion. Because gravity shrinks both the *inert* (nonburning) helium core and the surrounding *shell* of hydrogen, the hydrogen shell soon becomes hot enough to sustain hydrogen fusion (Figure 12.6). In fact, the shell becomes so hot that fusion proceeds at a higher rate than core hydrogen fusion did during the star's main-sequence life. This fusion of hydrogen in a shell around the core, which we call **hydrogen shell burning**, generates enough energy and pressure to push the surrounding layers of gas outward. Thus, hydrogen shell burning will cause the Sun's outer layers to swell in size even as its core shrinks.

Over the next billion years or so, the Sun will grow about 100-fold in radius and even more in luminosity as it expands to become a **red giant** star. The process of expansion will continue as long as the helium core remains inert. Recall that, today, a self-correcting feedback process called the *solar thermostat* regulates the Sun's fusion rate [Section 10.2]: For example, a rise in the Sun's current fusion rate would cause the core to inflate and cool until the fusion rate dropped back down. In contrast, thermal energy generated in the hydrogen-burning shell of a red giant cannot do anything to inflate the inert core that lies underneath. Instead, newly produced helium keeps adding to the mass of the helium core, amplifying its gravitational pull and shrinking it further. The hydrogen-burning shell shrinks along with the core, growing hotter and denser. The fusion rate in the shell consequently rises, feeding even more helium ash to the core. The star is caught in a vicious circle with a broken thermostat.

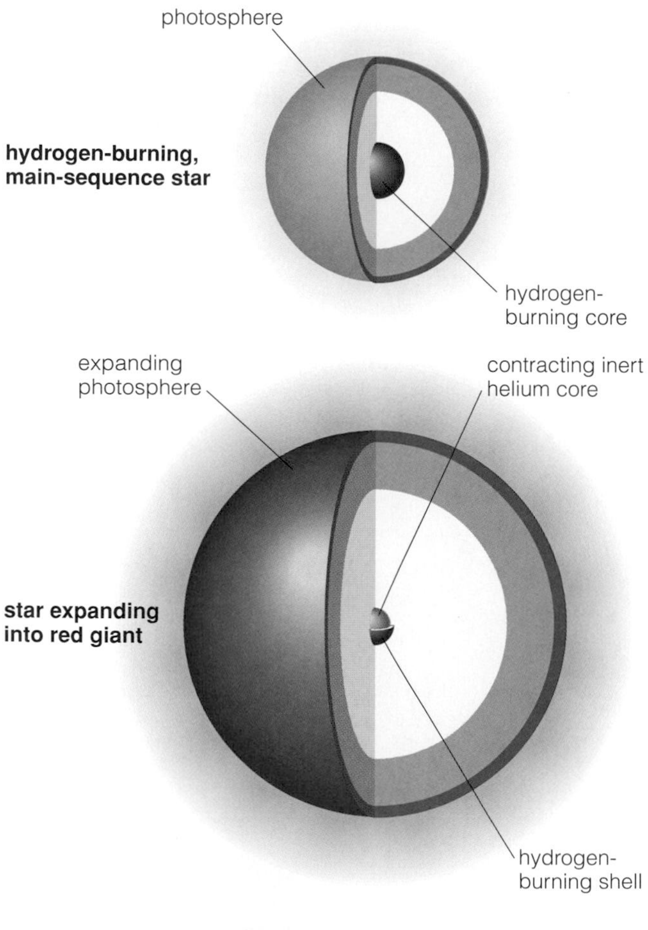

photosphere
hydrogen-burning, main-sequence star
hydrogen-burning core
expanding photosphere
contracting inert helium core
star expanding into red giant
hydrogen-burning shell

Figure 12.6 Interactive Figure

After a star ends its main-sequence life, its inert helium core contracts while hydrogen shell burning begins. The high rate of fusion in the hydrogen shell forces the star's upper layers to expand outward.

The Sun's core will continue to shrink and hydrogen shell burning will continue to intensify as the Sun grows into a red giant.

The core and shell will continue to shrink in size, while an ever-growing fusion rate in the shell pushes the Sun's upper layers outward. Eventually the temperature in the inert helium core will reach about 100 million K, which is hot enough for helium nuclei to fuse together. Meanwhile, the Sun's increasing radius will weaken the pull of gravity at its surface, allowing large amounts of mass to escape in a *stellar wind.* Observation of winds from red giants show that they carry away much more matter than the solar wind carries away from the Sun today.

Other low-mass stars expand into red giants in the same way, though the time it takes depends on their precise masses. Stars more massive than the Sun reach the red giant phase in less than 10 billion years, and proceed through that phase faster than the Sun. Lower-mass stars take longer to go through these stages than the Sun. In a very low-mass star, the inert helium core may never become hot enough to fuse helium. The core collapse will instead be halted by degeneracy pressure, ultimately leaving the star a *helium white dwarf.*

THINK ABOUT IT Before you read on, briefly summarize why a star grows larger and brighter after it exhausts its core hydrogen. When does the growth of a red giant finally halt, and why? How would a star's red giant stage be different if the temperature required for helium fusion were around 200 million K, rather than 100 million K? Why?

Helium Burning Recall that fusion occurs only when two nuclei come close enough together for the attractive *strong force* to overcome electromagnetic repulsion [Section 10.2]. Helium nuclei have two protons (and two neutrons) and hence a greater positive charge than the single proton of a hydrogen nucleus. The greater charge means that helium nuclei repel one another more strongly than hydrogen nuclei. **Helium fusion** therefore occurs only when nuclei slam into one another at much higher speeds than those needed for hydrogen fusion, which means that helium fusion requires much higher temperatures than hydrogen fusion.

The helium fusion process (often called the "triple-alpha" reaction because helium nuclei are sometimes called *alpha particles*) converts three helium nuclei into one carbon nucleus:

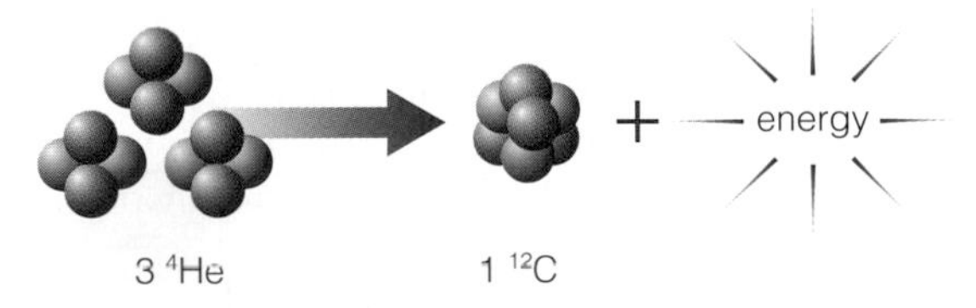

Energy is released because the carbon-12 nucleus has a slightly lower mass than the three helium-4 nuclei, and the lost mass becomes energy in accord with $E = mc^2$.

The ignition of helium burning in a low-mass star like the Sun has one subtlety. Theoretical models show that the thermal pressure in the inert helium core is too low to counteract gravity. Instead, the models show that the pressure fighting against gravity is *degeneracy pressure*—the same strange type of pressure that supports brown dwarfs. Because degeneracy pressure does *not* increase with temperature, the onset of helium fusion heats the core rapidly without causing it to inflate. The rising temperature causes the helium fusion rate to rocket upward in what is called a **helium flash**.

The sudden onset of helium fusion in the Sun's core will stop the core shrinkage, and the Sun will actually become smaller and less luminous than it was as a red giant.

The helium flash releases an enormous amount of energy into the core. In a matter of seconds, the temperature rises so much that thermal pressure again becomes dominant, and degeneracy pressure is no longer important. In fact, the thermal pressure becomes strong enough to push back against gravity, and the core actually begins to expand. This core expansion pushes the hydrogen-burning shell outward, lowering its temperature and its fusion rate. The result is that, even though the star now has core helium fusion and hydrogen shell burning taking place simultaneously (Figure 12.7), the total energy production falls from its peak during the red giant phase. The reduced total energy output of the star reduces its luminosity, allowing its outer layers to contract from their peak size during the red giant phase. As the outer layers contract, the star's surface temperature also increases somewhat. Thus, after the Sun spends about a billion years expanding into a luminous red giant, its size and luminosity will decline as it becomes a *helium-burning star.*

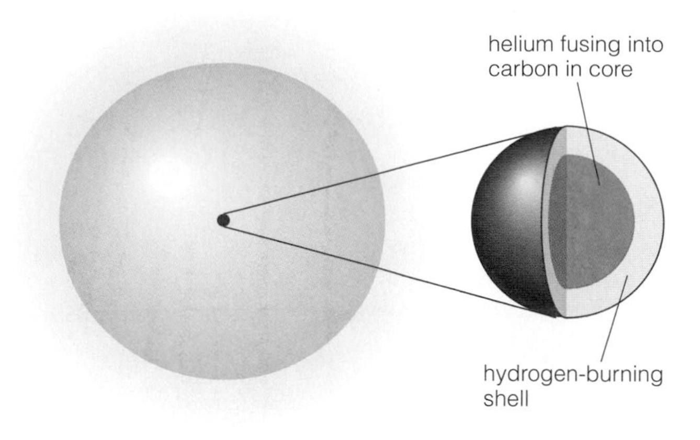

Figure 12.7

Core structure of a helium-burning star. Helium fusion causes the core and hydrogen-burning shell to expand and slightly cool, thereby reducing the overall energy generation rate in comparison to the rate during the red giant stage. The outer layers shrink back, so a helium-burning star is smaller than a red giant of the same mass.

We can see examples of low-mass stars in all the life stages we have discussed so far in the H–R diagram of a globular cluster (Figure 12.8). Stars that are still on the main sequence (below the main-sequence turnoff point) are in the hydrogen-burning phase that lasts most of a star's life. Just to the right of the main-sequence turnoff point we see subgiants—stars that have just begun their expansion into red giants as their cores have shut down and hydrogen shell burning has begun. The longer a star undergoes hydrogen shell burning, the larger and more luminous it becomes, which is why we see a continuous line of stars right up to the most luminous red giants. These are the red giants on the verge of helium flash. The stars that have already undergone the helium flash and become helium-burning stars appear below and to the left of the red giants, because they are somewhat smaller, hotter, and less luminous than they were at the moment of helium flash. The helium cores of all low-mass stars fuse helium into carbon at about the same rate, so these stars all have about the same luminosity. In an H–R diagram of a globular cluster they therefore trace out a horizontal line known as the *horizontal branch.*

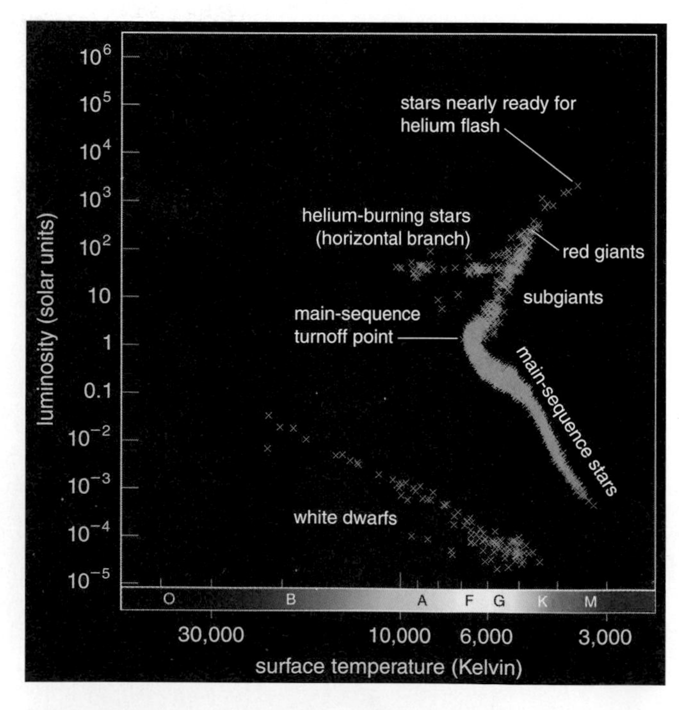

Figure 12.8

An H–R diagram of a globular cluster showing low-mass stars in several different life stages.

Last Gasps It is only a matter of time until a helium-burning star fuses all its core helium into carbon. In the Sun, the core helium will run out after about a hundred million years of burning. When the core helium is exhausted, fusion will again cease. The core, now made of the carbon "ash" from helium fusion, will begin to shrink once again under the crush of gravity.

Core shrinkage will resume after core helium burning ends, while both helium-burning and hydrogen-burning shells make the Sun bigger and more luminous than ever.

The exhaustion of core helium will cause the Sun to expand once again, just as it did in becoming a red giant. This time, the trigger for the expansion will be helium fusion in a shell around the inert carbon core. Meanwhile, the hydrogen shell will still burn atop the helium layer. Thus, the Sun will have become a *double-shell burning star.* Both shells will contract along with the inert core, driving their temperatures and fusion rates so high that the Sun will expand to an even greater size and luminosity than it had in its previous red giant phase.

The furious burning in the helium and hydrogen shells cannot last long—maybe a few million years or less. The Sun's only hope of extending its life will then lie with its carbon core, but this is a false hope for a low-mass star like the Sun. Carbon fusion is possible only at temperatures above about 600 million K, and the Sun's inert carbon core will never get that hot. Remember that degeneracy pressure can come into play in an inert core. For a star like the Sun, degeneracy pressure will halt the gravitational collapse of the core before it shrinks enough to reach the temperature needed for carbon fusion. With nothing left to fuse, the Sun will finally have reached the end of its life.

Suppose the universe contained only low-mass stars. Would elements heavier than carbon exist? Why or why not?

• How does a low-mass star die?

The huge size of the dying Sun will mean that it has a very weak grip on its outer layers. As the Sun's luminosity and radius keep rising, its wind will grow stronger. Observations of other stars already in this late stage of life show that their winds are an important source of the *interstellar dust grains* found in star-forming clouds. These dust grains form because the wind cools as it flows away from the star. At the point where the gas temperature has dropped to 1,000–2,000 K, some of the heavier elements in the wind begin to condense into microscopic clusters, forming small, solid particles of dust. This process of dust formation is much like the condensation that occurred in the solar nebula before the planets formed [Section 6.4]. The dust particles drift with the stellar wind into interstellar space, where they mix with other gas and dust in the galaxy.

The Sun's final end will be beautiful to those who witness it, as long as they are plenty far away. Through winds and other processes, the Sun will eject its outer layers into space, creating a huge shell of gas expanding away from the inert carbon core. The exposed core will still be very hot and will therefore emit intense ultraviolet radiation. This radiation will ionize the gas in the expanding shell, making it glow brightly as a **planetary nebula**. We have photographed many examples of planetary nebulae around other low-mass stars that have just died in this very same way (Figure 12.9). Note that, despite their name, planetary nebulae have

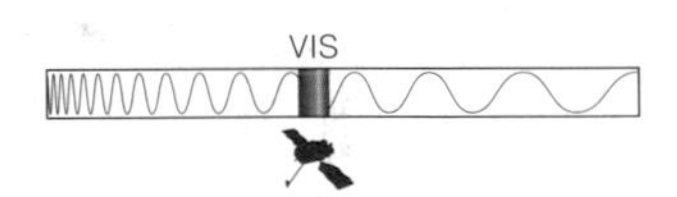

Figure 12.9

Hubble Space Telescope photos of planetary nebulae, which form when low-mass stars in their final death throes cast off their outer layers of gas. The central white dots are the remaining hot cores of the stars that ejected the gas. These hot cores ionize and energize the shells of gas that surround them. As the nebula gas disperses into space, the hot core remains as a white dwarf.

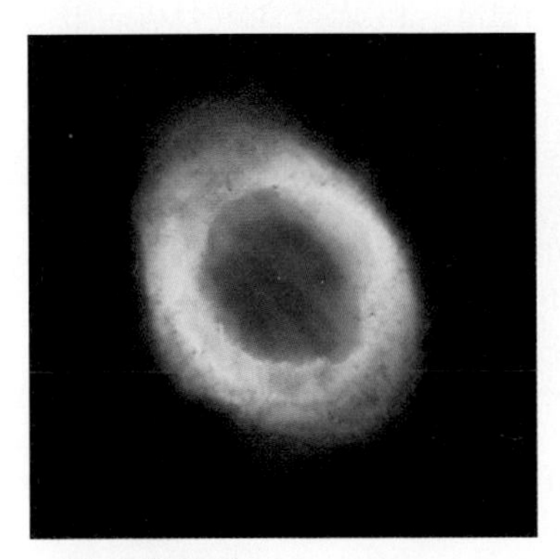

a Ring Nebula

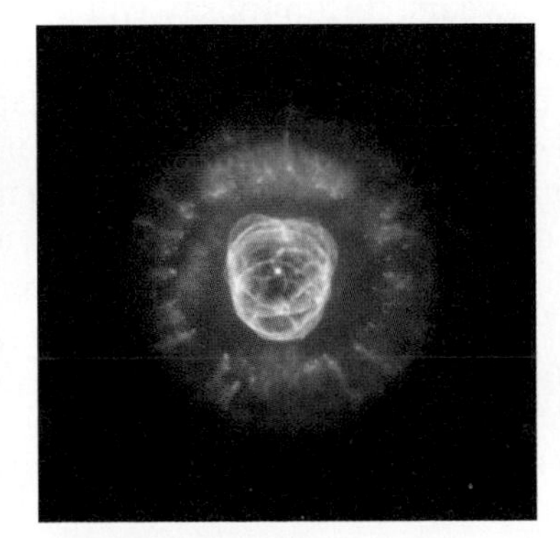

b Eskimo Nebula

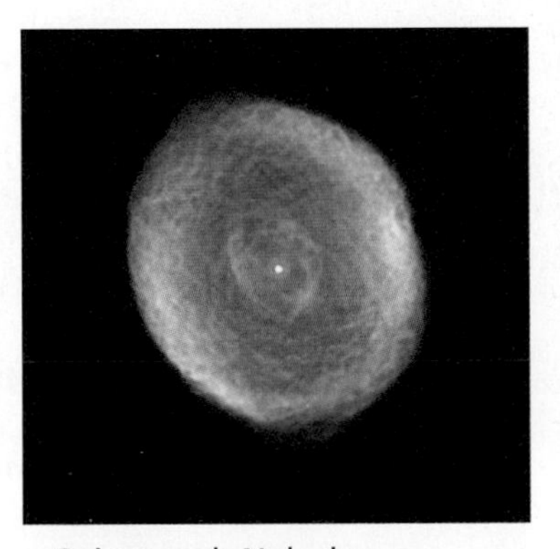

c Spirograph Nebula

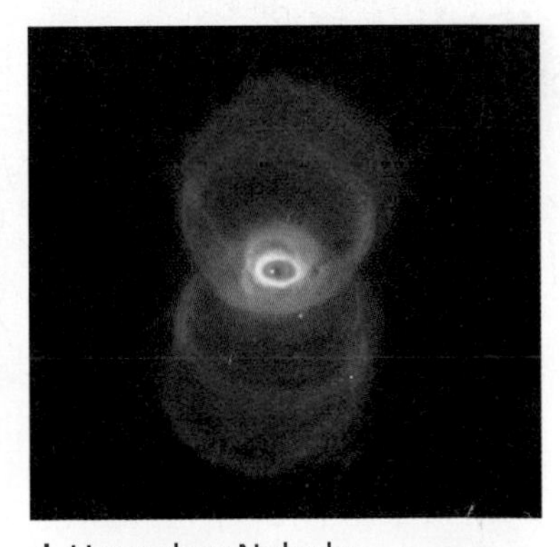

d Hourglass Nebula

nothing to do with planets. The name comes from the fact that nearby planetary nebulae look much like planets through small telescopes, appearing as simple disks.

When it dies, the Sun will eject its outer layers into space as a planetary nebula, leaving its exposed core behind as a white dwarf.

The glow of the planetary nebula will fade as the exposed core cools and the ejected gas disperses into space. The nebula will disappear within a million years, leaving the Sun's cooling carbon core behind as a *white dwarf*. Recall from Chapter 11 that white dwarfs are small in radius but high in mass and temperature. We can now understand why: They are small in radius but high in mass because they are the exposed cores of dead stars, supported against the crush of gravity by degeneracy pressure. They are hot because they only recently were in the center of a star, and have not yet had time to cool much. Thus, a white dwarf is little more than a decaying corpse that will cool for the indefinite future, eventually disappearing from view as it becomes too cold to emit any more visible light.

The Fate of the Earth The death of the Sun will obviously have consequences for Earth, and some of these consequences will begin even before the Sun enters the final stages of its life. Although the Sun will shine steadily for the next 5 billion years of its hydrogen-burning life, theoretical models show that it will become slightly more luminous with time. Although this rise in luminosity will be small compared to what happens in red giant stage, it will probably be enough to cause a *runaway greenhouse effect* [Section 7.4] on Earth between about 1 and 4 billion years from now, making Earth's oceans boil away. Temperatures on Earth will rise even more dramatically when the Sun finally exhausts its core supply of hydrogen, somewhere around the year A.D. 5,000,000,000.

Things will get steadily worse as the Sun grows into a red giant over the next several hundred million years. Just before helium flash, the Sun will be more than 1,000 times as luminous as it is today, and this huge luminosity will heat Earth's surface to more than 1,000 K. Clearly, any surviving humans will need to have found a new home. Saturn's moon Titan [Section 11.3] might not be a bad choice. Its surface temperature will have risen from well below freezing today to about the present temperature of Earth.

The Sun will shrink and cool somewhat after helium flash turns it into a helium-burning star, providing a temporary lull in the incineration of Earth. However, this respite will last only 100 million years or so, and then Earth will suffer one final disaster.

After exhausting its core helium, the Sun will expand again during its last million years. Its luminosity will soar to thousands of times what it is today, and its radius will grow to nearly the present radius of Earth's orbit—so large that solar prominences might lap at Earth's surface. Finally, the Sun will eject its outer layers as a planetary nebula that will engulf Jupiter and Saturn and drift on past Pluto into interstellar space. If Earth is not destroyed, its charred surface will be cold and dark in the faint, fading light of the white dwarf that the Sun has become.

The Sun's Life on an H–R Diagram We have discussed the Sun's life stages in general terms, but with computer models astronomers can determine how the Sun's surface temperature and luminosity will change during each stage of its life. We can plot the results on an H–R diagram,

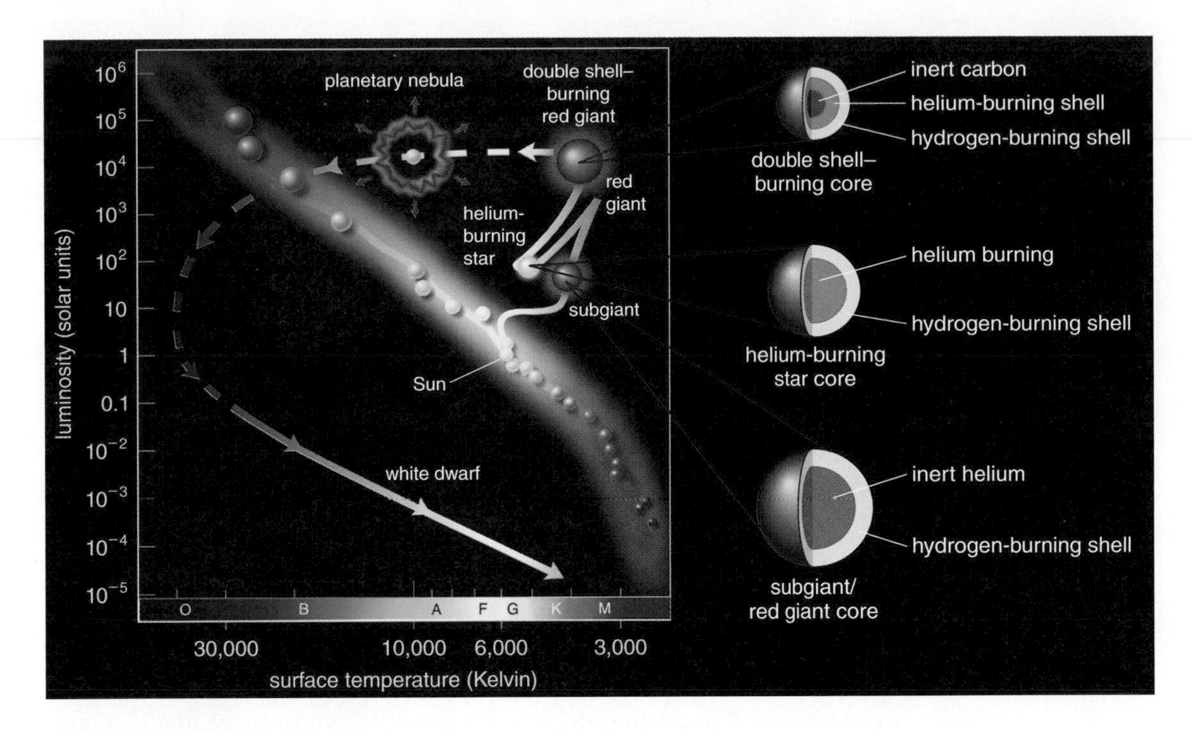

Figure 12.10 Interactive Figure

The life track of our Sun from its hydrogen-burning, main-sequence stage to the white dwarf stage. Core structure is shown at key stages.

producing a **life track** from which we can read the Sun's luminosity and surface temperature at any stage of its life (Figure 12.10).

A star's life track in an H–R diagram shows how its properties change with time.

You should study the Sun's life track carefully to make sure you understand each of the phases shown, because the life tracks of all low-mass stars are similar to the Sun's. Remember that the Sun spends most of its life as a hydrogen-burning, main-sequence star. The life track then shows what happens once the Sun exhausts its core hydrogen supply. Because the Sun expands in size and luminosity while it undergoes hydrogen shell burning, the life track goes generally upward as the Sun be-

SPECIAL TOPIC: HOW LONG IS FIVE BILLION YEARS?

The Sun's demise in about 5 billion years might at first seem worrisome, but 5 billion years is a very long time. It is longer than Earth has yet existed, and human time scales pale by comparison. A single human lifetime, if we take it to be about 100 years, is only 2×10^{-8}, or two hundred-millionths, of 5 billion years. Because 2×10^{-8} of a human lifetime is about 1 minute, we can say that a human lifetime compared to the life expectancy of the Sun is roughly the same as 60 heartbeats compared to a human lifetime.

What about human creations? The Egyptian pyramids have often been described as "eternal," but they are slowly eroding due to wind, rain, air pollution, and the impact of tourists. All traces of them will have vanished within a few hundred thousand years. While a few hundred thousand years may seem like a long time, the Sun's remaining lifetime is more than 1,000 times longer.

On a more somber note, we can gain perspective on 5 billion years by considering evolutionary time scales. During the past century, our species has acquired sufficient technology and power to destroy human life totally, if we so choose. However, even if we make that unfortunate choice, some species (including many insects) are likely to survive.

Would another intelligent species ever emerge on Earth? We have no way to know [Section 18.4], but we can look to the past for guidance. Many species of dinosaurs were biologically quite advanced, if not truly intelligent, when they were suddenly wiped out about 65 million years ago. Some small rodentlike mammals survived, and here we are 65 million years later. We therefore might guess that another intelligent species could evolve some 65 million years after a human extinction. If these beings also destroyed themselves, another species could evolve 65 million years after that, and so on.

Even at 65 million years per shot, Earth would have *nearly 80* more chances for an intelligent species to evolve in 5 billion years (5 billion $\div$ 65 million $\approx$ 77). Perhaps one of those species will not destroy itself, and future generations might move on to other star systems by the time the Sun finally dies. Perhaps this species will be our own.

comes a subgiant and then a red giant. The track also goes slightly to the right during this phase, because its surface temperature falls a bit. The tip of the red giant phase represents the moment of helium flash, after which the Sun contracts back in size, shown by the track going down and left as the Sun becomes a helium-burning star. Finally, after the Sun exhausts its core helium, it expands again as a double-shell burning red giant, and then ejects its planetary nebula into space and leaves a white dwarf behind. (The life track is dashed between its double-shell burning stage and white dwarf stage, because the ejection of the Sun's outer layers changes what we mean by the "surface" of the Sun.)

Stellar Evolution Tutorial, Lesson 3

12.3 LIFE AS A HIGH-MASS STAR

Human life would be impossible without both low-mass stars and high-mass stars. The long lives of low-mass stars allow evolution to proceed for billions of years, but only high-mass stars produce the full array of elements on which life depends. Low-mass stars can't produce these elements because their cores never get hot enough to fuse elements heavier than helium. The greater electric charges of heavier nuclei create greater repulsive forces that prevent fusion from happening except at extremely high temperatures. Only the cores of high-mass stars reach high enough temperatures, and that happens only near the very end of a high-mass star's life, when the immense weight of its overlying layers bears down on a core that has already exhausted its hydrogen fuel.

The early stages of a high-mass star's life are similar to the early stages of the Sun's life, except they proceed much more rapidly. However, in the final stages of life, the highest-mass stars proceed to fuse increasingly heavy elements until they have exhausted all possible fusion sources. When fusion finally stops for good, gravity causes the core to implode suddenly. As we will soon see, the implosion of the core causes the star to self-destruct in the titanic explosion we call a *supernova*. The fast-paced life and cataclysmic death of a high-mass star is surely among the great dramas of the universe.

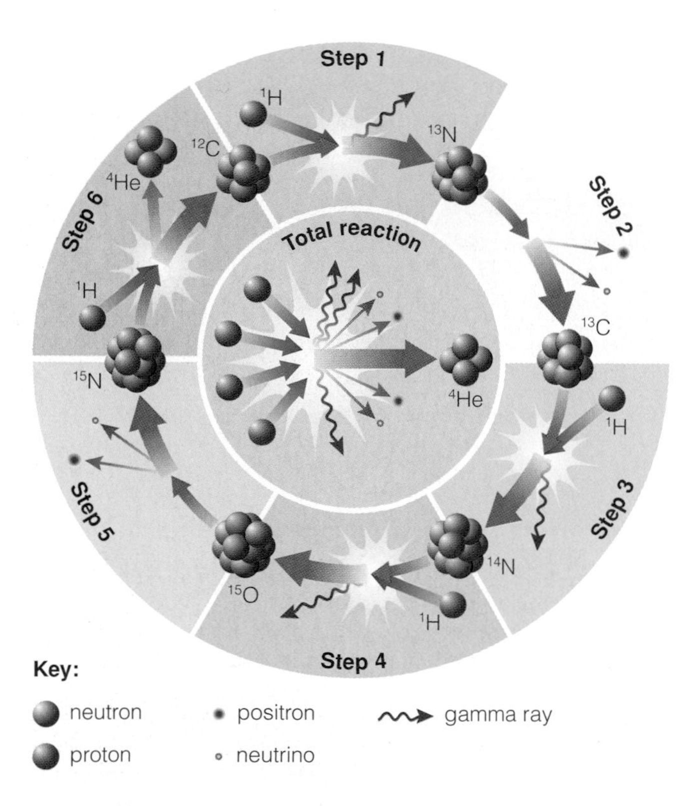

Figure 12.11

This diagram illustrates the six steps of the CNO cycle by which massive stars fuse hydrogen into helium. Note that the overall result is the same as that of the proton–proton chain: Four hydrogen nuclei fuse to make one helium nucleus. The carbon, nitrogen, and oxygen nuclei help the cycle proceed, but these nuclei are neither consumed nor created in the overall cycle.

• What are the life stages of a high-mass star?

To illustrate the life stages of a high-mass star, let's look at the life of a star born with 25 times the mass of the Sun ($25M_{\text{Sun}}$). Like all other stars, it forms out of a cloud fragment forced to contract into a protostar by gravity. Hydrogen burning begins when the gravitational potential energy released by the contracting protostar makes the core hot enough for fusion. However, hydrogen fusion inside a high-mass star proceeds through a different set of steps than hydrogen fusion in a low-mass star, which is part of the reason why high-mass stars live such brief but brilliant lives.

Hydrogen Fusion in a High-Mass Star Recall that a low-mass star like our Sun fuses hydrogen into helium through the *proton–proton chain* (see Figure 10.6). In a high-mass star, the strong gravity compresses the hydrogen core to a higher temperature than we find in lower-mass stars. The hotter core temperature makes it possible for protons to slam into carbon, oxygen, or nitrogen nuclei as well as into other protons. As a result, hydrogen fusion in high mass stars occurs through a chain of reactions called the **CNO cycle** (the letters *CNO* stand for carbon, nitrogen, and oxygen, respectively). Figure 12.11 shows the six steps of the CNO cycle.

A high-mass star lives a short life, rapidly fusing its core hydrogen into helium via the CNO cycle.

Notice that the overall reaction of the CNO cycle is the same as that of the proton–proton chain: Four hydrogen nuclei fuse into one helium-4 nucleus. The amount of energy generated in each reaction cycle therefore is also the same —it is equal to the difference in mass between the four hydrogen nuclei and the one helium nucleus multiplied by c^2. However, the CNO cycle allows hydrogen fusion to proceed at a far higher rate than would be possible by the proton-proton chain alone. That is why the luminosities of high-mass stars are so much higher than those of low-mass stars and their lives are so much shorter.

Becoming a Supergiant Our 25-solar-mass star will begin to run low on hydrogen fuel after only a few million years. As the core hydrogen runs out, the star responds much like a low-mass star, but much faster. It develops a hydrogen-burning shell, and its outer layers begin to expand outward, turning it into a giant star. At the same time, the core contracts, and this gravitational contraction releases energy that raises the core temperature until it becomes hot enough to fuse helium into carbon. However, there is no helium flash in high-mass stars. Their core temperatures are so high that thermal pressure remains strong, preventing degeneracy pressure from being a factor. Helium burning therefore ignites gradually, just as hydrogen burning did at the beginning of the star's life.

Our high-mass star fuses helium into carbon so rapidly that it is left with an inert carbon core after just a few hundred thousand years. Once again, the absence of fusion leaves the core without an energy source to fight off the crush of gravity. The inert carbon core shrinks, the crush of gravity intensifies, and the core pressure, temperature, and density all rise. Meanwhile, a helium-burning shell forms between the inert core and the hydrogen-burning shell. The shrinking core gets hotter and hotter, and in our high-mass star it soon becomes hot enough for carbon fusion. As we'll discuss in more detail shortly, the core goes through several more phases of fusion of increasingly heavy elements, but each phase lasts a much shorter time than the previous one.

Near the end of its life, a high-mass star expands to become a supergiant as fusion proceeds furiously in its core and surrounding shells.

Despite the dramatic events taking place in its interior, the high-mass star's outer appearance changes only slowly. As each stage of core fusion ceases, the surrounding shell burning intensifies and further inflates the star's outer layers, turning the star into a supergiant. Each time the core flares up, the outer layers contract somewhat but the star's overall luminosity remains about the same. The result is that the star's life track zigzags across the top of the H–R diagram (Figure 12.12). In the most massive stars of all, the core changes happen so quickly that the outer layers don't have time to respond, and the star progresses steadily toward becoming a red supergiant.

One of these massive, red supergiant stars happens to be relatively nearby: Betelgeuse, the upper-left shoulder of Orion. We are not sure exactly how much mass Betelgeuse was born with, because by this stage of

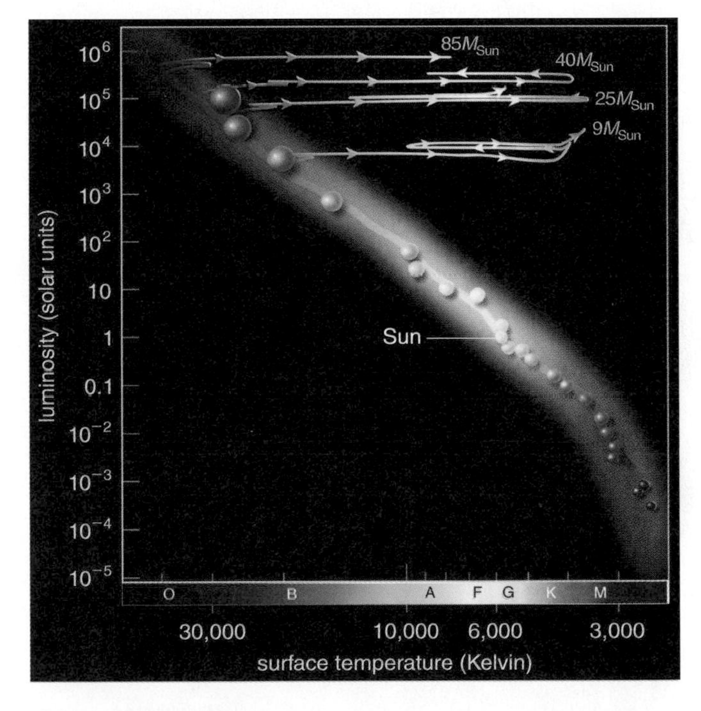

Figure 12.12

Life tracks on the H–R diagram from main-sequence star to red supergiant for selected high-mass stars. Labels on the tracks give the star's mass at the beginning of its main-sequence life. Because of the strong wind from such a star, its mass can be considerably smaller when it leaves the main sequence. (Based on models from A. Maeder and G. Meynet.)

its life it has lost considerable mass through powerful stellar winds. However, it is certainly a high-mass star like the one we have been discussing. The fact that Betelgeuse is now a red supergiant tells us that it is in the late stages of its life, although we have no way of knowing exactly what stage of nuclear burning is now taking place in its core. Betelgeuse may have a few thousand years of nuclear burning still ahead, or we may be seeing it during its final stages of life. If the latter is the case, then sometime soon we will witness one of the most dramatic events that ever occurs in the universe—a supernova explosion. To understand why a supernova happens, we need to look at how high-mass stars make elements heavier than carbon.

• How do high-mass stars make the elements necessary for life?

A low-mass star can't make elements heavier than carbon because degeneracy pressure halts the contraction of its inert carbon core before it can get hot enough for fusion. A high-mass star has no such problem. The crush of gravity in a high-mass star keeps its carbon core so hot that degeneracy pressure never comes into play. After helium fusion stops, the gravitational contraction of the carbon core continues until it reaches the 600 million K required to fuse carbon into heavier elements.

Carbon fusion provides the core with a new source of energy that restores gravitational equilibrium, but only temporarily. In the highest-mass stars, carbon burning may last only a few hundred years. When the core carbon is depleted, the core again begins to collapse, shrinking and heating once more until it can fuse a still-heavier element. The star is engaged in the final phases of a desperate battle against the ever-strengthening crush of gravity. The star will ultimately lose the battle, but it will be a victory for life in the universe: In the process of its struggle against gravity, the star will produce the heavy elements of which Earth-like planets and life are made.

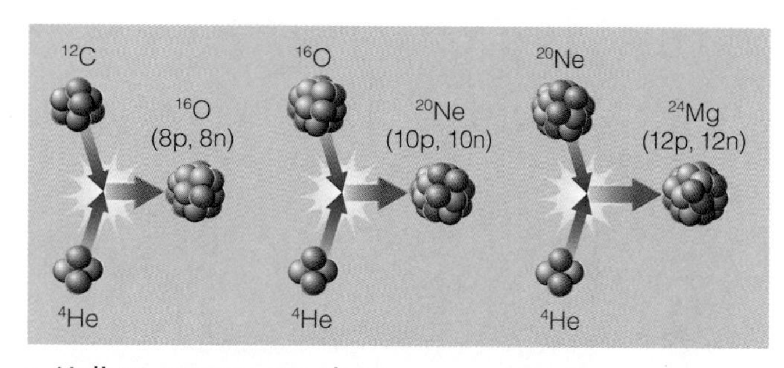

a Helium-capture reactions

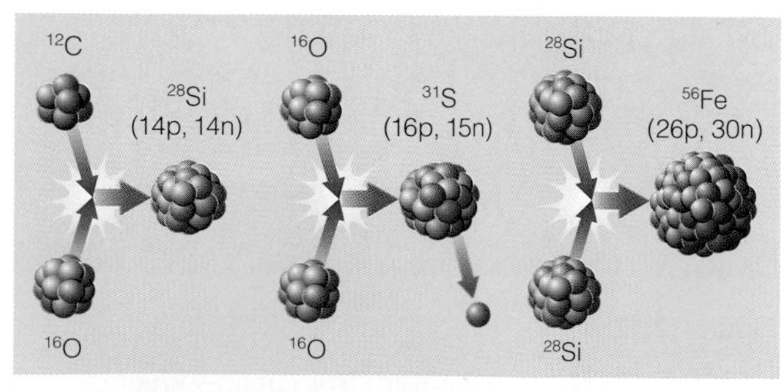

b Other reactions. (Note: Fusion of two silicon nuclei actually first produces nickel-56, but this decays rapidly to cobalt-56 and then to iron-56.)

Figure 12.13

A few of the many nuclear reactions that occur in the final stages of a high-mass star's life.

Advanced Nuclear Burning The nuclear reactions in a high-mass star's final stages of life become quite complex, and many different reactions may take place simultaneously. The simplest sequence of fusion stages occurs through successive **helium capture** reactions—reactions in which a helium nucleus fuses into some other nucleus (Figure 12.13a). (Some helium nuclei still remain in the core, but not enough to continue helium fusion efficiently.) Helium capture can fuse carbon into oxygen, oxygen into neon, neon into magnesium, and so on.

At high enough temperatures, a star's core plasma can fuse heavy nuclei to one another. For example, fusing carbon to oxygen creates silicon, fusing two oxygen nuclei creates sulfur, and fusing two silicon nuclei generates iron (Figure 12.13b). Some of these heavy-element reactions release free neutrons, which may fuse with heavy nuclei to make still rarer elements. The star is forging the variety of elements that, in our solar system at least, became the stuff of life.

The core of a high-mass star eventually becomes hot enough for fusion to produce the elements of which we and Earth are made.

Each time the core depletes the elements it is fusing, it shrinks and heats until it becomes hot enough for other fusion reactions. Meanwhile, a new type of shell burning ignites between the core and the overlying shells of fusion. Near the end, the star's central region resembles the inside of an onion, with layer upon layer of shells burning different elements (Figure 12.14). During the star's final few days, iron begins to pile up in the silicon-burning core.

Iron: Bad News for the Stellar Core The core continues shrinking and heating while iron piles up from nuclear burning in the surrounding shells. If iron were like the other elements in prior stages of nuclear burning, this core contraction would stop when iron fusion ignited. However, iron is unique among the elements in a very important way: It is the one element from which it is *not* possible to generate any kind of nuclear energy.

To understand why iron is unique, remember that only two basic processes can release nuclear energy: *fusion* of light elements into heavier ones, and *fission* of very heavy elements into not-so-heavy ones (see Figure 10.4). Recall that hydrogen fusion converts four protons (hydrogen nuclei) into a helium nucleus that consists of two protons and two neutrons. Thus, the total number of *nuclear particles* (protons and neutrons combined) does not change. However, this fusion reaction generates energy (in accord with $E = mc^2$) because the *mass* of the helium nucleus is less than the combined mass of the four hydrogen nuclei that fused to create it—despite the fact that the *number* of nuclear particles is unchanged.

In other words, fusing hydrogen into helium generates energy because helium has a lower *mass per nuclear particle* than hydrogen. Similarly, fusing three helium-4 nuclei into one carbon-12 nucleus generates energy because carbon has a lower mass per nuclear particle than helium, which means that some mass disappears and becomes energy in this fusion reaction. In fact, the decrease in mass per nuclear particle from hydrogen to helium to carbon is part of a general trend shown in Figure 12.15.

The mass per nuclear particle tends to decrease as we go from light elements to iron, which means that fusion of light nuclei into heavier nuclei generates energy. This trend reverses beyond iron: The mass per nuclear particle tends to *increase* as we look to still heavier elements. As a result, elements heavier than iron can generate nuclear energy only through fission into lighter elements. For example, uranium has a greater mass per nuclear particle than lead, so uranium fission (which ultimately leaves lead as a by-product) must convert some mass into energy.

A high-mass star's death is imminent when iron piles up in its core, because no energy can be released by fusion of iron.

Iron has the lowest mass per nuclear particle of all nuclei and therefore cannot release energy by either fusion or fission. Thus, once the matter in a stellar core turns to iron, it can generate no further energy. The iron core's only hope of resisting the crush of gravity lies with degeneracy pressure, but iron keeps piling up until even degeneracy pressure cannot support the core. What ensues is the ultimate nuclear-waste catastrophe. The star explodes as a supernova, scattering all the newly made elements into interstellar space.

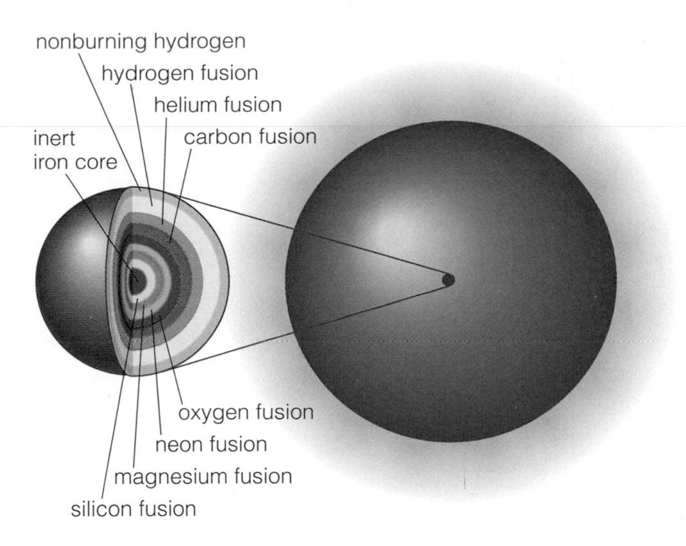

Figure 12.14 Interactive Figure

The multiple layers of nuclear burning in the core of a high-mass star during the final days of its life.

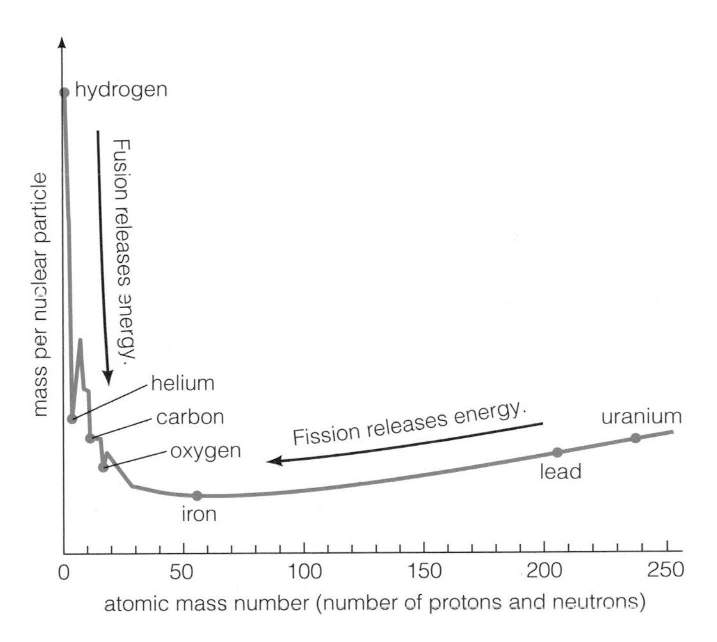

Figure 12.15

Overall, the average mass per nuclear particle declines from hydrogen to iron and then increases. Selected nuclei are labeled to provide reference points. (This graph shows the most general trends only. A more detailed graph would show numerous up-and-down bumps superimposed on the general trends. The vertical scale is arbitrary, but shows the general idea.)

How would the universe be different if hydrogen, rather than iron, had the lowest mass per nuclear particle? Why?

Evidence for the Origin of Elements Before we look at how a supernova happens, let's consider the evidence that indicates we actually understand the origin of the elements. We cannot see inside stars, so we cannot

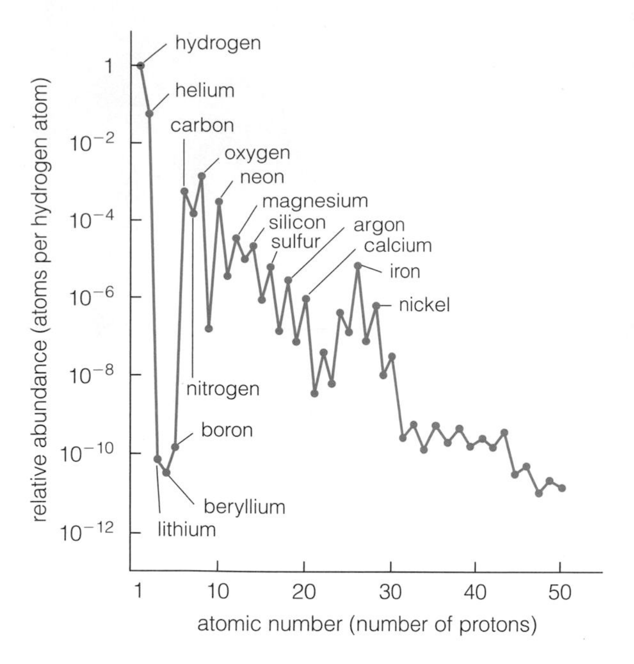

Figure 12.16

This graph shows the observed relative abundances of elements in the galaxy in comparison to the abundance of hydrogen (given as 1 in this comparison). For example, the abundance of nitrogen is about 10^{-4}, which means that there are about $10^{-4} = 0.0001$ times as many nitrogen atoms in the galaxy as hydrogen atoms.

directly observe elements being created in the ways we've discussed. However, the signature of nuclear reactions in massive stars is written in the patterns of elemental abundances across the universe.

For example, if massive stars really produce heavy elements (that is, elements heavier than hydrogen and helium) and scatter these elements into space when they die, the total amount of these heavy elements in interstellar gas should gradually increase with time (because additional massive stars have died). We should expect stars born recently to contain a greater proportion of heavy elements than stars born in the distant past, because they formed from interstellar gas that contained more heavy elements.

Stellar spectra confirm this prediction: Older stars do indeed contain smaller amounts of heavy elements than younger stars. For very old stars in globular clusters, elements besides hydrogen and helium typically make up as little as 0.1% of the total mass. In contrast, young stars that formed in the recent past contain about 2–3% of their mass in the form of heavy elements.

Measurements of element abundances in the cosmos confirm our models of how high-mass stars produce heavy elements.

We gain even more confidence in our model of elemental creation when we compare the abundances of various elements in the cosmos. For example, because helium-capture reactions add two protons (and two neutrons) at a time, we expect nuclei with even numbers of protons to outnumber those with odd numbers of protons that fall between them. Sure enough, even-numbered nuclei such as carbon, oxygen, and neon are relatively abundant (Figure 12.16). Similarly, because elements heavier than iron are made only by rare fusion reactions shortly before and during a supernova, we expect these elements to be extremely rare. Again, observations verify this prediction made by our model of nuclear creation.

• How does a high-mass star die?

Let's return now to our high-mass star, with iron piling up in its core. As we've discussed, it has no hope of generating any energy by fusion of this iron. After shining brilliantly for a few million years, the star will not live to see another day.

The Supernova Explosion The degeneracy pressure that briefly supports the inert iron core arises because the laws of quantum mechanics prohibit electrons from getting too close together. Once gravity pushes the electrons past the quantum mechanical limit, however, they can no longer exist freely. In an instant, the electrons disappear by combining with protons to form neutrons, releasing neutrinos in the process (Figure 12.17). The degeneracy pressure provided by the electrons instantly vanishes, and gravity has free rein.

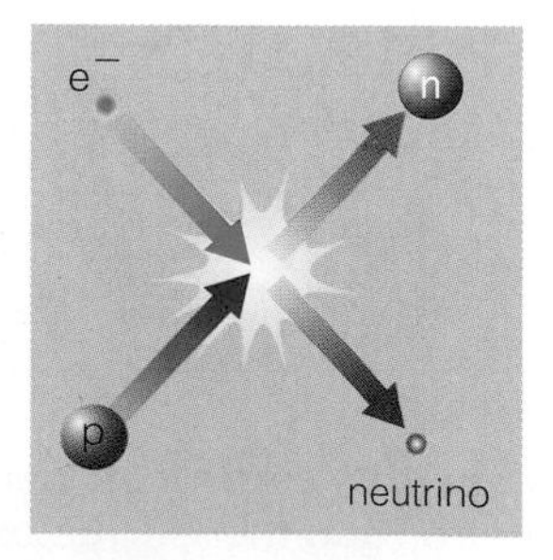

Figure 12.17

During the final, catastrophic collapse of a high-mass stellar core, electrons and protons combine to form neutrons, accompanied by the release of neutrinos.

In a fraction of a second, an iron core with a mass comparable to that of our Sun and a size larger than that of Earth collapses into a ball of neutrons just a few kilometers across. The collapse halts only because the neutrons have a degeneracy pressure of their own. The entire core then resembles a giant atomic nucleus. If you recall that ordinary atoms are made almost entirely of empty space [Section 5.1] and that almost all their mass is in their nuclei, you'll realize that a giant atomic nucleus must have an astoundingly high density.

The gravitational collapse of the core releases an enormous amount of energy—more than a hundred times what the Sun will radiate over its entire 10-billion-year lifetime. Where does this energy go? It drives the outer layers off into space in a titanic explosion called a **supernova**. The ball of neutrons left behind is called a **neutron star**. In some cases, the remaining mass may be so large that gravity also overcomes neutron degeneracy pressure, and the core continues to collapse until it becomes a *black hole* [Section 13.3].

When gravity overcomes degeneracy pressure in the iron core, the core collapses into a ball of neutrons and the star explodes in a supernova.

Theoretical models of supernovae successfully reproduce the observed energy outputs of real supernovae, but the precise mechanism of the explosion is not yet clear. Two general processes could contribute to the explosion. In the first process, neutron degeneracy pressure halts the gravitational collapse, causing the core to rebound slightly and ram into overlying material that is still falling inward. Until recently, most astronomers thought that this *core-bounce* process ejected the star's outer layers. Current models of supernovae, however, suggest that the more important process involves the neutrinos formed when electrons and protons combine to make neutrons. Although these ghostly particles rarely interact with anything [Section 10.2], so many are produced when the core implodes that they drive a shock wave that propels the star's upper layers into space.

The supernova scatters the elements produced by the star into space and leaves behind a neutron star or a black hole.

The shock wave sends the star's former surface zooming outward at a speed of 10,000 kilometers per second—fast enough to travel the distance from the Sun to Earth in only about 4 hours. The heat of the explosion makes the gas shine with dazzling brilliance. For about a week, a supernova blazes as powerfully as 10 billion Suns, rivaling the luminosity of a moderate-size galaxy. The ejected gases slowly cool and fade in brightness over the next several months, but they continue to expand outward until they eventually mix with other gases in interstellar space. The scattered debris from the supernova carries with it the variety of elements produced in the star's nuclear furnace, as well as additional elements created when some of the neutrons produced during the core collapse slam into other nuclei. Millions or billions of years later, this debris may be incorporated into a new generation of stars. We are truly "star stuff," because we and our planet were built from the debris of stars that exploded long ago.

Supernova Observations The study of supernovae owes a great debt to astronomers of many different epochs and cultures. Careful scrutiny of the night skies allowed ancient people to identify several supernovae whose remains can still be seen. The most famous example is the Crab Nebula in the constellation Taurus. The Crab Nebula is a **supernova remnant**—an expanding cloud of debris from a supernova explosion (Figure 12.18). A spinning neutron star lies at the center of the Crab Nebula, providing evidence that supernovae really do create neutron stars. Photographs taken years apart show that the nebula is growing larger at a rate of several thousand kilometers per second. Calculating backward from its present size, we can trace the nebula's birth to somewhere near A.D. 1100. Chinese observers recorded a "guest star" near this location on July 4, 1054, undoubtedly the supernova that created this nebula.

No supernova has been seen in our own galaxy since 1604, but today astronomers routinely discover supernovae in other galaxies. The nearest of these extragalactic supernovae, and the only one near enough to be

Figure 12.18 VIS

The Crab Nebula is the remnant of the supernova observed in A.D. 1054. This photograph was taken with the Very Large Telescope at the European Southern Observatory in Chile.

Figure 12.19

Before-and-after photos of the location of Supernova 1987A.

Before. The arrow points to the star observed to explode in 1987.

After. The supernova actually appeared as a bright point of light. It appears larger than a point in this photograph only because of overexposure.

visible to the naked eye, burst into view in 1987. Because it was the first supernova detected that year, it was given the name **Supernova 1987A**. Supernova 1987A was the explosion of a star in the *Large Magellanic Cloud,* a small galaxy that orbits the Milky Way and is visible only from southern latitudes. The Large Magellanic Cloud is about 150,000 light-years away, so the star really exploded some 150,000 years ago.

As the nearest supernova witnessed in four centuries, Supernova 1987A provided a unique opportunity to study a supernova and its debris in detail. Astronomers from all over the planet traveled to the Southern Hemisphere to observe it, and several orbiting spacecraft added observations in many different wavelengths of light.

Observations of supernova remnants confirm many aspects of our models of how high-mass stars live and die.

Older photographs of the Large Magellanic Cloud allowed astronomers to determine precisely which star had exploded (Figure 12.19). It turned out to be a blue star, not the red supergiant expected when core fusion has ceased. The most likely explanation is that the star's outer layers were unusually thin and warm near the end of its life, changing its appearance from that of a red supergiant to a blue one. The surprising color of the pre-explosion star demonstrates that we still have much to learn about supernovae. Reassuringly, most other theoretical predictions of stellar life cycles were well matched by observations of Supernova 1987A.

One of the most remarkable findings from Supernova 1987A was a burst of neutrinos, recorded by neutrino detectors in Japan and Ohio. The neutrino data confirmed that the explosion released most of its energy in the form of neutrinos, suggesting that we are correct in believing that the stellar core undergoes sudden collapse to a ball of neutrons.

THINK ABOUT IT When Betelgeuse explodes as a supernova, it will be more than 10 times as bright as the full moon in our sky. If our ancestors had seen Betelgeuse explode a few hundred or a few thousand years ago, do you think it could have had any effect on human history? How do you think our modern society would react if we saw Betelgeuse explode tomorrow?

12.4 SUMMARY OF STELLAR LIVES

Much of this chapter has been devoted to telling the life stories of two different stars, one with the mass of our Sun and another that began life with a mass 25 times as great. Now we will complete the picture of star life and star death by showing how these two basic life stories apply to stars of all masses. We will also see that some stars can have more complicated life stories, but only if they happen to be members of close binary systems.

• How does a star's mass determine its life story?

We have seen that the primary factor determining how a star lives its life is its mass. Low-mass stars live long lives and die in planetary nebulae, leaving behind white dwarfs. High-mass stars live short lives and die in supernovae, leaving behind neutron stars and black holes. Figure 12.20 summarizes the life cycles of the two stars we have focused on in this chapter.

Stars born with less than about $8M_{Sun}$ follow life stages similar to that of our Sun, while more massive stars live short but brilliant lives and die in supernova explosions.

Stars that begin life with less than about eight times the mass of the Sun ($8M_{Sun}$) have life stories similar to that of the $1M_{Sun}$ star in Figure 12.20. The main differences are in the overall life span and in what goes on inside the star. Stars less massive than $1M_{Sun}$ progress through their life stages more slowly, while more massive stars progress through their life stages more quickly. Stars with more than about two times the Sun's mass fuse hydrogen through the CNO cycle, and their cores remain so hot that they never undergo a helium flash. However, unless stars have masses above about $8M_{Sun}$, they never get hot enough to produce iron in their cores, and thus they die without reaching the point of supernova.

Stars that begin life with more than $8M_{Sun}$ have life stories similar to that of the $25M_{Sun}$ star in Figure 12.20. Again, stars with more mass fuse their elements more furiously and go through all these stages faster. The biggest difference in their life stories may only come at the very end—the most massive stars of all might become black holes when they die.

• How are the lives of stars with close companions different?

For the most part, stars in binary systems proceed from birth to death as if they were isolated and alone. The exceptions are close binary stars. Algol, the "demon star" in the constellation Perseus, consists of two stars that orbit each other closely: a $3.7M_{Sun}$ main-sequence star and a $0.8M_{Sun}$ subgiant.

A moment's thought reveals that something quite strange is going on. The stars of a binary system are born at the same time and therefore must both be the same age. We know that more massive stars live shorter lives, and therefore the more massive star must exhaust its core hydrogen and become a subgiant before the less massive star does. How, then, can Algol's less massive star be a subgiant while the more massive star is still burning hydrogen as a main-sequence star?

Continued on p. 324

Protostars: A star system forms when a cloud of interstellar gas collapses under gravity. The central protostar is surrounded by a protostellar disk in which planets may eventually form.

Blue main-sequence star: Star is fueled by hydrogen fusion in its core. In high-mass stars, hydrogen fusion proceeds by the series of reactions known as the CNO cycle.

Red supergiant: After core hydrogen is exhausted, the core shrinks and heats. Hydrogen shell burning begins around the inert helium core, causing the star to expand into a red supergiant.

Life of a $25M_{Sun}$ Star.
Main-sequence lifetime: 5 million years
Duration of later stages: <1 million years

Helium burning supergiant: Helium fusion begins when enough helium has collected in the core. The core then expands, slowing the fusion rate and allowing the star's outer layers to shrink somewhat. Hydrogen shell burning continues at a reduced rate.

Multiple shell–burning supergiant: After core helium is exhausted, the core shrinks until carbon fusion begins, while helium and hydrogen continue to burn in shells surrounding the core. Late in its life, the star fuses heavier elements like carbon and oxygen in shells while iron collects in the inert core.

Neutron star or black hole: The core collapse that initiates the supernova forms a ball of neutrons, which may remain behind as a neutron star or collapse further to make a black hole.

Supernova: Iron cannot provide fusion energy, so it accumulates in the core until degeneracy pressure can no longer support it. Then the core collapses, leading to the catastrophic explosion of the star.

Figure 12.20

Summary of stellar lives. The life stages of a high-mass star (on the left) and a low-mass star (on the right) are depicted in clockwise sequences beginning with the protostellar stage in the upper left corner. (Stars not drawn to scale.)

Yellow main-sequence star: Star is fueled by hydrogen fusion in its core, which converts four hydrogen nuclei into one helium nucleus. In low-mass stars, hydrogen fusion proceeds by the series of reactions known as the proton–proton chain.

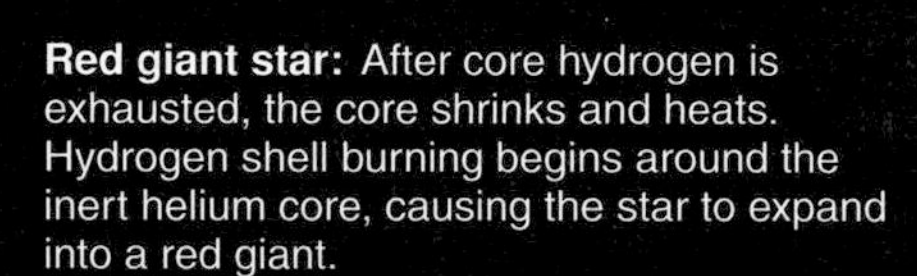

Red giant star: After core hydrogen is exhausted, the core shrinks and heats. Hydrogen shell burning begins around the inert helium core, causing the star to expand into a red giant.

Helium burning star: Helium fusion, in which three helium nuclei fuse to form a single carbon nucleus, begins when enough helium has collected in the core. The core then expands, slowing the fusion rate and allowing the star's outer layers to shrink somewhat. Hydrogen shell burning continues at a reduced rate.

Life of a 1M_{Sun} Star.
Main-sequence lifetime: 10 billion years
Duration of later stages: 1 billion years

Double shell–burning red giant: After core helium is exhausted, the core again shrinks and heats. Helium shell burning begins around the inert carbon core and the star enters its second red giant phase. Hydrogen shell burning continues.

Planetary nebula: The dying star expels its outer layers in a planetary nebula, leaving behind the exposed inert core.

White dwarf: The remaining white dwarf is made primarily of carbon and oxygen because the core never grew hot enough to fuse these elements into anything heavier.

Algol shortly after its birth. The higher-mass star (left) evolved more quickly than its lower-mass companion (right).

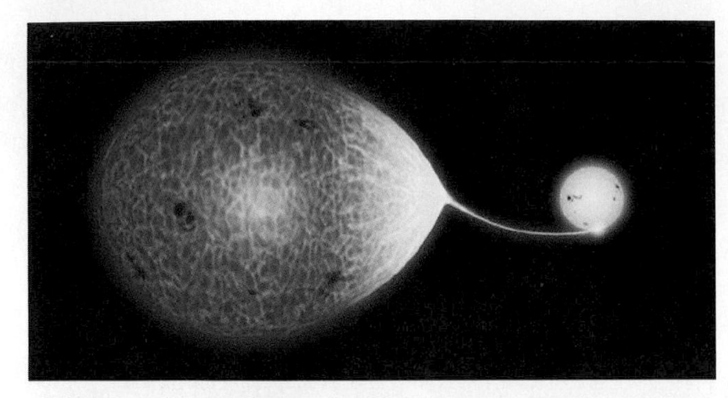

Algol at onset of mass transfer. When the more massive star expanded into a red giant, it began losing some of its mass to its normal, hydrogen-burning companion.

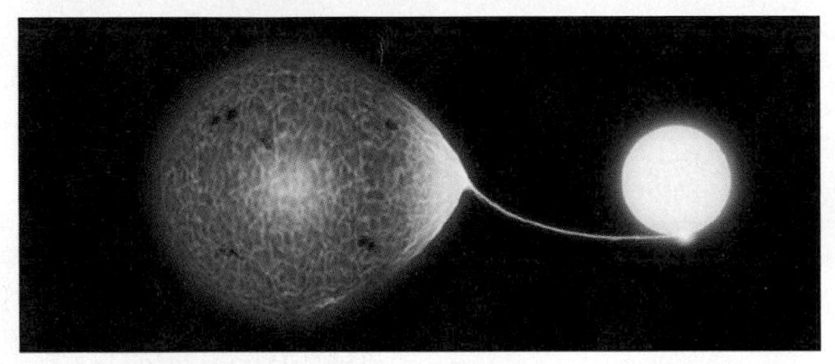

Algol today. As a result of the mass transfer, the red giant has shrunk to a subgiant, and the normal star on the right is now the more massive of the two stars.

Figure 12.21

Artist's conception of the development of the Algol close binary system.

This so-called *Algol paradox* reveals some of the complications in ordinary stellar life cycles that can arise in close binary systems. The two stars in a close binary are near enough to exert significant tidal forces on each other [Section 6.4]. The gravity of each star attracts the near side of the other star more strongly than it attracts the far side. The stars therefore stretch into football-like shapes rather than remaining spherical. In addition, the stars become *tidally locked* so that they always show the same face to each other, much as the Moon always shows the same face to Earth.

During the time that both stars are main-sequence stars, the tidal forces have little effect on their lives. However, when the more massive star (which exhausts its core hydrogen sooner) begins to expand into a red giant, gas from its outer layers can spill over onto its companion. This **mass exchange** occurs when the giant grows so large that its tidally distorted outer layers succumb to the gravitational attraction of the smaller companion star. The companion then begins to gain mass at the expense of the giant.

Stars in close binary systems can exchange mass with one another, altering their life histories.

The solution to the Algol paradox should now be clear (Figure 12.21). The $0.8M_{\text{Sun}}$ subgiant *used to be* much more massive. As the more massive star, it was the first to begin expanding into a red giant. As it expanded, however, so much of its matter spilled over onto its companion that it is now the less massive star.

The future may hold even more interesting events for Algol. The $3.7M_{\text{Sun}}$ star is still gaining mass from its subgiant companion. Thus, its life cycle is actually accelerating as its increasing gravity raises its core hydrogen fusion rate. Millions of years from now, it will exhaust its hydrogen and begin to expand into a red giant itself. At that point, it can begin to transfer mass *back* to its companion. Even stranger things can happen in other mass-exchange systems, particularly when one of the stars is a white dwarf or a neutron star. That is a topic for the next chapter.

THE BIG PICTURE

Putting Chapter 12 into Context

In this chapter, we answered the question of the origin of elements that we first discussed in Chapter 1. As you look back over this chapter, remember these "big picture" ideas:

- Virtually all elements besides hydrogen and helium were forged in the nuclear furnaces of stars and released into space from these stars. Thus, we and our planet are made of stuff produced in stars that lived and died long ago.
- Low-mass stars like our Sun live long lives and die with the ejection of a planetary nebula, leaving behind a white dwarf.
- High-mass stars live fast and die young, exploding dramatically as supernovae and leaving behind neutron stars or black holes.
- Close binary stars can exchange mass, altering the usual course of stellar evolution.

12.1 STAR BIRTH

- **How do stars form?**

Stars are born in cold, relatively dense molecular clouds. As a cloud fragment collapses under gravity, it becomes a protostar surrounded by a spinning disk of gas. The protostar may also fire jets of matter outward along its poles. Protostars rotate rapidly, and some may spin so fast that they split to form close binary star systems.

- **How massive are newborn stars?**

Newborn stars come in a range of masses but cannot be less massive than $0.08M_{Sun}$. Below this mass, degeneracy pressure prevents gravity from making the core hot enough for efficient hydrogen fusion, and the object becomes a "failed star" known as a brown dwarf.

12.2 LIFE AS A LOW-MASS STAR

- **What are the life stages of a low-mass star?**

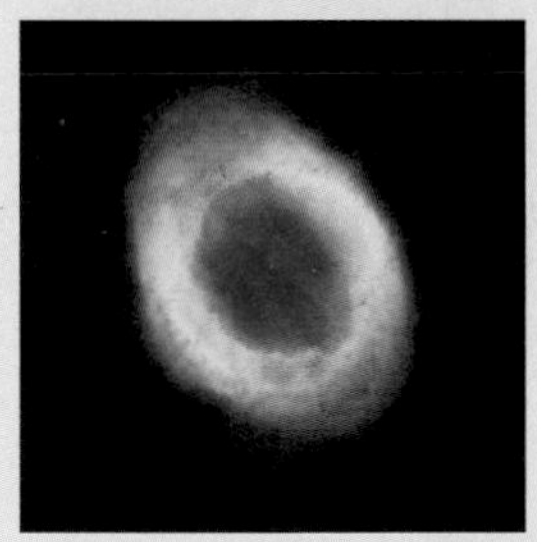

A low-mass star spends most of its life generating energy by fusing hydrogen in its core. Then it becomes a red giant, with a hydrogen shell burning around an inert helium core. Next comes helium core burning, followed by double-shell burning of hydrogen and helium shells around an inert carbon core.

- **How does a low-mass star die?**

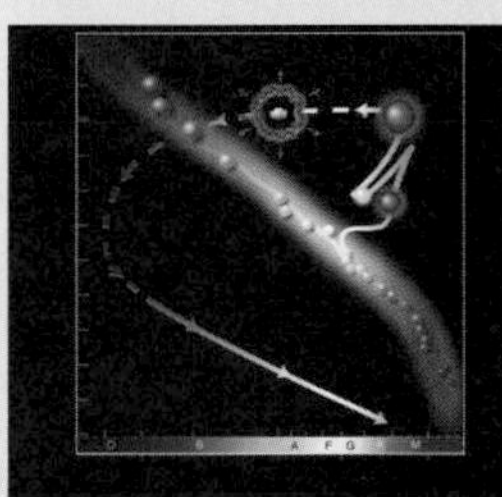

A low-mass star like the Sun never gets hot enough to fuse carbon in its core. It expels its outer layers into space as a planetary nebula, leaving behind a white dwarf.

12.3 LIFE AS A HIGH-MASS STAR

- **What are the life stages of a high-mass star?**

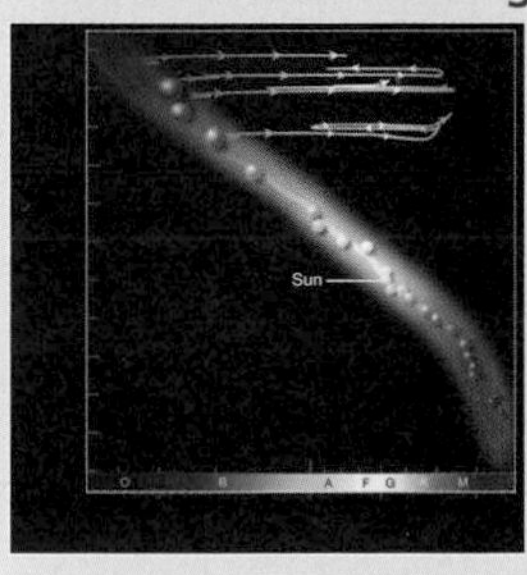

A high-mass star lives a much shorter life than a low-mass star, fusing hydrogen into helium via the CNO cycle. After exhausting its core hydrogen, a high-mass star begins hydrogen shell burning and then goes through a series of stages burning successively heavier elements. The furious rate of this fusion makes the star swell in size to become a supergiant.

- **How do high-mass stars make the elements necessary for life?**

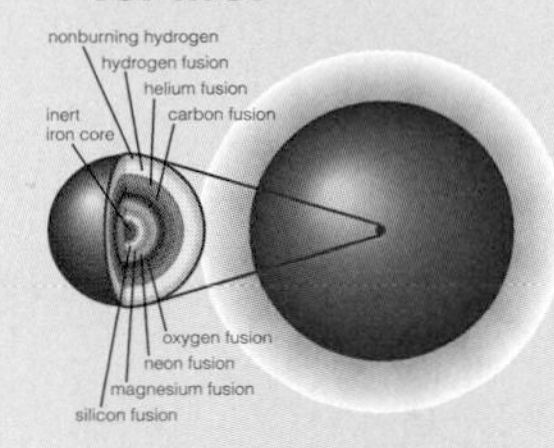

In its final stages of life, a high-mass star's core becomes hot enough to fuse carbon and other heavy elements. The variety of different fusion reactions produces a wide range of elements—including all the elements necessary for life—that are then released into space when the star dies.

- **How does a high-mass star die?**

A high-mass star dies in the cataclysmic explosion of a supernova, scattering newly produced elements into space and leaving a neutron star or black hole behind. The supernova occurs after fusion begins to pile up iron in the high-mass star's core. Because iron fusion cannot release energy, the core cannot hold off the crush of gravity for long. In the instant that gravity overcomes degeneracy pressure, the core collapses and the star explodes.

12.4 SUMMARY OF STELLAR LIVES

- **How does a star's mass determine its life story?**

A star's mass determines how it lives its life. Low-mass stars never get hot enough to fuse carbon or heavier elements in their cores, and end their lives by expelling their outer layers and leaving a white dwarf behind. High-mass stars live short but brilliant lives, ultimately dying in supernova explosions.

- **How are the lives of stars with close companions different?**

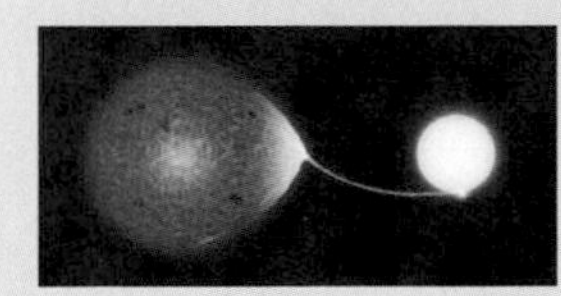

When one star in a close binary system begins to swell in size at the end of its hydrogen-burning life, it can begin to transfer mass to its companion. This mass exchange can then change the remaining life histories of both stars.

EXERCISES AND PROBLEMS

REVIEW QUESTIONS

1. What is a *molecular cloud?* Briefly describe the process by which a protostar forms from gas in a molecular cloud.
2. Why do protostars rotate rapidly? How can a close binary star system form?
3. Why does a spinning disk of gas surround a protostar? Describe some of the phenomena seen among protostars, such as strong winds and *jets.*
4. What is *degeneracy pressure,* and how does it differ from *thermal pressure?* Explain why degeneracy pressure can support a stellar core against gravity even when the core becomes very cold.
5. What is the minimum mass for a star, and why can't objects with lower masses be true stars? What is a *brown dwarf?*
6. Briefly explain the changes that the Sun will go through after it exhausts its core hydrogen. Be sure to explain both the changes occurring in the Sun's core and the changes visible from outside the Sun. What do we mean by the stages we call *hydrogen shell burning, helium burning,* and *double-shell burning?*
7. Why does helium fusion require much higher temperatures than hydrogen fusion? Briefly explain why helium fusion in the Sun will begin with a *helium flash.*
8. What is a *planetary nebula?* What happens to the core of a star after a planetary nebula occurs?
9. What will happen to Earth as the Sun changes in the future?
10. What do we mean by a star's *life track* on an H–R diagram? Summarize the stages of life that we see on the Sun's life track in Figure 12.10.
11. In broad terms, explain how the life of a high-mass star differs from that of a low-mass star.
12. Describe some of the nuclear reactions that can occur in high-mass stars after they exhaust their core helium. Why does this continued nuclear burning occur in high-mass stars but not in low-mass stars?
13. Why can't iron be fused to release energy?
14. Summarize some of the observational evidence supporting our ideas about how the elements form in massive stars.
15. What event initiates a supernova? Explain what happens during the explosion, and why a neutron star or black hole is left behind. What observational evidence supports our understanding of supernovae?
16. What is the Algol paradox and its resolution? Why can the lives of close binary stars differ from those of single stars?

SENSIBLE STATEMENTS?

Decide whether each of the following statements is sensible and explain why it is or is not.

17. The iron in my blood came from a star that blew up more than 4 billion years ago.
18. Humanity will eventually have to find another planet to live on, because one day the Sun will blow up as a supernova.
19. I sure am glad hydrogen has a higher mass per nuclear particle than many other elements. If it had the lowest mass per nuclear particle, none of us would be here.
20. I just discovered a $3.5M_{Sun}$ main-sequence star orbiting a $2.5M_{Sun}$ red giant. I'll bet that red giant was more massive than $3M_{Sun}$ when it was a main-sequence star.
21. If the Sun had been born as a high-mass star some 4.6 billion years ago, rather than as a low-mass star, the planet Jupiter would probably have Earth-like conditions today, while Earth would be hot like Venus.
22. If you could look inside the Sun today, you'd find that its core contains a much higher proportion of helium and a lower proportion of hydrogen than it did when the Sun was born.

PROBLEMS

Homes to Civilization? We do not yet know how many stars have Earth-like planets, nor do we know the likelihood that such planets might harbor advanced civilizations like our own. However, some stars can probably be ruled out as candidates for advanced civilizations. For example, given that it took a few billion years for humans to evolve on Earth, it seems unlikely that advanced life would have had time to evolve around a star that is only a few million years old. For each of the following stars, decide whether you think it is possible that it could harbor an advanced civilization. Explain your reasoning in one or two paragraphs.

23. A $10M_{Sun}$ hydrogen-burning star.
24. A $0.05M_{Sun}$ brown dwarf.
25. A $1.5M_{Sun}$ hydrogen-burning star.
26. A $1.5M_{Sun}$ red giant.
27. A $1M_{Sun}$ helium-burning star.
28. A red supergiant.
29. *Rare Elements.* Lithium, beryllium, and boron are elements with atomic numbers 3, 4, and 5, respectively. Despite their being three of the five simplest elements, Figure 12.16 shows that they are rare compared to many heavier elements. Suggest a reason for their rarity. (*Hint:* Consider the process by which helium fuses into carbon.)
30. *Research: Historical Supernovae.* Historical accounts exist for supernovae in the years 1006, 1054, 1572, and 1604. Choose one of these supernovae and learn more about historical records of the event. Did the supernova influence human history in any way? Write a two- to three-page summary of your research findings.

DISCUSSION QUESTIONS

31. *Connections to the Stars.* In ancient times, many people believed that our lives were somehow influenced by the patterns of the stars in the sky. Modern science has not found any evidence to support this belief, but instead has found that we have a connection to the stars on a much deeper level: In the words of Carl Sagan, we are "star stuff." Discuss in some detail our real connections to the stars as established by modern astronomy. Do you think these connections have any philosophical implications in terms of how we view our lives and our civilization? Explain.
32. *Humanity in A.D. 5,000,000,000.* Do you think it is likely that humanity will survive until the Sun begins to expand into a red giant 5 billion years from now? Why or why not?

For a complete list of media resources available, go to **www.astronomyplace.com**, and choose Chapter 12 from the pull-down menu.

ASTRONOMY PLACE WEB TUTORIALS

Tutorial Review of Key Concepts

Use the interactive **Tutorial** at **www.astronomyplace.com** to review key concepts from this chapter.

Stellar Evolution Tutorial

Lesson 1 Main-Sequence Lifetimes
Lesson 2 Evolution of a Low-Mass Star
Lesson 3 Late Stages of a High-Mass Star

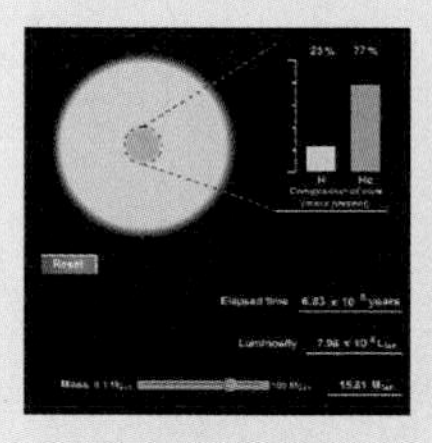

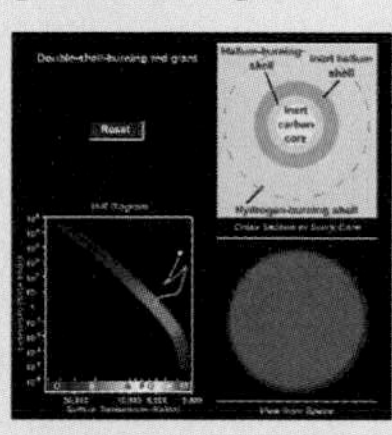

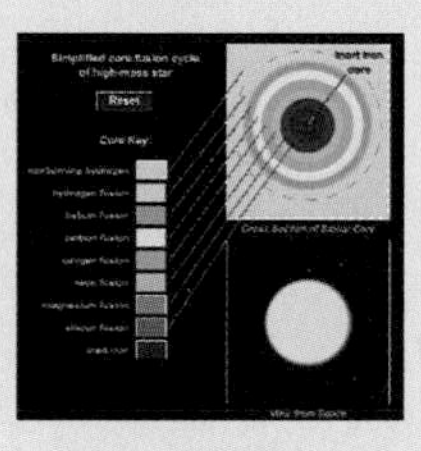

Supplementary Tutorial Exercises

Use the interactive **Tutorial Lessons** to explore the following questions.

Stellar Evolution Tutorial, Lesson 1

1. Use the tool for calculating stellar lifetimes to estimate the lifetimes of 10 stars of different mass. Record the mass and lifetime for each of your 10 stars.
2. Make a graph of your results from question 1, plotting mass on the x-axis and lifetime on the y-axis.
3. Based on your graph from question 2, briefly describe in words how lifetime depends on mass for a main-sequence star.

Stellar Evolution Tutorial, Lessons 2, 3

1. Study the animations for both the low-mass (Lesson 2) and high-mass (Lesson 3) stars. How are the lives of low-mass and high-mass stars similar? How are they different?
2. How do low-mass stars "move" on the H–R diagram as they go through their various stages of life?
3. How do high-mass stars "move" on the H–R diagram as they go through their various stages of life?

EXPLORING THE SKY AND SOLAR SYSTEM

Of the many activities available on the ***Voyager: SkyGazer*** **CD-ROM** accompanying your book, use the following files to observe key phenomena covered in this chapter.

Go to the **File: Demo** folder for the following demonstrations:

1. Crab from Finland

MOVIES

Check out the following narrated and animated short documentaries available on **www.astronomyplace.com** for a helpful review of key ideas covered in this chapter.

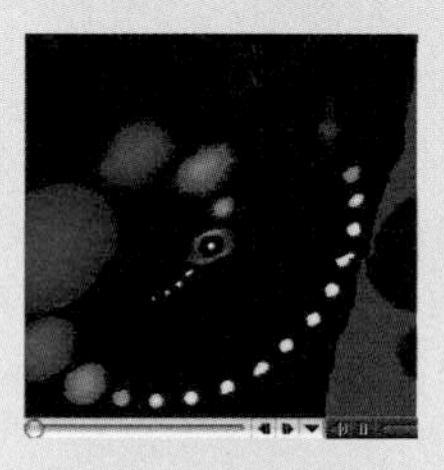

Lives of Stars Movie

Double Stars Movie

WEB PROJECTS

Take advantage of the useful Web links on **www.astronomyplace.com** to assist you with the following projects.

1. *Coming Fireworks in Supernova 1987A.* Astronomers believe that the show from Supernova 1987A is not yet over. In particular, sometime between now and about 2010, the expanding cloud of gas from the supernova is expected to ram into surrounding material, and the heat generated by the impact is expected to create a new light show. Learn more about how Supernova 1987A is changing and what we might expect to see from it in the future. Summarize your findings in a one- to two-page report.
2. *Picturing Star Birth and Death.* Photographs of stellar birthplaces (i.e., molecular clouds) and death places (e.g., planetary nebulae and supernova remnants) can be strikingly beautiful, but only a few such photographs are included in this chapter. Search the Web for additional photographs of these types. Look not only for photos taken in visible light, but also for photographs made from observations in other wavelengths of light. Put each photograph you find into a personal on-line journal, along with a one-paragraph description of what the photograph shows. Try to compile a journal of at least 20 such photographs.

13
The Bizarre Stellar Graveyard

LEARNING GOALS

13.1 WHITE DWARFS
- What is a white dwarf?
- What can happen to a white dwarf in a close binary system?

13.2 NEUTRON STARS
- What is a neutron star?
- How were neutron stars discovered?
- What can happen to a neutron star in a close binary system?

13.3 BLACK HOLES: GRAVITY'S ULTIMATE VICTORY
- What is a black hole?
- What would it be like to visit a black hole?
- Do black holes really exist?

13.4 THE MYSTERY OF GAMMA-RAY BURSTS
- What causes gamma ray bursts?

Welcome to the afterworld of stars, the fascinating domain of white dwarfs, neutron stars, and black holes. To scientists, these dead stars are ideal laboratories for testing the most extreme predictions of general relativity and quantum theory. To most other people, the eccentric behavior of stellar corpses demonstrates that the universe is stranger than they ever imagined.

Dead stars behave in unusual and unexpected ways that challenge our minds and stretch the boundaries of what we believe is possible. Stars that have finished nuclear burning have only one hope of staving off the crushing power of gravity: the strange quantum mechanical effect of degeneracy pressure. Even this strange pressure cannot save the most massive stellar cores. In this chapter, we will study the bizarre properties and occasional catastrophes of the stellar corpses known as white dwarfs, neutron stars, and black holes. Prepare to be amazed by the eerie inhabitants of the stellar graveyard!

Stellar Evolution Tutorial, Lessons 1–2

13.1 WHITE DWARFS

In the previous chapter, we saw that stars of different masses leave different types of stellar corpses. Low-mass stars like the Sun leave behind white dwarfs when they die. Higher-mass stars die in the titanic explosions known as supernovae, leaving behind a neutron star or black hole. Let's begin our study of stellar corpses with white dwarfs.

• What is a white dwarf?

A **white dwarf** is essentially the exposed core of a star that has died and shed its outer layers in a planetary nebula [Section 12.2]. It is quite hot when it first forms, because it was the inside of a star, but slowly cools with time. A white dwarf is stellar in mass but small in size [Section 11.2], which is why it is quite dim compared to a normal star like the Sun even when its surface is still quite hot. However, its high temperature may make it shine brightly in high-energy light such as ultraviolet and X rays (Figure 13.1).

A white dwarf is the corpse of a low-mass star, supported against the crush of gravity by electron degeneracy pressure.

A white dwarf's combination of a starlike mass with a small size makes gravity very strong near its surface. This strong gravity essentially tries to crush the white dwarf to even smaller size, so some sort of pressure must be pushing back equally hard to keep the white dwarf stable. Because there is no fusion to maintain heat and pressure inside a white dwarf, the pressure that opposes gravity must come from some other source. As we discussed in Chapter 12, the source is *degeneracy pressure*—a strange type of pressure that arises when subatomic particles are packed as closely as the laws of quantum mechanics allow. More specifically, the degeneracy pressure in white dwarfs arises from closely packed electrons, so we call it **electron degeneracy**

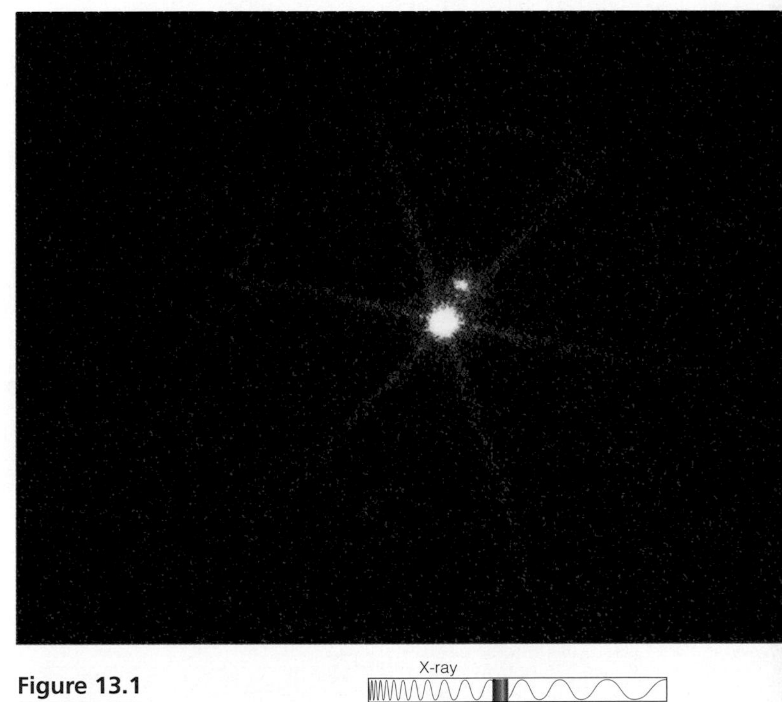

Figure 13.1 X-ray

The binary star system Sirius as seen by the Chandra X-Ray Observatory. Sirius A, to human eyes the brightest star in the night sky, is actually the dimmer of the two stars in this picture. Sirius B, its white dwarf companion, is much hotter and therefore much brighter in ultraviolet and X-ray light. (The spikes emanating from Sirius B are not real. They are artifacts created by the telescope's optics.)

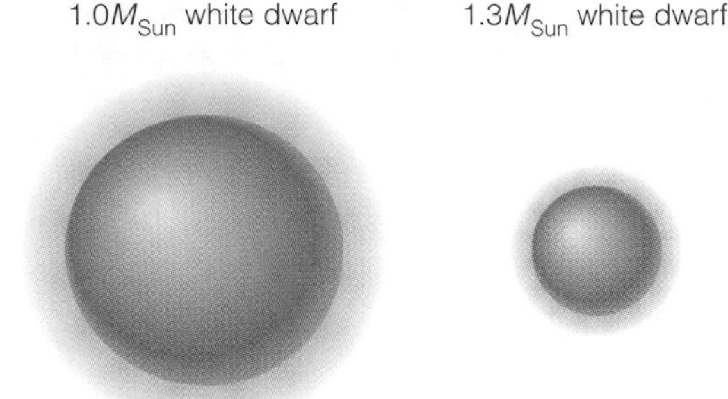

Figure 13.2

Contrary to what you might expect, more massive white dwarfs are actually *smaller* (and thus denser) than less massive white dwarfs.

pressure. Thus, a white dwarf exists in a state of balance between the inward crush of its gravity and the outward push of electron degeneracy pressure.

White Dwarf Composition, Density, and Size Because a white dwarf is the core left over after a star has ceased nuclear burning, its composition reflects the products of the star's final nuclear-burning stage. The white dwarf left behind by a $1M_{Sun}$ star like our Sun will be made mostly of carbon, since stars like the Sun fuse helium into carbon in their final stage of life. The cores of very low mass stars never become hot enough to fuse helium and thus end up as helium white dwarfs.

A teaspoon of white dwarf matter would weigh several tons.

Despite its ordinary-sounding composition, a scoop of matter from a white dwarf would be unlike anything ever seen on Earth. A typical white dwarf has the mass of the Sun ($1M_{Sun}$) compressed into an object the size of Earth. If you recall that Earth is smaller than a typical sunspot, it should be clear that packing the entire mass of the Sun into the volume of Earth is no small feat. The density of such a white dwarf is so high that a teaspoon of its material would weigh several tons if you could bring it to Earth.

More massive white dwarfs are actually smaller in size than less massive ones. For example, a $1.3M_{Sun}$ white dwarf is half the diameter of a $1.0M_{Sun}$ white dwarf (Figure 13.2). The more massive white dwarf is smaller because its greater gravity can compress its matter to a much greater density. According to the laws of quantum mechanics, the electrons in a white dwarf must respond to this compression by moving faster, thereby making the degeneracy pressure stronger. The most massive white dwarfs are therefore the smallest.

The White Dwarf Limit The fact that electron speeds are higher in more massive white dwarfs leads to a fundamental limit on the maximum mass of a white dwarf. Theoretical calculations show that the electron speeds would reach the speed of light in a white dwarf with a mass of about 1.4 times the mass of the Sun ($1.4M_{Sun}$). Because neither electrons nor anything else can travel faster than the speed of light (see Special Topic, Relativity and the Cosmic Speed Limit), no white dwarf can have a mass greater than this $1.4M_{Sun}$ **white dwarf limit.** (The white dwarf limit is often called the *Chandrasekhar limit,* after its discoverer.)

A white dwarf cannot have a mass greater than 1.4 times the mass of the Sun.

Strong observational evidence supports this theoretical limit on the mass of a white dwarf. Many known white dwarfs are members of binary systems, and hence we can measure their masses [Section 11.1]. In every observed case, the white dwarfs have masses below $1.4M_{Sun}$, just as we expect from theory.

• What can happen to a white dwarf in a close binary system?

Left to itself, a white dwarf will never again shine as brightly as the star it once was. With no source of fuel for fusion, it will simply cool with time into a cold, black dwarf. Its size will never change, because its electron degeneracy pressure will forever keep it stable against the crush of gravity. However, the situation can be quite different for a white dwarf in a close binary system.

A white dwarf in a close binary system can gradually gain mass if its companion is a main-sequence or giant star (Figure 13.3). When a clump of mass first spills over from the companion to the white dwarf, it has some small orbital velocity. The law of conservation of angular momentum dictates that the clump must orbit faster and faster as it falls toward the white dwarf's surface. The infalling matter therefore forms a whirlpool-like disk around the white dwarf. Because the process in which material falls onto another body is called *accretion* [Section 6.4], this rapidly rotating disk is called an **accretion disk**.

SPECIAL TOPIC: RELATIVITY AND THE COSMIC SPEED LIMIT

The idea that nothing can travel faster than the speed of light is a consequence of Einstein's *special theory of relativity,* published in 1905. Although relativity is often portrayed as being difficult, its basic ideas are easy to understand. Let's begin with an example that explains the "relative" part of relativity.

Imagine a supersonic plane trip from Nairobi, Kenya, to Quito, Ecuador, at a speed of 1,650 kilometers per hour. Since we've stated the airplane's speed, it might seem hard to argue with. But imagine that you could watch this trip from the Moon. Because both cities are near Earth's equator and Earth's equatorial rotation speed is 1,650 kilometers per hour from west to east, you'd see the airplane precisely matching Earth's rotation speed but in the opposite direction (see figure). Thus, from your point of view on the Moon, the airplane never moves at all (speed of zero)—it makes the trip from Quito to Nairobi because Earth rotates beneath it. So which viewpoint is correct: Did the plane travel at 1,650 kilometers per hour or did it remain stationary?

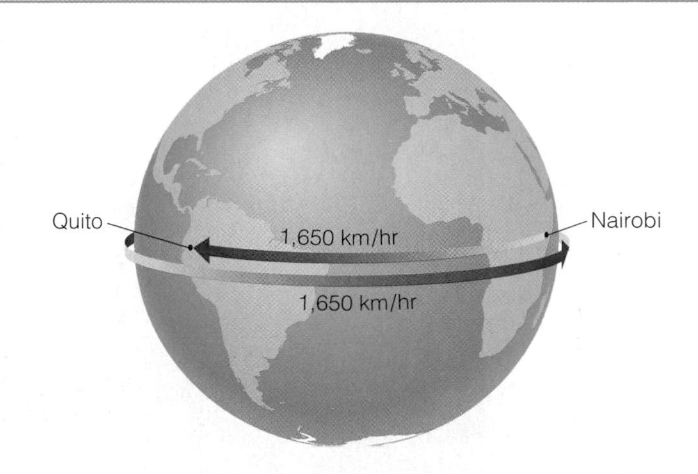

A plane flying at 1,650 km/hr from Nairobi to Quito travels precisely opposite Earth's rotation.

Einstein's theory tells us that both viewpoints are equally valid. That is, questions like "Who is really moving?" and "How fast are you going?" have no absolute answers, and a measurement of speed makes sense only if we measure it relative to some place or object. In this case, both observers would agree that the plane is traveling at a speed of 1,650 kilometers per hour *relative to* the surface of the Earth. The theory of relativity gets its name from the fact that motion is always relative.

Note that Einstein's theory does *not* say that "everything" is relative, only that motion is relative. In fact, it tell us that two things in the universe are absolute:

1. The laws of nature are the same for everyone.
2. The speed of light is the same for everyone.

The first absolute, that the laws of nature are the same for everyone, is a more general version of the idea that all viewpoints on motion are equally valid. If they weren't, different observers would disagree about the laws of physics. The second absolute, that the speed of light is the same for everyone, is much more surprising. Ordinarily, we expect speeds to add and subtract. If you watch someone throw a ball forward from a moving car, you'd see the ball traveling at the speed it is thrown *plus* the speed of the car. But if a person shines a light beam from a moving car, you'd see it moving at precisely the speed of light (about 300,000 kilometers per second), no matter how fast the car is going. This strange fact has been experimentally verified countless times.

The cosmic speed limit follows directly from this fact about the speed of light. To see why, imagine that you have just built the most incredible rocket possible, and you are taking it on a test ride. You push the acceleration button and just keep going faster and faster and faster. With enough fuel, you might expect that you'd eventually be moving faster than the speed of light. But we've left something important out: What is your speed being measured relative to? Remember that you will always measure light to be traveling at the speed of light, 300,000 kilometers per second. Thus, *you* will always find that the light from your rocket's headlights moves out ahead of you at the speed of light. This shouldn't be too surprising, as it's just another way of saying that you can't catch up with your own light. However, because everyone always measures the same speed of light, people back on Earth—or anyplace else—will *also* say that your headlight beams move through space at 300,000 kilometers per second. Thus, because you can't catch up with your own light, anyone who measures your speed will find that you are going *slower* than the speed of your headlight beams, which means slower than the speed of light.

These facts about the relativity of motion and the absoluteness of the speed of light also lead to several other famous consequences of Einstein's theory. For example, they tell us that if you observe a person moving past you at a speed close to the speed of light, you'll see her time running slower than yours, you'll measure her size to be compressed (in the direction of motion) from what you'd measure if she were stationary (relative to you), and you'd conclude that her mass is greater than it would be if she were stationary. The theory also predicts that mass and energy should be equivalent, as stated by Einstein's famous formula $E = mc^2$. All of these predictions of relativity have been experimentally tested and verified to high precision.

Figure 13.3

This artist's conception shows how mass spilling from a companion star (left) toward a white dwarf (right) forms an accretion disk around the white dwarf. The white dwarf itself is in the center of the accretion disk—too small to be seen on this scale. Matter streaming onto the disk creates a hot spot at the point of impact. The inset shows how the system looks from above rather than from the side.

In a close binary system, gas from a companion star can spill toward a white dwarf, forming a swirling accretion disk around it.

Accretion can provide a "dead" white dwarf with a new energy source. The inward-spiraling gas in the accretion disk should become quite hot as its gravitational potential energy is converted into thermal energy [Section 4.3], allowing it to shine with intense ultraviolet or X-ray radiation. More dramatic events can occur as fresh hydrogen gas from the companion star accumulates on the surface of a white dwarf.

Novae The hydrogen spilling toward the white dwarf from the companion gradually spirals inward through the accretion disk and eventually falls onto the surface of the white dwarf. The white dwarf's strong gravity compresses this hydrogen gas into a thin surface layer. Both the pressure and temperature rise as the layer builds up with more accreting gas. When the temperature at the bottom of the layer reaches about 10 million K, hydrogen fusion suddenly ignites.

The white dwarf blazes back to life as its hydrogen layer burns. This thermonuclear flash causes the binary system to shine for a few glorious weeks as a **nova** (Figure 13.4a). A nova can shine as brightly as 100,000 Suns. It generates heat and pressure, ejecting most of the material that has accreted onto the white dwarf. This material expands outward, creating a

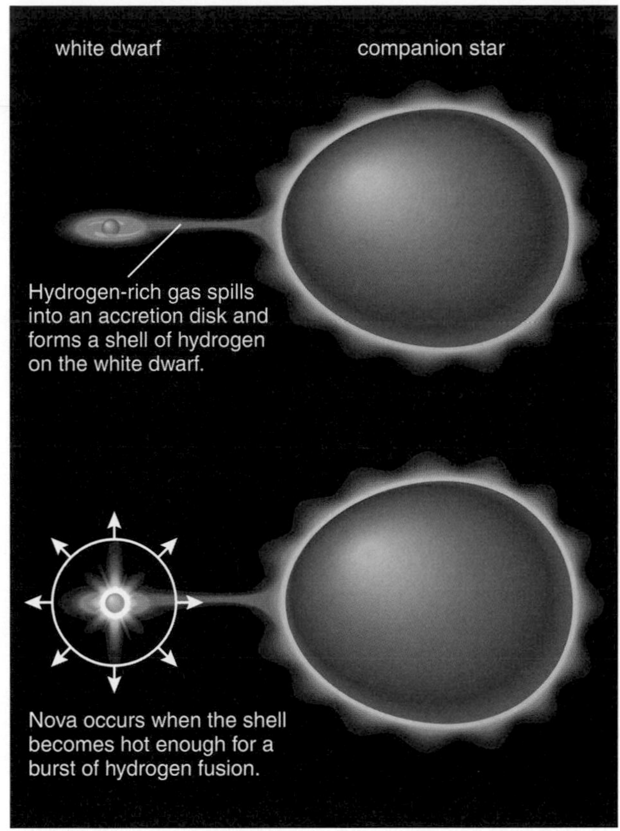

a Diagram of the nova process.

b Hubble Space Telescope image showing blobs of gas ejected from the nova T Pyxidis. The bright spot at the center of the blobs is the binary star system that generated the nova.

VIS

Figure 13.4

A nova occurs when hydrogen fusion ignites on the surface of a white dwarf in a binary star system.

nova remnant that sometimes remains visible years after the nova explosion (Figure 13.4b).

A nova is caused by hydrogen fusion on the surface of a white dwarf in a binary star system.

Accretion resumes after a nova explosion subsides, so the entire process can repeat itself. The time between successive novae in a particular system depends on the rate at which hydrogen accretes on the white dwarf surface and on how highly compressed this hydrogen becomes. The compression of hydrogen is greatest for the most massive white dwarfs, which have the strongest surface gravities. In some cases, novae have been observed to repeat after just a few decades. More commonly, accreting white dwarfs may have 10,000 years between nova outbursts.

White Dwarf Supernovae Each time a nova occurs, the white dwarf ejects some of its mass. Theoretical models cannot yet tell us whether the net result should be a gradual increase or decrease in the white dwarf's mass. Nevertheless, observations show that in at least some cases, accreting white dwarfs in binary systems continue to gain mass as time passes. If such a white dwarf gains enough mass, it can one day approach the $1.4M_{\text{Sun}}$ white dwarf limit. This day is the white dwarf's last.

Remember that most white dwarfs are made largely of carbon. As the white dwarf's mass approaches $1.4M_{\text{Sun}}$, its interior temperature rises high enough for carbon fusion to ignite. When carbon fusion begins, it ignites almost instantly throughout the star, and the white dwarf explodes completely in what we will call a **white dwarf supernova**.

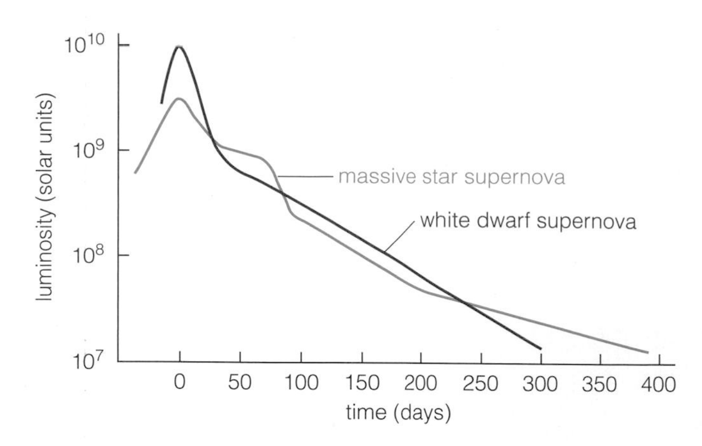

Figure 13.5

The curves on this graph show how the luminosities of the two types of supernovae fade with time. The white dwarf supernova fades steadily after its peak, while the massive star supernova fades in two distinct stages. (Notice that the luminosity scale goes in powers of 10.)

According to our understanding of novae and white dwarf supernovae, can either of these events ever occur with a white dwarf that is *not* a member of a binary star system? Explain.

If a white dwarf gains enough matter to exceed the 1.4 solar-mass white dwarf limit, it will explode completely in what we call a white dwarf supernova.

The "carbon bomb" detonation that creates a white dwarf supernova is quite different from the iron catastrophe that leads to a supernova ending the life of a high-mass star [Section 12.3], which we will call a **massive star supernova**.[1] However, both types of supernova shine brilliantly, with peak luminosities about 10 billion times that of the Sun ($10^{10} L_{Sun}$). Astronomers can distinguish between the two types by studying their light. Spectra of white dwarf supernovae lack hydrogen lines, because white dwarfs contain very little hydrogen. In contrast, massive stars usually have plenty of hydrogen in their outer layers when they explode, so hydrogen lines are prominent in the spectra of most massive star supernovae. In addition, the luminosities of white dwarf supernovae fade steadily with time, while massive star supernovae fade in two distinct stages (Figure 13.5).

Stellar Evolution Tutorial, Lesson 3

13.2 NEUTRON STARS

White dwarfs with densities of 5 tons per teaspoon may seem incredible, but neutron stars are stranger still. The possibility that neutron stars might exist was first proposed in the 1930s, but most astronomers thought it preposterous that nature could truly make anything so bizarre. Nevertheless, a vast amount of evidence now makes it quite clear that neutron stars really exist. In this section, we'll examine their properties, their discovery, and the strange things that can happen to them in close binary star systems.

• What is a neutron star?

A **neutron star** is the ball of neutrons created by the collapse of the iron core in a massive star supernova (Figure 13.6). Typically just 10 kilometers in radius yet more massive than the Sun, neutron stars are essentially giant atomic nuclei made almost entirely of neutrons and held together by gravity. Like white dwarfs, neutron stars resist the crush of gravity with the strange degeneracy pressure that arises when particles are packed as closely as nature allows. In the case of neutron stars, however, it is neutrons rather than electrons that are closely packed, so we say they are supported against the crush of gravity by **neutron degeneracy pressure**.

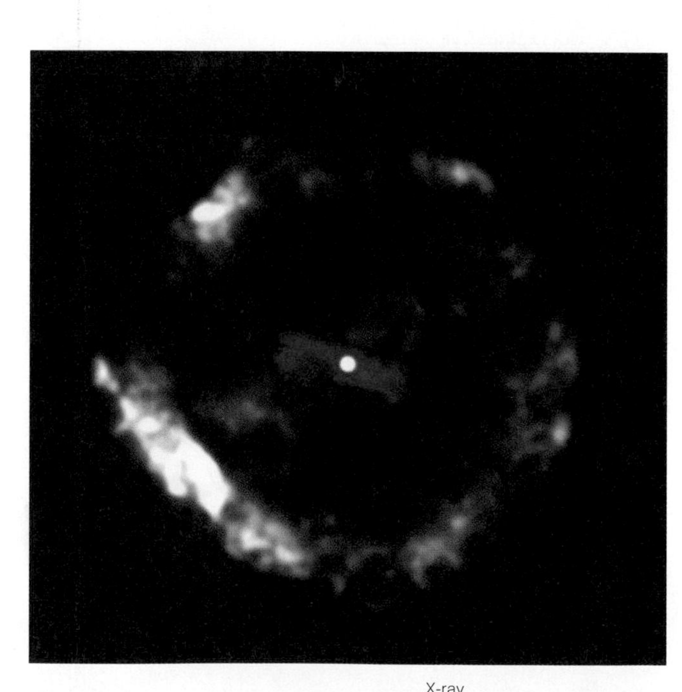

Figure 13.6

This X-ray image from the Chandra X-Ray Observatory shows the supernova remnant G11.2–03, the remains of a supernova observed by Chinese astronomers in A.D. 386. The white dot at the center is the neutron star left behind by the supernova. The different colors correspond to emission of X rays in different wavelength bands. The region pictured is about 23 light-years across.

[1]Observationally, astronomers classify supernovae as *Type II* if their spectra show hydrogen lines, and *Type I* otherwise. All Type II supernovae are assumed to be massive star supernovae. However, a Type I supernova can be either a white dwarf supernova or a massive star supernova in which the star blew away all its hydrogen before exploding. Type I supernovae appear in three classes whose light curves differ, called *Type Ia, Type Ib,* and *Type Ic.* Only Type Ia supernovae are thought to be white dwarf supernovae.

A neutron star is a ball of neutrons just a few kilometers in radius but with a mass like that of the Sun.

The force of gravity at the surface of a neutron star is truly awe-inspiring. Escape velocity is about half the speed of light. If you foolishly chose to visit a neutron star's surface, your body would be squashed immediately into a microscopically thin puddle of subatomic particles.

Things would be only slightly less troubling if a bit of neutron star could somehow come to visit you. A paper clip with the density of neutron star material would outweigh Mount Everest. If such a paper clip magically appeared in your hand, you could not prevent it from falling. Down it would plunge, passing through the Earth like a rock falling through air. It would gain speed until it reached the center of the Earth, and its momentum would carry it onward until it slowed to a stop on the other side of our planet. Then it would fall back down again. If it came in from space, each plunge of the neutron star material would drill a different hole through the rotating Earth. In the words of astronomer Carl Sagan, the inside of the Earth would "look briefly like Swiss cheese" (until the melted rock flowed to fill in the holes) by the time friction finally brought the piece of neutron star to rest at the center of the Earth.

A neutron star could fit in your hometown, but its gravity would quickly destroy our planet.

In the unfortunate event that an *entire* neutron star came to visit you, it would not fall at all. Because it would be only about 10 kilometers in size, the neutron star would probably fit in your hometown. Remember, however, that it would be 300,000 times as massive as Earth. As a result, the neutron star's immense surface gravity would quickly destroy your hometown and the rest of civilization. By the time the dust settled, the former Earth would be a shell no thicker than your thumb on the surface of the neutron star.

• How were neutron stars discovered?

The first observational evidence for neutron stars came in 1967, when a 24-year-old graduate student named Jocelyn Bell discovered a strange source of radio waves. The radio waves pulsed on and off at precise 1.337301-second intervals (Figure 13.7). At first, astronomers could not think of any natural process that could pulse in such a clockwork way, and they only half-jokingly called the new radio source "LGM," for Little Green Men. Today we refer to such rapidly pulsing radio sources as **pulsars**.

The mystery of pulsars was soon solved. By the end of 1968, astronomers had found two smoking guns: Pulsars sat at the centers of both the Crab Nebula and the Vela Nebula, the gaseous remnants of supernovae (Figure 13.8). The pulsars are neutron stars left behind by the supernova explosions.

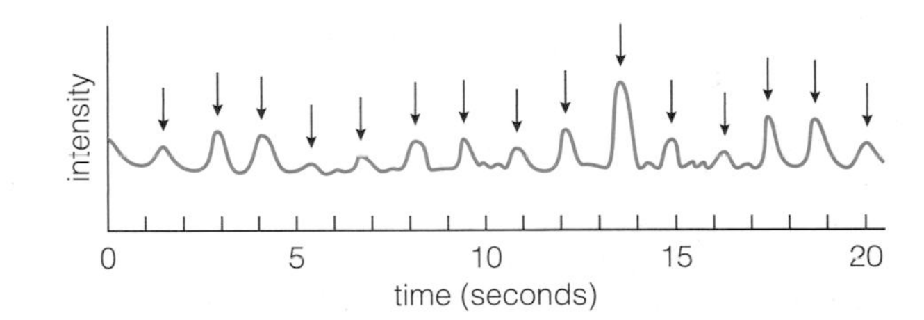

Figure 13.7

About 20 seconds of data from the first pulsar discovered by Jocelyn Bell in 1967. Arrows mark the pulses, which come precisely 1.337301 seconds apart.

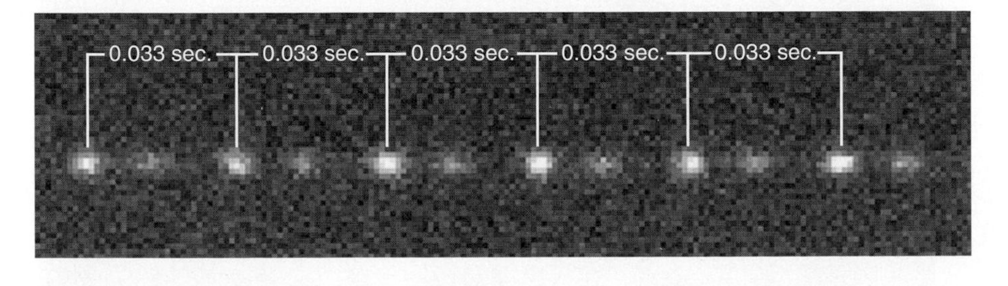

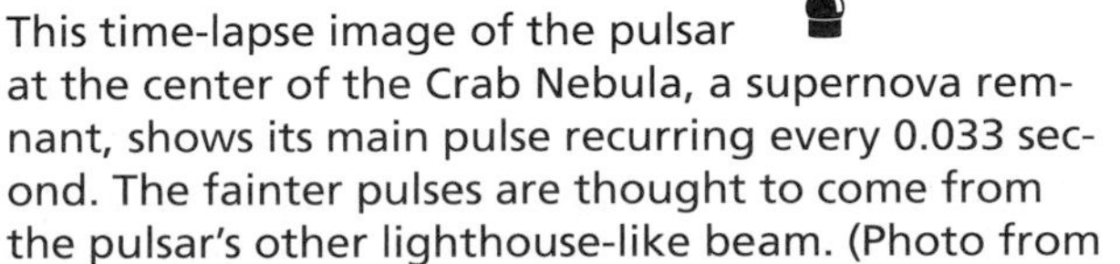

Figure 13.8

This time-lapse image of the pulsar at the center of the Crab Nebula, a supernova remnant, shows its main pulse recurring every 0.033 second. The fainter pulses are thought to come from the pulsar's other lighthouse-like beam. (Photo from the Very Large Telescope of the European Southern Observatory.)

Neutron stars can spin rapidly and emit beams of radiation along their magnetic poles, in which case we may see pulses of radiation as the beams sweep by Earth.

The pulsations arise because the neutron star is spinning rapidly as a result of the conservation of angular momentum [Section 4.3]: As an iron core collapses into a neutron star, its rotation rate must increase as it shrinks in size. The collapse also bunches the magnetic field lines running through the core far more tightly, greatly amplifying the strength of the magnetic field. These intense magnetic fields direct beams of radiation out along the magnetic poles, although we do not yet know exactly how. If a neutron star's magnetic poles are not aligned with its rotation axis, the beams of radiation sweep round and round (Figure 13.9). Like lighthouses, the neutron stars actually emit a fairly steady beam of light, but we see a pulse of light each time the beam sweeps past Earth.

Pulsars are not quite perfect clocks. The continual twirling of a pulsar's magnetic field generates electromagnetic radiation that carries away energy and angular momentum, causing the neutron star's rotation rate to slow gradually. The pulsar in the Crab Nebula, for example, currently spins about 30 times per second. Two thousand years from now, it will spin

Figure 13.9

Radiation from a neutron star can appear to pulse if the neutron star is rotating.

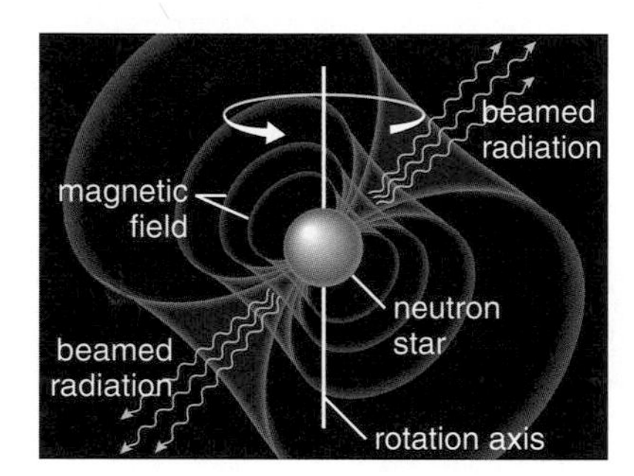

a A pulsar is a rapidly rotating neutron star that beams radiation along its magnetic axis.

b This artwork likens a pulsar (top) to a lighthouse (bottom). If a pulsar's radiation beams are not aligned with its rotation axis, they will sweep through space. Each time one of these beams sweeps across Earth, we see a pulse of radiation.

less than half as fast. Eventually, a pulsar's spin slows so much and its magnetic field becomes so weak that we can no longer detect it. In addition, some spinning neutron stars may be oriented so that their beams do not sweep past our location. Thus, we have the following rule: All pulsars are neutron stars, but not all neutron stars are pulsars.

Suppose we do *not* see pulses from a particular neutron star and hence do not call it a pulsar. Is it possible that a civilization living in some other star system would see this neutron star as a pulsar? Explain.

We know that pulsars must be neutron stars because no other massive object could spin so fast. A white dwarf, for example, can spin no faster than about once per second. An increase in spin would tear it apart because its surface would be rotating faster than the escape velocity. Pulsars have been discovered that rotate as fast as 625 times per second. Only an object as extremely small and dense as a neutron star could spin so fast without breaking apart.

• What can happen to a neutron star in a close binary system?

Like white dwarfs, neutron stars in close binary systems can brilliantly burst back to life. If the neutron star's companion has not yet died, gas from the companion star can flow toward the neutron star. Conservation of angular momentum demands that this gas form a swirling accretion disk around the neutron star, much like the accretion disk that can form around a white dwarf. However, the much stronger gravity of the neutron star makes its accretion disk much hotter and denser than the accretion disk around a white dwarf.

The high temperatures in the inner regions of the accretion disk make it radiate powerfully in X rays. Some close binaries with neutron stars emit 100,000 times more energy in X rays than our Sun emits in all wavelengths of light combined. Due to this intense X-ray emission, close binaries that contain accreting neutron stars are often called **X-ray binaries**. We have detected hundreds of X-ray binaries in the disk of the Milky Way Galaxy.

The emission from most X-ray binaries pulsates rapidly as the neutron star spins, much like the radio pulsations of the neutron stars we see as pulsars. However, the pulsation rates of X-ray binaries tend to accelerate with time, rather than slowing like radio pulsars. Apparently, matter accreting onto the neutron star adds angular momentum, speeding up the neutron star's rotation (Figure 13.10). Some of these neutron stars rotate so fast that they pulsate every few thousandths of a second (and hence are sometimes called *millisecond pulsars*).

The hot accretion disks around neutron stars in close binary systems shine brightly with X rays.

Like accreting white dwarfs that occasionally erupt into novae, accreting neutron stars sporadically erupt with enormous luminosities. Hydrogen-rich material from the companion star builds up on the surface of the neutron star, forming a layer about a meter thick. Pressures at the bottom of this hydrogen layer are high enough to maintain steady fusion, which produces a layer of helium beneath the hydrogen. Helium fusion suddenly ignites when the temperature in this layer builds to

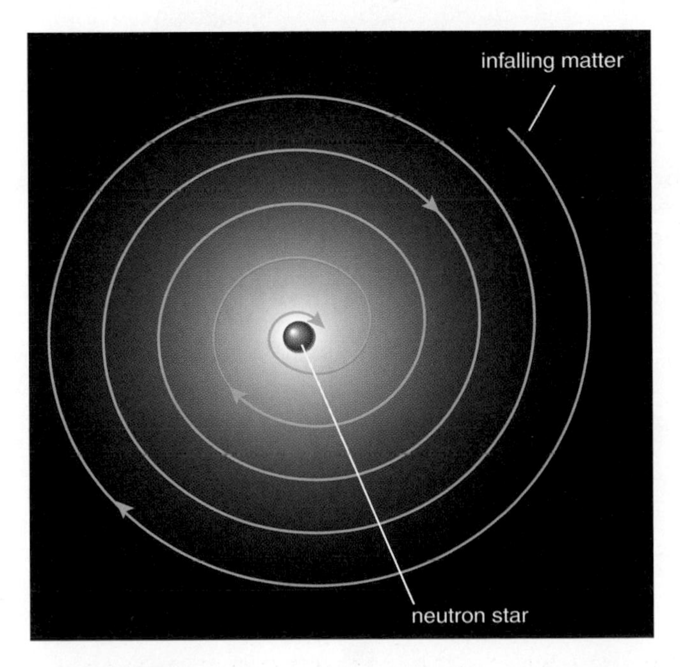

Figure 13.10

Matter accreting onto a neutron star adds angular momentum, increasing the neutron star's rate of spin.

100 million K. The helium burns rapidly to carbon and heavier elements, generating a burst of energy that flows from the neutron star in the form of X rays. These **X-ray bursters** typically flare every few hours to every few days. Each burst lasts only a few seconds, but during those seconds the system radiates 100,000 times as much power as the Sun, all in X rays. Within a minute after a burst, the X-ray burster cools back down and resumes accreting.

13.3 BLACK HOLES: GRAVITY'S ULTIMATE VICTORY

White dwarfs and neutron stars would be strange enough if the story ended here, but it does not. Sometimes, the gravity in a stellar corpse becomes so strong that nothing can prevent the corpse from collapsing under its own weight. The stellar corpse collapses without end, crushing itself out of existence and forming perhaps the most bizarre type of object in the universe: a *black hole.*

• What is a black hole?

The "black" in the name *black hole* comes from the fact that nothing—not even light—can escape from a black hole. The escape velocity of any object depends on the strength of its gravity, which depends on its mass and size [Section 4.4]. Making an object of a particular mass more compact makes its gravity stronger and hence raises its escape velocity. A black hole is so compact that it has an escape velocity greater than the speed of light. Because nothing can travel faster than the speed of light, neither light nor anything else can escape from within a black hole.

The "hole" part of the word *black hole* tells an even stranger story. Einstein discovered that space and time are not distinct, as we usually think of them, but instead are bound up together as four-dimensional **spacetime**. Moreover, in his general theory of relativity, Einstein showed that what we perceive as gravity arises from *curvature of spacetime* (see box, p. 339). The concept "curvature of spacetime" is not easily visualized, because we can see only three dimensions at once. However, we can understand the idea with a two-dimensional analogy.

A black hole is a place where gravity is so strong that nothing—not even light—can escape from within it.

Figure 13.11 represents all four dimensions of spacetime with a two-dimensional rubber sheet. In this analogy, the sheet is flat in a region far from any mass and its gravity (Figure 13.11a). Near a massive object with strong gravity, the sheet becomes curved (Figure 13.11b)—the stronger the gravity, the more curved it gets. In essence, a black hole is a bottomless pit in spacetime (Figure 13.11c). Because gravity gets stronger and stronger as we get closer to the black hole, the sheet curves more and more. Keep in mind that the illustration is only a two-dimensional analogy, and black holes actually are spherical in shape. Nevertheless, the analogy captures the key idea that a black hole really is a *hole* in the observable universe in the following sense: If you enter a black hole, you leave our observable universe and can never return.

The Event Horizon The boundary between the inside of the black hole and the universe outside is called the **event horizon**. The event horizon

Figure 13.11

We can use two-dimensional rubber sheets to show an analogy to curvature in four-dimensional spacetime.

a A two-dimensional representation of "flat" spacetime. The radial distance is the same between each of the circles shown.

b Gravity arises from curvature of spacetime, represented here by a mass pushing down on the rubber sheet. Notice how the circles become more widely separated near the mass, showing that the curvature is greater as we approach the mass on the sheet.

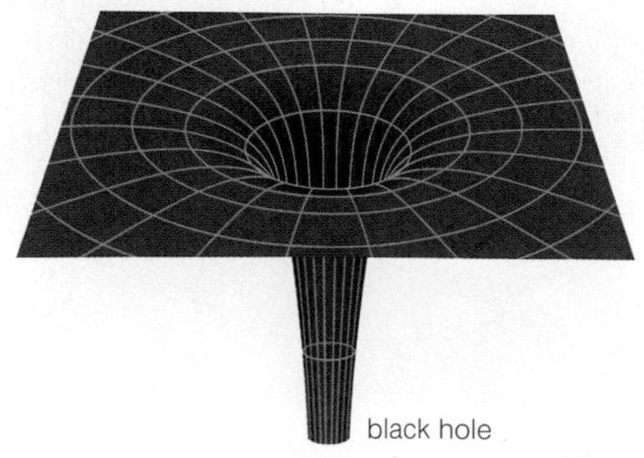

c The curvature of spacetime becomes greater and greater as we approach a black hole, and a black hole itself is a bottomless pit in spacetime.

essentially marks the point of no return for objects entering a black hole: It is the boundary around a black hole at which the escape velocity equals the speed of light. Nothing that passes within this boundary can ever escape. Thus, the event horizon gets its name because we have no hope of learning about any events that occur within it.

We usually think of the "size" of a black hole as the size of its event horizon. Our everyday understanding of "size" is hard to apply inside the event horizon because space and time are so distorted there. However, for someone outside the black hole the event horizon is a sphere with a well-defined size. We can therefore describe the size of the event horizon by its radius, which we call the **Schwarzschild radius** (after Karl Schwarzschild, who first computed it with Einstein's general theory of relativity).

The Schwarzschild radius of a black hole depends only on its mass. A black hole with the mass of the Sun has a Schwarzschild radius of about

SPECIAL TOPIC: GENERAL RELATIVITY AND CURVATURE OF SPACETIME

Einstein's special theory of relativity (see box, p. 331) solved a lot of problems that had troubled physicists prior to its 1905 publication. However, Einstein knew the theory was incomplete; in particular, it is called the *special* theory because it applies only to the special case of motion at constant velocity. It did not apply, for example, to objects whose velocities were accelerating under the influence of a gravitational force. Einstein therefore sought to generalize the theory to include gravity and acceleration.

In 1907, Einstein hit upon what he later called "the happiest thought of my life." He realized that gravity and acceleration have identical effects on the laws of physics. This idea is called the *equivalence principle,* and it states: *The effects of gravity are exactly equivalent to the effects of acceleration.*

To clarify the meaning of this principle, imagine that you are sitting inside with doors closed and window shades pulled down when your room is magically removed from Earth and sent hurtling through space with an acceleration of 1*g*—the acceleration of gravity on Earth, or 9.8 meters per second squared [Section 4.1]. According to the equivalence principle, you would have no way of knowing that you'd left Earth. Any experiment you performed, such as dropping balls of different weights, would yield the same results you'd get on Earth. Likewise, experiments performed in a freely falling elevator would yield the same results as those performed in an elevator drifting at constant velocity through empty space.

Einstein's new point of view on motion, acceleration, and gravity brought about a radical revision of how we think about space and time. Instead of thinking about the three dimensions of space and the one dimension of time as separate, we had to start thinking of them as a seamless four-dimensional entity known as *spacetime.* Einstein showed that a person would feel weightless, as though drifting through empty space, as long as the person's path through four-dimensional spacetime was as straight as possible.

So why do astronauts feel weightless even though they orbit Earth on a curved path? According to general relativity, they are still following the *straightest possible* path through four-dimensional spacetime. The path is curved only because spacetime itself is curved near the Earth. In other words, while Newton would have attributed the curved orbital path of the astronauts to the force of gravity, Einstein attributes it to curvature of spacetime. That is, gravity arises from curvature of spacetime.

The curvature of spacetime is caused by mass, and stronger gravity just means greater spacetime curvature. Thus, for example, the Sun's gravity curves spacetime more than Earth's gravity, and the strong gravity on the surface of a white dwarf curves spacetime more than the gravity on the surface of the Sun. Although we cannot visualize spacetime curvature, we can visualize two-dimensional analogies to it with rubber sheet diagrams like those shown in Figure 13.11. From this new perspective, planets orbit the Sun because of the way space is curved by the Sun: Each planet is going as straight as it can, but space is curved in a way that keeps it going round and round like a marble in a salad bowl (see figure).

Given that we cannot actually perceive all four dimensions of spacetime at once, you may wonder why scientists think spacetime curvature is real. The answer is that Einstein's theory predicts measurable effects from the curvature. For example, if gravity really does arise from curvature of spacetime, then light paths ought to bend when they pass near large masses. Sure enough, scientists have measured such bending of starlight as it passes near the Sun, and the amount of bending is precisely what general relativity predicts it should be. We have also observed such bending of light by distant galaxies, a phenomenon called *gravitatioinal lensing* [Section 16.1]. Another key prediction of general relativity is that time should run slower in regions of stronger gravity (remember that curvature affects both space and time)—again, numerous observations and experimental tests verify Einstein's predictions. Strange as it may seem, we live in a four-dimensional universe in which space and time are intertwined and can never be disentangled.

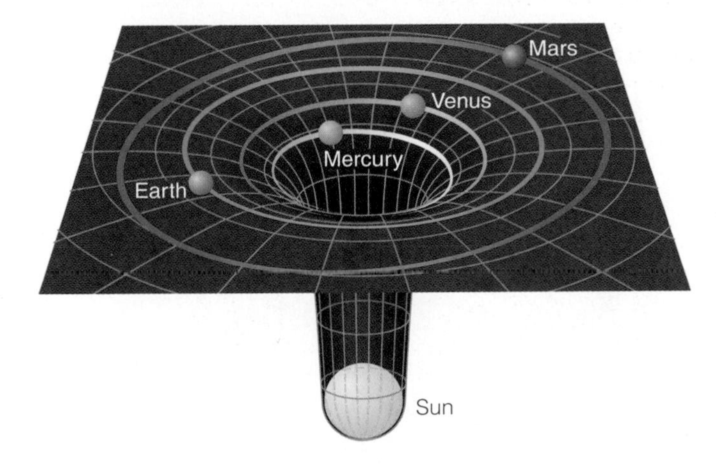

According to general relativity, planets orbit the Sun for much the same reason that you can make a marble go around in a salad bowl: The planet is going as straight as it can, but the curvature of spacetime causes its path through space to curve.

3 kilometers—only a little smaller than the radius of a neutron star of the same mass. More massive black holes have larger Schwarzschild radii. For example, a black hole with 10 times the mass of the Sun has a Schwarzschild radius of about 30 kilometers.

In essence, a collapsing stellar core becomes a black hole at the moment that it shrinks to a size smaller than its Schwarzschild radius. At that moment, the core disappears from view within its own event horizon. The black hole still contains all the mass and has the gravity associated with that mass, but its outward appearance tells us nothing about what fell in.

Singularity and the Limits to Knowledge We can't observe what happens to a stellar core once it collapses to make a black hole, because no information can ever emerge from within the event horizon. Nevertheless, we can use our understanding of the laws of physics to predict what must occur inside a black hole. Because nothing can stop the crush of gravity in a black hole, we might expect that all the matter that forms a black hole must ultimately be crushed to an infinitely tiny and dense point in the black hole's center. We call this point a **singularity**.

Unfortunately, this idea of a singularity pushes up against the limits to scientific knowledge today. The problem is that that two very successful theories give different answers about the nature of a singularity. Einstein's theory of general relativity, which seems to explain successfully how gravity works throughout the universe, predicts that spacetime should grow infinitely curved as it enters the pointlike singularity. Quantum physics, which successfully explains the nature of atoms and the spectra of light, predicts that spacetime should fluctuate chaotically near the singularity. These are clearly different claims, and no theory that can reconcile them has yet been found.

• What would it be like to visit a black hole?

Imagine that you are a pioneer of the future, making the first visit to a black hole. You've selected a black hole with a mass of $10M_{\text{Sun}}$ and a Schwarzschild radius of 30 km. As your spaceship approaches the black hole, you fire its engines to put the ship on a circular orbit a few thousand kilometers above the event horizon. This orbit will be perfectly stable—there is no need to worry about getting "sucked in."

Your first task is to test Einstein's general theory of relativity. This theory predicts that time should run more slowly as the force of gravity grows stronger. It also predicts that light coming out of a strong gravitational field should show a redshift, called a *gravitational redshift,* that is due to gravity rather than to the Doppler effect. You test these predictions with the aid of two identical clocks whose numerals glow with blue light. You keep one clock aboard the ship and push the other one, with a small rocket attached, directly toward the black hole (Figure 13.12). The small rocket automatically fires its engines just enough so that the clock falls only gradually toward the event horizon. Sure enough, the clock on the rocket ticks more slowly as it heads toward the black hole, and its light becomes increasingly redshifted. When the clock reaches a distance of about 10 kilometers above the event horizon, you see it ticking only half as fast as the clock on your spaceship, and its numerals are red instead of blue.

The rocket has to expend fuel rapidly to keep the clock hovering in the strong gravitational field, and it soon runs out of fuel. The clock plunges toward the black hole. From your safe vantage point inside the spaceship, you see the clock ticking more and more slowly as it falls. However, you soon need a radio telescope to "see" it, as the light from the clock face

Figure 13.12 Interactive Figure

Time runs more slowly on the clock nearer to the black hole, and gravitational redshift makes its glowing blue numerals appear red from your spaceship.

shifts from the red part of the visible spectrum, through the infrared, and on into the radio. Finally, its light is so far redshifted that no conceivable telescope could detect it. Just as the clock vanishes from view, you see that the time on its face has frozen to a stop.

Curiosity overwhelms the better judgment of one of your colleagues. He hurriedly climbs into a spacesuit, grabs the other clock, resets it, and jumps out of the airlock on a trajectory aimed straight for the black hole. Down he falls, clock in hand. He watches the clock, but because he and the clock are traveling together, its time seems to run normally and its numerals stay blue. From his point of view, time seems to neither speed up nor slow down. In fact, he'd say that *you* were the one with the strange time, as he would see your time running increasingly fast and your light becoming increasingly blueshifted. When his clock reads, say, 00:30, he and the clock pass through the event horizon. There is no barrier, no wall, no hard surface. The event horizon is a mathematical boundary, not a physical one. From his point of view, the clock keeps ticking. He is inside the event horizon, the first human being ever to leave our observable universe.

If you fell toward a black hole, you would rapidly accelerate and soon cross the event horizon. But to someone watching from afar, your fall would appear to take forever.

Back on the spaceship, you watch in horror as your overly curious friend plunges to his death. Yet, from your point of view, he will *never* cross the event horizon. You'll see time come to a stop for him and his clock just as he vanishes from view due to the huge gravitational redshift of light. When you return home, you can play a video for the judges at your trial, proving that your friend is still a part of our observable universe. Strange as it may seem, all of this is true according to Einstein's theory. From your point of view, your friend takes *forever* to cross the event horizon (even though he vanishes from view due to his ever-increasing redshift). From his point of view, it is but a moment's plunge before he passes into oblivion.

The truly sad part of this story is that your friend did not live to experience the crossing of the event horizon. The force of gravity he felt as he approached the black hole grew so quickly that it actually pulled much harder on his feet than on his head, simultaneously stretching him lengthwise and squeezing him from side to side (Figure 13.13). In essence, your friend was stretched in the same way the oceans are stretched by the tides, except that the *tidal force* near the black hole is trillions of times stronger than the tidal force of the Moon on Earth [Section 4.4]. No human could survive it.

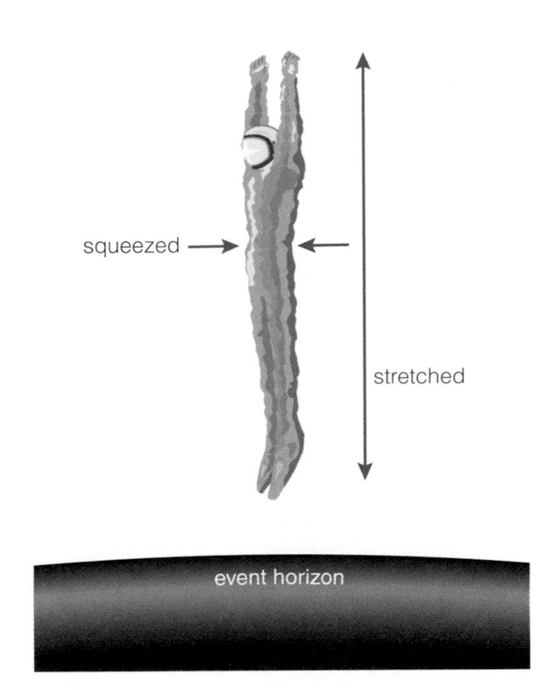

Figure 13.13

Tidal forces would be lethal near a black hole formed by the collapse of a star. The black hole would pull more strongly on the astronaut's feet than on his head, stretching him lengthwise and squeezing him from side to side.

COMMON MISCONCEPTIONS

BLACK HOLES DON'T SUCK

What would happen if our Sun suddenly became a black hole? For some reason, it has become part of our popular culture for most people to believe that Earth and the other planets would inevitably be "sucked in" by the black hole. That is not true. Although the sudden disappearance of light and heat from the Sun would be bad news for life, Earth's orbit would not change.

Newton's law of gravity tells us that the allowed orbits in a gravitational field are ellipses, hyperbolas, and parabolas [Section 4.4]. Note that "sucking" is not on the list! A spaceship would get into trouble only if it came so close to a black hole—within about three times its Schwarzschild radius—that the force of gravity would deviate significantly from what Newton's law predicts. Otherwise, a spaceship passing near a black hole would simply swing around it on an ordinary orbit (ellipse, parabola, or hyperbola). In fact, because most black holes are so small—typical Schwarzschild radii are far smaller than any star or planet—a black hole is actually one of the most difficult things in the universe to fall into by accident.

If he had thought ahead, your friend might have waited to make his jump until you visited a much larger black hole, like one of the *supermassive black holes* thought to reside in the centers of many galaxies [Section 15.4]. A 1 billion M_{Sun} black hole has a Schwarzschild radius of 3 billion kilometers—about the distance from our Sun to Uranus. Although the gravitational forces at the event horizon of all black holes are equally great, the larger size of a supermassive black hole makes its tidal forces much weaker and hence nonlethal. Your friend could safely plunge through the event horizon.

Again, from your point of view, the crossing would take forever, and you would see time come to a stop for him just as he vanished from sight because of the gravitational redshift. Again, he would experience time running normally and would see time in the outside universe running increasingly fast as he approached the event horizon. Unfortunately, anything he saw would do him little good as he plunged to oblivion inside the black hole.

• Do black holes really exist?

The idea of objects with escape velocities greater than the speed of light was first suggested in the late 1700s (by British philosopher John Mitchell and French physicist Pierre Laplace), though the bizarre implications of the idea were not understood until after Einstein published his general theory of relativity in 1915. As was the case for neutron stars, at first most astronomers who contemplated the idea of black holes thought them too strange to be true. Today, however, our understanding of physics gives us reason to think that black holes ought to be fairly common, and observational evidence strongly suggests that black holes really exist.

The Formation of a Black Hole The idea that black holes ought to exist comes from considering how they might form. Recall that white dwarfs cannot exceed $1.4M_{Sun}$, because gravity overcomes electron degeneracy pressure above that mass. Calculations show that the mass of a neutron star has a similar limit that lies somewhere between about 2 and 3 solar masses. Above this mass, neutron degeneracy pressure cannot hold off the crush of gravity in a collapsing stellar core.

A supernova occurs when the electron degeneracy pressure supporting the iron core of a massive star succumbs to gravity, causing the core to collapse catastrophically into a ball of neutrons [Section 12.3]. That is why most supernovae leave neutron stars behind. However, theoretical models show that the most massive stars might not succeed in blowing away all their upper layers. If enough matter falls back onto the neutron core, its mass may rise above the neutron star limit.

As soon as the core exceeds the neutron star limit, gravity overcomes the neutron degeneracy pressure and the core collapses once again. This time, no known force can keep the core from collapsing into oblivion as a black hole. Moreover, another effect of Einstein's theory of relativity makes it highly unlikely that any as-yet-unknown force could intervene either. Recall that Einstein's theory tells us that energy is equivalent to mass ($E = mc^2$) [Section 4.3]. Thus, like mass, energy must also exert some gravitational attraction. The gravity of pure energy usually is negligible, but not in a stellar core collapsing beyond the neutron star limit. Usually, the gravitational potential energy released as a star collapses boosts its temperature and pressure

According to the known laws of physics, nothing can stop the collapse of a stellar corpse with a mass greater than about 3 solar masses.

Figure 13.14 Interactive Figure

Artist's conception of the Cygnus X-1 system, which gets its name because it is the brightest X-ray source in the constellation Cygnus. The X rays come from the high-temperature gas in the accretion disk surrounding the black hole.

enough to fight off gravity. However, once a star collapses beyond the point where it could have been a neutron star, the energy associated with the enhanced temperature and pressure only makes gravity stronger. The more it collapses, the stronger gravity gets. To the best of our understanding, *nothing* can stop the star from becoming a black hole.

Observational Evidence for Black Holes The fact that black holes emit no light might make it seem as if they should be impossible to detect. However, a black hole's gravity can influence its surroundings in a way that reveals its presence. Astronomers have discovered many objects that show the telltale signs of an unseen gravitational influence with a large enough mass to suggest that it is black hole.

Strong observational evidence for black holes formed by supernovae comes from studies of X-ray binaries. Recall that the accretion disks around neutron stars in close binary systems can emit strong X-ray radiation, making an X-ray binary. The accretion disk forms because the neutron star's strong gravity pulls mass from the companion star. Because a black hole has even stronger gravity than a neutron star, a black hole in a close binary system should also be surrounded by a hot, X-ray emitting accretion disk. Thus, some X-ray binaries may contain black holes rather than neutron stars. The trick to learning which type of corpse resides in an X-ray binary depends on measuring the object's mass.

Some X-ray binaries probably contain accreting black holes rather than accreting neutron stars.

One of the most promising black hole candidates is in an X-ray binary called Cygnus X-1 (Figure 13.14). This system contains an extremely luminous star with an estimated mass of $18M_{Sun}$. Based on Doppler shifts of its spectral lines, astronomers have concluded that this star orbits a compact, unseen companion with a mass of about $10M_{Sun}$. Although there is some uncertainty in these mass estimates, the mass of the invisible accreting object clearly exceeds the $3M_{Sun}$ neutron star limit. Thus, it is too massive to be a neutron star, so by current knowledge it cannot be anything other than a black hole.

THINK ABOUT IT Recall that some X-ray binaries that contain neutron stars emit frequent X-ray bursts and are called X-ray bursters. Could an X-ray binary that contains a black hole exhibit the same type of X-ray bursts? Why or why not? (*Hint:* Where do the X-ray bursts occur in an X-ray binary with a neutron star?)

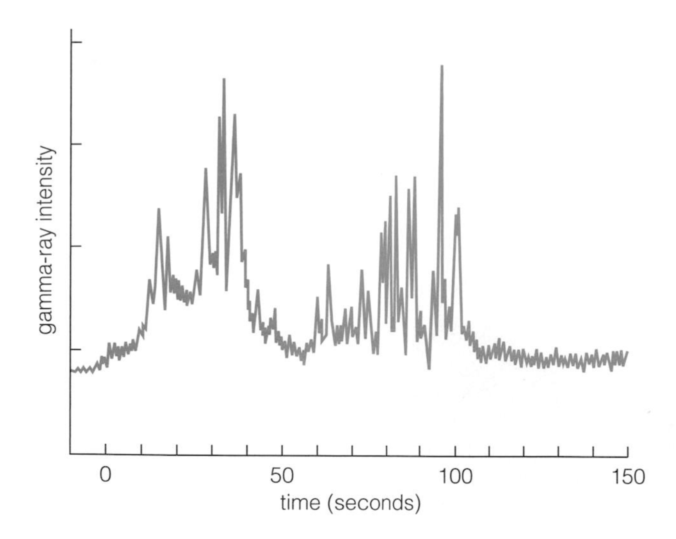

Figure 13.15

The intensity of a gamma-ray burst can fluctuate dramatically over time periods of just a few seconds.

13.4 THE MYSTERY OF GAMMA-RAY BURSTS

In the early 1960s, the United States began launching a series of top-secret satellites designed to look for gamma rays emitted by nuclear bomb tests. The satellites soon began detecting occasional bursts of gamma rays, typically lasting a few seconds (Figure 13.15). It took several years for military scientists to become convinced that these **gamma-ray bursts** were coming from space, not from some sinister human activity. They publicized the discovery in 1973. At first, astronomers assumed that these gamma ray bursts were just a variation on the X-ray bursts that can occur on neutron stars in close binary systems. But this idea was ruled out in 1991, leaving gamma ray bursts as one of the biggest mysteries in astronomy.

• What causes gamma ray bursts?

Gamma ray photons carry so much energy that they are very difficult to focus, since they tend to penetrate any mirror intended to deflect them. Thus, while astronomers before 1991 knew the bursts were occurring, they could not tell where they were coming from. That changed with the 1991 launch of NASA's *Compton Gamma Ray Observatory,* or *Compton* for short, which operated in Earth orbit for almost a decade.

Compton carried an array of eight detectors designed expressly to study gamma-ray bursts. By comparing the data recorded by all eight detectors, scientists could determine the direction of a gamma-ray burst within about 1°. The results were stunning. Compton detected a gamma-ray burst about once a day on average and soon had found enough bursts to show their distribution on a map of the sky. To almost everyone's surprise, the bursts were *not* concentrated like X-ray binaries in the disk of the Milky Way Galaxy, which ruled out the possibility that they came from the same types of systems as X-ray bursts. Instead, the gamma-ray bursts seemed to come randomly from all directions in space.

The even distribution across the sky of gamma-ray bursts suggested that they must come from far outside our own galaxy. If the bursts had been coming from objects distributed spherically about the Milky Way Galaxy, we would have seen a concentration of them in the direction of the galactic center. (Remember that we are located more than halfway out from the center of our galaxy.) Direct evidence for an extragalactic origin came in 1997, when astronomers first observed the afterglow of a gamma-ray burst in other wavelengths. The higher resolution possible in these other wavelengths allowed astronomers to pinpoint the burst's origin in a distant galaxy. Since then, numerous other bursts have also been traced to explosions in distant galaxies (Figure 13.16).

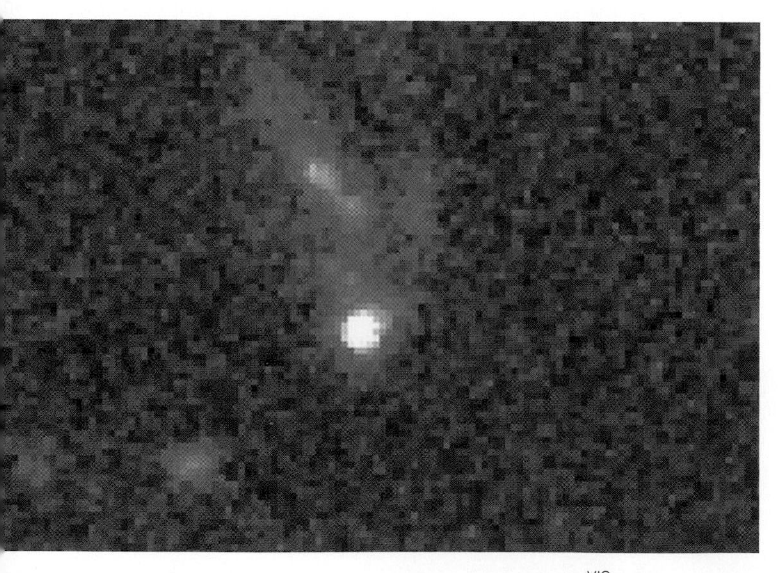

Figure 13.16

The bright dot near the center of this image is the visible-light afterglow of a gamma-ray burst, as seen by the Hubble Space Telescope. The elongated blob extending above the dot is the distant galaxy in which the burst occurred.

Learning that gamma-ray bursts come from outside our galaxy only deepened their mystery. The afterglows of some gamma-ray bursts can be seen with binoculars, even though they are coming from galaxies billions of light-years away—making them by far the most powerful bursts of energy that ever occur in the universe. If these bursts shine their light equally in all directions, like a light bulb, then the total luminosity of a burst can briefly exceed the combined luminosity of a million galaxies like our Milky Way! Because such a high luminosity is very difficult to explain, some scientists speculate that gamma-ray bursts channel their energy into narrow searchlight beams, like pulsars, so that only some are visible from Earth. But even in this case, a burst's luminosity would still temporarily surpass that of thousands of galaxies like our Milky Way.

What could cause such massive outbursts of energy? At least some gamma ray bursts appear to come from extremely powerful supernova explosions. An ordinary supernova that forms a neutron star does not release enough energy to explain the luminosity of the brightest gamma-ray bursts. However, a supernova that forms a black hole crushes even more matter into an even smaller radius, releasing many times more gravitational potential energy than one that forms a neutron star. This kind of event (sometimes called a *hypernova*) might be powerful enough to explain the most extreme gamma-ray bursts.

At least some gamma-ray bursts seem to come from unusually powerful supernovae, but no one knows exactly how the gamma rays are produced.

The primary evidence linking gamma-ray bursts to exploding stars comes from observing them in other wavelengths of light, including visible light and X rays. Several recently launched gamma-ray telescopes are capable of rapidly pinpointing the location of a gamma-ray burst in the sky, thereby allowing other types of telescopes to be pointed at the gamma-ray source almost as soon as the burst is detected. These observations have shown that numerous nearby gamma-ray bursts come from the locations of powerful supernova explosions. In addition, the extremely massive stars that could explode in this way are very short-lived and therefore should be found only in places where stars are actively forming. Observations also support this idea, because all the distant galaxies in which we have observed gamma-ray bursts to date also appear to be forming new stars.

The mystery of gamma-ray bursts is not yet fully solved, however. Even if they all come from powerful supernovae, no one knows precisely how these supernovae generate gamma-ray bursts. Moreover, careful study of gamma-ray bursts suggests that they come in at least two distinct types, and only one of these types has been linked with supernova explosions. The other type may turn out to have some completely different explanation. For example, one hypothesis suggests that some gamma-ray bursts occur when two neutron stars collide in a binary system. If two neutron stars orbit each other, Einstein's theory of general relativity predicts that they must gradually spiral in toward each other and eventually collide, presumably with catastrophic consequences. The energy of a neutron star collision would be great enough to produce a gamma-ray burst, but no one has come up with a mechanism by which the collision would actually generate gamma rays. Thus, we are left with one of the greatest mysteries in all of science: Gamma-ray bursts are the most powerful events in the universe, but we still don't know where they all come from or precisely how they are generated.

THE BIG PICTURE

Putting Chapter 13 into Context

We have now seen the mind-bending consequences of stellar death. As you think about the bizarre objects described in this chapter, try to keep in mind these "big picture" ideas:

- Despite the strange nature of stellar corpses, clear evidence exists for white dwarfs and neutron stars, and the case for black holes is very strong.
- White dwarfs, neutron stars, and black holes can all have close stellar companions from which they accrete matter. These binary systems produce some of the most spectacular events in the universe, including novae, white dwarf supernovae, and X-ray bursters.

- Black holes are holes in the observable universe that strongly warp space and time around them. The nature of black hole singularities remains beyond the frontier of current scientific understanding.
- Gamma-ray bursts were once thought to be related to neutron stars in our galaxy, but recent evidence indicates that this idea is wrong. At least some of them come from supernova explosions in distant galaxies.

SUMMARY OF KEY CONCEPTS

13.1 WHITE DWARFS

• What is a white dwarf?
A white dwarf is the core left over from a low-mass star, supported against the crush of gravity by electron degeneracy pressure. A white dwarf typically has the mass of the Sun compressed into a size no larger than Earth. No white dwarf can have a mass greater than $1.4M_{Sun}$.

• What can happen to a white dwarf in a close binary system?

A white dwarf in a close binary system can acquire hydrogen from its companion through an accretion disk that swirls toward its surface. As hydrogen builds up on the white dwarf's surface, it may ignite with nuclear fusion to make a nova that, for a few weeks, may shine as brightly as 100,000 Suns. In extreme cases, accretion may continue until the white dwarf's mass exceeds the white dwarf limit of $1.4M_{Sun}$, at which point it will explode as a white dwarf supernova.

13.2 NEUTRON STARS

• What is a neutron star?
A neutron star is the ball of neutrons created by the collapse of the iron core in a massive star supernova. It resembles a giant atomic nucleus 10 kilometers across but more massive than the Sun.

• How were neutron stars discovered?

Neutron stars spin rapidly when they are born, and their strong magnetic fields can direct beams of radiation that sweep through space as the neutron star spins. We see such neutron stars as pulsars, and these pulsars provided the first direct evidence for the existence of neutron stars.

• What can happen to a neutron star in a close binary system?
Neutron stars in close binary systems can accrete hydrogen from their companions, forming dense, hot accretion disks. The hot gas emits strongly in X rays, so we see these systems as X-ray binaries. In some of these systems, frequent bursts of helium fusion ignite on the neutron star's surface, emitting X-ray bursts.

13.3 BLACK HOLES: GRAVITY'S ULTIMATE VICTORY

• What is a black hole?

A black hole is a place where gravity has crushed matter into oblivion, creating a hole in the universe from which nothing can ever escape, not even light.

• What would it be like to visit a black hole?
You could orbit a black hole just like any other object of the same mass. However, you'd see strange effects for an object falling toward the black hole: Time would seem to run slowly for the object, and its light would be increasingly redshifted as it approached the black hole. The object would never quite reach the event horizon, but it would soon disappear from view as its light became so redshifted that no instrument could detect it.

• Do black holes really exist?

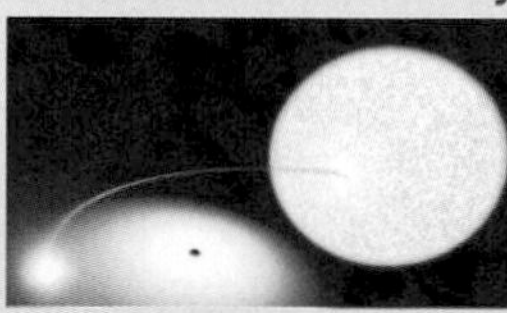

No known force can stop the collapse of a stellar corpse with a mass above the neutron star limit of 2 to 3 solar masses, and theoretical studies of supernovae suggest that such objects should sometimes form. Observational evidence supports this idea: Some X-ray binaries include compact objects far too massive to be neutron stars, making it likely that they are black holes.

13.4 THE MYSTERY OF GAMMA-RAY BURSTS

• What causes gamma ray bursts?
Gamma-ray bursts occur in distant galaxies and are the most powerful bursts of energy we observe anywhere in the universe. No one knows their precise cause, although at least some appear to come from unusually powerful supernovae.

EXERCISES AND PROBLEMS

REVIEW QUESTIONS

1. What is degeneracy pressure, and how is it important to white dwarfs and neutron stars? What is the difference between electron degeneracy pressure and neutron degeneracy pressure?
2. Describe the mass, size, and density of a typical white dwarf. How does the size of a white dwarf depend on its mass?
3. What happens to the electron speeds in a more massive white dwarf, and how does this idea lead to a limit on the mass of a white dwarf? What is the white dwarf limit?
4. What are *accretion disks,* and why do we find them only in close binary systems? Explain how the accretion disk provides a white dwarf with a new source of energy that we can detect from Earth.
5. What is a *nova?* Describe the process that creates a nova, and what a nova looks like.
6. What causes a *white dwarf supernova?* Observationally, how do we distinguish white dwarf and massive star supernovae?
7. Describe the mass, size, and density of a typical neutron star. What would happen if a neutron star came to your hometown?
8. How do we know that *pulsars* are neutron stars? Are all neutron stars also pulsars? Explain.
9. Explain how the presence of a neutron star can make a close binary star system appear to us as an *X-ray binary.* Why do some of these systems appear to us as *X-ray bursters?*
10. What do we mean when we say that a black hole really is a hole in *spacetime?* What is the *event horizon* of a black hole? How is it related to the *Schwarzschild radius?*
11. What do we mean by the *singularity* of a black hole? How do we know that our current theories are inadequate to explain what happens at the singularity?
12. Suppose you are falling into a black hole. How will you perceive the passage of your own time? How will you perceive the passage of time in the universe around you? Briefly explain why your trip is likely to be lethal.
13. Why do we think that black holes should sometimes be formed by supernovae? What observational evidence supports the existence of black holes?
14. How do we know that gamma-ray bursts do *not* come from the same sources as X-ray bursts? Summarize current ideas about the cause of gamma-ray bursts.

SENSIBLE STATEMENTS?

Decide whether each of the following statements is sensible and explain why it is or is not.

15. Most white dwarf stars have masses close to that of our Sun, but a few white dwarf stars are up to three times as massive as the Sun.
16. The radii of white dwarf stars in close binary systems gradually increase as they accrete matter.
17. If you want to find a pulsar, you might want to look near the remnant of a supernova described by ancient Chinese astronomers.
18. If a black hole 10 times as massive as our Sun were lurking just beyond Pluto's orbit, we'd have no way of knowing it was there.
19. If the Sun suddenly became a $1M_{\text{Sun}}$ black hole, the orbits of the nine planets would not change at all.
20. We can detect black holes with X-ray telescopes because matter falling into a black hole emits X rays after it smashes into the event horizon.

PROBLEMS

(Quantitative problems are marked with an asterisk.)

Life Stories of Stars. Write a one- to two-page life story for the following scenarios in problems 21 and 22. Each story should be detailed and scientifically correct but also creative. That is, it should be entertaining while at the same time proving that you understand stellar evolution. Be sure to state whether "you" are a member of a binary system.

21. You are a white dwarf of $0.8M_{\text{Sun}}$.
22. You are a neutron star of $1.5M_{\text{Sun}}$.

*23. *A Black Hole?* You've just discovered a new X-ray binary, which we will call *Hyp-X1* ("Hyp" for hypothetical). The system Hyp-X1 contains a bright, B2 main-sequence star orbiting an unseen companion. The separation of the stars is estimated to be 20 million kilometers, and the orbital period of the visible star is 4 days.
   a. Use Newton's version of Kepler's third law to calculate the sum of the masses of the two stars in the system. (*Hint:* See Problem 33 of Chapter 4 for the necessary formula.) Give your answer in both kilograms and solar masses ($M_{\text{Sun}} = 2.0 \times 10^{30}$ kg).
   b. Determine the mass of the unseen companion. Is it a neutron star or a black hole? Explain. (*Hint:* A main-sequence star with spectral type B2 has a mass of about $10M_{\text{Sun}}$.)

DISCUSSION QUESTIONS

24. *Too Strange to Be True?* Despite strong theoretical arguments for the existence of neutron stars and black holes, many scientists rejected the possibility that such objects could really exist until they were confronted with very strong observational evidence. Some people claim that this type of scientific skepticism demonstrates an unwillingness on the part of scientists to give up their deeply held scientific beliefs. Others claim that this type of skepticism is necessary for scientific advancement. What do you think? Defend your opinion.
25. *Black Holes in Popular Culture.* Phrases such as "it disappeared into a black hole" are now common in popular culture. Give a few examples in which the term *black hole* is used in popular culture but is not meant to be taken literally. In what ways are these uses correct in their analogies to real black holes? In what ways are they incorrect? Why do you think such an esoteric scientific idea as that of a black hole has so captured the public imagination?

MATH HELP AND EXERCISES

For additional help and practice with mathematical concepts applicable to this chapter, the Astronomy Place Web site has the following resources with detailed explanations, worked examples and practice problems:

- Density
- Orbital Separation and Newton's Version of Kepler's Third Law
- Formulas of Special Relativity

For a complete list of media resources available, go to **www.astronomyplace.com**, and choose Chapter 13 from the pull-down menu.

ASTRONOMY PLACE WEB TUTORIALS

Tutorial Review of Key Concepts

Use the interactive **Tutorials** at **www.astronomyplace.com** to review key concepts from this chapter.

Stellar Evolution Tutorial

Lesson 1 Main-Sequence Lifetimes
Lesson 2 Evolution of a Low-Mass Star
Lesson 3 Late Stages of a High-Mass Star

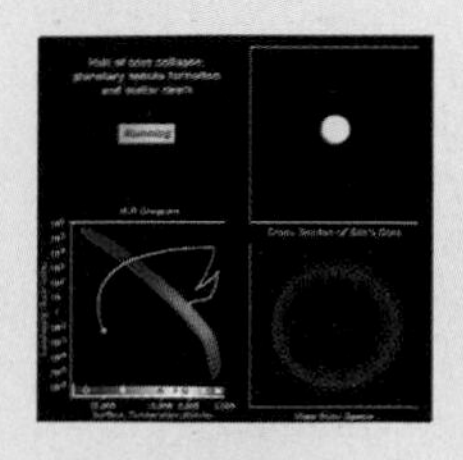

Black Holes Tutorial

Lesson 1 What Are Black Holes?
Lesson 2 The Search for Black Holes

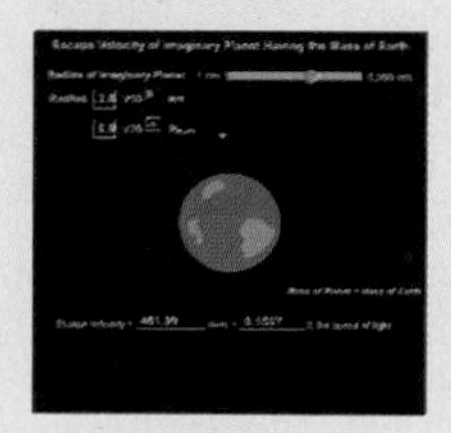
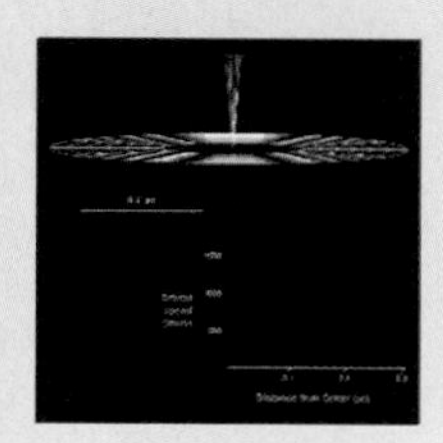
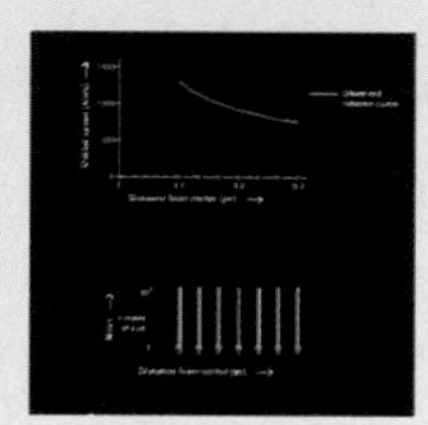

Supplementary Tutorial Exercises

Use the interactive **Tutorial Lessons** to explore the following questions.

Stellar Evolution Tutorial, Lessons 2, 3

1. What kind of "corpse" is left behind by a low-mass star? Why?
2. What kind of "corpse" is left behind by a high-mass star? Why?

Black Holes Tutorial, Lesson 1

1. What would have to happen to Earth to turn it into a black hole? Could this really happen?
2. Use the tool in the simulation to determine the Schwarzschild radii for 10 black holes of different masses. Record the mass and Schwarzschild radius for each black hole.
3. Make a graph of your results from question 2, plotting mass on the x-axis and Schwarzschild radius on the y-axis.
4. Based on your graph from question 3, briefly describe in words how the Schwarzschild radius depends on mass for a black hole.

Black Holes Tutorial, Lesson 2

1. Briefly explain how scientists can detect a black hole.
2. Why does the search for black holes generally begin with a search for X-ray sources in the sky?

WEB PROJECTS

Take advantage of the useful Web links on **www.astronomyplace.com** to assist you with the following projects.

1. *Gamma-Ray Bursts.* Go to the Web site for a mission studying gamma-ray bursts (such as HETE, INTEGRAL, or Swift) and find the latest information about gamma-ray bursts. Write a one- to two-page essay on recent discoveries and how they may shed light on the mystery of gamma-ray bursts.
2. *Black Holes.* Andrew Hamilton, a professor at the University of Colorado, maintains a Web site with a great deal of information about black holes and what it would be like to visit one. Visit his site and investigate some aspect of black holes that you find to be of particular interest. Write a short report on what you learn.

14
Our Galaxy

LEARNING GOALS

14.1 THE MILKY WAY REVEALED
- What does our galaxy look like?
- How do stars orbit in our galaxy?

14.2 GALACTIC RECYCLING
- How does our galaxy recycle gas into stars?
- Where do stars tend to form in our galaxy?

14.3 THE HISTORY OF THE MILKY WAY
- What clues to our galaxy's history do halo stars hold?
- How did our galaxy form?

14.4 THE MYSTERIOUS GALACTIC CENTER
- What lies in the center of our galaxy?

In previous chapters, we saw how stars forge new elements and expel them into space. We also studied how interstellar gas clouds enriched with these stellar by-products form new stars and planetary systems. These processes do not occur in isolation. Instead, they are part of a dynamic system that acts throughout our Milky Way Galaxy.

You are probably familiar with the idea that all living species on Earth interact with one another and with the land, water, and air to form a large, interconnected ecosystem. In a similar way, but on a much larger scale, our galaxy is a nearly self-contained system that cycles matter from stars into interstellar space and back into stars again. The birth of our solar system and the evolution of life on Earth would not have been possible without this "galactic ecosystem."

In this chapter, we will study our Milky Way Galaxy. We will investigate the galactic processes that maintain an ongoing cycle of stellar life and death, examine the structure and motion of the galaxy, probe its history, and explore the mysteries of the galactic center. Through it all, we will see that we are not only "star stuff" but "galaxy stuff"—the product of eons of complex recycling and reprocessing of matter and energy in the Milky Way Galaxy.

14.1 THE MILKY WAY REVEALED

On a dark night, you can see a faint band of light slicing across the sky through several constellations, including Sagittarius, Cygnus, Perseus, and Orion. This band of light looked like a flowing ribbon of milk to the ancient Greeks, and we now call it the *Milky Way* (see Figure 2.1). In the early 1600s, Galileo used his telescope to prove that the light of the Milky Way comes from myriad individual stars. Together these stars make up the kind of stellar system we call a *galaxy,* echoing the Greek word for "milk," *galactos.*

The true size and shape of our Milky Way Galaxy are hard to guess from the way it looks in our night sky. Because we live inside the galaxy, trying to determine its structure is somewhat like trying to draw a picture of your house without ever leaving your bedroom. Making the task even more difficult is the fact that much of our galaxy's visible light is hidden from our view. Only recently have we had the technology to observe the galaxy in other wavelengths of light. Nevertheless, by carefully observing our galaxy and comparing it to others that we see from the outside, we now have a good understanding of the processes that shape our galaxy. In this section, we'll begin our exploration of the galaxy by investigating its basic structure and orbital motion.

• What does our galaxy look like?

Our Milky Way Galaxy holds over 100 billion stars and is just one among tens of billions of galaxies in the observable universe [Section 1.1]. It is a vast **spiral galaxy**, so-named because of the spectacular **spiral arms** illustrated in Figure 14.1a. If we viewed it from the side, as shown in Figure 14.1b, we'd see that the spiral arms are part of a fairly flat **disk** of stars surrounding a bright central **bulge**. The entire disk is surrounded by a dimmer, rounder **halo**. Most of the galaxy's bright stars reside in its disk.

The most prominent stars in the halo are found in about two hundred **globular clusters** of stars [Section 11.3].

Our galaxy consists of a flat disk with spiral arms surrounding a central bulge, with a roughly spherical halo surrounding everything.

The entire galaxy is about 100,000 light-years in diameter, but the disk is only about 1,000 light-years thick. Our Sun is located in the disk about 28,000 light-years from the galactic center—a little more than halfway out from the center to the edge of the disk.

It took us a long time to learn these facts about the Milky Way's true size and shape. Clouds of interstellar gas and dust known collectively as the **interstellar medium** fill the galactic disk, obscuring our view when we try to peer directly through it. The dusty, smoglike nature of the interstellar medium hides most of our galaxy from us when we try to observe it in visible light, and as a result it long fooled astronomers into believing that we lived near our galaxy's center (see box below).

Astronomer Harlow Shapley finally proved otherwise in the 1920s, when he demonstrated that the Milky Way's globular clusters are centered on a point tens of thousands of light-years from our Sun. He concluded that this point, not our Sun, must be the center of the galaxy. Thus, it has been less than a century since we learned that our Sun holds no special place in the galaxy or the universe.

The Milky Way is a large galaxy. At least two smaller galaxies orbit the Milky Way.

The Milky Way is a relatively large galaxy. Within our Local Group of galaxies, only the Andromeda Galaxy is comparable in size (see Figure 1.1). The Milky Way's strong gravity influences smaller galaxies in its vicinity. For instance, two small galaxies, known as the *Large Magellanic Cloud* and *Small Magellanic Cloud,* orbit the Milky Way at distances

a Artist's conception of the Milky Way viewed from the outside.

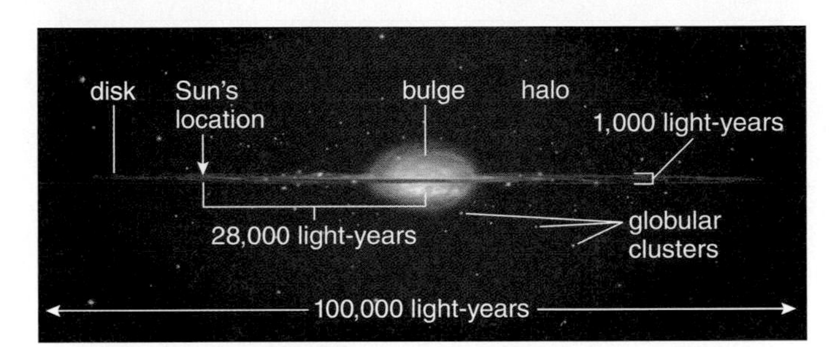

b Edge-on schematic view of the Milky Way.

Figure 14.1 

The Milky Way Galaxy.

SPECIAL TOPIC: HOW DID WE LEARN THE STRUCTURE OF THE MILKY WAY?

For most of human history, we knew the Milky Way only as an indistinct river of light in the sky. In 1610, Galileo used his telescope to discover that the Milky Way is made up of innumerable stars, but we still did not know the size or shape of our galaxy.

In the late 1700s, British astronomers William and Caroline Herschel (brother and sister) tried to determine the shape of the Milky Way more accurately by counting how many stars lay in each direction. Their approach suggested that the Milky Way's width was five times its thickness. More than a century later, in the early 1900s, Dutch astronomer Jacobus Kapteyn and his colleagues used a more sophisticated star-counting method to gauge the size and shape of the Milky Way. Their results seemed to confirm the general picture found by the Herschels and suggested that the Sun lay very near the center of the galaxy.

Kapteyn's results made astronomers with a sense of history slightly nervous. Only four centuries earlier, before Copernicus challenged the Ptolemaic system, astronomers had believed Earth was the center of the universe. Kapteyn's placement of the Sun near the Milky Way's center seemed to be giving Earth a central place again. Kapteyn knew that obscuring material could deceive us by hiding the rest of the galaxy like some kind of interstellar fog, but he found no evidence for such a fog.

While Kapteyn was counting stars, American astronomer Harlow Shapley was studying globular clusters. He found that these clusters appeared to be centered on a point tens of thousands of light-years from the Sun. (His original estimate was that the center of the galaxy was 45,000 light-years from the Sun, which was later refined to the current 28,000 light-years.) Shapley concluded that this point marked the true center of our galaxy and that Kapteyn must be wrong.

Today we know that Shapley was right. The Milky Way's interstellar medium is the "fog" that misled Kapteyn. Robert Trumpler, working at California's Lick Observatory in the 1920s, established the existence of this dusty gas by studying open clusters of stars. By assuming that all open clusters had about the same diameter, he estimated their distances from their apparent sizes in the sky, much as you might estimate the distances of cars at night from the apparent separation of their headlights. He found that stars in distant clusters appeared dimmer than expected based on their estimated distance, just as a car's headlights might appear in foggy weather. Trumpler concluded that light-absorbing material fills the spaces between the stars, partially obscuring the distant clusters and making them appear fainter than they would appear otherwise. Thus, we learned that interstellar material had been deceiving earlier astronomers, and that the stars visible in the night sky occupy a minuscule portion of the observable universe.

Subsequent studies established the spiral nature of our Milky Way Galaxy. Understanding the effects of interstellar dust allowed astronomers to take it into account in their observations, helping us learn the size and shape of our galaxy. Then, beginning in about the 1950s, careful studies of the motions of stars and gas clouds (made with radio observations of the 21-cm line from atomic hydrogen gas) gradually uncovered the detailed structure of the galactic disk, showing us the locations and motions of spiral arms.

of some 150,000 and 200,000 light-years, respectively. Both Magellanic Clouds are visible to the naked eye from the Southern Hemisphere.

Two other small galaxies lie even closer to the Milky Way than the Magellanic Clouds. These two galaxies are each in the process of colliding with the Milky Way's disk, and they will ultimately be ripped apart by our own galaxy's strong gravity. Both were discovered only recently, because their visible light is obscured from view by the gas and dust in the Milky Way's disk. One galaxy, known as the Sagittarius Dwarf, was discovered in 1994. The other, called the Canis Major Dwarf, was discovered in 2003.

Keep in mind that while these four nearby galaxies are quite small as galaxies go, they are still enormous, each containing perhaps 1 billion to a few billion stars. This makes them 1,000 or more times the size of typical globular clusters. These galaxies are small only when we compare them to the enormous size of our own Milky Way Galaxy, which is hundreds of times as large as its smaller companions.

• How do stars orbit in our galaxy?

Now that we have discussed the basic structure of our galaxy, let's turn our attention to our galaxy's orbital motion. A spiral galaxy like ours may look like it should rotate like a giant pinwheel, but the galaxy is not a solid structure. Instead, each individual star follows its own orbital path around the center of the galaxy.

Nearly all stars in the Milky Way follow one of two basic orbital patterns. Stars in the disk orbit in roughly circular paths that all go in the same direction in nearly the same plane. In contrast, stars in the bulge and halo soar high above and below the disk on randomly oriented orbits.

Orbits of Disk Stars If you could stand outside the Milky Way and watch it for a few billion years, the disk would resemble a huge merry-go-round. Like horses on a merry-go-round, individual stars bob up and down through the disk as they orbit. The general orbit of a star around the galaxy arises from its gravitational attraction toward the galactic center, while the bobbing arises from the localized pull of gravity within the disk itself (Figure 14.2). A star that is "too far" above the disk is pulled back into the disk by gravity. Because the density of interstellar gas is too low to slow the star, it flies through the disk until it is "too far" *below* the disk on the other side. Gravity then pulls it back in the other direction. This ongoing process produces the bobbing of the stars.

Disk stars orbit the galaxy's center in orderly circles that all go in the same direction, except for a slight up-and-down bobbing as they orbit.

The up-and-down motions of the disk stars give the disk its thickness of about 1,000 light-years—a great distance by human standards, but only about 1% of the disk's 100,000-light-year diameter. In the vicinity of our Sun, each star's orbit takes more than 200 million years, and each up-and-down "bob" takes a few tens of millions of years.

The galaxy's rotation is unlike that of a merry-go-round in one important respect: On a merry-go-round, horses near the edge move much faster than those near the center. But in our galaxy's disk, the orbital velocities of stars near the edge and those near the center are about the same. Stars closer to the center therefore complete each orbit in less time than stars farther out.

Orbits of Halo and Bulge Stars The orbits of stars in the halo and bulge are much less organized than the orbits of stars in the disk (see Fig-

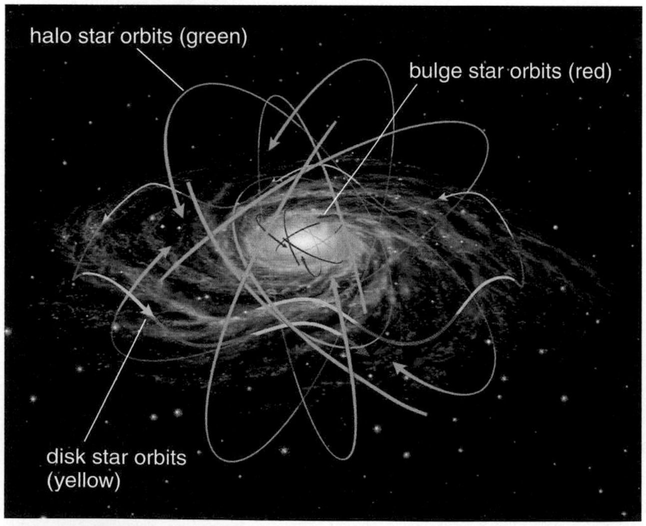

Figure 14.2

Characteristic orbits of disk stars (yellow), bulge stars (red), and halo stars (green) around the galactic center. (The yellow path exaggerates the up-and-down motion of the disk star orbits.)

ure 14.2). Individual bulge and halo stars travel around the galactic center on more or less elliptical paths, but the orientations of these paths are relatively random. Neighboring halo stars can circle the galactic center in opposite directions. They swoop from high above the disk to far below it and back again, plunging through the disk at velocities so high that the disk's gravity hardly alters their trajectories. Several fast-moving halo stars are currently passing through the disk not too far from our own solar system. One such star is Arcturus, the fourth-brightest star in the night sky.

Halo stars swoop high above and below the disk on randomly oriented orbits.

These swooping orbits explain why the bulge and halo are much puffier than the disk. Halo and bulge stars soar to heights above the disk far greater than the up-and-down bobbing of the disk stars. The great differences between orbits in the disk and orbits in the bulge and halo indicate that disk stars and halo stars have very different origins. As we'll discuss in Section 14.3, these orbital differences provide an important clue to how our galaxy formed.

THINK ABOUT IT Is there much danger that the Sun or Earth will someday be hit by a halo star swooping through the disk of the galaxy? Why or why not? (*Hint:* Think about the typical distances between stars, as illustrated by use of the 1-to-10-billion scale in Chapter 1.)

Detecting Dark Matter in a Spiral Galaxy Tutorial, Lessons 1–2

Using Stellar Orbits to Measure the Mass of the Galaxy The Sun's orbital path is fairly typical of disk stars located near our 28,000-light-year distance from the galactic center. By measuring the speeds of globular clusters relative to the Sun, we've determined that the Sun and its neighbors orbit the center of the Milky Way at a speed of about 220 kilometers per second (about 800,000 km/hr [Section 1.3]). Even at this speed, it takes the Sun about 230 million years to complete one orbit around the galactic center. Early dinosaurs were just emerging on Earth when our Sun last visited this side of the galaxy.

The orbital motion of the Sun and other stars gives us a way to determine the mass of the galaxy. Recall that Newton's law of gravity

SPECIAL TOPIC: HOW DO WE DETERMINE STELLAR ORBITS?

Astronomers learned how stars orbit the Milky Way by measuring the motions of many different stars relative to the Sun. Although these measurements are easy in principle, they can be difficult in practice.

Determining a star's precise motion relative to the Sun requires knowing its true velocity through space. However, our primary means of measuring speeds in the universe—the Doppler effect—can tell us only a star's *radial velocity,* the component of its velocity directed toward or away from us (see Figure 5.13). If we want to know the star's true velocity, we must also measure its *tangential velocity,* the component of its velocity directed across our line of sight.

Tangential velocity is difficult to measure because of the vast distances to stars. Over tens of thousands of years, the tangential velocities of stars cause their apparent positions in our sky to change, changing the shapes of the constellations. These changes are far too small for human eyes to notice. However, we can measure tangential velocities for many stars by comparing telescopic photographs taken years or decades apart. For example, if photographs taken 10 years apart show that a star has moved across our sky by an angle of 1 arcsecond, we know the star is moving at an angular rate of 0.1 arcsecond per year. We can convert this angular rate of motion (often called the star's *proper motion*) to a tangential velocity if we also know the star's distance. For a given angular rate, the tangential velocity is greater for more distant stars.

Because the earliest telescopic photographs date only to the late nineteenth century, we can measure tangential velocities only for objects that have moved noticeably since that time. In general, this limits us to stars within a few hundred light-years of Earth. Thus, we can determine precise stellar orbits only for relatively nearby stars. For more distant stars (and galaxies), we can usually measure only radial velocities. This is one reason why we have only limited knowledge of large-scale motion in the universe.

determines how quickly objects orbit one another. This fact, embodied in Newton's version of Kepler's third law [Section 4.4], allows us to determine the mass of a relatively large object when we know the period and average distance of a much smaller object in orbit around it.

We can calculate the mass of the galaxy *within* the Sun's orbit from the Sun's orbital properties using Newton's version of Kepler's third law.

For example, we can use the Sun's orbital velocity and its distance from the galactic center to determine the mass of our galaxy lying *within* the Sun's orbit. To understand why the Sun's orbital motion allows us to calculate only the mass within the Sun's orbit, rather than the total mass of the galaxy, we need to consider how the Sun is affected by mass within and beyond its orbit. Every part of the galaxy exerts gravitational forces on the Sun as it orbits, but the net force from matter outside the Sun's orbit is relatively small because the pulls from opposite sides of the galaxy virtually cancel one another. In contrast, the net gravitational forces from mass within the Sun's orbit all pull the Sun in the same direction—toward the galactic center. Thus, the Sun's orbital velocity responds almost exclusively to the gravitational pull of matter inside its orbit. By using the Sun's 28,000-light-year distance and 220-km/s orbital velocity in Newton's version of Kepler's third law, we find that the total amount of mass within the Sun's orbit is about 2×10^{41} kilograms, or about 100 billion times the mass of the Sun.

The orbits of the Milky Way's stars reveal that most of the galaxy's mass consists of invisible dark matter in the halo.

Similar calculations based on the orbits of more distant stars in the Milky Way have revealed one of the greatest mysteries in astronomy—one that we first encountered in Chapter 1 (see Figure 1.14). Photographs of spiral galaxies make it appear that most of their mass is concentrated near their centers. However, orbital motions tell us that just the opposite is true. If most of the mass were concentrated near the galaxy's center, the orbital speeds of more distant stars would be slower, just as the orbital speeds of the planets decline with distance from the Sun. Instead, we find that orbital speeds remain about the same out to great distances from the galactic center, telling us that most of the galaxy's mass resides far from the center and is distributed throughout the halo. Because we see very few stars and virtually no gas or dust in the halo, we conclude that most of the galaxy's mass must not give off any light that we can detect, and hence we refer to it as **dark matter**. We will discuss the evidence for dark matter in more detail in Chapter 16, along with ideas about its nature and its implications for the history and fate of the universe.

14.2 GALACTIC RECYCLING

The dusty gas that fills our galaxy's rotating disk turns out to be critically important to our existence. The birth of the Sun and the planets of our solar system could not have occurred without the galactic recycling that takes place within the disk of the galaxy and its interstellar medium.

Our galaxy is like a forest of stars whose ecology is shaped by the cycle of star birth and star death. Generations of stars continually recycle the same galactic matter through their cores, gradually raising the overall abundance of elements made by fusion. Recall that the universe was born with only the elements hydrogen and helium. All the elements heavier than helium, often called *heavy elements* by astronomers, have been made by fusion in stars. These elements are scattered into space when the stars die, and the galaxy then recycles them into new generations of stars.

Today, thanks to more than 10 billion years of galactic recycling, elements heavier than helium constitute about 2% of the galaxy's gaseous content by mass. The remaining 98% still consists of hydrogen (about 70%) and helium (about 28%).

THINK ABOUT IT Recall from Chapter 11 that stars in our galaxy's globular clusters are all very old, while stars in open clusters are relatively young. Based on this fact, which stars contain a higher proportion of heavy elements: stars in globular clusters or stars in open clusters? Explain.

Holding onto the elements released by stars is not an easy task. When a star explodes as a supernova, the ejected matter flies out at speeds of several thousand kilometers per second—much faster than the escape velocity from the galaxy. Thus, without the interstellar medium, the heavy elements released in the supernova would fly straight out of the Milky Way into intergalactic space. Instead, the matter expelled from the supernova collides with gas and dust in the interstellar medium, slows down, and eventually stops. It then blends in with the rest of the interstellar gas, where gravity can incorporate it into new generations of stars. In this section, we'll examine the galactic recycling process in more detail, so that we will understand why our existence owes as much to the functioning of our galaxy as it does to the creation of heavy elements by stars.

• How does our galaxy recycle gas into stars?

The galactic recycling process proceeds in several stages, making up what we will call the **star–gas–star cycle** and summarized in Figure 14.3. From Chapter 12, you are already familiar with the basic ideas in the lower three frames shown in the figure: Stars are born in molecular clouds, they shine and produce heavier elements through nuclear fusion, and they return much of their content back into the interstellar medium through

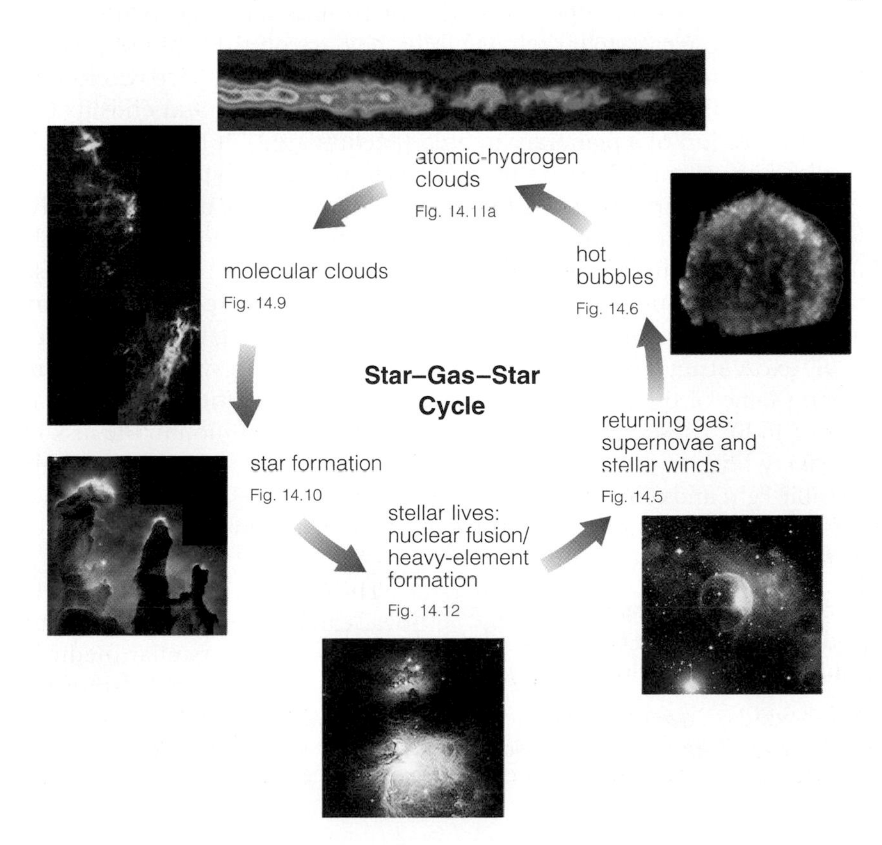

Figure 14.3

A pictorial representation of the star–gas–star cycle. These photos appear individually later in the chapter. Their figure numbers are indicated.

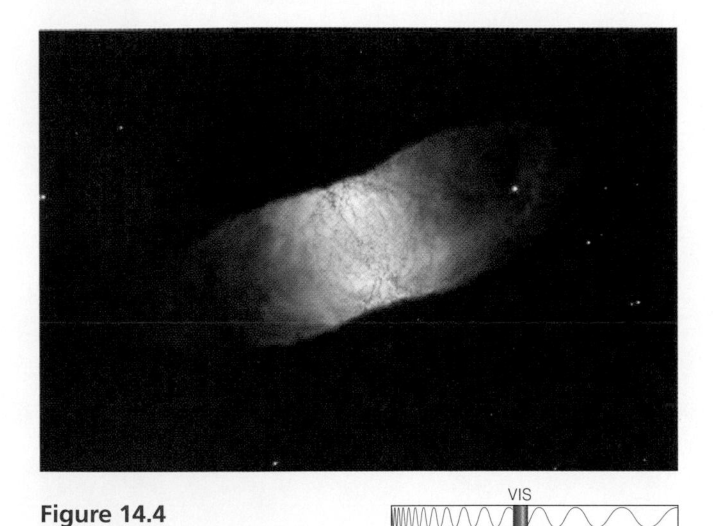

Figure 14.4 VIS

A dying low-mass star, like this one photographed by the Hubble Space Telescope, returns gas to the interstellar medium in a planetary nebula. This particular planetary nebula is known as the Retina Nebula. It is about 1 light-year across in the longer direction.

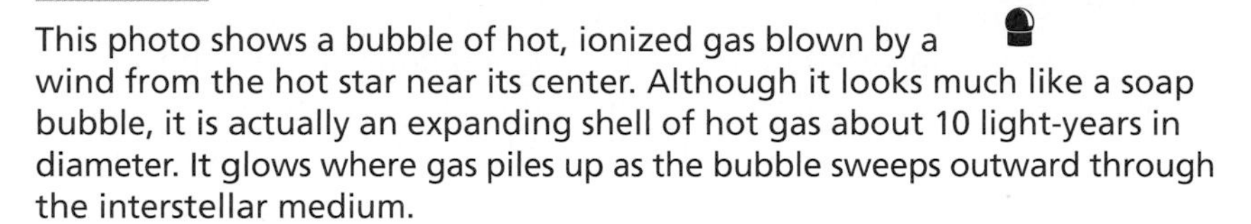

Figure 14.5 VIS

This photo shows a bubble of hot, ionized gas blown by a wind from the hot star near its center. Although it looks much like a soap bubble, it is actually an expanding shell of hot gas about 10 light-years in diameter. It glows where gas piles up as the bubble sweeps outward through the interstellar medium.

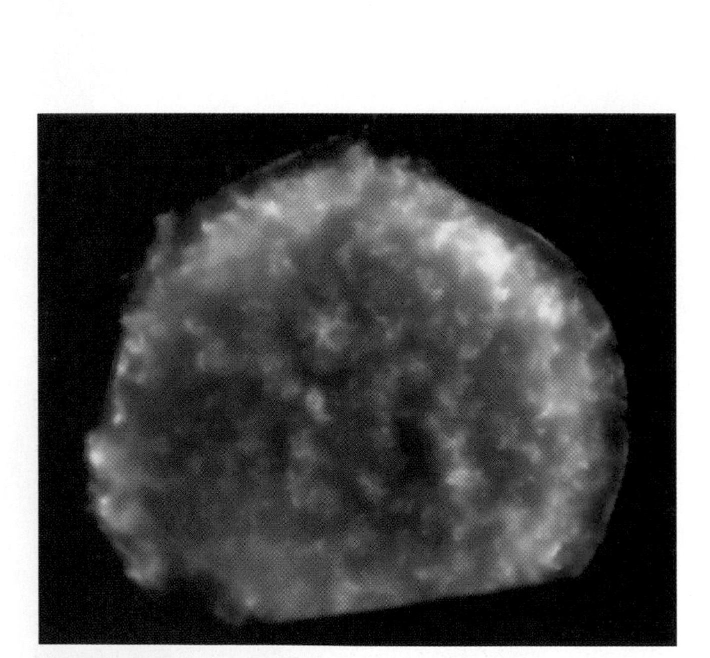

Figure 14.6 X-ray

This image from the Chandra X-Ray Observatory shows X-ray emission from hot gas in a young supernova remnant—the remnant from the supernova observed by Tycho Brahe in 1572. The most energetic X rays (blue), come from 20-million-degree gas just behind the expanding shock wave. Less energetic X rays (green and red) come from the 10-million-degree debris ejected by the exploded star. The remnant is about 20 light-years across. (The straight lower edge is an artifact of the X-ray detector's field of view.)

supernovae and stellar winds. Let's complete the loop, starting with the gas ejected by dying stars and finishing back at star birth.

Gas from Dying Stars All stars return much of their original mass to interstellar space in two basic ways: through stellar winds that blow throughout their lives, and through "death events" of planetary nebulae (for low-mass stars) or supernovae (for high-mass stars). Low-mass stars generally have weak stellar winds while they are on the main sequence. Their winds grow stronger and carry more material into space when they become red giants. By the time a low-mass star like the Sun ends its life with the ejection of a planetary nebula [Section 12.2], it has returned almost half its original mass to the interstellar medium (Figure 14.4).

High-mass stars lose mass much more dynamically and explosively. The powerful winds from supergiants and massive O and B stars recycle large amounts of matter into the galaxy. At the ends of their lives, these stars explode as supernovae. The high-speed gas ejected into space by supernovae or powerful winds sweeps up surrounding interstellar material, excavating a **bubble** of hot, ionized gas (gas in which atoms are missing some of their electrons [Section 5.2]) around the exploding star (Figure 14.5). These hot, tenuous bubbles are quite common in the disk of the galaxy, but they are not always easy to detect. While some emit strongly in visible light and others are hot enough to emit profuse amounts of X rays, many bubbles are evident only through radio emission from the shells of gas that surround them.

Supernovae and high-speed stellar winds can produce hot bubbles of gas in the interstellar medium.

The bubbles created by supernovae can have even more dramatic effects on the interstellar medium than those created by fast-moving stellar winds. Supernovae generate *shock waves*—waves of pressure that move faster than the speed of sound. A shock wave sweeps up surround-

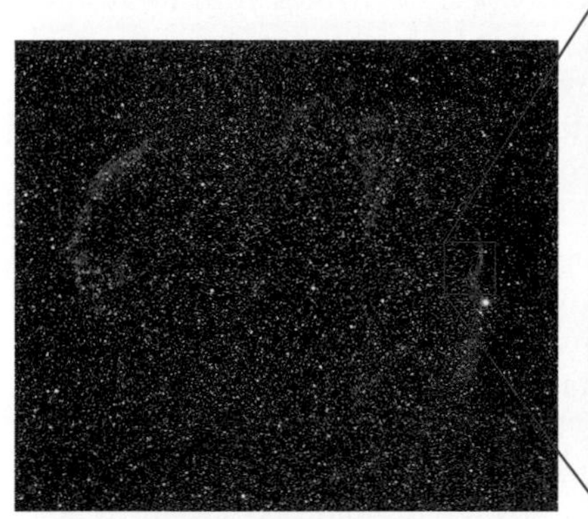

a This photograph shows the entire Cygnus Loop supernova remnant glowing in visible light. The angular size of this remnant in our sky is six times that of the Moon, and it is about 130 light-years across.

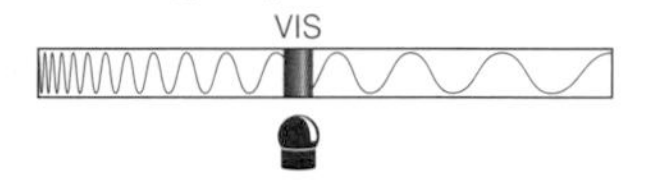

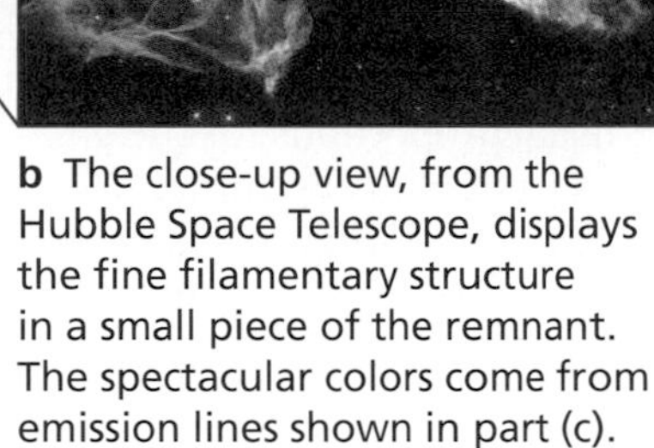

b The close-up view, from the Hubble Space Telescope, displays the fine filamentary structure in a small piece of the remnant. The spectacular colors come from emission lines shown in part (c).

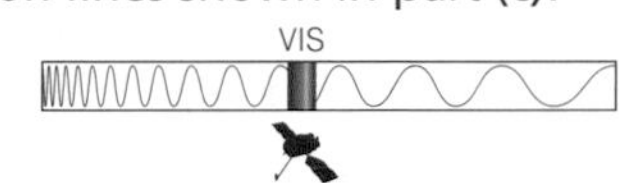

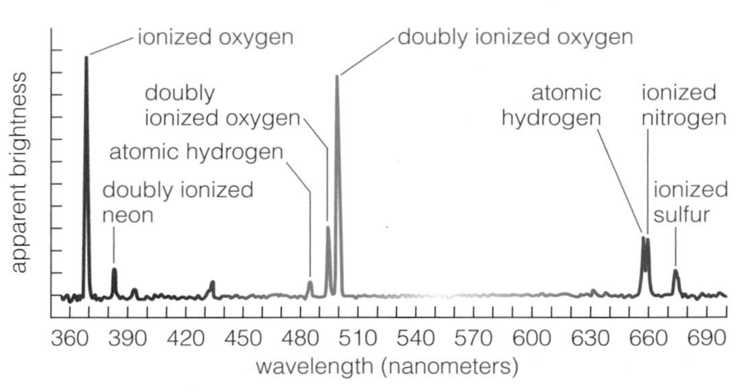

c A visible-light spectrum from the Cygnus Loop shows the strong emission lines that account for the distinct colors in the Hubble Space Telescope photo.

Figure 14.7

Emission of visible light from an older supernova remnant, the Cygnus Loop.

ing gas as it travels, creating a "wall" of fast-moving gas on its leading edge. When we observe a *supernova remnant* [Section 12.3], we are seeing the aftermath of its shock wave, which compresses, heats, and ionizes all the interstellar gas it encounters.

Figure 14.6 shows a young supernova remnant whose shocked gas is hot enough to emit X rays. In contrast, the older supernova remnant shown in Figure 14.7 is cooler because its shock wave has swept up more material, thereby distributing its energy more widely. Eventually, the shocked gas will radiate away most of its original energy, and the expanding wall of gas will slow to subsonic speeds and merge with the surrounding interstellar medium.

Giant bubbles of hot gas can erupt out of the disk, spreading their contents over a large region of the galaxy.

In regions of the galactic disk where many supernovae have recently exploded, we find giant, elongated bubbles extending from young clusters of stars to distances of 3,000 light-years or more above (or below) the disk. These probably are places where the bubbles from many individual supernovae have merged to make a giant bubble so large that it cannot be contained within the Milky Way's disk. Once the top of a giant bubble breaks out of the disk, where nearly all of the Milky Way's gas resides, nothing remains to slow its expansion except gravity. The result is a *blowout* that is in some ways similar to a volcanic eruption, but on a galactic scale: Hot gas erupts from the disk, spreading out as it shoots upward into the galactic halo (Figure 14.8). The gravity of the galactic disk slows the rise of the gas, eventually pulling it back down. Near the top of its trajectory, the ejected gas starts to cool and form clouds. These clouds can then rain back down into the disk, where their contents mix with the gas in a large region of the galaxy.

In addition to their effects on the interstellar medium, supernovae may also affect life by generating **cosmic rays** that can cause genetic mutations in living organisms. Cosmic rays are made of electrons, protons, and atomic nuclei that zip through interstellar space at close to the speed of light. Some cosmic rays penetrate Earth's atmosphere and reach Earth's

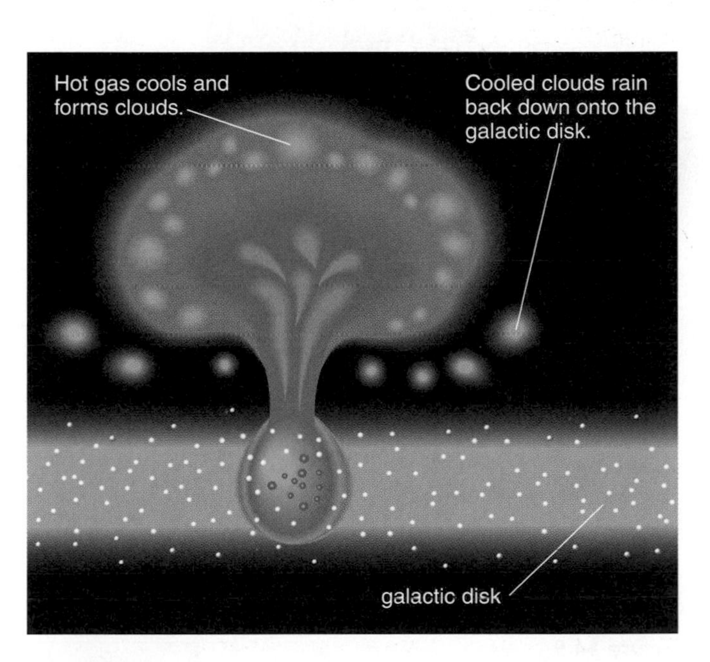

Figure 14.8

Hot gas erupting from a giant bubble out into the galactic halo eventually cools into gas clouds that rain back down on the disk. This process may be an important part of the galaxy-wide recycling system that incorporates the products of supernova explosions into new generations of stars and planets.

COMMON MISCONCEPTIONS

THE SOUND OF SPACE

In many science fiction movies, a thunderous sound accompanies the demolition of a spaceship. If the moviemakers wanted to be more realistic, they would silence the explosion. On Earth, we perceive sound when sound waves—which are waves of alternately rising and falling pressure—cause trillions of gas atoms to push our eardrums back and forth. Although sound waves can and do travel through interstellar gas, the extremely low density of this gas means that only a handful of atoms per second would collide with something the size of a human eardrum. As a result, it would be impossible for a human ear (or a similar-size microphone) to register any sound. Despite the presence of sound waves and shock waves in space, the sound of space is silence.

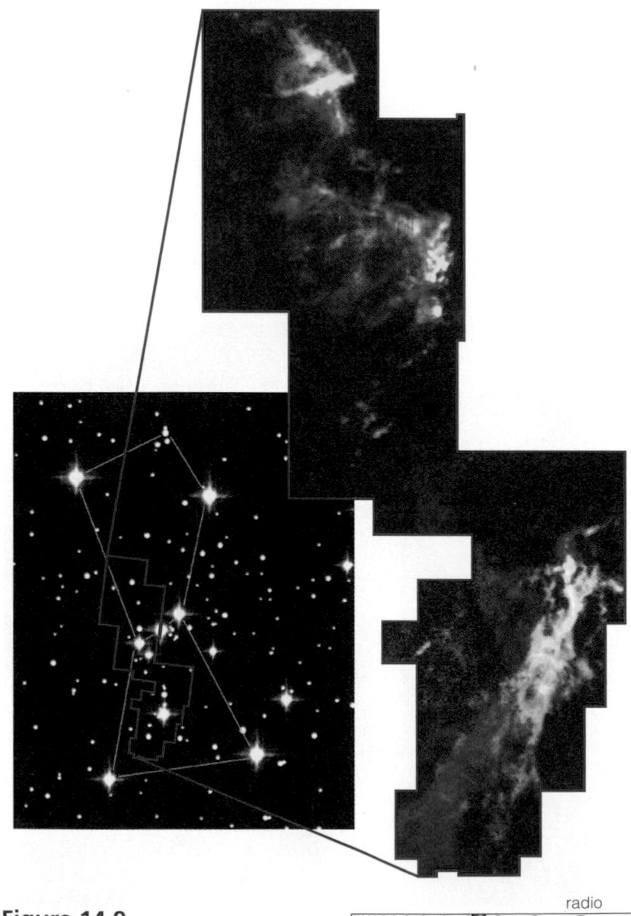

Figure 14.9

radio

This image shows the complex structure of a molecular cloud in the constellation Orion. The picture was made by measuring Doppler shifts of emission lines from carbon monoxide molecules, and the colors indicate gas motions: Bluer parts are moving toward us and redder parts are moving away from us (relative to the cloud as a whole). This enormous cloud is about 1,600 light-years distant and several hundred light-years across.

surface. On average, about one cosmic-ray particle strikes your body each second. The cosmic-ray bombardment rate is 100 times greater at the high altitudes at which jet planes fly, and even more cosmic rays funnel along magnetic field lines to Earth's magnetic poles. That is why some airlines restrict how often their flight crews cross the Earth's polar regions.

Cooling and Cloud Formation The hot, ionized gas in bubbles heated by supernovae is dynamic and widespread, but it is a relatively small fraction of the gas in the Milky Way. Most of the gas is much cooler—cool enough so that hydrogen atoms remain neutral rather than being ionized. We therefore refer to this gas as **atomic hydrogen gas**, although the hydrogen is mixed with neutral atoms of helium and heavy elements in the usual proportions for the galaxy (70% hydrogen, 28% helium, and 2% heavy elements by mass). Once the gas in bubbles cools, it becomes part of this widespread atomic hydrogen gas in the galaxy.

We can map the distribution of atomic hydrogen gas in the Milky Way with radio observations. Atomic hydrogen emits a spectral line with a wavelength of 21 centimeters, which lies in the radio portion of the electromagnetic spectrum (see Figure 5.2). We see the radio emission from this **21-centimeter line** coming from all directions, telling us that atomic hydrogen gas is distributed throughout the galactic disk. Based on the overall strength of the 21-centimeter emission, the total amount of atomic hydrogen gas in our galaxy must be about 5 billion solar masses, which is a few percent of the galaxy's total mass.

Matter remains in the atomic hydrogen stage of the star–gas–star cycle for millions of years. Gravity slowly draws blobs of this gas together into tighter clumps, which radiate energy more efficiently as they grow denser. The blobs therefore cool and contract, forming clouds of cooler and denser gas. The cooling and contraction of atomic hydrogen clouds is a very slow process, taking a much longer time than the other steps in the cycle from star death to star birth. That is why so much of the Milky Way's gas is in the atomic hydrogen stage of the star–gas–star cycle.

Clouds of atomic hydrogen also contain a small but important amount of interstellar dust. Interstellar **dust grains** are tiny, solid flecks of carbon and silicon minerals that resemble particles of smoke and form in the winds of red giant stars [Section 12.2]. Once formed, dust grains remain in the interstellar medium until they are heated and destroyed by a passing shock wave or incorporated into a protostar. Dust grains make up only about 1% of the mass of atomic hydrogen clouds, but they are responsible for the absorption of visible light that prevents us from seeing through the disk of the galaxy.

Gas heated by supernovae first cools into atomic hydrogen clouds, then cools further into clouds of molecular hydrogen.

As the temperature drops further in the center of a cool cloud of atomic hydrogen, hydrogen atoms combine into molecules, making a *molecular cloud* [Section 12.1]. Molecular clouds are the coldest, densest collections of gas in the interstellar medium. They often congregate into *giant molecular clouds* that hold up to a million solar masses of gas. The total mass of molecular clouds in the Milky Way is somewhat uncertain, but it is probably about the same as the total mass of atomic hydrogen gas—about 5 billion solar masses. Throughout much of this molecular gas, temperatures hover only a few degrees above absolute zero.

Molecular hydrogen (H_2) is by far the most abundant molecule in molecular clouds, but it is difficult to detect because temperatures are usually too cold for the gas to produce H_2 emission lines. Most of what

Figure 14.10

A portion of the Eagle Nebula, as seen by the Hubble Space Telescope. The dark columns of gas are molecular clouds, and stars are currently forming in the densest parts of these clouds. Arrows indicate two of the locations where dense knots of gas are giving birth to stars. The region pictured here is about 5 light-years across.

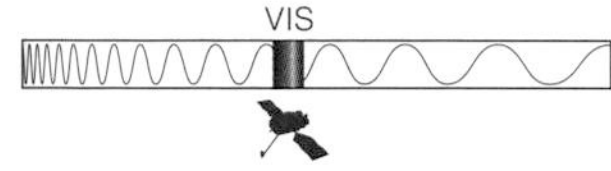

we know about molecular clouds comes from observing spectral lines of molecules that make up only a tiny fraction of a cloud's mass. Carbon monoxide (CO) is the most abundant of these molecules. It produces strong emission lines in the radio portion of the spectrum at the 10–30 K temperatures of molecular clouds (Figure 14.9). More than 120 other molecules have been identified in molecular clouds by their radio emission lines, including water (H_2O), ammonia (NH_3), and ethyl alcohol (C_2H_5OH).

Molecular clouds are heavy and dense compared to the rest of the interstellar gas and therefore tend to settle toward the central layers of the Milky Way's disk. This tendency creates a phenomenon you can see with your own eyes: the dark lanes running through the luminous band of light in our sky that we call the Milky Way (see Figure 2.1).

Completing the Cycle A large molecular cloud gives birth to a cluster of stars. Once a few stars form in a newborn cluster, their radiation begins to erode the surrounding gas in the molecular cloud. Ultraviolet photons from high-mass stars heat and ionize the gas, and winds and radiation pressure push the ionized gas away. This kind of feedback prevents much of the gas in a molecular cloud from turning into stars.

Ultraviolet radiation from newly forming stars can erode the molecular clouds that gave them birth.

The process of molecular cloud erosion is vividly illustrated in the Eagle Nebula, a complex of clouds where new stars are forming (Figure 14.10). The dark, lumpy columns are molecular clouds. To the upper right (outside the picture), newly formed massive stars glow with ultraviolet radiation. This radiation sears the surface of the molecular clouds,

TABLE 14.1 *Typical States of Gas in the Interstellar Medium*

	State of Gas		
	Hot Bubbles	*Atomic Hydrogen Clouds*	*Molecular Clouds*
Primary Constituent	Ionized hydrogen	Atomic hydrogen	Molecular hydrogen
Approximate Temperature	1,000,000 K	100–10,000 K	30 K
Approximate Density (atoms per cm^3)	0.01	1–100	300
Description	Pockets of gas heated by stellar winds or supernovae	The most common form of gas, filling much of the galactic disk	Regions of star formation

destroying molecules and stripping electrons from atoms. As a result, matter "evaporates" from the molecular clouds and joins the hotter ionized gas encircling them. Only the densest knots of gas resist evaporation. Stars are forming in some of these dense knots, which remain compact while the rest of the cloud erodes. These star-forming knots are the tips of the dark protrusions on the columns of molecular gas in the figure.

We have arrived back where we started in the star–gas–star cycle. The most massive stars now forming in the Eagle Nebula will explode within a few million years, filling the region with bubbles of hot gas and newly formed heavy elements. The expanding bubbles will slow and cool as their gas merges with the widespread atomic hydrogen gas in the galaxy. Farther in the future, this gas will cool further and coalesce into molecular clouds, forming new stars, new planets, and maybe even new civilizations.

Despite the recycling of matter from one generation of stars to the next, the star–gas–star cycle cannot go on forever. With each new generation of stars, some of the galaxy's gas becomes permanently locked away in brown dwarfs that never return material to space and in stellar corpses left behind when stars die (white dwarfs, neutron stars, and black holes). The interstellar medium therefore is slowly running out of gas, and the rate of star formation will gradually taper off over the next 50 billion years or so. Eventually, star formation will cease.

Putting It All Together: The Distribution of Gas in the Milky Way

Different regions of the galaxy are in different stages of the star–gas–star cycle. Because the cycle proceeds over such a long period of time compared to a human lifetime, each stage appears to us as a snapshot. We therefore see the interstellar medium in a wide variety of manifestations, ranging from the tenuous million-degree gas of bubbles to the cold, dense gas of molecular clouds. Table14.1 summarizes the different states in which we see interstellar gas in the galactic disk.

To get a complete picture of the star-gas-star cycle of the Milky Way, we need to observe its gas in many different kinds of light.

We can see how these different states of gas are arranged in our galaxy by observing the galaxy in many different wavelengths of light. Figure 14.11 shows seven views of the disk of the Milky Way Galaxy. Each view represents a panorama in a particular wavelength band, made by photographing the Milky Way's disk in every direction from Earth.

- Figure 14.11a shows variations in the intensity of radio emission from the 21-centimeter line of atomic hydrogen. Thus, it maps the distribution of atomic hydrogen gas, demonstrating that this gas fills much of the galactic disk.
- Figure 14.11b shows variations in the intensity of radio emission lines from carbon monoxide (CO) and therefore maps the distribution of molecular clouds. These cold, dense clouds are concentrated in a narrow layer near the midplane of the galactic disk.
- Figure 14.11c shows variations in the intensity of long-wavelength infrared emission from interstellar dust grains. Notice that the regions of strongest emission from dust correspond to the locations of the molecular clouds in Figure 14.11b.
- Figure 14.11d shows shorter-wavelength infrared light from stars at wavelengths that penetrate clouds of gas and dust. Thus, this

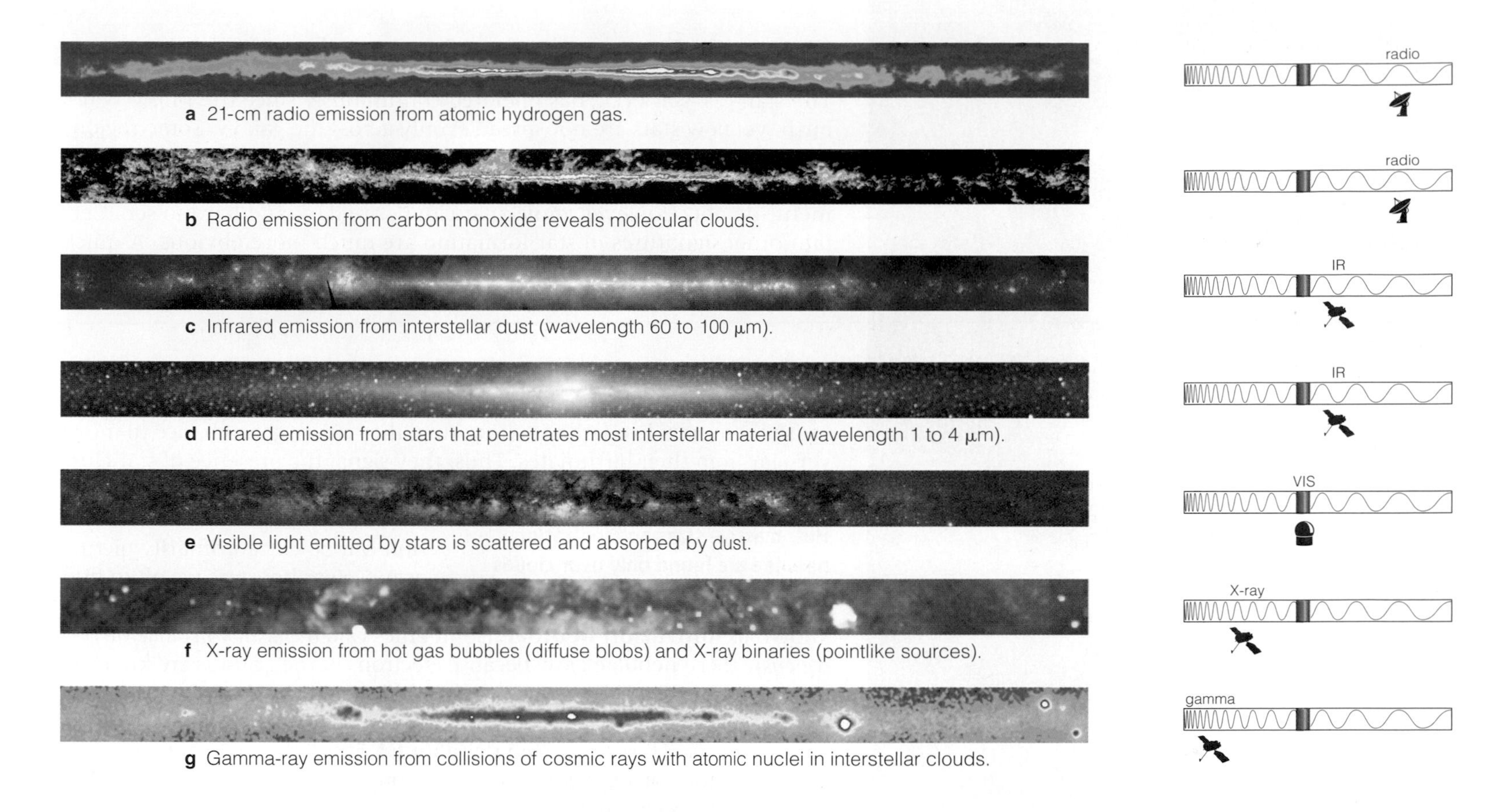

a 21-cm radio emission from atomic hydrogen gas.

b Radio emission from carbon monoxide reveals molecular clouds.

c Infrared emission from interstellar dust (wavelength 60 to 100 μm).

d Infrared emission from stars that penetrates most interstellar material (wavelength 1 to 4 μm).

e Visible light emitted by stars is scattered and absorbed by dust.

f X-ray emission from hot gas bubbles (diffuse blobs) and X-ray binaries (pointlike sources).

g Gamma-ray emission from collisions of cosmic rays with atomic nuclei in interstellar clouds.

Figure 14.11

Panoramic views of the Milky Way in different bands of the spectrum. The center of the galaxy, which lies in the direction of the constellation Sagittarius, is in the center of each strip. The rest of each strip shows all other directions in the Milky Way disk as seen from Earth. (Imagine attaching the left and right ends of each strip to form a circular band that corresponds to the 360° band of the Milky Way in our sky.)

image shows how our galaxy would look if there were no dust blocking our view. The galactic bulge is clearly evident at the center.

- Figure 14.11e shows the galactic disk in visible light, just as it appears in the night sky. (Of course, only part of the Milky Way is above the horizon at any one time.) Because visible light cannot penetrate interstellar dust, the dark blotches correspond closely to the bright patches of molecular radio emission and infrared dust emission in Figure 14.11b and c.
- Figure 14.11f shows X-ray light from the galactic disk. The pointlike blotches in this view are mostly X-ray binaries [Section 13.2]. The rest of the X-ray emission comes primarily from hot gas bubbles. Because hot gas tends to rise into the halo, it is less concentrated toward the midplane than the atomic and molecular gas.
- Figure 14.11g shows gamma-ray emission from the Milky Way. Most of the gamma-ray emission is produced by collisions between cosmic-ray particles and atomic nuclei in interstellar clouds. Such collisions happen most frequently where gas densities are highest, so the gamma-ray emission corresponds closely to the locations of molecular and atomic gas.

Carefully compare and contrast the views of the Milky Way's disk in Figure 14.11. Why do regions that appear dark in some views appear bright in others? What general patterns do you notice?

Figure 14.12

A photo of the Orion Nebula, an ionization nebula energized by ultraviolet photons from hot stars.

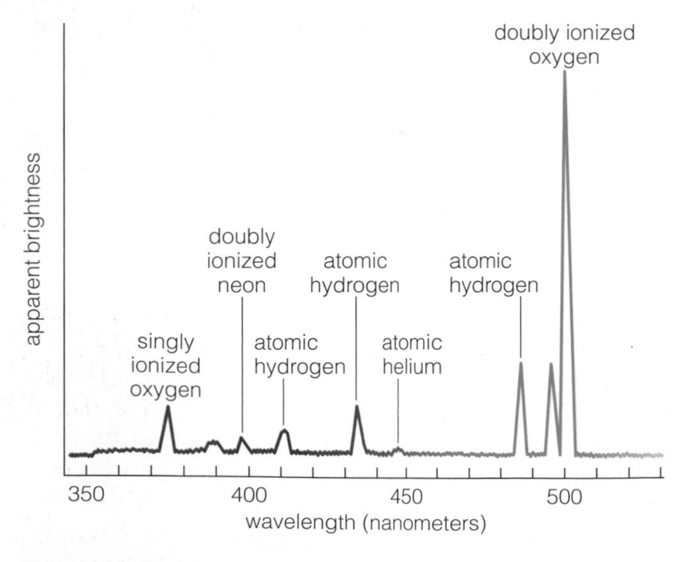

Figure 14.13

The spectrum of an ionization nebula in blue and green light. The prominent emission lines reveal the atoms and ions that emit most of the light. Through careful study of these lines, we can determine the nebula's chemical composition.

• Where do stars tend to form in our galaxy?

The star–gas–star cycle has operated continuously since the Milky Way's birth, yet new stars are not spread evenly across the galaxy. Some regions seem much more fertile than others. Galactic environments rich in molecular clouds tend to spawn new stars easily, while gas-poor environments do not. However, molecular clouds are dark and hard to see. Certain other signatures of star formation are much more obvious. A quick tour of some star-forming galactic environments will help you spot where the action is.

Hallmarks of Star-Forming Regions Wherever we see hot, massive stars, we know that we have spotted a region of active star formation. Because these stars live fast and die young, they never get a chance to move very far from their birthmates. Thus, they signal the presence of star clusters in which many of their lower-mass companions are still forming.

Hot, massive stars and ionization nebulae are found only near clouds that are actively forming stars.

Regions of active star formation can be extraordinarily picturesque. Near hot stars we often find colorful, wispy blobs of glowing gas known as **ionization nebulae** (sometimes called *emission nebulae* or *H II regions*). These nebulae glow because electrons in their atoms are knocked up to high energy levels [Section 5.2] or ionized by ultraviolet photons from the hot stars, so they emit light as the electrons fall back to lower energy levels. The Orion Nebula, about 1,500 light-years away in the "sword" of the constellation Orion, is among the most famous. Few astronomical objects can match its spectacular beauty (Figure 14.12).

Most of the striking colors in an ionization nebula come from particular spectral lines produced by particular atomic transitions. For example, the transition in which an electron falls from energy level 3 to energy level 2 in a hydrogen atom generates a red photon with a wavelength of 656 nanometers (see Figure 5.9). Ionization nebulae appear predominantly red in photographs because of all the red photons released by this particular transition. (Jumps from level 2 to level 1 are even more common, but they produce ultraviolet photons that can be studied only with ultraviolet telescopes in space.) Transitions in other elements produce other spectral lines of different colors (Figure 14.13).

The blue and black tints in some star-forming regions have a different origin. Starlight reflected from dust grains produces the blue colors, because interstellar dust grains scatter blue light much more readily than red light (Figure 14.14). These so-called *reflection nebulae* are always bluer in color than the stars supplying the light. (The effect is similar to the scattering of sunlight in our atmosphere that makes the sky blue [Section 7.1].) The black regions of nebulae are dark, dusty gas clouds that block our view of the stars beyond them. Figure 14.15 shows a multicolored nebula characteristic of a hot-star neighborhood.

In Figure 14.15, identify the red ionized regions, the blue reflecting regions, and the dark obscuring regions. Briefly explain the origin of the colors in each region.

Spiral Arms Taking a broader view of our galaxy, we can see that the spiral arms must be full of newly forming stars because they bear all the

Figure 14.14

The blue tints in this nebula in the constellation are produced by reflected light.

Figure 14.15

A photo of the Horsehead Nebula and its surroundings. (The region pictured is about 150 light-years across.)

hallmarks of star formation. They are home to both molecular clouds and numerous clusters of young, bright, blue stars surrounded by ionization nebulae. Detailed images of other spiral galaxies show these characteristics more clearly (Figure 14.16). Hot, blue stars and ionization nebulae trace out the arms while the stars between the arms are generally redder and older. We also see enhanced amounts of molecular and atomic gas in the spiral arms, and streaks of interstellar dust often obscure the inner sides of the arms themselves (Figure 14.17). Thus, spiral arms contain both young stars and the material necessary to make new stars.

At first glance, spiral arms look as if they ought to move with the stars, like the fins of a giant pinwheel in space. However, we know that spiral arms cannot be fixed patterns of stars that rotate along with the galaxy. The reason is that stars near the center of the galaxy complete an orbit in much less time than stars far from the center. If the spiral arms simply moved along with the stars, the central parts of the arms would complete several orbits around the galaxy as the outer parts orbited just once. This difference in orbital periods would eventually wind up the spiral arms into a tight coil. Because we generally don't see such tightly wound spiral

COMMON MISCONCEPTIONS

WHAT IS A NEBULA?

The term nebula means "cloud," but in astronomy it can refer to many different kinds of objects—a state of affairs that sometimes leads to misconceptions. Many astronomical objects look "cloudy" through small telescopes, and in past centuries astronomers called any such object a nebula as long as they were sure it wasn't a comet. For example, galaxies were called nebulae because they looked like either fuzzy round blobs or fuzzy spiral blobs.

Using the term nebula to refer to a galaxy now sounds somewhat dated, given the enormous differences between these distant star systems and the much smaller clouds of gas that populate the interstellar medium. Nevertheless, some people still refer to spiral galaxies as "spiral nebulae." Today, we generally use the term nebula to refer to true interstellar clouds, but be aware that the term is still sometimes used in other ways.

Figure 14.16

Galaxy M51, a spiral galaxy with two prominent spiral arms, as photographed by the Hubble Space Telescope. Notice that the spiral arms are much bluer in color than the central bulge and the regions of the disk between the arms. The blue arms also contain many red blotches, which are ionization nebulae. Because massive, blue stars live only for a few million years, the relative blueness of the spiral arms tells us that stars must be forming more actively within them than elsewhere in the galaxy. (This portion of M51 is roughly 30,000 light-years across.)

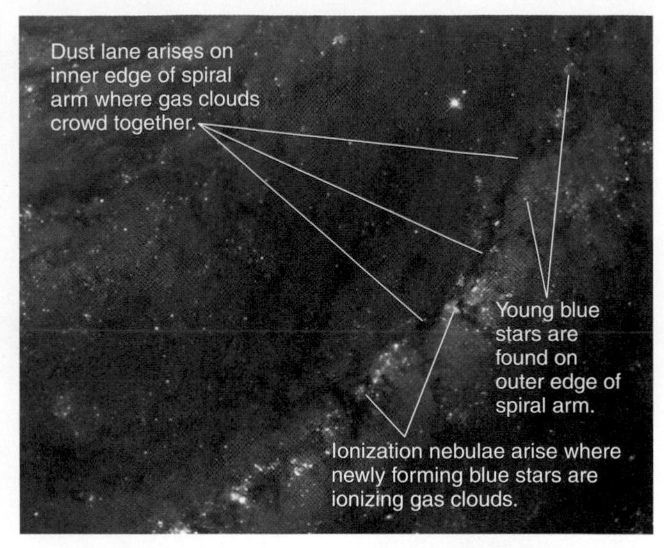

Figure 14.17

This photo showing a segment of a spiral arm in Galaxy M51 illustrates the relationship between dust, gas, and new stars in a spiral arm.

arms in galaxies, we conclude that spiral arms are more like swirling ripples in a whirlpool than like the fins of a giant pinwheel.

In fact, we now believe that spiral arms are enormous waves of star formation that propagate through the gaseous disk of a spiral galaxy like the Milky Way. Theoretical models suggest that disturbances called **spiral density waves** are responsible for the spiral arms. According to these models, spiral arms are places in a galaxy's disk where stars and gas clouds get more densely packed. Pushing the stars closer together has little effect on the stars themselves—they are still much too widely separated to collide with each other. However, the large gas clouds do collide, and packing the clouds closer together enhances the force of gravity within them, triggering the formation of many new stars.

Spiral arms are waves of star formation that spread through our galaxy's disk.

To visualize how spiral density waves propagate through a galaxy's disk, consider how traffic backs up behind a slow-moving tractor on a rural highway. Cars approaching the tractor slow down and bunch together. After cars pass the tractor, they speed up and spread out again. Thus, a pack of cars is always bunched up behind the tractor, even though the cars themselves are constantly flowing past it.

In a spiral density wave, gravity plays the role of the tractor, while stars and gas clouds play the role of the cars. The stars and gas clouds of a galaxy's disk are constantly flowing through its spiral arms, but the extra density of matter in the spiral arm alters that flow. The extra matter exerts a gravitational force that pulls stars and gas clouds into the arm and tries to halt their escape as they move out the other side. This gravitational pull is not strong enough to trap the stars and gas clouds. However, like the tractor, it temporarily slows them down, and this temporary slowdown produces a long-lasting pattern that gets stretched into spiral shape by the rotation of the disk.

14.3 THE HISTORY OF THE MILKY WAY

Now that we have discussed the basic properties of the Milky Way, including the star-gas-star cycle of the disk, we are ready to turn our attention to the history of our galaxy. All the galaxy's properties provide clues to its history. However, some of the most important clues come from a detailed comparison of disk stars with halo stars. We'll begin with this comparison and then discuss a basic model of galaxy formation that explains many of the differences between these two groups of stars.

• What clues to our galaxy's history do halo stars hold?

We have already seen how the disorderly orbits of halo stars differ from the generally circular orbits of disk stars. Two other differences between halo stars and disk stars give us further clues to their origins. First, we don't find any young stars in the halo, while in the disk we see stars of many different ages. Second, the spectra of halo stars show that they contain fewer heavy elements than do disk stars.

Halo stars are all old with a very low proportion of heavy elements, while disk stars come in all ages and contain higher proportions of heavy elements.

Because of these striking differences, astronomers divide the Milky Way's stars into two distinct populations. The *disk population* (sometimes called *Population I*) contains both young stars and old stars, all of which have heavy-element proportions of about 2%, like our Sun. The *spheroidal population* (or *Population II*) consists of stars in the halo and bulge, both of which are roughly spherical in shape. Stars in this population are always old and therefore low in mass, and those in the halo sometimes have heavy-element proportions as low as 0.02%—making heavy elements about 100 times rarer in these stars than in the Sun.

We can understand why bulge and halo stars differ by looking at how the Milky Way's gas is distributed. The halo does not contain the cold, dense molecular clouds required for star formation. In fact, the halo contains very little gas at all, and that small amount of gas is generally quite hot. Because star-forming molecular clouds are found only in the disk, new stars can be born only in the disk and not in the halo.

The relative lack of heavy elements in halo stars indicates that they must have formed early in the galaxy's history—before many supernovae had exploded, adding heavy elements to star-forming clouds. Thus, we conclude that the halo has lacked the gas needed for star formation for a very long time. Apparently, all the Milky Way's cool gas settled into the disk long ago. The only stars that still survive in the halo are long-lived, low-mass stars. Any more massive stars that were once born in the halo died long ago.

How does the halo of our galaxy resemble the distant future fate of the galactic disk? Explain.

• How did our galaxy form?

Any model for our galaxy's formation must account for the differences between disk stars and halo stars. The most basic model proposes that our

Figure 14.18 Interactive Figure

This four-picture sequence illustrates a simple schematic model of galaxy formation, showing how a spiral galaxy might develop from a protogalactic cloud of hydrogen and helium gas.

galaxy began as a giant **protogalactic cloud** containing all the hydrogen and helium gas that the galaxy eventually turned into stars. Gravity would have caused such a cloud to contract and fragment, just as in present-day star-forming clouds [Section 12.1].

According to this model, the stars of the spheroidal population (bulge and halo) formed first. Early on, the gravity associated with our protogalactic cloud drew in matter from all directions, creating a cloud that was blobby in shape and had little or no measurable rotation. The orbits of stars forming within such a cloud could have had any orientation, accounting for the randomly oriented orbits of stars in the spheroidal population.

Later, conservation of angular momentum caused the remaining gas to flatten into a spinning disk as it contracted under the force of gravity (Figure 14.18). This process was much like the process that leads to the formation of spinning disks of gas around young stars [Sections 6.3, 12.1], but on a much larger scale. Collisions among gas particles tended to average out their random motions, leading them to acquire orbits in the same direction and in the same plane. Stars that form within this spinning disk are born on orbits moving at the same speed and in the same direction as their neighbors and thus become members of the disk population of stars.

Halo stars formed when our galaxy's protogalactic cloud was still large and blobby, and disk stars formed after the gas settled into a spinning disk.

We can test this basic model by studying the "fossil record" written within the stars of the Milky Way. The clues found to date support the basic picture but suggest that the full story of galaxy formation may be somewhat more complex.

All available evidence confirms that the stars in the Milky Way's halo are indeed old. The main-sequence turnoff points in H–R diagrams of globular clusters show that their stars were born at least 12 billion years ago [Section 11.3]. Individual halo stars (i.e., those not in globular clusters) and some of the bulge stars appear similarly old. Furthermore, the proportions of heavy elements in halo stars are much lower than in the Sun, indicating that they formed before many generations of supernovae had a chance to enrich the Milky Way's interstellar medium.

However, careful study of heavy-element proportions suggests that our galaxy formed from a few different gas clouds. If the Milky Way had formed from a single protogalactic cloud, it would have steadily accumulated heavy elements during its inward collapse as stars formed and exploded within it. In that case, the outermost stars in the halo would be the oldest and the most deficient in heavy elements. Stars belonging to different globular clusters in the Milky Way's halo do indeed differ in age and heavy-element content, but these variations do not seem to depend on the stars' distance from the galactic center.

Our galaxy's halo stars may have formed in several small protogalactic clouds that later merged to form a single, larger protogalactic cloud.

The easiest way to account for the variations is to suppose that the Milky Way's earliest stars formed in relatively small protogalactic clouds, each with a few globular clusters, and that these clouds later collided and combined to create the full protogalactic cloud that became the Milky Way (Figure 14.19). In fact, similar processes may still be happening at some level. As we discussed earlier, the Sagittarius and Canis Major dwarf galaxies are currently crashing through the Milky Way's disk and are being torn apart in the process. A billion years from now, their stars will be indistinguishable from halo stars because they will all be circling the Milky Way on orbits that carry them high above the disk.

THINK ABOUT IT If the preceding scenario is true, then the Milky Way suffered several collisions early in its history. Explain why we should not be surprised that galaxy collisions (or collisions between protogalactic clouds) were rather common in the distant past. (*Hint:* How did the average separations of galaxies in the past compare to their average separations today?)

Once the Milky Way's full protogalactic cloud was in place, its collapse and heavy-element enrichment should have continued in a more orderly fashion than previously. Support for this scenario comes from a layer of stars intermediate between the disk and the halo. The heavy-element content of stars and globular clusters in this intermediate layer does indeed depend on their distances from the galactic center. These stars are nearly as old as halo stars but formed before the spinning protogalactic cloud finished flattening into a disk. Their proportions of heavy elements suggest that, at the time the disk formed, the Milky Way contained about 10% as much material in the form of heavy elements as it does today.

After the disk formed, generations of stars lived and died in the star–gas–star cycle that gradually increased the abundance of heavy elements. Thus, the ages of stars in the Milky Way's disk range from newly born to 10 billion or more years old. New stars will continue to be born as long as enough gas remains in the disk.

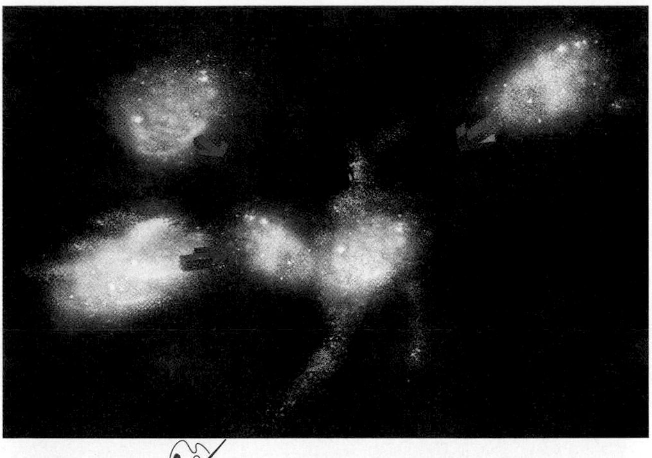

Figure 14.19

This painting shows a model of how the Milky Way's halo may have formed. The characteristics of stars in the Milky Way's halo suggest that several smaller gas clouds, already bearing some stars and globular clusters, may have merged to form the Milky Way's protogalactic cloud. These stars and star clusters remained in the halo while the gas settled into the Milky Way's disk.

Black Holes Tutorial, Lessons 1–2

14.4 THE MYSTERIOUS GALACTIC CENTER

The center of the Milky Way Galaxy lies in the direction of the constellation Sagittarius. This region of the sky does not look particularly special to our unaided eyes. However, if we could remove the interstellar dust that obscures our view, the galaxy's central bulge would be one of the night sky's most spectacular sights. Moreover, a gigantic black hole with more than 3 million times the mass of our Sun may lurk deep within the bulge, at the very center of our galaxy.

• What lies in the center of our galaxy?

Although the Milky Way's clouds of gas and dust prevent us from seeing visible light from the center of the galaxy, we can peer into the heart of our galaxy with radio, infrared, and X-ray telescopes. Figure 14.20 shows a series of infrared and radio views, looking ever-deeper into the galaxy's center. Within about 1,000 light-years of the center, we find swirling clouds of gas and a cluster of several million stars. Bright radio emission traces out the magnetic fields that thread this turbulent region. In the very center we find a source of radio emission named Sagittarius A* (pronounced "Sagittarius A-star"), or Sgr A* for short, that is quite unlike any other radio source in our galaxy. Several hundred stars crowd the region within about 1 light-year of Sgr A*.

The motions of stars and gas near Sgr A* indicate that it contains a few million solar masses within a very small region of space. Observations show that there are not nearly enough stars in this region to account for so much mass, even though stars here are much more crowded together than in our region of the galaxy. As a result, astronomers suspect that Sgr A* contains a very massive black hole. These suspicions received a big boost in 2002 when astronomers monitoring infrared light from the galactic

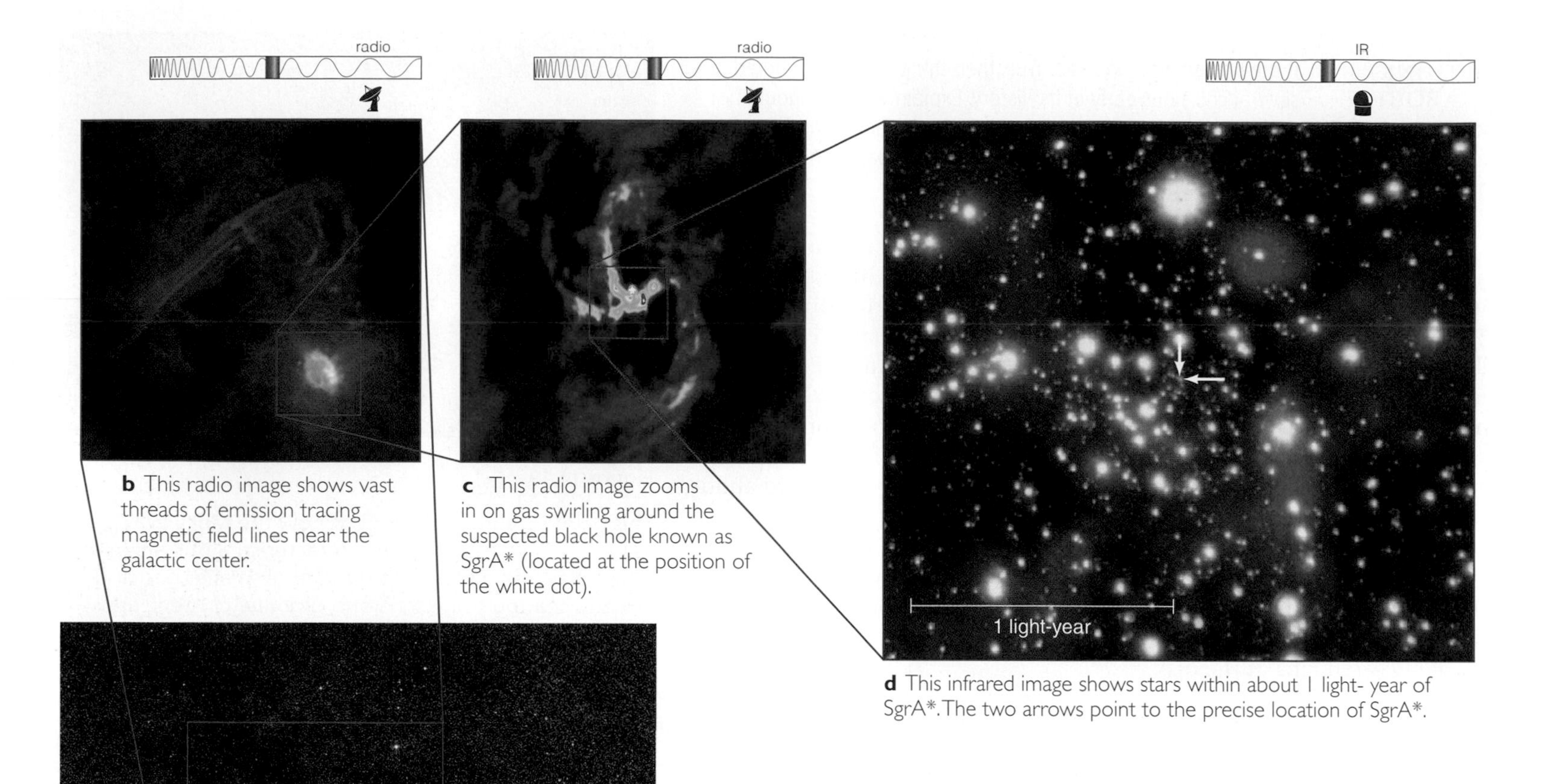

b This radio image shows vast threads of emission tracing magnetic field lines near the galactic center.

c This radio image zooms in on gas swirling around the suspected black hole known as SgrA* (located at the position of the white dot).

d This infrared image shows stars within about 1 light- year of SgrA*. The two arrows point to the precise location of SgrA*.

a This infrared image shows stars and gas clouds within 1,000 light-years of the center of the Milky Way.

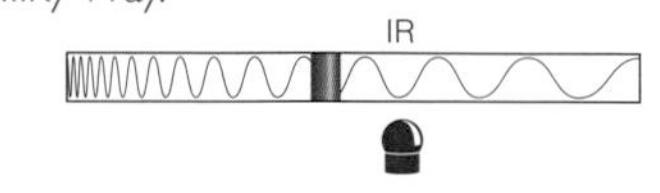

Figure 14.20

Zooming into the galactic center at infrared and radio wavelengths.

center observed stars swooping within a few light-hours of this massive object in their orbits (Figure 14.21). By applying Newton's version of Kepler's third law to the orbits of these stars, they concluded that the central object must have a mass of 3 to 4 million solar masses, all packed into a region of space just a little larger than our solar system. An object that massive within such a small space is almost certainly a black hole.

Stars very close to our galaxy's center orbit a nearly invisible and very small object more than 3 million times as massive as our Sun—probably a huge black hole.

However, the behavior of this suspected black hole is puzzling. Most other suspected black holes are thought to accumulate matter through accretion disks that radiate brightly in X rays. These include black holes in binary star systems like Cygnus X-1 [Section 13.3] and some giant black holes at the centers of other galaxies that we will discuss in Chapter 15. If the black hole at the center of our galaxy had an accretion disk like these others, its X-ray light would easily penetrate the dusty gas of our galaxy and it would appear fairly bright to our X-ray telescopes. Yet the X-ray emission from Sgr A* has usually been relatively faint.

Recent observations of Sgr A* made with the Chandra X-Ray Observatory are helping us understand this surprising behavior. In October 2000, an enormous X-ray flare lasting 3 hours was observed coming from the location of the suspected black hole (Figure 14.22). This sudden change in X-ray brightness probably came from energy released by a comet-size lump of matter that was torn apart by tidal forces just before it disappeared into the black hole. If we continue to observe similar X-ray flares from Sgr A*, then the explanation for its generally low X-ray brightness may be that matter falls into it in big chunks instead of in the smooth, swirling flow of an accretion disk. Until we better understand Sgr A*, it is sure to remain a favorite target for X-ray telescopes, infrared telescopes, and radio telescopes alike.

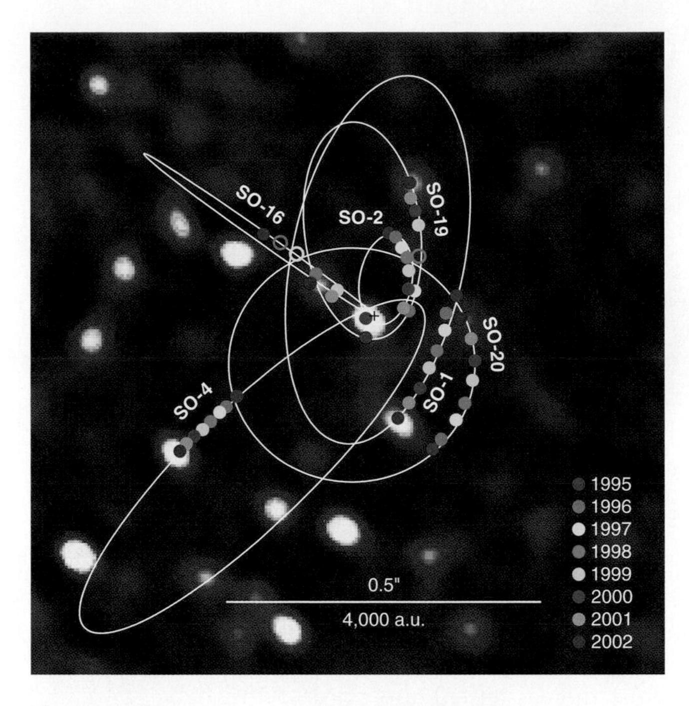

Figure 14.21

This diagram shows observed stellar positions (colored dots indicate year of observation) and calculated orbits for several stars near the very center of the galaxy. By applying Newton's version of Kepler's third law to these orbits, we infer that the central object has a mass 3 to 4 million times that of our Sun, packed into a space so small that it is almost certainly a black hole. (The 4,000 AU shown on the scale bar is equivalent to about 23 light-days.)

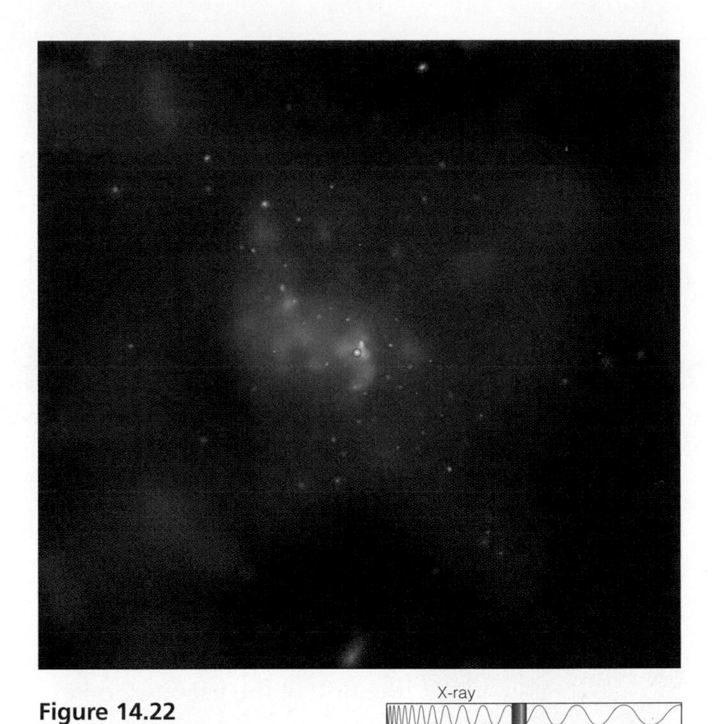

Figure 14.22

This X-ray image from the Chandra X-Ray Observatory shows the central 60 light-years of our galaxy. The circled white dot in the middle of the image is an X-ray flare from the suspected black hole at the Milky Way's center. The flare probably came from energy released by a comet-size lump of matter that was torn apart by tidal forces just before it disappeared into the black hole.

THE BIG PICTURE

Putting Chapter 14 into Context

In this chapter, we have explored the structure, motion, and history of our galaxy, along with the recycling of gas that has made our existence possible. When you review this chapter, pay attention to these "big picture" ideas:

- The inability of visible light to penetrate deeply through interstellar gas and dust concealed the true nature of our galaxy until recent times. Modern astronomical instruments reveal the Milky Way Galaxy to be a dynamic system of stars and gas that continually gives birth to new stars and planetary systems.
- Stellar winds and explosions make interstellar space a violent place. Hot gas tears through the atomic hydrogen gas that fills much of the galactic disk, leaving expanding bubbles and fast-moving clouds in its wake. All this violence might seem quite dangerous, but it performs the great service of mixing new heavy elements into the gas of the Milky Way.
- Although the elements from which we are made were forged in stars, we could not exist if stars were not organized into galaxies. The Milky Way Galaxy acts as a giant recycling plant, converting gas expelled from each generation of stars into the next and allowing some of the heavy elements to solidify into planets like our own.

SUMMARY OF KEY CONCEPTS

14.1 THE MILKY WAY REVEALED

• What does our galaxy look like?

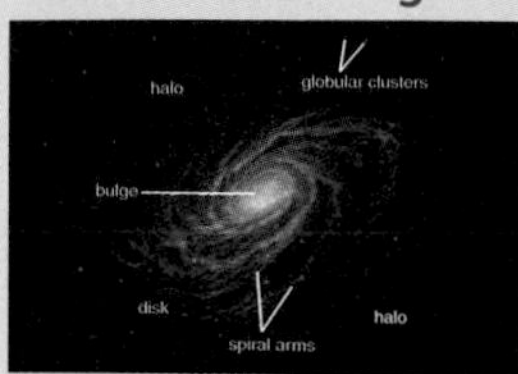

The Milky Way Galaxy consists of a thin **disk** about 100,000 light-years in diameter with a central bulge and a spherical region called the **halo** that surrounds the entire disk. The disk contains the gas and dust of the **interstellar medium**, while the halo contains very little gas.

• How do stars orbit in our galaxy?

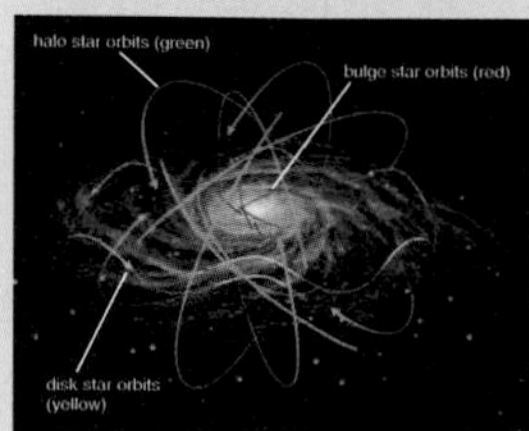

Stars in the disk all orbit the galactic center in about the same plane and in the same direction. Halo and bulge stars also orbit the center of the galaxy, but their orbits are randomly inclined to the disk of the galaxy. Orbital motions of stars allow us to determine the distribution of mass in our galaxy.

14.2 GALACTIC RECYCLING

• How does our galaxy recycle gas into stars?

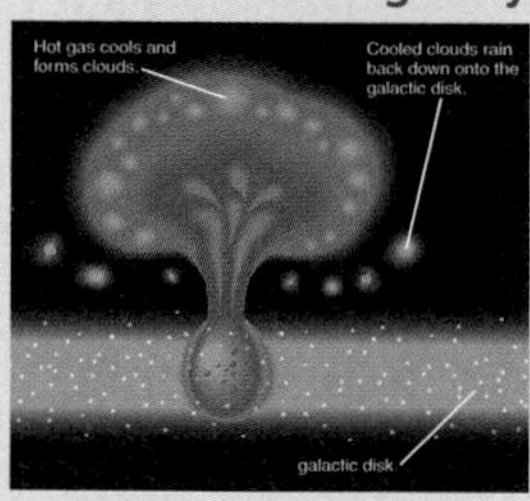

Stars are born from the gravitational collapse of gas clumps in **molecular clouds**. Massive stars explode as supernovae when they die, creating hot **bubbles** in the interstellar medium that contain the new elements made by these stars. Eventually, this gas cools and mixes into the surrounding interstellar medium, turning into atomic hydrogen and then cooling further, producing molecular clouds. These molecular clouds then form stars, completing the **star–gas–star cycle**.

• Where do stars tend to form in our galaxy?

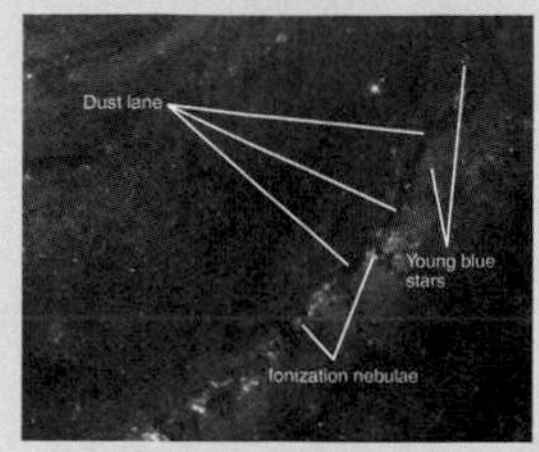

Active star-forming regions, marked by the presence of hot, massive stars and ionization nebulae, are found preferentially in **spiral arms**. The spiral arms represent regions where a **spiral density wave** has caused gas clouds to crash into each other, thereby compressing them and making star formation more likely.

14.3 THE HISTORY OF THE MILKY WAY

• What clues to our galaxy's history do halo stars hold?

The halo generally contains only old, low-mass stars with a much smaller proportion of heavy elements than stars in the disk. Thus, halo stars must have formed early in the galaxy's history, before the gas settled into a disk.

• How did our galaxy form?

The galaxy probably began as a huge blob of gas called a **protogalactic cloud**. Gravity caused the cloud to shrink in size, and conservation of angular momentum caused the gas to form the spinning disk of our galaxy. Stars in the halo formed before the gas finished collapsing into the disk.

14.4 THE MYSTERIOUS GALACTIC CENTER

• What lies in the center of our galaxy?

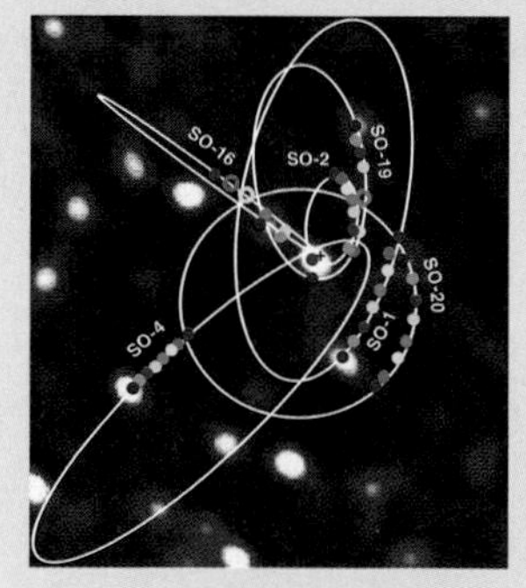

Motions of stars near the center of our galaxy suggest that it contains a black hole about 3 to 4 million times as massive as the Sun. The black hole appears to be powering a bright source of radio emission known as Sgr A*.

EXERCISES AND PROBLEMS

REVIEW QUESTIONS

1. Draw simple sketches of our galaxy as it would appear face-on and edge-on. Identify the *disk, bulge, halo,* and *spiral arms,* and indicate the galaxy's approximate dimensions.
2. What do we mean by the *interstellar medium?* Where is it found in the galaxy, and how does it affect our view of the galaxy?
3. What are the Large and Small Magellanic Clouds, and the Sagittarius and Canis Major dwarfs?

4. Describe the basic characteristics of stellar orbits in the bulge, disk, and halo of our galaxy. How can we use orbital properties to learn about the mass of the galaxy? What have we learned?
5. Summarize the stages of the star–gas–star cycle illustrated in Figure 14.3.
6. What creates a *bubble* of hot, ionized gas? What happens to the gas in the bubble over time?
7. What are *cosmic rays,* and where are they thought to come from?
8. What do we mean by *atomic hydrogen gas?* How common is it, and how do we map its distribution in the galaxy?
9. What are interstellar dust grains? What are they made of, and where are they found?
10. Briefly summarize the different types of gas present in the disk of the galaxy, and describe how they appear when we view the galaxy in different wavelengths of light.
11. What are ionization nebulae, and why are they found near hot, massive stars?
12. How do we know that spiral arms do not rotate like giant pinwheels? What makes spiral arms bright, and how do we think the spiral arms are maintained?
13. Briefly describe the characteristics that distinguish the galaxy's *disk population* of stars from its *spheroidal population* of stars.
14. How do the different ages of disk stars and halo stars support the idea that our galaxy formed from the gravitational collapse of a *protogalactic cloud?*
15. Why do we think that the Milky Way's full protogalactic cloud may have been formed from the merger of several smaller protogalactic clouds?
16. What is Sgr A*? What evidence suggests that it contains a massive black hole?

SENSIBLE STATEMENTS?

Decide whether each of these statements is sensible and explain why it is or is not.

17. We did not understand the true size and shape of our galaxy until NASA satellites were launched into the galactic halo, enabling us to see what the Milky Way looks like from the outside.
18. Planets like Earth probably didn't form around the very first stars because there were so few heavy elements back then.
19. If I could see infrared light, the galactic center would look much more impressive.
20. Many spectacular ionization nebulae are seen throughout the Milky Way's halo.
21. The carbon in my diamond ring was once part of an interstellar dust grain.
22. The Sun's velocity around the Milky Way tells us that most of our galaxy's dark matter lies in the galactic disk near the center of the galaxy.
23. We know that a black hole lies at our galaxy's center because numerous stars near it have vanished over the past several years, telling us that they've been sucked in.
24. If we could watch a time-lapse movie of a spiral galaxy over millions of years, we'd see many stars being born and dying within the spiral arms.

PROBLEMS

25. *Unenriched Stars.* Suppose you discovered a star made purely of hydrogen and helium. How old do you think it would be? Explain your reasoning.
26. *Enrichment of Star Clusters.* The gravitational pull of an isolated globular cluster is rather weak—a single supernova explosion can blow all the interstellar gas out of a globular cluster. How might this fact be related to observations indicating that stars ceased to form in globular clusters long ago? How might it be related to the fact that globular clusters are deficient in elements heavier than hydrogen and helium? Summarize your answers in one or two paragraphs.
27. *High-Velocity Star.* The average speed of stars relative to the Sun in the solar neighborhood is about 20 km/s (i.e., the speed at which we see stars moving toward or away from the Sun—*not* their orbital speed around the galaxy). Suppose you discover a star in the solar neighborhood that is moving relative to the Sun at a much higher speed, say 200 km/s. What kind of orbit does this star probably have around the Milky Way? In what part of the galaxy does it spend most of its time? Explain.
28. *Research: Discovering the Milky Way.* Humans have been looking at the Milky Way since long before recorded history, but only in the past century did we verify the true shape of the galaxy and our location within it. Learn more about how conceptions of the Milky Way developed through history. What names did different cultures give the band of light they saw? What stories did they tell about it? How have ideas about the galaxy changed in the past few centuries? Try to locate diagrams that illustrate these changes. Write a two- to three-page summary of your findings.

DISCUSSION QUESTIONS

29. *Galactic Ecosystem.* We have likened the star–gas–star cycle in our Milky Way to the ecosystem that sustains life on Earth. Here on our planet, water molecules cycle from the sea to the sky to the ground and back to the sea. Our bodies convert atmospheric oxygen molecules into carbon dioxide, and plants convert the carbon dioxide back into oxygen molecules. How are the cycles of matter on Earth similar to the cycles of matter in the galaxy? How do they differ? Do you think the term *ecosystem* is appropriate to discussions of the galaxy?
30. *Galaxy Stuff.* In the chapters on stars, we learned why we are "star stuff." Based on what you've learned in this chapter, explain why we are also "galaxy stuff." Does the fact that the entire galaxy was involved in bringing forth life on Earth change your perspective on Earth or on life in any way? If so, how? If not, why not?

For a complete list of media resources available, go to **www.astronomyplace.com** and choose Chapter 14 from the pull-down menu.

ASTRONOMY PLACE WEB TUTORIALS

Tutorial Review of Key Concepts

Use the interactive **Tutorials** at **www.astronomyplace.com** to review key concepts from this chapter.

Detecting Dark Matter in a Spiral Galaxy Tutorial

Lesson 1 Introduction to Rotation Curves
Lesson 2 Determining the Mass Distribution

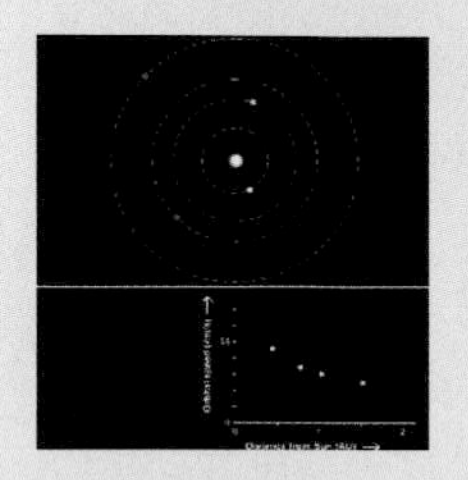
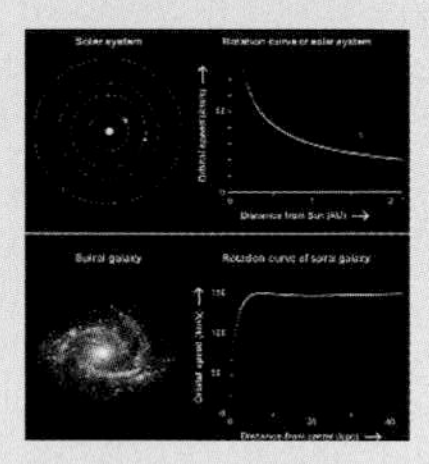
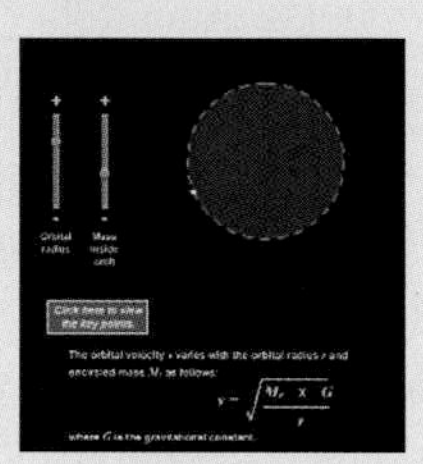

Black Holes Tutorial

Lesson 1 What Are Black Holes?
Lesson 2 The Search for Black Holes

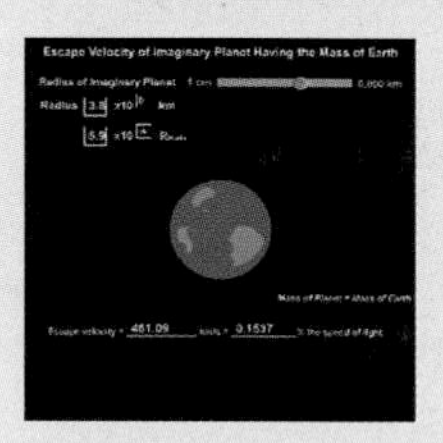
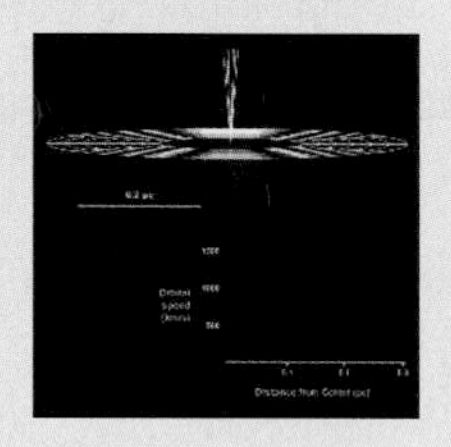
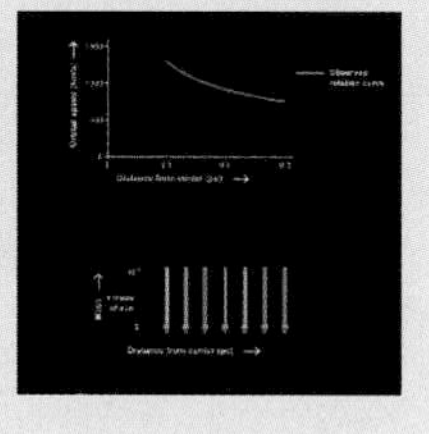

Supplementary Tutorial Exercises

Use the interactive **Tutorial Lessons** to explore the following questions.

Detecting Dark Matter in a Spiral Galaxy Tutorial, Lesson 1

1. What is a rotation curve? Use the tool in the simulation to determine how the rotation curve for a star system changes as the mass of the central star changes. Describe the change in words.
2. Compare the rotation curve of a galaxy with the rotation curve of the solar system.

Detecting Dark Matter in a Spiral Galaxy Tutorial, Lesson 2

1. How does orbital speed depend on the mass and distance of the central object?
2. What can we infer about the distribution of mass in our galaxy by observing the orbital speed of stars as they get farther from the center of our galaxy? Explain your answer.

Black Holes Tutorial, Lesson 2

1. Explain how we determine the mass of the central black hole in the Milky Way Galaxy.
2. Do other galaxies have giant black holes in their centers? Explain.

EXPLORING THE SKY AND SOLAR SYSTEM

Of the many activities available on the ***Voyager: SkyGazer* CD-ROM** accompanying your book, use the following files to observe key phenomena covered in this chapter.

Go to the **File: Basics** folder for the following demonstrations.

1. Milky Way
2. Wide Field Milky Way
3. Winter Milky Way
4. Lagoon Nebulae

Go to the **Explore** menu for the following demonstration.

1. Solar Neighborhood

MOVIES

Check out the following narrated and animated short documentary available on **www.astronomyplace.com** for a helpful review of key ideas covered in this chapter.

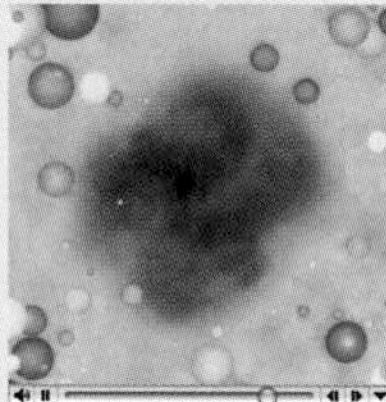

The Milky Way Galaxy Movie

WEB PROJECTS

Take advantage of the useful Web links on **www.astronomyplace.com** to assist you with the following projects.

1. *Images of the Star–Gas–Star Cycle.* Explore the Web to find pictures of nebulae and other forms of interstellar gas in different stages of the star–gas–star cycle. Assemble the pictures into a sequence that tells the story of interstellar recycling, with a one-paragraph explanation accompanying each image.
2. *The Galactic Center.* Search the Web for recent images of the galactic center, along with information about whether the center hides a massive black hole. Present a two- to three-page report, with pictures, giving an update on current knowledge about the center of the Milky Way Galaxy.

15
A Universe of Galaxies

LEARNING GOALS

15.1 ISLANDS OF STARS
- What are the three major types of galaxies?
- How are galaxies grouped together?

15.2 DISTANCES OF GALAXIES
- How do we measure the distances to galaxies?
- What is Hubble's Law?
- How do distance measurements tell us the age of the universe?

15.3 GALAXY EVOLUTION
- How do we observe the life histories of galaxies?
- How did galaxies form?
- Why do galaxies differ?

15.4 QUASARS AND OTHER ACTIVE GALACTIC NUCLEI
- What are quasars?
- What is the power source for quasars and other active galactic nuclei?
- Do supermassive black holes really exist?

Far beyond the Milky Way, we see billions of other galaxies scattered throughout space. Some look similar to our own galaxy, while others look quite different. The sight of all these galaxies inspires us not only to wonder how they came to be, but also to ask fundamental questions about our universe: How old is the universe? How big is it? How is the universe changing with time? Such questions might have seemed ridiculously speculative a century ago. Today, we believe we know the answers to these questions with respectable accuracy.

Edwin Hubble, the man for whom the Hubble Space Telescope is named, provided the key discovery when he proved conclusively that galaxies exist beyond the Milky Way. The distances he measured revealed an astonishing fact: The more distant a galaxy is, the faster it moves away from us. Hubble's discovery dealt a mortal blow to the traditional belief in a static, eternal, and unchanging universe. The motions of the galaxies away from one another imply instead that the entire universe is expanding and that its age is finite.

In this chapter, we will get acquainted with these islands in space, the galaxies of stars. We will discuss how we measure their distances, and how learning their distances has helped us learn about the age and size of our universe. Finally, we will study the galaxies themselves, looking for clues to how they have evolved through time and in the process learning about the history of our universe as a whole.

15.1 ISLANDS OF STARS

Figure 15.1 shows an amazing image of a tiny patch of the sky taken by the Hubble Space Telescope. The telescope pointed in a single direction in the sky and collected all the light it could for 10 days. If you held a grain of sand at arm's length, the angular size of the grain would match the angular size of everything in this picture. Almost every blob of light in this image is a galaxy—an island of stars bound together by gravity. Like our own Milky Way, each galaxy is a dynamic system that has cycled hydrogen gas through stars for billions of years, producing new elements for future generations of stars. Counting the galaxies in this patch of the sky and multiplying by the number of such patches it would take to make a montage of the entire sky, we find that the observable universe contains more than 80 billion galaxies. (The Hubble Space Telescope has since taken an even deeper photo of a different patch of the sky, shown as the opener to Chapter 1 on page 1. This photo, called the *Hubble Ultra Deep Field*, was made with an 11-day exposure using a newer camera installed by Shuttle astronauts in 2002.)

Even without knowing the distances to the individual galaxies, we can see that galaxies come in many sizes, colors, and shapes. Some look large, some small. Some are reddish, some whitish. Some appear round, and some appear flat. We would like to understand why galaxies differ in these ways, but it is not easy to learn their histories. Just as with stars, our observations capture only the briefest instant in any galaxy's life, leaving us to piece together the life story of a typical galaxy from pictures of different galaxies at various life stages. The task is made even more difficult because we see young galaxies only at great distances—distances at which we are looking far back into the universe's past (see Figure 1.4). Our task therefore begins with an attempt to categorize the galaxies we see nearby, whose details are easier to observe.

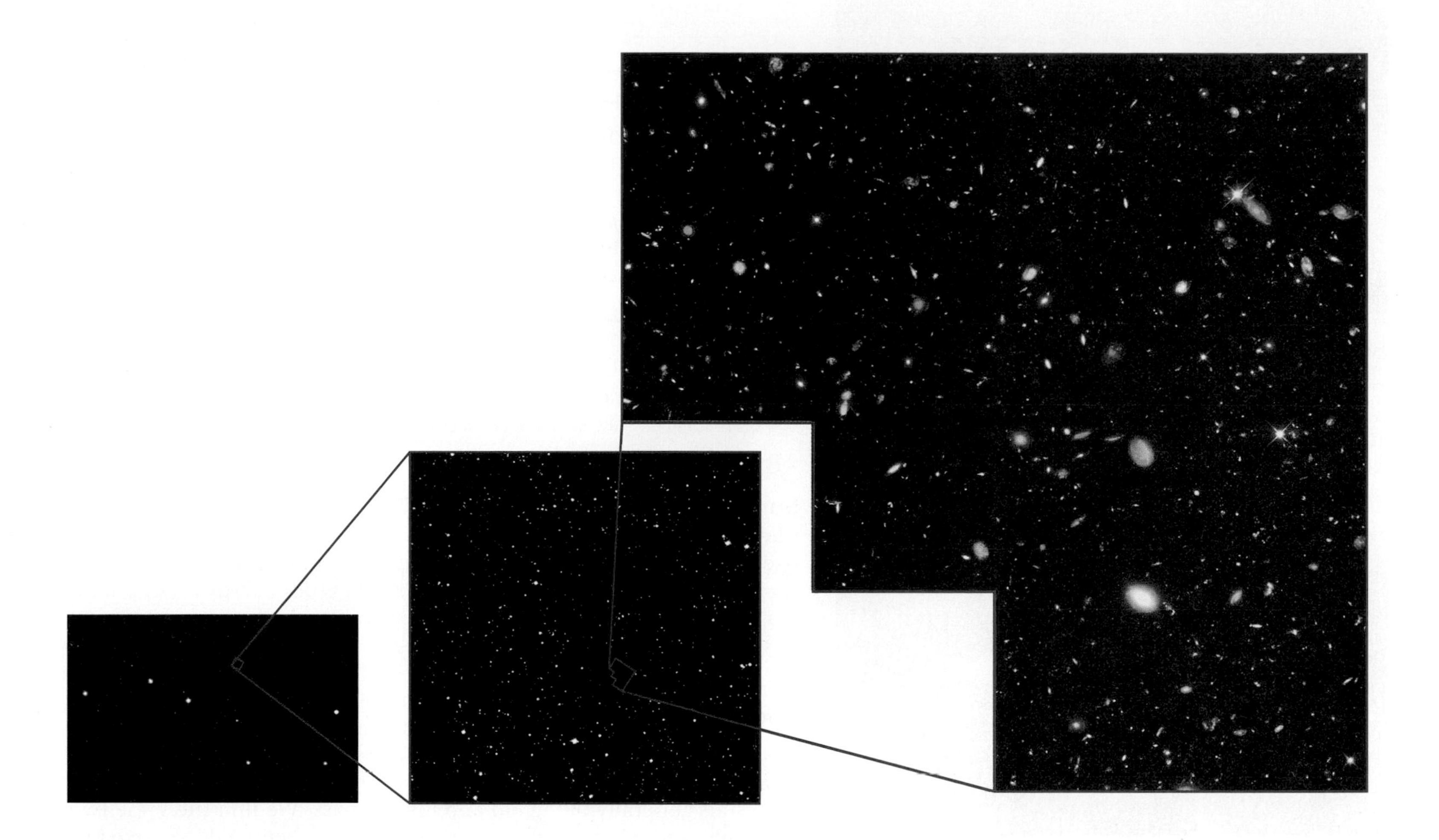

Figure 15.1

VIS

The Hubble Deep Field, an image composed of 10 days of exposures taken with the Hubble Space Telescope. Some of the galaxies pictured are three-quarters of the way across the observable universe. The zoom-out sequence shows the location of the field within the Big Dipper, recognizable in the lower left frame.

• What are the three major types of galaxies?

Astronomers classify galaxies into three major categories:

- **Spiral galaxies,** such as our own Milky Way, look like flat white disks with yellowish bulges at their centers. The disks are filled with cool gas and dust, interspersed with hotter ionized gas , and usually display beautiful spiral arms.
- **Elliptical galaxies** are redder, more rounded, and often longer in one direction than in the other, like a football. Compared with spiral galaxies, elliptical galaxies contain very little cool gas and dust, though they often contain very hot, ionized gas.
- **Irregular galaxies** appear neither disklike nor rounded.

Figure 15.2

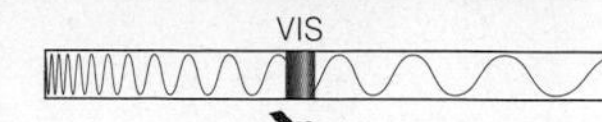

The spiral galaxy NGC 4414, whose disk is somewhat tilted to our line of sight. It is about 100,000 light-years in diameter. (NGC stands for the New General Catalog, a listing of more than 7,000 objects published in 1888.)

Figure 15.3

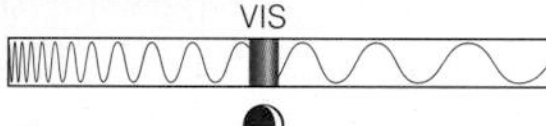

NGC 4594 (the Sombrero Galaxy) is a spiral galaxy with a large bulge and a dusty disk that we see almost edge-on. A much larger but nearly invisible halo surrounds the entire galaxy. The bulge and halo together make up the spheroidal component of the galaxy. This image shows a region of the galaxy about 82,000 light years across.

Figure 15.4

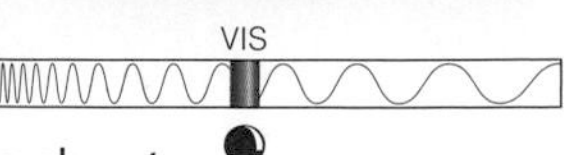

NGC 1300, a barred spiral galaxy about 150,000 light-years in diameter.

Galaxies come in three major types: spiral, elliptical, and irregular.

The differing colors of galaxies arise from the different kinds of stars that populate them: Spiral and irregular galaxies look white because they contain stars of all different colors and ages, while elliptical galaxies look redder because old, reddish stars produce most of their light. Galaxies also come in a wide range of sizes, from *dwarf galaxies* containing as few as 100 million (10^8) stars to *giant galaxies* with more than 1 trillion (10^{12}) stars. Let's take a closer look at galaxies of each type.

THINK ABOUT IT

Take a moment and try to classify the larger galaxies in Figure 15.1. How many appear spiral? Elliptical? Irregular? Do the colors of galaxies seem related to their shapes?

Spiral Galaxies

Like the Milky Way, other spiral galaxies also have a thin *disk* extending outward from a central *bulge* (Figure 15.2). The bulge itself merges smoothly into a *halo* that can extend to a radius of more than 100,000 light-years. However, the halo is difficult to see in photographs because its stars are generally dim and spread over a large volume of space.

Spiral galaxies have a disk, bulge, and halo like the Milky Way.

Recall that the Milky Way is made up of two distinct populations of stars [Section 14.3]. The *disk population* includes stars of all ages and masses that orbit in the disk of the galaxy. The *spheroidal population* consists of halo and bulge stars, with the halo stars generally being old and low in mass. We find the same two populations in other spiral galaxies, and we use them to define two primary components of galaxies:

- The **disk component** is the flat disk in which stars follow orderly, nearly circular orbits around the galactic center. The disk component always contains an *interstellar medium* of gas and dust, but the amounts and proportions of the interstellar medium in molecular, atomic, and ionized forms differ from one spiral galaxy to the next.
- The bulge and halo together make up the **spheroidal component**, which gets its name from its rounded shape. Stars in the spheroidal component have orbits with many different inclinations, and the spheroidal component generally contains little cool gas and dust. Figure 15.3 shows a spiral galaxy with an unusually large bulge that illustrates the general shape of the spheroidal component.

All spiral galaxies have both a disk and a spheroidal component, but there are some variations on the general theme. Some spiral galaxies appear to have a straight bar of stars cutting across the center, with spiral arms curling away from the ends of the bar. Such galaxies are known as *barred spiral galaxies* (Figure 15.4). Astronomers suspect that the Milky Way itself is a barred spiral galaxy, because our galaxy's bulge appears to be somewhat elongated.

Other galaxies have disk and spheroidal components like spiral galaxies but appear to lack spiral arms. These so-called *lenticular galaxies* (*lenticular* means "lens-shaped") are sometimes considered an intermediate class between spirals and ellipticals, because they tend to have less cool gas than normal spirals but more than ellipticals.

Among large galaxies in the universe, most (75–85%) are spiral or lenticular. Galaxies with obvious disks are much rarer among small galaxies.

Elliptical Galaxies The major difference between elliptical and spiral galaxies is that ellipticals lack a significant disk component. Thus, an elliptical galaxy has only a spheroidal component and looks much like the bulge and halo of a spiral galaxy. (In fact, elliptical galaxies are sometimes called *spheroidal galaxies.*) Although most large galaxies in the universe are spiral, some of the largest galaxies in the universe are giant elliptical galaxies (Figure 15.5). Nevertheless, the vast majority of elliptical galaxies are small, and these small elliptical galaxies are the most common type of galaxy in the universe. Particularly small ellipticals with less than about a billion stars, known as *dwarf elliptical galaxies,* are often found near larger spiral galaxies. For example, at least 10 dwarf elliptical galaxies belong to the Local Group.

Elliptical galaxies differ from spiral galaxies in that they do not have significant disks.

Elliptical galaxies usually contain very little dust or cool gas, although some have relatively small and cold gaseous disks rotating at their centers. However, some large elliptical galaxies contain substantial amounts of very hot gas. This low-density, X ray–emitting gas is much like the gas in the hot bubbles created by supernovae and powerful stellar winds in the Milky Way [Section 14.2].

The lack of cool gas in elliptical galaxies means that, like the halo of the Milky Way, they generally have little or no ongoing star formation. Thus, elliptical galaxies tend to look red or yellow in color because they do not have any of the hot, young, blue stars found in the disks of spiral galaxies.

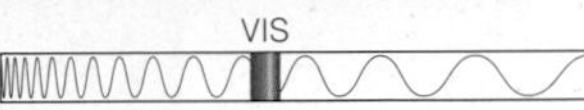

Figure 15.5

M 87, a giant elliptical galaxy in the Virgo Cluster. The region shown is more than 120,000 light-years across.

Irregular Galaxies Some of the galaxies we see nearby fall into neither of the two major categories. This *irregular* class of galaxies is a miscellaneous class, encompassing small galaxies such as the Magellanic Clouds (Figure 15.6) and larger "peculiar" galaxies that appear to be in disarray. These blobby star systems are usually white and dusty, like the disks of spirals. Their colors tell us that they contain young, massive stars.

Irregular galaxies appear to be in disarray.

Among nearby galaxies, only a small percentage of galaxies as large as the Milky Way are irregular. Telescopic observations probing deeper into the universe show that distant galaxies are more likely to be irregular in shape than nearby galaxies. Because the light of more distant galaxies has taken longer to reach us, these observations tell us that irregular galaxies were more common when the universe was younger.

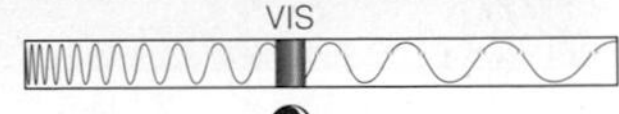

Figure 15.6

The Large Magellanic Cloud, an irregular galaxy that is a small companion to the Milky Way. It is about 30,000 light-years across.

Hubble's Galaxy Classes Edwin Hubble invented a system for classifying galaxies that remains widely used. It assigns the letter *E,* followed by a number, to each elliptical galaxy. The larger the number, the flatter the galaxy. An E0 galaxy is round, and an E7 galaxy is highly elongated. A spiral galaxy is assigned an uppercase *S,* or *SB* if it has a bar, and a lowercase *a, b,* or *c* that indicates the size of the bulge and the dustiness of the disk. The bulge size decreases from *a* to *c,* while the amount of dusty gas increases. Lenticular galaxies are designated S0, and irregular galaxies are designated Irr. Figure 15.7 shows Hubble's scheme.

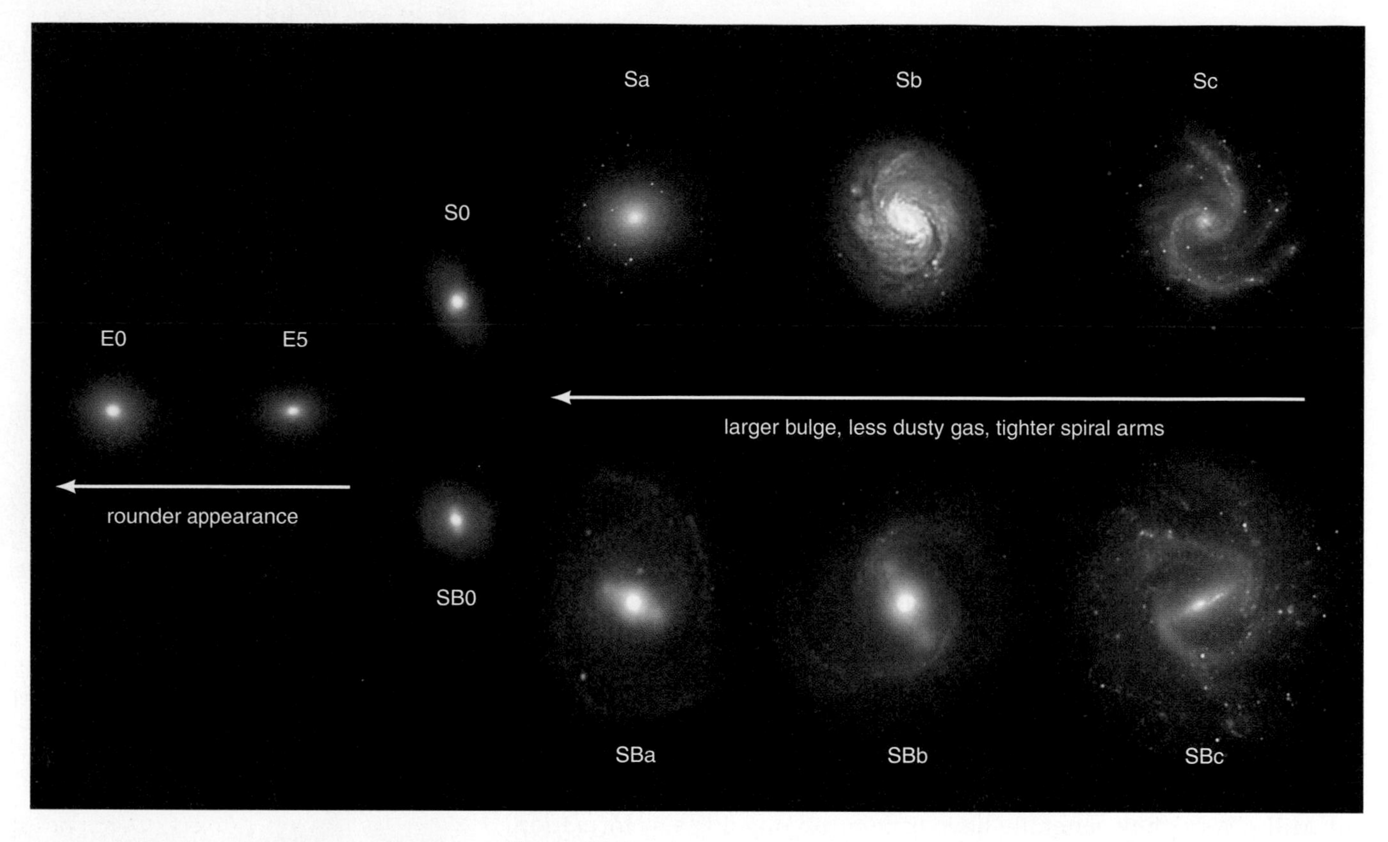

Figure 15.7

This "tuning fork" diagram illustrates Hubble's galaxy classes. An elliptical galaxy is classified with the letter *E* and a number indicating how round it is (a lower number means a rounder galaxy). A spiral galaxy is classified with an *S*, or *SB* for a barred spiral, and a lowercase letter based on its appearance. Lenticular galaxies, an intermediate class between spiral and elliptical galaxies, are classified S0 or SB0, depending on whether their bulge is bar-shaped.

Figure 15.8

Hickson Compact Group 87, a small group of galaxies consisting of a large edge-on spiral galaxy at the bottom of this photo, two smaller spiral galaxies at the center and upper left, and an elliptical galaxy to the right. The whole group is about 170,000 light-years in diameter. (The other objects in this photograph are foreground stars in our own galaxy.)

Astronomers had once hoped that the classification of galaxies might yield deep insights, just as the classification of stars did in the early twentieth century. The Hubble classification scheme itself was suspected for a time to be an evolutionary sequence in which galaxies flattened and spread out as they aged. Unfortunately for astronomers, galaxies turn out to be far more complex than stars, and classification schemes like this one have not led to easy answers about their nature.

• How are galaxies grouped together?

Although some galaxies travel solo through the universe, many are gravitationally bound together with neighboring galaxies. Spiral galaxies are often found in loose collections of up to a few dozen galaxies, called **groups**. Our Local Group is one example (see Figure 1.1). Figure 15.8 shows another galaxy group.

Spiral galaxies tend to congregate in small groups, while elliptical galaxies are primarily found in large clusters.

Elliptical galaxies appear to be more social than spiral galaxies. They are particularly common in **clusters** of galaxies, which can contain hundreds and sometimes thousands of galaxies extending over more than 10 million light-years (Figure 15.9). Elliptical galaxies make up about half the large galaxies in the central regions of clusters, while they represent only a small minority (about 15%) of the large galaxies found outside clusters.

Measuring Cosmic Distances Tutorial, Lessons 1–4

15.2 DISTANCES OF GALAXIES

To learn more about galaxies than just their shape, color, and type, we need to know how far away they are. Measuring the distances to galaxies is one of the most challenging tasks we face when trying to understand galaxies and the universe as a whole, but the payoff is enormous. Besides telling us where galaxies are located, such measurements also reveal the size and age of the observable universe.

• How do we measure the distances to galaxies?

Our determinations of astronomical distances depend on a chain of methods in which each step allows us to measure greater distances in the universe. We have already discussed the measurement of distances to nearby stars by parallax [Section 11.1]. Because parallax involves apparent shifts in stellar positions as Earth orbits the Sun, measuring distances by parallax requires knowing the precise Sun–Earth distance, or astronomical unit (AU; see Figure 1.12). Astronomers measure the AU with a technique called **radar ranging**, in which radio waves are transmitted from Earth and bounced off Venus. Because radio waves travel at the speed of light, the round-trip travel time for the radar signals tells us Venus's distance from Earth. We can then use Kepler's laws and a little geometry to calculate the length of an AU. Thus, radar ranging to measure the AU represents the first link in the distance chain, and parallax measurements of distances to nearby stars represent the second link. We will now follow the rest of this chain, link by link, to the outermost reaches of the observable universe.

Standard Candles Once we have measured distances to nearby stars through parallax, we can begin to measure distances to other stars in the same way that you might estimate the distance to a street lamp at night. If the street lamp does not look very bright, then it's probably far away. If it looks very bright, then it's probably quite close.

We can determine the lamp's distance more accurately if we can measure its apparent brightness. For example, suppose we see a distant street lamp and know that all street lamps of its type put out 1,000 watts of light. If we then measure the street lamp's apparent brightness, we can calculate its distance by using the inverse square law for light [Section 11.1].

We can determine distance by measuring the apparent brightness of an object whose luminosity we already know and applying the inverse square law for light.

An object such as a street lamp, for which we are likely to know the true luminosity, represents what astronomers call a **standard candle**. The term *standard candle* is meant to suggest a light source of a known, standard luminosity. Unlike light bulbs, astronomical objects do not come marked with wattage. An astronomical object can serve as a

Figure 15.9 VIS

Central part of the galaxy cluster Abell 1689. Almost every object in this photograph is a galaxy belonging to the cluster. Yellowish elliptical galaxies outnumber the whiter spiral galaxies. The region pictured is about 2 million light-years across. (A few stars from our own galaxy appear in the foreground, looking like white dots with four spikes in the form of a cross.)

standard candle only if we have some way of knowing its true luminosity without first measuring its apparent brightness and distance. Fortunately, many astronomical objects meet this requirement. For example, any star that is a twin of our Sun—that is, a main-sequence star with spectral type G2—should have about the same luminosity as the Sun. Thus, if we measure the apparent brightness of a Sun-like star, we can assume it has the same luminosity as the Sun (3.8×10^{26} watts) and then use the inverse square law for light to estimate its distance.

Beyond the few hundred light-years for which we can measure distances by parallax, we use standard candles for most cosmic distance measurements. These distance measurements always have some uncertainty, because no astronomical object is a perfect standard candle. The challenge of measuring astronomical distances comes down to the challenge of finding the objects that make the best standard candles. The more confidently we know an object's true luminosity, the more certain we are of its distance.

Main-Sequence Fitting We can use Sun-like stars as standard candles because we know that they are similar to the Sun and because we can measure the Sun's luminosity quite easily. However, Sun-like stars are relatively dim, and we cannot detect them at great distances. To measure distances beyond 1,000 light-years or so, we need brighter standard candles.

An obvious first choice is to use brighter main-sequence stars. However, before we can use any main-sequence star as a standard candle, we must first have some way of knowing its true luminosity. Thus, we must follow two steps to use bright main-sequence stars as standard candles:

1. We identify a star cluster that is close enough for us to determine its distance by parallax and plot its H–R diagram. Because we know the distances to the cluster stars, we can use the inverse square law for light to establish their true luminosities from their apparent brightnesses.
2. We can look at stars in other clusters that are too far away for parallax measurements and measure their apparent brightnesses. If we assume that main-sequence stars in other clusters have the same true luminosities as their counterparts in the nearby cluster, we can calculate their distances from the inverse square law for light.

Twentieth-century astronomers laid the groundwork for this technique by calibrating the luminosities on a "standard" H–R diagram. This calibration relied largely on a single, nearby star cluster—the Hyades Cluster in the constellation Taurus, whose distance is now known from its parallax. We can find the distances to other star clusters by comparing the apparent brightnesses of their main-sequence stars with those in the Hyades Cluster and assuming that all main-sequence stars of the same color have the same luminosity (Figure 15.10). This technique of determining distances by comparing main sequences in different star clusters is called **main-sequence fitting**.

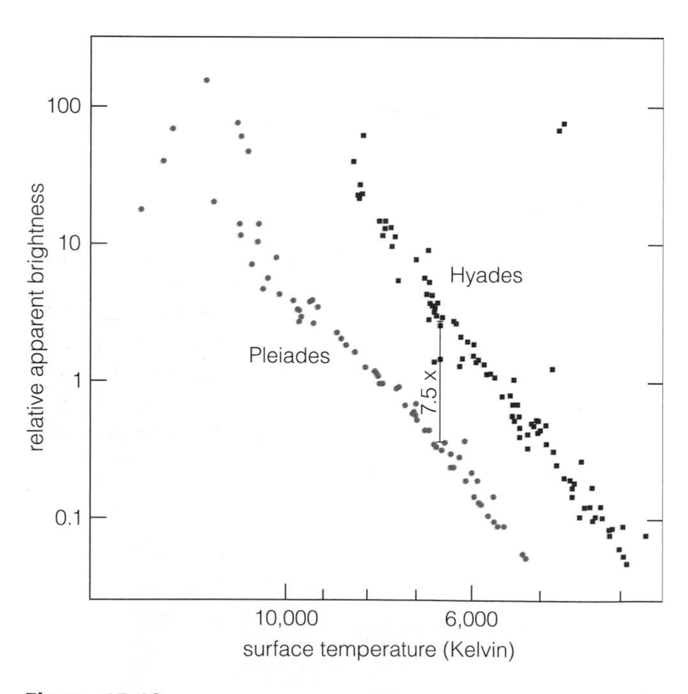

Figure 15.10

Comparison of the apparent brightness of stars in the Hyades Cluster with those in the Pleiades Cluster shows that the Pleiades are about 2.75 times farther away because they are $2.75^2 \approx 7.5$ times dimmer.

Cepheid Variables Main-sequence fitting works well for measuring distances to star clusters throughout the Milky Way, but not for measuring distances to other galaxies. Most main-sequence stars are too faint to be seen in other galaxies, even with our largest telescopes. Instead, we need very bright stars to serve as standard candles for distance measurements beyond the Milky Way.

The most useful bright stars for measuring the distances to galaxies are called **Cepheid variable stars,** or **Cepheids** for short. These stars vary in brightness in our sky, alternately becoming dimmer and brighter with periods ranging between a few days and a few months. Each Cepheid has its own particular time period between peaks in luminosity. For example, Figure 15.11 shows the brightness variations of a Cepheid with a period of about 50 days.

In 1912, Henrietta Leavitt discovered that the periods of Cepheids are very closely related to their luminosities: The longer the period, the more luminous the star (Figure 15.12). We say that Cepheids obey a **period-luminosity relation** that allows us to determine (within about 10%) a Cepheid's luminosity simply by measuring its period of brightness variation. For example, Figure 15.12 shows that a Cepheid variable whose brightness peaks every 30 days is effectively screaming out, "Hey, everybody, my luminosity is 10,000 times that of the Sun!" Thus, once we measure a Cepheid's period, we know its luminosity and we can use the inverse square law for light to determine its distance.

Leavitt discovered the period-luminosity relation with careful observations of Cepheids in a nearby galaxy (the Large Magellanic Cloud), but she did not know why Cepheids vary in this special way. We now know that Cepheids vary because they have a peculiar problem in matching the amount of energy their surfaces radiate with the amount welling up from the core. In a futile quest for a steady equilibrium, the upper layers of a Cepheid variable star alternately expand and contract, causing the star's luminosity to rise and fall. The period-luminosity relation holds because larger (and hence more luminous) Cepheids take longer to pulsate in and out in size.

Cepheid variable stars are useful for measuring distances because we can determine a Cepheid's luminosity from the period between its peaks of brightness.

Cepheids have been used for almost a century to measure distances to nearby galaxies. As we'll discuss shortly, they played a critical role in Edwin Hubble's discoveries. More recently, one of the main missions of the Hubble Space Telescope has been to measure accurate distances to galaxies up to 100 million light years away by studying Cepheids within them. This distance may sound very large, but it is still quite small compared with the distances of galaxies in Figure 15.1. In order to go further, we use the distances determined with these Cepheids to measure the luminosities of even brighter standard candles, such as supernovae.

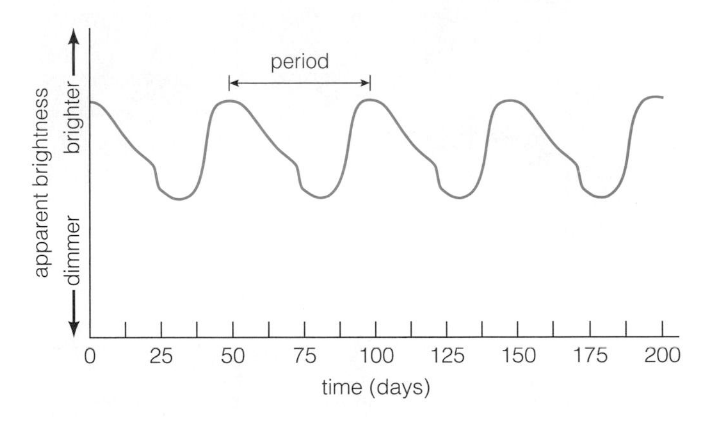

Figure 15.11

This graph shows how the brightness of a Cepheid varies with time. The period is the time from one peak of brightness to the next. This Cepheid has a period of about 50 days.

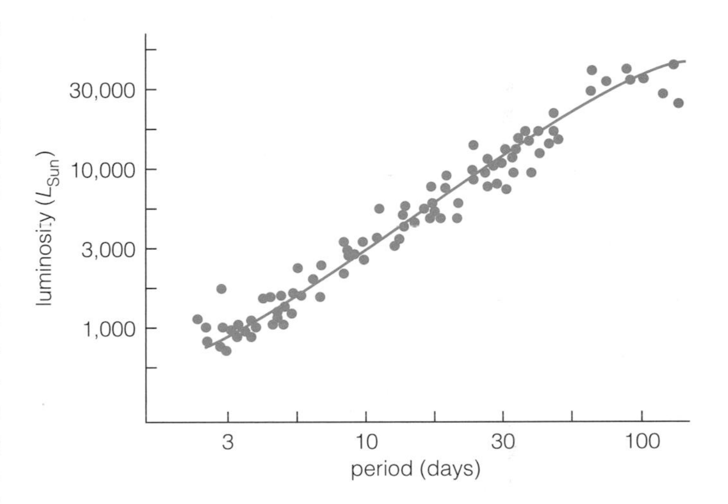

Figure 15.12

Cepheid period-luminosity relation. The data show that all Cepheids of a particular period have very nearly the same luminosity. Thus, by measuring a Cepheid's period, we can determine its luminosity and hence its distance. (Cepheids actually come in two types with two different period-luminosity relations. The relation here is for Cepheids with heavy-element content similar to that of our Sun, or "Type I Cepheids.")

Distant Standard Candles Recall that white dwarf supernovae are thought to be exploding white dwarf stars that have reached the 1.4-solar-mass limit [Section 13.1]. These supernovae should all have nearly the same luminosity, because they all come from stars of the same mass that explode in the same way. Although white dwarf supernovae are quite rare in any individual galaxy, several have been detected during the past century in galaxies within about 50 million light-years of the Milky Way. Astronomers kept careful records of those events, so that today we can determine the true luminosities of these supernovae by using Cepheids to measure the distances to the galaxies in which they occurred. These measurements confirm that the luminosities of all white dwarf supernovae are about the same.

Because white dwarf supernovae are so bright—about 10 billion solar luminosities at their peak—we can detect them even when they occur

Figure 15.13

Edwin Hubble at the Mount Wilson Observatory.

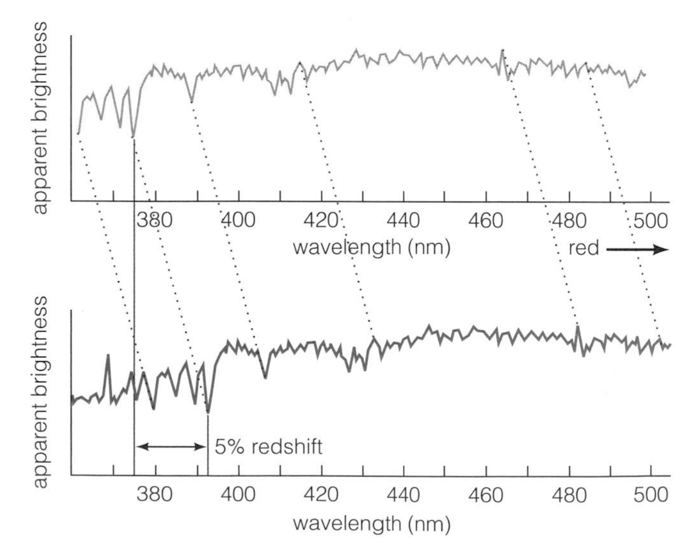

Figure 15.14

Redshifted galaxy spectrum. The gray line shows the spectrum of light originally emitted by the galaxy. The blue line shows the spectrum we observe, which is shifted by 5% to longer (redder) wavelengths, indicating that this galaxy is moving away from ours at 5% of the speed of light.

in galaxies billions of light-years away. Thus, white dwarf supernovae are standard candles that allow us to measure the distances of galaxies in the far reaches of the observable universe, completing the distance chain.

White dwarf supernovae are useful for measuring large distances because they are bright and all have about the same peak luminosity.

Unfortunately, the number of galaxies whose distances we can measure with this technique is relatively small because white dwarf supernovae occur only once every few hundred years in a typical galaxy. We use a different technique for most distant galaxies, and that technique relies on the expansion of the universe discovered by Edwin Hubble.

Hubble's Law Tutorial, Lessons 1–3

• What is Hubble's Law?

The ability to measure distances to galaxies is the key to much of our modern understanding of the size and age of the universe. We can trace the beginning of this understanding directly back to discoveries made by Edwin Hubble. Let's explore how Hubble discovered the famous law that bears his name, and how this law has helped us answer fundamental questions about the universe in which we live.

Edwin Hubble and the Andromeda Galaxy Before Edwin Hubble's groundbreaking work in the 1920s, no one knew for certain whether the spiral-shaped objects they saw in the sky were merely clouds of gas within the Milky Way—and therefore that the Milky Way represented the entire universe—or distant and distinct galaxies. The opinions of astronomers were quite split on this issue, which became a subject of great debate. The problem was that neither side could prove its case, because the techniques available at the time could not distinguish objects within the Milky Way from those beyond it.

Hubble put the debate to rest in 1924. Using the new, 100-inch telescope atop southern California's Mount Wilson (Figure 15.13)—the largest telescope in the world at the time—he discovered Cepheid variables in the Andromeda Galaxy by comparing photographs of the galaxy taken days apart. He then used those observations to estimate the galaxy's distance, proving once and for all that the Andromeda Galaxy sat far beyond the outer reaches of stars in the Milky Way.

Edwin Hubble used Cepheids to prove that the Andromeda Galaxy lies beyond the Milky Way.

This single scientific discovery dramatically changed our view of the universe. Rather than thinking the Milky Way was the entire universe, we suddenly knew that it is just one among many galaxies in an enormous universe. The stage was set for an even greater discovery.

Distance and Redshift Astronomers had known since the 1910s that the spectra of most spiral galaxies tended to be *redshifted* (Figure 15.14). Recall that redshifts occur when the object emitting the radiation is moving away from us [Section 5.2]. Because Hubble had not yet proved that the spiral galaxies were in fact separate from the Milky Way, no one understood the true significance of their motions.

Following his discovery of Cepheids in Andromeda, Hubble and his coworkers spent the next few years busily measuring the redshifts of galaxies and estimating their distances. Because even Cepheids were too dim to be seen in most of these galaxies, Hubble needed brighter standard candles for his distance estimates. One of his favorite techniques was to use the brightest object he could see in each galaxy as a standard candle, because he assumed these objects to be very bright stars that would always have about the same luminosity.

A galaxy's redshift tells us how fast it is moving away from us, and the relationship between redshift and distance shows that the universe is expanding.

In 1929, Hubble announced his conclusion: The more distant a galaxy, the greater its redshift and hence the faster it moves away from us. As we discussed in Chapter 1 (see Figure 1.15), this discovery implies that the entire universe is expanding. Thus, Hubble had discovered the expansion of the universe.

Hubble's original assertion was based on an amazingly small sample of galaxies. Even more incredibly, he had grossly underestimated the luminosities of his standard candles. The "brightest stars" he had been using as standard candles were really entire *clusters* of bright stars. Fortunately, Hubble was both bold and lucky. Subsequent studies of much larger samples of galaxies showed that they are indeed receding from us, but they are even farther away than Hubble thought.

Hubble's Law We express the idea that more distant galaxies move away from us faster with a very simple formula, now known as **Hubble's law**:

$$v = H_0 \times d,$$

where v stands for a galaxy's velocity away from us (sometimes called a *recession velocity*), d stands for its distance, and H_0 (pronounced "H-naught") is a number called **Hubble's constant**. We usually write Hubble's law in this form to express the idea that galaxy speeds depend on their distances. However, astronomers more often use the law in reverse—measuring a galaxy's speed from its redshift and then using Hubble's law to estimate its distance.

Hubble's Law expresses a relationship between galaxy speeds and distances and hence allows us to determine a galaxy's distance from its speed.

Because Hubble's law in principle applies to all distant galaxies, it does not have the limitations of other distance measurement techniques (such as white dwarf supernovae). Nevertheless, we encounter two important practical difficulties when we try to use Hubble's law to measure galactic distances:

1. Galaxies do not obey Hubble's law perfectly. Hubble's law would give an exact distance only for a galaxy whose speed is determined solely by the expansion of the universe. In reality, nearly all galaxies experience gravitational tugs from other galaxies, and these tugs alter their speeds from the "ideal" values predicted by Hubble's law.
2. Even when galaxies obey Hubble's law well, the distances we find with it are only as accurate as our best measurement of Hubble's constant.

The first problem is most serious for nearby galaxies. Within the Local Group, for example, Hubble's law does not work at all: The galaxies in the Local Group are gravitationally bound together with the Milky Way and therefore are *not* moving away from us in accord with Hubble's law. However, Hubble's law works fairly well for more distant galaxies. The recession speeds of galaxies at large distances are so great that any motions caused by the gravitational tugs of neighboring galaxies are tiny in comparison.

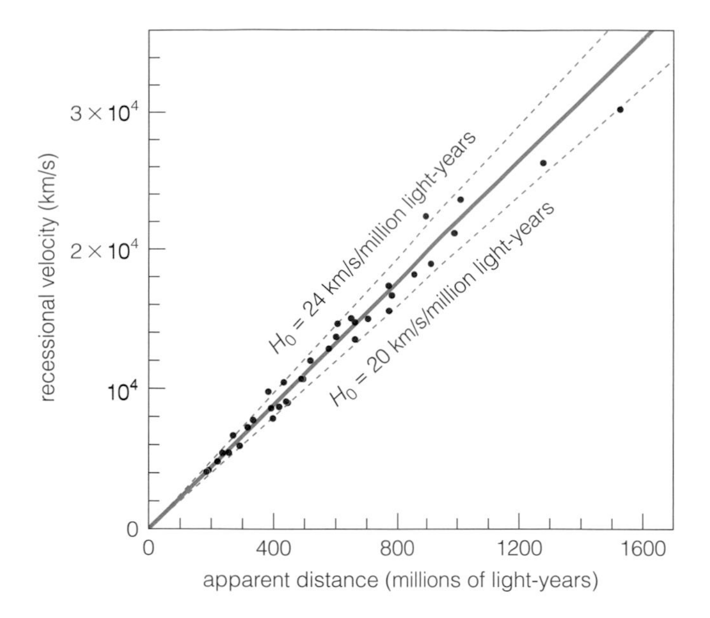

Figure 15.15 Interactive Figure

White dwarf supernovae can be used as standard candles to establish Hubble's law out to very large distances. The points on this figure show the apparent distances of white dwarf supernovae and the recessional velocities of the galaxies in which they exploded. The fact that these points all fall close to a straight line demonstrates that these supernovae are good standard candles.

The second problem means that, even for distant galaxies, we can know only *relative* distances until we pin down the true value of H_0. For example, Hubble's law tells us that a galaxy moving away from us at 20,000 km/s is twice as far away as one moving at 10,000 km/s, but we can determine the actual distances of the two galaxies only if we know H_0. Until recently, estimates of the value of H_0 were quite uncertain.

The quest to measure Hubble's constant was one of the main missions of the Hubble Space Telescope.

One of the main missions of the Hubble Space Telescope has been to obtain an accurate value of H_0. Astronomers used the telescope to discover Cepheid variables in galaxies out to about 60 million light-years and then used those distances to determine the luminosities of distant standard candles such as white dwarf supernovae. Plotting galactic distances measured with those distant standard candles against the velocities indicated by their redshifts has pinned down the value of H_0 to somewhere between 20 and 24 *kilometers per second per million light-years* (Figure 15.15). In other words, for every million light-years of distance away from us, a galaxy's speed away from us is between 20 and 24 kilometers per second. For example, with this range of values for Hubble's constant, Hubble's law predicts that a galaxy located 100 million light-years away would be moving away from us at a speed between 2,000 and 2,400 kilometers per second.

Distance Chain Summary: From Radar Ranging to Hubble's Law

Figure 15.16 summarizes the chain of measurements that allows us to determine ever-greater distances. With each link in the distance chain, however, uncertainties become somewhat greater. Thus, although we know the Earth–Sun distance at the beginning of the chain extremely accurately, distances to the farthest reaches of the observable universe remain uncertain by about 20%.

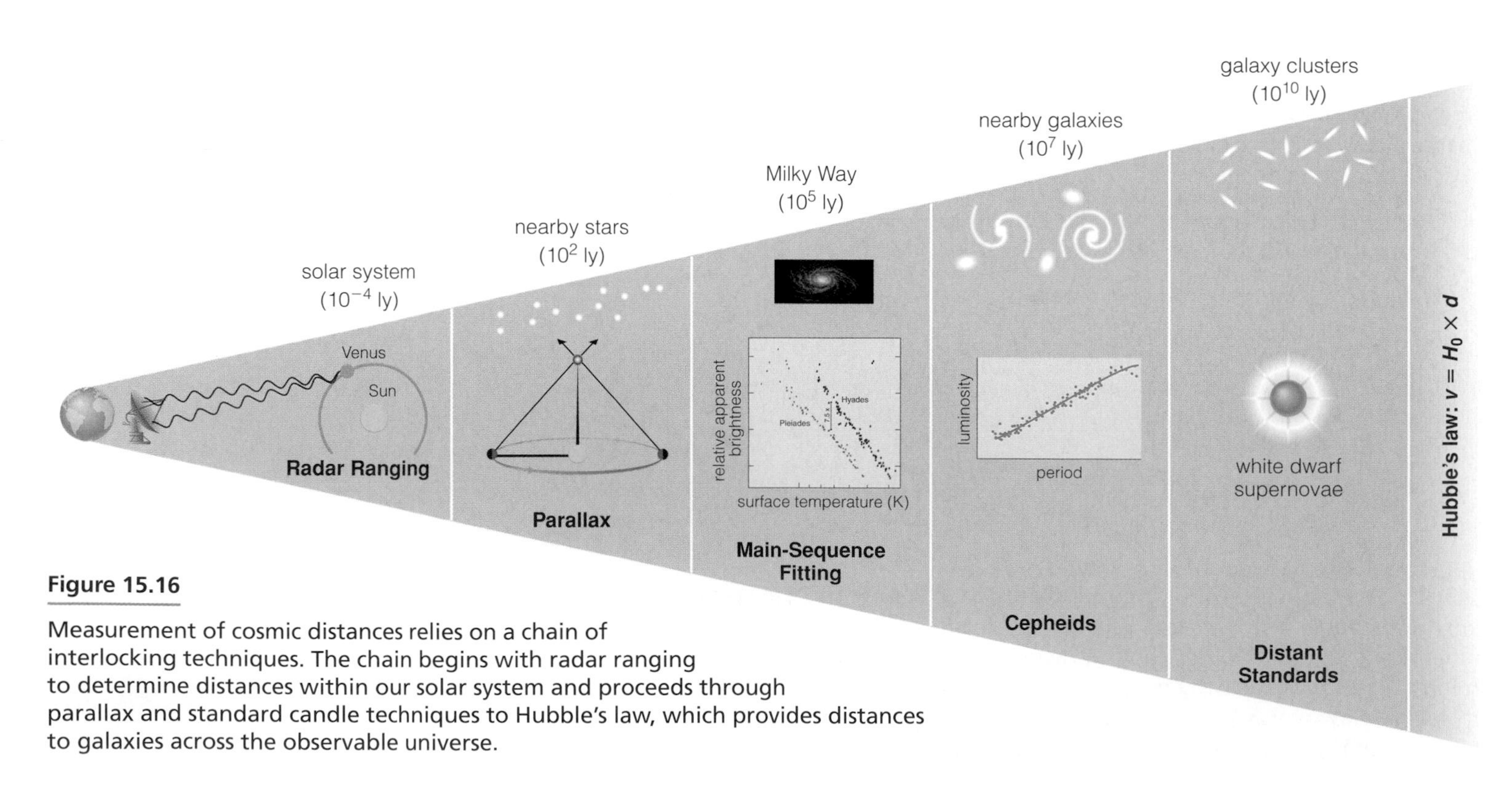

Figure 15.16

Measurement of cosmic distances relies on a chain of interlocking techniques. The chain begins with radar ranging to determine distances within our solar system and proceeds through parallax and standard candle techniques to Hubble's law, which provides distances to galaxies across the observable universe.

- ***Radar ranging:*** We measure the Earth–Sun distance by bouncing radio waves off planets and using some geometry.
- ***Parallax:*** We measure the distances to nearby stars by observing how their positions change, relative to the background stars, as Earth orbits the Sun. These distances thus rely on our knowledge of the Earth–Sun distance, determined with radar ranging to planets.
- ***Main-sequence fitting:*** We know the distance to the Hyades star cluster in our Milky Way Galaxy through parallax. Comparing the apparent brightnesses of its main-sequence stars to those of stars in other clusters gives us the distances to these other star clusters in our galaxy.
- ***Cepheid variables:*** By studying Cepheids in star clusters with distances measured by main-sequence fitting, we learn the precise period-luminosity relation for Cepheids. When we find a Cepheid in a more distant star cluster or galaxy, we can determine its true luminosity by measuring the period between its peaks in brightness and then use this true luminosity to determine the distance.
- ***Distant standards:*** By measuring distances to relatively nearby galaxies with Cepheids, we learn the true luminosities of white dwarf supernovae and other distant standard candles, enabling us to measure great distances through the universe.
- ***Hubble's law:*** Distances measured to galaxies with white dwarf supernovae and other distant standards allow us to measure Hubble's constant, H_0. Once we know H_0, we can use Hubble's law to determine a galaxy's distance from its redshift.

• How do distance measurements tell us the age of the universe?

Hubble's law is a remarkably powerful tool for understanding the universe. Not only does it tell us that the universe is expanding and give us a way to measure galactic distances, but it also helps us determine the age and size of the observable universe. To see how, we must first consider the expansion of the universe in a little more detail.

Universal Expansion We first discussed the expansion of the universe in Chapter 1. Galaxies all across the universe are moving away from one another, and this fact implies that galaxies must have been closer together in the past. Tracing this convergence back in time, we reason that all the matter in the observable universe started very close together and that the entire universe came into being at a single moment.

The expansion of the universe implies that the universe came into being at a single moment in time.

It may be tempting to think of the expanding universe as a ball of galaxies expanding into a void, but this impression is mistaken. To the best of our knowledge, the universe is not expanding *into* anything. As far as we can tell, there is no edge to the distribution of galaxies in the universe. On very large scales, the distribution of galaxies appears to be relatively smooth, meaning that the overall appearance of the universe around you would look more or less the same no matter where you are located. The idea that the matter in the universe is evenly distributed, without a center or an edge, is often called the **cosmological principle**. Although we cannot prove it to be true, it is completely consistent with all our observations of the universe [Section 16.3].

So how can the universe be expanding if it's not expanding *into* anything? Back in Chapter 1, we likened the expanding universe to a raisin cake baking [Section 1.3], but a cake has a center and edges that grow into empty space as it bakes. A better analogy is something that can expand but that has no center and no edges. The surface of a balloon can fit the bill, as can an infinite surface such as a sheet of rubber that extends to infinity in all directions.

Because it's hard to visualize infinity, let's use the surface of a balloon as our analogy to the expanding universe (Figure 15.17). Note that this analogy uses the balloon's two-dimensional *surface* to represent all three dimensions of space. Thus, the surface of the balloon represents the entire universe, and the spaces inside and outside the balloon have no meaning in this analogy. Aside from the reduced number of dimensions, the analogy works well because the balloon's spherical surface has no center and no edges, just as no city is the center of Earth's surface and no edges exist where you could walk or sail off the Earth. We can represent galaxies with plastic polka dots attached to the balloon, and we can make our model universe expand by inflating the balloon.

Figure 15.17 Interactive Figure

As the balloon expands, dots move apart in the same way that galaxies move apart in our expanding universe.

The Age of the Universe

To see how Hubble's law leads us to an age for the universe, let's imagine that some miniature scientists are living on dot B in Figure 15.17. Suppose that, three seconds after the balloon begins to expand, they measure the following:

Dot A is 3 cm away and moving at 1 cm/s.

Dot C is 3 cm away and moving at 1 cm/s.

Dot D is 6 cm away and moving at 2 cm/s.

They could summarize these observations as follows: *Every dot is moving away from our home with a speed that is 1 cm/s for each 3 cm of distance.* Because the expansion of the balloon is uniform, scientists living on any other dot would come to the same conclusion. Each scientist living on the balloon would determine that the following formula relates the distances and velocities of other dots on the balloon:

$$v = \left(\frac{1}{3}\text{s}\right) \times d$$

where v and d are the velocity and distance of any dot, respectively.

THINK ABOUT IT Confirm that this formula gives the correct values for the speeds of dots C and D, as seen from dot B, 3 seconds after the balloon began expanding. How fast would a dot located 9 cm from dot B move, according to the scientists on dot B?

If the miniature scientists think of their balloon as a bubble, they might call the number relating distance to velocity—the term $\frac{1}{3}$s in the preceding formula—the "bubble constant." An especially insightful miniature scientist might flip over the "bubble constant" and find that it is exactly equal to the time since the balloon started expanding. That is, the "bubble constant" $\frac{1}{3}$s tells them that the balloon has been expanding for 3 seconds. Perhaps you see where we are heading.

Just as the inverse of the "bubble constant" tells the miniature scientists that their balloon has been expanding for 3 seconds, the inverse of the Hubble constant, or $1/H_0$, tells us something about how long our universe has been expanding. The "bubble constant" for the balloon depends on when it is measured, but it is always equal to 1 divided by the time since the balloon started expanding. Similarly, the Hubble constant actually changes with time but stays roughly equal to 1 divided by the age of the universe. We call it a constant because it is the same at all locations in the universe, and because its value does not change noticeably on the time scale of human civilization.

The rate at which the universe expands tells us how old it is—about 14 billion years.

Current estimates based on the value of Hubble's constant put the age of the universe between about 12 and 15 billion years. To derive a more precise value for the universe's age, we need to know whether the expansion has been speeding up or slowing down over time, a question we will examine more closely in Chapter 16. If the gravitational pull of each galaxy on every other galaxy has significantly slowed the expansion rate, then the universe's age is somewhat less than $1/H_0$. If some mysterious force has accelerated the expansion rate, then the universe's age is somewhat more than $1/H_0$. The best available evidence as of 2004 suggests that the universe is about 14 billion years old.

Lookback Time The expansion of the universe leads to a complication in discussing galaxy distances that we have ignored up to this point. To understand the complication, imagine observing a supernova in a distant galaxy. Suppose the supernova occurred 400 million years ago but we are only just now seeing it. The supernova's light must have traveled 400 million light-years to reach Earth, but this simple statement about how far the light has traveled does not translate easily into a distance for the galaxy in which the supernova occurred. The problem is that the universe is expanding, making the distance between Earth and the supernova greater today than it was at the time of the supernova event. Thus, if we simply say that the galaxy is "400 million light-years away," it's not clear whether we mean its distance now, its distance at the time of the supernova, or something in between.

An object's lookback time is the time it took for the object's light to reach us.

Because distances between galaxies are always changing, it is easier to speak about faraway objects in terms of how much time their light takes to reach us—400 million years in the case of the supernova. We call this the **lookback time** to the supernova. In other words, a distant object's lookback time is the difference between the current age of the universe and the age of the universe when the light left the object. Unlike a statement about distance, the meaning of a lookback time is unambiguous: If the lookback time is 400 million years, it means the light really traveled through space for a period of 400 million years to reach us.

An object's lookback time is directly related to its redshift. Recall that the redshifts of galaxies tell us how quickly they are moving away from us. In the context of an expanding universe, redshifts have an additional, more fundamental interpretation. Let's return again to the universe on the balloon. Suppose you draw wavy lines on the balloon's surface to represent light waves. As the balloon inflates, these wavy lines stretch out, and their wavelengths increase (Figure 15.18). This stretching closely resembles what happens to photons in an expanding universe. The expansion of the universe stretches out all the photons within it, shifting them to longer, redder wavelengths. We call this effect a **cosmological redshift**.

Figure 15.18

As the universe expands, photon wavelengths stretch like the wavy lines on this expanding balloon.

COMMON MISCONCEPTIONS

BEYOND THE HORIZON

Perhaps you're thinking there must be something beyond the cosmological horizon. After all, we can see farther and farther with each passing second. The new matter we see had to come from somewhere, didn't it? The problem with this reasoning is that the cosmological horizon, unlike a horizon on Earth, is a boundary in time, not in space.

At any one moment, in whatever direction we look, the cosmological horizon lies at the beginning of time and encompasses a certain volume of the universe. At the next moment, the cosmological horizon still lies at the beginning of time, but it encompasses a slightly larger volume. When we peer into the distant universe, we are looking back in both space and time. We cannot look "past the horizon" because we cannot look back to a time before the universe began.

In a sense, we have a choice when we interpret the redshift of a distant galaxy: We can think of the redshift as being caused either by the Doppler effect as the galaxy moves away from us or by a photon-stretching, cosmological redshift. However, as we look to very distant galaxies, the ambiguity in the meaning of *distance* also makes it difficult to specify precisely what we mean by a galaxy's *speed*. Thus, it becomes preferable to interpret the redshift as being due to photon stretching in an expanding universe.

From this perspective, it is better to think of space itself as expanding, carrying the galaxies along for the ride, than to think of the galaxies as projectiles flying through a static universe. The cosmological redshift of a galaxy thus tells us how much space has expanded during the time since light from the galaxy left on its journey to us.

The Horizon of the Universe When we began our discussion of the expanding universe, we stressed that the universe as a whole does not seem to have an edge. Yet the universe does have a horizon, a place beyond which we cannot see. The **cosmological horizon** that marks the limits of the observable universe is a boundary in time, not in space. It exists because we cannot see back to a time before the universe began. For example, if the universe is 14 billion years old, then no object can have a lookback time greater than 14 billion years (see Figure 1.4). Thus, the age of the universe fundamentally limits the size of our observable universe.

The size of the observable universe is determined by the age of the universe.

Our quest to study galaxies and measure their distances has brought us to the very limits of the observable universe. The galaxies in Figure 15.1 at the beginning of this chapter extend from relatively nearby almost all the way to the cosmological horizon.

15.3 GALAXY EVOLUTION

Now that we understand how galactic distances are measured, we are ready to turn our attention back to understanding galaxies themselves. We understand less about the formation and development of galaxies, or **galaxy evolution**, than we do about the lives of stars. Nevertheless, we've made great progress in understanding galaxy evolution in recent decades, thanks largely to the same types of powerful telescopes that have allowed us to measure distances more accurately. In this section, we'll investigate the current state of research into the lives of galaxies.

• How do we observe the life histories of galaxies?

Powerful telescopes can be used like time machines to observe the life histories of galaxies—the farther we look into the universe, the further back we look in time [Section 1.1]. The most distant galaxies we can see already had some stars in place about 13 billion years ago, about the same time that the oldest stars in our own galaxy formed. It therefore seems safe to assume that most galaxies began to form at about this time, in which case most galaxies would be roughly the same age today. Thus, when we look to such great distances that we are looking deep into the past, we see galaxies as they were when they were much younger than their current age of about 13 billion years.

Images of the deep universe allow us to study galaxies at many different distances and therefore many different ages.

The linkage between a galaxy's distance and its age gives us a remarkable ability: Simply by photographing galaxies at different distances, we can assemble "family albums" showing galaxies in different stages of development. Pictures of the farthest galaxies show galaxies in their childhood, and pictures of the nearest show mature galaxies as they are today. Figure 15.19 shows partial family albums for elliptical, spiral, and irregular galaxies. Each individual photograph shows a single galaxy at a single stage in its life. Grouping these photographs by galaxy type allows us to see how galaxies of a particular type have changed through time.

THINK ABOUT IT In the preceding paragraph, we use the term *today* in a very broad sense. For example, if we look at a relatively nearby galaxy—one located, say, 20 million light-years away—we see it as it was 20 million years ago. In what sense is this "today"? (*Hint:* How does 20 million years compare to the age of the universe? To the lifetime of a massive star?)

Figure 15.19 VIS

Family albums for elliptical, spiral, and irregular galaxies of different ages, plus some very young galaxies shown on the far left. These photos are all zoom-outs of galaxies shown in Figure 15.1. We see more distant galaxies as they were when they were younger, indicated by the approximate age of the universe along the horizontal axis. The younger galaxies appear smaller because they are more distant. (The time scale runs to 12 billion years, rather than to the current age of the universe of 14 billion years, because the nearest galaxies in Figure 15.1 are about 2 billion light-years away.)

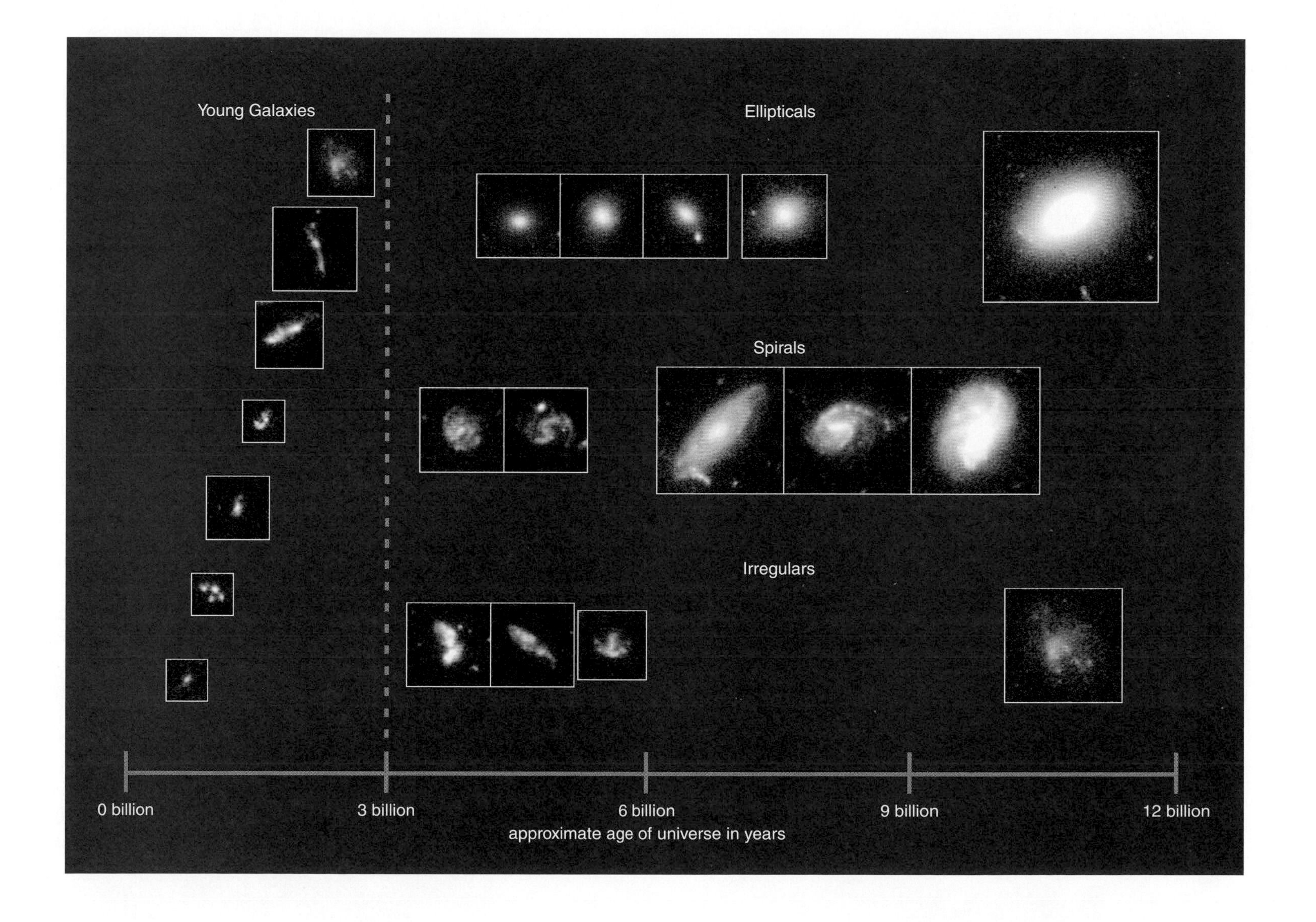

The photographs in Figure 15.19, taken by the Hubble Space Telescope, show galaxies extending back to a time when the universe was just 1 or 2 billion years old. However, we suspect that the first stars and galaxies formed even earlier than this. Observing these first stars and galaxies is a challenge not even Hubble can meet. Detecting such faraway galaxies will require larger telescopes, and the extreme redshifts expected for such galaxies mean the telescopes will need to be particularly sensitive to infrared light. Within a decade or so, NASA hopes to launch a much larger, infrared-sensitive successor to the Hubble Space Telescope (called the James Webb Space Telescope), but for now we have little direct information about galaxy birth.

• How did galaxies form?

Because our telescopes cannot yet see back to the time when galaxies formed their first stars, we must use theoretical modeling to study the earliest stages in galaxy evolution. The most successful models for galaxy formation assume the following:

- Hydrogen and helium gas filled all of space more or less uniformly when the universe was very young—say, in the first million years after its birth.
- The distribution of matter in the universe was not perfectly uniform—certain regions of the universe started out ever so slightly denser than others.

Beginning from these assumptions, which are supported by a mounting body of evidence (discussed in Chapter 17), we can model galaxy formation using well-established laws of physics to trace how the denser regions in the early universe grew into galaxies. The models show that the regions of enhanced density originally expanded along with the rest of the universe. However, the slightly greater pull of gravity in these regions gradually slowed their expansion. Within about a billion years, the expansion of these denser regions halted and reversed, and the material within them began to contract into *protogalactic clouds* like the cloud of matter that eventually formed our Milky Way [Section 14.3].

Our most successful models of galaxy formation suggest that protogalactic clouds formed in regions of slightly enhanced density in the early universe.

According to the models, the protogalactic clouds that eventually formed spiral galaxies initially cooled as they contracted, radiating away their thermal energy, and the first generation of stars grew from the densest, coldest clumps of gas. These first-generation stars were probably quite massive, living and dying within just a few million years—a short time compared to the time required for the collapse of a protogalactic cloud into a mature galaxy. The supernovae of these and later generations of massive stars seeded the galaxy with its first smattering of heavy elements, generating shock waves that heated the surrounding interstellar gas. This heating slowed the collapse of young galaxies and the rate at which stars formed within them, allowing time for the rest of the gas to settle into a rotating disk.

This picture explains many of the basic features of galaxies, but it leaves at least two major questions unanswered. First, our models assume that galaxies formed in regions of slightly enhanced density, but they do not tell us where these density enhancements came from. The nature and origin of density enhancements in the early universe is one of the major puzzles in astronomy, and we'll revisit it in Chapters 16 and 17. Second, our basic picture explains the origin of spiral galaxies quite well, but it does not tell us why some galaxies are elliptical and others irregular. This is the question to which we turn our attention next.

• Why do galaxies differ?

All protogalactic clouds are thought to begin in the same basic way, with gravity collecting matter around a pre-existing density enhancement. Thus, our models predict that the differences between the galaxy types must arise either from differing conditions in their protogalactic clouds or from later interactions with other galaxies. It now seems likely that both factors play a role. To help differentiate the role of conditions in the protogalactic cloud from that of subsequent interactions, we will explore a single key question: Why do spiral galaxies have gas-rich disks, while elliptical galaxies do not?

Conditions in the Protogalactic Cloud Two plausible explanations for the differences between spiral galaxies and elliptical galaxies trace a galaxy's type back to the protogalactic cloud from which it formed:

- *Protogalactic spin.* A galaxy's type might be determined by the spin of the protogalactic cloud from which it formed. If the original cloud has a significant amount of angular momentum, it will rotate quickly as it collapses. The galaxy it produces will therefore tend to form a disk, and the resulting galaxy would be a spiral galaxy. If the protogalactic cloud has little or no angular momentum, its gas might not form a disk at all, and the resulting galaxy would be elliptical.
- *Protogalactic density.* A galaxy's type might be determined by the density of the protogalactic cloud from which it formed. A protogalactic cloud with relatively high gas density would radiate energy more effectively and cool more quickly, thereby allowing more rapid star formation. If the star formation proceeds fast enough, all the gas could be turned into stars before any of it has time to settle into a disk. The resulting galaxy would therefore lack a disk, making it an elliptical galaxy. In contrast, a lower-density cloud would form stars more slowly, leaving plenty of gas to form the disk of a spiral galaxy.

Elliptical galaxies may have formed from protogalactic clouds that were spinning more slowly or were more dense than those that formed spiral galaxies.

Evidence for the latter scenario comes from a few giant elliptical galaxies at very great distances. These galaxies look very red even after we have accounted for their large redshifts. They apparently have no blue or white stars at all, indicating that new stars no longer form within these galaxies—even though we are seeing them as they were when the universe was only a few billion years old. This finding suggests that all the stars in these elliptical galaxies formed almost simultaneously, which is consistent with the idea that all the stars formed before a disk could develop.

Galactic Collisions The two previous scenarios, in which the formation of a gas-rich disk depends on the angular momentum or density of the protogalactic cloud, probably describe important parts of the overall

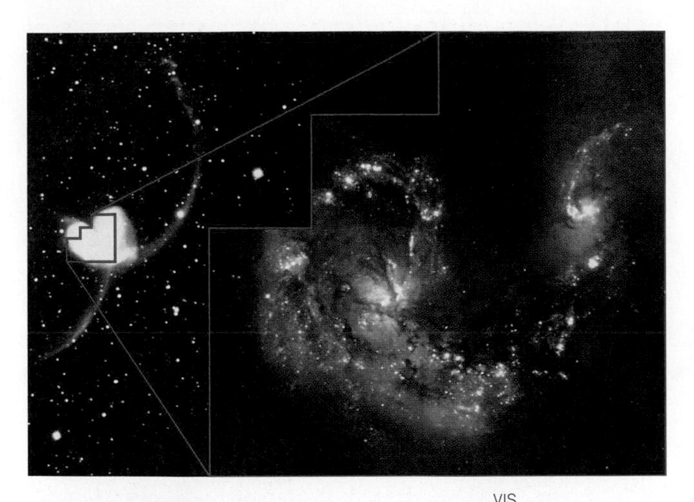

Figure 15.20 VIS

A pair of colliding spiral galaxies known as the Antennae (also known as NGC4038/4039). The image taken from the ground (left) reveals their vast tidal tails, and the close-up from the Hubble Space Telescope shows the fine details of the burst of star formation at the center of the collision.

story. However, they ignore one key fact: Galaxies rarely evolve in perfect isolation.

Think back to our scale-model solar system in Chapter 1. Using a scale on which the Sun was the size of a grapefruit, the nearest star was like another grapefruit a few thousand kilometers away. Because the average distances between stars are so huge compared to the sizes of stars, collisions between stars are extremely rare. However, if we rescale the universe so that our *galaxy* is the size of a grapefruit, the Andromeda Galaxy is like another grapefruit only about 3 meters away, and a few smaller galaxies lie considerably closer. Thus, the average distances between galaxies are not much larger than the sizes of galaxies, and collisions between galaxies are inevitable.

Collisions between galaxies are spectacular events that unfold over hundreds of millions of years (Figure 15.20). In our short lifetimes, we can at best see a snapshot of a collision in progress, distorting the shapes of the colliding galaxies. Galactic collisions must have been even more frequent in the distant past, when the universe was smaller and galaxies were even closer together. Observations confirm that distorted-looking galaxies—probably galaxy collisions in progress—were more common in the early universe than they are today.

Computer models show that a collision between two spiral galaxies can form an elliptical galaxy.

We can learn much more about galactic collisions with the aid of computer simulations that allow us to "watch" collisions that in nature take hundreds of millions of years to unfold. These computer models show that a collision between two spiral galaxies can create an elliptical galaxy (Figure 15.21). Tremendous tidal forces between the colliding galaxies tear apart the two disks, randomizing the orbits of their stars. Meanwhile, a large fraction of their gas sinks to the center of the collision and rapidly forms new stars. Supernovae and stellar winds eventually blow away the rest of the gas. When the cataclysm finally settles down, the merger of the two spirals has produced a single elliptical galaxy. Little gas is left for a disk, and the orbits of the stars have random orientations.

Observations support the idea that at least some elliptical galaxies result from collisions and subsequent mergers. Elliptical galaxies dominate the galaxy populations at the cores of dense clusters of galaxies, where collisions should be most frequent. This fact may mean that any spirals once present became ellipticals through collisions.

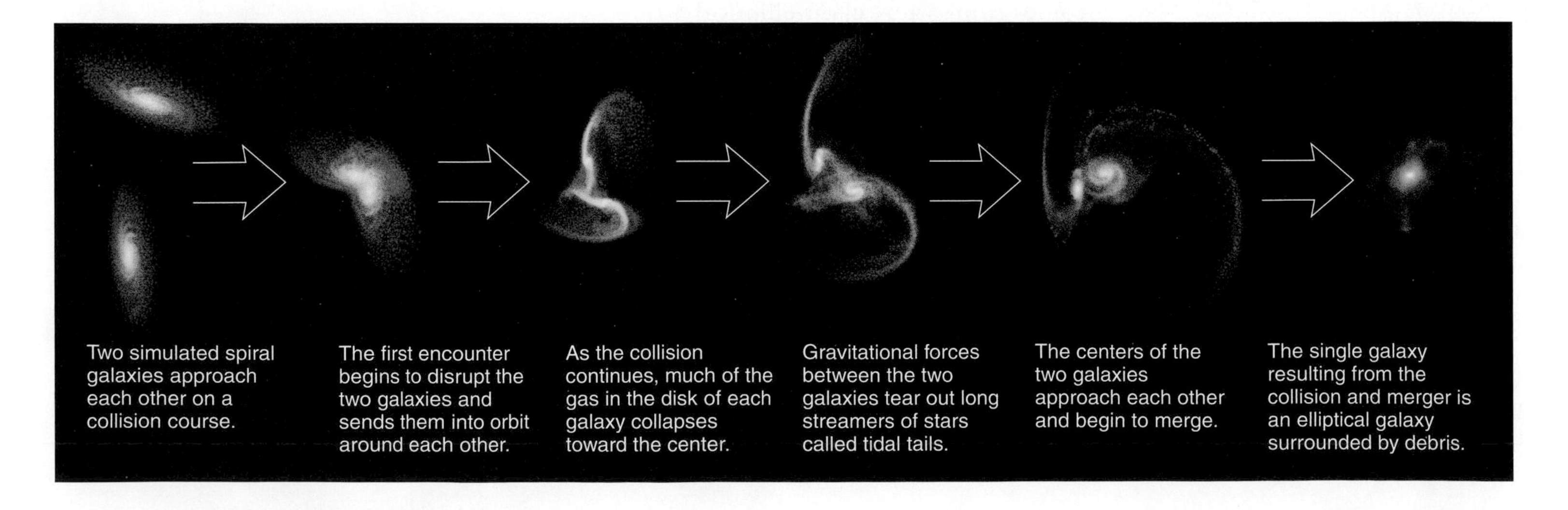

Figure 15.21

Several stages in a supercomputer simulation of a collision between two spiral galaxies that results in an elliptical galaxy. At least some of the elliptical galaxies in the present-day universe formed in this way. The whole sequence spans about 1.5 billion years.

Stronger evidence comes from structural details of elliptical galaxies, which often attest to a violent past. Some elliptical galaxies have stars and gas clouds with orbits suggesting that they are leftover pieces of galaxies that merged in a past collision. Others are surrounded by shells of stars that may have formed in gas stripped out of a galaxy during a past collision.

Observations show that at least some elliptical galaxies have experienced past collisions.

The most decisive evidence that collisions affect the evolution of elliptical galaxies comes from observations of the **central dominant galaxies** found at the centers of many dense clusters. Central dominant galaxies are gigantic elliptical galaxies that apparently grew to a huge size by consuming other galaxies through collisions. They frequently contain several tightly bound clumps of stars that probably were the centers of individual galaxies before being swallowed by the giant (Figure 15.22). This process of *galactic cannibalism* can create central dominant galaxies more than 10 times as massive as the Milky Way, making them the largest galaxies in the universe.

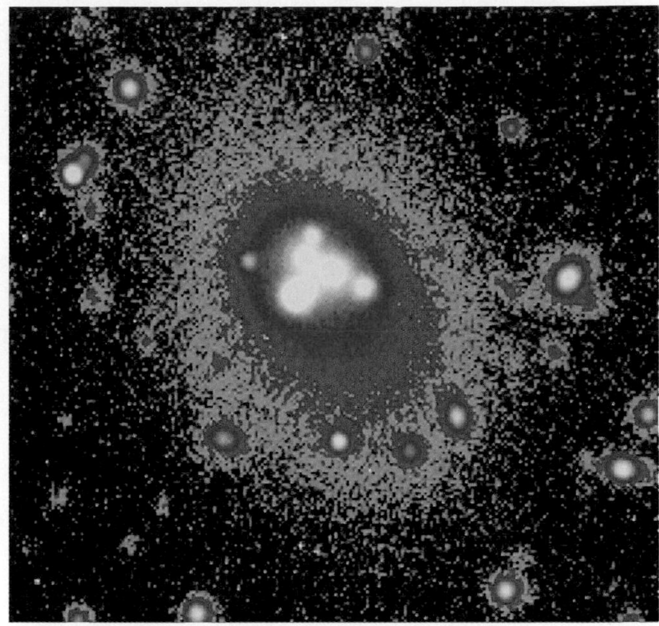

Figure 15.22

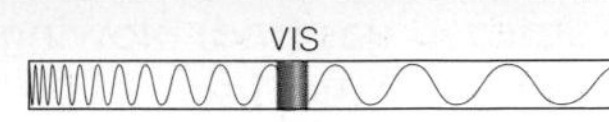

The central dominant galaxy of the cluster Abe113827 contains multiple clumps of stars that probably once were the centers of individual galaxies. The image has been specially processed so that different colors represent different levels of brightness. This procedure makes the brightest clumps (yellow) easier to see. This giant elliptical galaxy has apparently grown partly by consuming smaller galaxies that have collided with it.

Starbursts We have discussed two general ways in which elliptical galaxies may have formed: (1) as a result of birth in an unusually slow-rotating or high-density protogalactic cloud; and (2) as a result of later collisions and mergers of spiral galaxies. The birth scenarios make it easy to account for the lack of hot, young stars and ongoing star formation in elliptical galaxies, since they envision all the cool gas needed for star formation being used up before a spiral disk can form. But how can we account for the lack of young stars and cool gas in ellipticals that formed from more recent collisions?

Computer models suggest that galaxy collisions should ignite huge bursts of rapid star formation—so rapid that all the cool gas would be quickly transformed into stars, leaving little left over to form a disk. Observational support for this idea comes from a relatively small number of galaxies that appear to be in the midst of such bursts of star formation. We call these objects **starburst galaxies**.

Most galaxies with ongoing star formation appear comfortably settled, steadily forming new stars at modest rates. The Milky Way Galaxy produces an average of about one new star per year. At this rate, the Milky Way won't exhaust the interstellar gas in its disk until long after the Sun has died. Starburst galaxies are not so thrifty. Some form new stars at rates exceeding 100 stars per year, more than 100 times the star formation rate in the Milky Way. These galaxies cannot possibly sustain such a torrid pace of star formation for long. At their current rates of star formation, starburst galaxies would consume *all* their interstellar gas in just a few hundred million years. Because this is a relatively short time compared to the 10 billion or more years since galaxies first formed, a starburst must be only a temporary phase in a galaxy's life.

Starburst galaxies are forming stars so quickly that they will run out of star-forming clouds in just a few hundred million years.

A star formation rate 100 times that of the Milky Way also means that supernovae will occur at 100 times the Milky Way's rate. The bubbles of hot gas blown out by these many individual supernovae merge into a gigantic bubble so large that it rapidly bursts through any gaseous disk. The hot gas then erupts into intergalactic space, flowing outward in a **galactic wind** (Figure 15.23).

a This visible-light photograph (from the Subaru telescope) shows violently disturbed gas (red) blowing out both above and below the disk.

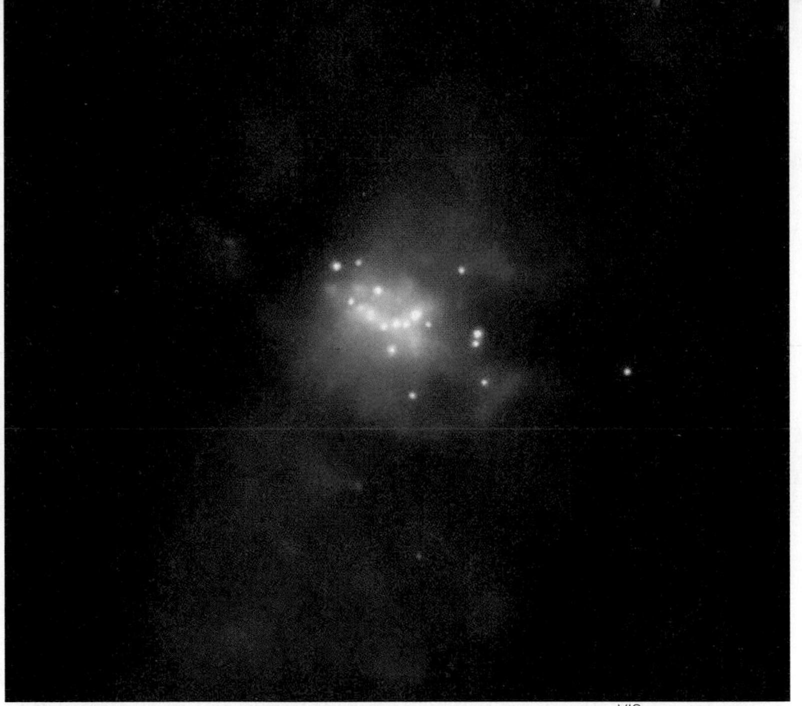

b This X-ray image from the Chandra X-Ray Observatory shows the same region as the visible-light photograph in (a). The reddish region represents X-ray emission from very hot gas blowing out of the disk. The bright dots in the galactic disk probably represent X-ray emission from accretion disks around black holes or neutron stars produced by recent supernovae.

Figure 15.23

Visible-light and X-ray views of a starburst galaxy called M82, showing its galactic wind. Both images show the same region, which is about 16,000 light-years across.

Many starbursts do indeed appear to be occurring in colliding galaxies. For example, the colliding galaxy pair shown in Figure 15.20 is undergoing a starburst. Thus, starbursts may be the answer to the mystery of why elliptical galaxies lack young stars and cool gas. The starburst uses up most of the cool gas; the galactic wind blows away what remains; and all the hot, massive stars die out within just a few hundred million years after the starburst ends. By the time the collision ends and the merger into an elliptical galaxy is complete, there is simply no cool gas left to support ongoing star formation.

At least some elliptical galaxies formed in collisions followed by starbursts, but it's not yet clear whether all of them do. Likewise, at least some irregular galaxies look irregular because they are currently undergoing collisions and starbursts, but not all irregular galaxies are colliding. For example, the Large Magellanic Cloud (which orbits the Milky Way) is an irregular galaxy that is also undergoing a period of rapid star formation. The starburst leading to this galaxy's irregular appearance might have been triggered not by a collision but rather by a close encounter with the Milky Way. No matter what their exact causes turn out to be, starbursts clearly represent an important piece in the overall puzzle of galaxy evolution.

Black Holes Tutorial, Lessons 1, 2

15.4 QUASARS AND OTHER ACTIVE GALACTIC NUCLEI

Starbursts may be spectacular, but some galaxies display even more incredible phenomena: extreme amounts of radiation, and sometimes powerful jets of material, emanating from deep in their centers (Figure 15.24).

These unusually bright galactic centers are usually called **active galactic nuclei**. The very brightest active galactic nuclei are known as **quasars**. The brightest quasars shine more powerfully than 1,000 galaxies the size of the Milky Way.

The glory days of quasars are long past. We find quasars primarily at great distances, telling us that these blazingly luminous objects were most common billions of years ago, when galaxies were in their youth. Because we find no nearby quasars and relatively few nearby galaxies with any type of active galactic nucleus, we conclude that the objects that shine as quasars in young galaxies must become dormant as the galaxies age. Thus, many nearby galaxies that now look quite normal must have centers that once shone brilliantly as quasars.

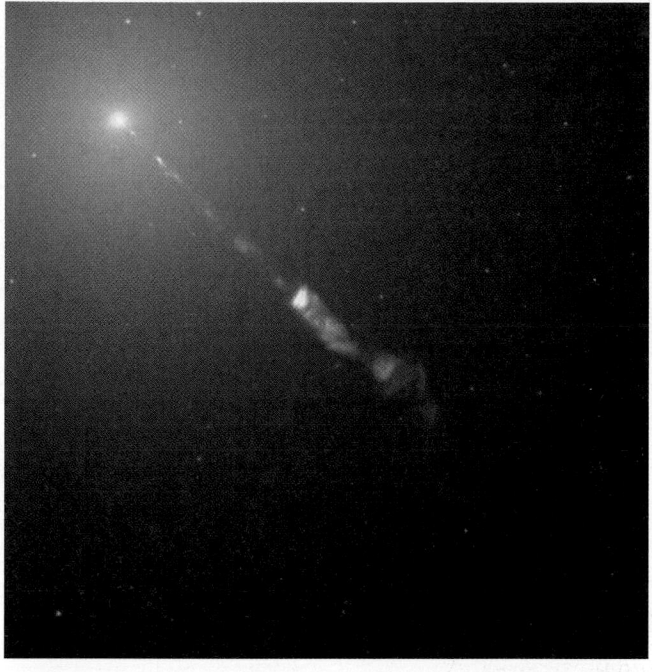

Figure 15.24

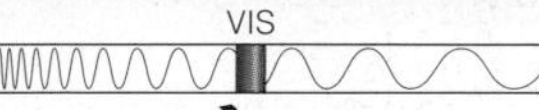

The active galactic nucleus in the elliptical galaxy M87. The bright yellow spot is the active nucleus, and the blue streak is a jet of particles shooting outward from the nucleus at nearly the speed of light.

• What are quasars?

What could possibly be the source of the incredible luminosities of quasars, and why did quasars fade away? A growing body of evidence points to a single answer: The energy output of quasars comes from gigantic accretion disks surrounding **supermassive black holes**—black holes with masses millions to billions of times that of our Sun. Before we study how these incredible powerhouses work, let's investigate the evidence that points to their existence.

The Discovery of Quasars In the early 1960s, a young professor at the California Institute of Technology named Maarten Schmidt was busy identifying cosmic sources of radio-wave emission. Radio astronomers would tell him the coordinates of newly discovered radio sources, and he would try to match them with objects seen through visible-light telescopes. Usually the radio sources turned out to be normal-looking galaxies, but one day he discovered a major mystery: A radio source called 3C273 looked like a blue star through a telescope but had strong emission lines at wavelengths that did not appear to correspond to any known chemical element. (The designation 3C273 stands for 3rd Cambridge Radio Catalogue, object 273.)

After months of puzzlement, Schmidt suddenly realized that the emission lines were not coming from an unfamiliar element at all. Instead, they were emission lines of hydrogen that were hugely redshifted from their normal wavelengths. Schmidt calculated that the expansion of the universe was carrying 3C273 away from us at 17% of the speed of light.

Schmidt computed the distance to 3C273 using Hubble's law. Then he used the distance and apparent brightness of 3C273 to estimate its luminosity [Section 11.1]. What he found was astonishing: 3C273 has a luminosity of about 10^{39} watts, or well over a trillion (10^{12}) times that of our Sun—making it hundreds of times more luminous than the entire Milky Way Galaxy. Discoveries of similar but even more distant objects soon followed. Because the first few of these objects were strong sources of radio emission that looked like stars through visible-light telescopes, they were named "quasi-stellar radio sources," or *quasars* for short. Later, astronomers learned that most quasars are not such powerful radio emitters, but the name has stuck.

Quasars can be more than a trillion times as luminous as the Sun.

For many years, a debate raged among astronomers over whether Hubble's law could really be used to determine quasar distances. Some argued that quasars might have high redshifts for other reasons and therefore might be much nearer to us than Hubble's law would suggest. The vast majority of astronomers now consider

Figure 15.25 VIS

The bright, central object in this photo is a quasar. Notice that it has spikes like those that appear in telescopic photographs of bright stars, which is what led to the "quasi-stellar" name. Nevertheless, this photo clearly shows that the quasar is located in the middle of a distant galaxy: We can see the rest of the galaxy surrounding the quasar, as well as other galaxies in the same galaxy cluster. Spectra confirm that all these objects have the same redshift, proving that the quasar is an extremely bright active galactic nucleus.

this debate settled. Improved images show that quasars are indeed the centers of extremely distant galaxies and often are members of very distant galaxy clusters (Figure 15.25).

Most quasars lie more than halfway to the cosmological horizon. The lines in typical quasar spectra are shifted to more than three times their rest wavelengths, which tells us that the light from these quasars emerged when the universe was less than a third of its present age. The light we see from the farthest known quasars began its journey when the universe was only about 6% of its present age.

Evidence from Nearby Active Galactic Nuclei Quasars are difficult to study in detail because they are so far away. Luckily, some quasar-like objects are much closer to home. About 1% of present-day galaxies—that is, galaxies we see nearby—have active galactic nuclei that look very much like quasars, except that they are less powerful. (These galaxies are often called *Seyfert galaxies* after astronomer Carl Seyfert, who in 1943 grouped galaxies with active galactic nuclei into a special class.)

The light-emitting regions of active galactic nuclei are so small that even the sharpest images do not resolve them. Our best visible-light images show only that active galactic nuclei must be smaller than 100 light-years across. Radio-wave images made with the aid of *interferometry* [Section 5.3] show that these nuclei are even smaller: less than 3 light-years across. Rapid changes in the luminosities of some active galactic nuclei point to an even smaller size.

The immense luminosities of quasars and other active galactic nuclei come from regions no larger than our solar system.

To understand how variations in luminosity give us clues about an object's size, imagine that you are a master of the universe and you want to signal one of your fellow masters a billion light-years away. An active galactic nucleus would make an excellent signal beacon, because it is so bright. However, suppose the smallest nucleus you can find is 1 light-year across. Each time you flash it on, the photons from the front end of the source reach your fellow master a full year before the photons from the back end. Thus, if you flash it on and off more than once a year, your signal will be smeared out. Similarly, if you find a source that is 1 light-day across, you can transmit signals that flash on and off no more than once a day. If you want to send signals just a few hours apart, you need a source no more than a few light-hours across.

Occasionally, the luminosity of an active galactic nucleus doubles in a matter of hours. The fact that we see a clear signal indicates that the source must be less than a few light-hours across. In other words, the incredible luminosities of active galactic nuclei and quasars are apparently being generated in a volume of space not much bigger than our solar system.

Radio Galaxies and Jets Another clue to the nature of quasars came in the early 1950s, a decade before the discovery of quasars. Radio astronomers noticed that certain galaxies, now called **radio galaxies**, emit unusually strong radio waves. Upon closer inspection, they learned that much of the radio emission comes not from the galaxies themselves but rather from pairs of huge *radio lobes*, one on either side of the galaxy (Figure 15.26).

Active galactic nuclei in radio galaxies can propel huge jets of particles moving at nearly the speed of light.

Today, radio telescopes resolve the structure of radio galaxies in vivid detail. The power source is an active galactic nucleus less than

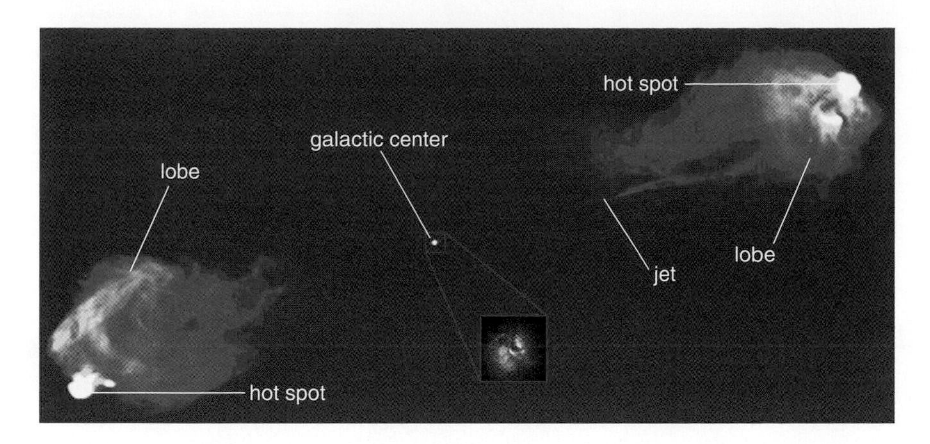

Figure 15.26

radio

This image, made with the Very Large Array in New Mexico, shows radio wave emission from a radio galaxy called Cygnus A. The emission comes from two long jets of particles shot out from the center of the galaxy at nearly the speed of light. The hot spots at the end of each jet occur where the particles ram into surrounding intergalactic gas. The particles are then deflected to make the large radio lobes, much like the spray of water from a fire hose is deflected when it hits a wall. The inset photo shows the galaxy in visible light; notice that the radio lobes lie far beyond the bounds of the visible galaxy. (The two lobes are about 400,000 light-years apart.)

a few light-years across at the center of the galaxy. We often see two gigantic **jets** of plasma shooting out of the active galactic nucleus in opposite directions. (Sometimes, only the jet tilted in our direction is visible.) Using time-lapse radio images taken several years apart, we can track the motions of various plasma blobs in the jets. Some of these blobs move at close to the speed of light. The lobes lie at the ends of the jets, which are sometimes as far as a million light-years from the galactic center.

We now suspect that quasars and radio galaxies are the same types of objects viewed in slightly different ways. We have discovered that many quasars have jets and radio lobes like those seen in radio galaxies (Figure 15.27). Moreover, the active galactic nuclei of many radio galaxies seem to be concealed beneath donut-shaped rings of dark molecular clouds (Figure 15.28). Such structures may look like quasars when they are oriented so that we can see the active galactic nucleus at the center and look like the nuclei of radio galaxies when the ring of dusty gas dims our view of the central object. (A subset of active galactic nuclei called *BL Lac objects* probably represents the centers of radio galaxies whose jets happen to point directly at us.)

• What is the power source for quasars and other active galactic nuclei?

Astronomers have worked hard to envision physical processes that might explain how radio galaxies, quasars, and other active galactic nuclei release so much energy within such small central volumes. Only one explanation seems to fit: The energy comes from matter falling into a supermassive black hole. The gravitational potential energy of the infalling matter is converted into kinetic energy and collisions between infalling particles convert the kinetic energy into thermal energy. The resulting heat causes this matter to emit the intense radiation we observe. As in X-ray binaries, we expect that the infalling matter swirls through an accretion disk before it disappears beneath the event horizon of the black hole [Section 13.3].

Quasars are probably powered by matter falling into supermassive black holes.

If the supermassive black hole model is correct, it should be able to explain the major observed features of quasars and other active galactic nuclei. In particular, it should explain their extreme luminosities, the fact that they emit radiation over a broad range of wavelengths from radio waves to X-rays, and the presence of their powerful jets.

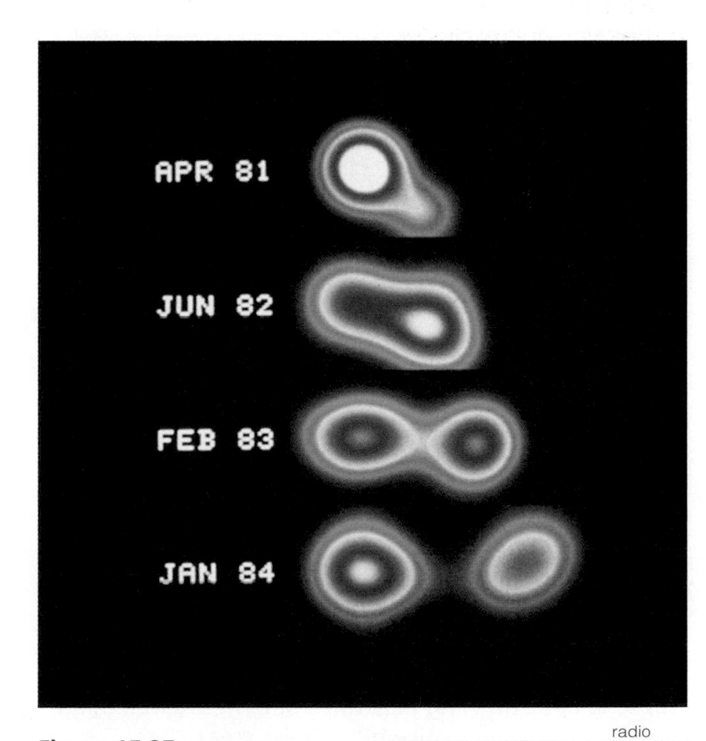

Figure 15.27

radio

These images show a jet from a quasar known as 3C345. Taken over a period of several years, they show a blob of plasma (right) moving away from the quasar (left) at close to the speed of light.

Figure 15.28

Artist's conception of the central few hundred light-years of a radio galaxy. The active galactic nucleus, obscured by a ring of dusty molecular clouds, lies at the point from which the jets emerge. If viewed along a direction closer to the jet axis, the active nucleus would not be obscured and would look more like a quasar.

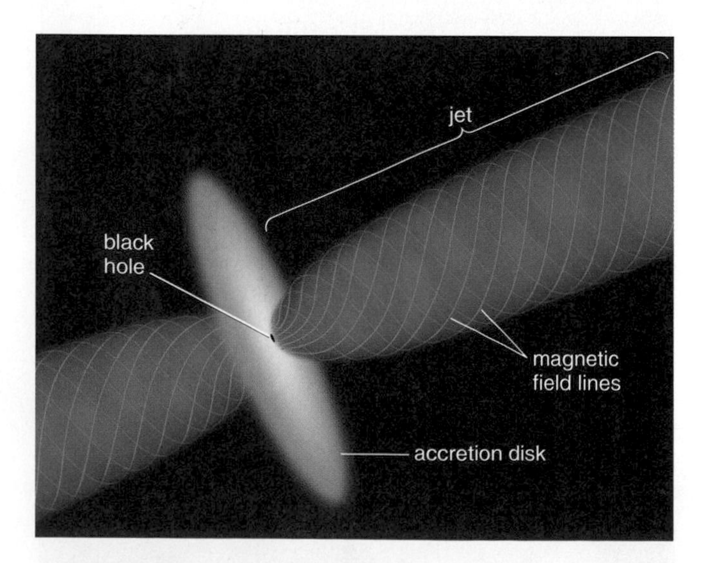

Figure 15.29

This schematic drawing illustrates one theory that might explain how supermassive black holes create jets. The theory relies on the magnetic field lines thought to thread the accretion disk surrounding a black hole. As the accretion disk spins, it twists the magnetic field lines. Charged particles at the accretion disk's surface can then fly outward along the twisted magnetic field lines.

Explaining the extreme luminosities of quasars was the main motivation for the supermassive black hole model. Matter falling into a black hole can generate awesome amounts of energy. During its fall to the event horizon of a black hole, as much as 10–40% of the mass-energy ($E = mc^2$) of a chunk of matter can be converted into thermal energy and ultimately to radiation. Thus, accretion by black holes can produce light far more efficiently than nuclear fusion, which converts less than 1% of mass-energy into photons. Remember that the light is coming not from the black hole itself but rather from the hot gas in the accretion disk that surrounds it.

The environment surrounding a supermassive black hole explains why active galactic nuclei emit light across such a broad wavelength range. Hot gas in and above the accretion disk produces enormous amounts of ultraviolet and X-ray photons. This radiation ionizes surrounding interstellar gas, creating ionization nebulae that emit visible light. (The emission lines produced by these nebulae are the same ones Maarten Schmidt used to measure the first quasar redshifts.) Dust grains in the molecular clouds that encircle the active galactic nucleus (see Figure 15.28) absorb high-energy light and reemit it as infrared light. Finally, the fast-moving electrons that we sometimes see jetting from these nuclei at nearly the speed of light can produce the radio emission from active galactic nuclei.

The powerful jets emerging from active galactic nuclei are more difficult to explain, but certainly there is plenty of energy available for flinging material outward at nearly the speed of light. One plausible model for jet production relies on the twisted magnetic fields thought to accompany accretion disks (Figure 15.29). As an accretion disk spins, it pulls the magnetic field lines that thread it around in circles. Charged particles fly outward along the field lines like beads on a twirling string, forming a jet that shoots out into space.

In summary, the supermassive black hole model seems to explain the major observed features of quasars and other active galactic nuclei. However, several important mysteries remain. For example, we do not yet know what would create such giant black holes, nor do we know why quasars eventually run out of gas to accrete and stop shining brightly. The preponderance of quasars during the first several billion years of the universe suggests that the formation of supermassive black holes is somehow linked to galaxy formation. This suggestion is supported by recent observations showing that the mass of a galaxy's central black hole is closely related to the mass of its spheroidal component, which was probably assembled during the time that quasars were most luminous. Some scientists have suggested that clusters of neutron stars resulting from extremely dense starbursts at the centers of galaxies might somehow coalesce to form an enormous black hole, but these speculations are still unverified. The origins of supermassive black holes remain mysterious.

• Do supermassive black holes really exist?

Do supermassive black holes really drive the tremendous activity of radio galaxies, quasars, and other active galactic nuclei? The idea of such monster black holes has been hard for some astronomers to swallow. Proving that we have found a black hole is tricky. Black holes themselves do not emit any light, so we need to infer their existence from the ways in which they alter their surroundings. In the vicinity of a black hole, matter should be orbiting at high speed around something invisible. We have already examined the evidence for a black hole at the center of our Milky Way [Section 14.4], but what about other galaxies?

Detailed observations of matter orbiting at the centers of nearby galaxies suggest that supermassive black holes are quite common. In fact, it is possible that *all* galaxies might contain supermassive black holes at their centers. One prominent example is the relatively nearby galaxy M87, which features a bright, active galactic nucleus and a jet that emits both radio and visible light (see Figure 15.24). Thus, it was already a prime black-hole suspect when astronomers pointed the Hubble Space Telescope at its core in 1994 (Figure 15.30). The spectra they gathered showed blueshifted emission lines on one side of the nucleus and redshifted emission lines on the other. This pattern of Doppler shifts is the characteristic signature of orbiting gas: On one side of the orbit the gas is coming toward us and hence is blueshifted, while on the other side it is moving away from us and is redshifted. The magnitude of these Doppler shifts shows that the gas, located up to 60 light-years from the center, is orbiting something invisible at a speed of hundreds of kilometers per second. This high-speed orbital motion indicates that the central object has a mass some 2–3 billion times that of our Sun.

Observations of rapidly orbiting material at the centers of galaxies indicate that at least some galaxies, and perhaps all of them, harbor supermassive black holes.

Observations of NGC4258, another galaxy with a visible jet, have delivered even more persuasive evidence. A ring of molecular clouds orbits the nucleus of this galaxy in a circle less than 1 light-year in radius. We can pinpoint these clouds because they amplify the microwave emission lines of water molecules, generating beams of microwaves very similar to laser beams. (The word *laser* stands for "*l*ight *a*mplification by *s*timulated *e*mission of *r*adiation." These clouds contain *water masers*. The word *maser* stands for "*m*icrowave *a*mplification by *s*timulated *e*mission of *r*adiation.") The Doppler shifts of these emission lines allow us to determine the orbits of the clouds very precisely. Their orbital motion tells us that the clouds are circling a single, invisible object with a mass of 36 million solar masses. A supermassive black hole is the only thing we know of that could be so massive while remaining unseen.

We may never be 100% certain that we have discovered black holes in other galaxies. The best we can do is rule out all other possibilities. However, the hypothesis that gigantic black holes lie at the cores of quasars, nearby active galactic nuclei, and radio galaxies is so far withstanding the tests of time and thousands of observations.

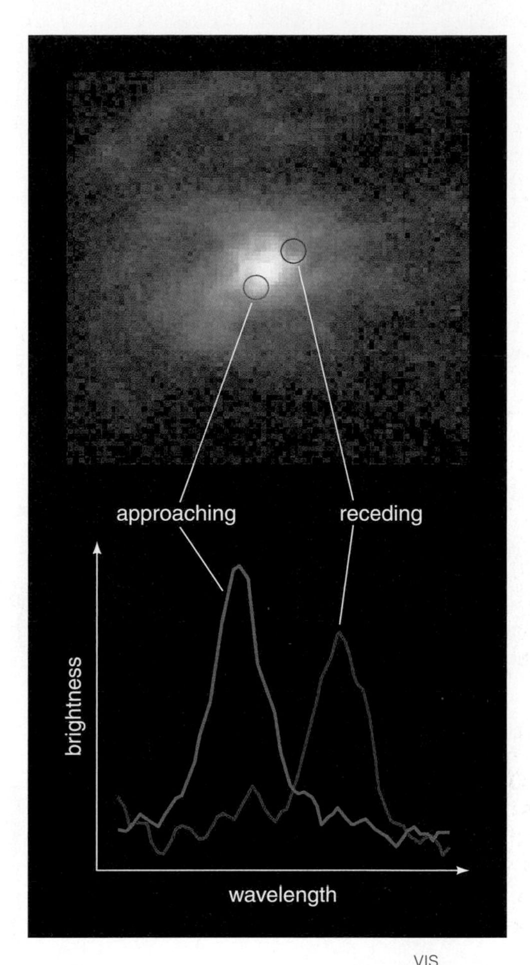

Figure 15.30

This Hubble Space Telescope photo shows gas near the center of the galaxy M87, and the graph shows Doppler shifts of spectra from gas on either side of the center (the circled regions in the photo). Notice that light from the gas on one side is redshifted (red curve), indicating that this gas is rotating away from us, while light from the other side (blue curve) is blueshifted, indicating that this gas is rotating toward us. Precise measurement of the Doppler shifts tells us that the gas orbits the galactic center at about 800 km/s, and applying Newton's laws tells us that the central object must have a mass 2–3 billion times that of the Sun.

THE BIG PICTURE

Putting Chapter 15 into Context

The picture could hardly get any bigger than it has in this chapter. Looking back through both space and time, we have seen a wide variety of galaxies extending nearly to the limits of the observable universe. As you look back, keep sight of these "big picture" ideas:

- The universe is filled with galaxies that come in a variety of shapes and sizes. The most fundamental distinctions in galaxies are between *disk components,* with stars of many ages and abundant gas for new star formation, and *spheroidal components,* which contain old stars and little gas. Both components are present in spiral galaxies, while elliptical galaxies generally lack significant disks.

- Our measurements of the distances to faraway galaxies rely on a chain of techniques, each of which builds upon the ones that come before it. This chain allowed Hubble to prove that the universe is far larger than the Milky Way and to discover the expansion of the universe. Today, refinements of the distance chain have allowed us to pin down the expansion rate fairly precisely, telling us that we live in a universe that is about 14 billion years old.
- When we say that the universe is expanding, we mean that *space itself* is expanding. The universe is not expanding "into" anything, and the universe has no center and no edges.
- Although we do not yet know the complete story of galaxy evolution, we are rapidly learning more. We know that galaxies grow from protogalactic clouds of gas, but collisions with neighboring galaxies have probably affected many galaxies, sometimes igniting starbursts and turning spiral galaxies into elliptical galaxies.
- The tremendous energy outputs of quasars and other active galactic nuclei, including those of radio galaxies, are probably powered by gas accreting onto supermassive black holes. The centers of many present-day galaxies must still contain the supermassive black holes that once enabled them to shine as quasars.

15.1 ISLANDS OF STARS

- **What are the three major types of galaxies?**

(1) **Spiral galaxies** have prominent disks and spiral arms.

(2) **Elliptical galaxies** are rounder and redder than spiral galaxies and contain less cool gas and dust.

(3) **Irregular galaxies** are neither disklike nor rounded in appearance.

- **How are galaxies grouped together?**

Spiral galaxies tend to collect in **groups** of galaxies, which contain up to several dozen galaxies. Elliptical galaxies are more common in **clusters** of galaxies, which contain hundreds to thousands of galaxies, all bound together by gravity.

15.2 DISTANCES OF GALAXIES

- **How do we measure the distances to galaxies?**

Our measurements of galaxy distances depend on a chain of methods. The chain begins with **radar ranging** in our own solar system and parallax measurements of distances to nearby stars, then relies on **standard candles** to measure greater distances.

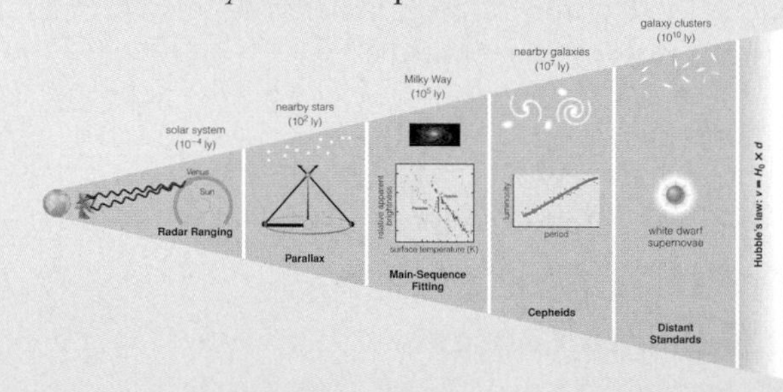

- **What is Hubble's law?**

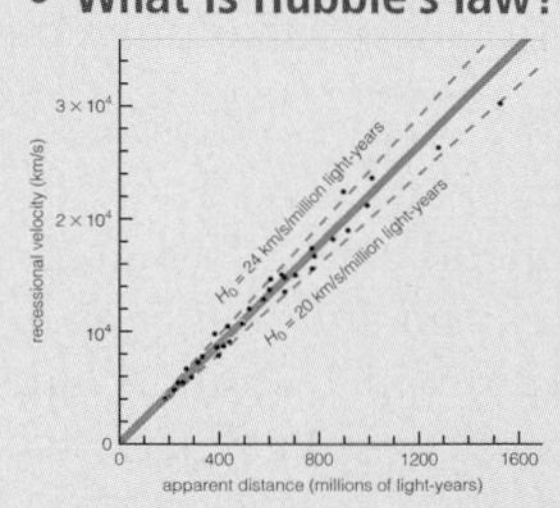

Hubble's law tells us that more-distant galaxies are moving away faster: $v = H_0 \times d$, where H_0 is **Hubble's constant**. It allows us to determine a galaxy's distance from the speed at which it is moving away from us, which we can measure from its Doppler shift.

• How do distance measurements tell us the age of the universe?
Combining distance measurements with velocity measurements tells us Hubble's constant, and the inverse of Hubble's constant tells us how long it would have taken the universe to reach its present size *if* the expansion rate had never changed. Based on Hubble's constant and estimates of how it has changed with time, we now estimate the age of the universe at about 14 billion years, which restricts our view of the universe to **lookback times** smaller than that age.

15.3 GALAXY EVOLUTION

• How do we observe the life histories of galaxies?

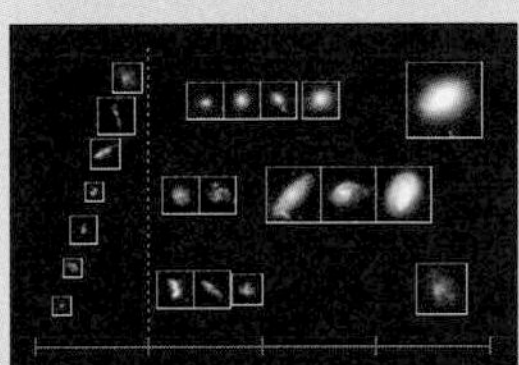

Today's telescopes enable us to observe galaxies of many different ages because they are powerful enough to detect light from objects with lookback times almost as large as the age of the universe. We can therefore assemble "family albums" of galaxies at different distances and lookback times.

• How did galaxies form?
The most successful models of galaxy formation assume that galaxies formed as gravity pulled together regions of the universe that were ever so slightly denser than their surroundings. Gas collected in protogalactic clouds, and stars began to form as the gas cooled.

• Why do galaxies differ?
Differences between present-day galaxies probably can arise both from conditions in their protogalactic clouds and from collisions with other galaxies. Slowly rotating or high-density protogalactic clouds may form elliptical rather than spiral galaxies. Ellipticals may also form through the collision and merger of two spiral galaxies.

15.4 QUASARS AND OTHER ACTIVE GALACTIC NUCLEI

• What are quasars?
Some galaxies have unusually bright centers known as **active galactic nuclei**. A **quasar** is a particularly bright active galactic nucleus. Quasars are generally found at very great distances, telling us that they were much more common early in the history of the universe.

• What is the power source for quasars and other active galactic nuclei?

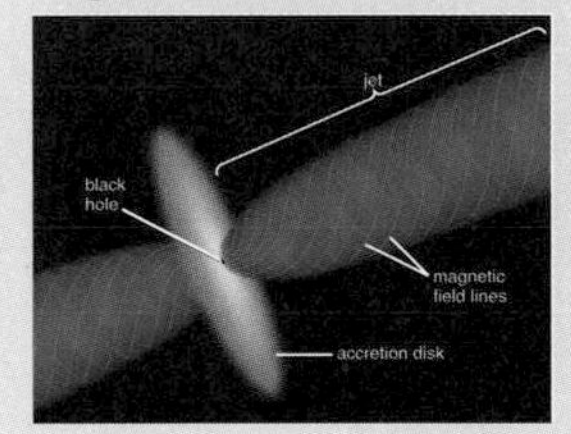

Supermassive black holes are thought to be the power sources for active galactic nuclei. As matter falls into a supermassive black hole through an accretion disk, its gravitational potential energy is transformed into thermal energy and then into light with enormous efficiency.

• Do supermassive black holes really exist?
Observations of orbiting stars and gas clouds in the nuclei of galaxies suggest that *all* galaxies may harbor supermassive black holes at their centers.

EXERCISES AND PROBLEMS

REVIEW QUESTIONS

1. What are the three major types of galaxies, and how do they look different from one another?
2. Describe the differences between normal spiral galaxies, barred spiral galaxies, and lenticular galaxies.
3. Distinguish between the *disk component* and the *spheroidal component* of a spiral galaxy. Which component includes cool gas and active star formation?
4. What is the major difference between spiral and elliptical galaxies? Answer in terms of the presence or absence of disk and spheroidal components. How does this difference explain the lack of hot, young stars in elliptical galaxies?
5. How are galaxy types different in clusters of galaxies than among smaller groups and isolated galaxies?
6. What do we mean by a *standard candle?* Explain how we can use standard candles to measure distances.
7. Summarize each of the major links in the distance chain. Why are *Cepheid variable stars* so important? Why are white dwarf supernovae so useful, even though they are quite rare?
8. Explain how Hubble used Cepheid variable stars to prove that the Andromeda Galaxy lies beyond the bounds of the Milky Way.
9. What is *Hubble's law?* What is *Hubble's constant?* Explain what we mean when we say that Hubble's constant is between 20 and 24 *kilometers per second per million light-years.*
10. What is the cosmological principle, and how is it important to our understanding of the universe?
11. In what ways is the surface of an expanding balloon a good analogy to the universe? In what ways is this analogy limited? Use the analogy to explain why Hubble's constant is related to the age of the universe.
12. What do we mean by the *lookback time* to a distant galaxy? Briefly explain why lookback times are less ambiguous than distances when discussing objects very far away.
13. What is the *cosmological horizon,* and what determines how far away it lies?
14. What do we mean by a *cosmological redshift?* How does our interpretation of a distant galaxy's redshift differ if we think of it as a cosmological redshift rather than as a Doppler shift?

15. What do we mean by *galaxy evolution?* How do telescopic observations allow us to study galaxy evolution?
16. Briefly describe the starting assumptions for galaxy formation, and how models suggest that these starting conditions can lead to the formation of a spiral galaxy.
17. Describe two ways in which conditions in a protogalactic cloud might lead to the birth of an elliptical rather than a spiral galaxy. How might elliptical galaxies form from later collisions of spiral galaxies?
18. Briefly explain why we expect that collisions between galaxies should be relatively common, while collisions between stars are extremely rare. Why should galaxy collisions have been more common in the past than they are today? What evidence supports the idea that galaxy collisions sometimes occur?
19. What is a *starburst galaxy?* How do observations of starbursts help us understand the old ages of stars in elliptical galaxies?
20. Briefly describe the discovery of *quasars.* What evidence convinced astronomers that the high redshifts of quasars really do imply great distances? Why can we learn more about quasars by studying nearby *active galactic nuclei* and *radio galaxies?*
21. Briefly explain how we can use variations in luminosity to set limits on the size of an object's emitting region. For example, if an object doubles its luminosity in 1 hour, how big can it be?
22. Summarize the supermassive black hole model for the energy output of quasars and other active galactic nuclei. What evidence suggests that such black holes really exist?

SENSIBLE STATEMENTS?

Decide whether each of the following statements is sensible and explain why it is or is not.

23. After measuring the galaxy's redshift, I used Hubble's law to estimate its distance.
24. The center of the universe is more crowded with galaxies than any other place in the universe.
25. I'd love to live in one of the galaxies near our cosmological horizon, because I want to see the black void into which the universe is expanding.
26. If someone in a galaxy with a lookback time of 4.6 billion years had a superpowerful telescope, they could see our solar system in the process of its formation.
27. Galaxies that are more than 10 billion years old are too far away to see even with our most powerful telescopes.
28. If the Andromeda Galaxy someday collides and merges with the Milky Way, the resulting galaxy may be elliptical.
29. NGC9645 is a starburst galaxy that has been forming stars at the same furious pace for some 10 billion years.
30. Astronomers proved that quasar 3C473 contains a supermassive black hole when they discovered that its center is completely dark.

PROBLEMS

(Quantitative problems are marked with an asterisk.)

31. *Life Story of a Spiral.* Imagine that you are a spiral galaxy. Describe your life history from birth to the present day. Your story should be detailed and scientifically consistent, but also creative. That is, it should be entertaining while at the same time incorporating current scientific ideas about the formation of spiral galaxies.
32. *Life Story of an Elliptical.* Imagine that you are an elliptical galaxy. Describe your life history from birth to the present. There are several possible scenarios for the formation of elliptical galaxies, so choose one and stick to it. Be creative while also incorporating scientific ideas that demonstrate your understanding.
33. *Cepheids as Standard Candles.* Suppose you are observing Cepheids in a nearby galaxy. You observe one Cepheid with a period of 8 days between peaks in brightness, and another with a period of 35 days. Estimate the luminosity of each star. Explain how you arrived at your estimate. (*Hint:* See Figure 15.12.)

*34. *Counting Galaxies.* Estimate how many galaxies are pictured in Figure 15.1. Explain the method you used to arrive at this estimate. This picture shows about $\frac{1}{30{,}000{,}000}$ of the sky, so multiply your estimate by 30,000,000 to obtain an estimate of how many galaxies like these fill the entire sky.

*35. *Your Last Hurrah.* Suppose you fell into an accretion disk that swept you into a supermassive black hole. On your way down, the disk radiates 10% of your mass-energy, $E = mc^2$.
   a. What is your mass in kilograms? (Recall that 1 kg = 2.2 pounds.) Calculate how much radiative energy will be produced by the accretion disk as a result of your fall into the black hole.
   b. Calculate approximately how long a 100-watt light bulb would have to burn to radiate this same amount of energy.

DISCUSSION QUESTIONS

36. *Cosmology and Philosophy.* One hundred years ago, many scientists believed that the universe was infinite and eternal, with no beginning and no end. When Einstein first developed his general theory of relativity, he found that it predicted that the universe should be either expanding or contracting. He believed so strongly in an eternal and unchanging universe that he modified the theory, a modification he would later call his "greatest blunder." Why do you think Einstein and others assumed that the universe had no beginning? Do you think that a universe with a definite beginning in time, some 14 billion or so years ago, has any important philosophical implications? Explain.
37. *The Case for Supermassive Black Holes.* The evidence for supermassive black holes at the center of galaxies is strong. However, it is very difficult to prove absolutely that they exist because the black holes themselves emit no light. We can only infer their existence from their powerful gravitational influences on surrounding matter. How compelling do you find the evidence? Do you think astronomers have proved the case for black holes beyond a reasonable doubt? Defend your opinion.

MATH HELP AND EXERCISES

For additional help and practice with mathematical concepts applicable to this chapter, the Astronomy Place Web site has the following resources with detailed explanations, worked examples, and practice problems.

- The Inverse Square Law of Light
- Redshift and Hubble's Law
- Feeding a Black Hole
- Weighing Supermassive Black Holes

For a complete list of media resources available, go to **www.astronomyplace.com** and choose Chapter 15 from the pull-down menu.

ASTRONOMY PLACE WEB TUTORIALS

Tutorial Review of Key Concepts

Use the interactive **Tutorials** at **www.astronomyplace.com** to review key concepts from this chapter.

Measuring Cosmic Distances Tutorial

Lesson 1 Radar
Lesson 2 Stellar Parallax
Lesson 3 Standard Candles: Main-Sequence Stars and Cepheid Variables
Lesson 4 Standard Candles: White Dwarf Supernovae and Spiral Galaxies

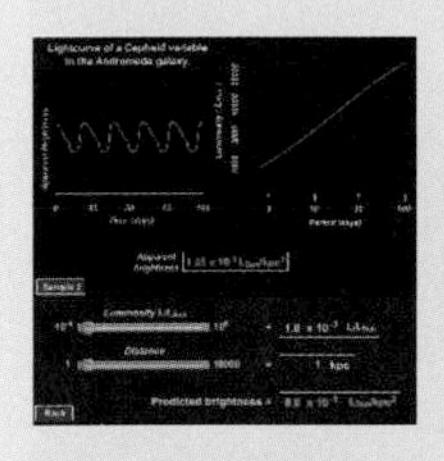

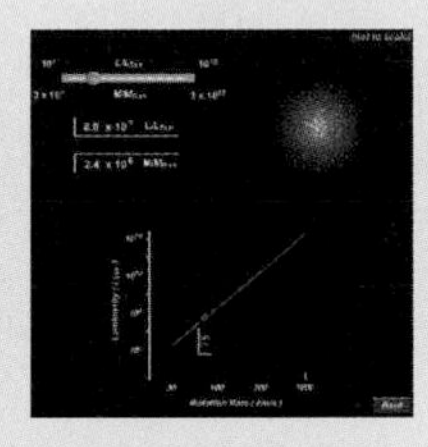

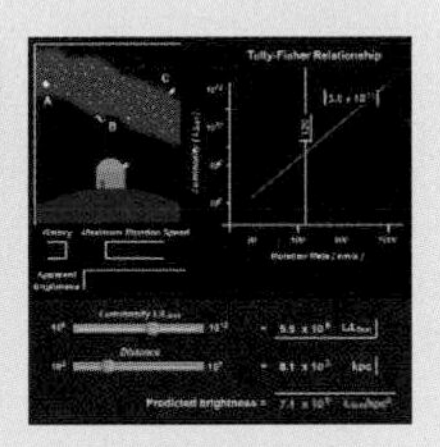

Hubble's Law Tutorial

Lesson 1 Hubble's Law
Lesson 2 The Expansion of the Universe
Lesson 3 The Age of the Universe

Black Holes Tutorial

Lesson 1 What Are Black Holes?
Lesson 2 The Search for Black Holes

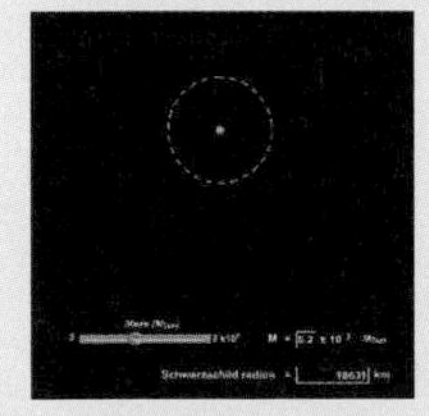

Supplementary Tutorial Exercises

Use the interactive **Tutorial Lessons** to explore the following questions.

Measuring Cosmic Distances Tutorial, Lessons 3–4

1. What makes it possible for a streetlight to appear much brighter than a star?
2. How can we measure the distance to a faraway galaxy in which we observe a white dwarf supernova?
3. How can we measure the distance to a faraway spiral galaxy in which we do *not* observe a supernova?

Hubble's Law Tutorial, Lesson 1

1. How do we determine the motion of distant galaxies toward or away from us?
2. How can we use Hubble's law to determine the distance to a far-away galaxy?

Black Holes Tutorial, Lessons 1, 2

1. What is the Schwarzschild radius for a black hole of 1 million solar masses? For black holes of 10 million and 100 million solar masses?
2. Describe how scientists determine whether the center of a galaxy contains a supermassive black hole.

EXPLORING THE SKY AND SOLAR SYSTEM

Of the many activities available on the ***Voyager: SkyGazer*** **CD-ROM** accompanying your book, use the following files to observe key phenomena covered in this chapter.

Go to the **File: Basics** folder for the following demonstration.

1. Galaxies in Coma

MOVIES

Check out the following narrated and animated short documentary available on **www.astronomyplace.com** for a helpful review of key ideas covered in this chapter.

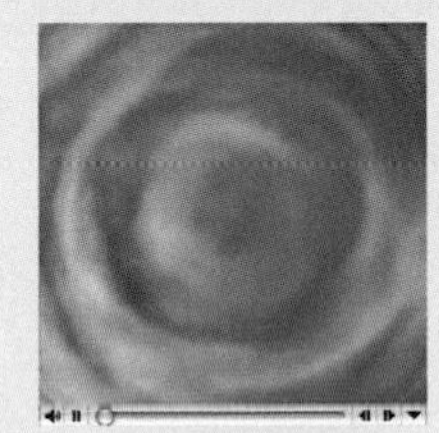

From the Big Bang to the Galaxies Movie

WEB PROJECTS

Take advantage of the useful Web links on **www.astronomyplace.com** to assist you with the following projects.

1. *Galaxy Gallery.* Many fine images of galaxies are available on the Web. Collect several images of each major type and build a galaxy gallery of your own. Supply a descriptive paragraph about each galaxy.
2. *Greatest Lookback Time.* Look for recent discoveries of objects with the largest lookback times (or redshifts). What is the current record for the most distant known object? What kind of object is it?
3. *Future Observatories.* The subject of galaxy evolution is a very active area of research. Look for information on current and future observatories involved in investigating galaxy evolution (such as the James Webb Space Telescope). How big are the planned telescopes? What wavelengths will they look at? When will they be built? Write a short summary of a proposed mission.

16
Dark Matter, Dark Energy, and the Fate of the Universe

16.1 UNSEEN INFLUENCES IN THE COSMOS
- What do we mean by dark matter and dark energy?

16.2 EVIDENCE FOR DARK MATTER
- What is the evidence for dark matter in galaxies?
- What is the evidence for dark matter in clusters of galaxies?
- Does dark matter really exist?
- What might dark matter be made of?

16.3 STRUCTURE FORMATION
- What is the role of dark matter in galaxy formation?
- What are the largest structures in the universe?

16.4 THE UNIVERSE'S FATE
- Will the universe continue expanding forever?
- Is the expansion of the universe accelerating?

In Chapter 15, we explored current understanding of the evolution of galaxies in an expanding universe. However, we left at least two crucial questions unaddressed: First, what is the source of the gravity that causes galaxies to form and holds them together? Second, what will happen to the expansion of the universe in the future? Both questions would be interesting in their own right, but they've taken on even greater importance because our attempts to answer them have led us to two of the greatest mysteries in science.

If we are interpreting the data correctly, then the dominant source of gravity in the universe is an unidentified type of mass, known as *dark matter*, that is completely invisible to our eyes and telescopes. The gravity of dark matter must therefore be the "glue" that holds galaxies like our own together. Moreover, the very fate of the universe seems to hinge on the total amount of dark matter and on the existence of an even more mysterious force or energy—often called *dark energy*—that may counteract the effects of gravity on large scales.

In this chapter, we will explore the evidence for dark matter and dark energy and discuss why their nature remains so mysterious. We'll also investigate how dark matter determines the current structure of the universe and the implications of dark energy for the fate of the universe, if dark matter and dark energy indeed prove to be real.

16.1 UNSEEN INFLUENCES IN THE COSMOS

What is the universe made of? Ask an astronomer this seemingly simple question, and you might see a professional scientist blush with embarrassment. Based on all the available evidence today, the answer to this simple question is "We do not know."

It might seem incredible that we still do not know the composition of most of the universe, but you might also wonder why we should be so clueless. After all, astronomers can measure the chemical composition of distant stars and galaxies from their spectra, and we've said over and over that stars and gas clouds are made almost entirely of hydrogen and helium, with small amounts of heavier elements mixed in. But notice the key words "chemical composition." When we say these words, we are talking about the composition of material built from atoms of elements such as hydrogen, helium, carbon, and iron. While it is true that all familiar objects—including people, planets, and stars—are built from atoms, the same may not be true of the universe as a whole.

In fact, we now have good reason to think that the vast majority of the universe is *not* composed of atoms. Instead, the universe may consist largely of a mysterious form of mass known as *dark matter* and a mysterious form of energy known as *dark energy.* As we'll discuss in Chapter 17, some recent observations apparently tell us the precise percentages of the universe that consist of dark energy, dark matter, and ordinary atoms. Nevertheless, the actual nature of dark matter and dark energy remains completely unknown.

• What do we mean by dark matter and dark energy?

It's easy to talk about dark matter and dark energy, but what do these terms really mean? They are nothing more than names given to unseen influences in the cosmos. In both cases we have been led to think that there is something out there, even though we cannot identify it, because we have observed phenomena that otherwise do not make sense.

Dark matter is the name given to mass that we infer to exist through its gravitational effects but that emits no detectable radiation.

We might naively think that the major source of gravity that holds together galaxies is the same gas and dust that makes up their stars. However, decades of observations suggest otherwise: By carefully observing gravitational effects on matter that we can see, such as stars or glowing clouds of gas, we've learned that there must be far more matter than meets the eye. Because this matter apparently gives off little or no light, we call it **dark matter**. Thus, dark matter is simply a name we give to whatever unseen influence is causing the observed gravitational effects. We've already discussed dark matter briefly in Chapters 1 and 14, noting that studies of the Milky Way's rotation suggest that most of our galaxy's mass is distributed throughout its halo while most of the galaxy's light comes from stars and gas clouds in the thin galactic disk (see Figure 1.14).

The second unseen influence is inferred to exist from careful studies of the expansion of the universe. From the time that Hubble first discovered the expansion, it was generally assumed that gravity must slow the expansion with time. In just the past few years, however, mounting evidence has suggested that the expansion of the universe is actually accelerating [Section 16.4]. If so, some mysterious type of force or energy must be able to counteract the effects of gravity on very large scales.

Dark energy is the name given to an unseen influence that may be causing the expansion of the universe to accelerate with time.

Dark energy is the most common name given to whatever it is that may be causing the expansion to accelerate, but it is not the only name; you may occasionally hear the same unseen influence attributed to *quintessence* or to a *cosmological constant*. The term *dark energy* has become popular because it echoes the term *dark matter*, but there's nothing unusually "dark" about it—after all, we don't expect to see light from the mere presence of a force or energy field. Thus, despite the similarity in their names, dark matter and dark energy are thought to exist for completely different reasons.

Scientists are also far more confident in the existence of dark matter than in the existence of dark energy. Scientific confidence in dark matter has been building for decades and is now at the point where dark matter seems almost indispensable to explaining the current structure of the universe. That is why we will devote most of this chapter to a discussion of dark matter and its presumed role as the dominant source of gravity in our universe. In contrast, the idea of dark energy has only recently gained observational support, and the data remain quite limited. Moreover, while we have at least some reasonable hypotheses about the possible nature of dark matter, we are virtually clueless about the potential nature of dark energy. Thus, we cannot say much about dark energy other than that it would explain some recent observations of the universal expansion, and for that reason we will defer further discussion of it to the final section of this chapter.

Before we continue, it's important to think about dark matter and dark energy in the context of science. Upon first hearing of these ideas, you might be tempted to think that astronomers have "gone medieval," arguing about unseen influences in the same way that people in medieval times supposedly argued about the number of angels that could dance on the head of a pin. However, strange as the ideas of dark matter and dark energy may seem, they have emerged from careful scientific study conducted in accordance with the hallmarks of science that we discussed in Chapter 3 (see Figure 3.23). Dark matter and dark energy were each proposed to exist because they seemed the simplest way to explain observed motions in the universe. They've each gained credibility because models of the universe that assume their existence make testable predictions and, at least so far, the predictions have been borne out by further observations. Thus, even if we someday conclude that we were wrong to infer the existence of dark matter and/or dark energy, we will still need alternative explanations for the observations made to date. One way or the other, our view of the universe will be forever changed by what we learn as we explore the mysteries of these unseen influences.

Detecting Dark Matter in a Spiral Galaxy Tutorial, Lessons 1–3

16.2 EVIDENCE FOR DARK MATTER

We are now ready to begin investigating dark matter in greater detail. In this section, we'll examine the evidence for the existence of dark matter and what the evidence indicates about the nature of dark matter.

• What is the evidence for dark matter in galaxies?

Let's begin our discussion of the evidence for dark matter by examining the case for dark matter in our own Milky Way Galaxy. We'll then proceed to other galaxies and clusters of galaxies. Along the way, we'll see why dark matter appears to be the dominant form of matter in our universe.

Distribution of Mass in the Milky Way In Chapter 14 we saw how the Sun's motion around the galaxy reveals the total amount of mass within its orbit. Similarly, we can use the orbital motion of any other star to measure the mass of the Milky Way within that star's orbit. In principle, we could determine the complete distribution of mass in the Milky Way by doing the same thing with the orbits of stars at every different distance from the galactic center.

In practice, interstellar dust obscures our view of disk stars more than a few thousand light-years away from us, making it very difficult to measure stellar velocities. However, radio waves penetrate this dust, and clouds of atomic hydrogen gas emit a spectral line at the radio wavelength of 21 cm [Section 14.2]. Measuring the Doppler shift of this 21-cm line tells us a cloud's velocity toward or away from us. With the help of a little geometry, we can then determine the cloud's orbital velocity.

A rotation curve plots the orbital velocities of objects in a galaxy against their distances from the center of the galaxy.

A diagram called a **rotation curve**, which plots the *rotational velocity* of stars or gas clouds against their distance from the center of the galaxy, summarizes the results of these orbital velocity measurements. As a simple example of the concept, let's construct a rotation curve for a merry-go-round. Every object on a merry-go-round goes around the center with the same rotational period, but objects farther from the center move in larger circles. Thus, objects

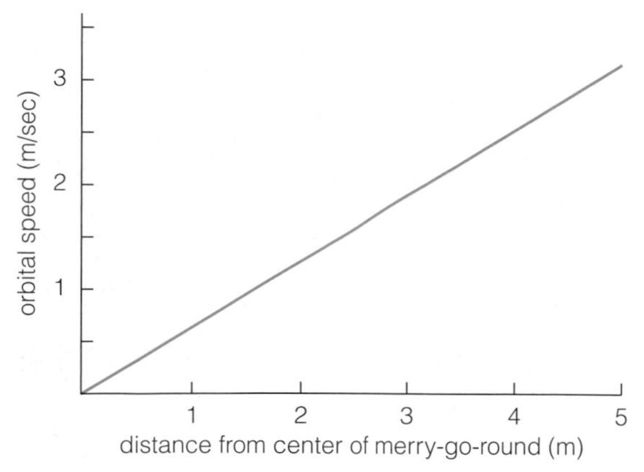

a A rotation curve for a merry-go-round.

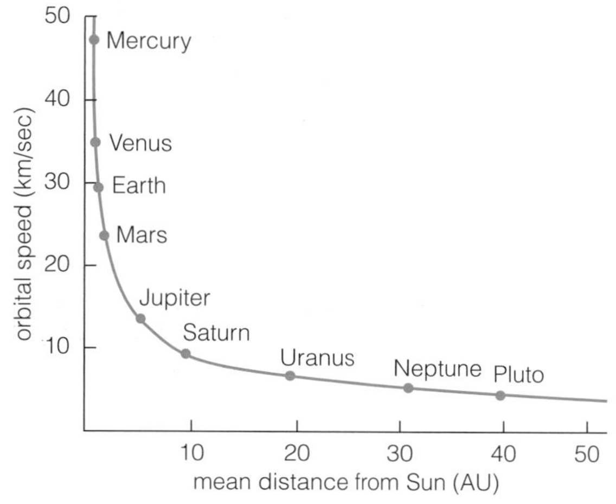

b The rotation curve for the planets in our solar system.

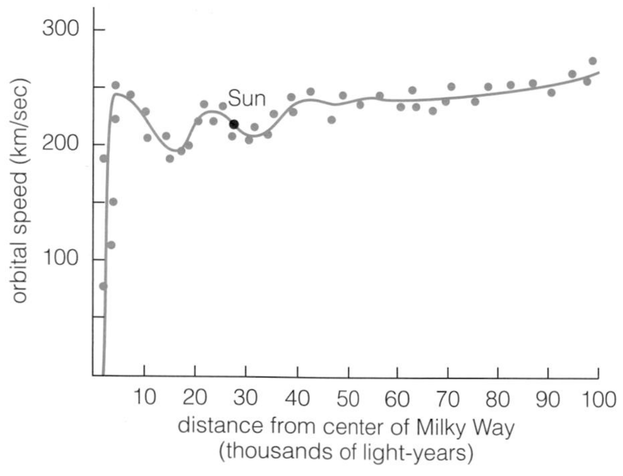

c The rotation curve for the Milky Way Galaxy. Dots represent stars or gas clouds whose rotational speeds have been measured.

Figure 16.1 Interactive Figure

Rotation curves show how the orbital speed of a system depends on distance from its center. The solar system's rotation curve declines with radius because its mass is concentrated at the center. The Milky Way's rotation curve is flat, indicating that the Milky Way's mass extends well beyond the Sun's orbit.

farther from the center move at faster speeds, and the rotation curve for a merry-go-round is a straight line that rises steadily outward (Figure 16.1a).

In contrast, the rotation curve for our solar system drops off with distance from the Sun, because inner planets orbit at faster speeds than outer planets (Figure 16.1b). This drop-off in speed with distance occurs because virtually all the mass of the solar system is concentrated in the Sun. The gravitational force holding a planet in its orbit decreases with distance from the Sun, and a smaller force means a lower orbital speed. The rotation curve of any astronomical system whose mass is concentrated toward the center must drop similarly.

Figure 16.1c shows the rotation curve for the Milky Way Galaxy. Each individual dot represents the distance from the galactic center and the orbital speed of a particular star or gas cloud. The curve running through the dots represents a "best fit" to the data. Notice that the orbital velocities remain approximately constant beyond the inner few thousand light-years, making the rotation curve look flat. This behavior contrasts sharply with the steeply declining rotation curve of the solar system. Thus, unlike the case for the solar system, most of the mass of the Milky Way must *not* be concentrated at its center. Instead, the orbits of progressively more distant gas clouds must encircle more and more mass. The Sun's orbit encompasses about 100 billion solar masses, but a circle twice as large surrounds twice as much mass, and a larger circle surrounds even more mass. Because of the difficulty of finding clouds to measure on the outskirts of the galaxy, we have not yet found the "edge" of this mass distribution.

The flatness of the Milky Way's rotation curve indicates that a large amount of dark matter lies beyond our galaxy's visible regions.

The flatness of the Milky Way's rotation curve therefore implies that most of our galaxy's mass lies well beyond our Sun, tens of thousands of light-years from the galactic center. A more detailed analysis suggests that most of this mass is located in the spherical halo that surrounds the disk of our galaxy, and that the total amount of this mass might be *10 times* the total mass of all the stars in the disk. Because we have detected very little radiation coming from this enormous amount of mass, it qualifies as dark matter. Thus, if we are interpreting the evidence correctly, the luminous part of the Milky Way's disk must be rather like the tip of an iceberg, marking only the center of a much larger clump of mass (Figure 16.2).

Dark Matter in Other Spiral Galaxies Other galaxies also seem to contain vast quantities of dark matter. We can determine the amount of dark matter in a galaxy by comparing the galaxy's mass to its luminosity. (More formally, astronomers calculate the galaxy's *mass-to-light ratio.*) The procedure is fairly simple in principle. First, we use the galaxy's luminosity to estimate the amount of mass that the galaxy contains in the form of stars. Next, we determine the galaxy's total mass by applying the law of gravity to observations of the orbital velocities of stars and gas clouds. If this total mass is larger than the mass that we can attribute to stars, then we infer that the excess mass must be dark matter.

Measuring a galaxy's luminosity is relatively easy, as long as we can determine its distance with one of the techniques discussed in Chapter 15 (see Figure 15.16). We simply point a telescope at the galaxy in question, measure its apparent brightness, and calculate its luminosity from its distance and the inverse square law of light [Section 11.1]. Measuring the galaxy's total mass requires measuring orbital speeds of stars or gas clouds as far from the galaxy's center as possible. Atomic hydrogen gas clouds can be found in spiral galaxies at greater distances from the center than stars, so most of our data come from radio observations of the 21-cm line from these clouds. We use Doppler shifts of the 21-cm line to determine how fast a cloud is moving toward us or away from us (Figure 16.3). We then combine observations for clouds at varying orbital distances to construct the galaxy's rotation curve.

The flat rotation curves of spiral galaxies other than the Milky Way indicate that they, too, harbor lots of dark matter.

The rotation curves of most spiral galaxies turn out to be remarkably flat as far out as we can see (Figure 16.4). Just as in the Milky Way, these flat rotation curves imply that a great deal of matter lies far out in the haloes of these other spiral galaxies. A detailed analysis tells us that these other spiral galaxies also have 10 times as much mass in dark matter (or more) as they do in stars. In other words, typical spiral galaxies are made 90% or more of dark matter, and 10% or less of matter in stars.

Figure 16.2

The dark matter associated with a spiral galaxy like the Milky Way occupies a much larger volume than the galaxy's luminous matter. The radius of this dark-matter halo may be 10 times as large as the galaxy's halo of stars.

Dark Matter in Elliptical Galaxies

We must use a different technique to weigh elliptical galaxies, because most of them contain very little atomic hydrogen gas and hence do not produce detectable 21-cm radiation. We generally weigh the inner parts of elliptical galaxies by observing the motions of the stars themselves.

The motions of stars in an elliptical galaxy are disorganized, so we cannot assemble their velocities into a sensible rotation curve. Nevertheless, the velocity of each individual star still depends on the mass inside the star's orbit. At any particular distance from an elliptical galaxy's center, some stars are moving toward us and some are moving away from us. As a result, every star has a slightly different Doppler shift. When we look at spectral lines from the galaxy as a whole, we see the combined effect of all these Doppler shifts. Together, they change the spectral line from a nice sharp line at a particular wavelength to a *broadened* line spanning a range of wavelengths (Figure 16.5). The greater the broadening of the spectral line, the faster the stars must be moving.

The widths of the absorption lines in an elliptical galaxy's spectrum tell us how much matter it contains, and much of that matter is dark.

When we compare spectral lines from different regions of an elliptical galaxy, we find that the speeds of the stars remain fairly constant as we look farther from the galaxy's center. Thus, just as in spirals, most of the matter in elliptical galaxies must lie beyond the distance where the light trails off and hence must be dark matter. The evidence for dark matter becomes even more

Figure 16.3

Measuring the rotation of a spiral galaxy with the 21-cm line of atomic hydrogen. Blueshifted lines on the left side of the disk show how fast that side is rotating toward us. Redshifted lines on the right side show how fast that side is rotating away from us.

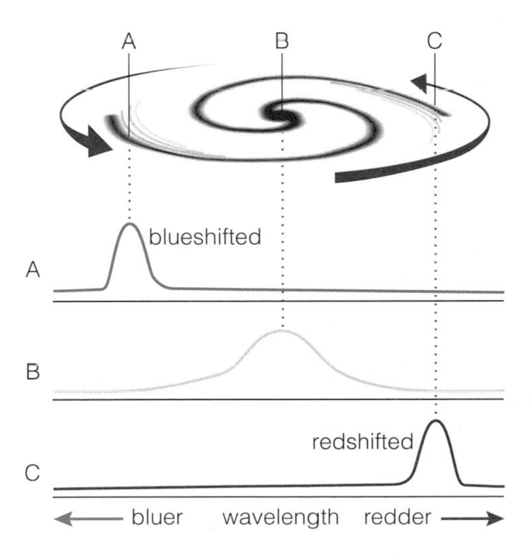

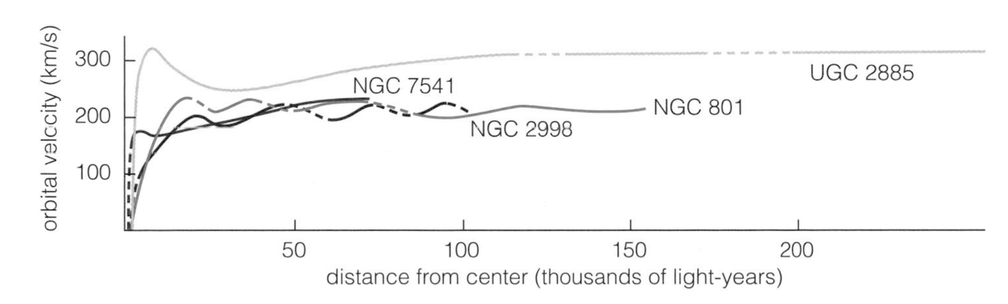

Figure 16.4

Actual rotation curves of four spiral galaxies. They are all nearly flat over a wide range of distances from the center, indicating that dark matter is common in spiral galaxies.

convincing in cases in which we can measure the speeds of globular star clusters orbiting at large distances from the center of an elliptical galaxy.

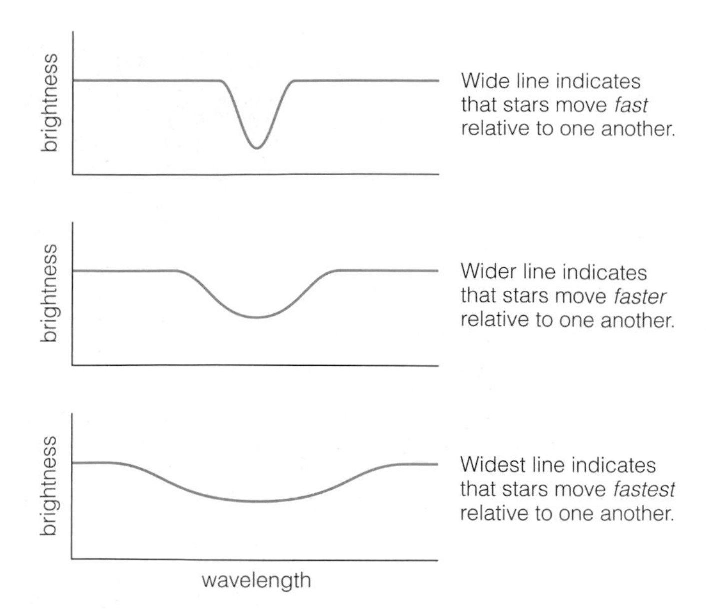

Figure 16.5

The broadening of absorption lines in an elliptical galaxy's spectrum tells us how fast its stars move relative to one another.

Figure 16.6

Fritz Zwicky, discoverer of dark matter in clusters of galaxies. Zwicky had an eccentric personality, but some of his ideas that seemed strange in the 1930s proved correct many decades later.

• What is the evidence for dark matter in clusters of galaxies?

The evidence we have discussed so far suggests that stars make up only about 10% of a galaxy's mass—the remaining mass consists of dark matter. Observations of galaxy clusters suggest that the total proportion of dark matter is even greater. The mass of dark matter in clusters appears to be as much as 50 times the mass in stars.

The evidence for dark matter in clusters comes from three different ways of measuring cluster masses: measuring the speeds of galaxies orbiting the center of the cluster, studying the X-ray emission from hot gas between the cluster galaxies, and observing how the clusters bend light as *gravitational lenses*. Let's investigate each of these techniques more closely.

Orbits of Galaxies in Clusters The problem of dark matter in astronomy is not particularly new. In the 1930s, astronomer Fritz Zwicky was already arguing that clusters of galaxies held enormous amounts of this mysterious stuff (Figure 16.6). Few of his colleagues paid attention, but later observations supported Zwicky's claims.

Zwicky was one of the first astronomers to think of galaxy clusters as huge swarms of galaxies bound together by gravity. It seemed natural to him that galaxies clumped closely in space should all be orbiting one another, just like the stars in a star cluster. He therefore assumed that he could measure cluster masses by observing galaxy motions and applying Newton's laws of motion and gravitation.

Armed with a spectrograph, Zwicky measured the redshifts of the galaxies in a particular cluster and used these redshifts to calculate the velocities at which the individual galaxies are moving away from us. He determined the velocity of the cluster as a whole by averaging the velocities of its individual galaxies. He then estimated the speed of a galaxy around the cluster—that is, its orbital speed—by subtracting this average velocity from the individual galaxy's velocity. Finally, he used these orbital speeds to estimate the cluster's mass and compared this mass to the luminosity of the cluster.

The orbits of galaxies in clusters tell us that galaxy clusters contain huge amounts of dark matter.

To his surprise, Zwicky found that clusters of galaxies have much greater masses than their luminosities would suggest. That is, when he estimated the total mass of stars necessary to account for the overall luminosity of a cluster, he found that it was far less than the mass he measured by studying galaxy speeds. He concluded that most of the matter within these clusters must not be in the form of stars and instead must be almost entirely dark. Many astronomers disregarded Zwicky's result, believing that he must have done something wrong to arrive at such a strange answer. Today, far more sophisticated measurements of galaxy orbits in clusters confirm Zwicky's original finding.

Hot Gas in Clusters A second method for measuring the mass of a cluster of galaxies relies on X-ray observations of the hot gas that fills the space between the galaxies in the cluster (Figure 16.7). This gas (sometimes called the *intracluster medium*) is so hot that it emits primarily X rays and therefore went undetected until the 1960s, when X-ray telescopes were finally launched above Earth's atmosphere. The temperature of this gas is tens of millions of degrees in many clusters and can exceed 100 million degrees in the largest clusters. This hot gas can also contain a great deal of mass. Some large clusters have up to seven times as much mass in the form of X-ray emitting gas than they do in the form of stars.

The hot gas can tell us about dark matter because its temperature depends on the total mass of the cluster. The gas in most clusters is nearly in a state of *gravitational equilibrium*—that is, the outward gas pressure balances gravity's inward pull [Section 10.1]. In this state of balance, the average kinetic energies of the gas particles are determined primarily by the strength of gravity and hence by the amount of mass within the cluster. Because the temperature of a gas reflects the average kinetic energies of its particles, the gas temperatures we measure with X-ray telescopes tell us the average speeds of the X-ray-emitting particles. We can then use these particle speeds to determine the cluster's total mass.

Temperature measurements of hot gas can also tell us the amount of dark matter in clusters, and give results that agree with those we infer from galaxy velocities.

The results obtained with this method agree well with the results found by studying the orbital motions of the cluster's galaxies. Even after we account for the mass of the hot gas, we find that the amount of dark matter in clusters of galaxies is up to 50 times that of the combined mass of the stars in the cluster's galaxies. In other words, the gravity of dark matter seems to be binding the galaxies of a cluster together in much the same way that it helps bind individual galaxies together.

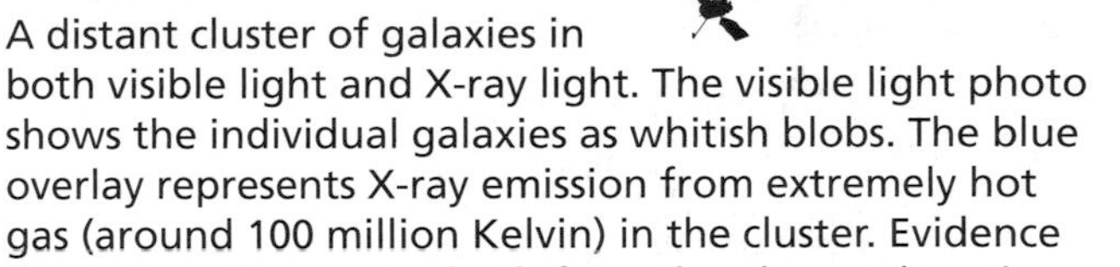

Figure 16.7

A distant cluster of galaxies in both visible light and X-ray light. The visible light photo shows the individual galaxies as whitish blobs. The blue overlay represents X-ray emission from extremely hot gas (around 100 million Kelvin) in the cluster. Evidence for dark matter comes both from the observed motions of the visible galaxies and from the temperature of the hot gas. (The photo shows a region about 8 million light-years across.)

Gravitational Lensing Until very recently, astronomers relied exclusively on methods based on Newton's laws to measure galaxy and cluster masses. These laws keep telling us that the universe holds far more matter than we can see. Can we trust these laws? One way to check is to measure masses in a different way. Today, astronomers have another tool for measuring masses: *gravitational lensing*.

SPECIAL TOPIC: PIONEERS OF SCIENCE

Scientists always take a risk when they publish what they think are ground-breaking results. If their results turn out to be in error, their reputations may suffer. In the case of dark matter, the pioneers in its discovery risked their entire careers. A case in point is Fritz Zwicky, with his proclamations in the 1930s about dark matter in clusters of galaxies. Most of his colleagues considered him an eccentric who leapt to premature conclusions.

Another pioneer in the discovery of dark matter was Vera Rubin, an astronomer at the Carnegie Institution. Working in the 1960s, she became the first woman to observe under her own name at California's Palomar Observatory, then the largest telescope in the world. (Another woman, Margaret Burbidge, was permitted to observe at Palomar earlier but was required to apply for time under the name of her husband, also an astronomer.) Rubin first saw the gravitational signature of dark matter in spectra she recorded of stars in the Andromeda Galaxy. She noticed that stars in the outskirts of Andromeda moved at surprisingly high speeds, suggesting a stronger gravitational attraction than could be explained by the mass of the galaxy's stars alone. In other words, she found that the rotation curve for Andromeda is relatively flat to great distances from the center, just as we now know is also the case for the Milky Way.

Working with a colleague, Kent Ford, Rubin constructed rotation curves for the hydrogen gas in many other spiral galaxies (by studying Doppler shifts in the spectra of hydrogen gas) and discovered that flat rotation curves are common. Although Rubin and Ford did not immediately recognize the significance of the results, they were soon arguing that the universe must contain substantial quantities of dark matter.

For a while, many other astronomers had trouble believing the results. Some astronomers suspected that the bright galaxies studied by Rubin and Ford were unusual for some reason. So Rubin and Ford went back to work, obtaining rotation curves for fainter galaxies. They found flat rotation curves—a signature of dark matter—even in these galaxies. By the 1980s, the evidence compiled by Rubin, Ford, and other astronomers measuring rotation curves was so overwhelming that even the critics came around. Either the theory of gravity was wrong, or they had discovered dark matter in spiral galaxies.

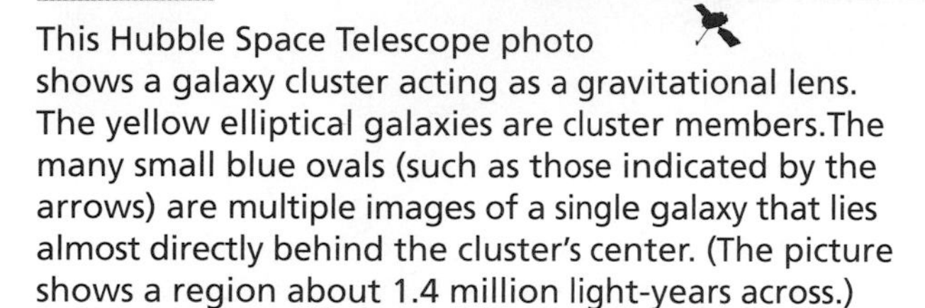

Figure 16.8

This Hubble Space Telescope photo shows a galaxy cluster acting as a gravitational lens. The yellow elliptical galaxies are cluster members.The many small blue ovals (such as those indicated by the arrows) are multiple images of a single galaxy that lies almost directly behind the cluster's center. (The picture shows a region about 1.4 million light-years across.)

Gravitational lensing occurs because masses distort the space around them [Section 13.3]. Massive objects can therefore act as **gravitational lenses** that bend light beams passing nearby. This prediction of Einstein's general theory of relativity was first verified in 1919 during an eclipse of the Sun. Because the light-bending angle of a gravitational lens depends on the mass of the object doing the bending, we can measure the masses of objects by observing how strongly they distort light paths.

Gravity's light-bending effects distort the images of galaxies lying behind a cluster, enabling us to measure the cluster's mass without relying on Newton's laws.

Figure 16.8 shows a striking example of how a cluster of galaxies can act as a gravitational lens. Many of the yellow elliptical galaxies concentrated toward the center of the picture belong to the cluster, but at least one of the galaxies pictured does not. At several positions on various sides of the central clump of yellow galaxies you will notice multiple images of the same blue galaxy. Each one of these images, whose sizes differ, looks like a distorted oval with an off-center smudge.

The blue galaxy seen in these multiple images lies almost directly behind the center of the cluster, at a much greater distance. We see multiple images of this single galaxy because photons do not follow straight paths as they travel from the galaxy to Earth. Instead, the cluster's gravity bends the photon paths, allowing light from the galaxy to arrive at Earth from a few slightly different directions (Figure 16.9). Each alternative path produces a separate, distorted image of the blue galaxy.

Multiple images of a gravitationally lensed galaxy are rare. They occur only when a distant galaxy lies directly behind the lensing cluster. However, single, distorted images of gravitationally lensed galaxies are quite common. Figure 16.10 shows a typical example. This picture shows numerous normal-looking galaxies and several arc-shaped galaxies. The oddly curved galaxies are not members of the cluster, nor are they really curved. They are normal galaxies lying far beyond the cluster whose images have been distorted by the cluster's gravity.

Cluster masses measured through gravitational lensing agree with those measured from galaxy velocities and gas temperatures.

Careful analyses of the distorted images created by clusters enable us to measure cluster masses without resorting to Newton's laws. Instead, Einstein's theory of general relativity tells us how massive these clusters must be to generate the observed distortions. It is reassuring that cluster masses derived in this way generally agree with those derived from galaxy velocities and X-ray temperatures. The three different methods all indicate that clusters of galaxies hold very substantial amounts of dark matter.

• Does dark matter really exist?

Astronomers have made a strong case for the existence of dark matter, but could they be completely off base? Is it possible that dark matter is a figment of human imagination, and that there's a completely different explanation for the observations we've discussed?

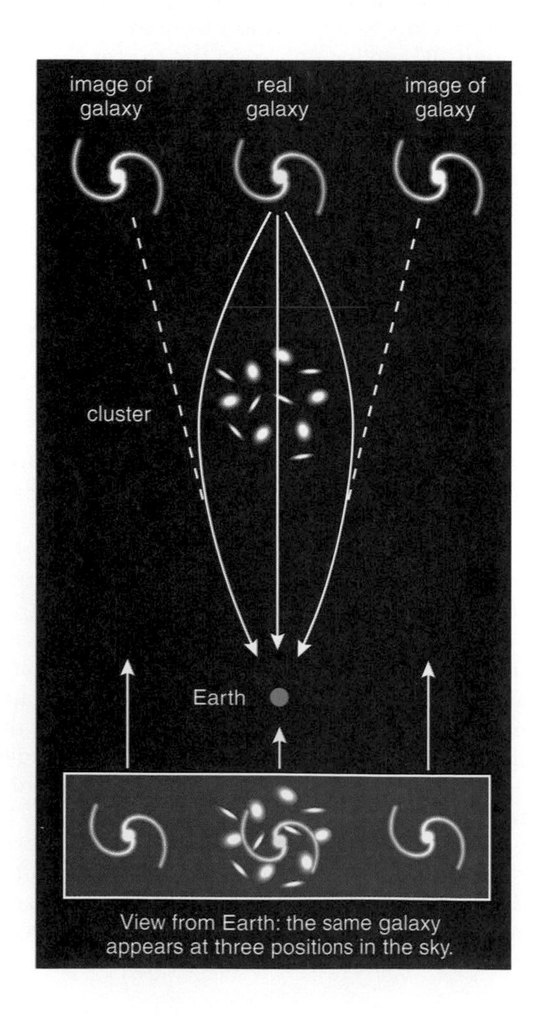

Figure 16.9 Interactive Figure

A cluster's powerful gravity bends light paths from background galaxies to Earth. If light can arrive from several different directions, we see multiple images of the same galaxy.

Figure 16.10

Hubble Space Telescope photo of the cluster Abell 2218. The thin, elongated galaxies around the main clump of galaxies on the left side of the picture are the images of background galaxies distorted by the cluster's gravity. By measuring these distortions, astronomers can determine the total amount of mass in the cluster. (The region pictured is about 1.4 million light-years from side to side.)

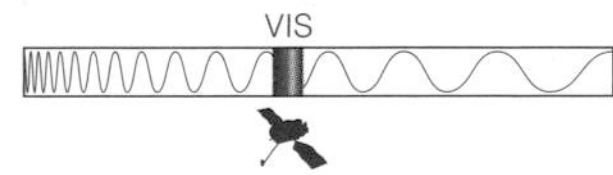

All the evidence for dark matter rests on our understanding of gravity. For individual galaxies, the case for dark matter rests primarily on applying Newton's laws of motion and gravity to observations of the orbital speeds of stars and gas clouds. We've used the same laws to make the case for dark matter in clusters, along with additional evidence based on gravitational lensing predicted by Einstein's general theory of relativity. Thus, it seems that one of the following must be true:

1. Dark matter really exists, and we are observing the effects of its gravitational attraction.
2. There is something wrong with our understanding of gravity, which is causing us to mistakenly infer the existence of dark matter.

We cannot yet rule out the second possibility, but most astronomers consider it very unlikely. Newton's laws of motion and gravity are among the most trustworthy tools in science. We have used them time and again to measure masses of celestial objects from their orbital properties. We found the masses of Earth and the Sun by applying Newton's version of Kepler's third law to objects that orbit them [Section 5.3]. We used this same law to calculate the masses of stars in binary star systems, revealing the general relationships between the masses of stars and their outward appearances. Newton's laws have also told us the masses of things we can't see directly, such as the masses of orbiting neutron stars in X-ray binaries and of black holes in active galactic nuclei. Einstein's general theory of relativity likewise stands on solid ground, having been repeatedly tested and verified to high precision in many observations and experiments. Thus, we have good reason to trust our current understanding of gravity.

Moreover, many scientists have already made valiant efforts to come up with alternate theories of gravity that could account for the observations without assuming the existence of dark matter. (After all, there's a Nobel Prize waiting for anyone who can substantiate a new theory of gravity.) So far, no one has succeeded in doing so in a way that can still explain the many other observations accounted for by our current theories of gravity.

Either dark matter exists or our current understanding of gravity is incorrect.

In essence, our high level of confidence in our current understanding of gravity gives us equally high confidence that dark matter really exists. Thus, while we should always keep an open mind about the possibility of future changes in our understanding, we will proceed for now under the assumption that dark matter is real. With that assumption in mind, let's turn our attention to the nature of what seems to be the most common form of matter in the universe.

Should the fact that we have three different ways of measuring cluster masses give us greater confidence that we really do understand gravity and that dark matter really does exist? Why or why not?

• What might dark matter be made of?

What is all this dark stuff in galaxies and clusters of galaxies? The answer is that we don't yet know. Nevertheless, we can make at least some educated guesses.

At least some of the dark matter is likely to be *ordinary*, made of protons, neutrons, and electrons. The only unusual thing about this dark matter is that it doesn't emit much detectable radiation, because it is made of the same stuff as all the "bright matter" that we can see. However, as we'll discuss shortly, it's also likely that some of the dark matter is *extraordinary*, made of particles that we have yet to discover.

A bit of terminology will be useful. Because the protons and neutrons that make up most of the mass of ordinary matter belong to a category of particles called **baryons**, ordinary matter is sometimes called **baryonic matter**. By extension, extraordinary matter is called **nonbaryonic matter**.

Ordinary Dark Matter Matter need not be extraordinary to be dark. In astronomy, "dark" merely means not as bright as a normal star and therefore not visible across vast distances of space. Your body is dark matter, because you would be far too dim for our telescopes to detect if you were somehow flung into the halo of our galaxy. Everything you own is dark matter. Earth and the rest of the planets are dark matter as well. In fact, the "failed stars" known as *brown dwarfs* [Section 12.1] and even faint red main-sequence stars of spectral type M [Section 11.2] are too dim for current telescopes to see in the halo and therefore qualify as dark matter. Thus, it's conceivable that trillions of faint red stars, brown dwarfs, and Jupiter-size objects left over from the Milky Way's formation still roam our galaxy's halo, providing much of its mass. These objects are sometimes called **MACHOs**, for *massive compact halo objects*, although they are better thought of as dim, starlike (or planetlike) objects.

MACHOs such as dim, red stars and brown dwarfs are too faint for us to see directly, but there are other ways to search for them. One innovative technique takes advantage of gravitational lensing on a much smaller scale than that we studied for clusters of galaxies. If trillions of these dim stars and similar objects really roam the halo of the galaxy, every once in a while one of them should drift across our line of sight to a more distant star. When the object lies almost directly between us and the farther star, its gravity will focus the star's light directly toward the Earth. The distant star will appear much brighter than usual for several days or weeks as the

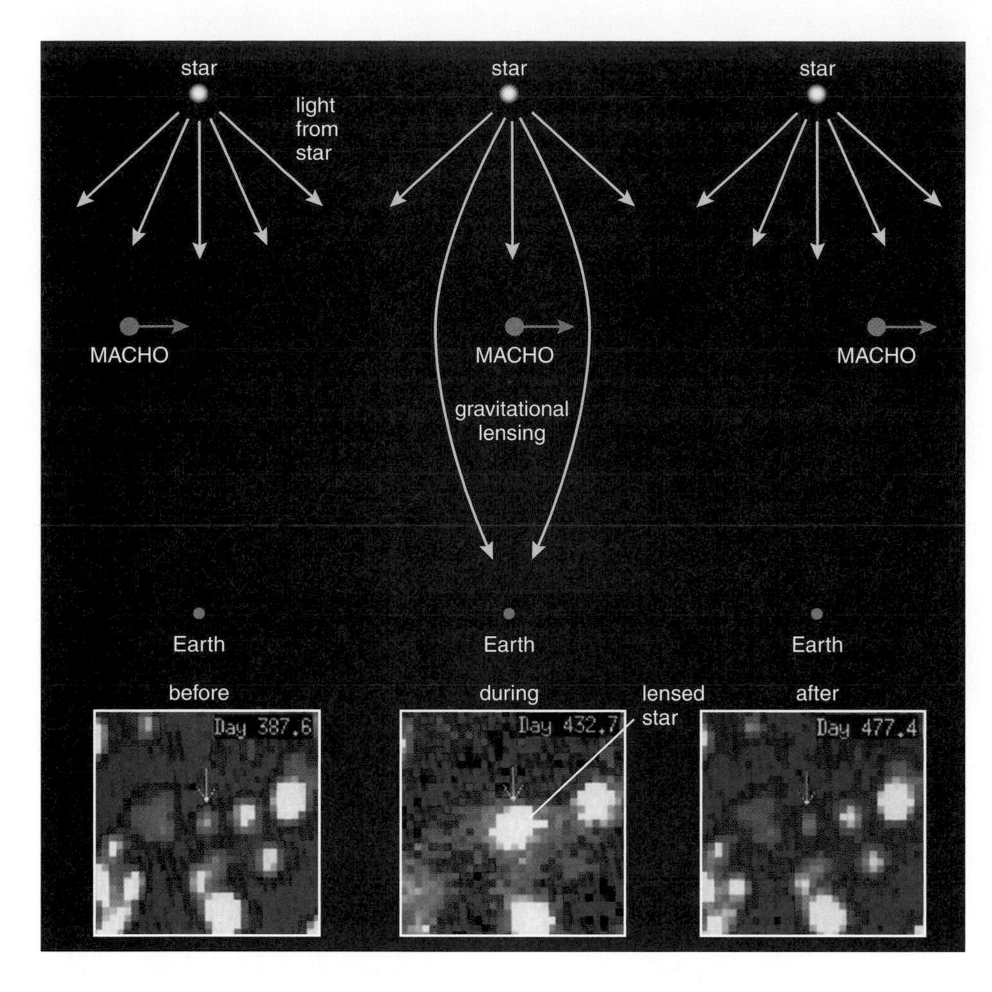

Figure 16.11 Interactive Figure

When a small, starlike object (MACHO) passes in front of a more distant star, gravitational lensing temporarily makes the star appear brighter. Searches for such events show that our galaxy's halo does indeed contain dim, starlike objects, but that these objects do not constitute the majority of the galaxy's dark matter.

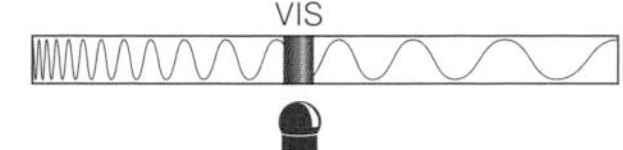

lensing object passes in front of it (Figure 16.11). We cannot see the object itself, but the duration of the lensing event reveals its mass.

Gravitational lensing by starlike objects in our galaxy's halo does not turn up enough objects to account for all the dark matter.

Gravitational lensing events such as these are rare, happening to about one star in a million each year. To detect such lensing events, we therefore must monitor huge numbers of stars. Current large-scale monitoring projects now record numerous lensing events annually. These events demonstrate that dim starlike objects (MACHOs) do indeed populate our galaxy's halo, but not in large enough numbers to account for all the Milky Way's dark matter. Similar measurements rule out black holes with masses like our Sun's. Something else lurks unseen in the outer reaches of our galaxy.

Extraordinary Dark Matter A more exotic possibility is that most of the dark matter in galaxies and clusters of galaxies is not made of ordinary, baryonic matter at all. Let's begin to explore this possibility by taking another look at those nonbaryonic particles we discussed in Section 10.2: neutrinos. These unusual particles are dark by their very nature, because they have no electrical charge and hence cannot emit electromagnetic radiation of any kind. Moreover, they are never bound together with charged particles in the way that neutrons are bound in atomic nuclei, so their presence cannot be revealed by associated light-emitting particles.

Particles like neutrinos interact with other forms of matter only through the force of gravity and a force called the *weak force*, which governs some nuclear reactions. For this reason, they are said to be *weakly interacting particles*. If you recall that trillions of neutrinos from the Sun are passing through your body at this very moment without doing any damage, you'll see why the name *weakly interacting* fits well.

The dark matter in galaxies cannot be made of neutrinos, because these very low mass particles travel through the universe at enormous speeds and can easily escape a galaxy's gravitational pull. (However, neutrinos make up a small amount of the dark matter outside galaxies.) What if other weakly interacting particles exist that are similar to neutrinos but considerably heavier? They too would evade direct detection, but they would move more slowly so that a large collection of them could be held together by their mutual gravity. Such hypothetical particles are called **WIMPs,** for *weakly interacting massive particles*. Note that WIMPs are subatomic particles, so the "massive" in their name means only that they have higher masses than some other subatomic particles. (They are also often called *cold dark matter* to set them apart from the faster-moving neutrinos.) WIMPs could make up most of the mass of a galaxy or cluster or galaxies, but they would be completely invisible in all wavelengths of light. Most astronomers now consider it likely that WIMPs make up the vast majority of dark matter, and hence the majority of all matter in the universe.

Scientists suspect that dark matter is mostly made up of weakly interacting particles that are like neutrinos but more massive.

It might surprise you that scientists suspect the universe to be filled with particles they haven't yet discovered. However, this hypothesis would also explain why dark matter seems to be distributed throughout spiral galaxy halos rather than concentrated in flattened disks like the visible matter. Recall that galaxies are thought to have formed as gravity pulled together matter in regions of slightly enhanced density in the early universe [Section 15.3]. This matter would have consisted mostly of dark matter mixed with some ordinary (hydrogen and helium) gas. The ordinary gas could collapse to form a rotating disk because individual gas particles could lose orbital energy: Collisions among many gas particles can cause some of their orbital energy to be radiated away in the form of light. In contrast, WIMPs cannot radiate energy and they rarely interact and exchange energy with other particles. Thus, as the gas collapsed to form a disk, WIMPs would have remained stuck on orbits far out in the galactic halo—just where most dark matter seems to be located.

By itself, the agreement between the measured distribution of dark matter in galaxies and what we'd expect from dark matter made of WIMPs doesn't prove that extraordinary dark matter exists. However, as we'll discuss in the next chapter, there are additional reasons why many astronomers believe that baryons represent only a minority of the universe's mass and hence that WIMPs are the most common form of matter in the universe.

THINK ABOUT IT What do you think of the idea that much of the universe is made of as-yet-undiscovered particles? Can you think of other instances in the history of science in which the existence of something was predicted before it was discovered?

16.3 STRUCTURE FORMATION

Dark matter remains enigmatic, but every year we are learning more about its role in the universe. Because galaxies and clusters of galaxies seem to contain much more dark matter than luminous matter, dark matter's gravitational pull must be the primary force holding these structures together. Thus, we strongly suspect that the gravitational attraction of dark matter is what pulled galaxies and clusters together in the first place.

• What is the role of dark matter in galaxy formation?

Stars, galaxies, and clusters of galaxies are all *gravitationally bound systems*—their gravity is strong enough to hold them together. In most of the gravitationally bound systems we have discussed so far, gravity has completely overwhelmed the expansion of the universe. That is, while the universe as a whole is expanding, space is *not* expanding within our solar system, our galaxy, or our Local Group of galaxies.

Our best guess at how galaxies formed, outlined in Section 15.3, envisions them growing from slight density enhancements that were present in the very early universe. During the first few million years after the Big Bang, the universe expanded everywhere. Gradually, the stronger gravity in regions of enhanced density pulled in matter until these regions stopped expanding and became protogalactic clouds, even as the universe as a whole continued (and still continues) to expand.

The gravity of dark matter was probably the main force that caused protogalactic clouds to become galaxies and galaxies to group into clusters of galaxies.

If dark matter is indeed the most common form of mass in galaxies, it must have provided most of the gravitational attraction responsible for creating the protogalactic clouds. The hydrogen and helium gas in the protogalactic clouds collapsed inward and gave birth to stars, while weakly interacting dark matter remained in the outskirts because of its inability to radiate away its orbital energy. According to this model, the luminous matter in each galaxy must still be nestled inside the larger cocoon of dark matter that initiated the galaxy's formation (see Figure 16.2), just as observational evidence seems to suggest.

The formation of galaxy clusters probably echoes the formation of galaxies. Early on, all the galaxies that will eventually constitute a cluster are flying apart with the expansion of the universe, but the gravity of the dark matter associated with the cluster eventually reverses the trajectories of these galaxies. The galaxies ultimately fall back inward and start orbiting each other randomly, like the stars in the halo of our galaxy.

Some clusters of galaxies apparently have not yet finished forming, because their immense gravity is still drawing in new members. For example, the large, nearby Virgo Cluster of galaxies appears to be drawing in the Milky Way and other galaxies of the Local Group. The evidence comes from careful study of galaxy speeds. Plugging the Virgo Cluster's distance into Hubble's law tells us the speed at which the Milky Way and the Virgo Cluster should be drifting apart due to the universal expansion [Section 15.2]. However, the measured speed is about 400 km/s slower than the speed we predict from Hubble's law alone. We conclude that this 400 km/s discrepancy (sometimes called a *peculiar velocity*) arises because the cluster's gravity is pulling us back against the flow of universal expansion. In other words, while the Milky Way and other galaxies of our

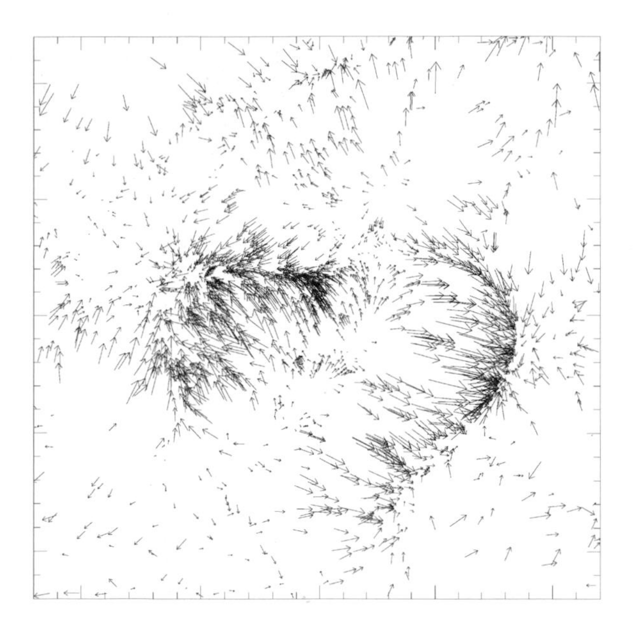

Figure 16.12

This graph represents the motions of galaxies attributable to effects of gravity. Each arrow represents the amount by which a galaxy's actual velocity (inferred from a combination of observations and modeling) differs from the velocity we'd expect it to have from Hubble's law alone. The Milky Way is at the center of the picture, which shows an area about 600 million light-years from side to side. (Only a representative sample of galaxies is shown.) Note how the galaxies tend to flow into regions where the density of galaxies is already high. These vast, high-density regions are probably superclusters in the process of formation.

Local Group are still moving away from the Virgo Cluster with the expansion of the universe, the rate at which we are separating from the cluster is slowing with time. Eventually, the cluster's gravity may stop the separation altogether, at which point the cluster will begin pulling in the galaxies of our Local Group, ultimately making them members of the cluster.

Similar processes are taking place on the outskirts of other large clusters of galaxies, where we see many galaxies whose speeds indicate that the cluster's gravity is pulling on them. Eventually, some or perhaps all of these galaxies will fall into the cluster. Thus, many clusters are still attracting galaxies, adding to the hundreds they already contain. On even larger scales, clusters themselves seem to be tugging on one another, hinting that they are parts of even bigger gravitationally bound systems, called **superclusters,** that are just beginning to form (Figure 16.12). But some structures are even larger than superclusters.

• What are the largest structures in the universe?

Beyond about 300 million light-years from Earth, deviations from Hubble's law owing to gravitational tugs are insignificant compared with the universal expansion, so Hubble's law becomes our primary method for measuring galaxy distances [Section 15.2]. Using this law, astronomers can make maps of the distribution of galaxies in space. Such maps reveal **large-scale structures** much vaster than clusters of galaxies.

Mapping Large-Scale Structures Making maps of galaxy locations requires an enormous amount of data. A long-exposure photo showing galaxy positions is not enough, because it does not tell us the galaxy distances. Instead, we must measure the redshift of each individual galaxy so that we can estimate its distance by applying Hubble's law. These measurements used to require intensive labor, and up until about a decade ago it took years of effort to map the locations of just a few hundred galaxies. However, astronomers have recently developed technology that allows redshift measurements for hundreds of galaxies during a single night of telescopic observation. As a result, we now have redshift measurements—and hence estimated distances—for many thousands of distant galaxies. We are rapidly acquiring more data, allowing astronomers to create ever-better and more detailed three-dimensional maps of the universe.

Galaxies appear to be arranged in immense structures hundreds of millions of light years across.

Figure 16.13 shows the distribution of galaxies in three slices of the universe, each extending farther out in distance. Our Milky Way Galaxy is located at the vertex at the far left, and each dot represents another entire galaxy of stars. The slice at the left comes from one of the first surveys of large-scale structures performed in the 1980s. This map, which required years of effort by many astronomers, dramatically revealed the complex structure of our corner of the universe. It showed that galaxies are not scattered randomly through space but instead are arranged in huge chains and sheets that span many millions of light-years. Clusters of galaxies are located at the intersections of these chains. Between these chains and sheets of galaxies lie giant empty regions called **voids**. The other two slices show data from the more recent Sloan Digital Sky Survey. Scheduled for completion in 2005, the Sloan survey is measuring redshifts for nearly a million galaxies spread across about one-fourth of the sky.

Some of the structures in these pictures are amazingly large. The so-called Sloan Great Wall, clearly visible in the center slice, extends more

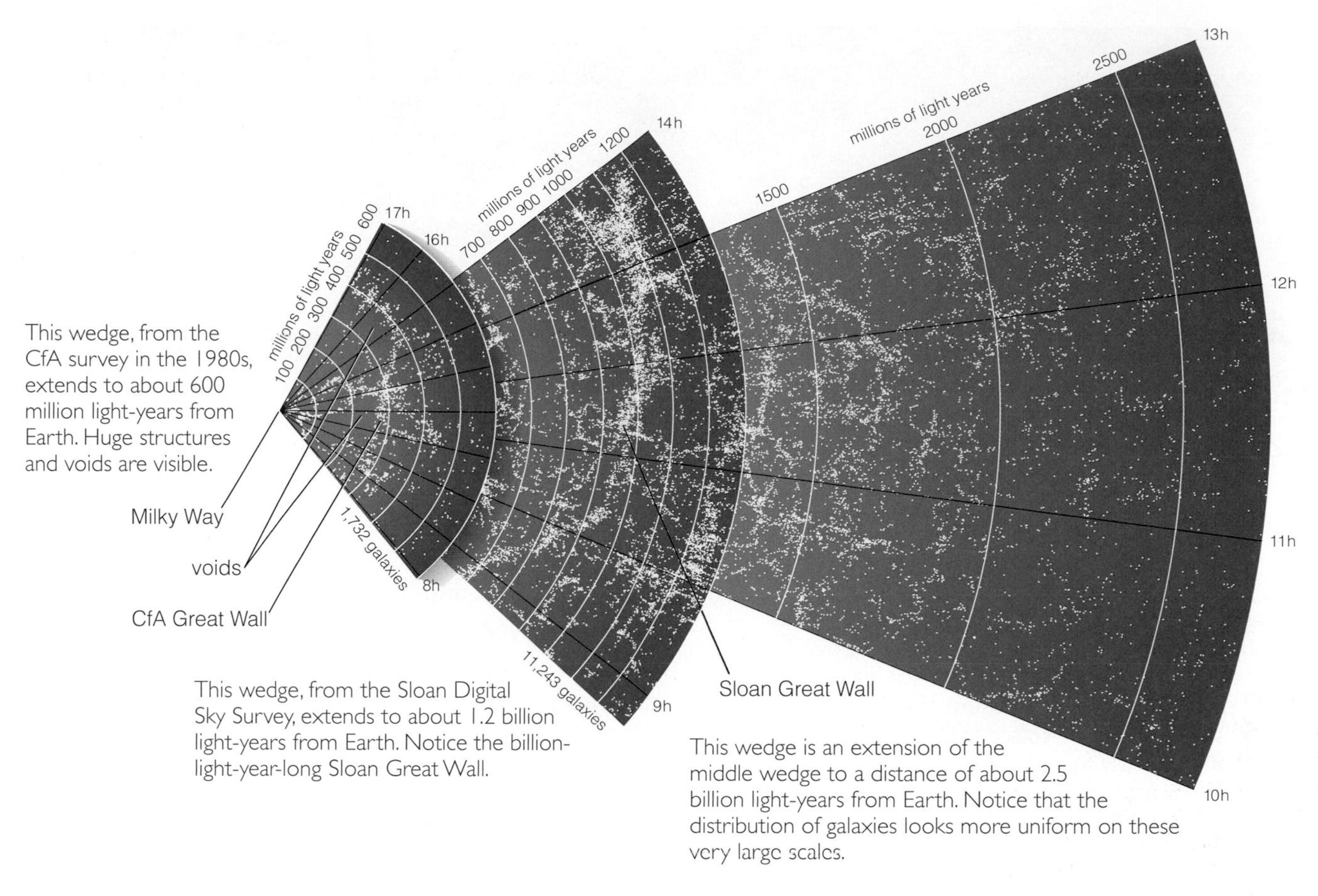

Figure 16.13

Each of these three wedges shows a "slice" of the universe extending outward from our own Milky Way Galaxy. The dots represent galaxies, shown at their measured distances from Earth. We see that galaxies are not scattered randomly but instead trace out long chains and sheets surrounded by huge voids containing very few galaxies. (The wedges are shown flat but actually are a few angular degrees in thickness; the CFA wedge at left does not actually line up with the two Sloan wedges.)

than one billion light-years from end to end. Immense structures such as these apparently have not yet collapsed into randomly orbiting, gravitationally bound systems.

The universe may still be growing structures on these very large scales. However, there seems to be a limit to the size of the largest structures. If you look closely at the right-most slice in Figure 16.13, you'll notice that the overall distribution of galaxies appears nearly uniform on scales larger than about a billion light-years. In other words, on very large scales the universe looks much the same everywhere, in agreement with what we expect from the *Cosmological Principle* [Section 15.2].

The Origin of Large Structures Why is gravity collecting matter on such enormous scales? Just as we suspect that galaxies formed from regions of slightly enhanced density in the early universe, we suspect that these larger structures were also regions of enhanced density. Galaxies, clusters, superclusters, and the Sloan Great Wall probably all started as mildly high-density regions of different sizes. The voids in the distribution of galaxies probably started as mildly low-density regions.

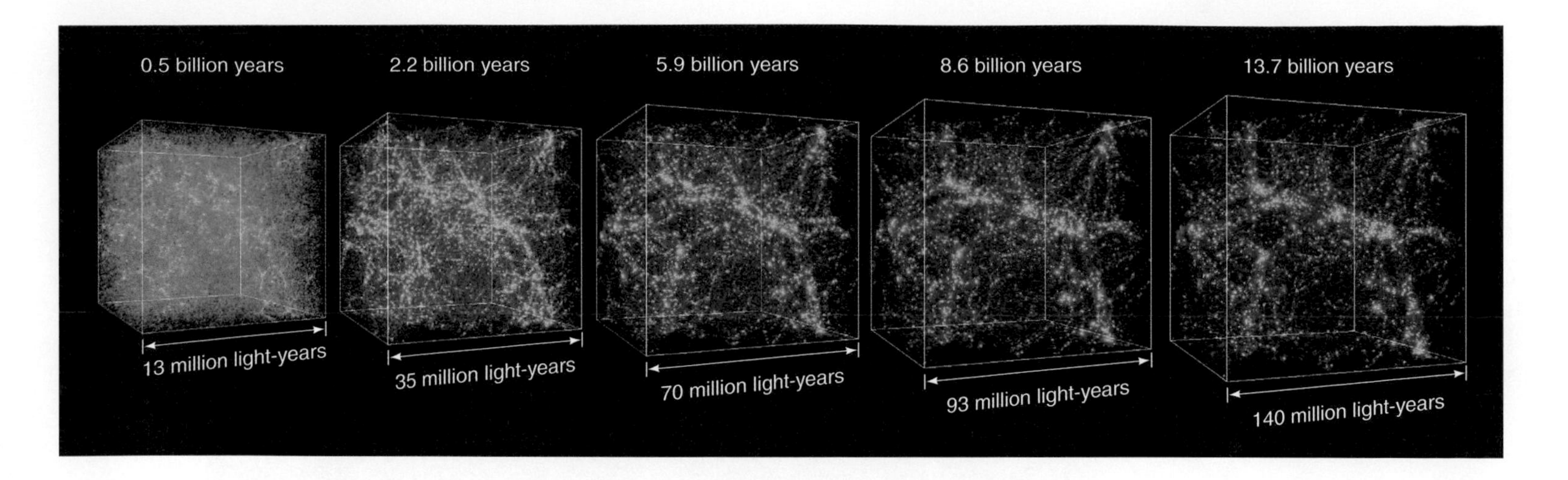

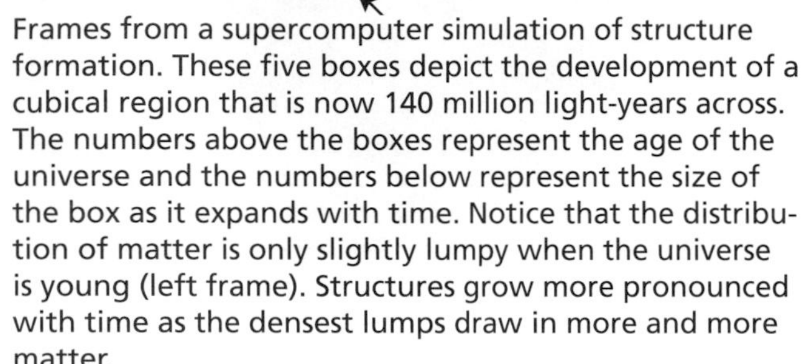

Figure 16.14 Interactive Figure

Frames from a supercomputer simulation of structure formation. These five boxes depict the development of a cubical region that is now 140 million light-years across. The numbers above the boxes represent the age of the universe and the numbers below represent the size of the box as it expands with time. Notice that the distribution of matter is only slightly lumpy when the universe is young (left frame). Structures grow more pronounced with time as the densest lumps draw in more and more matter.

The structure we see in today's universe probably mirrors the distribution of dark matter when the universe was very young.

If this picture of structure formation is correct, then the structures we see in today's universe mirror the original distribution of dark matter very early in time. Supercomputer models of structure formation in the universe can now simulate the growth of galaxies, clusters, and larger structures from tiny density enhancements as the universe evolves (Figure 16.14). The results of these models look remarkably similar to the slices of the universe in Figure 16.13, bolstering our confidence in this scenario. However, the models do not tell us *why* the universe started with these slight density enhancements—that is a topic for the next chapter. Nevertheless, it seems increasingly clear that these "lumps" in the early universe were the seeds of all the marvelous structures we see in the universe today.

Fate of the Universe Tutorial, Lessons 1–3

16.4 THE UNIVERSE'S FATE

Some say the world will end in fire,
Some say in ice.
From what I've tasted of desire
I hold with those who favor fire.
But if it had to perish twice,
I think I know enough of hate
To say that for destruction ice
Is also great
And would suffice.

Robert Frost

We now arrive at one of the ultimate questions in astronomy: How will the universe end? Edwin Hubble's work established that the galaxies in the universe are rapidly flying away from one another [Section 15.2], but the gravitational pull of each galaxy on every other galaxy acts to slow the expansion. The possible outcomes would appear to fall into two general categories. If gravity is strong enough, the expansion will someday halt and the universe will begin collapsing, eventually ending in a cataclysmic crunch. Alternatively, if the expansion can overcome the pull of gravity, the universe will continue to expand forever, growing ever colder as its galaxies grow ever farther apart. The fate of the universe thus seems to boil down to a simple question: Is the universe expanding fast enough to escape its own gravitational pull and keep on expanding forever?

• Will the universe continue expanding forever?

Let's begin by considering the fate of the universe as it seemed just a few years ago, before the discoveries that suggested the presence of dark energy in the universe. In the absence of dark energy, we would expect gravity to be slowing the expansion of the universe with time. In that case, the fate of the universe hinges on the overall strength of the universe's gravitational pull. The strength of this pull depends on the density of matter in the universe: The greater the density, the greater the overall strength of gravity and the higher the likelihood that gravity will someday halt the expansion.

Precise calculations show that gravity can win out over expansion if the current density of the universe exceeds a seemingly minuscule 10^{-29} gram per cubic centimeter, which is roughly equivalent to a few hydrogen atoms in a volume the size of a closet. The precise density marking the dividing line between eternal expansion and eventual collapse is called the **critical density**. (Remember that for the moment we are assuming a universe without dark energy.)

Observations of the luminous matter in galaxies show that the mass contained in stars falls far short of the critical density. The visible parts of galaxies contribute about 0.5% of the matter density needed to halt the universe's expansion. The fate of the universe would thus seem to rest with the dark matter. Is there enough dark matter to halt the expansion of the universe?

Because stars seem to contribute about 0.5% of the matter density needed to halt the expansion, the expansion could halt only if the total mass of dark matter were at least 200 times that of the mass in stars. Our studies of galaxies suggest that they contain at least 10 times as much dark matter as matter in stars, and studies of clusters of galaxies raise that number further to about 50 times as much dark matter as matter in stars. However, this is still only about a quarter of the amount of dark matter needed to halt the expansion. Thus, if the proportion of dark matter in the universe at large is similar to that in clusters, the universe seems destined to expand forever. For gravity to reverse the expansion and pull the universe back together, even more dark matter would have to lie beyond the boundaries of clusters.

The universe does not appear to contain enough dark matter to prevent it from expanding forever.

If large-scale structures contain a higher proportion of dark matter than do clusters, the influence of that extra dark matter should show up in the velocities of galaxies near those large-scale structures: Larger amounts of dark matter should cause greater deviations from Hubble's law. As of 2004, however, most studies of galaxy velocities hold the line near the value we infer from clusters, which is about 25% of the critical density required to reverse the expansion. If that is the case, the universe seems destined to expand forever, even if there is no dark energy affecting the rate of expansion.

• Is the expansion of the universe accelerating?

In the past few years, observations of distant white dwarf supernovae have enabled us to probe the fate of the universe in an entirely new way.

Because white dwarf supernovae are such good standard candles [Section 15.2], we can use them to determine whether gravity has been slowing the universe's expansion, as it must if the universe is destined to end in a cataclysmic crunch. However, the astronomers who set out to measure gravity's influence over the universe by observing supernovae discovered something quite unexpected.

Instead of slowing because of gravity, the expansion of the universe appears to be speeding up, suggesting that some mysterious repulsive force—the so-called *dark energy*—is pushing all the universe's galaxies apart. This discovery, if it holds up to further scrutiny, has far-reaching implications for both the fate of the universe and our understanding of the forces that govern its behavior on large scales. To understand the evidence for an accelerating expansion, we must become more familiar with the possible futures of the universe.

Four Expansion Patterns Astronomers subdivide the two general possibilities for the fate of the universe, expanding forever or someday collapsing, into four broad categories. Each represents a particular pattern of change in the future expansion rate (Figure 16.15). We will call these four possible expansion patterns *recollapsing, critical, coasting,* and *accelerating.* The first three possibilities assume that gravity is the only force that affects the expansion rate of the universe, while the fourth adds a repulsive force (dark energy) that opposes gravity:

The expansion rate of the universe depends on the balance between gravity, which acts to slow the expansion, and dark energy, which acts to accelerate it.

- If there is no dark energy and the matter density of the universe is *larger* than the critical density, the collective gravity of all its matter will eventually halt the universe's expansion and reverse it. The gal-

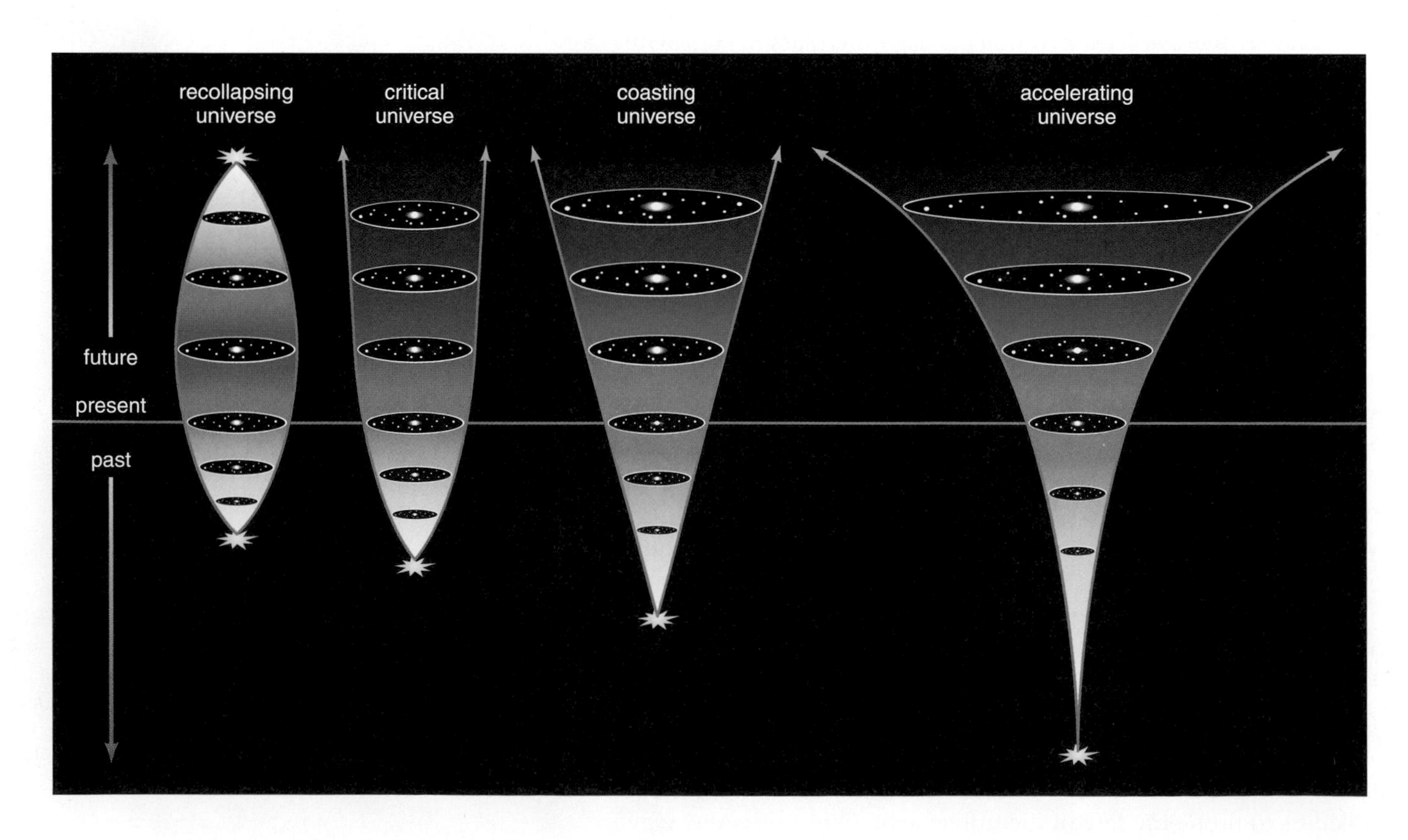

Figure 16.15

Four models for the fate of the universe. Each diagram shows how the size of a circular slice of the universe changes with time in a particular model. The slices are the same size at the present time, marked by the red line, but the models make different predictions about the sizes of the slices in the past and future. The first three cases assume that there is no dark energy, so that the fate of the universe depends only on how its actual density compares to the critical density. The last case assumes that a repulsive force—the so-called dark energy—is accelerating the expansion with time. (The diagram assumes continuous acceleration, but it is also possible that the universe initially slowed before the acceleration began.)

axies will come crashing back together, and the entire universe will end in a fiery "Big Crunch." If this is the fate of our universe, then we live in a **recollapsing universe**. (A recollapsing universe is sometimes called a *closed universe*, because mathematical calculations show that it must have an overall geometry that closes upon itself like the surface of a sphere, but in more dimensions [Section 17.4].)

- If there is no dark energy and the matter density of the universe *equals* the critical density, the collective gravity of all its matter is exactly the amount needed to balance the expansion. The universe will never collapse but will expand more and more slowly as time progresses. If this is the fate of our universe, then we live in a **critical universe**. (Mathematically speaking, a critical universe stops expanding after infinite time, and its overall geometry is "flat"—like the surface of a table but in more dimensions. Thus, a critical universe is one example of what astronomers call a *flat universe*.)
- If there is no dark energy and the matter density of the universe is *smaller* than the critical density, the collective gravity of all its matter cannot halt the expansion. The universe will keep expanding forever, with little change in its rate of expansion. If this is the fate of our universe, then we live in a **coasting universe**. (A coasting universe is sometimes called an *open universe*, because its overall geometry is more like the open surface of a saddle than like the closed surface of a sphere.)
- If dark energy exerts a repulsive force that causes the expansion of the universe to accelerate with time, then we live in an **accelerating universe**. Its galaxies will recede from one another increasingly faster, and it will become cold and dark more quickly than a coasting universe. (Depending on the strength of gravity relative to the repulsive force, the overall geometry of an accelerating universe could be flat, open, or closed. As we'll discuss in Chapter 17, current evidence suggests a flat geometry for this case.)

THINK ABOUT IT Do you think that one of the potential fates of the universe is preferable to the others? If so, why? If not, why not?

Evidence for Acceleration

Figure 16.16 illustrates how the average distance between galaxies should change with time for each possibility. The lines for the accelerating, coasting, and critical universes always continue upward as time increases, because in these cases the universe is always expanding. The steeper the slope, the faster the expansion. In the recollapsing case, the line begins on an upward slope but eventually turns around and declines as the universe contracts. All the lines pass through the same point and have the same slope at the moment labeled "now," because the current separation between galaxies and the current expansion rate in each case must agree with observations of the present-day universe.

Note that the age that we infer for the universe from its expansion rate differs in each case. A recollapsing universe requires the least amount of time to arrive at the current separation between galaxies—the example in Figure 16.16 goes from zero separation to the current separation in less than 5 billion years. The cases for which gravity is less important require more time to achieve the current separation between galaxies.

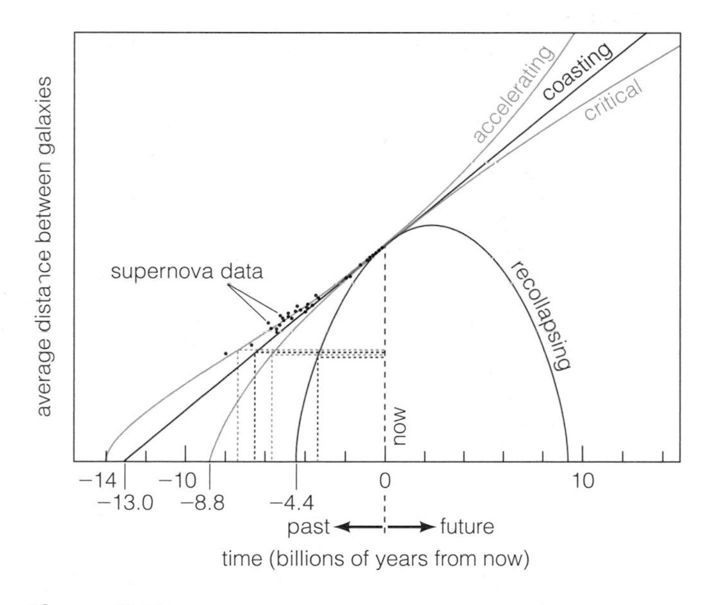

Figure 16.16

Data from white dwarf supernovae are shown along with four possible models for the expansion of the universe. Each curve shows how the average distance between galaxies changes with time for a particular model. A rising curve means that the universe is expanding, and a falling curve means that the universe is contracting. Notice that the supernovae data fit the accelerating universe better than the other models.

The ages we would infer from the examples in Figure 16.16 are 8.8 billion years for a critical universe, 13 billion years for a coasting universe, and around 14 billion years for an accelerating universe.

This relationship between the age of the universe and its expansion pattern enables us to determine the expansion pattern from observations of white dwarf supernovae. Because these supernovae are so bright and make such excellent standard candles, we can identify them and measure their distances even when they are more than halfway across the observable universe. Measuring the distances to supernovae that exploded at different times in the past therefore tells us how the rate of expansion of the universe has changed with time.

Distances measured to faraway white dwarf supernovae indicate that the expansion of the universe is now speeding up.

Observations of such distant supernovae are still very difficult, but we have some data that are plotted as dots in Figure 16.16. Although there is some scatter in these data, they appear to fit the curve for an accelerating universe better than any of the other models, and they do not fit either a critical or a recollapsing universe. In other words, the observations to date seem to favor an accelerating universe.

Exactly why the expansion of the universe might be accelerating remains a deep mystery. No known force would act to push the universe's galaxies apart, and an enormous amount of energy would be required. Of course, our lack of understanding does not stop us from giving a name to the repulsive force that is causing the acceleration, and we have already discussed why it is often dubbed *dark energy.* Keep in mind, however, that we do not yet have any idea of what the dark energy might actually be. Nevertheless, if dark energy really exists, it is the most prevalent form of energy in the universe, outstripping the total mass-energy of all the matter in the universe—including the dark matter. Only continued observations will tell us whether the dark energy is real or an artifact of our still-limited data.

A Never-Ending Expansion? Whether or not the dark energy exists, and whatever dark energy might turn out to be, it now seems likely that

SPECIAL TOPIC: WHAT DID EINSTEIN CONSIDER HIS GREATEST BLUNDER?

Shortly after Einstein completed his general theory of relativity in 1915, he found that it predicted that the universe could not be standing still: The mutual gravitational attraction of all the matter would make the universe collapse. Because Einstein thought at the time that the universe should be eternal and static, he decided to alter his equations. In essence, he inserted a "fudge factor" called the *cosmological constant* that acted as a repulsive force to counteract the attractive force of gravity.

Had he not been so convinced that the universe should be standing still, Einstein might instead have come up with the correct explanation for why the universe is not collapsing: because it is still expanding from the event of its birth. After Hubble discovered the universal expansion, Einstein called his invention of the cosmological constant "the greatest blunder" of his career.

Recently, astronomers have begun to take the idea of a cosmological constant more seriously. In the mid-1990s, a few observations suggested the ages of the oldest stars were slightly older than the age of the universe derived from Hubble's constant under the assumption that gravity is the only force affecting the universe's expansion. Clearly, stars cannot be older than the universe. If these observations were being interpreted correctly, the universe had to be older than Hubble's constant implies. If the expansion rate has accelerated, so that the universe is expanding faster today than it was in the past, then the age of the universe would be greater than that ordinarily found from Hubble's constant (see Figure 16.16). What could cause the expansion of the universe to accelerate over time? The repulsive force represented by a cosmological constant, of course.

Further study of the troubling observations has shown that the oldest stars probably are *not* older than the age of the universe derived from Hubble's constant. However, measurements of distances to high-redshift galaxies using white dwarf supernovae as standard candles now suggest that the expansion *is* accelerating. A cosmological constant arising from dark energy could account for this startling finding, but we'll need more observations before we can be sure it is correct. Einstein's greatest blunder, it seems, just won't go away.

the universe is indeed doomed to expand forever, its galaxies receding ever more quickly into an icy, empty future. After all, our examination of the strength of gravity showed it to be too weak to stop the expansion even without dark energy, and the acceleration due to dark energy would only seem to seal this fate. Some scientists even hypothesize that the dark energy could eventually cause galaxies, stars, and planets to break apart and disperse with the never-ending expansion.

Based on current data, the universe seems destined to expand forever.

However, before we convince ourselves that we now know the fate of the universe, we should bear in mind that forever is a very long time. The universe may hold other surprises that we haven't yet discovered, surprises that might force us to rethink what might happen between now and the end of time.

This is the way the world ends
This is the way the world ends
This is the way the world ends
Not with a bang but a whimper.

From* The Hollow Men *by T. S. Eliot

THE BIG PICTURE

Putting Chapter 16 into Context

We have found that there may be much more to the universe than meets the eye. Dark matter too dim for us to see seems to far outweigh the stars, and a mysterious dark energy may be even more prevalent. Here are some key "big picture" points to remember about this chapter:

- Dark matter and dark energy sound very similar, but they are each hypothesized to explain different observations. Dark matter is thought to exist because we detect its gravitational influence. Dark energy is a term given to whatever strange, repulsive force may be accelerating the expansion of the universe.
- Either dark matter exists, or we do not understand how gravity operates across galaxy-size distances. There are many reasons to be confident in our understanding of gravity, leading most astronomers to conclude that dark matter is real.
- Dark matter seems to be by far the most abundant form of mass in the universe. We still do not know what it is, but we suspect it is largely made up of some type of as-yet-undiscovered subatomic particles.
- If dark matter is indeed the dominant source of gravity in the universe, then it is the glue that binds together galaxies, clusters, superclusters, and other large-scale structures in the universe. All this structure has probably grown from regions where the density of dark matter in the early universe was slightly enhanced.
- The fate of the universe depends on whether gravity can ever halt the present expansion. The total strength of gravity seems too weak to do so even when we account for dark matter, and the evidence suggesting that the expansion is accelerating only reinforces the suggestion that expansion will never cease.

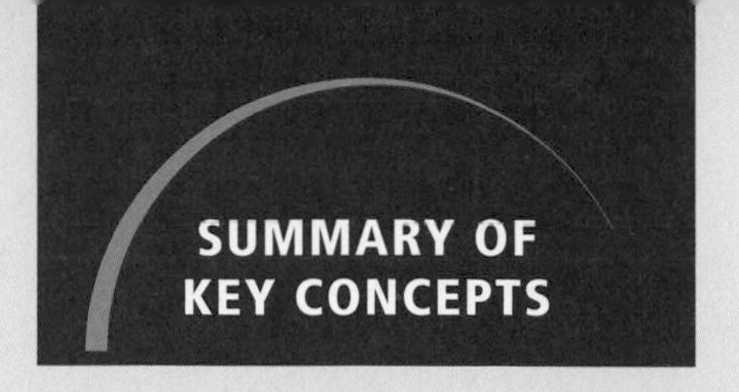

16.1 UNSEEN INFLUENCES IN THE COSMOS

- **What do we mean by dark matter and dark energy?**

Dark matter and **dark energy** have never been directly observed, but each has been proposed to exist because it seems the simplest way to explain a set of observed motions in the universe. Dark matter is the name given to the unseen mass whose gravity governs the observed motions of stars and gas clouds. Dark energy is the name given to whatever may be causing the expansion of the universe to accelerate.

16.2 EVIDENCE FOR DARK MATTER

- **What is the evidence for dark matter in galaxies?**

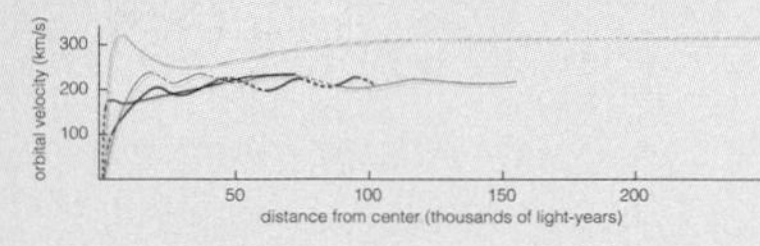

The orbital velocities of stars and gas clouds in galaxies do not change much with distance from the center of the galaxy. Applying Newton's laws of gravitation and motion to these orbits leads to the conclusion that the total mass of a galaxy is far larger than the mass of its stars. Because no detectable visible light is coming from this matter, we call it **dark matter**.

- **What is the evidence for dark matter in clusters of galaxies?**

We have three different ways of measuring the amount of dark matter in clusters of galaxies: from galaxy orbits, from the temperature of the hot gas in clusters, and from the **gravitational lensing** predicted by Einstein. All of these methods agree that the total mass of a cluster is about 50 times the mass of its stars, implying huge amounts of dark matter.

- **Does dark matter really exist?**

We infer that dark matter exists from its gravitational influence on the matter we can see, leaving two possibilities: Either dark matter exists, or there is something wrong with our understanding of gravity. We cannot rule out the latter possibility, but we have good reason to be confident of our current understanding of gravity and the idea that dark matter is real.

- **What might dark matter be made of?**

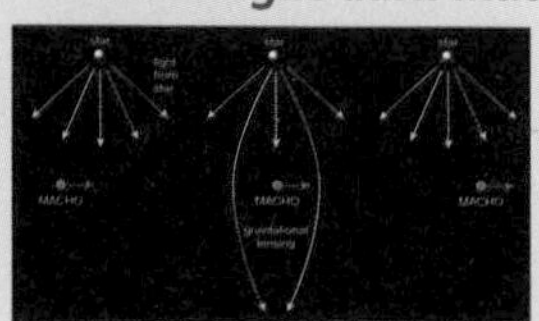

Some of the dark matter could be ordinary or **baryonic matter** in the form of dim stars or planetlike objects, but there does not appear to be enough ordinary matter to account for all the dark matter. Most of it is probably extraordinary or **nonbaryonic matter** consisting of undiscovered particles that we call **WIMPs**.

16.3 STRUCTURE FORMATION

- **What is the role of dark matter in galaxy formation?**

Because most of a galaxy's mass is in the form of dark matter, the gravity of that dark matter is probably what formed protogalactic clouds and then galaxies from slight density enhancements in the early universe.

- **What are the largest structures in the universe?**

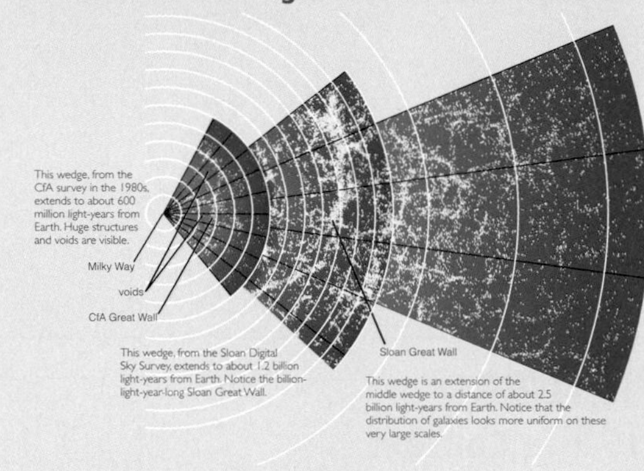

Galaxies appear to be distributed in gigantic chains and sheets that surround great voids. These giant structures trace their origin directly back to regions of slightly enhanced density early in time.

16.4 THE UNIVERSE'S FATE

- **Will the universe continue expanding forever?**

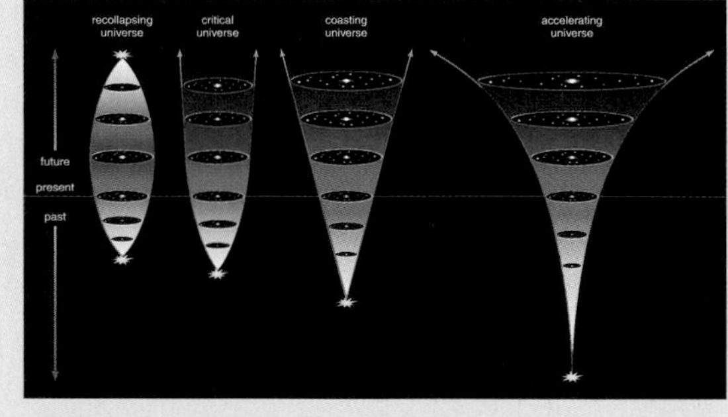

Even before we consider the possibility of a mysterious dark energy, the evidence points to eternal expansion. The **critical density** is the average matter density the universe would need for the strength of gravity to eventually halt the expansion. The overall matter density of the universe appears to be only about 25% of the critical density.

- **Is the expansion of the universe accelerating?**

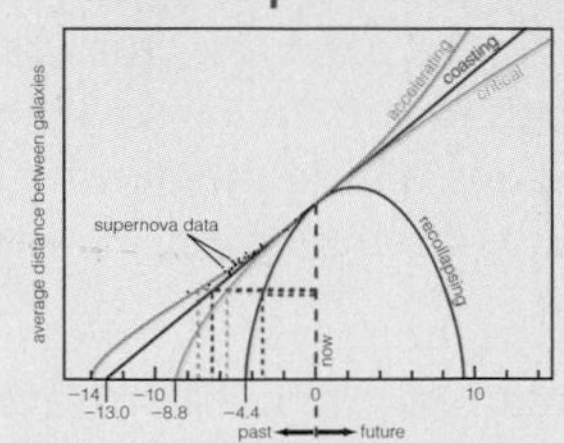

Observations of distant supernovae indicate that the expansion of the universe is speeding up. No one knows the nature of the mysterious force (dark energy) that could be causing this acceleration.

EXERCISES AND PROBLEMS

REVIEW QUESTIONS

1. Define *dark matter* and *dark energy*, and clearly distinguish between them. What types of observations have led scientists to propose the existence of each of these unseen influences?
2. What is a *rotation curve?* Describe the rotation curve of the Milky Way, and explain how it indicates the presence of large amounts of dark matter.
3. How do we construct rotation curves for other spiral galaxies? What do they tell us about the galaxy masses and dark matter?
4. How do we measure the masses of elliptical galaxies? What do these masses lead us to conclude about dark matter in elliptical galaxies?
5. Briefly describe the three different ways of measuring the mass of a cluster of galaxies. Do the results from the different methods agree? What do they tell us about dark matter in galaxy clusters?
6. What is *gravitational lensing?* Why does it occur? How can we use it to estimate the masses of lensing objects?
7. Briefly explain why the conclusion that dark matter exists rests on assuming that we understand gravity correctly. Is it possible that our understanding of gravity is not correct? Explain.
8. In what sense is dark matter "dark"? Briefly explain why objects like you, planets, and even dim stars could qualify as dark matter.
9. What do we mean by *MACHOs?* How can we search for them? Briefly describe why these searches suggest that starlike (or planetlike) objects and black holes *cannot* account for all the dark matter in the halo of our galaxy.
10. Explain what we mean when we say that a neutrino is a *weakly interacting particle.* Why can't the dark matter in galaxies be made of neutrinos?
11. What do we mean by *WIMPs?* Why does it seem likely that dark matter consists of these particles, even though we do not yet know what they are?
12. Briefly explain why dark matter is thought to have played a major role in the formation of galaxies and larger structures in the universe. What evidence suggests that large structures are still forming?
13. What does the large-scale structure of the universe look like? Explain why we think this structure reflects the density of the early universe.
14. What do we mean by the *critical density* of the universe? According to current evidence, how does the actual density of matter in the universe compare to the critical density?
15. Describe and compare four possible patterns to the expansion of the universe: recollapsing, critical, coasting, and accelerating. Explain why our estimate of the current age of the universe depends on which model is correct.
16. Observationally, how can we decide which of the four possible expansion models is the right one? Based on the current evidence, which model is favored?
17. Assuming the acceleration is real, what does it imply for the fate of the universe? What does current evidence suggest for the fate if the acceleration is not real? Explain.

TRUE STATEMENTS?

Decide whether each of the following statements is true and explain why it is or is not.

18. Strange as it may sound, most of both the mass and energy in the universe may take forms that we are unable to detect directly.
19. A cluster of galaxies is held together by the mutual gravitational attraction of all the stars in the cluster's galaxies.
20. We can estimate the total mass of a cluster of galaxies by studying the distorted images of galaxies whose light passes through the cluster.
21. Clusters of galaxies are the largest structures that we have so far detected in the universe.
22. The primary evidence for an accelerating universe comes from observations of young stars in the Milky Way.
23. There is no doubt remaining among astronomers that the fate of the universe is to expand forever.

PROBLEMS

24. *Dark Matter.* Overall, how convincing do you consider the case for the existence of dark matter? Write a short essay in which you describe what we mean by dark matter, describe the evidence for its existence, and discuss your opinion about the strength of the evidence.
25. *Dark Energy.* Overall, how convincing do you consider the case for the existence of dark energy? Write a short essay in which you describe what we mean by dark energy, describe the evidence for its existence, and discuss your opinion about the strength of the evidence.
26. *The Future Universe.* Based on current evidence concerning the growth of structure in the universe, briefly describe what you would expect the universe to look like on large scales about 10 billion years from now.

DISCUSSION QUESTIONS

27. *Dark Matter or Revised Gravity.* One possible explanation for the evidence we find for dark matter is that we are currently using the wrong law of gravity to measure the masses of very large objects. If we really do misunderstand gravity, then many fundamental theories of physics, including Einstein's theory of general relativity, will need to be revised. Which explanation for our observations do you find more appealing, dark matter or revised gravity? Explain why. Why do you suppose most astronomers find dark matter more appealing?
28. *Our Fate.* Scientists, philosophers, and poets alike have speculated on the fate of the universe. How would you prefer the universe as we know it to end, in a "Big Crunch" or through eternal expansion? Explain the reasons behind your preference.

MATH HELP AND EXERCISES

The Astronomy Place Web site has additional mathematical topics, with worked examples and additional practice problems. For this chapter, you'll find the following:

- The Orbital Velocity Law
- Mass–to–Light Ratio
- Masses of Clusters

For a complete list of media resources available, go to **www.astronomyplace.com** and choose Chapter 16 from the pull-down menu.

ASTRONOMY PLACE WEB TUTORIALS

Tutorial Review of Key Concepts

Use the following interactive **Tutorials** at **www.astronomyplace.com** to review key concepts from this chapter.

Detecting Dark Matter in a Spiral Galaxy Tutorial

Lesson 1 Introduction to Rotation Curves
Lesson 2 Determining the Mass Distribution
Lesson 3 Where Is the Dark Matter?

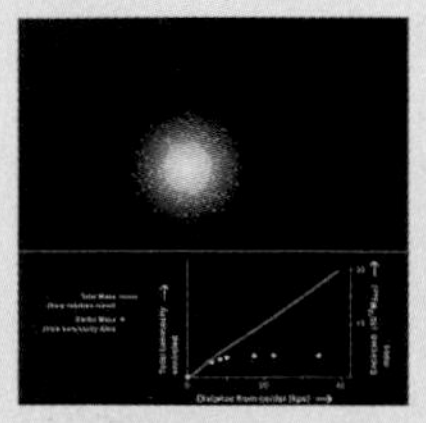
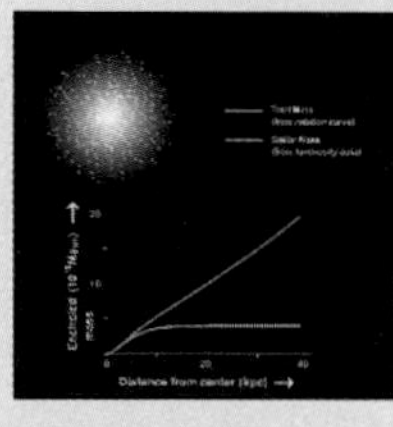
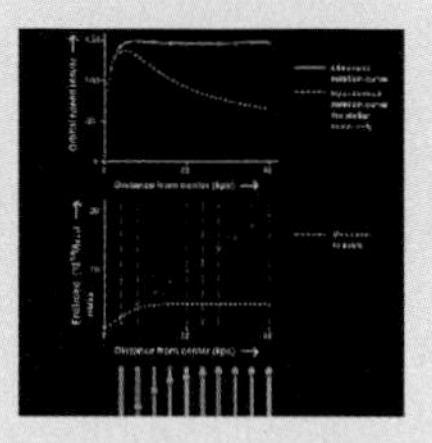

Fate of the Universe Tutorial

Lesson 1 The Role of Gravity
Lesson 2 The Role of Dark Energy
Lesson 3 Fate and History of the Universe

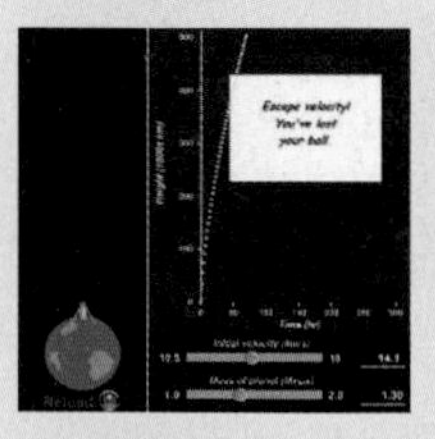
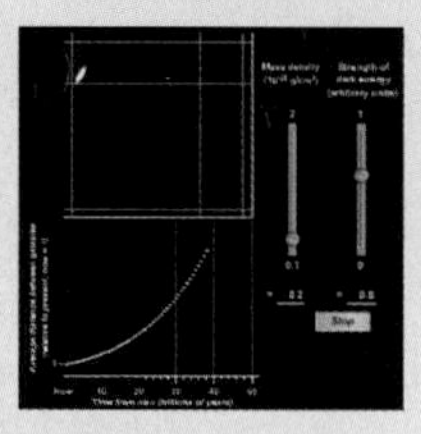
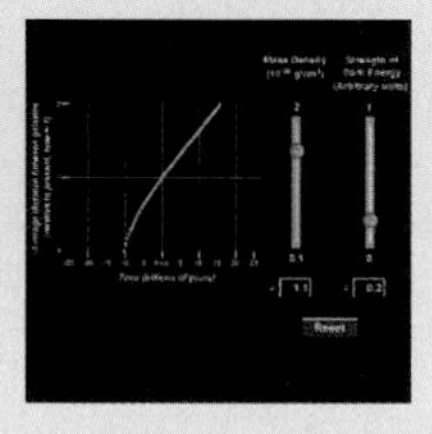

Supplementary Tutorial Exercises

Use the interactive **Tutorial Lessons** to explore the following questions.

Detecting Dark Matter in Spiral Galaxies Tutorial, Lesson 3

1. Explain how stellar rotation curves suggest the existence of dark matter.
2. How does the amount of dark matter change as you move away from the center of a galaxy?

Fate of the Universe Tutorial, Lesson 1

1. If gravity were the only factor affecting the fate of the universe, what would the possible fates be?
2. How would the mass density of the universe determine which fate would be in store for the universe?

Fate of the Universe Tutorial, Lesson 2

1. What does dark energy do to the expansion of the universe?
2. How would the existence of dark energy change the fate of the universe from what it would be if only gravity played a role in its fate?

Fate of the Universe Tutorial, Lesson 3

1. How would the existence of dark energy change estimates of the age of the universe?
2. What evidence suggests that dark energy really exists?

MOVIES

Check out the following narrated and animated short documentary available on **www.astronomyplace.com** for a helpful review of key ideas covered in this chapter.

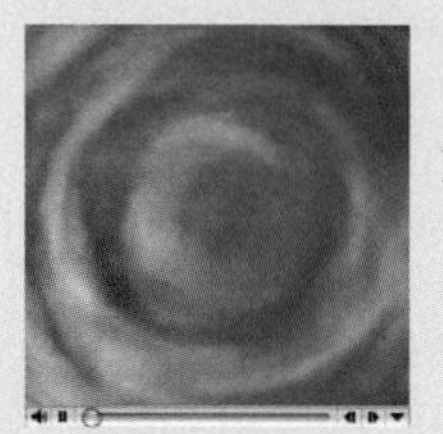

From the Big Bang to the Galaxies Movie

WEB PROJECTS

Take advantage of the useful Web links on **www.astronomyplace.com** to assist you with the following projects.

1. *Gravitational Lenses.* Gravitational lensing occurs in numerous astronomical situations. Compile a catalog of examples from the Web. Try to find pictures of lensed stars, quasars, and galaxies. Give a one-paragraph explanation of what is going on in each picture.
2. *Accelerating Universe.* Search for the most recent information about possible acceleration of the expansion of the universe. Write a one- to three-page report on your findings.
3. *The Nature of Dark Matter.* Find and study recent reports on new ideas about the possible nature of dark matter. Write a one- to three-page report that summarizes the latest ideas about what dark matter is made of.

17
The Beginning of Time

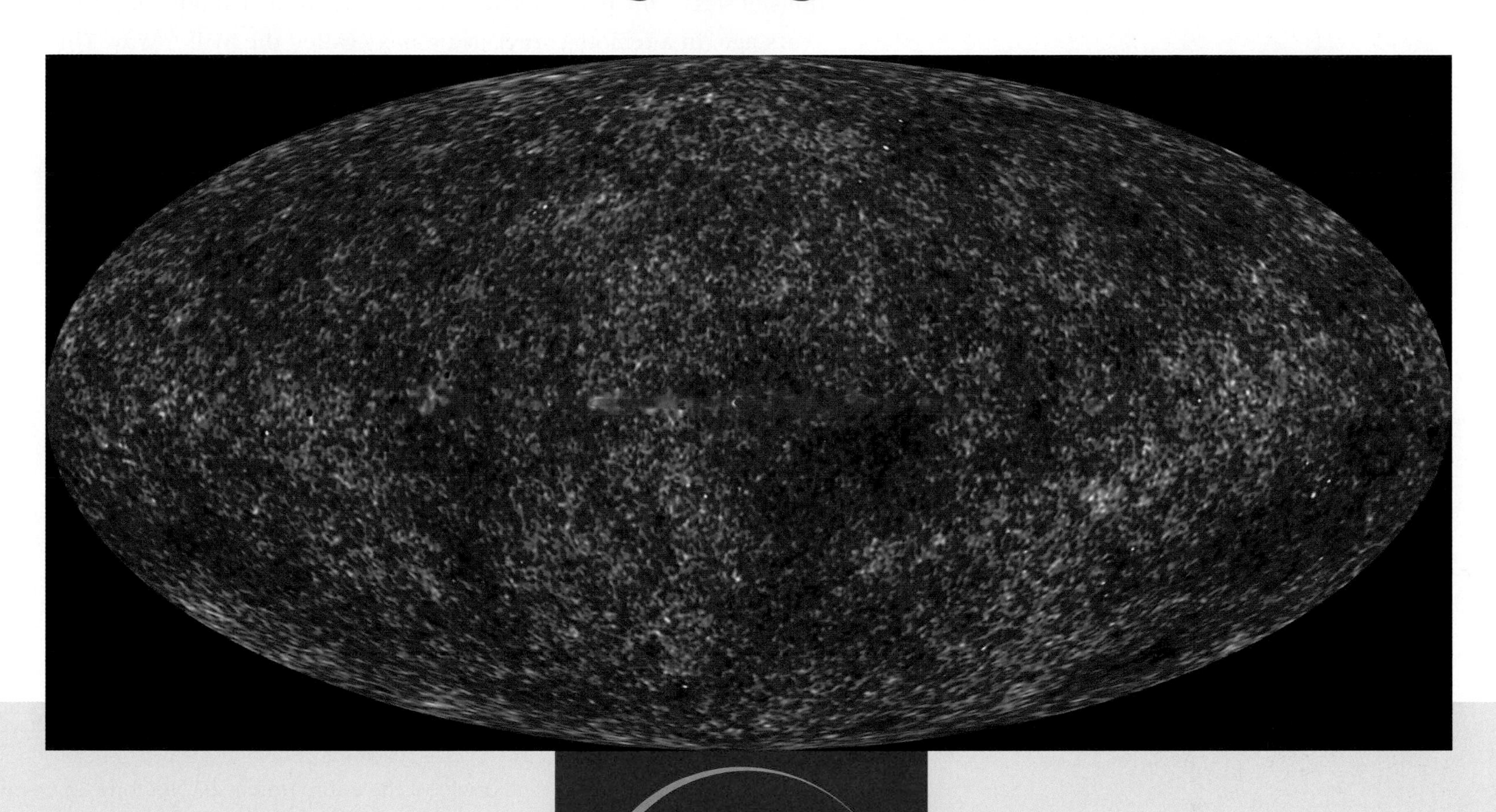

LEARNING GOALS

17.1 THE BIG BANG
- What were conditions like in the early universe?
- What is the history of the universe according to the Big Bang theory?

17.2 EVIDENCE FOR THE BIG BANG
- How do we observe the radiation left over from the Big Bang?
- How do the abundances of elements support the Big Bang?

17.3 THE BIG BANG AND INFLATION
- What aspects of the universe were originally unexplained by the Big Bang model?
- How does inflation explain these features of the universe?
- How can we test the idea of inflation?

17.4 OBSERVING THE BIG BANG FOR YOURSELF
- Why is the darkness of the night sky evidence for the Big Bang?

The universe has been expanding for about 14 billion years. During that time, matter collected into galaxies. Stars formed in those galaxies, producing heavy elements that were recycled into later generations of stars. One of these late-coming stars formed about 4.6 billion years ago, in a remote corner of a galaxy called the Milky Way. This star was born with a host of planets that formed in a flattened disk surrounding it. One of these planets soon became covered with life that gradually evolved into ever more complex forms. Today, the most advanced life-form on this planet, human beings, can look back on this series of events and marvel at how the universe created conditions suitable for life.

To this point in the book, we have discussed how the matter produced in the early universe gradually assembled into planets, stars, and galaxies. However, we have not yet answered one big question: Where did the *matter itself* come from? To answer this ultimate question, we must go beyond the most distant galaxies and even beyond what we can see near the horizon of the universe. We must go back not only to the origins of matter and energy but to the beginning of time itself.

17.1 THE BIG BANG

Is it really possible to study the origin of the entire universe? Not long ago, questions about the origin of everything we see were considered unfit for scientific study. That attitude began to change with Hubble's discovery that the universe is expanding. This discovery led to the insight that all things very likely sprang into being at a single moment in time, in an event that we have come to call the *Big Bang*. Today, powerful telescopes allow us to view how galaxies have changed over the past 14 billion years, and at great distances we see young galaxies still in the process of forming [Section 15.3]. These observations confirm that the universe is gradually aging, just as we should expect if the entire universe really was born some 14 billion years ago.

Unfortunately, we cannot see back to the very beginning of time. Light from the most distant galaxies shows us what the universe looked like when it was one or two billion years old. Beyond these galaxies, we have not yet found any objects shining brightly enough for us to see them. Ultimately, we face an even more fundamental problem. The universe is filled with a faint glow of radiation that appears to be the remnant heat of the Big Bang. This faint glow is light that has traveled freely through space since the universe was about 380,000 years old, which is when the universe first became transparent to light. Before that time, light could not pass freely through the universe, so there is no possibility of seeing light from earlier times. Thus, just as we must rely on mathematical modeling to determine what the Sun is like on the inside, we must use modeling to investigate what the universe was like during its earliest moments.

• What were conditions like in the early universe?

Scientific models of the conditions that prevailed in the early universe are based on fundamental principles of physics. The universe is cooling and becoming less dense as it expands, so it must have been hotter and denser in the past. Calculating exactly how hot and dense the universe must have been when it was more compressed is similar to calculating the temperature and density of gas in a balloon when you squeeze it, except that the conditions become much more extreme. Figure 17.1 shows just how hot the universe was during its earliest moments, according to such calculations.

The universe was so hot during the first few seconds that photons could transform themselves into matter, and vice versa, in accordance with Einstein's formula $E = mc^2$ [Section 4.3]. Reactions that create and destroy matter are now relatively rare in the universe at large, but physicists can reproduce many such reactions in their laboratories.

One such reaction is the creation or destruction of an *electron–antielectron pair* (Figure 17.2). When two photons collide with a total energy greater than twice the mass-energy of an electron (the electron's mass times c^2), they can create two brand-new particles: a negatively charged electron and its positively charged twin, the *antielectron* (also known as a *positron*). The electron is a particle of **matter**, and the antielectron is a particle of **antimatter**. The reaction that creates an electron–antielectron pair also runs in reverse. When an electron and an antielectron meet, they *annihilate* each other totally, transforming all their mass-energy back into photon energy.

The very early universe was so hot that energy could be transformed into matter and vice versa.

Similar reactions can produce or destroy any particle–antiparticle pair, such as a proton and antiproton or a neutron and antineutron. The early universe therefore was filled with an extremely hot and dense blend of photons, matter, and antimatter, converting furiously back and forth. Despite all these vigorous reactions, describing conditions in the early universe is straightforward, at least in principle. We simply need to use the laws of physics to calculate the proportions of the various forms of radiation and matter at each moment in the universe's early history. The only difficulty is our incomplete understanding of the laws of physics.

To date, physicists have investigated the behavior of matter and energy at temperatures as high as those that existed in the universe just *one ten-billionth* (10^{-10}) of a second after the Big Bang, giving us confidence that we actually understand what was happening at that early moment in the history of the universe. Our understanding of physics is less certain under the more extreme conditions that prevailed even earlier, but we do have some ideas about what the universe was like when it was a mere 10^{-38} second old, and perhaps a glimmer of what it was like at the age of just 10^{-43} second. These tiny fractions of a second are so small that, for all practical purposes, we are studying the very moment of creation—the Big Bang itself.

• What is the history of the universe according to the Big Bang theory?

The **Big Bang theory**—the scientific theory of the universe's earliest moments—is based on applying known and tested laws of physics to the idea that all we see today, from Earth to the cosmic horizon, began as an

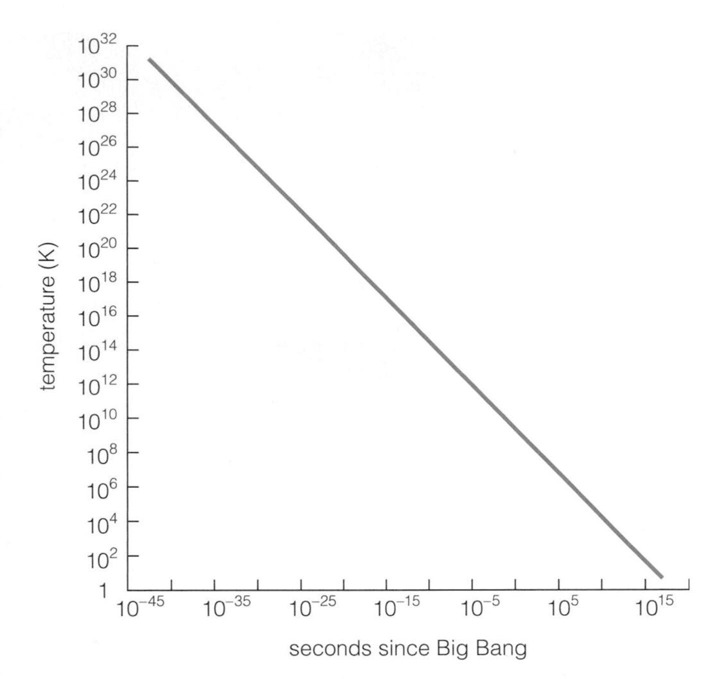

Figure 17.1

The universe cools as it expands. By using the laws of physics and the current temperature of the universe (about 3 K), we can calculate how hot the universe must have been in the past. This graph shows the results. Notice that both axis scales use powers of 10. (The graph extends to the present; 10^{10} years $\approx 3 \times 10^{17}$ seconds.)

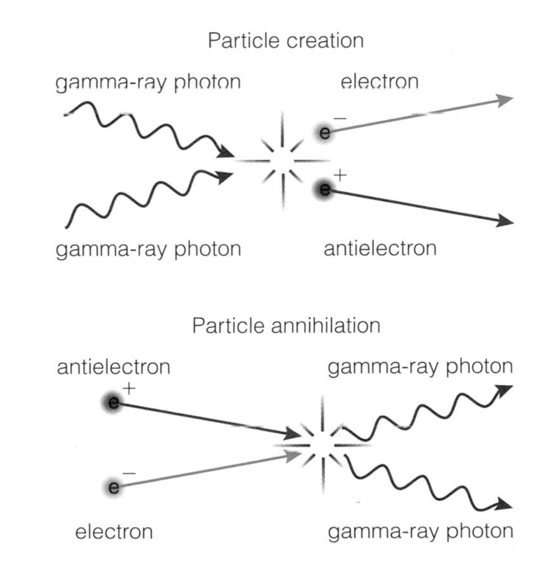

Figure 17.2

Electron–antielectron creation and annihilation. Reactions like these constantly converted photons to particles and vice versa in the early universe.

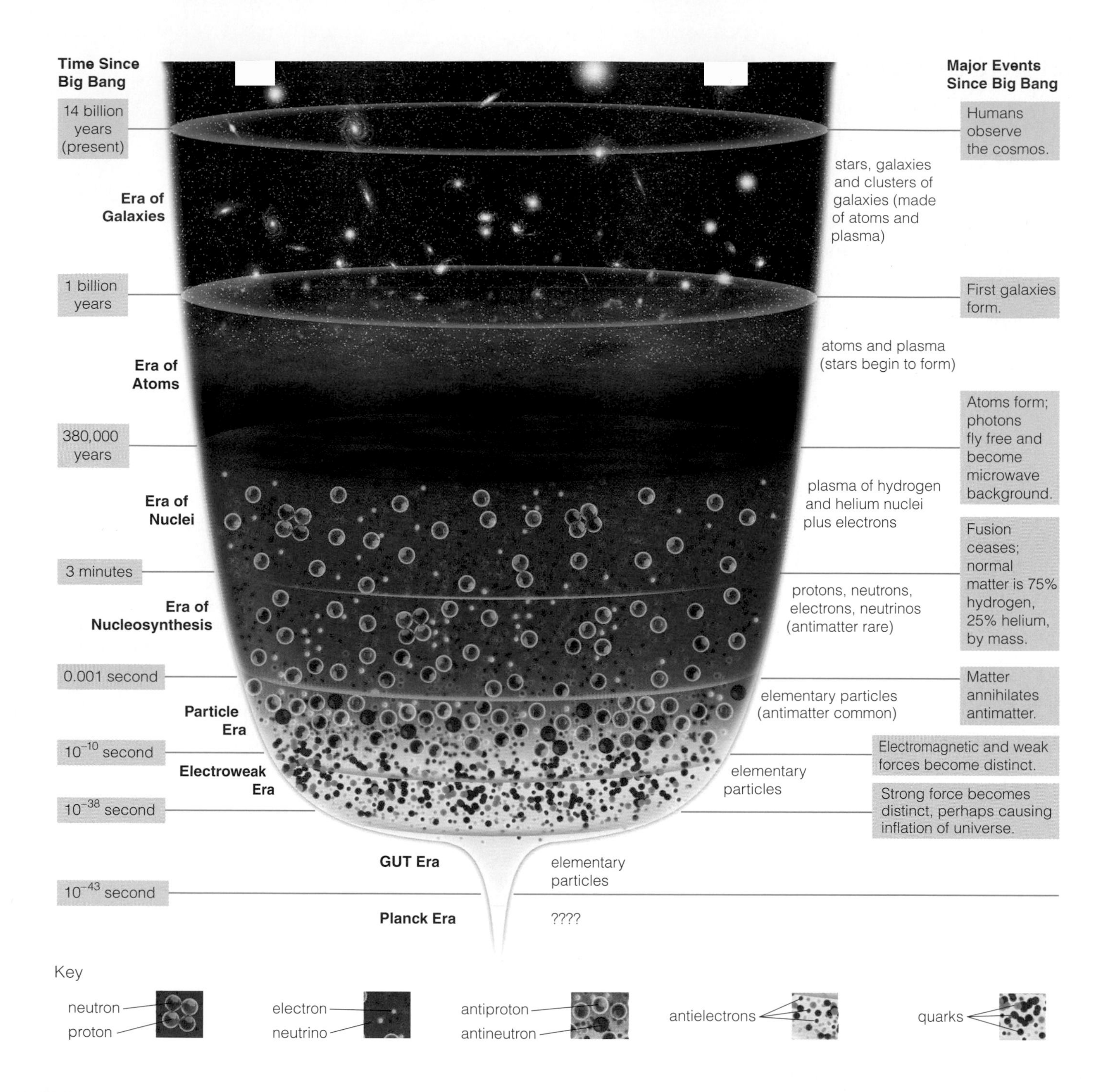

Figure 17.3

This diagram summarizes the eras of the universe. The names of the eras and their ending times are indicated on the left, and the state of matter during each era and the events marking the end of each era are indicated on the right.

incredibly tiny, hot, and dense collection of matter and radiation. The Big Bang theory describes how expansion and cooling of this unimaginably intense mixture of particles and photons could have led to the present universe of stars and galaxies, and it explains several aspects of today's universe with impressive accuracy. We will discuss the evidence supporting the Big Bang theory later in this chapter. First, in order to help you understand the significance of the evidence, we'll examine the history of the universe according to this theory.

Figure 17.3 summarizes the story by dividing the overall history of the universe into a series of *eras,* or time periods. (Some scientists further subdivide the eras described here.) Each era is distinguished from the next by some major change in the conditions of the universe. You'll find it easiest to keep track of the various eras if you refer back to this figure as we discuss each era in detail. Notice that most of the key events in the history of the universe occurred in a very short period of time. It will take you longer to read this chapter than it took the universe to progress through the first five eras we will discuss, by which point the chemical composition of the universe had already been determined.

The Planck Era As we work our way back through time, we ultimately reach the limit of our current scientific ability to understand the physical conditions when the universe was an incomprehensibly young 10^{-43} second old. This instant in time is called the *Planck time* after physicist Max Planck, one of the founders of the science of quantum mechanics. We refer to all times prior to the Planck time as the **Planck era**; that is, the Planck era represents the first 10^{-43} second in the history of the universe.

We do not yet understand the physics of the universe well enough to describe what it was like during the Planck era.

Current theories cannot adequately describe the extreme conditions that must have existed during the Planck era. According to the laws of quantum mechanics, there must have been substantial energy fluctuations from point to point in the very early universe. Because energy and mass are equivalent, Einstein's theory of general relativity tells us that these energy fluctuations must have generated a rapidly changing gravitational field that randomly warped space and time. During the Planck era, these random energy fluctuations were so large that our current theories are powerless to describe what might have been happening. In essence, the problem is that we do not yet have a theory that links quantum mechanics (our successful theory of the very small) and general relativity (our successful theory of the very big). Perhaps someday we will be able to merge these theories of the very small and the very big into a single "theory of everything." Until that happens, science cannot describe the universe before the Planck time.

The GUT Era Understanding the transition that marked the beginning of the next era requires thinking in terms of the *forces* that operate in the universe. Everything that happens in the universe today is governed by four distinct forces: *gravity, electromagnetism,* the *strong force,* and the *weak force.* We have already encountered each of these forces individually.

Gravity is the most familiar of the four forces, providing the "glue" that holds planets, stars, and galaxies together. The electromagnetic force, which depends on the electrical charge of a particle instead of its mass, is far stronger than gravity. It is therefore the dominant force between particles in atoms and molecules, responsible for all chemical and biological reactions. However, the existence of both positive and negative electrical charge causes the electromagnetic force to lose out to gravity on large scales, even though both forces decline with distance by an inverse square law. Most large astronomical objects (such as planets and stars) are electrically neutral overall, making the electromagnetic force unimportant on that scale. Gravity therefore becomes the dominant force for such objects, because more mass always means more gravity.

The strong and weak forces operate only over extremely short distances, making them important within atomic nuclei but not on larger scales. The strong force binds atomic nuclei together [Section 10.2]. The weak force plays a crucial role in nuclear reactions such as fission or fusion, and it is the only force besides gravity that affects weakly interacting particles such as neutrinos or WIMPs (weakly interacting massive particles [Section 16.2]).

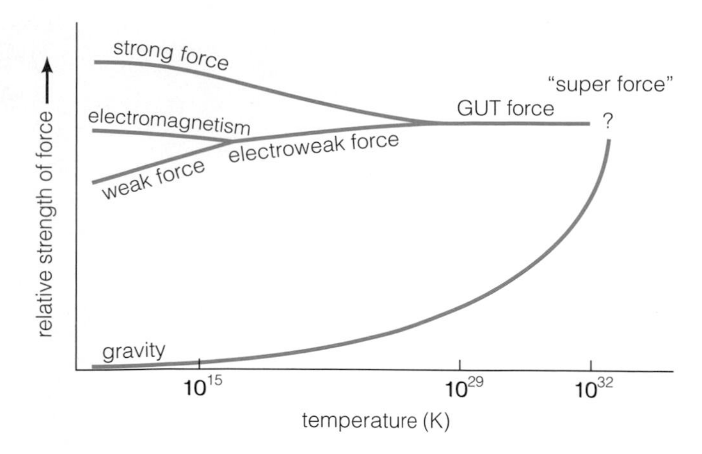

Figure 17.4

The four forces are distinct at low temperatures but may merge at very high temperatures.

Although the four forces behave quite differently from one another, we now believe that they are actually just different aspects of a smaller number of more fundamental forces, probably only one or two (Figure 17.4). At the high temperatures that prevailed in the early universe, the four forces were not so distinct as they are today.

As an analogy, think about ice, liquid water, and water vapor. These three substances are quite different from one another in appearance and behavior, yet they are just different phases of the single substance H_2O. In a similar way, experiments have shown that the electromagnetic and weak forces lose their separate identities under conditions of very high temperature or energy and merge together into a single **electroweak force**. At even higher temperatures and energies, the electroweak force may merge with the strong force and ultimately with gravity. Theories that predict the merger of the electroweak and strong forces are called **grand unified theories**, or **GUTs** for short. Thus, the merger of the strong, weak, and electromagnetic forces is often called the *GUT force*. Many physicists suspect that at even higher energies the GUT force and gravity merge into a single "super force" that governs the behavior of everything. (Among the names you may hear for theories linking all four forces are *supersymmetry, superstrings,* and *supergravity.*)

Calculations from general relativity and quantum mechanics suggest that this unified "super force" may have reigned in the universe during the Planck era. If so, the Planck time (10^{-43} s) marks the instant when gravity became distinct from the other three forces, which were still merged as the GUT force. By analogy to ice crystals forming as a liquid cools, we say that gravity "froze out" at the Planck time. The universe subsequently entered the **GUT era**, when two forces operated in the universe: gravity and the GUT force.

Energy released near the end of the GUT era may have caused a rapid expansion of the universe known as inflation.

The GUT era is thought to have lasted but a tiny fraction of a second, coming to an end when the universe had cooled to 10^{29} K at an age of about 10^{-38} second. Grand unified theories predict that the strong force froze out from the GUT force at this point, leaving three forces operating in the universe: gravity, the strong force, and the electroweak force. The freezing out of the strong force may have released an enormous amount of energy. This energy release may have caused the universe to undergo a sudden and dramatic expansion that we call **inflation**. In the space of a mere 10^{-36} second, pieces of the universe the size of an atomic nucleus may have grown to the size of our solar system. Inflation sounds bizarre, but as we will discuss later, it explains several puzzling features of today's universe.

The Electroweak Era The end of the GUT era marks the beginning of the **electroweak era**, so named because the electromagnetic and weak forces were still unified in the electroweak force. Intense radiation filled all of space, as it had since the Planck era, spontaneously producing matter and antimatter particles that almost immediately annihilated and turned back into photons. The universe continued to expand and cool throughout the electroweak era, dropping to a temperature of 10^{15} K when it reached an age of 10^{-10} second. This temperature is still 100 million times hotter than the temperature in the core of the Sun, but it was low enough for the electromagnetic and weak forces to freeze out from the electroweak force. After this instant (10^{-10} s), all four forces were forever distinct in the universe.

The end of the electroweak era marks an important transition not only in the physical universe, but also in human understanding of the universe. The theory that unified the weak and electromagnetic forces, developed in the 1970s, predicted the emergence of new types of particles (called the W and Z bosons, or *weak bosons*) at temperatures above the 10^{15} K that pervaded the universe when it was 10^{-10} second old. In 1983, experiments performed in a huge particle accelerator near the French/Swiss border (CERN) reached energies equivalent to such high temperatures for the first time. The new particles showed up just as predicted, produced from the extremely high energy in accord with $E = mc^2$.

Experimental evidence supports current theory about the universe since it was 10^{-10} second old, at the end of the electroweak era, but we have no direct evidence about earlier times.

Thus, we have direct experimental evidence concerning the conditions in the universe at the end of the electroweak era. We do *not* have any direct experimental evidence of conditions prior to that time. Our theories concerning the earlier parts of the electroweak era and the GUT era consequently are much more speculative than our theories describing the universe from the end of the electroweak era to the present.

The Particle Era As long as it was hot enough for the spontaneous creation and annihilation of particles to continue, the total number of particles was roughly in balance with the total number of photons. Once it became too cool for this spontaneous exchange of matter and energy to continue, photons became the dominant form of energy in the universe. We refer to the time between the end of the electroweak era and the moment when spontaneous particle production ceased as the **particle era**, to emphasize the importance of subatomic particles during this period.

During the early parts of the particle era (and earlier eras), photons turned into all sorts of exotic particles that we no longer find freely existing in the universe today, including *quarks*—the building blocks of protons and neutrons. By the end of the particle era, all quarks had combined into protons and neutrons, which shared the universe with other particles such as electrons, neutrinos, and perhaps WIMPs. The particle era came to an end when the universe reached an age of 1 millisecond (0.001 s) and the temperature had fallen to 10^{12} K. At this point, it was no longer hot enough to produce protons and antiprotons (or neutrons and antineutrons) spontaneously from pure energy.

Protons must have slightly outnumbered antiprotons during the particle era, or we would not be here today.

If the universe had contained equal numbers of protons and antiprotons (or neutrons and antineutrons) at the end of the particle era, all of the pairs would have annihilated each other, creating photons and leaving essentially no matter in the universe. From the obvious fact that the universe contains a significant amount of matter, we conclude that protons must have slightly outnumbered antiprotons at the end of the particle era.

We can estimate the size of the imbalance between matter and antimatter by comparing the present numbers of protons and photons in the universe. The two numbers should have been similar in the very early universe, but today photons outnumber protons by about a billion to one. This ratio indicates that for every billion antiprotons in the early universe, there must have been about a billion and one protons. Thus, for each 1 billion protons and antiprotons that annihilated each other at the end of the particle era, a single proton was left over. This seemingly slight excess of matter over antimatter makes up all the ordinary matter in the present-day universe. Some of those protons (and neutrons) left over from when the universe was 0.001 second old are the very ones that make up our bodies.

The Era of Nucleosynthesis So far, everything we have discussed occurred within the first 0.001 second of the universe's existence—a time span shorter than the time it takes you to blink an eye. At this point, the protons and neutrons left over after the annihilation of antimatter began to fuse into heavier nuclei. However, the heat of the universe remained so high that most nuclei broke apart as fast as they formed. This dance of fusion and breakup marked the **era of nucleosynthesis**.

Most of the helium in the universe was made during the first 3 minutes.

The era of nucleosynthesis ended when the universe was about 3 minutes old. After this time, the density in the expanding universe had dropped so much that fusion no longer occurred, even though the temperature was still about a billion Kelvin (10^9 K)—much hotter than the temperature at the center of the Sun today. When fusion ceased, about 75% of the mass of the ordinary (baryonic) matter in the universe remained as individual protons, or hydrogen nuclei. The other 25% of this mass had fused into helium nuclei, with trace amounts of deuterium (hydrogen with a neutron) and lithium (the next heaviest element after hydrogen and helium). Except for the relatively small amount of matter that stars later forged into heavier elements, the chemical composition of the universe remains the same today.

The Era of Nuclei At the end of the era of nucleosynthesis, the universe consisted of a very hot plasma of hydrogen nuclei, helium nuclei, and free electrons. This basic picture held for about the next 380,000 years as the universe continued to expand and cool. The fully ionized nuclei moved independently of electrons during this period (rather than being bound with electrons in neutral atoms), which we call the **era of nuclei**. Throughout this era, photons bounced rapidly from one electron to the next, just as they do deep inside the Sun today [Section 10.2], never managing to travel far between collisions. Any time a nucleus managed to capture an electron to form a complete atom, one of the photons quickly ionized it.

Photons began to travel freely through the universe about 380,000 years after the Big Bang, when electrons first combined with nuclei to make atoms.

The era of nuclei came to an end when the expanding universe was about 380,000 years old. At this point the temperature had fallen to about 3,000 K—roughly half the temperature of the Sun's surface today. Hydrogen and helium nuclei finally captured electrons for good, forming stable, neutral atoms for the first time. With electrons now bound into atoms, the universe suddenly became transparent, as if a thick fog had suddenly lifted. Photons, formerly trapped among the electrons, began to stream freely across the universe. We still see these photons today as the *cosmic microwave background,* which we will discuss shortly.

The Era of Atoms and the Era of Galaxies We've already discussed the rest of the story in earlier chapters. The end of the era of nuclei marked the beginning of the **era of atoms**, when the universe consisted of a mixture of neutral atoms and plasma (ions and electrons), along with a large number of photons. Thanks to the slight density enhancements present in the universe at this time and the gravitational attraction of dark matter, the atoms and plasma slowly assembled into protogalactic clouds. Stars formed in these clouds, transforming the gas clouds into galaxies. The first full-fledged galaxies had formed by the time the universe was about 1 billion years old, beginning what we call the **era of galaxies**.

The first galaxies formed by the time the universe was a billion years old.

The era of galaxies continues to this day. Generation after generation of star formation in galaxies steadily builds elements heavier than helium and incorporates them into new star systems. Some of these star systems develop planets, and on at least one of these planets life burst into being a few billion years ago. Now here we are, thinking about it all. Describing both the follies and the achievements of the human race, Carl Sagan once said, "These are the things that hydrogen atoms do—given 15 billion years of cosmic evolution." (Sagan died in 1996, before we refined the age of the universe to 14 billion years.)

17.2 EVIDENCE FOR THE BIG BANG

Like any scientific theory, the Big Bang theory is a model of nature designed to explain a set of observations. If it is close to the truth, it should be able to make predictions about the real universe that we can verify through more observations or experiments. The Big Bang model has gained wide scientific acceptance for two key reasons:

- The Big Bang model predicts that the radiation that began to stream across the universe at the end of the era of nuclei should still be present today. Sure enough, we find that the universe is filled with what we call the **cosmic microwave background**. Its characteristics precisely match what we expect according to the Big Bang model.
- The Big Bang model predicts that some of the original hydrogen in the universe should have fused into helium during the era of nucleosynthesis. Observations of the actual helium content of the universe closely match the amount of helium predicted by the Big Bang model.

Let's take a closer look at this evidence, starting with the cosmic microwave background.

• How do we observe the radiation left over from the Big Bang?

The first major piece of evidence supporting the Big Bang theory was announced in 1965. Arno Penzias and Robert Wilson, two physicists working at Bell Laboratories in New Jersey, were calibrating a sensitive microwave antenna designed for satellite communications (Figure 17.5). (*Microwaves* fall within the radio portion of the electromagnetic spectrum; see Figure 5.2.) Much to their chagrin, they kept finding unexpected "noise" in every measurement they made with the antenna.

Fearing that they were doing something wrong, Penzias and Wilson worked frantically to discover and eliminate all possible sources of background noise. They even climbed up on their antenna to scrape off pigeon droppings, on the off-chance that these were somehow causing the noise.

Figure 17.5

Arno Penzias and Robert Wilson, discoverers of the cosmic microwave background, with the Bell Labs microwave antenna.

No matter what they did, the microwave noise wouldn't go away. The noise was the same no matter where they pointed their antenna, indicating that it came from all directions in the sky and ruling out the possibility that it came from any particular astronomical object or from any place on Earth. Embarrassed by their inability to explain the noise, Penzias and Wilson prepared to "bury" their discovery about the noise at the end of a long scientific paper about their antenna.

Meanwhile, physicists at nearby Princeton University were busy calculating the expected characteristics of the radiation left over from the heat of the Big Bang. They concluded that, if the Big Bang had really occurred, this radiation should be permeating the entire universe and should be detectable with a microwave antenna. On a fateful airplane trip home from an astronomical meeting, Penzias sat next to an astronomer who told him of the Princeton calculations. The Princeton group soon met with Penzias and Wilson to compare notes. The "noise" in the Bell Labs antenna was not an embarrassment after all. Instead, it was the cosmic microwave background—and the first strong evidence that the Big Bang had really happened. Penzias and Wilson received the 1978 Nobel Prize in physics for their discovery of the cosmic microwave background.[1]

The cosmic microwave background is radiation left over from the Big Bang.

The cosmic microwave background consists of photons arriving at Earth directly from the end of the era of nuclei, when the universe was about 380,000 years old. Because neutral atoms finally could remain stable, they captured most of the electrons in the universe. With no more free electrons to block them, the photons from that epoch have flown unobstructed through the universe ever since (Figure 17.6). Thus, when we observe the cosmic microwave background, we essentially are seeing back to a time when the universe was only 380,000 years old. In that sense, we are seeing light from the most distant regions of the observable universe—only 380,000 light-years from our cosmological horizon [Section 15.2].

Surprisingly, it does not take a particularly powerful telescope to "see" this radiation. In fact, you can pick it up with an ordinary television antenna. If you set an antenna-fed television (i.e., *not* cable or satellite TV) to a channel for which there is no local station, you will see a screen full of static "snow." About 1% of this static is due to photons in the cosmic microwave background. Try it. If your friends ask why you are watching nothing, tell them that you are actually watching the most incredible sight ever seen on a television screen: the Big Bang, or at least as close to it as we'll ever get.

The cosmic microwave background came from the heat of the universe itself and therefore should have an essentially perfect thermal radiation spectrum [Section 5.2]. When this radiation first broke free 380,000 years after the Big Bang, the temperature of the universe was about 3,000 K, not too different from that of a red giant star's surface. Thus, the spectrum of the cosmic microwave background originally peaked in visible light, just like the thermal radiation from a red star, with wavelengths of a few hundred nanometers. However, the universe has expanded by a factor of about 1,000 since that time, stretching the wavelengths of these photons by the same amount [Section 15.2]. Their wavelengths have therefore shifted to about a millimeter, squarely in the microwave por-

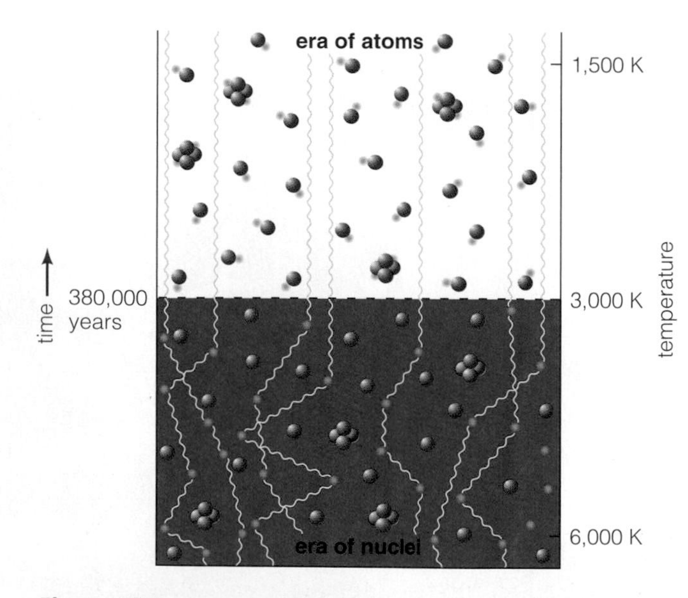

Figure 17.6 Interactive Figure

Photons (yellow squiggles) frequently collided with free electrons during the era of nuclei and thus could travel freely only after electrons became bound into atoms. This transition was something like the transition from a dense fog to clear air. The photons released at the end of the era of nuclei, when the universe was about 380,000 years old, make up the cosmic microwave background. Precise measurements of these microwaves tell us what the universe was like at this moment in time.

[1]The dramatic story of the discovery of the cosmic microwave background is told in greater detail, along with much more scientific history, in Timothy Ferris, *The Red Limit* (New York: Quill, 1983).

tion of the spectrum and corresponding to a temperature of a few degrees above absolute zero.

The cosmic microwave background—the heat of the universe itself—has a perfect thermal radiation spectrum corresponding to a temperature of 2.73 K.

In the early 1990s, a NASA satellite called the *Cosmic Background Explorer (COBE)* was launched to test these ideas about the cosmic microwave background. The results were a stunning success for the Big Bang theory. As shown in Figure 17.7, the cosmic microwave background does indeed have a perfect thermal radiation spectrum, with a peak corresponding to a temperature of 2.73 K. In a very real sense, the temperature of the night sky is a frigid 3 degrees above absolute zero.

Figure 17.7

This graph shows the spectrum of the cosmic microwave background recorded by NASA's COBE satellite. A theoretically calculated thermal radiation spectrum (smooth curve) for a temperature of 2.73 K perfectly fits the data (dots). This excellent fit is important evidence in favor of the Big Bang theory.

THINK ABOUT IT Suppose the cosmic microwave background did not really come from the heat of the universe itself but instead came from many individual stars and galaxies. Explain why, in that case, we would not expect it to have a perfect thermal radiation spectrum. How does the spectrum of the cosmic microwave background lend support to the Big Bang theory?

COBE achieved an even greater success mapping the temperature of the cosmic microwave background in all directions. It was already known that the cosmic microwave background is extraordinarily uniform throughout the universe. Conditions in the early universe must have been extremely uniform to produce such a smooth radiation field. For a time, this uniformity was considered a strike against the Big Bang theory because, as we discussed in Chapters 15 and 16, the universe must have contained some regions of enhanced density in order to explain the formation of galaxies. The COBE measurements restored confidence in the Big Bang theory because they showed that the cosmic microwave background is *not quite* perfectly uniform. Instead, its temperature varies very slightly from one place to another by a few parts in 100,000.

More recently, NASA's Wilkinson Microwave Anisotropy Probe (WMAP) has provided even more dramatic confirmation of the small temperature variations, with a map of the cosmic microwave background released in 2003 (Figure 17.8). These variations in temperature indicate that the density of the early universe really did differ slightly from place to place. The seeds of structure formation were indeed present during the era of nuclei.

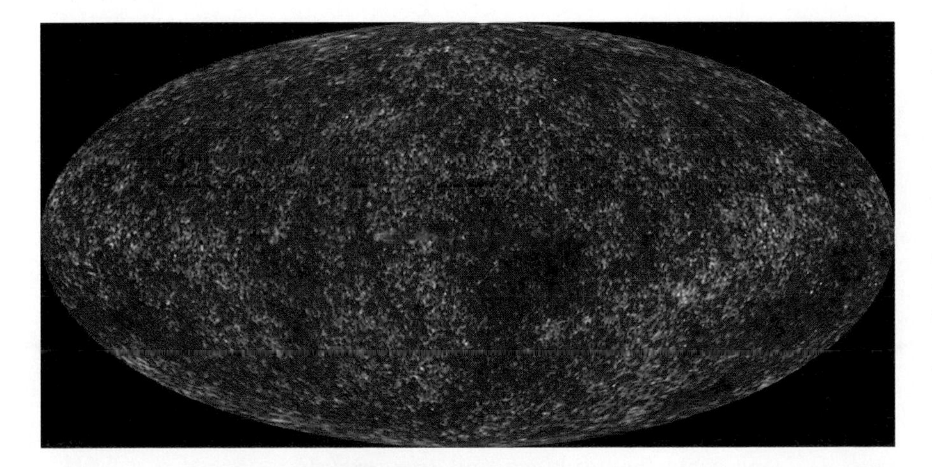

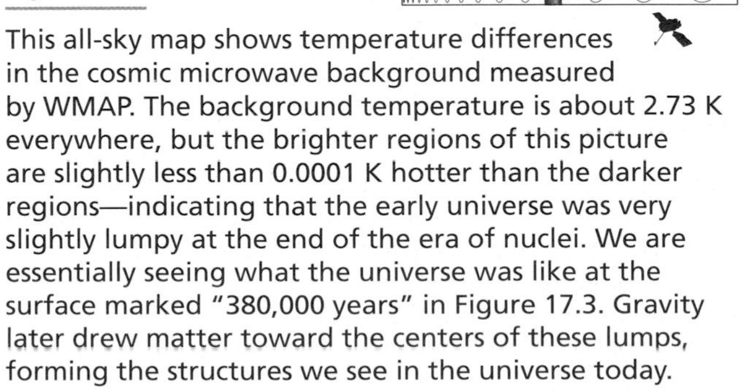

Figure 17.8

This all-sky map shows temperature differences in the cosmic microwave background measured by WMAP. The background temperature is about 2.73 K everywhere, but the brighter regions of this picture are slightly less than 0.0001 K hotter than the darker regions—indicating that the early universe was very slightly lumpy at the end of the era of nuclei. We are essentially seeing what the universe was like at the surface marked "380,000 years" in Figure 17.3. Gravity later drew matter toward the centers of these lumps, forming the structures we see in the universe today.

Maps of the cosmic microwave background reveal the density enhancements from which galaxies and larger structures later formed.

The discovery of density enhancements bolstered the idea that some of the dark matter consists of WIMPs that we have not yet identified [Section 16.2] and that the gravity of this dark matter drove the formation of structure in the universe. Regions of enhanced density can grow into galaxies because the extra gravity in these regions draws matter together even while the rest of the universe expands. The greater the density enhancements, the faster matter should have collected into galaxies.

Detailed calculations show that, to explain the fact that galaxies formed within a few billion years, the density enhancements at the end of the era of nuclei must have been significantly greater than the few parts in 100,000 suggested by the temperature variations in the cosmic microwave background. Because WIMPs are weakly interacting and do not interact with photons, we do not expect them to influence the temperature of the cosmic microwave background directly. However, the gravity of the WIMPs can collect ordinary baryonic matter into clumps that *do* interact with photons. Thus, the small density enhancements detected by microwave telescopes appear to echo much larger density enhancements made up of WIMPs. Careful modeling of these temperature variations shows that they are consistent with dark-matter density enhancements large enough to account for the structure we see in the universe today.

• How do the abundances of elements support the Big Bang?

The discovery of the cosmic microwave background in 1965 quickly solved another long-standing astronomical problem: the origin of cosmic helium. Everywhere in the universe, about one-quarter of the mass of ordinary matter (i.e., not dark matter) is helium. The Milky Way's helium fraction is about 28%, and no galaxy has a helium fraction lower than 25%. A small proportion of this helium comes from hydrogen fusion in stars, but most does not: Fusion of hydrogen to helium in stars could have produced only about 10% of the observed helium.

The majority of the helium in the universe must already have been present in the protogalactic clouds that preceded the formation of galaxies. In other words, the universe itself must once have been hot enough to fuse hydrogen into helium. The current microwave background temperature of 2.73 K tells us precisely how hot the universe was in the dis-

SPECIAL TOPIC: THE STEADY STATE UNIVERSE

Although the Big Bang theory enjoys wide acceptance among scientists today, alternative ideas have been proposed and considered. One of the cleverest alternatives, developed in the late 1940s, was called the *steady state* universe. This hypothesis accepted the fact that the universe is expanding but rejected the idea of a Big Bang, instead postulating that the universe is infinitely old. The steady state hypothesis may seem paradoxical at first: If the universe has been expanding forever, shouldn't every galaxy be infinitely far away from every other galaxy? Proponents of the steady state universe answered by claiming that new galaxies continually form in the gaps that open up as the universe expands, thereby keeping the same average distance between galaxies at all times. In a sense, the steady state hypothesis said that the creation of the universe is an ongoing and eternal process rather than having happened all at once with a Big Bang.

Two key discoveries caused the steady state universe to lose favor. First, the 1965 discovery of the cosmic microwave background matched a prediction of the Big Bang theory but was not adequately explained by the steady state hypothesis. Second, a steady state universe should look about the same at all times, which is inconsistent with observations showing that galaxies at great distances look younger than nearby galaxies. As a result of these predictive failures, most astronomers no longer take the steady state universe seriously.

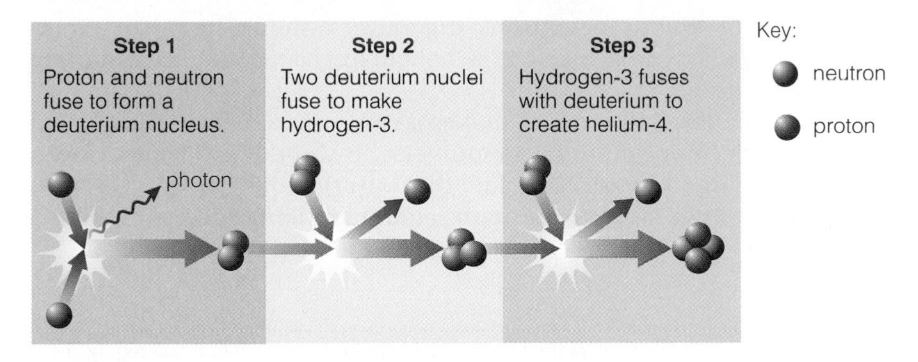

Figure 17.9

During the 3-minute-long era of nucleosynthesis, virtually all the neutrons in the universe fused with protons to form helium-4. This figure illustrates one of several possible reaction pathways.

tant past and exactly how much helium it should have made. The result—25% helium—is another impressive success of the Big Bang theory.

Helium Formation in the Early Universe In order to see why 25% of ordinary matter became helium, we need to understand what protons and neutrons were doing during the 3 minutes that marked the era of nucleosynthesis. Early in this era, when the universe's temperature was 10^{11} K, nuclear reactions could convert protons into neutrons, and vice versa. As long as the universe remained hotter than 10^{11} K, these reactions kept the numbers of protons and neutrons nearly equal. But as the universe cooled, neutron–proton conversion reactions began to favor protons.

Neutrons are slightly more massive than protons, and therefore reactions that convert protons to neutrons require energy to proceed (in accordance with $E = mc^2$). As the temperature fell below 10^{11} K, the required energy for neutron production was no longer readily available, so the rate of these reactions slowed. In contrast, reactions that convert neutrons into protons release energy and thus are unhindered by cooler temperatures. By the time the temperature of the universe fell to 10^{10} K, protons began to outnumber neutrons because the conversion reactions ran only in one direction. Neutrons changed into protons, but the protons didn't change back.

For about the next 3 minutes, the universe was still hot and dense enough for nuclear fusion to operate. Protons and neutrons constantly combined to form *deuterium*—the rare form of hydrogen nuclei that contains a neutron in addition to a proton—and deuterium nuclei fused to form helium (Figure 17.9). However, during the early part of the era of nucleosynthesis, the helium nuclei were almost immediately blasted apart by one of the many gamma rays that filled the universe.

The Big Bang theory predicts that the universe should have a chemical composition of 75% hydrogen and 25% helium by mass, which agrees with the observed composition.

Fusion began to create long-lasting helium nuclei when the universe was about 1 minute old and had cooled to a temperature at which it contained few destructive gamma rays. Calculations show that the proton-to-neutron ratio at this time should have been about 7 to 1. Moreover, almost all the available neutrons should have become incorporated into nuclei of helium-4. Figure 17.10 shows that, based on the

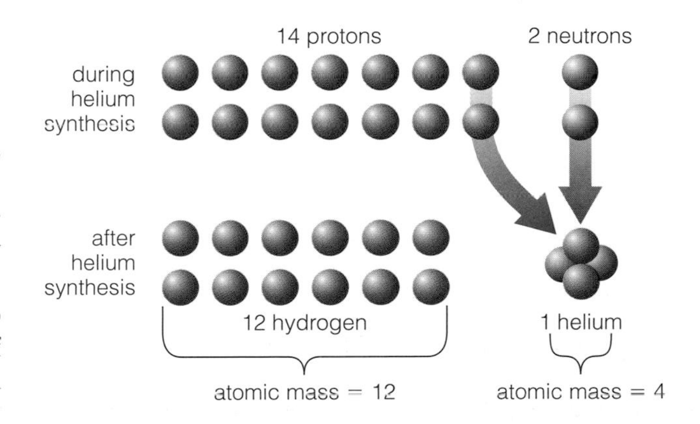

Figure 17.10

Calculations show that protons outnumbered neutrons 7 to 1, which is the same as 14 to 2, during the era of nucleosynthesis. The result was 12 hydrogen nuclei (individual protons) for each helium nucleus. Thus, the hydrogen-to-helium mass ratio is 12 to 4, which is the same as 75% to 25%. The agreement between this prediction and the observed abundance of helium is important evidence in favor of the Big Bang theory.

7-to-1 ratio of protons to neutrons, the universe should have had a composition of 75% hydrogen and 25% helium by mass at the end of the era of nucleosynthesis.

Thus, the Big Bang theory makes a very concrete prediction about the chemical composition of the universe: It should be 75% hydrogen and 25% helium by mass. The fact that observations confirm this predicted ratio of hydrogen to helium is another striking success of the Big Bang theory.

THINK ABOUT IT Briefly explain why it should not be surprising that some galaxies contain a little more than 25% helium, but it would be very surprising if some galaxies contained less. (*Hint:* Think about how the relative amounts of hydrogen and helium in the universe are affected by fusion in stars.)

Abundances of Other Light Elements Why didn't the Big Bang produce heavier elements? By the time stable helium nuclei formed, when the universe was about a minute old, the temperature and density of the rapidly expanding universe had already dropped too far for a process like carbon production (three helium nuclei fusing into carbon [Section 12.2]) to occur. Reactions between protons, deuterium nuclei, and helium were still possible, but most of these reactions led nowhere. In particular, fusing two helium-4 nuclei results in a nucleus that is unstable and falls apart in a fraction of a second, as does fusing a proton to a helium-4 nucleus.

Rapid cooling and expansion of the early universe shut off nuclear fusion before the universe could produce elements much heavier than helium.

A few reactions involving hydrogen-3 (also known as *tritium*) or helium-3 can create long-lasting nuclei. For example, fusing helium-4 and hydrogen-3 produces lithium-7. However, the contributions of these reactions to the overall composition of the universe were minor because hydrogen-3 and helium-3 were so rare. Models of element production in the early universe show that, before the cooling of the universe shut off fusion entirely, such reactions generated only trace amounts of lithium, the next lightest element after helium. Thus, aside from hydrogen, helium, and lithium, all other elements were forged much later in the nuclear furnaces of stars. (Beryllium and boron, which are heavier than lithium but lighter than carbon, were created later when high-energy particles broke apart heavier nuclei that formed in stars.)

The Density of Ordinary Matter Calculations made with the Big Bang model allow scientists to estimate the density of ordinary (baryonic) matter in the universe from the observed amount of deuterium in the universe today. Remember that, during the era of nucleosynthesis, protons and neutrons first fused into deuterium and the deuterium nuclei then fused into helium. The fact that some deuterium nuclei still exist in the universe indicates that this process stopped before all the deuterium nuclei were used up. The amount of deuterium in the universe today therefore tells us about the density of protons and neutrons (baryons) during the era of nucleosynthesis: The higher the density, the more efficiently fusion would have proceeded. Thus, a higher density in the early universe would have left less deuterium in the universe today, and a lower density would have left more deuterium.

Observations show that about one out of every 40,000 hydrogen atoms contains a deuterium nucleus—that is, a nucleus with a neutron in addition to its proton. Calculations based on this deuterium abundance show that the density of ordinary (baryonic) matter in the universe is about 4% of the critical density (Figure 17.11). (Recall that the critical density is the density required if the expansion of the universe is to stop and reverse someday [Section 16.4].) Similar calculations based on the observed abundance of lithium and helium-3 lead to the same conclusion, adding to our confidence in the Big Bang model.

The abundances of light elements indicate that the overall amount of ordinary matter is much less than the total amount of dark matter.

These results also lead to an astonishing prediction about the nature of dark matter. Recall that the overall density of the universe appears to be close to 25% of the critical density [Section 16.4]. Because this is about six times as large as the 4% of critical density that we find for ordinary matter, we conclude that the universe contains about six times as much extraordinary (nonbaryonic) dark matter as it does of ordinary (baryonic) matter. Thus, unless we are missing something fundamental in our understanding of all these issues, the Big Bang model predicts that extraordinary (nonbaryonic) dark matter such as WIMPs constitutes the majority of the universe's mass. That is why most astronomers think that dark matter consists mostly of WIMPs, and why many scientists are actively trying to find ways to detect WIMPs and learn about their properties.

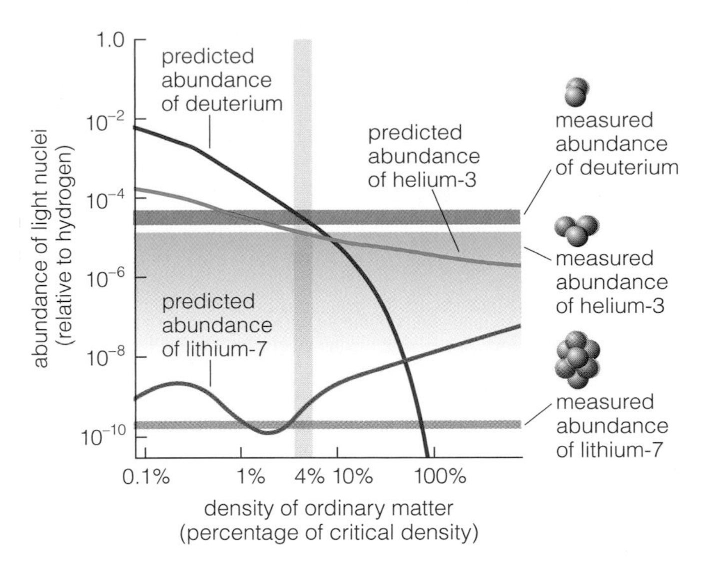

Figure 17.11

This graph shows how the measured abundances of deuterium, helium-3, and lithium-7 lead to the conclusion that the density of ordinary matter is about 4% of the critical density. The three horizontal swaths show measured abundances; the thickness of each swath represents the range of uncertainty in the measurements. (The upper edge of the blue swath indicates the upper limit on the helium-3 abundance; a lower limit has not yet been established.) The three curves represent models based on the Big Bang theory; these curves show how the abundance of each type of nucleus is expected to depend on the density of ordinary matter in the universe. Notice that the predictions (curves) match up with all three measurements (horizontal swaths) only in the gray vertical strip, which represents a density of about 4% of the critical density.

THINK ABOUT IT The ideas just discussed point to a rather amazing fact: Although we have yet to discover any WIMPs, we suspect they dominate the total mass of the universe. Briefly explain how this is possible, and comment on how confident we can be that weakly interacting particles make up the bulk of dark matter.

17.3 THE BIG BANG AND INFLATION

The Big Bang model relies heavily on our knowledge of particle physics, which has been tested to temperatures of 10^{15} K, corresponding to temperatures at the end of the electroweak era, when the universe was a mere 10^{-10} second old. Our knowledge of earlier times rests on a weaker foundation because we are less certain of the physical laws at work. In fact, the best laboratory for studying the laws of physics at such high temperatures is the Big Bang itself.

Different models of how matter might behave at such high energies predict different outcomes for the universe we see today. If a particular model predicts that our universe should look different from the way it really does, then that model must be wrong. On the other hand, if a new model of particle physics explains some previously unexplained aspects of the universe, then it may be on the right track. The grand unified theories of particle physics discussed in Section 17.1 have not yet been proved, but many scientists believe that these theories are on the right track for just this reason. In particular, the grand unified theories suggest that the universe at a very early age underwent the dramatic burst of expansion that we call *inflation*, and inflation seems to explain several key aspects of our universe that are otherwise left unexplained by the Big Bang theory.

• What aspects of the universe were originally unexplained by the Big Bang model?

The Big Bang theory has gained wide acceptance because of the strong evidence from the cosmic microwave background and the measured abundances of light elements in the universe. However, without the addition of more speculative physics such as that of the grand unified theories and their prediction of inflation, the Big Bang theory leaves several major aspects of our universe unexplained. Three of the most pressing questions are the following:

The original Big Bang model left several important questions unanswered.

- *Where does structure come from?* Recall that our models of the formation of galaxies and larger structures all assume that gravity collected matter around regions of slightly enhanced density in the early universe. Thus, explaining the origin of structure requires that the Big Bang must have somehow produced these slight density enhancements. The subtle temperature differences seen in the cosmic microwave background (see Figure 17.8) tell us that regions of enhanced density did indeed exist at the end of the era of nuclei, when the universe was 380,000 years old. But we still need to explain where the density enhancements came from.
- *Why is the large-scale universe so uniform?* Although the slight temperature variations in the cosmic microwave background show that the universe is not *perfectly* uniform on large scales, the overall smoothness is nonetheless remarkable. Observations of the cosmic microwave background tell us that the density of the universe at the end of the era of nuclei varied from place to place by no more than about 0.01%, and this uniformity explains why distant reaches of the universe look so similar today. Without inflation, the Big Bang theory does not explain why distant reaches of the universe look so similar.
- *Why is the density of the universe close to the critical density?* The total density of dark matter plus dark energy in the universe [Section 16.4] appears to be remarkably close to the critical density—so close that it is difficult to consider it a coincidence. After all, there is no obvious reason why the density could not have turned out to be anything, such as 1,000 times the critical density or 0.0000001 times the critical density. But without inflation, the Big Bang model is unable to explain the near-critical density of the universe as anything other than luck.

• How does inflation explain these features of the universe?

Physicist Alan Guth realized in 1981 that grand unified theories could potentially answer all three questions. These theories predict that the separation of the strong force from the GUT force should have released enormous energy, causing the universe to expand dramatically, perhaps by a factor of 10^{30} in less than 10^{-36} second. This dramatic expansion is what we call *inflation*. While it sounds outrageous to talk about something that happened in the first trillion-trillion-trillionth of a second in the history of the universe, inflation may have shaped the way the universe looks today.

Structure: Giant Quantum Fluctuations To understand how inflation explains the origin of structure, we need to recognize a special feature of energy fields. Laboratory-tested principles of quantum mechanics tell us that on very small scales the energy fields at any point in space are always fluctuating. Thus, the distribution of energy through space is very slightly irregular, even in a complete vacuum. The tiny quantum "ripples" that make up the irregularities can be characterized by a wavelength that corresponds roughly to their size. In principle, quantum ripples in the very early universe could have been the seeds for density enhancements that later grew into galaxies. However, the wavelengths of the original ripples were far too small to explain density enhancements like those we see imprinted on the cosmic microwave background.

Inflation would have stretched tiny quantum ripples to enormous sizes, allowing them to become the density enhancements around which galaxies later formed.

Inflation would have dramatically increased the wavelengths of these quantum fluctuations. The fantastic growth of the universe during the period of inflation would have stretched tiny ripples from sizes smaller than an atomic nucleus to the size of our solar system (Figure 17.12), making them large enough to become the density enhancements from which galaxies and larger structures later formed. Amazingly, the structure of today's universe may have started as tiny quantum fluctuations just before the period of inflation.

Figure 17.12

During inflation, ripples in spacetime would have stretched by a factor of perhaps 10^{30}. The peaks of these ripples then would have become the density enhancements that produced all the structure we see in the universe today.

Uniformity: Equalizing Temperatures and Densities We'll next consider how inflation explains the overall uniformity of the universe on large scales, but first let's look more closely at the question of why the uniformity is surprising. The idea that different parts of the universe were very similar shortly after the Big Bang may seem quite natural, but on further inspection the uniformity of the universe becomes difficult to explain.

Imagine observing the cosmic microwave background in a certain part of the sky. You are seeing microwaves that have traveled through the universe since the end of the era of nuclei, just 380,000 years after the Big Bang. Thus, you are seeing a region of the universe as it was some 14 billion years ago, when the universe was only 380,000 years old. Now imagine turning around and looking at the background radiation coming from the opposite direction. You are also seeing this region at an age of 380,000 years, and it looks virtually identical in temperature and density. The surprising part is this: The two regions are billions of light-years apart on opposite sides of our observable universe but we are seeing them as they were when they were only 380,000 years old. They can't possibly have exchanged light or any other information. A signal traveling at the speed of light from one to the other would barely have started its journey. So how did they come to have the same temperature and density?

Inflation explains large-scale uniformity by saying that distant regions of our observable universe today were once close enough to exchange radiation.

Inflation answers this question by saying that even though the two regions cannot have had any contact *since* the time of inflation, they were in contact prior to that time. Before the onset of inflation when the universe was 10^{-38} second old, the two regions were less than 10^{-38} light-second away from one another. Thus, radiation traveling at the speed of light would have had time to bounce between the two regions, and this exchange of energy equalized their temperatures and densities. Inflation then pushed these equalized regions to much greater distances, far out of contact with one another. Like criminals getting their stories straight before being locked in separate jail cells, the two regions (and all other parts of the observable universe) came to the same temperature and density before inflation spread them far apart.

Because inflation caused different regions of the universe to separate so vastly in such a short period of time, many people wonder whether it violates Einstein's theories saying that nothing can move faster than the speed of light. It does not, because nothing actually *moves* through space as a result of inflation or the ongoing expansion of the universe. Instead, the expansion of the universe is the expansion of *space itself.* Objects may be separating from one another at a speed faster than the speed of light, but no matter or radiation is able to travel between them during that time. In essence, inflation opens up a huge gap in space between objects that were once close together. The objects get very far apart, but nothing ever travels between them at a speed that exceeds the speed of light.

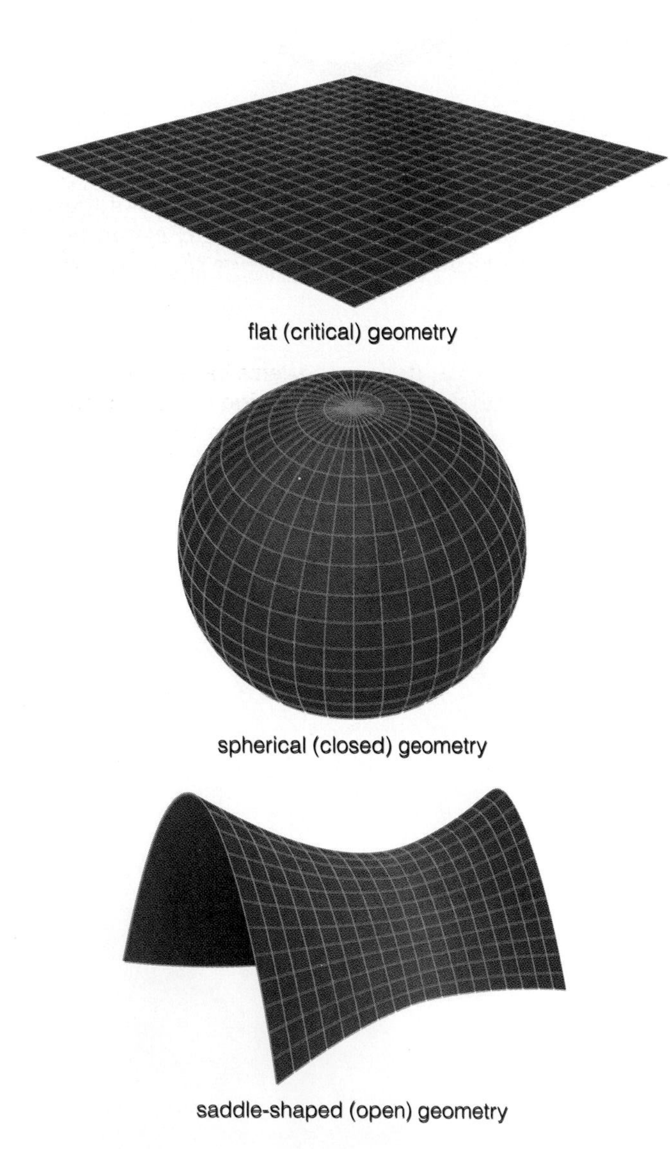

Figure 17.13

Analogies to the three possible categories of overall geometry for the universe. Keep in mind that the real universe has these "shapes" in more dimensions than we can see.

Density: Balancing the Universe The third question answered by inflation asks why the matter density of the universe is so close to the critical density. Another way to say that the universe's density is close to critical is to say that the overall geometry of the universe is remarkably "flat." To understand this idea, we must consider the overall geometry of the universe in a little more detail.

Recall that Einstein's general theory of relativity tells us that gravity can curve the structure of spacetime [Section 13.3]. Although we cannot visualize this curvature in all three dimensions of space (or all four dimensions of spacetime), we can detect its presence by its effects on light. For example, observations of gravitational lensing [Section 16.2] tell us that we are seeing light that has passed through a curved region of space. Although the curvature of the universe can vary from place to place, the universe as a whole must have some overall shape. Almost any shape is possible, but all the possibilities fall into just three general categories (Figure 17.13). By analogy to objects that we can see in three dimensions, scientists refer to these three categories of shape as *spherical* (or closed), *flat* (or critical), and *saddle-shaped* (or open).

According to general relativity, the overall geometry would be flat if the matter density of the universe were precisely equal to the critical density, in which case the kinetic energy of expansion would precisely balance the universe's overall gravitational pull. In the absence of dark energy, any imbalance in these energies causes curvature of spacetime, and deviations from precise balance grow more severe as the universe evolves. For example, if the universe had been 10% denser at the end of the era of nuclei, it would have collapsed long ago. On the other hand, if it had been 10% less dense at that time, galaxies would never have formed before expansion spread all the matter too thin. Thus, the universe had to start out remarkably balanced to be even remotely close to flat today.

Inflation predicts that the overall geometry of the universe should appear flat, and hence that the overall density of matter and energy should equal the critical density.

Inflation can explain this precise balance by its effects on the geometry of the universe. In terms of Einstein's theory, the effect of inflation on spacetime curvature is similar to the flattening of a balloon's surface when you blow into it (Figure 17.14). The flattening of space during the period of inflation would have been so enormous that any curvature the universe might have had previously would be noticeable only on size scales much larger than the observable universe. Thus, inflation predicts that the overall geometry of the universe should appear perfectly flat, in which case the overall density of matter plus energy should be precisely equal to the critical density.

The fine balance predicted by inflation turns out to be both a success and a potential pitfall of the inflation theory. On one hand, it explains how the universe managed to have a density just right to allow the birth of galaxies. On the other hand, its prediction of a perfectly flat universe seems to disagree with observations showing that the total density of matter (including dark matter) is only about 25% of the critical density.

Dark energy might explain this shortfall in the density of dark matter. Remember that Einstein's theory of relativity tells us that mass can be transformed into energy and vice versa, which means that energy must be able to curve spacetime in the same way that mass can curve spacetime (see box General Relativity and Curvature of Spacetime, p. 339). Thus, the dark energy associated with a large-scale repulsive force could compensate for the shortfall in the matter density, making the present-day universe flatter than it would otherwise be. Remarkably, the supernova measurements indicating that the expansion of the universe is accelerating [Section 16.4] also show that the strength of the implied repulsive force is about right to render the universe perfectly flat, just as inflation predicts. In other words, the total density of matter and dark energy together may indeed be precisely equal to the critical density.

Figure 17.14

As a balloon expands, its surface seems increasingly flat to an ant crawling along it. Inflation is thought to have made the universe seem flat in a similar way.

• How can we test the idea of inflation?

We've seen that inflation answers some outstanding mysteries about the universe, but did it really happen? We cannot directly observe the universe at the very early time when inflation is thought to have occurred. Nevertheless, we can test the idea of inflation by exploring whether its predictions are consistent with our observations of the universe at later times. Scientists are only beginning to make observations that test inflation, but the findings to date are consistent with the idea that an early inflationary episode made the universe uniform and flat while planting the seeds of structure formation.

Observed patterns in the temperature of the cosmic microwave background are consistent with the idea of inflation and reveal details about the universe's geometry, composition, and age.

The strongest tests of inflation to date come from detailed studies of the cosmic microwave background, and in particular of the map made by the WMAP satellite (see Figure 17.8). Remember that this map shows tiny temperature differences corresponding to density variations in the universe at the end of the era of nuclei, when the universe was about 380,000 years old. However, according to the theory of inflation, these density enhancements were actually created much earlier, when inflation caused tiny quantum ripples to expand into seeds of structure. Thus, careful observations of the temperature variations in the microwave background can tell us about the structure of the universe during its first instant of existence.

Figure 17.15 shows an analysis of the temperature variations observed by WMAP in the cosmic microwave background, along with additional data from other microwave telescopes. The graph shows how the typical temperature differences between patches of sky depend on their angular size on the celestial sphere. The dots represent data from the ob-

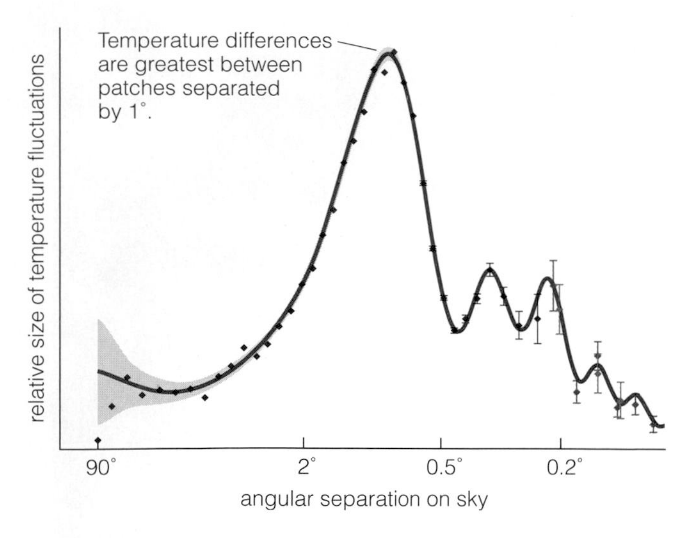

Figure 17.15

This graph shows how detailed analysis of temperature differences in the cosmic microwave background supports the idea of inflation. The data points indicate how the typical temperature differences between patches of sky depend on their angular size on the celestial sphere. (Black dots come from the WMAP data shown in Figure 17.8, and the blue and green points represent data from other telescopes.) The red curve shows the prediction of a model that relies on inflation to produce a universe whose ordinary matter density, dark matter density, and expansion rate are all similar to their observed values. Close agreement between the data points and the model indicates that the "genetic code" predicted by inflation is very similar to the true "genetic code" of our universe.

servations, and the red curve shows the inflation-based model that best fits the observations. This model makes specific predictions not only about the data shown in the figure, but also about other characteristics of our universe such as its overall geometry, composition, and age. Thus, in a sense, these new observations of the cosmic microwave background are revealing the "genetic code" of the universe. To the extent that we have been able to read this code, it aligns reassuringly well with the universe we observe about us at the present time. The "genetic code" according to this model corresponds to a universe with the following features:

- The overall geometry is flat, implying that the total mass-energy of the universe must be equivalent to the critical density.
- The density of ordinary (baryonic) matter is 4.4% of the critical density, in agreement with observations of deuterium in the universe [Section 17.2].
- The total matter density is 27% of the critical density. Subtracting the 4.4% for ordinary matter, we conclude that extraordinary (nonbaryonic) dark matter makes up about 23% of the critical density, in agreement with what we infer from measurements of the masses of clusters of galaxies [Section 16.2].
- The combination of a flat geometry and a matter density lower than the critical density implies the existence of a repulsive "dark energy" that currently accelerates the expansion, in agreement with observations of distant supernovae [Section 16.4]. Because the total mass-energy of the universe is the critical density, and matter accounts for only 27% of this, dark energy must account for the remaining 73% of the mass-energy of the universe.
- The universe's age should be about 13.7 billion years at the current microwave temperature of 2.73 K, in agreement with what we infer from Hubble's constant [Section 15.2] and the ages of the oldest stars [Section 11.3].

This close correspondence between the genetic code inherent in the universe at an age of 380,000 years and our observations of the present-day universe, some 14 billion years later, is persuasive evidence in favor of the Big Bang in general and inflation in particular. The bottom line is that, all things considered, inflation does a remarkable job of explaining features of our universe that are otherwise unaccounted for in the Big Bang theory. Many astronomers and physicists therefore suspect that some process akin to inflation did affect the early universe, but the details of the interaction between high-energy particle physics and the evolving universe remain unclear. If these details can be worked out successfully, we face an amazing prospect—a breakthrough in our understanding of the very smallest particles, achieved by studying the universe on the largest observable scales.

17.4 OBSERVING THE BIG BANG FOR YOURSELF

You might occasionally read an article in a newspaper or a magazine questioning whether the Big Bang really happened. We will never be able to prove with absolute certainty that the Big Bang theory is correct. However, no one has come up with any other model of the universe that so successfully explains so much of what we see. As we have discussed, the Big Bang model makes at least two specific predictions that we have observationally verified: the characteristics of the cosmic microwave background and the composition of the universe. It also explains quite naturally many other features of the universe. So far, at least, we know of nothing that is inconsistent with the Big Bang model.

The Big Bang theory's very success has also made it a target for respected scientists, skeptical nonscientists, and crackpots alike. The nature of scientific work requires that we test established wisdom to make sure it is valid. A sound scientific disproof of the Big Bang theory would be a discovery of great importance. However, stories touted in the news media as disproofs of the Big Bang usually turn out to be disagreements over details rather than fundamental problems that threaten to bring down the whole theory. Yet scientists must keep refining the theory and tracking down disagreements, because once in a while a small disagreement blossoms into a full-blown scientific revolution.

SPECIAL TOPIC: HOW WILL THE UNIVERSE END?

According to the Big Bang theory, time and space have a beginning in the Big Bang. Do they also have an end? If we live in a recollapsing universe [Section 16.4], the answer seems to be a clear yes. Sometime in the distant future, the universal expansion will cease and reverse, and the universe will eventually come to an end in a fiery "Big Crunch." However, it now appears unlikely that we live in a recollapsing universe, in which case the universe seems destined to expand forever.

If dark energy is real, our Local Group of galaxies will become increasingly more lonely as the acceleration of the universe carries the more distant galaxies ever faster away from us. Some scientists speculated that the repulsive force of dark energy might even strengthen with time. In that case, the galaxies of the Local Group will someday separate, and the growing repuslive force could eventually tear apart our galaxy, our solar system, and even matter itself in a catastrophic event sometimes called the "Big Rip." However, the Big Rip is quite speculative; a more plausible scenario for the end of a perpetually expanding universe proceeds as follows.

The star–gas–star cycle in galaxies cannot continue forever, because not all the material is recycled. With each generation of stars, more mass becomes locked up in planets, brown dwarfs, white dwarfs, neutron stars, and black holes. Eventually, about a trillion years from now, even the longest-lived stars will burn out, and the galaxies will fade into darkness.

At this point, the only new action in the universe will occur on the rare occasions when two objects—such as two brown dwarfs or two white dwarfs—collide within a galaxy. The vast distances separating star systems in galaxies make such collisions extremely rare. For example, the probability of our Sun (or the white dwarf that it will become) colliding with another star is so small that it would be expected to happen only once in a quadrillion (10^{15}) years. Forever is a long time, however, and even low-probability events will eventually happen many times. If a star system experiences a collision once in a quadrillion years, it will experience about 100 collisions in 100 quadrillion (10^{17}) years. By the time the universe reaches an age of 10^{20} years, star systems will have suffered an average of 100,000 collisions each, making a time-lapse history of any galaxy look like a cosmic game of pinball.

These multiple collisions will severely disrupt galaxies. As in any gravitational encounter, some objects lose energy in such collisions and some gain energy. Objects that gain enough energy will be flung into intergalactic space, where they will drift ever farther from all other objects with the expansion of the universe. Objects that lose energy will eventually fall to the galactic center, forming a gigantic black hole. The remains of the universe will consist of black holes with masses as great as a trillion solar masses widely separated from a few scattered planets, brown dwarfs, and stellar corpses. If Earth somehow survives, it will be a frozen chunk of rock in the darkness of the expanding universe, billions of light-years from any other solid object.

If grand unified theories are correct, Earth still cannot last forever. These theories predict that protons will eventually fall apart. The predicted lifetime of protons is extremely long: a half-life of at least 10^{32} years. However, if protons really do decay, then by the time the universe is 10^{40} years old Earth and all other atomic matter will have disintegrated into radiation and subatomic particles, such as electrons and neutrinos.

The final phase may come through a mechanism proposed by Stephen Hawking. According to Hawking's theory, even black holes cannot last forever. Instead, they slowly "evaporate," their mass-energy turning into radiation. The process is so slow that we do not expect to be able to see it from any existing black holes, but if Hawking is correct, then black holes in the distant future will disappear in brilliant bursts of radiation. The largest black holes last longest, but even trillion-solar-mass black holes will evaporate sometime after the universe reaches an age of 10^{100} years. From then on, the universe will consist only of individual photons and subatomic particles, each separated by enormous distances from the others. Nothing new will ever happen, and no events will ever occur that would allow an omniscient observer to distinguish past from future. In a sense, the universe will finally have reached the end of time.

Lest any of this sound depressing, keep in mind that 10^{100} years is an indescribably long time. As an example, imagine that you wanted to write on a piece of paper a number that consisted of a 1 followed by 10^{100} zeros (that is, the number $10^{10^{100}}$). It sounds easy, but a piece of paper large enough to fit all those zeros *would not fit in the observable universe* today. If that still does not alleviate your concerns, you may be glad to know that a few creative thinkers speculate about ways in which the universe might undergo rebirth, even after the end of time.

You don't need to accept all you have read without question. The next time you are musing on the universe's origins, try an experiment for yourself. Go outside on a clear night, look at the sky, and ask yourself why it is dark.

• Why is the darkness of the night sky evidence for the Big Bang?

If the universe were infinite, unchanging, and everywhere the same, then the entire night sky would blaze as brightly as the Sun. Johannes Kepler [Section 3.3] was one of the first people to reach this conclusion, but we now refer to the idea as **Olbers' paradox** after Heinrich Olbers, a German astronomer of the 1800s.

If the universe were infinite, unchanging, and everywhere the same, then the entire night sky would be bright, because it would be completely covered with stars.

To understand how Olbers' paradox comes about, imagine that you are in a dense forest on a flat plain. If you look in any direction, you'll likely see a tree. If the forest is small, you might be able to see through some gaps in the trees to the open plains, but larger forests have fewer gaps (Figure 17.16). An infinite forest would have no gaps at all—a tree trunk would block your view along any line of sight.

The universe is like a forest of stars in this respect. In an unchanging universe with an infinite number of stars, we would see a star in every di-

Figure 17.16

Olbers' paradox is similar to the view through a forest.

a In a large forest, a tree will block your view no matter where you look. Similarly, in an unchanging universe with an infinite number of stars, we would expect to see stars in every direction, making the sky bright even at night.

b In a small forest, you can see open spaces beyond the trees. Because the night sky is dark, the universe must similarly have spaces in which we see nothing beyond the stars, which means either that the number of stars is finite or that the universe changes in a way that prevents us from seeing an infinite number of them.

rection, making every point in the sky as bright as the Sun's surface. Even the presence of obscuring dust would not change this conclusion. The intense starlight would heat the dust over time until it too glowed like the Sun or evaporated away.

There are only two ways out of this dilemma. Either the universe has a finite number of stars, in which case we would not see a star in every direction, or it changes over time in some way that prevents us from seeing an infinite number of stars. For several centuries after Kepler first recognized the dilemma, astronomers leaned toward the first option. Kepler himself preferred to believe that the universe had a finite number of stars because he thought it had to be finite in space, with some kind of dark wall surrounding everything. Astronomers in the early twentieth century preferred to believe that the universe was infinite in space but that we lived inside a finite collection of stars. They thought of the Milky Way as an island floating in a vast black void. However, subsequent observations showed that galaxies fill all of space more or less uniformly. We are therefore left with the second option: The universe changes over time.

The Big Bang theory is the best explanation we have for why the night sky is dark.

The Big Bang theory solves Olbers' paradox in a particularly simple way. It tells us that we can see only a finite number of stars because the universe began at a particular moment. While the universe may contain an infinite number of stars, we can see only those that lie within the observable universe, inside our cosmological horizon [Section 15.2]. There are other ways in which the universe could change over time and prevent us from seeing an infinite number of stars, so Olbers' paradox does not *prove* that the universe began with a Big Bang. However, we must have some explanation for why the sky is dark at night, and no explanation besides the Big Bang also explains so many other observed properties of the universe so well.

THE BIG PICTURE

Putting Chapter 17 into Context

Our "big picture" is now about as complete as it gets. We've discussed the universe from Earth outward, and from the beginning to the end. When you think back on this chapter, keep in mind the following ideas:

- Predicting conditions in the early universe is straightforward, as long as we know how matter and energy behave under such extreme conditions.
- Our current understanding of physics allows us to reconstruct the conditions that prevailed in the universe all the way back to the first 10^{-10} second. Our understanding is less certain back to 10^{-38} second. Beyond 10^{-43} second, we run up against the present limits of human knowledge.
- Although it may sound strange to talk about the universe during its first fraction of a second, our ideas about the Big Bang rest on a solid foundation of observational, experimental, and theoretical evidence. We cannot say with absolute certainty that the Big Bang really happened, but no other model ever proposed has so successfully explained how our universe came to be as it is.

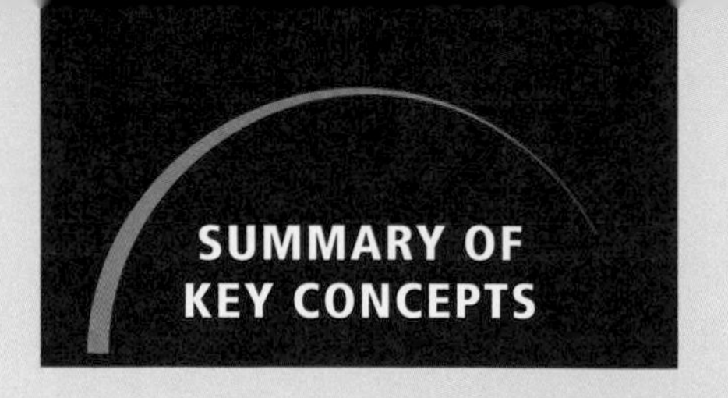

17.1 THE BIG BANG

- **What were conditions like in the early universe?**

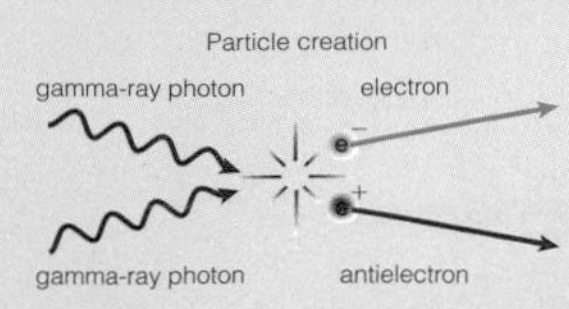

The early universe was filled with radiation and elementary particles. It was so hot and dense that the energy of radiation could turn into particles of **matter** and **antimatter**, which then collided and turned back into radiation.

- **What is the history of the universe according to the Big Bang theory?**

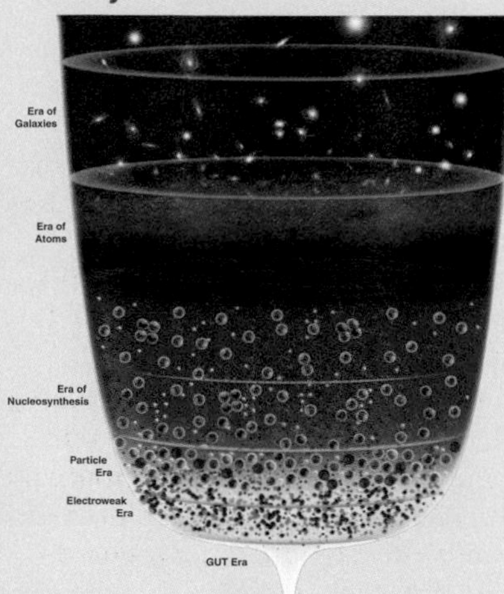

The universe has been through a series of eras, each marked by unique physical conditions. We know little about the **Planck era**, when the four forces may have all behaved as one. Gravity became distinct at the start of the **GUT era**, which may have ended with the rapid expansion called **inflation**. Electromagnetism and the weak force became distinct at the end of the **electroweak era**. Matter particles annihilated all the antimatter particles at the end of the **particle era**. Fusion of protons and neutrons into helium ceased at the end of the **era of nucleosynthesis**. Hydrogen nuclei captured all the free electrons, forming hydrogen atoms at the end of the **era of nuclei**. Galaxies began to form at the end of the **era of atoms**. The **era of galaxies** continues to this day.

17.2 EVIDENCE FOR THE BIG BANG

- **How do we observe the radiation left over from the Big Bang?**

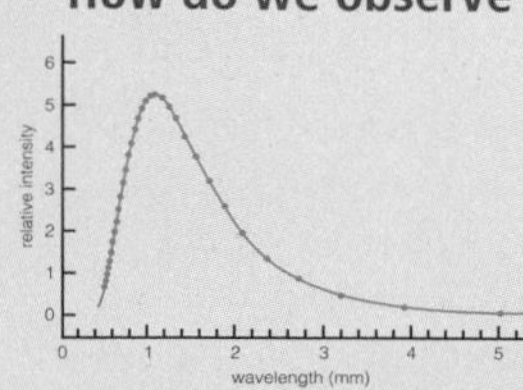

Telescopes that can detect microwaves allow us to observe the **cosmic microwave background**—radiation left over from the Big Bang. Its spectrum matches the characteristics expected of the radiation released at the end of the era of nuclei, spectacularly confirming a key prediction of the Big Bang theory.

- **How do the abundances of elements support the Big Bang?**

The Big Bang theory predicts the ratio of protons to neutrons during the era of nucleosynthesis, and from this predicts that the chemical composition of the universe should be about 75% hydrogen and 25% helium (by mass). This matches observations of the cosmic abundances, another spectacular confirmation of the Big Bang theory.

17.3 THE BIG BANG AND INFLATION

- **What aspects of the universe were originally unexplained by the Big Bang model?**

(1) The origin of the density enhancements that turned into galaxies and larger structures. (2) The overall smoothness of the universe on large scales. (3) The fact that the actual density of matter is close to the critical density.

- **How does inflation explain these features of the universe?**

(1) The episode of inflation stretched tiny, random quantum fluctuations to sizes large enough for them to become the density enhancements around which structure later formed. (2) The universe is smooth on large scales because, prior to inflation, everything we can observe today was close enough together for temperatures and densities to equalize. (3) Inflation caused the universe to expand so much that the observable universe appears geometrically flat, implying that its overall density of mass plus energy equals the critical density.

- **How can we test the idea of inflation?**

Models of inflation make specific predictions about the temperature patterns we should observe in the cosmic microwave background. The observed patterns seen in recent observations by microwave telescopes match those predicted by inflation.

17.4 OBSERVING THE BIG BANG FOR YOURSELF

- **Why is the darkness of the night sky evidence for the Big Bang?**

Olbers' paradox tells us that if the universe were infinite, unchanging, and filled with stars, the sky would be everywhere as bright as the surface of the Sun, and it would not be dark at night. The Big Bang theory solves this paradox by telling us that the night sky is dark because the universe has a finite age, which means we can see only a finite number of stars in the sky.

EXERCISES AND PROBLEMS

REVIEW QUESTIONS

1. What is *antimatter?* How were particle–antiparticle pairs created in the early universe? How were they destroyed?
2. Explain what we mean by the *Big Bang theory.*
3. Make a list of the major eras in the history of the universe, summarizing the important events thought to have occurred during each era.
4. Why can't our current theories describe the history of the universe during the *Planck era?*
5. What are the four forces that operate in the universe today? Why do we think there were fewer forces operating in the early universe?
6. What are *grand unified theories?* According to these theories, how many forces operated during the *GUT era?* How were these forces related to the four forces that operate today?
7. What do we mean by *inflation,* and when do we think it occurred?
8. Why do we think there was a slight imbalance between matter and antimatter in the early universe? What happened to all the antimatter, and when?
9. How long did the *era of nucleosynthesis* last? Explain why this era was so important in determining the chemical composition of the universe forever after.
10. When we observe the *cosmic microwave background,* at what age are we seeing the universe? How long have the photons in the background been traveling through space? Explain.
11. Briefly describe how the cosmic microwave background was discovered. How does the existence and nature of this radiation support the Big Bang theory?
12. How does the chemical abundance of helium in the universe support the Big Bang theory? Explain.
13. How do measurements of deuterium and lithium tell us about the density of the universe, and why do they suggest that most dark matter consists of WIMPs?
14. Describe each of the three major questions left unanswered by the Big Bang theory without inflation, and explain how inflation answers each of them.
15. How can observations of the cosmic microwave background—radiation released when the universe was 380,000 years old—tell us about the universe at the much earlier time when inflation occurred? Summarize the geometry, composition, and age of the universe according to observations made to date.
16. What is *Olbers' paradox,* and how is it resolved by the Big Bang theory?

TRUE STATEMENTS?

Decide whether each of the following statements is true and explain why it is or is not.

17. Although the universe today appears to be made mostly of matter and not antimatter, the Big Bang theory suggests that the early universe had nearly equal amounts of matter and antimatter.
18. According to the Big Bang theory, the cosmic microwave background was created when energetic photons ionized the neutral hydrogen atoms that originally filled the universe.
19. While the existence of the cosmic microwave background is consistent with the Big Bang theory, it is also easily explained by assuming that it comes from individual stars and galaxies.
20. According to the Big Bang theory, most of the helium in the universe was created by nuclear fusion in the cores of stars.
21. The theory of inflation suggests that the structure in the universe today may have originated as tiny quantum fluctuations.
22. The fact that the night sky is dark tells us that the universe cannot be infinite, unchanging, and everywhere the same.

PROBLEMS

(Quantitative problems are marked with an asterisk.)

23. *Life Story of a Proton.* Tell the life story of a proton from its formation shortly after the Big Bang to its presence in the nucleus of an oxygen atom you have just inhaled. Your story should be creative and imaginative, but it should also demonstrate your scientific understanding of as many stages in the proton's life as possible. You can draw on material from the entire book, and your story should be three to five pages long.
24. *Creative History of the Universe.* The story of creation as envisioned by the Big Bang theory is quite dramatic, but it is usually told in a fairly straightforward, scientific way. Write a more dramatic telling of the story, in the form of a short story, play, or poem. Be as creative as you wish, but be sure to remain accurate according to the science as it is understood today.

*25. 10^{100} *Years.* In the box "How Will the Universe End?" we found that the final stage in the history of a perpetually expanding universe will come about 10^{100} years from now. Such a large number is easy to write but difficult to understand. This problem investigates some of the incredible properties of very large numbers.
   a. The current age of the universe is around 10^{10} years. How much longer is a trillion years than this current age? How much longer is 10^{15} years? 10^{20} years?
   b. Suppose protons decay with a half-life of 10^{32} years. When will the number of remaining protons be half the current amount? When will it be a quarter of its current amount? How many half-lives will have gone by when the universe reaches an age of 10^{34} years? What fraction of the original protons will remain at this time? Based on your answers, is it reasonable to conclude that *all* protons in today's universe will be gone by the time the universe is 10^{40} years old? Explain.
   c. Suppose you were trying to write 10^{100} zeros on a piece of paper and could write microscopically, so that each zero (including the thickness of the pencil mark) occupied a volume of 1 cubic micrometer—about the size of a bacterium. Could 10^{100} zeros of this size fit in the observable universe? Explain. (*Hints:* Calculate the volume of the observable universe in cubic micrometers by assuming it is a sphere with a radius of 14 billion light-years. The volume of a sphere is $\frac{4}{3} \times \pi \times (\text{radius})^3$; 1 light-year $\approx 10^{15}$ meters; 1 cubic meter $= 10^{18}$ cubic micrometers.)

DISCUSSION QUESTIONS

26. *The Moment of Creation.* You've probably noticed that, in discussing the Big Bang theory, we never quite talk about the first instant. Even our most speculative theories at present take us back only to within 10^{-43} second of creation. Do you think it will *ever* be possible for science to consider the moment of creation itself? Will we ever be able to answer questions such as *why* the Big Bang happened? Defend your opinions.
27. *The Big Bang.* How convincing do you find the evidence for the Big Bang model of the universe's origin? What are the strengths of the theory? What does it fail to explain? Overall, do *you* think the Big Bang really happened? Defend your opinion.

MATH HELP AND EXERCISES

The Astronomy Place Web site has additional mathematical topics, with worked examples and additional practice problems. For this chapter, you'll find the following:

- Laws of Thermal Radiation

For a complete list of media resources available, go to **www.astronomyplace.com** and choose Chapter 17 from the pull-down menu.

ASTRONOMY PLACE WEB TUTORIALS

Tutorial Review of Key Concepts

Use the following interactive **Tutorial** at **www.astronomyplace.com** to review key concepts from this chapter.

Hubble's Law Tutorial

Lesson 1 Hubble's Law
Lesson 2 The Expansion of the Universe
Lesson 3 The Age of the Universe

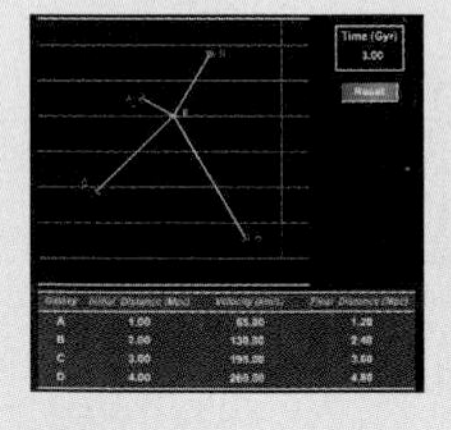

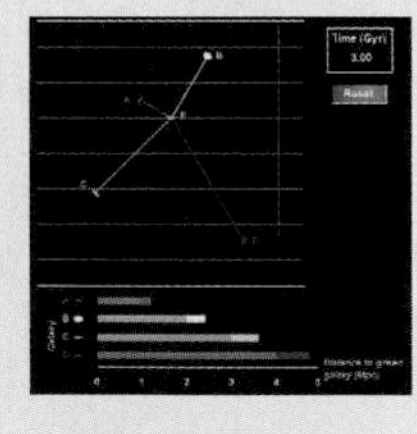

Supplementary Tutorial Exercises

Use the interactive **Tutorial Lesson** to explore the following questions.

Hubble's Law Tutorial, Lesson 2

1. How does the observed rate of expansion allow us to determine distances between galaxies in the distant past?
2. Suppose we have an uncertainty of 10% in the rate of expansion. Would this cause significant changes in what we conclude about temperatures in the early universe? Why or why not?

MOVIES

Check out the following narrated and animated short documentary available on **www.astronomyplace.com** for a helpful review of key ideas covered in this chapter.

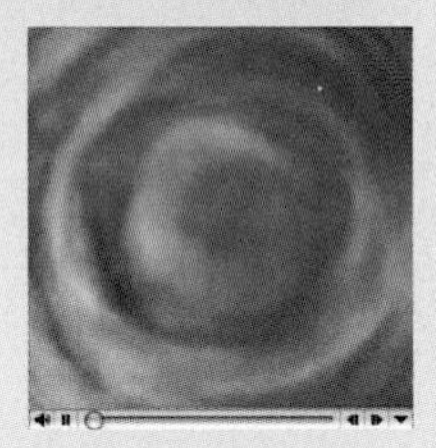

From the Big Bang to the Galaxies Movie

WEB PROJECTS

Take advantage of the useful Web links on **www.astronomyplace.com** to assist you with the following projects.

1. *New Tests of the Big Bang Theory.* The Cosmic Background Explorer (COBE) satellite provided striking confirmation of several predictions of the Big Bang theory. Newer satellites have been designed to test the Big Bang theory further, primarily by observing the subtle variations in the cosmic microwave background with a much higher sensitivity than COBE. Use the Web to gather pictures and information about the COBE mission and its successors, such as WMAP or Planck. Write a one- to two-page report about the strength of the evidence compiled by COBE and WMAP and how much more we might learn from upcoming missions.
2. *Decay of the Proton.* One of the most startling predictions of grand unified theories is that protons will eventually decay, albeit with a half-life of more than 10^{32} years. If this is true, it may be possible to observe an occasional proton decay, despite the extraordinarily long half-life. Several experiments to look for proton decay are under way or being planned. Find out about one or more of these experiments, and write a one- to two-page summary in which you describe the experiment(s), any results to date, and what these results (or potential results) mean to the grand unified theories.
3. *New Ideas in Inflation.* The idea of inflation solves many of the puzzles associated with the standard Big Bang theory, but we are still a long way from finding evidence confirming that inflation really occurred. Find recent articles that discuss some of the latest ideas about inflation and how we might test these ideas. Write a two- to three-page summary of your findings.

18
Life in the Universe

LEARNING GOALS

18.1 LIFE ON EARTH
- When did life arise on Earth?
- How did life arise on Earth?
- What are the necessities of life?

18.2 LIFE IN THE SOLAR SYSTEM
- Could there be life on Mars?
- Could there be life on Europa or other jovian moons?

18.3 LIFE AROUND OTHER STARS
- Are habitable planets likely?
- Are Earth-like planets rare or common?

18.4 THE SEARCH FOR EXTRATERRESTRIAL INTELLIGENCE
- How many civilizations are out there?
- How does SETI work?

18.5 INTERSTELLAR TRAVEL AND ITS IMPLICATIONS TO CIVILIZATION
- How difficult is interstellar travel?
- Where are the aliens?

We have covered a lot of ground in this book, discussing fundamental questions about the nature and origin of our planet, our star, our galaxy, and our universe. But we have not yet discussed one of the most profound questions of all: Are we alone? The universe seems to be filled with worlds beyond imagination—more than 100 billion star systems in our galaxy alone, and some 100 billion galaxies in the observable universe—yet we do not know whether any other world has ever been home to life.

In this chapter, we will discuss the possibility of life beyond Earth. We'll begin by considering the history of life on Earth, which should help us understand the prospects for finding life elsewhere. We'll then consider the possibility of finding microbial life elsewhere in our solar system or beyond and examine efforts to search for extraterrestrial intelligence (SETI). Finally, we'll discuss the astonishing implications that the search for life may hold for the future of our own civilization.

18.1 LIFE ON EARTH

It may seem that aliens are everywhere. Popular shows and movies, such as *Star Trek* and *Star Wars,* show aliens from many different worlds traveling easily among the stars. Closer to home, supermarket tabloids routinely carry headlines about the latest alien atrocities or about alien corpses hidden by the government at "Area 51."

Despite their media popularity, any alien visitors have been sadly negligent in leaving scientific evidence of their trespass. Decades of scientific observation and study have not turned up a single piece of undeniable evidence that aliens have been here (see the box "Are Aliens Already Here?," p. 483). Why, then, do so many scientists believe it is worth the effort to search for life beyond Earth?

For a long time, the answer was simply that it seemed natural for other worlds to be inhabited. Belief in life on other worlds was common even among ancient Greek philosophers, and many famous scientists of the past few centuries took it as a given that intelligent beings exist on other planets. For example, Kepler suggested that the Moon was inhabited, and William Herschel (co-discoverer of Uranus with his sister Caroline) spoke of life on virtually all the planets in our solar system. Most famously, in the late nineteenth century Percival Lowell claimed to see networks of canals on Mars, which he argued were the mark of an advanced civilization. Lowell's ideas became the basis for H. G. Wells's novel *The War of the Worlds.*

Hopes of finding life on our planetary neighbors were dashed by the bleak images of Mars returned by spacecraft and the discovery of the runaway greenhouse effect on Venus. For a few decades, scientific interest in extraterrestrial life waned. However, scientific interest in extraterrestrial life has been on the upswing again during the past couple of decades. Part of the renewed interest has come from the discovery of planets around other stars [Section 6.5], which confirms that there are plenty of places to look for life beyond our own solar system. But new discoveries about life on Earth have played an even more important role. In particular, three recent developments in the study of life on Earth have made it seem much more likely that we could find life elsewhere:

- We have learned that life arose quite early in Earth's history, suggesting that life might also form quickly on other worlds with the right conditions.
- Laboratory experiments have shown that the chemical constituents thought to have been common on the young Earth combine readily into complex organic molecules. We may never know exactly how life arose on Earth, but these experiments suggest that life might arise through naturally occurring chemistry—in which case the same chemistry could have given rise to life on many other worlds.
- We have discovered microscopic living organisms that can survive in conditions similar to those found on at least some other worlds in our solar system, suggesting that the basic necessities of life may be quite common in the universe.

Because these three ideas are so important to understanding the modern science of life in the universe, let's examine each of them in greater detail. Along the way, we'll also discuss a little bit about the nature and history of life on Earth. We will then be prepared to consider how we might actually search for life on other worlds.

• When did life arise on Earth?

Our planet was born about 4.6 billion years ago, but we would not expect life to have had much of a chance to survive during the first several hundred million years of Earth's history. This was the time of the *heavy bombardment* [Section 6.4], when Earth should have been repeatedly struck by large asteroids or comets. At least a few of these impacts probably had sufficient energy to vaporize the early oceans completely, killing off any life that might already have been present on Earth. The ancestors of living organisms today could not have arisen much before the end of the heavy bombardment.

Life could not have taken a permanent hold on Earth until the end of the heavy bombardment, between about 4.2 and 3.9 billion years ago.

Studies of craters on the Moon suggest that the last major impacts of the heavy bombardment occurred between about 4.2 and 3.9 billion years ago. Remarkably, we now have evidence suggesting that life was thriving prior to 3.85 billion years ago. If we are interpreting the evidence correctly, it means that life arose in a geological blink of the eye from the moment when conditions first allowed it. To understand this evidence and its importance, we need to discuss how scientists study the history of life on Earth.

The Geological Time Scale We learn about the history of life on Earth through the study of **fossils**, relics of organisms that lived and died long ago. Most fossils form when dead organisms fall to the bottom of a sea (or other body of water) and are gradually buried by layers of sediment. The sediments are produced by erosion on land and carried by rivers to the sea. Over millions of years, sediments pile up on the seafloor, and the weight of the upper layers compresses underlying layers into rock. Erosion or tectonic activity can later expose the fossils (Figure 18.1). In some

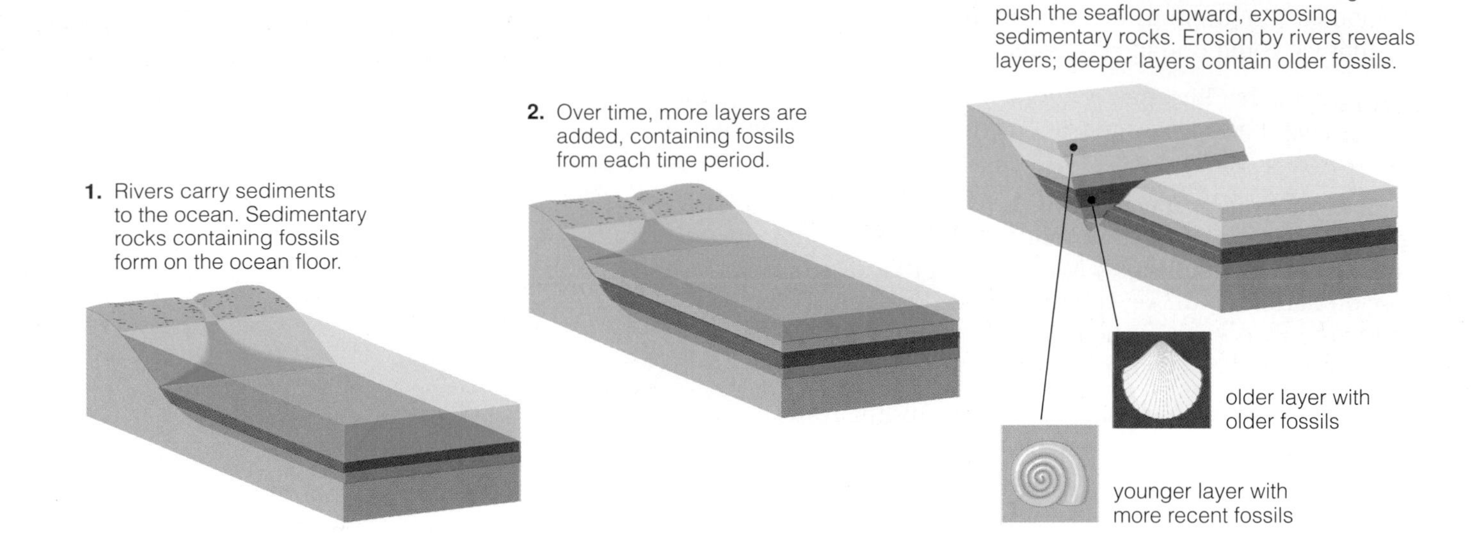

Figure 18.1

Formation of sedimentary rock. Each layer represents a particular time and place in Earth's history and is characterized by fossils of organisms that lived in that time and place.

places, such as the Grand Canyon, the sedimentary layers record billions of years of Earth's history (Figure 18.2).

The key to reconstructing the history of life is to determine the dates at which fossil organisms lived. The *relative* ages of fossils found in different layers are easy to determine: Deeper layers formed earlier and contain more ancient fossils. Radiometric dating [Section 6.4] confirms these relative ages and gives us fairly precise absolute ages for fossils. Based on the layering of rocks and fossils, geologists divide Earth's history into a set of distinct intervals that make up what we call the **geological time scale**. Figure 18.3 shows the names of the various intervals on a timeline, along with numerous important events in Earth's history.

THINK ABOUT IT Study Figure 18.3. How does the length of time during which animals and plants have lived on land compare to the length of time during which life has existed? How does the length of time during which humans have existed compare to the length of time since mammals and dinosaurs first arose?

Figure 18.2

The rock layers of the Grand Canyon record 2 billion years of Earth's history.

Fossil Evidence for the Early Origin of Life You might wonder why the geological time scale shows so much more detail in the last few hundred million years than it does for earlier times. The answer is that fossils become increasingly difficult to find as we look deeper into Earth's history, for three major reasons: First, older rocks are much rarer than younger rocks, because most of Earth's surface is geologically young. Second, even when we find very old rocks, they often turn out to have been subject to transformations (caused by heat and pressure) that would have destroyed any fossil evidence they may have contained. Third, all life prior to a few hundred million years ago was microscopic, and microscopic fossils are much more difficult to identify than dinosaur bones.

Fossil evidence suggests that life on Earth was already thriving by 3.5 billion years ago.

Despite these difficulties, geologists have found a few very old rocks that suggest life was already thriving on Earth by 3.5 billion years ago, and possibly for several hundred million years before that. One

Figure 18.3

The geological time scale. Be sure to notice that the lower timeline is an expanded view of the last portion of the upper timeline. The eons, eras, and periods are defined by changes observed in the fossil record. The absolute ages come from radiometric dating. (The K–T event is the geological term for the impact linked to the mass extinction of the dinosaurs [Section 9.4].)

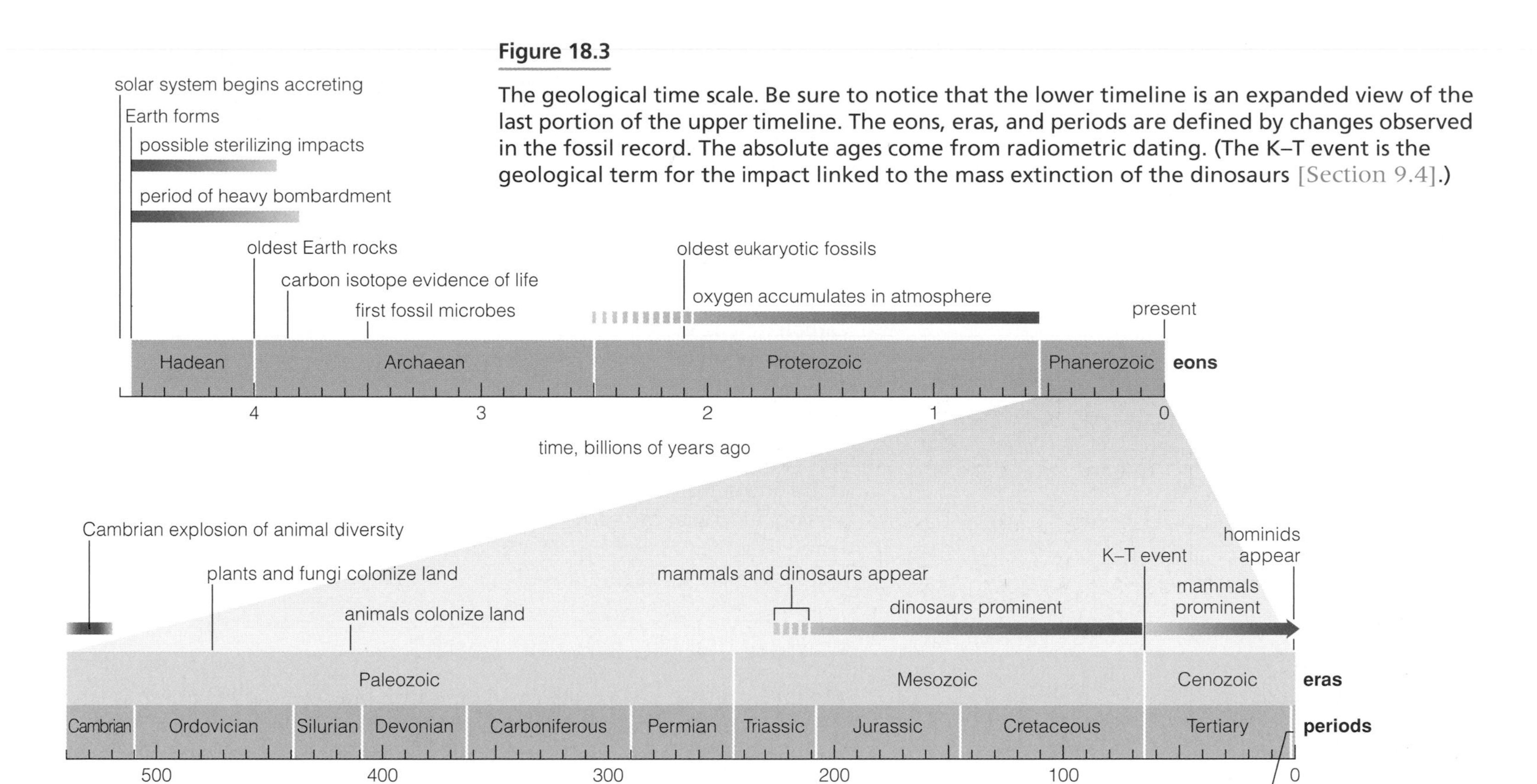

strong line of evidence comes from rocks called *stromatolites*. These rocks show structure nearly identical to that found in large bacterial mats today (Figure 18.4). Radiometric dating shows that some stromatolites are 3.5 billion years old.

The fact that organisms were already advanced enough to build stromatolites by 3.5 billion years ago suggests that more primitive organisms must have lived even earlier. Some evidence supports this idea. More ancient rocks that have undergone too much change to leave fossils intact may still hold carbon that was once part of living organisms.

Figure 18.4

Rocks called stromatolites offer evidence of photosynthetic life as early as 3.5 billion years ago.

a These large mats photographed at Shark Bay, Western Australia, are colonies of microbes known as "living stromatolites."

b The bands visible in this section of a modern-day mat are formed by layers of sediment adhering to different types of microbes.

c This section of a 3.5-billion-year-old stromatolite shows a structure nearly identical to that of a living mat. Thus, it offers strong evidence of having been made by microbes, including some photosynthetic ones, that lived 3.5 billion years ago.

Carbon has two stable isotopes: carbon-12, with six protons and six neutrons in its nucleus, and carbon-13, which has one extra neutron (see Figure 5.5). Living organisms incorporate carbon-12 slightly more easily than carbon-13. As a result, the fraction of carbon-13 is always a bit lower in fossils than in rock samples that lack fossils. All life and all fossils tested to date show the same characteristic ratio of the two carbon isotopes. Dating very ancient rocks can be difficult, but some rocks that are more than 3.85 billion years old show the same ratio of carbon-12 to carbon-13, suggesting that these rocks contain remnants of life.

Carbon isotope evidence suggests that life arose more than 3.85 billion years ago—quite soon after the end of the heavy bombardment.

To summarize, fossil evidence points to life already thriving by 3.5 billion years ago and carbon isotope evidence suggests that life was present more than 3.85 billion years ago. Thus, even under the conservative assumption that the last major impact of the heavy bombardment occurred 4.2 billion years ago, we conclude that life arose in no more than about three hundred million years. If the last major impact occurred 3.9 billion years ago, then life arose within a few tens of millions of years or less.

• How did life arise on Earth?

We have discussed when life arose on Earth, but how did life arise? To answer this question, we must first consider how life today differs from life at the time of its origin on Earth. Then we can ask how the first organisms may have originated.

The Theory of Evolution The fossil record clearly shows that life has gone through great changes over time (see Figure 18.3). Thus, if we are going to understand how life arose, we must understand what causes this change, so that we can trace life back to its origin. The unifying theory through which scientists understand the history of life on Earth is the **theory of evolution**, proposed by Charles Darwin (1809–1882).

The fossil record shows us how life has changed through time, and the theory of evolution explains how these changes have occurred.

Evolution simply means "change with time," and strong fossil evidence for evolution had already been gathered before Darwin published his theory. Thus, most scientists of Darwin's time had already accepted the reality of evolution, but no one prior to Darwin could explain how evolution occurs. In essence, the fossil record provides strong evidence that evolution *has* occurred, while Darwin's theory of evolution explains *how* it occurs.

Darwin's theory tells us that evolution proceeds through a process called **natural selection**. The many individuals of any species always differ in certain small ways. If an individual possesses a trait that gives it an advantage in survival and reproduction, then the trait is likely to be passed on to future generations. We say that nature "selects" the advantageous trait, which is why the process is called natural selection. Over time, natural selection can help individuals of a species become better able to compete for scarce resources. If enough small individual variations accumulate, natural selection can even give rise to an entirely new species.

Darwin compiled a tremendous body of evidence supporting the idea that natural selection is the primary mechanism by which evolution proceeds. More than 100 years of ongoing research has only given further support to Darwin's idea, which is why we call it the *theory* of evolution

[Section 3.4]. Perhaps the strongest support for the theory has come from the discovery of DNA (short for *deoxyribonucleic acid*), which has allowed scientists to understand how evolution occurs on a molecular level.

You have probably heard of DNA, which is the genetic material of all life on Earth. Living organisms reproduce by copying DNA molecules and passing them on to their descendants. A molecule of DNA consists of two long strands—somewhat like the interlocking strands of a zipper—wound together in the spiral shape known as a double helix (Figure 18.5). The instructions for assembling a living organism are written in the precise order of four chemical bases that make up the interlocking portions of the DNA "zipper." (The four chemical bases are represented in Figure 18.5 by the letters A, T, G, and C, which stand for the first letters of their chemical names.) These bases pair up in a way that ensures that each of the two strands of a DNA molecule contains the same genetic information. Thus, by unwinding and allowing new strands to form alongside the original ones (with the new strands made from chemicals floating around inside a cell), a single DNA molecule can give rise to two identical copies of itself. This is how genetic material is copied and passed on to future generations.

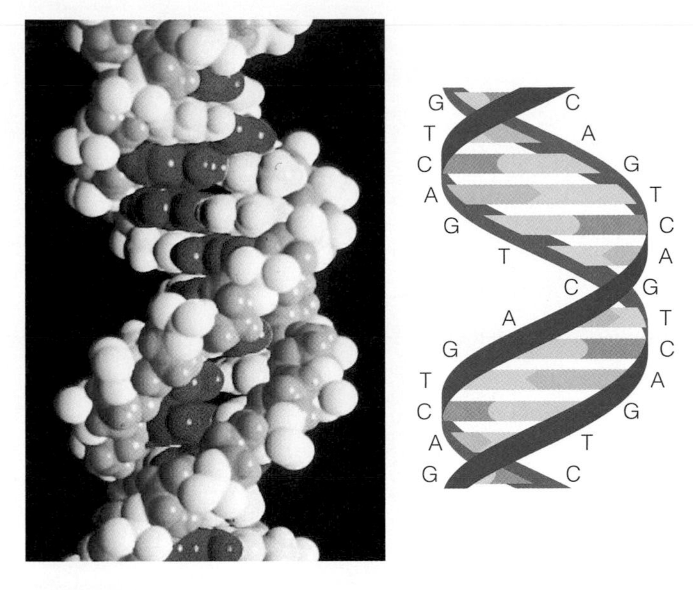

Figure 18.5

DNA is the genetic material of all life on Earth. Left: A model of a small piece of a DNA molecule. Right: This diagram shows that a DNA molecule is made from two interlocking strands wound in the shape of a double helix. The interlocking pieces (which look like ladder steps in the figure) are composed of four chemical bases, labeled A, T, G, and C. Note that A always connects to T, and G always connects to C. Thus, each strand contains the same basic information, so when the two strands separate and "helper molecules" bring new chemical bases to attach to each one, the result is two new molecules each identical to the original one that separated. This ability of DNA to copy itself is the key to heredity.

Evolution occurs because the passing of genetic information from one generation to the next is not always perfect. An organism's DNA may be altered by occasional copying errors or by external influences, such as ultraviolet light from the Sun or exposure to toxic or radioactive chemicals. Any change in the sequence of the chemical bases in an organism's DNA is called a **mutation**. Many mutations are lethal, killing the cell in which the mutation occurs. Some, however, may improve a cell's ability to survive and reproduce. The cell then passes on this improvement to its offspring.

This molecular understanding of the mechanism of natural selection has put the theory of evolution on a stronger foundation than ever. While no theory can ever be proven true beyond all doubt, the theory of evolution today is as solid as almost any theory in science, including the theory of gravity and the theory of atoms. Indeed, biologists today routinely witness evolution occurring before their eyes among laboratory microorganisms, or over periods of just a few decades among plants and animals subjected to some kind of environmental stress.

The First Living Organisms The basic chemical nature of DNA (as well as of many other important biological molecules) is virtually identical among all living organisms, telling us that all life on Earth today can trace its origins to a common ancestor. In other words, sometime before about 3.85 billion years ago, a single organism arose and gradually evolved into every other organism that has ever lived.

All living organisms today evolved from a common ancestor that lived long ago.

What did the first living organism look like? Although we may never have fossils of the earliest organisms, in principle we can learn about early life through careful studies of the DNA of living organisms. Biologists can determine the evolutionary relationships among living species by comparing the sequences of bases in their DNA. For example, two organisms whose DNA sequences differ in five places for a particular gene are probably more distantly related than two organisms whose gene sequences differ in only one place.

Many such DNA comparisons suggest that all living organisms are related in a way depicted schematically by the "tree of life" in Figure 18.6. Although details in the structure of this tree remain uncertain, it confirms that living organisms share a common ancestry. As it is currently

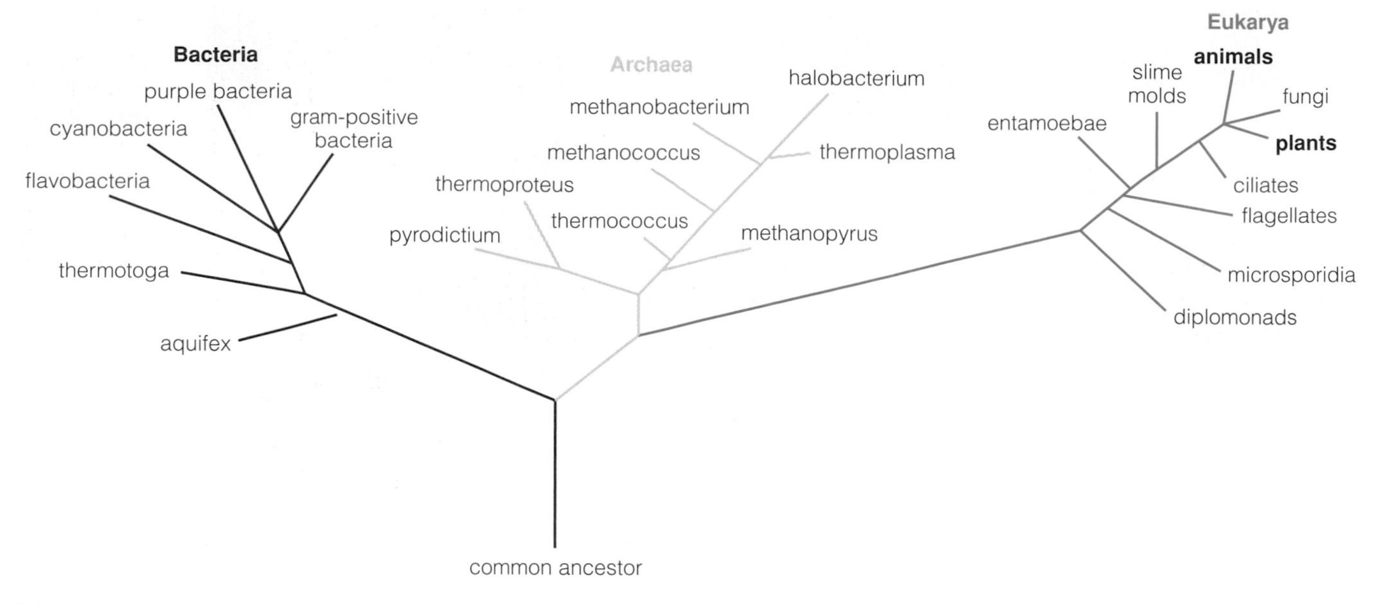

Figure 18.6

The tree of life, showing evolutionary relationships determined by comparison of DNA sequences in different organisms. Just two small branches represent *all* plant and animal species.

understood, the tree of life shows that life on Earth is divided into three major groupings (known as the domains Bacteria, Archaea, and Eukarya). Notice that plants and animals represent only two tiny branches of the great diversity of life on Earth.

Comparison of DNA sequences also allows at least reasonable guesses as to which organisms found on Earth today most resemble the common ancestor. The answer appears to be microscopic organisms living in the deep oceans around seafloor volcanic vents called *black smokers* (after the dark, mineral-rich water that flows out of them) and in hot springs in places such as Yellowstone (Figure 18.7). These organisms thrive in temperatures as high as 110°C (230°F). (The high pressures at the seafloor prevent the water from boiling despite the high temperature.) Unlike most life at the Earth's surface, which depends on sunlight, these organisms get energy from chemical reactions in water heated volcanically by Earth itself.

DNA studies suggest that the earliest organisms probably lived in hot water near sources of volcanic heat.

That early organisms might have lived in such "extreme" conditions may at first seem surprising, but it makes sense when we think about it more deeply. Sunlight can provide energy only to organisms capable of photosynthesis, which is a very complex molecular process. The chemical pathways for extracting energy from the heat of Earth itself are actually much simpler and therefore far more likely to have been used by the earliest organisms. Moreover, chemical reactions tend to proceed faster and with greater diversity in hotter water than cooler water, so the hot water near volcanic vents may have offered a natural laboratory in which chemistry "experiments" were constantly underway.

The Transition from Chemistry to Biology The theory of evolution tells us how a single organism can have evolved into the great diversity of life on Earth today. But where did the first organism come from? We may never know for sure, because it is unlikely that any fossil evidence could have recorded the transition from nonlife to life. However, over the past several decades, scientists have conducted many laboratory

(continued on p. 464)

a This photograph shows a black smoker—a volcanic vent on the ocean floor that spews out hot, mineral-rich water.

b This aerial photo shows a hot spring in Yellowstone National Park. The different colors are from different microbes that survive in water of different temperatures. For a sense of scale, note the walkway winding along the lower right.

Figure 18.7

DNA evidence suggests that the ancestor of all life on Earth resembled organisms that live today in hot water near volcanic vents or hot springs.

SPECIAL TOPIC: WHAT IS LIFE?

You may have noticed that while we've been talking about life, we haven't actually defined the term. In fact, it's surprisingly difficult to draw a clear boundary between life and nonlife. Life can be so difficult to define that we may be tempted to fall back on the famous words of Supreme Court Justice Potter Stewart who, in avoiding the difficulty of defining pornography, wrote: "I shall not today attempt further to define [it] ... But I know it when I see it." If living organisms on other worlds turn out to be much like those found on Earth, it may prove true that we'll know them when we see them. But if the organisms are fairly different from those on Earth, we'll need clearer guidelines to decide whether they are truly "living."

One way to seek distinguishing features of life is to study living organisms, looking for common characteristics. Given the difficulty of defining life, you probably won't be surprised to learn that there are exceptions to almost any "rule" we think of. Nevertheless, biologists have identified at least six key properties that appear to be shared by most or all living organisms on Earth:

1. *Order:* Living organisms are not random collections of molecules but rather have molecules arranged in orderly patterns that form cell structures.
2. *Reproduction:* Organisms are capable of reproducing.
3. *Growth and development:* Living organisms grow and develop in patterns determined at least in part by heredity.
4. *Energy utilization:* Living organisms use energy to fuel their many activities.
5. *Response to the environment:* Life actively responds to changes in its surroundings. For example, simple organisms may alter their chemistry in the presence of a food source, while warm-blooded mammals may sweat, pant, or adjust blood flow to maintain a constant internal temperature.
6. *Evolutionary adaptation:* Life evolves through natural selection, as organisms pass on traits that make them better adapted to survival in their local environments.

These six properties are all important, but biologists today regard evolution as the most fundamental and unifying of them. Evolutionary adaptation is the only property that can explain the great diversity of life on Earth. Moreover, understanding how evolution works allows us to understand how all the other properties came to be. As a result, the simplest definition of life might be "something that can reproduce and evolve through natural selection."

For most practical purposes, this definition of life would probably suffice. However, some cases may still challenge this definition. For example, computer scientists can now write programs (that is, lines of computer code) that can reproduce themselves (that is, create additional sets of identical lines of code). By adding to the programs instructions that allow random changes, they can even make "artificial life" that evolves on a computer. Should this "artificial life," which consists of nothing but electronic signals processed by computer chips, be considered alive?

The fact that we have such difficulty distinguishing the living from the nonliving on Earth suggests that we should be very cautious about constraining our search for life elsewhere. No matter what definition of life we choose, there's always the possibility that we'll someday encounter something that challenges our definition. Nevertheless, the properties of reproduction and evolution seem very likely to be shared by most if not all life in the universe and therefore provide a useful starting point as we consider how to explore the possibility of extraterrestrial life.

Figure 18.8

This photograph shows chemist Stanley Miller with a reproduction of the experimental setup he first used in the 1950s to study pathways to the origin of life. (He worked with Harold Urey, so the experiment is called the Miller-Urey experiment.) The flasks are filled with liquids or gases thought to have been present on the early Earth, and an electrical discharge supplies energy. The result is the production of many complex organic molecules. Scientists now think the early atmosphere was different from that used in Miller's original experiment, but similar experiments with the correct gases still yield similar results.

experiments designed to mimic the conditions that existed on the young Earth. These experiments suggest that life could have arisen through natural chemical reactions.

The first such experiments were performed in the 1950s (Figure 18.8), and they have been greatly refined and improved since that time. In essence, the experiments simply mix chemical ingredients thought to have been present on the early Earth and then "spark" the chemicals with electricity to simulate lightning or other energy sources. The chemical reactions that follow have produced all the major molecules of life, including all the amino acids and DNA bases. Many of these same molecules are found today in meteorites, suggesting that impacts may also have been an important source of organic molecules on the early Earth.

Laboratory experiments also show that mixing a warm, dilute solution of organic molecules with naturally occurring sand or clay allows the molecules to assemble themselves into much more complex molecules. Strands of RNA, a molecule that looks much like a single strand of DNA, have been produced in the laboratory with lengths of up to nearly 100 bases. Because some RNA molecules are capable of self-replication, many biologists now presume that RNA was the original genetic material of life on Earth, with DNA coming later.

Laboratory experiments show that the molecules of life form easily under conditions that existed on the early Earth.

Other experiments have shown that microscopic, enclosed membranes also form under conditions expected on the early Earth (Figure 18.9). Self-replicating RNA molecules may have become enclosed in such membranes, making what are sometimes called "pre-cells." Those pre-cells in which RNA replicated faster and more accurately were more likely to spread, leading to a type of positive feedback that would have encouraged even faster and more accurate replication. Given millions of years of chemical reactions occurring all over the Earth, RNA might eventually have evolved into DNA, making true living organisms.

Figure 18.10 summarizes the steps by which chemistry might have become biology on Earth. We may never know whether life really arose in this way, but it certainly seems possible.

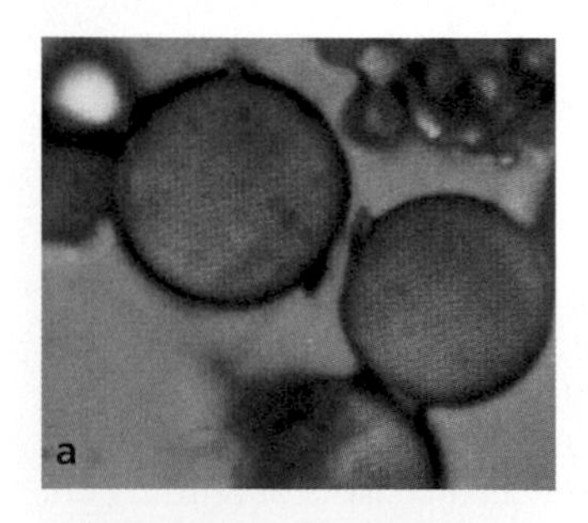

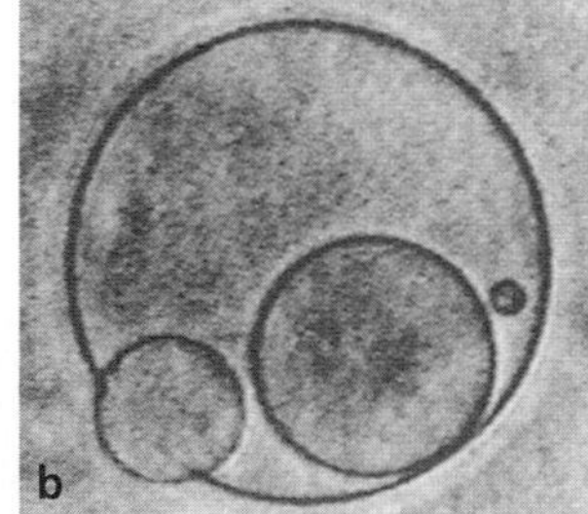

Figure 18.9

These microscopic photos show enclosed membranes that self-assemble in laboratory experiments mimicking conditions on the early Earth. **a** These microscopic spheres were made by cooling a warm water solution of amino acids. They are not alive, but they exhibit many life-like properties. **b** These microscopic membranes are made from lipids that, when mixed with water, spontaneously form enclosed droplets.

Could Life Have Migrated to Earth?

Our scenario suggests that life could have arisen naturally here on Earth. However, an alternative possibility is that life arose somewhere else first—perhaps on Venus or Mars—and then migrated to Earth on meteorites. Remember that we have collected meteorites that were blasted by impacts from the surfaces of the Moon and Mars [Section 9.1]. Calculations suggest that Venus, Earth, and Mars all should have exchanged many tons of rock, especially in the early days of the solar system when impacts were more common.

The idea that life could travel through space to land on Earth once seemed outlandish. After all, it's hard to imagine a more forbidding environment than that of space, with no air, no water, and constant bombardment by dangerous radiation from the Sun and stars. However, the presence of organic molecules in meteorites and comets tells us that the building blocks of life can indeed survive in space, and tests have shown that some microbes can survive in space for years.

In a sense, Earth, Venus, and Mars have been "sneezing" on each other for billions of years. Life could conceivably have originated on any of these three planets and been transported to the others. It's an intriguing thought, but it does not change our basic scenario for the origin of life—it simply moves it from one planet to another.

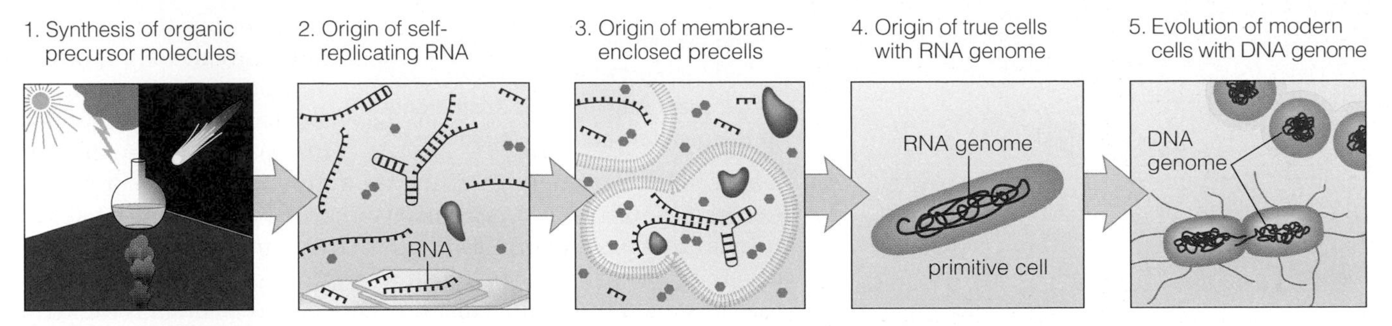

Figure 18.10

A summary of the steps by which chemistry on the early Earth may have led to the origin of life.

A Brief History of Life on Earth We are now ready to take a quick look at the history of life on Earth, as summarized in Figure 18.3. We will focus on events that may help us understand what we are looking for when we search for life on other worlds.

Chemical reactions probably began producing organic molecules not long after Earth's formation nearly 4.6 billion years ago. Some mineral evidence suggests that Earth already had oceans by 4.3 to 4.4 billion years ago, and the early oceans would have been natural laboratories for chemistry that could have led to life. However, even if life arose at such early times, it was almost certainly extinguished by major impacts of the heavy bombardment. The common ancestor of life today arose after the impacts subsided.

Whether through chemical reactions in the oceans or through migration from Venus or Mars, living organisms quickly took hold on Earth after the end of the heavy bombardment. Starting from the first organism, evolution rapidly diversified life on Earth.

Despite the rapid pace of evolution, the most complex life-forms remained single-celled for at least a billion years after life first arose. Some 2 billion years ago, the land was still inhospitable because of the lack of a protective ozone layer [Section 7.1]. Continents much like those of today were surrounded by oceans teeming with life, but, despite pleasant temperatures and plentiful rainfall, the land itself was probably as barren as Mars is now. Things began to change only when oxygen started building up in Earth's atmosphere.

Nearly all the oxygen in our atmosphere was originally released through photosynthesis by single-celled organisms known as cyanobacteria (Figure 18.11). These microbes may have been producing oxygen through photosynthesis as early as 3.5 billion years ago. However, the oxygen did not immediately begin to accumulate in the atmosphere. For more than a billion years, chemical reactions with surface rocks pulled oxygen back out of the atmosphere as fast as the cyanobacteria could produce it. But these tiny organisms were abundant and persistent, and eventually the surface rock was so saturated by oxygen that the rate of oxygen removal slowed down. At that point, some 2 billion years ago, oxygen began to accumulate in the atmosphere, though it may not have reached a level that we could have breathed until just a few hundred million years ago.

Today, we often think of oxygen as a necessity for life. However, oxygen was probably poisonous to most organisms living before about 2 billion years ago (and remains a poison to many microbes still living today). The rise of atmospheric oxygen therefore caused tremendous evolutionary pressure and may well have been a major factor in the eventual evolution of complex plants and animals.

There were undoubtedly many crucial changes as primitive microbes gradually evolved into multicellular organisms and early plants and

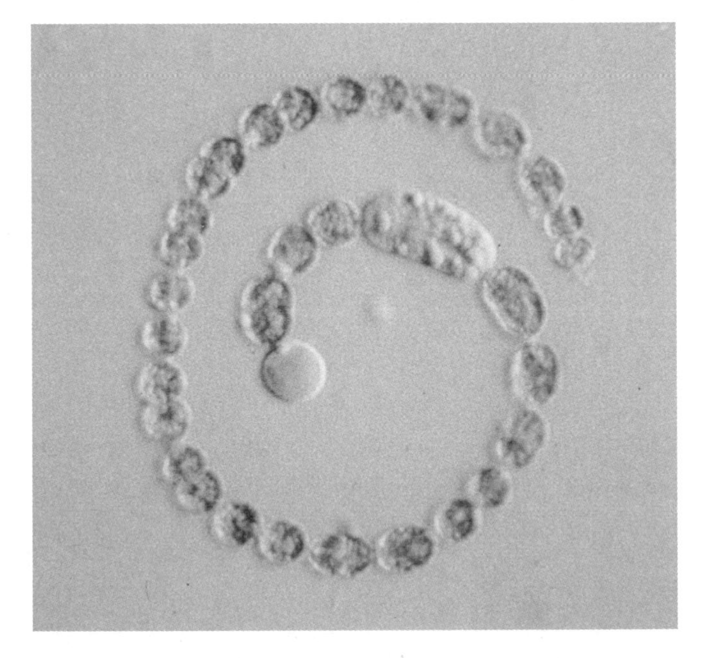

Figure 18.11

This microscopic photo shows a chain of modern cyanobacteria. The ancestors of these living organisms produced essentially all the oxygen in Earth's atmosphere.

Living organisms remained single-celled for most of Earth's history, with larger plants and animals arising only in the past few hundred million years.

animals, but the fossil record does not allow us to pinpoint the times at which all these changes occurred. However, we see a dramatic change in the fossil record dating to about 540 million years ago. Over the next 40 million years or so, animal life evolved from tiny and primitive organisms into all the basic body plans (phyla) that we find on Earth today. This remarkable flowering of animal diversity occurred in such a short time relative to the history of Earth that it is often called the *Cambrian explosion.* (*Cambrian* is the name geologists give to the period from about 540 million to 500 million years ago; see Figure 18.3.)

Early dinosaurs and mammals arose around the same time, some 225 to 250 million years ago, but dinosaurs at first proved more successful and dominated the landscape for well over 100 million years. Their sudden demise 65 million years ago [Section 9.4] paved the way for the evolution of large mammals—including humans. The earliest humans appeared on the scene only a few million years ago, or after 99.9% of Earth's history to date had already gone by. Our few centuries of industry and technology have come after 99.99999% of Earth's history.

• What are the necessities of life?

If we are going to look for other worlds that might harbor life, a first step is to understand the necessities of life. While it is always possible that life on other worlds could be quite different from life on Earth, it's easiest to begin the search for life by looking for conditions that would be comfortable for organisms from Earth. So what exactly does Earth life need to survive?

If we think about ourselves, the requirements for life seem fairly stringent: We need abundant oxygen in an atmosphere that is otherwise not poisonous, we need temperatures in a fairly narrow range of conditions, and we need abundant and varied food sources. However, the discovery of life in "extreme" environments—such as in the hot water near black smokers and hot springs—shows that many microbes (often called *extremophiles*) can survive in a much wider range of conditions.

The organisms living in hot water prove that at least some microbes can survive in much higher temperatures than we would have guessed. Other organisms live in other extremes. In the freezing cold but very dry valleys of Antarctica, the surface appears barren but microbes have been found living *inside* rocks, surviving on tiny droplets of liquid water and energy from sunlight. Other microscopic organisms have been found living up to several kilometers underground in water that fills pores within the subterranean rock. We have found life thriving in environments so acidic, alkaline, or salty that we humans would be poisoned almost instantly. We have even found microbes that can survive high doses of radiation, making it possible for them to survive for many years in the radiation-filled environment of space.

If we compare all the different forms of life on Earth, we find that life as a whole has only three basic requirements:

- A source of nutrients (elements and molecules) from which to build living cells.
- Energy to fuel the activities of life, either from sunlight, from chemical reactions, or from the heat of Earth itself.
- Liquid water.

These requirements give us a basic road map for the search for life elsewhere. If we want to find life on another world, it makes sense to start by searching for worlds that offer these basic necessities.

Life on Earth requires nutrients, energy, and liquid water. Of the three requirements, liquid water is the only one that is not common on other worlds.

Interestingly, only the third requirement (liquid water) seems to pose much of a constraint. Organic molecules are present almost everywhere—even on meteorites and comets. Many worlds are large enough to retain internal heat that could provide energy for life, and virtually all worlds have sunlight (or starlight) bathing their surfaces, although the inverse square law of light [Section 11.1] means that light provides less energy to worlds farther from their sun or star. Thus, nutrients and energy should be available to at least some degree on almost every planet and moon. In contrast, liquid water seems to be relatively rare, and the search for liquid water therefore drives the search for life in our universe.

18.2 LIFE IN THE SOLAR SYSTEM

If we assume that life elsewhere would be at least a little bit like life on Earth, then the search for life begins with a search for **habitable worlds**—worlds that contain the basic necessities for life as we know it, including liquid water. The liquid water requirement rules out most of the worlds in our solar system. For example, Mercury and the Moon are barren and dry, Venus is too hot for liquid water, and Pluto and most of the small moons of the outer solar system are too cold. The jovian planets may have droplets of liquid water in some of their clouds, but the strong vertical winds on these planets make their clouds unlikely homes for life. That leaves two major possibilities besides Earth for habitability in our solar system: (1) Mars, and (2) a few of the large moons orbiting jovian planets, most notably Europa [Section 8.2]. Let's discuss the current evidence about the potential habitability of these worlds.

Is it possible for a world to be habitable but not actually have life? Is it possible for a world to have life but not be habitable? Explain.

• Could there be life on Mars?

We now have sufficiently detailed images of Mars to be quite confident that no civilizations have ever existed there. The canals seen by Percival Lowell were some type of mirage, formed by a combination of real Martian features and his own imagination (and perhaps retinal features of his own eye). Nevertheless, we have good reason to believe that liquid water flowed on Mars in the past [Section 7.3], making it at least seem possible that life could once have found a home there. Moreover, since Mars today contains subsurface ice, it's conceivable that life might still survive near sources of volcanic heat, where pockets of liquid water might persist underground.

Missions to Mars Mars is not only the best candidate in our solar system for life beyond Earth but also the only place where we've begun an actual search for life. Our first attempt to search for life on Mars came

Figure 18.12

The Opportunity rover studies a rock on Mars during its mission in 2004. Such studies are helping scientists understand whether and when Mars may have been habitable.

with the Viking missions to Mars in the 1970s. Two Viking landers arrived on the surface of Mars in 1976. Each was equipped with a robotic arm for scooping up soil samples, which were fed into several on-board, robotically controlled experiments.

Three of the Viking experiments were designed expressly to look for signs of life. None of these experiments could actually "see" life but rather looked for chemical changes that could be attributed to living organisms. Although all three experiments gave results that initially seemed consistent with life, further study suggested that chemical reactions could have produced the same results. Moreover, a fourth experiment, which analyzed the content of Martian soil, found no measurable level of organic molecules—the opposite of what we would expect if life were present. As a result, most scientists have concluded that no life was present at the locations where Viking sampled the soil.

More recent missions to Mars, including the Pathfinder lander in 1997 and the Spirit and Opportunity rovers that landed in 2004, have taken a more measured approach to the search for life (Figure 18.12). Rather than conducting direct experiments like Viking, these missions have been designed to help us better understand Martian conditions, so that we'll have a better idea of how and where to search for life with future missions.

Mars may have been habitable in the past and could have underground liquid water today, making it a prime target for space missions searching for life.

Within a decade or so, NASA hopes to launch a mission to Mars that will bring back surface samples for study in laboratories on Earth. Later, we may send more advanced robots or humans to Mars, where they could search for fossils or living organisms in deep canyons like Valles Marineris, in ancient valley bottoms and dried-up lake beds, or in underground pockets of water near not-quite-dead volcanoes. The search will not be easy, but we should eventually learn whether life has ever existed on Mars.

The Debate over Martian Meteorites An entirely different approach to searching for life on Mars relies on studies of the couple of dozen known meteorites whose chemical composition suggests they came from Mars. One of these meteorites, designated ALH84001, generated particular excitement when a group of scientists claimed that it contained evidence of past life on Mars.

ALH84001 was found on the Antarctic ice in 1984 (Figure 18.13). Careful study of the meteorite shows that it landed in Antarctica about 13,000 years ago, following a 16-million-year journey through space after being blasted from Mars by an impact. The rock itself dates to 4.5 billion years ago, indicating that it solidified shortly after Mars formed and therefore resided on Mars throughout the times when Mars may have been warmer and wetter.

Figure 18.13

The Martian meteorite ALH84001, before it was cut open for detailed study. The small block shown for scale to the right measures 1 cubic centimeter, about the size of a typical sugar cube.

Painstaking analysis of the meteorite reveals several lines of evidence that could indicate the past presence of life on Mars. For example, the rock contains layered carbonate minerals and complex organic molecules (polycyclic aromatic hydrocarbons, or PAHs) that are associated with life when found in Earth rock, as well as microscopic chains of magnetite crystals quite similar to chains made in Earth rocks by living bacteria. Most intriguingly, highly magnified images reveal rod-shaped structures that look much like recently discovered *nanobacteria* on Earth (Figure 18.14). The terrestrial nanobacteria are about $\frac{1}{100}$ the size of ordinary bacteria, and some biologists question whether they truly are living organisms.

However, they appear to contain DNA, which suggests that they are indeed a form of life. If so, could the similar-looking structures in the Martian meteorite be fossil life from Mars?

A meteorite from Mars shows tantalizing evidence of past life, but the evidence can also be explained in a nonbiological way.

It's possible, but each of the tantalizing hints of Martian life can also be explained in a nonbiological way. Subsequent studies have shown that chemical and geological processes can produce structures very similar to those found in the Martian meteorite. In addition, terrestrial bacteria have been found living inside the meteorite, indicating that it was contaminated by Earth life during the 13,000 years it resided in Antarctica. This contamination may explain the presence of the complex molecules found in the rock.

On balance, most scientists now doubt that the Martian meteorite shows true evidence of Martian life. Nevertheless, studies of ALH84001 and other Martian meteorites are continuing, and they may yet turn up surprises.

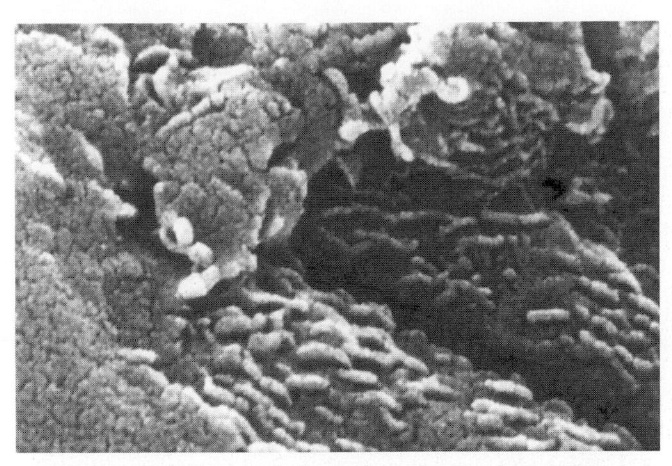

a This photo shows rod-shaped structures found in a highly magnified slice of ALH84001. They measure about 100 nanometers in length and are as small as 10–20 nanometers in width.

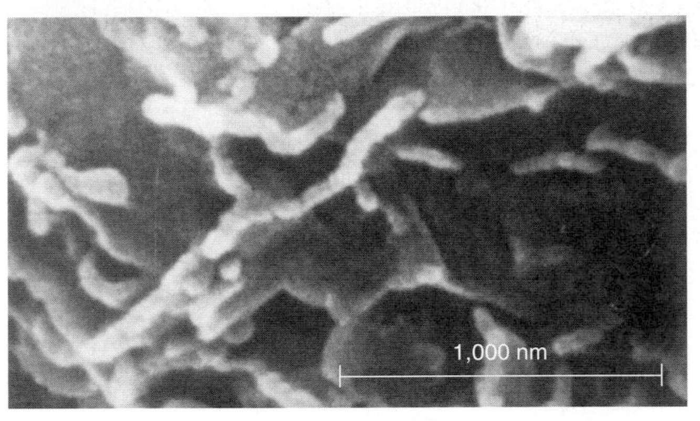

b This photo shows terrestrial nanobacteria in a sample of volcanic rock from Sicily. They are close in size to the structures seen in ALH84001. The scale bar at the bottom is 1 micrometer, or 1,000 nanometers.

Figure 18.14

Does Martian meteorite ALH84001 contain fossils of Martian organisms?

• Could there be life on Europa or other jovian moons?

After Mars, the next most likely candidates for life in our solar system are some of the moons of the jovian planets—especially Jupiter's moons Europa, Ganymede, and Callisto and Saturn's moon Titan [Section 8.2]. These moons are all large enough that they would be considered planets if they orbited the Sun independently.

The strongest of these candidates for harboring life is Jupiter's moon Europa, on which tidal heating probably creates a deep ocean beneath an icy crust (see Figure 8.17). The ice and rock from which Europa formed undoubtedly included the necessary chemical ingredients for life, and Europa's internal heating (primarily due to tidal heating) is strong enough to power volcanic vents on the sea bottom. Thus, it's easy to imagine places on Europa's ocean floor that look much like black smokers on Earth (see Figure 18.7a). If life on Earth first really arose near such undersea volcanic vents, Europa would seem to have everything needed for an origin of life.

If Europa's ocean really exists, it may have seafloor vents much like the black smokers where life on Earth may have first arisen.

The possibility of life on Europa is especially interesting because, unlike any potential life on Mars, it would not necessarily have to be microscopic. After all, the several kilometers of surface ice that hide Europa's ocean (if it exists) could also hide large creatures swimming within it. However, the potential energy sources for life on Europa are far more limited than the energy sources for life on Earth (mainly because sunlight could not fuel photosynthesis in the subsurface ocean). As a result, most scientists suspect that any life that might exist on Europa would probably be quite small and primitive.

As we discussed in Chapter 8, some evidence suggests that Jupiter's moons Ganymede and Callisto may also have subsurface oceans. However, these moons have much less internal heat and therefore would have even less energy for life than Europa. If they have life at all, it is almost certainly small and primitive. Nevertheless, Europa, Ganymede, and Callisto offer the astonishing possibility that Jupiter alone could be orbited by more worlds with life than we find in all the rest of the solar system.

Other possible homes to life in our solar system include Jupiter's moons Ganymede and Callisto and Saturn's moon Titan.

Titan offers another enticing place to look for life. Its surface is far too cold for liquid water, but it may have lakes or oceans of liquid ethane and methane—and we should find out by early 2005, after a robotic spacecraft (the Huygens probe) descends to Titan's surface. Although many biologists think it unlikely, it is possible that these liquids could play the same role for life as does water on Earth. Titan also appears to have many other organic molecules on its surface, and, at the very least, Titan probably offers an incredible natural laboratory of interesting organic chemistry.

If you are reading this book after the Huygens probe descent to Titan, take a couple minutes to search the Web and learn what it found. How have the results altered our views of Titan?

Moreover, we can't completely rule out water-based life on Titan, even if it seems improbable. While liquid water does not exist on Titan's surface today, pockets of liquid water may have existed in the past following impacts of asteroids or comets. The heat of a large impact could have melted water ice and kept it liquid for as long as a few thousand years. If water-based life arose in such temporary pockets of water, it might have found a way to survive as the water froze, perhaps by migrating deep underground where Titan's internal heat might still keep some water liquid.

Detecting Extrasolar Planets Tutorial, Lessons 1–3

18.3 LIFE AROUND OTHER STARS

We already know of planets orbiting more than 100 stars besides our Sun, and it's quite likely that billions of planetary systems inhabit our galaxy alone. These numbers might immediately make prospects for life elsewhere seem quite good, but numbers alone don't tell the whole story. In this section, we'll consider the prospects for life on worlds orbiting other stars.

Before we begin, we must distinguish between *surface* life like that on Earth and *subsurface* life like that we envision as a possibility on Mars or Europa. While large telescopes could in principle allow us to discover surface life on distant planets, no foreseeable technology will allow us to find life that is hidden deep underground in other star systems (unless the subsurface life has a noticeable effect on the planet's atmosphere). We therefore will focus on the search for life on planets with habitable surfaces—surfaces with temperatures and pressures that could allow liquid water to exist.

• Are habitable planets likely?

We have not yet discovered any extrasolar planets that seem likely to be habitable. All the extrasolar planets found to date are much more massive than Earth, suggesting that they are more like the jovian planets of our solar system than the terrestrial planets. Thus, they are unlikely candidates for life (though they could conceivably have moons with life).

However, the existence of other jovian planets makes it seem quite reasonable to suppose that terrestrial planets are also common. Our technology is just not yet up to the task of finding them. Thus, if we hope to find other habitable worlds, the first task is to decide whether we should expect them to exist, and if so, to determine where to look.

Constraints on Star Systems Before we consider planets themselves, it's useful to ask how many stars have any chance of having planets with life. In other words, which stars would make good "Suns," providing heat and light to the surfaces of terrestrial planets that happen to orbit them?

The first requirement for a star to have life-bearing worlds is that it be old enough that life could have arisen. More massive stars live shorter lives, and the most massive stars live no more than a few million years [Section 12.3]. Given that life on Earth did not arise for hundreds of millions of years after our solar system was born, we can rule out any star with more than a few times the mass of our Sun. However, because less massive stars are far more common than more massive stars, the lifetime constraint rules out only about 1% of all stars.

A second requirement is that the star allow planets to have stable orbits. About half of all stars are in binary or multiple star systems, in which stable planetary orbits are less likely than around single stars. If life is not possible in such systems, then we can rule out half the stars in our galaxy as potential homes to life. Of course, the other half—still some 100 billion stars or more—remain possible homes for life. Moreover, under some circumstances stable planetary orbits are possible in multiple star systems, so we shouldn't entirely rule out life in such systems.

A third constraint on the likelihood of finding habitable planets is the size of a star's **habitable zone**—the region in which a terrestrial planet of the right size could have a surface temperature that might allow for liquid water and life. Figure 18.15 shows the approximate sizes, to scale, of the habitable zones around our Sun, around a star with about half the mass of the Sun (spectral type K), and around a star with about $\frac{1}{10}$ the mass of our Sun (spectral type M). Although habitable planets seem possible

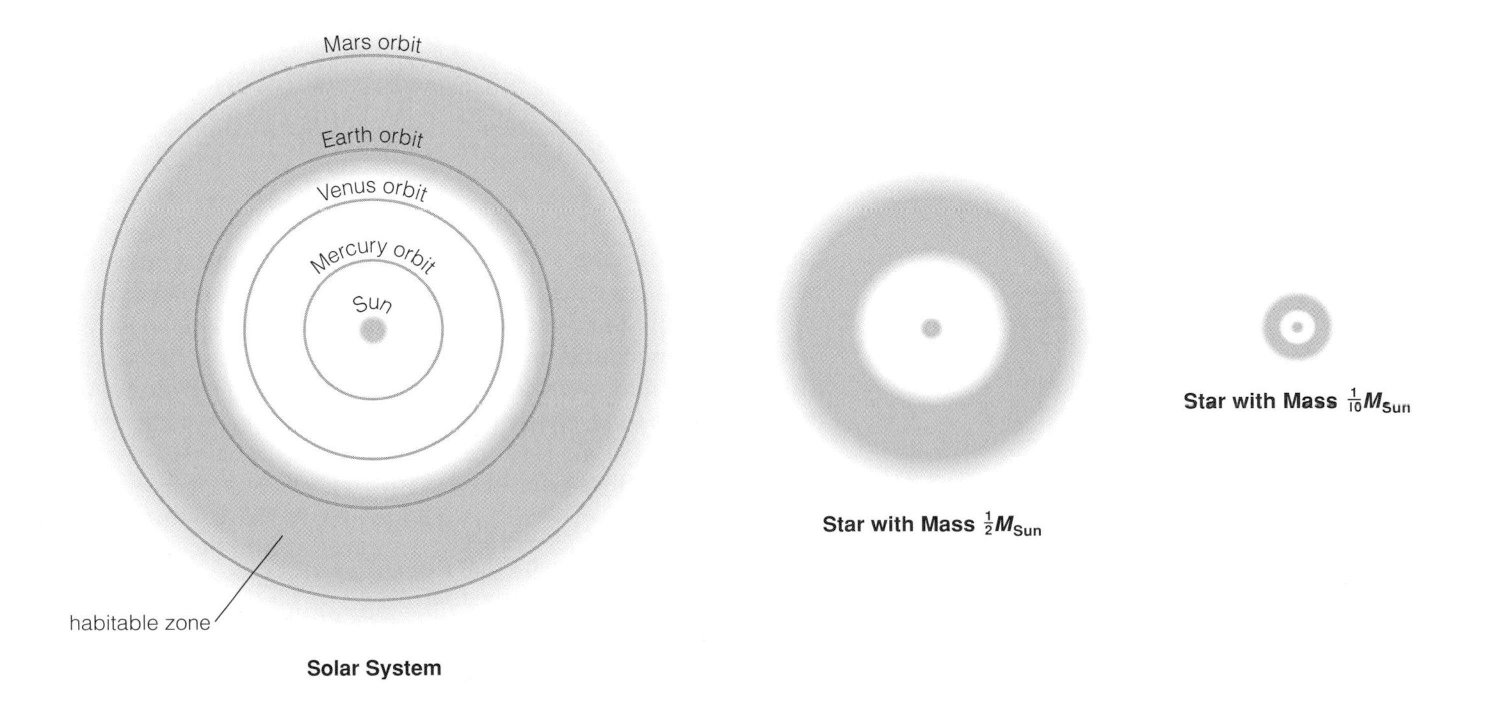

Figure 18.15

The approximate habitable zones around our Sun, a star with half the mass of the Sun (spectral type K), and a star with $\frac{1}{10}$ the mass of the Sun (spectral type M), shown to scale. The habitable zone becomes increasingly smaller and closer-in for stars of lower mass and luminosity.

in all three cases, the smaller size of the habitable zones around the less massive stars makes it less likely that suitable planets would have formed in these regions.

Although not all star systems are potential homes for life, many billions of stars in our galaxy could be orbited by habitable planets.

All in all, it seems that the vast majority of stars are at least potentially capable of having life-bearing planets. Moreover, even very conservative assumptions suggest enormous numbers of possibilities in our galaxy alone. For example, limiting the search for habitable planets to stars very similar to our Sun (that is, spectral type G) would still mean billions of potential other Suns in the Milky Way.

Finding Habitable Planets Finding Earth-size planets is a daunting technological challenge [Section 6.5]. Recall that looking for an Earth-like planet around a nearby star is like standing on the East Coast of the United States and looking for a pinhead on the West Coast. Nevertheless, advancing technology should soon put Earth-like planets within our telescopic reach. The Kepler mission, scheduled for launch in 2007, will look for transits of Earth-size planets in front of their stars (see Figure 6.30). Scientists hope that Kepler will detect hundreds of Earth-size planets, and orbital properties measured by Kepler will tell us whether these planets lie within their stars' habitable zones. The Space Interferometry Mission (SIM), scheduled for launch in 2009, may also be capable of detecting Earth-size planets around some nearby stars.

We will need images or spectra to determine whether the planets really are habitable or have life. Scientists are actively working on technologies that may provide such data. If all goes well, within about a decade NASA hopes to follow SIM with the *Terrestrial Planet Finder (TPF)*, an orbiting interferometer (several telescopes working together as one [Section 5.3]). It will be capable of obtaining low-resolution spectra and crude images (a few pixels) of Earth-like planets around nearby stars.

Astronomers hope to deploy even more powerful interferometers in later decades, either in space or on the Moon. Within the lifetimes of today's college students, we could conceivably have optical interferometers with dozens of telescopes spread across hundreds of kilometers. Such telescopes will be able not only to detect Earth-like planets around nearby stars but also to obtain fairly clear images and high-resolution spectra of those planets.

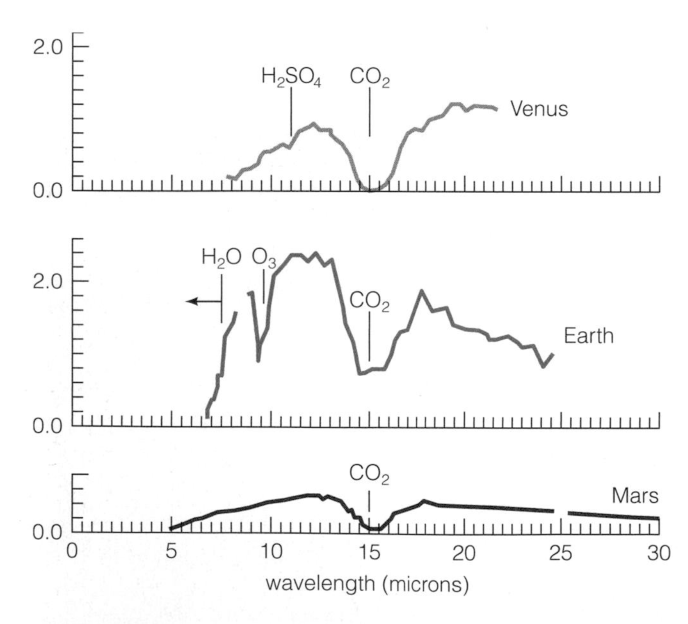

Figure 18.16

The infrared spectra of Venus, Earth, and Mars, as they might be seen from afar, showing absorption features that point to the presence of carbon dioxide (CO_2), ozone (O_3), and sulfuric acid (H_2SO_4) in their atmospheres. While carbon dioxide is present in all three spectra, only our own planet has appreciable oxygen (and hence ozone)—a product of photosynthesis. If we could make similar spectral analyses of distant planets, we might detect atmospheric gases that would indicate life.

Signatures of Life The images from future telescopes may tell us whether the planets have continents and oceans like Earth and perhaps will even allow us to monitor seasonal changes. The spectra from future telescopes should prove even more important to the search for life. Moderate-resolution infrared spectra can reveal the presence and abundance of many atmospheric gases, including carbon dioxide, ozone, methane, and water vapor (Figure 18.16). Careful analysis of atmospheric makeup might tell us whether a planet has life.

Future telescopes may be capable of obtaining images and spectra that will tell us whether distant terrestrial worlds are habitable or have life.

On Earth, for example, the large abundance of oxygen (21% of our atmosphere) is a direct result of photosynthetic life. Abundant oxygen in the atmosphere of a distant world might similarly indicate the presence of life, since we know of no nonbiological way to produce an oxygen abundance as high as Earth's.

Other evidence might come from the ratio of oxygen to the other detected gases. Scientists involved in the search for life in other planetary systems are working to improve our understanding of how life influences atmospheric chemistry so we can recognize the particular gas combinations that are unmistakable signatures of life.

• Are Earth-like planets rare or common?

We will not know for certain whether habitable planets exist or how common such planets may be until we survey many star systems with telescopes capable of detecting such small planets. Nevertheless, the existence of more than one Earth-size planet in our own solar system and our understanding of planetary formation make it seem reasonable to imagine that many Earth-size planets exist within the habitable zones of other stars. But should we expect these planets to have Earth-like conditions in which life could arise and evolve?

Most scientists think so, but a few have raised some very interesting questions. In essence, these scientists suggest that Earth's hospitality is the result of several rare kinds of planetary luck. According to this idea, sometimes called the "rare Earth hypothesis," the specific circumstances that have allowed life on Earth to survive and evolve into complex forms (such as oak trees and people) might be so rare that ours could be the only planet in the galaxy that harbors anything but the simplest life. Let's briefly examine some of the key issues involved in the rare Earth hypothesis.

Galactic Constraints Proponents of the rare Earth hypothesis suggest that Earth-like planets can form in only a relatively small region of the Milky Way Galaxy, making the number of potential homes for life far smaller than we might otherwise expect it to be. In essence, they argue that there is a fairly narrow ring at about our solar system's distance from the center of the Milky Way Galaxy that makes up a *galactic habitable zone* analogous to the habitable zone around an individual star (Figure 18.17).

According to the arguments for a galactic habitable zone, outer regions of our galaxy are unlikely to have terrestrial planets because of a low abundance of heavy elements (elements other than hydrogen and helium). Recall that the fraction of heavy elements varies among different stars from less than 0.1% among the old stars in globular clusters to more than 2% among young stars in the galactic disk [Section 14.3]. Even within the galactic disk, the abundance tends to decline with distance from the center of the galaxy. Because terrestrial planets are made almost entirely of heavy elements, a lower abundance of these elements might lessen the chance of forming terrestrial planets. The inner regions of the galaxy are ruled out primarily through an argument concerning supernova rates. Supernovae are more common in the more crowded, inner regions of the galactic disk, making it more likely that a terrestrial planet would be exposed to the intense radiation from a nearby supernova. By assuming that this radiation would be detrimental to life, the proponents of a galactic habitable zone argue against finding habitable planets in the inner regions of the galaxy. Together, the constraints on finding Earth-like planets in the inner and outer regions of the galaxy leave the galactic habitable zone as a relatively narrow ring encompassing no more than about 10% of the stars in the galactic disk.

However, other scientists offer counterarguments to both sets of galactic constraints. For the heavy element abundance, they note that Earth's mass is less than $\frac{1}{100{,}000}$ of the mass of the Sun. Thus, even a very small

Figure 18.17

The green ring in this diagram of the Milky Way Galaxy represents what some scientists believe is a galactic habitable zone—the only region of the galaxy in which Earth-like planets are likely to be found. However, other scientists doubt the claims that underlie the galactic habitable zone, in which case Earth-like planets could be far more widespread.

heavy-element abundance could be enough to make one or more Earth-like planets. Unless there is something about the planetary accretion process that prevents terrestrial planets from forming in systems with low heavy-element abundances, then we might find terrestrial planets around most any star. Regarding the radiation danger from supernovae, we do not really know that such radiation would be detrimental to life. A planet's atmosphere might protect life against the effects of the radiation. It is even possible that the radiation could be beneficial to life by increasing the rate of mutations and thereby accelerating the pace of evolution. If these counterarguments are correct, then Earth-like planets might be found throughout much or all of the galaxy.

Impact Rates and Jupiter Another issue raised by rare Earth proponents is the impact rate on planets in other star systems. We have seen that Earth was probably subjected to numerous large impacts—some large enough to vaporize the oceans and sterilize the planet—during the heavy bombardment that went on during the first half-billion years or so after our planet's birth. In our solar system, the impact rate lessened dramatically after that. Might the impact rate remain high much longer in other planetary systems?

The most numerous small objects in our solar system are the trillion or so comets of the distant Oort cloud [Section 9.2]. Fortunately for us, nearly all these myriad objects are essentially out of reach, posing no threat to our planet. However, the reason they are out of reach can be traced directly to Jupiter: Recall that the Oort cloud's comets are thought to have formed among the jovian planets, later being "kicked out" to their current orbits by close encounters with these planets, especially Jupiter (see Figure 9.20). Thus, if Jupiter did not exist, many of the comets might have remained in regions of the solar system where they could pose a danger to Earth. In that case, the heavy bombardment might never have ended, and huge impacts would continue to this day. From this viewpoint, our existence on Earth has been possible only because of the "luck" of having Jupiter as a planetary neighbor.

The primary question in this case is just how "lucky" this situation might be. Our discoveries of extrasolar planets so far suggest that Jupiter-size planets are in fact quite common. However, we've also found that many large planets migrate inward [Section 6.5], perhaps disrupting terrestrial planet orbits along the way. Whether having a Jupiter in the right place for ejecting comets is lucky or typical remains an open question.

Climate Stability Another issue affecting the rarity of Earth-like planets is climate stability. Earth's climate has been stable enough for liquid water to exist throughout the past 4 billion years. This climate stability has almost certainly played a major role in allowing complex life to evolve on our planet. If our planet had frozen over like Mars or overheated like Venus, we would not be here today. Advocates of the rare Earth hypothesis point to at least two pieces of "luck" related to Earth's stable climate.

The first piece of "luck" is the existence of plate tectonics. Recall that plate tectonics plays a major role in regulating Earth's climate through the carbon dioxide cycle [Section 7.5]. Plate tectonics probably was not necessary to the origin of life, but it seems to have been very important in keeping the climate stable enough for the subsequent evolution of plants and animals. But are we really "lucky" to have plate tectonics, or should this geological process be common on similar-size planets elsewhere? We do not yet know. The lack of plate tectonics on Venus, which is quite similar in size to Earth, might seem to argue for plate tectonics being rare. On the other hand, some scientists suspect that the lack of plate tectonics on Venus can be traced to its runaway greenhouse effect, which may have baked out subsurface water and caused Venus's lithosphere to thicken too much to allow plate movements [Section 7.4]. Venus suffered its runaway greenhouse effect because it is not quite far enough from the Sun to be within the Sun's habitable zone. In that case, it's possible that any Earth-size planet within a star's habitable zone would have plate tectonics.

The second piece of "luck" in climate stability is the existence of Earth's relatively large Moon. Models of Earth's rotation and orbit show that if the Moon did not exist, gravitational tugs from other planets would cause large swings in Earth's axis tilt over periods of tens to hundreds of thousands of years. (Such swings in axis tilt are thought to occur on Mars.) Changes in axis tilt would affect the severity of the seasons, which in turn could cause deeper ice ages and more intense periods of warmth. Given that the Moon formed as a result of a random, giant impact [Section 6.4], we might seem to be very lucky to have the Moon and the climate stability it brings.

Again, however, there are other ways to look at the issue. Changes in axis tilt might warm or cool different parts of the planet dramatically, but the changes would probably occur slowly enough for life to adapt or migrate as the climate changed. In addition, our Moon's presumed formation in a random giant impact does not necessarily mean that large moons will be rare. At least a few giant impacts should be expected in any planetary system. Indeed, Earth may not be the only planet in our own solar system that ended up with a large moon through a giant impact. Pluto's moon (Charon) may have formed in the same way [Section 9.3]. Thus, while luck was certainly involved in Earth's having a large moon, it might not be a very rare kind of luck.

The Bottom Line The bottom line is that while the rare Earth hypothesis offers some intriguing arguments, it is too early to say whether any of them will hold up over time. For each potential argument that Earth has been lucky, we've seen counterarguments suggesting otherwise. There's no doubt that our solar system and our world have "personality"—they exhibit properties that might be found only occasionally in other star systems—but were such properties truly essential for our existence or merely a help?

The rare Earth hypothesis suggests that Earth has been the beneficiary of several instances of rare planetary luck, but it remains quite controversial.

Our solar system has no properties obviously essential to complex or even intelligent life that other star systems would never have. Indeed, it may be that we have missed out on some helpful phenomena that could have sped evolution on Earth. We might be less lucky than we recognize, and creatures on other worlds might regard the nature of our planet with disappointment. Until we learn much more about other planets in the universe, we cannot know whether Earth-like planets and complex life are common or rare.

18.4 THE SEARCH FOR EXTRATERRESTRIAL INTELLIGENCE

So far, we have focused on search strategies for microbial or other nonintelligent life. However, if intelligent beings and civilizations exist elsewhere, we might be able to find them with a completely different type of search strategy. Instead of searching for hard-to-find spectroscopic signs of life, we might simply listen for signals that intelligent beings are sending into interstellar space, either in deliberate attempts to contact other civilizations or as a means of communicating among themselves. The search for signals from other civilizations is generally known as the **search for extraterrestrial intelligence**, or **SETI** for short.

• How many civilizations are out there?

SETI efforts have a chance to succeed only if other advanced civilizations are broadcasting signals that we could potentially receive. Thus, in order to judge the chances of SETI success, we'd need to know how many civilizations are broadcasting such signals right now.

Given that we do not even know whether microbial life exists anywhere beyond Earth, it should be clear that we cannot yet know whether other civilizations exist, let alone how many there might be. Nevertheless, for the purposes of planning a search for extraterrestrial intelligence, it is useful to have an organized way of thinking about the number of civilizations that might be out there. To keep our discussion simple, let's consider only the number of potential civilizations in our own galaxy. We can always extend our estimate to the rest of the universe by simply multiplying the result we find for our galaxy by 100 billion, the approximate number of galaxies in our universe.

The Drake Equation The first scientific conference on the search for extraterrestrial intelligence was held in 1961 in Green Bank, West Virginia, where a pioneering search for an alien signal had recently been conducted. During the meeting, astronomer Frank Drake wrote a simple equation designed to summarize the factors that would determine the number of civilizations we might contact (Figure 18.18).

The Drake equation summarizes the factors that determine the number of civilizations in our galaxy with whom we could potentially communicate.

His equation is now known as the **Drake equation**, and in principle it gives us a simple way to calculate the number of civilizations capable of interstellar communication that are currently sharing the Milky Way Galaxy with us. In a form slightly modified from the original, the Drake equation looks like this:

$$\text{Number of civilizations} = N_{\text{HP}} \times f_{\text{life}} \times f_{\text{civ}} \times f_{\text{now}}$$

This equation will make sense once you understand the meaning of each factor:

- N_{HP} is the number of habitable planets in the galaxy; that is, the number of planets that are at least capable of having life.
- f_{life} is the fraction of habitable planets that actually *have* life. For example, if $f_{\text{life}} = 1$ it would mean that all habitable planets have life, and if $f_{\text{life}} = 1/1{,}000{,}000$ it would mean that only 1 in a million habitable planets have life. Thus, the product $N_{\text{HP}} \times f_{\text{life}}$ tells us the number of life-bearing planets in the galaxy.
- f_{civ} is the fraction of the life-bearing planets upon which a civilization capable of interstellar communication *has at some time* arisen. For example, if $f_{\text{civ}} = 1/1{,}000$ it would mean that such a civilization has existed on 1 out of 1,000 planets with life, while the other 999 out of 1,000 have not had a species that learns to build radio trans-

Figure 18.18

Astronomer Frank Drake, with the equation he first wrote in 1961. (With Dr. Drake's approval, we use a slightly modified form of his equation in this book.)

mitters, high-powered lasers, or other devices for interstellar conversation. When we multiply this factor by the first two factors to form the product $N_{HP} \times f_{life} \times f_{civ}$, we get the total number of planets upon which intelligent beings have evolved and developed a communicating civilization at some time in the galaxy's history.

- f_{now} is the fraction of these civilization-bearing planets that happen to have a civilization *now*, as opposed to, say, millions or billions of years in the past. This factor is important because we can hope to contact only civilizations that are broadcasting signals we could receive at present (assuming we take into account the light-travel time for signals from other stars).

Because the product of the first three factors tells us the total number of civilizations that have *ever* arisen in the galaxy, multiplying by f_{now} tells us how many civilizations we could potentially make contact with today. Thus, the result of the Drake equation is the number of civilizations that we might hope to contact.

We do not know the values of any of the factors in the Drake equation, but it is still a useful tool for organizing our thinking.

Unfortunately, we don't yet know the value of any of the factors in the Drake equation, so we cannot actually calculate the number of civilizations in our galaxy. Nevertheless, the equation is a useful way of organizing our thinking, as we can see by thinking about the potential values for each of its factors.

THINK ABOUT IT Try the following sample numbers in the Drake equation. Suppose there are 1,000 habitable planets in our galaxy, that 1 in 10 habitable planets has life, that 1 in 4 planets with life has at some point had an intelligent civilization, and that 1 in 5 civilizations that have ever existed is in existence now. How many civilizations would exist at present? Explain.

The Number of Life-Bearing Planets Let's begin with the first two factors in the Drake equation, whose product ($N_{HP} \times f_{life}$) tells us the number of life-bearing planets in our galaxy. We can make a reasonably educated guess only about the first factor, the number of habitable planets (N_{HP}). Current understanding of solar system formation (see Figure 6.27) suggests that terrestrial planets ought to form fairly easily and, as we discussed earlier, there ought to be many billions of stars in our galaxy with habitable zones large enough to have Earth-like planets. Thus, unless some of the "rare Earth" ideas prove to be correct, it seems entirely reasonable to suppose our galaxy has billions of habitable planets.

Our galaxy is likely to have billions of habitable planets, but we have no reliable way to estimate the fraction of them that actually have life.

The factor f_{life} presents more difficulty. For the moment, we have no reliable way to estimate the fraction of habitable planets upon which life actually arose. The problem is that we cannot generalize when we have only one example to study—our own Earth. Still, the fact that life arose rapidly on Earth suggests that the origin of life was fairly "easy." In that case, we might expect most or all habitable planets to also have life, making the fraction f_{life} close to 1. Of course, until we have solid evidence that life arose anywhere else, such as on Mars, it is also possible that Earth might really have been very lucky and that f_{life} might be so close to zero that life has never arisen on any other planet in our galaxy.

The Question of Intelligence Even if life-bearing planets are very common, civilizations capable of interstellar communication might not be. The fraction of life-bearing planets that at some time have such civilizations, f_{civ}, depends on at least two things: First, a planet would have to have a species evolve with sufficient intelligence to develop interstellar communication. In other words, the planet needs a species at least as smart as we are. Second, that species would have to develop a civilization with technology at least as great as ours.

Although we cannot really be sure, most scientists suspect that only the first requirement is difficult to meet. A fundamental assumption in nearly all of science today is that we are not "special" in any particular way. We live on a fairly typical planet orbiting an ordinary star in a normal galaxy, and we assume that living creatures elsewhere—whether they prove to be rare or common—would be subjected to evolutionary pressures quite similar to those that have operated on Earth. Thus, if species with intelligence similar to ours have arisen elsewhere, we assume that they would have similar sociological drives that would eventually lead them to develop the technology necessary for interstellar communication.

If this assumption is correct, then the fraction f_{civ} depends primarily on the question of whether sufficient intelligence is rare or common among life-bearing planets. As with the question of life of any kind, the short answer to this question is we just don't know, but we can get at least some insight by considering what happened on Earth.

Life on Earth arose quickly, but it took nearly 4 billion years to get a species intelligent enough to talk about it.

Look again at Figure 18.3. While life arose quite quickly on Earth, all life remained microbial until just a few hundred million years ago, and it took nearly 4 billion years for us to arrive on the scene. This slow progress toward intelligence might suggest that producing a civilization is very difficult even when life is present. On the other hand, roughly half the stars in the Milky Way are older than our Sun, so if Earth's case is typical then plenty of planets have existed long enough for intelligence to arise.

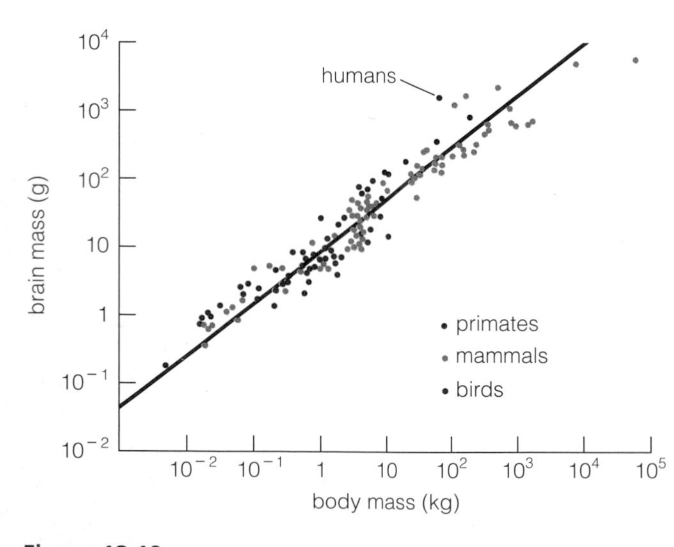

Figure 18.19

This graph shows how brain mass compares to body mass for some mammals (including primates) and birds. The straight line represents an "average" of the ratio of brain mass to body mass, so that animals that fall above the line are "smarter" than average and animals that fall below the line are "intellectually challenged." Note that the scale uses powers of 10 on both axes. (Adapted from Sagan, 1977.)

Another way to address the question is by considering our level of intelligence in comparison to that of other animals on Earth. It's difficult to get animals to take IQ tests, but we can get a rough measure of intelligence by comparing brain mass to total body mass (a measure sometimes called the *encephalization quotient,* or EQ). Figure 18.19 shows the brain weights for a sampling of birds and mammals (including primates) plotted against their body weight. There is a clear and expected trend in that heavier animals have heavier brains. By drawing a straight line that fits these data, we can define an average value of brain mass for each body mass. Animals whose brain mass falls above the line are smarter than average, while animals whose brain mass falls below the line are less mentally agile. Keep in mind that it is the *vertical* distance above the line that tells us how much smarter a species is than the average, and that the scale goes in powers of 10 on both axes. If you look closely, you'll see that the data point for humans lies significantly farther above the line than the data point for any other species. Thus, by this measure of intelligence, we are far smarter than any other species that has ever existed on Earth.

Some people use this fact to argue that even on a planet with complex life, a species as intelligent as we are would be very rare. They say that even if there is an evolutionary drive toward intelligence in general, it takes extreme luck to reach our level of intelligence. After all, while it's evolutionarily useful to have enough intelligence to capture prey and evade other predators, it's not clear why natural selection would lead to brains big enough to build spacecraft. However, the same data can be used to reach an opposite conclusion. The scatter in the levels of intelligence among different animals tells us that some variation should be expected, and statistical analysis shows that we are not unreasonably far above the average. It might therefore be inevitable that some species would develop our level of intelligence on any planet with complex life.

Technological Lifetimes For the sake of argument, let's assume that life and intelligence are at least reasonably likely, so that thousands or millions of planets in our galaxy have at some time given birth to a civilization. In that case, the final factor in the Drake equation, f_{now}, determines the likelihood of there being someone whom we could contact now. The value of this factor depends on the survivability of civilizations.

Consider our own example. In the roughly 12 billion years during which our galaxy has existed, we have been capable of interstellar communication via radio for only about 60 years. Thus, if we were to destroy ourselves tomorrow (saving you the unpleasantness of a final exam), then other civilizations could have received signals from us during only 60 years out of the galaxy's 12-billion-year existence, equivalent to 1 part in 200 million of the galaxy's history. If such a short technological lifetime is typical of civilizations, then f_{now} would be only $\frac{1}{200{,}000{,}000}$, and some 200 million civilization-bearing planets would need to have existed at one time or another in the Milky Way in order for us to have a good chance of finding another civilization out there now.

Even if civilizations arise on many planets, they are unlikely to exist today unless they avoid early self-destruction.

However, we'd expect f_{now} to be so small only if we are on the brink of self-destruction—after all, the fraction will grow larger for as long as our civilization survives. Thus, if civilizations are at all common, survivability is the key factor in whether any are out there now. If most civilizations self-destruct shortly after achieving the technology for interstellar communication, then we are almost certainly alone in the galaxy at present. But if most survive and thrive for thousands or millions of years, the Milky Way may be brimming with civilizations—most of them far more advanced than our own.

THINK ABOUT IT Describe a few reasons why a civilization capable of interstellar communication would also be capable of self-destruction. Overall, do you believe our civilization can survive for thousands or millions of years? Defend your opinion.

• How does SETI work?

If there are indeed other civilizations out there, then in principle we ought to be able to make contact with them. Based on our current understanding of physics, it seems likely that even very advanced civilizations would communicate much as we do—by encoding signals in radio waves or other forms of light. Most SETI researchers use large radio telescopes to search for alien radio signals (Figure 18.20). A few researchers are beginning to check other parts of the electromagnetic spectrum. For example, some scientists use visible light telescopes to search for communications encoded as laser pulses. Of course, advanced civilizations may well have invented

Figure 18.20

The 64-meter Parkes radio telescope in New South Wales, Australia. A SETI experiment "piggybacks" on this telescope while it is engaged in other astronomical research.

communication technologies that we cannot even imagine. In that case, SETI efforts will not detect them.

In principle, aliens within about 50 light-years could watch past television broadcasts. Our own current SETI efforts could pick up only much stronger signals.

A good way to think about our chances of picking up an alien signal is to imagine what aliens would need to do to pick up signals from us. We have been sending relatively high-power transmissions into space since about the 1950s in the form of television broadcasts. Thus, in principle, anyone within about 50 light-years of Earth could watch our old television shows (perhaps a frightening thought). However, in order to detect our broadcasts, they would need far larger and more sensitive radio telescopes than we have today. If their technology were at the same level as ours, they could receive a signal from us only if we deliberately broadcast an unusually high-powered transmission.

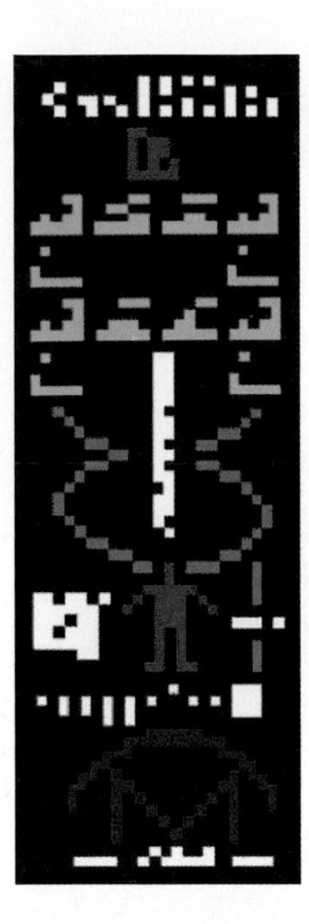

Figure 18.21

In 1974, a short message was broadcast to the globular cluster M13 using the Arecibo radio telescope. The picture shown here was encoded by using two different radio frequencies, one for "on" and one for "off" (the colors shown here are arbitrary). To decode the message, the aliens would need to realize that the bits are meant to be arranged in a rectangular grid as shown, but that should not be difficult: The grid has 73 rows and 23 columns, and aliens would presumably know that these are both prime numbers. The picture represents the Arecibo radio dish, our solar system, a human stick figure, and a schematic of DNA and the eight simple molecules used in its construction.

To date, humans have made only a few attempts to broadcast our existence in this way. The most powerful of these occasional transmissions was made in 1974 and lasted only 3 minutes. The powerful planetary radar transmitter on the Arecibo radio telescope (see Figure 5.17) was used to send a simple pictorial message to the globular cluster M13 (Figure 18.21). This target was chosen in part because it contains a few hundred thousand stars, seemingly offering a good chance that at least one has a civilization around it. However, M13 is about 21,000 light-years from Earth, so it will take some 21,000 years for our signal to get there and another 21,000 years for any response to make its way back to Earth.

Several SETI projects under way or in development would be capable of detecting signals like the one we broadcast from Arecibo if they came from civilizations within a few hundred light-years. However, we could detect the signal only if we had the receiver tuned to the frequency of the broadcast—just as you can listen to your favorite radio station only by calling up the correct frequency on your radio dial. What radio frequency would aliens use? In the past, some astronomers made guesses about popular alien frequencies based on things such as frequencies emitted by common molecules in interstellar space. Today, SETI efforts generally seek to bypass this question by scanning millions of frequency bands simultaneously. Thus, if anyone nearby is deliberately broadcasting on an ongoing basis, we have a good chance of detecting the signals.

THINK ABOUT IT SETI efforts are often controversial because of their cost and uncertain chances of success. However, SETI supporters say the cost is justified because contact with an extraterrestrial intelligence would be one of the most important discoveries in human history. Do you agree? Defend your opinion.

18.5 INTERSTELLAR TRAVEL AND ITS IMPLICATIONS TO CIVILIZATION

So far, we have discussed ways of detecting distant civilizations without ever leaving the comfort of our own planet. Could we ever actually visit other worlds in other star systems? A careful analysis of this question turns out to have profound implications for the future of our civilization. To see why, we first need to consider the prospects for achieving interstellar travel.

• How difficult is interstellar travel?

In many science fiction movies, our descendants travel among the stars as routinely as we jet about the Earth in airplanes. They race around the galaxy in starships of all sizes and shapes, circumventing nature's prohibition on faster-than-light travel by entering hyperspace, wormholes, or warp drive. They witness incredible cosmic phenomena firsthand, such as stars and planets in all stages of development, accretion disks around white dwarfs and neutron stars, and the distortion of spacetime near black holes. Along the way they encounter numerous alien species, most of which happen to look and act a lot like us.

Unfortunately, real interstellar travel is likely to be limited by the speed of light, and even approaching that speed will require overcoming huge technological hurdles. Nevertheless, we have already sent out our first emissaries to the stars, and there's no reason to believe that we won't develop better technologies in the future.

The Challenge of Interstellar Travel To date, we have launched four spacecraft that will leave our solar system and eventually travel among the stars: the planetary probes *Pioneer 10, Pioneer 11, Voyager 1,* and *Voyager 2.* These spacecraft are traveling about as fast as anything ever built by humans, but their speeds are still less than $\frac{1}{10,000}$ the speed of light. It would take each of them some 100,000 years just to reach the next nearest star system (Alpha Centauri), but their trajectories won't take them anywhere near it. Instead, they will simply continue their journey without passing close to any nearby stars, wandering the Milky Way for millions or even billions of years to come. Each carries a greeting from Earth, just in case someone comes across one of them someday (Figure 18.22).

If we want to make interstellar journeys within human lifetimes, we will need starships that can travel at speeds close to the speed of light. We will need entirely new types of engines to reach such high speeds.

High-speed interstellar travel would require many thousands of times as much energy as the entire world currently uses each year.

The energy requirements of interstellar spacecraft may pose an even more daunting challenge. For example, the energy needed to accelerate a single ship the size of *Star Trek*'s *Enterprise* to just half the speed of light would be more than 2,000 times the total annual energy use of the world today. Clearly, interstellar travel will require vast new sources of energy. In addition, fast-moving starships will require new types of shielding to protect crew members from instant death. As a starship travels through interstellar gas at near-light speed, ordinary atoms and ions will hit it like a deadly flood of high-energy cosmic rays.

If we succeed in building starships capable of traveling at speeds close to the speed of light, the crews will face significant social challenges. According to well-tested principles of Einstein's theory of relativity (see box, Relativity and the Cosmic Speed Limit, p. 331), time will run much slower on a spaceship that travels at high speed to the stars than it does here on Earth. For example, in a ship traveling at an average speed of 99.9% of the speed of light, the 50-light-year round trip to the star Vega would take the travelers aboard only about 2 years—but more than 50 years would pass on Earth while they were gone. Thus, the crew will need only 2 years' worth of provisions and will age only 2 years during the voyage, but they will return to a world quite different from the one they left. Family and friends will be older or deceased, new technologies might have made their knowledge and skills obsolete, and many political and social changes

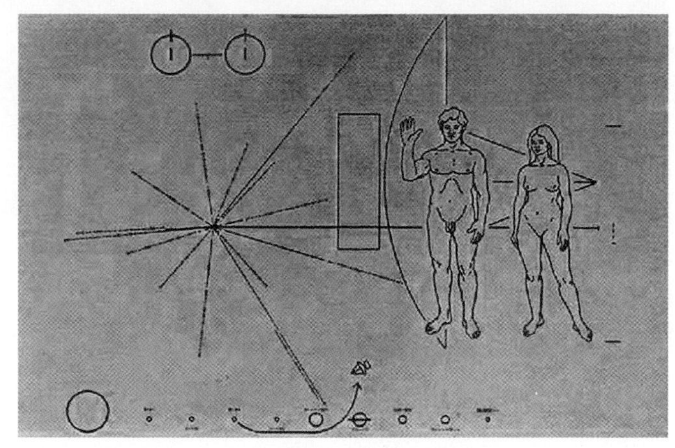

a The Pioneer plaque, about the size of an automobile license plate. The human figures are shown in front of a drawing of the spacecraft to give them a sense of scale. The "prickly" graph to their left shows the Sun's position relative to nearby pulsars, and Earth's location around the Sun is shown below. Binary code indicates the pulsar periods; because pulsars slow with time, the periods will allow someone reading the plaque to determine when the spacecraft was launched.

b *Voyagers 1* and *2* carry a phonograph record—a 12-inch gold-plated copper disk containing music, greetings, and images from Earth.

Figure 18.22

Messages aboard the Pioneer and Voyager spacecraft, which are bound for the stars.

Figure 18.23

Artist's conception of the Project Orion starship, showing one of the small hydrogen-bomb detonations that propel it. Debris from the detonation strikes the flat disk, called the pusher plate, at the back of the spaceship. The central sections (enclosed in a lattice) hold the bombs, and the front sections house the crew.

Figure 18.24

Artist's conception of a spaceship powered by an interstellar ramjet. The giant scoop in the front (left) collects interstellar hydrogen for use as fusion fuel.

may have occurred in their absence. Clearly, the crew will face a difficult adjustment when they come home to Earth.

Starship Design Despite all the difficulties, some scientists and engineers have already proposed designs that could in principle take us to nearby stars. In the 1960s a group of scientists proposed an approach, called *Project Orion,* that envisioned accelerating a spaceship with repeated detonations of relatively small hydrogen bombs. Each explosion would take place a few tens of meters behind the spaceship and would propel the ship forward as the vaporized debris impacted a "pusher plate" on the back of the spacecraft (Figure 18.23). Calculations showed that a spaceship accelerated by the rapid-fire detonation of a million H-bombs could reach Alpha Centauri in just over a century. In principle, we could build an Orion spacecraft with existing technology, though it would be very expensive and would require an exception to the international treaty banning nuclear detonations in space.

While a century to the nearest star is not bad, it still wouldn't make interstellar travel easy. Unfortunately, no available technology could go much faster. The problem is weight: Making a rocket faster requires more fuel, but adding fuel adds weight and makes it more difficult for the rocket to accelerate. Calculations show that even in the best case, rockets carrying nuclear fuel could achieve speeds no more than a few percent of the speed of light. Nevertheless, we can envision some possible future technologies that might get around this problem.

One idea suggests powering starships with engines that generate energy through matter–antimatter annihilation. Whereas nuclear fusion converts less than 1% of the mass of atomic nuclei into energy, matter–antimatter annihilation [Section 17.1] converts *all* the annihilated mass into energy. Starships with matter–antimatter engines could probably reach speeds of 90% or more of the speed of light. At these speeds, the slowing of time predicted by relativity becomes noticeable, putting many nearby stars within a few years' journey for the crew members. However, because no natural reservoirs of antimatter exist, we would have to be able to manufacture many tons of antimatter and then store it safely for the trip—capabilities that are far beyond our present means.

High-speed interstellar travel remains well beyond our current capabilities, but we can envision future technologies that could make it possible.

An even more speculative and futuristic design, know as an *interstellar ramjet,* would collect interstellar hydrogen with a gigantic scoop, using the collected gas as fuel for its nuclear engines (Figure 18.24). By collecting fuel along the way, the ship would not need to carry the weight of fuel on-board. However, because the density of interstellar gas is so low, the scoop would need to be enormous. As astronomer Carl Sagan said, we are talking about "spaceships the size of worlds."

The bottom line is that while we face enormous obstacles to achieving interstellar travel, there's no reason to think it's impossible. If we can avoid self-destruction and we continue to explore space, it seems quite likely that our descendants will make journeys to the stars.

SPECIAL TOPIC: ARE ALIENS ALREADY HERE?

In this chapter, we have discussed contact with intelligent aliens as a possibility, not a reality. However, public opinion polls suggest that up to half the American public believes that aliens are already visiting us. What can science say about this remarkable notion?

The bulk of the claimed evidence for alien visitation consists of sightings of UFOs—unidentified flying objects. Many thousands of UFOs are reported each year, and no one doubts that unidentified objects are being seen. The question is whether they are alien spacecraft.

Aliens have long been a staple of science fiction, but modern interest in UFOs began with a widely reported sighting in 1947. While flying a private plane near Mount Rainier in Washington State, businessman Kenneth Arnold saw nine mysterious objects streaking across the sky. He told a reporter that the objects "flew erratic, like a saucer if you skip it across the water." (In fact, he may have seen meteors skipping across the atmosphere, though no one knows for sure.) He did *not* say that the objects were saucer-shaped, but the reporter nevertheless wrote up Arnold's experience as a sighting of "flying saucers." The story was front-page news throughout America, and within a decade "flying saucers" had invaded popular culture, if not our planet.

The flying saucer reports also interested the U.S. Air Force, largely out of concern that the UFOs might represent new types of aircraft developed by the Soviet Union. For two decades, the air force hired teams of academics to study UFO reports. In the overwhelming majority of cases, these experts were able to specify a plausible identification of the UFO. The explanations included bright stars and planets, aircraft and gliders, rocket launches, balloons, birds, ball lightning, meteors, atmospheric phenomena, and the occasional hoax. For a minority of the sightings, the investigators could not deduce what was seen, but their overall conclusion was that there was no reason to believe the UFOs were either highly advanced Soviet craft or visitors from other worlds. The air force ultimately dropped its investigations of the UFO phenomenon.

Believers discounted the air force denials, claiming to have other evidence of alien visitation. So far, none of this evidence has ever withstood scientific scrutiny. Photographs and film clips are nearly always too fuzzy to clearly show alien spacecraft, except in cases that are obviously faked. UFO witnesses are frequently credible (they include seasoned pilots), but generally there are several possible explanations for what they've seen besides alien spacecraft. Crop circles are easily made by pranksters. Stories of alien abductions are dramatic but cannot be verified. Pieces of metal that "UFO experts" say could not have been made by humans have turned out to be pieces of cars or refrigerators. Champions of alien visitation generally explain away the lack of clear evidence in one of two ways: government cover-ups or a failure of the mainstream scientific community to take the relevant phenomena seriously. Neither explanation is compelling.

It's certainly conceivable that a secretive government might *try* to put the lid on evidence of alien visits, though the motivation for doing so is unclear. The usual explanations are that the public couldn't handle the news or that the government is taking secret advantage of the alien materials to design new military hardware (via "reverse engineering"). Both explanations are silly. Half the population already believes in alien visitors and would hardly be shocked if newspapers announced that aliens were stacked up in government warehouses. As for reverse-engineering extraterrestrial spacecraft, we should keep in mind how difficult it is to travel from star to star. Any society that could do so routinely would be technologically far beyond our own. Reverse-engineering their spaceships is as unlikely as expecting Neanderthals to construct personal computers just because a laptop somehow landed in their cave. In addition, while a government might successfully hide evidence for a short time, does it really seem possible that evidence could remain secret for decades? After all, talk show fame and riches would await anyone who uncovered the conspiracy. Moreover, unless the aliens landed only in the United States, can we seriously believe that *every* government has cooperated in hiding the evidence?

Alleged disinterest on the part of the scientific community is an equally unimpressive claim. Scientists are constantly competing with one another to be the first with a great discovery, and clear evidence of alien visitors would certainly rank high on the all-time list. Countless researchers would work evenings and weekends, without pay, if they thought they could make such a discovery. The fact that few scientists are engaged in such study reflects not a lack of interest, but a lack of evidence worthy of study.

Of course, absence of evidence is not evidence of absence. Most scientists are open to the possibility that we might someday find evidence of alien visits, and many would welcome aliens with open arms. So far, however, we have no hard evidence to support the belief that aliens are already here.

• Where are the aliens?

Imagine that we survive and become interstellar travelers and that we begin colonizing habitable planets around nearby stars. As the colonies grow at each new location, some of the people may decide to set out for other star systems. Even if our starships traveled at relatively low speeds—say, a few percent of the speed of light—we could have dozens of outposts around nearby stars within a few centuries. In 10,000 years, our descendants would be spread among stars within a few hundred light-years of Earth. In a few million years, we could have outposts throughout the Milky Way Galaxy. We will have become a true galactic civilization.

Now, if we take the idea that *we* could develop a galactic civilization within a few million years and combine it with the reasonable (though unproved) idea that civilizations ought to be common, we are led to an astonishing conclusion: Someone else should already have created a galactic civilization. In fact, it should have been done a long time ago.

If *we* can develop interstellar travel in the future, then it would seem that other civilizations should have developed the capability long ago.

To see why, let's take some sample numbers. For example, suppose the overall odds of a civilization arising around a star are about the same as your odds of winning the lottery, or 1 in a million. Taking a low estimate of 100 billion stars in the Milky Way Galaxy, this means some 100,000 civilizations in our galaxy alone. Further, suppose we are a fairly typical civilization, so that civilizations generally arise when their stars are approaching 5 billion years old. Given that the galaxy is some 12 billion years old, the first of these 100,000 civilizations would have arisen at least 7 billion years ago. Others would have arisen, on average, about every 70,000 years. Under these assumptions, the youngest civilization besides ourselves would be some 70,000 years ahead of us technologically, and most would be millions or billions of years ahead of us.

Thus, we encounter a strange paradox: Plausible arguments suggest that a galactic civilization should already exist, yet we have so far found no evidence of such a civilization. This paradox is often called *Fermi's paradox,* after the Nobel Prize–winning physicist Enrico Fermi. During a 1950 conversation with other scientists about the possibility of extraterrestrial intelligence, Fermi responded to speculations by asking, "So where is everybody?"

This paradox has many possible solutions, but broadly speaking we can group them into three categories:

1. We are alone. There is no galactic civilization because civilizations are extremely rare—so rare that we are the first to have arisen on the galactic scene, perhaps even the first in the universe.
2. Civilizations are common, but no one has colonized the galaxy. There are at least three possible reasons why this might be the case. Perhaps interstellar travel is much harder or vastly more expensive than we have guessed, and civilizations are unable to venture far from their home worlds. Perhaps the desire to explore is unusual, and other societies either never leave their home star systems or stop exploring before they've colonized much of the galaxy. Most ominously, perhaps many civilizations have arisen, but they have all destroyed themselves before achieving the ability to colonize the stars.
3. There *is* a galactic civilization, but it has not yet revealed its existence to us.

We do not know which, if any, of these explanations is the correct solution to the question "Where are the aliens?" However, each category of solution has astonishing implications for our own species.

Consider the first solution—that we are alone. If this is true, then our civilization is a remarkable achievement. It implies that through all of cosmic evolution, among countless star systems, we are the first piece of our galaxy or the universe ever to know that the rest of the universe exists. Through us, the universe has attained self-awareness. Some philosophers and many religions argue that the ultimate purpose of life is to become truly self-aware. If so, and if we are alone, then the destruction of our civilization and the loss of our scientific knowledge would represent an inglorious end to something that took the universe some 14 billion years to achieve. From this point of view, humanity becomes all the more precious, and the collapse of our civilization would be all the more tragic.

The second category of solutions has much more terrifying implications. If thousands of civilizations before us have all failed to achieve interstellar travel on a large scale, what hope do we have? Unless we somehow think differently than all other civilizations, this solution says that we will never go far in space. Because we have always explored when the opportunity arose, this solution almost inevitably leads to the conclusion that failure will come about because we destroy ourselves. We can only hope that this answer is wrong.

The third solution is perhaps the most intriguing. It says that we are newcomers on the scene of a galactic civilization that has existed for millions or billions of years before us. Perhaps this civilization is deliberately leaving us alone for the time being and will someday decide the time is right to invite us to join it.

No matter what the answer turns out to be, learning it is sure to mark a turning point in the brief history of our species. Moreover, this turning point is likely to be reached within the next few decades or centuries. We already have the ability to destroy our civilization. If we do so, then our fate is sealed. But if we survive long enough to develop technology that can take us to the stars, the possibilities seem almost limitless.

THE BIG PICTURE

Putting Chapter 18 into Context

Throughout our study of astronomy, we have taken the "big picture" view of trying to understand how we fit into the universe. Here, at last, we have returned to Earth and examined the role of our own generation in the big picture of human history. Tens of thousands of past human generations have walked this Earth. Ours is the first generation with the technology to study the far reaches of our universe, to search for life elsewhere, and to travel beyond our home planet. It is up to us to decide whether we will use this technology to advance our species or to destroy it.

Imagine for a moment the grand view, a gaze across the centuries and millennia from this moment forward. Picture our descendants living among the stars, having created or joined a great galactic civilization. They will have the privilege of experiencing ideas, worlds, and discoveries far beyond our wildest imagination. Perhaps, in their history lessons, they will learn of our generation—the generation that history placed at the turning point and that managed to steer its way past the dangers of self-destruction and onto the path to the stars.

18.1 LIFE ON EARTH

- **When did life arise on Earth?**

Fossil evidence puts the origin of life at least 3.5 billion years ago, and carbon isotope evidence pushes this date to more than 3.85 billion years ago. Thus, life arose within a few hundred million years after the last major impact of the heavy bombardment, and possibly in a much shorter time.

- **How did life arise on Earth?**

Genetic evidence suggests that all life on Earth evolved from a common ancestor, and this ancestor was probably similar to microbes that live today in hot water near undersea volcanic vents or hot springs. We do not know how this first organism arose, but laboratory experiments suggest that it may have been the result of natural chemical processes on the early Earth.

- **What are the necessities of life?**

Life on Earth thrives in a wide range of environments, and in general seems to require only three things: a source of nutrients, a source of energy, and liquid water.

18.2 LIFE IN THE SOLAR SYSTEM

- **Could there be life on Mars?**

Mars once had conditions that may have been conducive to an origin of life. If life arose, it might still survive in pockets of liquid water underground.

- **Could there be life on Europa or other jovian moons?**

Europa probably has a subsurface ocean of liquid water, and may have undersea volcanoes on its ocean floor. If so, it has conditions much like those in which life on Earth probably arose, making it a good candidate for life beyond Earth. Ganymede and Callisto might have oceans as well. Titan may have other liquids on its surface, though it is too cold for liquid water. Perhaps life can survive in these other liquids, or perhaps Titan has liquid water deep underground.

18.3 LIFE AROUND OTHER STARS

- **Are habitable planets likely?**

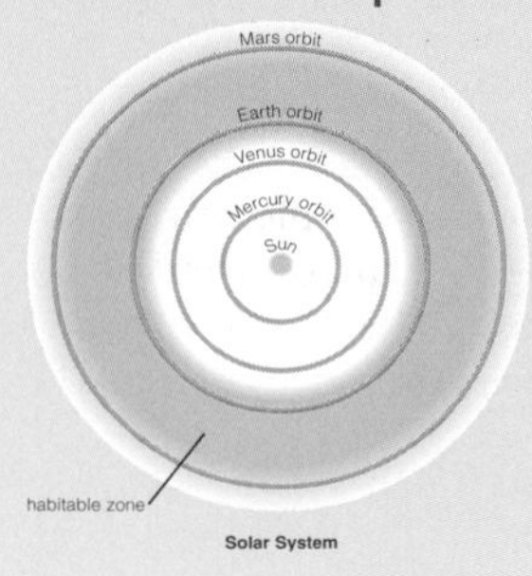

Billions of stars have at least moderate-size **habitable zones** in which life-bearing planets might exist. We do not yet have the technology to search for habitable planets directly, but several planned missions should be able to begin the search soon.

- **Are Earth-like planets rare or common?**

We don't know. Arguments can be made on both sides of the question, and we lack the data to determine their validity at present.

18.4 THE SEARCH FOR EXTRATERRESTRIAL INTELLIGENCE

- **How many civilizations are out there?**

We don't know, but the **Drake equation** gives us a way to organize our thinking about the question. The equation (in a modified form) says that the number of civilizations in the Milky Way Galaxy with whom we could potentially communicate is $N_{HP} \times f_{life} \times f_{civ} \times f_{now}$, where N_{HP} is the number of habitable planets in the galaxy, f_{life} is the fraction of habitable planets that actually have life on them, f_{civ} is the fraction of life-bearing planets upon which a civilization capable of interstellar communication has at some time arisen, and f_{now} is the fraction of all these civilizations that exist now.

- **How does SETI work?**

SETI, the search for extraterrestrial intelligence, generally refers to efforts to detect signals—such as radio or laser communications—coming from civilizations on other worlds.

18.5 INTERSTELLAR TRAVEL AND ITS IMPLICATIONS TO CIVILIZATION

- **How difficult is interstellar travel?**

Convenient interstellar travel remains well beyond our technological capabilities, because of the technological requirements for engines, the enormous energy needed to accelerate spacecraft to speeds near the speed of light, and the difficulties of shielding the crew from radiation. Nevertheless, people have proposed ways around all these difficulties, and it seems reasonable to think that we will someday achieve interstellar travel if we survive long enough.

- **Where are the aliens?**

It seems that we should be capable of colonizing the galaxy in a few million years or less, and the galaxy was around for at least 7 billion years before Earth was even born. Thus, it seems that someone should have colonized the galaxy long ago—yet we have no evidence of other civilizations. Every possible category of explanation for this surprising fact has astonishing implications for our species and our place in the universe.

EXERCISES AND PROBLEMS

REVIEW QUESTIONS

1. Describe three recent developments in the study of life on Earth that make it seem much more likely that we could find life elsewhere.
2. How do we study the history of life on Earth? Describe the *geological time scale* and a few of the major events along it.
3. Summarize the evidence pointing to an early origin of life on Earth. How far back in Earth's history did life exist?
4. Why is the theory of evolution so critical to our understanding of the history of life on Earth? Explain how evolution proceeds by *natural selection*, and what happens to DNA that allows species to evolve.
5. Give a brief overview of the history of life on Earth. What evidence points to a common ancestor for all life? How and when did oxygen accumulate in Earth's atmosphere? When did larger animals diversify on Earth?
6. Where do we think life on Earth first arose, and why?
7. Describe how laboratory experiments are helping us study the origin of life on Earth.
8. Is it possible that life migrated to Earth from elsewhere? Explain.
9. Describe the range of environments in which life thrives on Earth. What three basic requirements apply to life in all these environments?
10. What is a *habitable world?* Which worlds in our solar system seem potentially habitable, and why?
11. Briefly summarize the debate over possible fossil evidence of life in a meteorite from Mars.
12. What do we mean by a star's *habitable zone?* Do we expect many stars to be capable of having habitable planets? Explain.
13. What is the *rare Earth hypothesis?* Summarize the arguments on both sides regarding the validity of this hypothesis.
14. What is the *Drake equation?* Define each of its factors, and describe the current state of understanding about the potential values of each factor.
15. What is SETI? Describe the capabilities of current SETI efforts.
16. Summarize the factors that make interstellar travel difficult, and describe a few technologies that might someday make it possible nonetheless.
17. What is Fermi's paradox? Describe several potential solutions to the paradox, and the implications of each to the future of our civilization.

FANTASY OR SCIENCE FICTION?

Each of the following describes some futuristic scenario that, while perhaps common and entertaining, may or may not be plausible. In each case, decide whether the scenario is plausible according to our present understanding of science or whether it is unlikely to be possible. Explain your reasoning.

18. The first human explorers on Mars discover that the surface is littered with the ruins of an ancient civilization, including remnants of tall buildings and temples.
19. The first human explorers on Mars drill a hole into a Martian volcano to collect a sample of soil from several meters underground. Upon analysis of the soil, they discover that it holds living microbes resembling terrestrial bacteria but with a different biochemistry.
20. In 2020, a spacecraft lands on Europa and melts its way through the ice into the Europan ocean. It finds numerous strange, living microbes, along with a few larger organisms that feed on the microbes.
21. It's the year 2075. A giant telescope on the Moon, consisting of hundreds of small telescopes linked together across a distance of 500 kilometers, has just captured a series of images of a planet around a distant star that clearly show seasonal changes in vegetation.
22. A century from now, after completing a careful study of planets around stars within 100 light-years of Earth, we've discovered that the most diverse life exists on a planet orbiting a young star that formed just 100 million years ago.
23. In 2030, a brilliant teenager working in his garage builds a coal-powered rocket that can travel at half the speed of light.
24. In the year 2750, we receive a signal from a civilization around a nearby star telling us that the *Voyager 2* spacecraft recently crash-landed on its planet.
25. Crew members of the matter–antimatter spacecraft *Star Apollo,* which left Earth in the year 2165, return to Earth in the year 2450, looking only a few years older than when they left.
26. Aliens from a distant star system invade Earth with intent to destroy us and occupy our planet, but we successfully fight them off when their technology proves no match for ours.
27. A single, great galactic civilization exists. It originated on a single planet long ago but is now made up of beings from many different planets, each of which was assimilated into the galactic culture in turn.

PROBLEMS

28. *Most Likely to Have Life.* Suppose you were asked to vote in a contest to name the world in our solar system (besides Earth) "most likely to have life." Which world would you cast your vote for? Explain and defend your choice in a one-page essay.
29. *Are Earth-like Planets Common?* Based on what you have learned in this book, form an opinion as to whether you think Earth-like planets will ultimately prove to be rare, common, or something in between. Write a one- to two-page essay explaining and defending your opinion.
30. *Aliens in the Movies.* Choose a science fiction movie (or television show) that involves an alien species. Do you think aliens like this could really exist? Do you think they are portrayed in a realistic way? Write a one- to two-page critical review of the movie, focusing primarily on the question of how well the movie addresses the aliens in light of current scientific knowledge.
31. *Solution to the Fermi Paradox.* Among the various possible solutions to the question "Where are the aliens?" which do you think is most likely? (If you have no opinion on their likelihood, which do you like best?) Write a one- to two-page essay in which you explain why you favor this solution.

DISCUSSION QUESTIONS

32. *Funding the Search for Life.* Imagine that you are a member of Congress, so that your job includes deciding how much government funding goes to research in different areas of science. How much would you allot to the search for life in the universe compared to the amount allotted to research in other areas of astronomy and planetary science? Why?
33. *Distant Dream or Near-Reality?* Considering all the issues surrounding interstellar flight, when (if ever) do you think we are likely to begin traveling among the stars? Why?
34. *The Turning Point.* Discuss the idea that our generation has acquired a greater responsibility for the future than any previous generation. Do you agree with this assessment? If so, how should we deal with this responsibility? Defend your opinions.

For a complete list of media resources available, go to **www.astronomyplace.com** and choose Chapter 18 from the pull-down menu.

ASTRONOMY PLACE WEB TUTORIALS

Tutorial Review of Key Concepts

Use the following interactive **Tutorial** at **www.astronomyplace.com** to review key concepts from this chapter.

Detecting Extrasolar Planets Tutorial

Lesson 1 Taking a Picture of a Planet
Lesson 2 Stars' Wobbles and Properties of Planets
Lesson 3 Planetary Transits

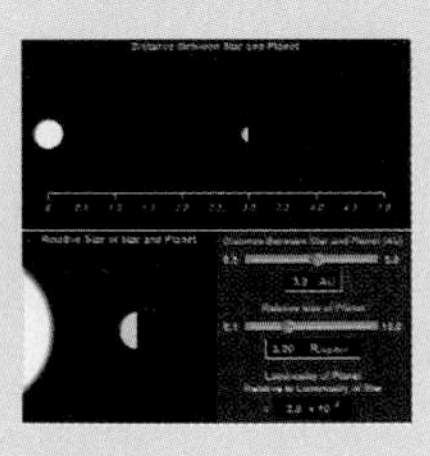

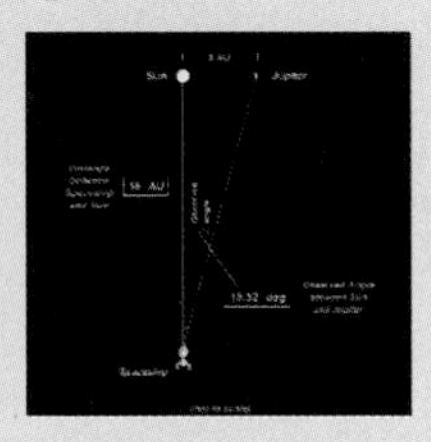

Supplementary Tutorial Exercises

Use the interactive **Tutorial Lesson** to explore the following questions.

Detecting Extrasolar Planets Tutorial, Lessons 1–3

1. Give two reasons why visual detection of planets orbiting other stars is extremely difficult.
2. As you move away from two objects what happens to the apparent angle between them?
3. How can we detect extrasolar planets?
4. How might transits allow us to detect Earth-size planets around other stars?

MOVIES

Check out the following narrated and animated short documentary available on **www.astronomyplace.com** for a helpful review of key ideas covered in this chapter.

The Search for Extraterrestrial Life Movie

WEB PROJECTS

Take advantage of the useful Web links on **www.astronomyplace.com** to assist you with the following projects.

1. *Astrobiology News.* Go to NASA's Astrobiology home page and read some of the recent news from the search for life in the universe. Choose one recent news article and write a one- to two-page summary of the research and how it relates to the question of life in the universe in general.
2. *Martian Meteorites.* Find information about the latest discoveries concerning Martian meteorites and whether the meteorites contain evidence of life. Choose one recent discovery that seems important, and write a short summary of how you think it alters the debate about the habitability of Mars or about life on Mars.
3. *The Search for Extraterrestrial Intelligence.* Go to the home page for the SETI Institute. Learn more about how SETI is funded and carried out. In one page or less, describe the SETI Institute and its work.
4. *Starship Design.* Find more details about a proposal for starship propulsion or design. How would the starship work? What new technologies would be needed, and what existing technologies could be applied? Summarize your findings in a one- to two-page report.
5. *Advanced Spacecraft Technologies.* NASA supports many efforts to incorporate new technologies into spaceships. Although few of them reach the level of being suitable for interstellar colonization, most are innovative and fascinating. Learn about one such NASA project, and write a short summary of your findings.

Appendixes

A | Useful Numbers

ASTRONOMICAL DISTANCES

1 AU $\approx 1.496 \times 10^8$ km

1 light-year $\approx 9.46 \times 10^{12}$ km

1 parsec (pc) $\approx 3.09 \times 10^{13}$ km $\approx$ 3.26 light-years

1 kiloparsec (kpc) = 1,000 pc $\approx 3.26 \times 10^3$ light-years

1 megaparsec (Mpc) = 10^6 pc $\approx 3.26 \times 10^6$ light-years

ASTRONOMICAL TIMES

1 solar day (average) = 24^{h}

1 sidereal day $\approx 23^{h}\,56^{m}\,4.09^{s}$

1 synodic month (average) $\approx$ 29.53 solar days

1 sidereal month (average) $\approx$ 27.32 solar days

1 tropical year $\approx$ 365.242 solar days

1 sidereal year $\approx$ 365.256 solar days

UNIVERSAL CONSTANTS

Speed of light:	$c = 3 \times 10^5$ km/s $= 3 \times 10^8$ m/s
Gravitational constant:	$G = 6.67 \times 10^{-11} \dfrac{\text{m}^3}{\text{kg} \times \text{s}^2}$
Planck's constant:	$h = 6.626 \times 10^{-34}$ joule $\times$ s
Stefan–Boltzmann constant:	$\sigma = 5.7 \times 10^{-8} \dfrac{\text{watt}}{\text{m}^2 \times \text{Kelvin}^4}$
mass of a proton:	$m_p = 1.67 \times 10^{-27}$ kg
mass of an electron:	$m_e = 9.1 \times 10^{-31}$ kg

USEFUL SUN AND EARTH REFERENCE VALUES

Mass of the Sun: $1\,M_{\text{Sun}} \approx 2 \times 10^{30}$ kg

Radius of the Sun: $1\,R_{\text{Sun}} \approx 696{,}000$ km

Luminosity of the Sun: $1\,L_{\text{Sun}} \approx 3.8 \times 10^{26}$ watts

Mass of the Earth: $1\,M_{\text{Earth}} \approx 5.97 \times 10^{24}$ kg

Radius (equatorial) of the Earth: $1\,R_{\text{Earth}} \approx 6{,}378$ km

Acceleration of gravity on Earth: $g = 9.8$ m/s^2

Escape velocity from surface of Earth: $v_{\text{escape}} = 11$ km/s $= 11{,}000$ m/s

ENERGY AND POWER UNITS

Basic unit of energy: 1 joule $= 1 \dfrac{\text{kg} \times \text{m}^2}{\text{s}^2}$

Basic unit of power: 1 watt = 1 joule/s

Electron-volt: 1 eV $= 1.60 \times 10^{-19}$ joule

B | Useful Formulas

- Universal law of gravitation for the force between objects of mass M_1 and M_2, distance d between their centers:

$$F = G\frac{M_1 M_2}{d^2}$$

- Newton's version of Kepler's third law; p and a are period and semimajor axis, respectively, of either orbiting mass:

$$p^2 = \frac{4\pi^2}{G(M_1 + M_2)} a^3$$

- Escape velocity at distance R from center of object of mass M:

$$v_{\text{escape}} = \sqrt{\frac{2GM}{R}}$$

- Relationship between a photon's wavelength (λ), frequency (f), and the speed of light (c):

$$\lambda \times f = c$$

- Energy of a photon of wavelength λ or frequency f:

$$E = hf = \frac{hc}{\lambda}$$

- Stefan–Boltzmann law for thermal radiation at temperature T (in Kelvin):

$$\text{emitted power per unit area} = \sigma T^4$$

- Wien's law for the peak wavelength (λ_{max}) thermal radiation at temperature T (in Kelvin):

$$\lambda_{\text{max}} = \frac{2{,}900{,}000}{T}\,\text{nm}$$

- Doppler shift (radial velocity is positive if the object is moving away from us and negative if it is moving toward us):

$$\frac{\text{radial velocity}}{\text{speed of light}} = \frac{\text{shifted wavelength} - \text{rest wavelength}}{\text{rest wavelength}}$$

- Angular separation (α) of two points with an actual separation s, viewed from a distance d (assuming d is much larger than s):

$$\alpha = \frac{s}{2\pi d} \times 360^\circ$$

- Inverse Square Law for Light:

$$\text{apparent brightness} = \frac{\text{luminosity}}{4\pi d^2}$$

(where d is the distance to the object)

- Parallax formula (distance d to a star with parallax angle p in arcseconds):

$$d\ (\text{in parsecs}) = \frac{1}{p\ (\text{in arcseconds})}$$

- The orbital velocity law, to find the mass M_r contained within the circular orbit of radius r for an object moving at speed v:

$$M_r = \frac{r \times v^2}{G}$$

C | A Few Mathematical Skills

This appendix reviews the following mathematical skills: powers of 10, scientific notation, working with units, the metric system, and finding a ratio. You should refer to this appendix as needed while studying the textbook.

C.1 POWERS OF 10

Powers of 10 simply indicate how many times to multiply 10 by itself. For example:

$$10^2 = 10 \times 10 = 100$$

$$10^6 = 10 \times 10 \times 10 \times 10 \times 10 \times 10 = 1{,}000{,}000$$

Negative powers are the reciprocals of the corresponding positive powers. For example:

$$10^{-2} = \frac{1}{10^2} = \frac{1}{100} = 0.01$$

$$10^{-6} = \frac{1}{10^6} = \frac{1}{1{,}000{,}000} = 0.000001$$

Table C.1 lists powers of 10 from 10^{-12} to 10^{12}. Note that powers of 10 follow two basic rules:

1. A positive exponent tells how many zeros follow the 1. For example, 10^0 is a 1 followed by no zeros, and 10^8 is a 1 followed by eight zeros.
2. A negative exponent tells how many places are to the right of the decimal point, including the 1. For example, $10^{-1} = 0.1$ has one place to the right of the decimal point; $10^{-6} = 0.000001$ has six places to the right of the decimal point.

Multiplying and Dividing Powers of 10

Multiplying powers of 10 simply requires adding exponents, as the following examples show:

$$10^4 \times 10^7 = \underbrace{10{,}000}_{10^4} \times \underbrace{10{,}000{,}000}_{10^7} = \underbrace{100{,}000{,}000{,}000}_{10^{4+7} = 10^{11}} = 10^{11}$$

$$10^5 \times 10^{-3} = \underbrace{100{,}000}_{10^5} \times \underbrace{0.001}_{10^{-3}} = \underbrace{100}_{10^{5+(-3)} = 10^2} = 10^2$$

$$10^{-8} \times 10^{-5} = \underbrace{0.00000001}_{10^{-8}} \times \underbrace{0.00001}_{10^{-5}} = \underbrace{0.0000000000001}_{10^{-8+(-5)} = 10^{-13}} = 10^{-13}$$

TABLE C.1 *Powers of 10*

Zero and Positive Powers			*Negative Powers*		
Power	**Value**	**Name**	**Power**	**Value**	**Name**
10^0	1	One			
10^1	10	Ten	10^{-1}	0.1	Tenth
10^2	100	Hundred	10^{-2}	0.01	Hundredth
10^3	1,000	Thousand	10^{-3}	0.001	Thousandth
10^4	10,000	Ten thousand	10^{-4}	0.0001	Ten thousandth
10^5	100,000	Hundred thousand	10^{-5}	0.00001	Hundred thousandth
10^6	1,000,000	Million	10^{-6}	0.000001	Millionth
10^7	10,000,000	Ten million	10^{-7}	0.0000001	Ten millionth
10^8	100,000,000	Hundred million	10^{-8}	0.00000001	Hundred millionth
10^9	1,000,000,000	Billion	10^{-9}	0.000000001	Billionth
10^{10}	10,000,000,000	Ten billion	10^{-10}	0.0000000001	Ten billionth
10^{11}	100,000,000,000	Hundred billion	10^{-11}	0.00000000001	Hundred billionth
10^{12}	1,000,000,000,000	Trillion	10^{-12}	0.000000000001	Trillionth

Dividing powers of 10 requires subtracting exponents, as in the following examples:

$$\frac{10^5}{10^3} = \underbrace{100{,}000}_{10^5} \div \underbrace{1{,}000}_{10^3} = \underbrace{100}_{10^{5-3} = 10^2} = 10^2$$

$$\frac{10^3}{10^7} = \underbrace{1{,}000}_{10^3} \div \underbrace{10{,}000{,}000}_{10^7} = \underbrace{0.0001}_{10^{3-7} = 10^{-4}} = 10^{-4}$$

$$\frac{10^{-4}}{10^{-6}} = \underbrace{0.0001}_{10^{-4}} \div \underbrace{0.000001}_{10^{-6}} = \underbrace{100}_{10^{-4-(-6)} = 10^2} = 10^2$$

Powers of Powers of 10

We can use the multiplication and division rules to raise powers of 10 to other powers or to take roots. For example:

$$(10^4)^3 = 10^4 \times 10^4 \times 10^4 = 10^{4+4+4} = 10^{12}$$

Note that we can get the same end result by simply multiplying the two powers:

$$(10^4)^3 = 10^{4\times3} = 10^{12}$$

Because taking a root is the same as raising to a fractional power (e.g., the square root is the same as the $\frac{1}{2}$ power, the cube root is the same as the $\frac{1}{3}$ power, etc.), we can use the same procedure for roots, as in the following example:

$$\sqrt{10^4} = (10^4)^{1/2} = 10^{4\times(1/2)} = 10^2$$

Adding and Subtracting Powers of 10

Unlike with multiplication and division, there is no shortcut for adding or subtracting powers of 10. The values must be written in longhand notation. For example:

$$10^6 + 10^2 = 1{,}000{,}000 + 100 = 1{,}000{,}100$$
$$10^8 + 10^{-3} = 100{,}000{,}000 + 0.001 = 100{,}000{,}000.001$$
$$10^7 - 10^3 = 10{,}000{,}000 - 1{,}000 = 9{,}999{,}000$$

Summary

We can summarize our findings using n and m to represent any numbers:

- To *multiply* powers of 10, *add* exponents: $10^n \times 10^m = 10^{n+m}$
- To *divide* powers of 10, *subtract* exponents: $\dfrac{10^n}{10^m} = 10^{n-m}$
- To *raise* powers of 10 to other powers, multiply exponents: $(10^n)^m = 10^{n \times m}$

C.2 SCIENTIFIC NOTATION

When we are dealing with large or small numbers, it's generally easier to write them with powers of 10. For example, it's much easier to write the number 6,000,000,000,000 as 6×10^{12}. This format, in which a number *between* 1 and 10 is multiplied by a power of 10, is called **scientific notation**.

Converting a Number to Scientific Notation

We can convert numbers written in ordinary notation to scientific notation with a simple two-step process:

1. Move the decimal point to come after the *first* nonzero digit.
2. The number of places the decimal point moves tells you the power of 10; the power is *positive* if the decimal point moves to the left and *negative* if it moves to the right.

   Examples:

$$3{,}042 \xrightarrow[\text{3 places to left}]{\text{decimal needs to move}} 3.042 \times 10^3$$

$$0.00012 \xrightarrow[\text{4 places to right}]{\text{decimal needs to move}} 1.2 \times 10^{-4}$$

$$226 \times 10^2 \xrightarrow[\text{2 places to left}]{\text{decimal needs to move}} (2.26 \times 10^2) \times 10^2 = 2.26 \times 10^4$$

Converting a Number from Scientific Notation

We can convert numbers written in scientific notation to ordinary notation by the reverse process:

1. The power of 10 indicates how many places to move the decimal point; move it to the *right* if the power of 10 is positive and to the *left* if it is negative.

2. If moving the decimal point creates any open places, fill them with zeros.

   Examples:

$$4.01 \times 10^2 \xrightarrow[\text{2 places to right}]{\text{move decimal}} 401$$

$$3.6 \times 10^6 \xrightarrow[\text{6 places to right}]{\text{move decimal}} 3{,}600{,}000$$

$$5.7 \times 10^{-3} \xrightarrow[\text{3 places to left}]{\text{move decimal}} 0.0057$$

Multiplying or Dividing Numbers in Scientific Notation

Multiplying or dividing numbers in scientific notation simply requires operating on the powers of 10 and the other parts of the number separately.

Examples:

$$(6 \times 10^2) \times (4 \times 10^5) = (6 \times 4) \times (10^2 \times 10^5) = 24 \times 10^7 = (2.4 \times 10^1) \times 10^7 = 2.4 \times 10^8$$

$$\frac{4.2 \times 10^{-2}}{8.4 \times 10^{-5}} = \frac{4.2}{8.4} \times \frac{10^{-2}}{10^{-5}} = 0.5 \times 10^{-2-(-5)} = 0.5 \times 10^3 = (5 \times 10^{-1}) \times 10^3 = 5 \times 10^2$$

Note that, in both these examples, we first found an answer in which the number multiplied by a power of 10 was *not* between 1 and 10. We therefore followed the procedure for converting the final answer to scientific notation.

Addition and Subtraction with Scientific Notation

In general, we must write numbers in ordinary notation before adding or subtracting.

Examples:

$$(3 \times 10^6) + (5 \times 10^2) = 3{,}000{,}000 + 500 = 3{,}000{,}500 = 3.0005 \times 10^6$$

$$(4.6 \times 10^9) - (5 \times 10^8) = 4{,}600{,}000{,}000 - 500{,}000{,}000 = 4{,}100{,}000{,}000 = 4.1 \times 10^9$$

When both numbers have the *same* power of 10, we can factor out the power of 10 first.

Examples:

$$(7 \times 10^{10}) + (4 \times 10^{10}) = (7 + 4) \times 10^{10} = 11 \times 10^{10} = 1.1 \times 10^{11}$$

$$(2.3 \times 10^{-22}) - (1.6 \times 10^{-22}) = (2.3 - 1.6) \times 10^{-22} = 0.7 \times 10^{-22} = 7.0 \times 10^{-23}$$

C.3 WORKING WITH UNITS

Showing the units of a problem as you solve it usually makes the work much easier and also provides a useful way of checking your work. If an answer does not come out with the units you expect, you probably did something wrong. In general, working with units is very similar to working with numbers, as the following guidelines and examples show.

Five Guidelines for Working with Units

Before you begin any problem, think ahead and identify the units you expect for the final answer. Then operate on the units along with the numbers as you solve the problem. The following five guidelines may be helpful when you are working with units:

1. Mathematically, it doesn't matter whether a unit is singular (e.g., meter) or plural (e.g., meters); we can use the same abbreviation (e.g., m) for both.

2. You cannot add or subtract numbers unless they have the *same* units. For example, 5 apples + 3 apples = 8 apples, but the expression 5 apples + 3 oranges cannot be simplified further.

3. You *can* multiply units, divide units, or raise units to powers. Look for key words that tell you what to do.

   - *Per* suggests division. For example, we write a speed of 100 kilometers per hour as:

   $$100\frac{\text{km}}{\text{hr}} \quad \text{or} \quad 100\frac{\text{km}}{1\text{ hr}}$$

   - *Of* suggests multiplication. For example, if you launch a 50-kg space probe at a launch cost *of* $10,000 per kilogram, the total cost is:

   $$50\,\cancel{\text{kg}} \times \frac{\$10{,}000}{\cancel{\text{kg}}} = \$500{,}000$$

   - *Square* suggests raising to the second power. For example, we write an area of 75 square meters as 75 m^2.

   - *Cube* suggests raising to the third power. For example, we write a volume of 12 cubic centimeters as 12 cm^3.

4. Often the number you are given is not in the units you wish to work with. For example, you may be given that the speed of light is 300,000 km/s but need it in units of m/s for a particular problem. To convert the units, simply multiply the given number by a *conversion factor:* a fraction in which the numerator (top of the fraction) and denominator (bottom of the fraction) are equal, so that the value of the fraction is 1; the number in the denominator must have the units that you wish to change. In the case of changing the speed of light from units of km/s to m/s, you need a conversion factor for kilometers to meters. Thus, the conversion factor is:

   $$\frac{1{,}000\text{ m}}{1\text{ km}}$$

   Note that this conversion factor is equal to 1, since 1,000 meters and 1 kilometer are equal, and that the units to be changed (km) appear in the denominator. We can now convert the speed of light from units of km/s to m/s simply by multiplying by this conversion factor:

   $$\underbrace{300{,}000\frac{\cancel{\text{km}}}{\text{s}}}_{\substack{\text{speed of light}\\ \text{in km/s}}} \times \underbrace{\frac{1{,}000\text{ m}}{1\,\cancel{\text{km}}}}_{\substack{\text{conversion from}\\ \text{km to m}}} = \underbrace{3 \times 10^8\frac{\text{m}}{\text{s}}}_{\substack{\text{speed of light}\\ \text{in m/s}}}$$

   Note that the units of km cancel, leaving the answer in units of m/s.

5. It's easier to work with units if you replace division with multiplication by the reciprocal. For example, suppose you want to know how many minutes are represented by 300 seconds. We can find the answer by dividing 300 seconds by 60 seconds per minute:

$$300 \text{ s} \div 60 \frac{5}{\text{min}}$$

However, it is easier to see the unit cancellations if we rewrite this expression by replacing the division with multiplication by the reciprocal (this process is easy to remember as "invert and multiply"):

$$300 \text{ s} \div 60 \frac{\text{s}}{\text{min}} = \underbrace{300 \cancel{\text{s}} \times \underbrace{\frac{1 \text{ min}}{60 \cancel{\text{s}}}}_{\text{invert}}}_{\text{and multiply}} = 5 \text{ min}$$

We now see that the units of seconds (s) cancel in the numerator of the first term and the denominator of the second term, leaving the answer in units of minutes.

More Examples of Working with Units

Example 1. How many seconds are there in 1 day?

Solution: We can answer the question by setting up a *chain* of unit conversions in which we start with 1 *day* and end up with *seconds*. We use the facts that there are 24 hours per day (24 hr/day), 60 minutes per hour (60 min/hr), and 60 seconds per minute (60 s/min):

$$\underbrace{1 \cancel{\text{day}}}_{\substack{\text{starting} \\ \text{value}}} \times \underbrace{\frac{24 \cancel{\text{hr}}}{\cancel{\text{day}}}}_{\substack{\text{conversion} \\ \text{from} \\ \text{day to hr}}} \times \underbrace{\frac{60 \cancel{\text{min}}}{\cancel{\text{hr}}}}_{\substack{\text{conversion} \\ \text{from} \\ \text{hr to min}}} \times \underbrace{\frac{60 \text{ s}}{\cancel{\text{min}}}}_{\substack{\text{conversion} \\ \text{from} \\ \text{min to s}}} = 86{,}400 \text{ s}$$

Note that all the units cancel except *seconds*, which is what we want for the answer. There are 86,400 seconds in 1 day.

Example 2. Convert a distance of 10^8 cm to km.

Solution: The easiest way to make this conversion is in two steps, since we know that there are 100 centimeters per meter (100 cm/m) and 1,000 meters per kilometer (1,000 m/km):

$$\underbrace{10^8 \text{ cm}}_{\substack{\text{starting} \\ \text{value}}} \times \underbrace{\frac{1 \text{ m}}{100 \text{ cm}}}_{\substack{\text{conversion} \\ \text{from} \\ \text{cm to m}}} \times \underbrace{\frac{1 \text{ km}}{1{,}000 \text{ m}}}_{\substack{\text{conversion} \\ \text{from} \\ \text{m to km}}} = 10^8 \cancel{\text{cm}} \times \frac{1 \cancel{\text{m}}}{10^2 \cancel{\text{cm}}} \times \frac{1 \text{ km}}{10^3 \cancel{\text{m}}} = 10^3 \text{ km}$$

Alternatively, if we recognize that the number of kilometers should be smaller than the number of centimeters (because kilometers are larger), we might decide to do this conversion by dividing as follows:

$$10^8 \text{ cm} \div \frac{100 \text{ cm}}{\text{m}} \div \frac{1{,}000 \text{ m}}{\text{km}}$$

In this case, before carrying out the calculation, we replace each division with multiplication by the reciprocal:

$$10^8 \text{ cm} \div \frac{100 \text{ cm}}{\text{m}} \div \frac{1{,}000 \text{ m}}{\text{km}} = 10^8 \text{ cm} \times \frac{1 \text{ m}}{100 \text{ cm}} \times \frac{1 \text{ km}}{1{,}000 \text{ m}}$$

$$= 10^8 \cancel{\text{cm}} \times \frac{1 \cancel{\text{m}}}{10^2 \cancel{\text{cm}}} \times \frac{1 \text{ km}}{10^3 \cancel{\text{m}}}$$

$$= 10^3 \text{ km}$$

Note that we again get the answer that 10^8 cm is the same as 10^3 km, or 1,000 km.

Example 3. Suppose you accelerate at 9.8 m/s^2 for 4 seconds, starting from rest. How fast will you be going?

Solution: The question asked "how fast?" so we expect to end up with a speed. Therefore, we multiply the acceleration by the amount of time you accelerated:

$$9.8 \frac{\text{m}}{\text{s}^2} \times 4 \text{ s} = (9.8 \times 4) \frac{\text{m} \times \cancel{\text{s}}}{\text{s}^{\cancel{2}}} = 39.2 \frac{\text{m}}{\text{s}}$$

Note that the units end up as a speed, showing that you will be traveling 39.2 m/s after 4 seconds of acceleration at 9.8 m/s^2.

Example 4. A reservoir is 2 km long and 3 km wide. Calculate its area, in both square kilometers and square meters.

Solution: We find its area by multiplying its length and width:

$$2 \text{ km} \times 3 \text{ km} = 6 \text{ km}^2$$

Next we need to convert this area of 6 km^2 to square meters, using the fact that there are 1,000 meters per kilometer (1,000 m/km). Note that we must square the term 1,000 m/km when converting from km^2 to m^2:

$$6 \text{ km}^2 \times \left(1{,}000 \frac{\text{m}}{\text{km}}\right)^2 = 6 \text{ km}^2 \times 1{,}000^2 \frac{\text{m}^2}{\text{km}^2} = 6 \cancel{\text{km}^2} \times 1{,}000{,}000 \frac{\text{m}^2}{\cancel{\text{km}^2}}$$

$$= 6{,}000{,}000 \text{ m}^2$$

The reservoir area is 6 km^2, which is the same as 6 million m^2.

C.4 THE METRIC SYSTEM (SI)

The modern version of the metric system, known as *Système Internationale d'Unites* (French for "International System of Units") or **SI**, was formally established in 1960. Today, it is the primary measurement system in nearly every country in the world with the exception of the United States. Even in the United States, it is the system of choice for science and international commerce.

The basic units of length, mass, and time in the SI are:

- The **meter** for length, abbreviated m
- The **kilogram** for mass, abbreviated kg
- The **second** for time, abbreviated s

Multiples of metric units are formed by powers of 10, using a prefix to indicate the power. For example, *kilo* means 10^3 (1,000), so a kilometer is

TABLE C.2 *SI (Metric) Prefixes*

Small Values			*Large Values*		
Prefix	**Abbreviation**	**Value**	**Prefix**	**Abbreviation**	**Value**
Deci	d	10^{-1}	Deca	da	10^{1}
Centi	c	10^{-2}	Hecto	h	10^{2}
Milli	m	10^{-3}	Kilo	k	10^{3}
Micro	μ	10^{-6}	Mega	M	10^{6}
Nano	n	10^{-9}	Giga	G	10^{9}
Pico	p	10^{-12}	Tera	T	10^{12}

1,000 meters; a microgram is 0.000001 gram, because *micro* means 10^{-6}, or one millionth. Some of the more common prefixes are listed in Table C.2.

Metric Conversions

Table C.3 lists conversions between metric units and units used commonly in the United States. Note that the conversions between kilograms and pounds are valid only on Earth, because they depend on the strength of gravity.

TABLE C.3 *Metric Conversions*

To Metric	*From Metric*
1 inch = 2.540 cm	1 cm = 0.3937 inch
1 foot = 0.3048 m	1 m = 3.28 feet
1 yard = 0.9144 m	1 m = 1.094 yards
1 mile = 1.6093 km	1 km = 0.6214 mile
1 pound = 0.4536 kg	1 kg = 2.205 pounds

Example 1. International athletic competitions generally use metric distances. Compare the length of a 100-meter race to that of a 100-yard race.

Solution: Table C.3 shows that 1 m = 1.094 yd, so 100 m is 109.4 yd. Note that 100 meters is almost 110 yards; a good "rule of thumb" to remember is that distances in meters are about 10% longer than the corresponding number of yards.

Example 2. How many square kilometers are in 1 square mile?

Solution: We use the square of the miles-to-kilometers conversion factor:

$$(1\ \text{mi}^2) \times \left(\frac{1.6093\ \text{km}}{1\ \text{mi}}\right)^2 = (1\ \cancel{\text{mi}^2}) \times \left(1.6093^2 \frac{\text{km}^2}{\cancel{\text{mi}^2}}\right) = 2.5898\ \text{km}^2$$

Therefore, 1 square mile is 2.5898 square kilometers.

C.5 FINDING A RATIO

Suppose you want to compare two quantities, such as the average density of the Earth and the average density of Jupiter. The way we do such a comparison is by dividing, which tells us the *ratio* of the two quantities. In this case, the Earth's average density is 5.52 grams/cm^3 and Jupiter's average density is 1.33 grams/cm^3 (see Figure 8.1), so the ratio is:

$$\frac{\text{average density of Earth}}{\text{average density of Jupiter}} = \frac{5.52\ \cancel{\text{g/cm}^3}}{1.33\ \cancel{\text{g/cm}^3}} = 4.15$$

Notice how the units cancel on both the top and bottom of the fraction. We can state our result in two equivalent ways:

- The ratio of the Earth's average density to Jupiter's average density is 4.15.
- The Earth's average density is 4.15 times Jupiter's average density.

Sometimes, the quantities that you want to compare may each involve an equation. In such cases, you could, of course, find the ratio by first calculating each of the two quantities individually and then dividing. However, it is much easier if you first express the ratio as a fraction, putting the equation for one quantity on top and the other on the bottom. Some of the terms in the equation may then cancel out, making any calculations much easier.

Example 1. Compare the kinetic energy of a car traveling at 100 km/hr to that of a car traveling at 50 km/hr.

Solution: We do the comparison by finding the ratio of the two kinetic energies, recalling that the formula for kinetic energy is $\frac{1}{2}mv^2$. Since we are not told the mass of the car, you might at first think that we don't have enough information to find the ratio. However, notice what happens when we put the equations for each kinetic energy into the ratio, calling the two speeds v_1 and v_2:

$$\frac{\text{K.E. car at } v_1}{\text{K.E. car at } v_2} = \frac{\cancel{\frac{1}{2}}\cancel{m}_{\text{car}} {v_1}^2}{\cancel{\frac{1}{2}}\cancel{m}_{\text{car}} {v_2}^2} = \frac{{v_1}^2}{{v_2}^2} = \left(\frac{v_1}{v_2}\right)^2$$

All the terms cancel except those with the two speeds, leaving us with a very simple formula for the ratio. Now we put in 100 km/hr for v_1 and 50 km/hr for v_2:

$$\frac{\text{K.E. car at 100 km/hr}}{\text{K.E. car at 50 km/hr}} = \left(\frac{100 \ \cancel{\text{km/hr}}}{50 \ \cancel{\text{km/hr}}}\right)^2 = 2^4 = 4$$

The ratio of the car's kinetic energies at 100 km/hr and 50 km/hr is 4. That is, the car has four times as much kinetic energy at 100 km/hr as it has at 50 km/hr.

Example 2. Compare the strength of gravity between the Earth and the Sun to the strength of gravity between the Earth and the Moon.

Solution: We do the comparison by taking the ratio of the Earth–Sun gravity to the Earth–Moon gravity. In this case, each quantity is found from the equation of Newton's law of gravity. (See Section 4.4.) Thus, the ratio is:

$$\frac{\text{Earth–Sun gravity}}{\text{Earth–Moon gravity}} = \frac{\cancel{G}\dfrac{\cancel{M_{\text{Earth}}} M_{\text{Sun}}}{(d_{\text{Earth–Sun}})^2}}{\cancel{G}\dfrac{\cancel{M_{\text{Earth}}} M_{\text{Moon}}}{(d_{\text{Earth–Moon}})^2}} = \frac{M_{\text{Sun}}}{(d_{\text{Earth–Sun}})^2} \times \frac{(d_{\text{Earth–Moon}})^2}{M_{\text{Moon}}}$$

Note how all but four of the terms cancel; the last step comes from replacing the division with multiplication by the reciprocal (the "invert and multiply" rule for division). We can simplify the work further by rearranging the terms so that we have the masses and distances together:

$$\frac{\text{Earth–Sun gravity}}{\text{Earth–Moon gravity}} = \frac{M_{\text{Sun}}}{M_{\text{Moon}}} \times \frac{(d_{\text{Earth–Moon}})^2}{(d_{\text{Earth–Sun}})^2}$$

Now it is just a matter of looking up the numbers (see Appendix E) and calculating:

$$\frac{\text{Earth–Sun gravity}}{\text{Earth–Moon gravity}} = \frac{1.99 \times 10^{30} \ \cancel{\text{kg}}}{7.35 \times 10^{22} \ \cancel{\text{kg}}} \times \frac{(384.4 \times 10^3 \ \cancel{\text{km}})^2}{(149.6 \times 10^6 \ \cancel{\text{km}})^2} = 179$$

In other words, the Earth–Sun gravity is 179 times stronger than the Earth–Moon gravity.

D | The Periodic Table of the Elements

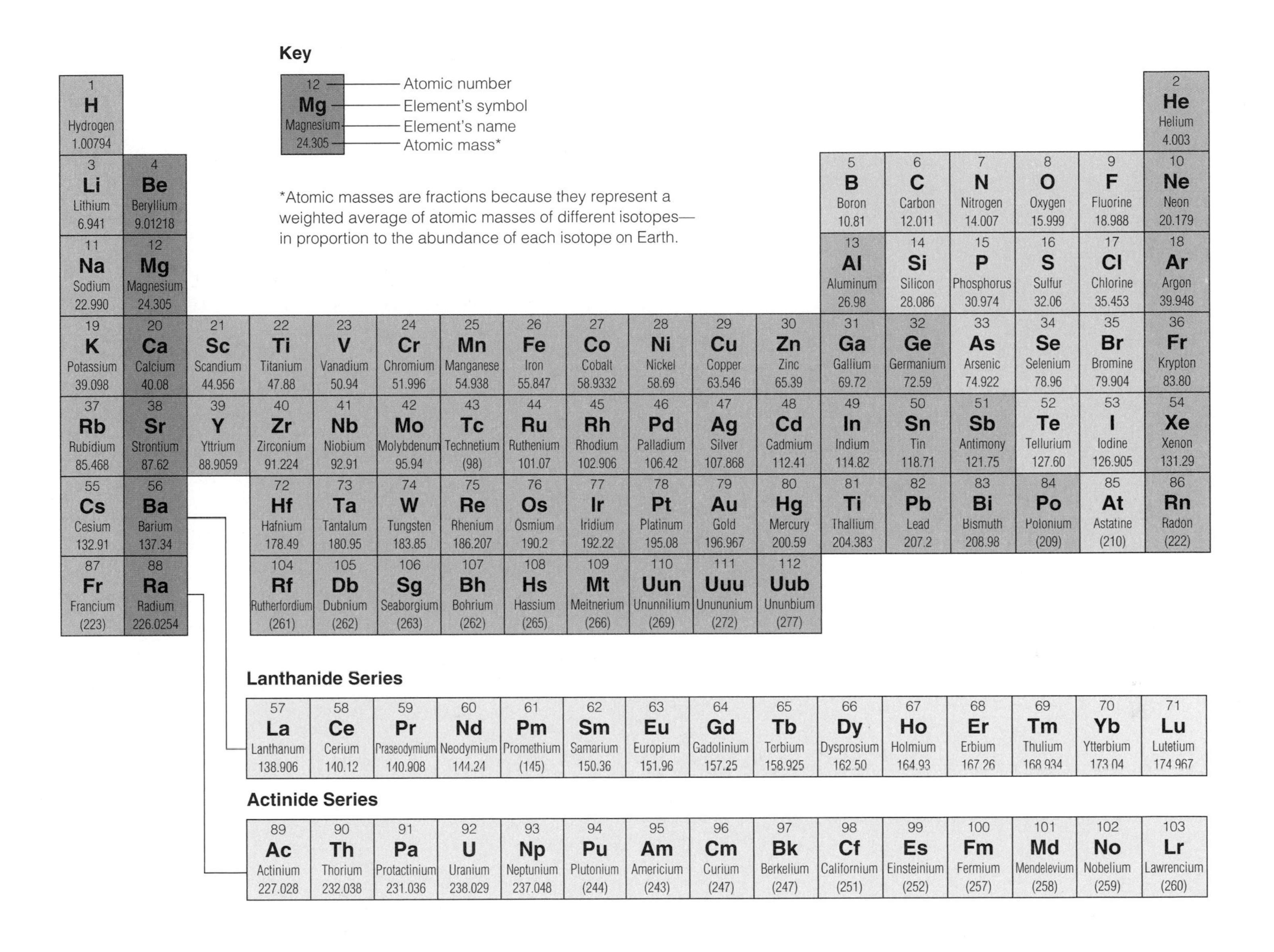

E | Planetary Data

TABLE E.1 *Physical Properties of the Sun and Planets*

Name	Radius (Eq[a]) (km)	Radius (Eq) (Earth units)	Mass (kg)	Mass (Earth units)	Average Density (g/cm^3)	Surface Gravity (Earth = 1)
Sun	695,000	109	1.99×10^{30}	333,000	1.41	27.5
Mercury	2,440	0.382	3.30×10^{23}	0.055	5.43	0.38
Venus	6,051	0.949	4.87×10^{24}	0.815	5.25	0.91
Earth	6,378	1.00	5.97×10^{24}	1.00	5.52	1.00
Mars	3,397	0.533	6.42×10^{23}	0.107	3.93	0.38
Jupiter	71,492	11.19	1.90×10^{27}	317.9	1.32	2.36
Saturn	60,268	9.46	5.69×10^{26}	95.18	0.70	0.92
Uranus	25,559	3.98	8.66×10^{25}	14.54	1.22	0.91
Neptune	24,764	3.81	1.03×10^{26}	17.13	1.64	1.14
Pluto	1,160	0.181	1.31×10^{22}	0.0022	2.05	0.07

[a]Eq = equatorial.

TABLE E.2 *Orbital Properties of the Sun and Planets*

Name	Distance from Sun[a] (AU)	Distance from Sun[a] (10^6 km)	Orbital Period (years)	Orbital Inclination[b] (degrees)	Orbital Eccentricity	Sidereal Rotation Period (Earth days)[c]	Axis Tilt (degrees)
Sun	—	—	—	—	—	25.4	7.25
Mercury	0.387	57.9	0.2409	7.00	0.206	58.6	0.0
Venus	0.723	108.2	0.6152	3.39	0.007	−243.0	177.3
Earth	1.00	149.6	1.0	0.00	0.017	0.9973	23.45
Mars	1.524	227.9	1.881	1.85	0.093	1.026	25.2
Jupiter	5.203	778.3	11.86	1.31	0.048	0.41	3.08
Saturn	9.539	1,427	29.42	2.48	0.056	0.44	26.73
Uranus	19.19	2,870	84.01	0.77	0.046	−0.72	97.92
Neptune	30.06	4,497	164.8	1.77	0.010	0.67	29.6
Pluto	39.48	5,906	248.0	17.14	0.248	−6.39	112.5

[a] Semimajor axis of the orbit.

[b] With respect to the ecliptic.

[c] A negative sign indicates rotation is backward relative to other planets.

TABLE E.3 *Satellites of the Solar System (as of 2003)*[a]

Planet Satellite	*Radius or Dimensions*[b] *(km)*	*Distance from Planet (10^3 km)*	*Orbital Period*[c] *(Earth days)*	*Mass*[d] *(kg)*	*Density*[d] *(g/cm^3)*	*Notes About the Satellite*
Earth						**Earth**
Moon	1,738	384.4	27.322	7.349×10^{22}	3.34	*Moon:* Probably formed in giant impact.
Mars						**Mars**
Phobos	13 × 11 × 9	9.38	0.319	1.3×10^{16}	2.2	*Phobos, Deimos:* Probable captured asteroids.
Deimos	8 × 6 × 5	23.5	1.263	1.8×10^{15}	1.7	
Jupiter						**Jupiter**
Small inner moons (4 moons)	10 to 135 × 82 × 75	128–222	0.295–0.6745	—	—	*Metis, Adrastea, Amalthea, Thebe:* Small moonlets within and near Jupiter's ring system.
Io	1,821	421.6	1.769	8.933×10^{22}	3.57	*Io:* Most volcanically active object in the solar system.
Europa	1,565	670.9	3.551	4.797×10^{22}	2.97	*Europa:* Possible oceans under icy crust.
Ganymede	2,634	1,070.0	7.155	1.482×10^{23}	1.94	*Ganymede:* Largest satellite in solar system; unusual ice geology.
Callisto	2,403	1,883.0	16.689	1.076×10^{23}	1.86	*Callisto:* Cratered iceball.
Irregular group 1 (7 moons)	4–85	7,500–17,100	30–457	—	—	*Themisto, Leda, Himalia, Lysithea, Elara, and 2 others:* Probable captured moons with inclined orbits.
Irregular group 2 (46 moons)	1–30	18,300–23,100	−854 to −901 −504 to 1,312	—	—	*Ananke, Carme, Pasiphae, Sinope, and 44 others:* Probable captured moons in inclined backward orbits.
Saturn						**Saturn**
Small inner moons (6)	10 to 97 × 95 × 77	134–151	0.574–0.695	—	—	*Pan, Atlas, Prometheus, Pandora, Epimetheus, Janus:* Small moonlets within and near Saturn's ring system.
Mimas	199	185.52	0.942	3.70×10^{19}	1.17	*Mimas, Enceladus, Tethys:* Small and medium-size iceballs, many with interesting geology.
Enceladus	249	238.02	1.370	1.2×10^{20}	1.24	
Tethys	530	294.66	1.888	6.17×10^{20}	1.26	
Calypso	15 × 8 × 8	294.66	1.888	4×10^{15}	—	*Calypso, Telesto:* Small moonlets sharing Tethys's orbit.
Telesto	15 × 13 × 8	294.67	1.888	6×10^{15}	—	
Dione	559	377.4	2.737	1.08×10^{21}	1.44	*Dione:* Medium-size iceball, with interesting geology.
Helene	18 × ? × 15	377.4	2.737	1.6×10^{16}	—	*Helene:* Small moonlet sharing Dione's orbit.
Rhea	764	527.04	4.518	2.31×10^{21}	1.33	*Rhea:* Medium-size iceball, with interesting geology.
Titan	2,575	1,221.85	15.945	1.3455×10^{23}	1.88	*Titan:* Dense atmosphere shrouds surface; ongoing geological activity possible.
Hyperion	180 × 140 × 112	1,481.1	21.277	2.8×10^{19}	—	*Hyperion:* Only satellite known not to rotate synchronously.

Iapetus	718	3,561.3	79.331	1.59×10^{21}	1.21	*Iapetus:* Bright and dark hemispheres show greatest contrast in the solar system.
Phoebe	110	12,952	−550.4	1×10^{19}	—	*Phoebe:* Very dark; material ejected from Phoebe may coat one side of Iapetus.
Irregular group 1 (4 moons)	7–22	11,400–17,100	453–829	—	—	*2000 S2, S3, S5, S6:* Probable captured moons with highly inclined orbits.
Irregular group 2 (3 moons)	5–15	17,400–18,000	854–901	—	—	*2000 S4, S10, S11:* Probable captured moons in inclined orbits.
Irregular group 3 (5 moons)	4–10	15,600–23,400	−723 to −1,325	—	—	*2000 S1, S7, S8, S9, S12, and 2003 S1:* Probable captured moons in inclined backward orbits.
Uranus						**Uranus**
Small inner moons (13 moons)	10 to 97 × 95 × 77	134–151	0.574–0.695	—	—	*Cordelia, Ophelia, Bianca, Cressida, Desdemona, Juliet, Portia, Rosalind, Belinda, Puck, 1986 U10, 2003 U1, 2003 U3:* Small moonlets within and near Uranus's ring system.
Miranda	236	129.8	1.413	6.6×10^{19}	1.26	*Miranda, Ariel, Umbriel, Titania, Oberon:* Small and medium-size iceballs, with some interesting geology.
Ariel	579	191.2	2.520	1.35×10^{21}	1.65	
Umbriel	584.7	266.0	4.144	1.17×10^{21}	1.44	
Titania	788.9	435.8	8.706	3.52×10^{21}	1.59	
Oberon	761.4	582.6	13.463	3.01×10^{21}	1.50	
Irregular group (9 moons)	???–60	7,170–25,000	580–2,280	—	—	*Caliban, Sycorax, Stephano, Prospero, Setebos, Trinculo, 2001 U2, 2001 U3, 2003 U3:* Too recently discovered for accurate determination of their properties; several in backward orbits.
Neptune						**Neptune**
Small inner moons (5 moons)	29 to 104 × ? × 89	48–74	0.296–0.554	—	—	*Naiad, Thalassa, Despina, Galatea, Larissa:* Small moonlets within and near Neptune's ring system.
Proteus	218 × 208 × 201	117.6	1.121	6×10^{19}	—	
Triton	1,352.6	354.59	−5.875	2.14×10^{22}	2.0	*Triton:* Probable captured Kuiper belt object—largest captured object in solar system.
Nereid	170	5,588.6	360.125	3.1×10^{19}	—	*Nereid:* Small, icy moon; very little known.
Irregulars (5 moons)	15–20	20,200–21,900	2,520–2,870	—	—	*2002 N1, N2, N3, N4, 2003 N1:* Possible captured moons in inclined or backward orbit.
Pluto						**Pluto**
Charon	593	19.6	6.38718	1.56×10^{21}	1.6	*Charon:* Unusually large compared to its planet; may have formed in giant impact.

[a] *Note:* Authorities differ substantially on many of the values in this table.

[b] a × b × c values for the Dimensions are the approximate lengths of the axes (center to edge) for irregular moons.

[c] Negative sign indicates backward orbit.

[d] Masses and densities are most accurate for those satellites visited by a spacecraft on a flyby. Masses for the smallest moons have not been measured but can be estimated from the radius and an assumed density.

F | Stellar Data

TABLE F.1 *Stars Within 12 Light-Years*

Star	*Distance (ly)*	*Spectral Type*		*RA* h	*RA* m	*Dec* °	*Dec* ′	*Luminosity (L/L_{Sun})*
Sun	0.000016	G2	V	—	—	—	—	1.0
Proxima Centauri	4.2	M5.5	V	14	30	−62	41	0.0006
α Centauri A	4.4	G2	V	14	40	−60	50	1.6
α Centauri B	4.4	K0	V	14	40	−60	50	0.53
Barnard's Star	6.0	M4	V	17	58	+04	42	0.005
Wolf 359	7.8	M6	V	10	56	+07	01	0.0008
Lalande 21185	8.3	M2	V	11	03	+35	58	0.03
Sirius A	8.6	A1	V	06	45	−16	42	26.0
Sirius B	8.6	DA2	—	06	45	−16	42	0.002
Luyten 726-8A	8.7	M5.5	V	01	39	−17	57	0.0009
Luyten 726-8B	8.7	M6	V	01	39	−17	57	0.0006
Ross 154	9.7	M3.5	V	18	50	−23	50	0.004
Ross 248	10.3	M5.5	V	23	42	+44	11	0.001
ε Eridani	10.5	K2	V	03	33	−09	28	0.37
Lacaille 9352	10.7	M1.5	V	23	06	−35	51	0.05
Ross 128	10.9	M4	V	11	48	+00	49	0.003
EZ Aquarii A	11.3	M5	V	22	39	−15	18	0.0006
EZ Aquarii B	11.3	M6	V	22	39	−15	18	0.0004
EZ Aquarii C	11.3	M6.5	V	22	39	−15	18	0.0003
61 Cygni A	11.4	K5	V	21	07	+38	42	0.15
61 Cygni B	11.4	K7	V	21	07	+38	42	0.09
Procyon A	11.4	F5	IV–V	07	39	+05	14	7.4
Procyon B	11.4	DA	—	07	39	+05	14	0.0005
Gliese 725 A	11.4	M3	V	18	43	+59	38	0.02
Gliese 725 B	11.4	M3.5	V	18	43	+59	38	0.01
Gliese 15 A	11.6	M1.5	V	00	18	+44	01	0.03
Gliese 15 B	11.6	M3.5	V	00	18	+44	01	0.003
DX Cancri	11.8	M6.5	V	08	30	+26	47	0.0003
ε Indi	11.8	K5	V	22	03	−56	45	0.26
τ Ceti	11.9	G8	V	01	44	−15	57	0.59
GJ 1061	11.9	M5.5	V	03	36	−44	31	0.0009

Note: These data were provided by the RECONS project, courtesy of Dr. Todd Henry. The luminosities are all total (bolometric) luminosities. The DA stellar types are white dwarfs. The coordinates are for the year 2000.

TABLE F.2 *Twenty Brightest Stars*

Star	Constellation	RA h	RA m	Dec °	Dec ′	Distance (ly)	Spectral Type		Apparent Magnitude	Luminosity (L/L_{Sun})
Sirius	Canis Major	6	45	−16	42	8.6	A1	V	−1.46	26
Canopus	Carina	6	24	−52	41	313	F0	Ib-II	−0.72	13,000
α Centauri	Centaurus	14	40	−60	50	4.4	G2	V	−0.01	1.6
							K0	V	1.3	0.53
Arcturus	Boötes	14	16	+19	11	37	K2	III	−0.06	170
Vega	Lyra	18	37	+38	47	25	A0	V	0.04	60
Capella	Auriga	5	17	+46	00	42	G0	III	0.75	70
							G8	III	0.85	77
Rigel	Orion	5	15	−08	12	772	B8	Ia	0.14	70,000
Procyon	Canis Minor	7	39	+05	14	11.4	F5	IV–V	0.37	7.4
Betelgeuse	Orion	5	55	+07	24	427	M2	Iab	0.41	38,000
Achernar	Eridanus	1	38	−57	15	144	B5	V	0.51	3,600
Hadar	Centaurus	14	04	−60	22	525	B1	III	0.63	100,000
Altair	Aquila	19	51	+08	52	17	A7	IV–V	0.77	10.5
Acrux	Crux	12	27	−63	06	321	B1	IV	1.39	22,000
							B3	V	1.9	7,500
Aldebaran	Taurus	4	36	+16	30	65	K5	III	0.86	350
Spica	Virgo	13	25	−11	09	260	B1	V	0.91	23,000
Antares	Scorpio	16	29	−26	26	604	M1	Ib	0.92	38,000
Pollux	Gemini	7	45	+28	01	34	K0	III	1.16	45
Fomalhaut	Piscis Austrinus	22	58	−29	37	25	A3	V	1.19	18
Deneb	Cygnus	20	41	+45	16	2,500	A2	Ia	1.26	170,000
β Crucis	Crux	12	48	−59	40	352	B0.5	IV	1.28	37,000

Note: Three of the stars on this list, Capella, α Centauri, and Acrux, are binary systems with members of comparable brightness. They are counted as single stars because that is how they appear to the naked eye. All the luminosities given are total (bolometric) luminosities. The coordinates are for the year 2000.

G | Galaxy Data

TABLE G.1 *Galaxies of the Local Group*

Galaxy Name	*Distance (millions of ly)*	*Type*[a]	*RA* *h*	*RA* *m*	*Dec* °	*Dec* ′	*Luminosity (millions of* L_{Sun}*)*
Milky Way	—	Sbc	—	—	—	—	15,000
WLM	3.0	Irr	00	02	−15	30	50
NGC 55	4.8	Irr	00	15	−39	13	1,300
IC 10	2.7	dIrr	00	20	+59	18	160
NGC 147	2.4	dE	00	33	+48	30	131
And III	2.5	dE	00	35	+36	30	1.1
NGC 185	2.0	dE	00	39	+48	20	120
NGC 205	2.7	E	00	40	+41	41	370
M 32	2.6	E	00	43	+40	52	380
M 31	2.5	Sb	00	43	+41	16	21,000
And I	2.6	dE	00	46	+38	00	4.7
SMC	0.19	Irr	00	53	−72	50	230
Sculptor	0.26	dE	01	00	−33	42	2.2
LGS 3	2.6	dIrr	01	04	+21	53	1.3
IC 1613	2.3	Irr	01	05	+02	08	64
And II	1.7	dE	01	16	+33	26	2.4
M 33	2.7	Sc	01	34	+30	40	2,800
Phoenix	1.5	dIrr	01	51	−44	27	0.9
Fornax	0.45	dE	02	40	−34	27	15.5
EGB0427 + 63	4.3	dIrr	04	32	+63	36	9.1
LMC	0.16	Irr	05	24	−69	45	1,300
Carina	0.33	dE	06	42	−50	58	0.4
Leo A	2.2	dIrr	09	59	+30	45	3.0
Sextans B	4.4	dIrr	10	00	+05	20	41
NGC 3109	4.1	Irr	10	03	−26	09	160
Antlia	4.0	dIrr	10	04	−27	19	1.7
Leo I	0.82	dE	10	08	+12	18	4.8
Sextans A	4.7	dIrr	10	11	−04	42	56
Sextans	0.28	dE	10	13	−01	37	0.5
Leo II	0.67	dE	11	13	+22	09	0.6
GR 8	5.2	dIrr	12	59	+14	13	3.4
Ursa Minor	0.22	dE	15	09	+67	13	0.3
Draco	2.7	dE	17	20	+57	55	0.3
Sagittarius	0.08	dE	18	55	−30	29	18
SagDIG	3.5	dIrr	19	30	−17	41	6.8
NGC 6822	1.6	Irr	19	45	−14	48	94
DDO 210	2.6	dIrr	20	47	−12	51	0.8
IC 5152	5.2	dIrr	22	03	−51	18	70
Tucana	2.9	dE	22	42	−64	25	0.5
UKS2323-326	4.3	dE	23	26	−32	23	5.2
Pegasus	3.1	dIrr	23	29	+14	45	12

[a] Types beginning with S are spiral galaxies classified according to Hubble's system (see Chapter 15). Type E galaxies are elliptical or spheroidal. Type Irr galaxies are irregular. The prefix d denotes a dwarf galaxy.

TABLE G.2 *Nearby Galaxies in the Messier Catalog*[a,b]

Galaxy Name (M / NGC)[c]	*RA* *h*	*m*	*Dec* °	′	RV_{hel}[d]	RV_{gal}[e]	*Type*[f]	*Nickname*
M31 / NGC224	00	43	+41	16	−300 ± 4	−122	Spiral	Andromeda
M32 / NGC221	00	43	+40	52	−145 ± 2	32	Elliptical	
M33 / NGC598	01	34	+30	40	−179 ± 3	−44	Spiral	Triangulum
M49 / NGC4472	12	30	+08	00	997 ± 7	929	Elliptical/ Lenticular/Seyfert	
M51 / NGC5194	13	30	+47	12	463 ± 3	550	Spiral/Interacting	Whirlpool
M58 / NGC4579	12	38	+11	49	1,519 ± 6	1,468	Spiral/Seyfert	
M59 / NGC4621	12	42	+11	39	410 ± 6	361	Elliptical	
M60 / NGC4649	12	44	+11	33	1,117 ± 6	1,068	Elliptical	
M61 / NGC4303	12	22	+04	28	1,566 ± 2	1,483	Spiral/Seyfert	
M63 / NGC5055	13	16	+42	02	504 ± 4	570	Spiral	Sunflower
M64 / NGC4826	12	57	+21	41	408 ± 4	400	Spiral/Seyfert	Black Eye
M65 / NGC3623	11	19	+13	06	807 ± 3	723	Spiral	
M66 / NGC3627	11	20	+12	59	727 ± 3	643	Spiral/Seyfert	
M74 / NGC628	01	37	+15	47	657 ± 1	754	Spiral	
M77 / NGC1068	02	43	−00	01	1,137 ± 3	1,146	Spiral/Seyfert	
M81 / NGC3031	09	56	+69	04	−34 ± 4	73	Spiral/Seyfert	
M82 / NGC3034	09	56	+69	41	203 ± 4	312	Irregular/Starburst	
M83 / NGC5236	13	37	−29	52	516 ± 4	385	Spiral/Starburst	
M84 / NGC4374	12	25	+12	53	1,060 ± 6	1,005	Elliptical	
M85 / NGC4382	12	25	+18	11	729 ± 2	692	Spiral	
M86 / NGC4406	12	26	+12	57	−244 ± 5	−298	Elliptical/Lenticular	
M87 / NGC4486	12	30	+12	23	1,307 ± 7	1,254	Elliptical/Central Dominant/Seyfert	Virgo A
M88 / NGC4501	12	32	+14	25	2,281 ± 3	2,235	Spiral/Seyfert	
M89 / NGC4552	12	36	+12	33	340 ± 4	290	Elliptical	
M90 / NGC4569	12	37	+13	10	−235 ± 4	−282	Spiral/Seyfert	
M91 / NGC4548	12	35	+14	30	486 ± 4	442	Spiral/Seyfert	
M94 / NGC4736	12	51	+41	07	308 ± 1	360	Spiral	
M95 / NGC3351	10	44	+11	42	778 ± 4	677	Spiral/Starburst	
M96 / NGC3368	10	47	+11	49	897 ± 4	797	Spiral/Seyfert	
M98 / NGC4192	12	14	+14	54	−142 ± 4	−195	Spiral/Seyfert	
M99 / NGC4254	12	19	+14	25	2,407 ± 3	2,354	Spiral	
M100 / NGC4321	12	23	+15	49	1,571 ± 1	1,525	Spiral	
M101 / NGC5457	14	03	+54	21	241 ± 2	360	Spiral	
M104 / NGC4594	12	40	−11	37	1,024 ± 5	904	Spiral/Seyfert	Sombrero
M105 / NGC3379	10	48	+12	35	911 ± 2	814	Elliptical	
M106 / NGC4258	12	19	+47	18	448 ± 3	507	Spiral/Seyfert	
M108 / NGC3556	11	09	+55	57	695 ± 3	765	Spiral	
M109 / NGC3992	11	55	+53	39	1,048 ± 4	1,121	Spiral	
M110 / NGC205	00	38	+41	25	−241 ± 3	−61	Elliptical	

[a]Galaxies identified in the catalog published by Charles Messier in 1781; these galaxies are relatively easy to observe with small telescopes.

[b]Data obtained from NED: NASA/IPAC Extragalactic Database (http://ned.ipac.caltech.edu). The original Messier list of galaxies was obtained from SED, and the list data were updated to 2001 and M 102 was dropped.

[c]The galaxies are identified by the Messier number (M followed by a number) and by their NGC numbers, which come from the *New General Catalog* published in 1888.

[d]Radial velocity in km/s, with respect to the Sun (heliocentric). Positive values mean motion away from the Sun, and negative values are toward the Sun.

[e]Radial velocity in km/s, with respect to the Milky Way Galaxy, calculated from the RV_{hel} values with a correction for the Sun's motion around the galactic center.

[f]Galaxies are first listed by their primary type (spiral, elliptical, or irregular) and then by any other special categories that apply (see Chapter 15).

TABLE G.3 *Nearby, X-ray Bright Clusters of Galaxies*

Cluster Name	*Redshift*	*Distance*[a] *(billions of ly)*	*Temperature of Intracluster Medium (millions of K)*	*Average Orbital Velocity of Galaxies*[b] *(km/sec)*	*Cluster Mass*[c] *(10^{15} M_{Sun})*
Abell 2142	0.0907	1.20	101. ± 2	1,132 ± 110	1.8
Abell 2029	0.0766	1.07	100. ± 3	1,164 ± 98	1.8
Abell 401	0.0737	1.03	95.2 ± 5	1,152 ± 86	1.6
Coma	0.0233	0.34	95.1 ± 1	821 ± 49	1.6
Abell 754	0.0539	0.77	93.3 ± 3	662 ± 77	1.6
Abell 2256	0.0589	0.83	87.0 ± 2	1,348 ± 86	1.4
Abell 399	0.0718	1.01	81.7 ± 7	1,116 ± 89	1.3
Abell 3571	0.0395	0.57	81.1 ± 3	1,045 ± 109	1.3
Abell 478	0.0882	1.22	78.9 ± 2	904 ± 281	1.2
Abell 3667	0.0566	0.80	78.5 ± 6	971 ± 62	1.2
Abell 3266	0.0599	0.85	78.2 ± 5	1,107 ± 82	1.2
Abell 1651a	0.0846	1.17	73.1 ± 6	685 ± 129	1.2
Abell 85	0.0560	0.80	70.9 ± 2	969 ± 95	1.2
Abell 119	0.0438	0.63	65.6 ± 5	679 ± 106	0.94
Abell 3558	0.0480	0.69	65.3 ± 2	977 ± 39	0.94
Abell 1795	0.0632	0.89	62.9 ± 2	834 ± 85	0.88
Abell 2199	0.0314	0.46	52.7 ± 1	801 ± 92	0.68
Abell 2147	0.0353	0.51	51.1 ± 4	821 ± 68	0.65
Abell 3562	0.0478	0.68	45.7 ± 8	736 ± 49	0.55
Abell 496	0.0325	0.47	45.3 ± 1	687 ± 89	0.54
Centaurus	0.0103	0.15	42.2 ± 1	863 ± 34	0.49
Abell 1367	0.0213	0.31	41.3 ± 2	822 ± 69	0.47
Hydra	0.0126	0.19	38.0 ± 1	610 ± 52	0.42
C0336	0.0349	0.50	37.4 ± 1	650 ± 170	0.41
Virgo	0.0038	0.06	25.7 ± 0.5	632 ± 41	0.23

Note: This table lists the 25 brightest clusters of galaxies in the X-ray sky from a catalog by J. P. Henry (2000).

[a]Cluster distances were computed using a value for Hubble's constant of 65 km/sec/Mpc.

[b]The average orbital velocities of galaxies given in this column are the velocity dispersions of the clusters' galaxies.

[c]This column gives each cluster's mass within the largest radius at which the intracluster medium can be in gravitational equilibrium. Because our estimates of that radius depend on Hubble's constant, these masses are inversely proportional to Hubble's constant, which we have assumed to be 65 km/s/Mpc.

H | Selected Astronomical Web Sites

The Web contains a vast amount of astronomical information. For all your astronomical Web surfing, the best starting point is the Web site for this textbook:

Astronomy Place
www.astronomyplace.com

The following are some other sites that may be of particular use. In case any of the links change, you can always find live links to these sites, and many more, on the Astronomy Place Web site.

KEY MISSION SITES

The following table lists the Web pages for major current astronomy missions.

Site	*Description*	*Web Address*
NASA's Office of Space Science Missions Page	**Direct links to all past, present, and planned NASA space science missions**	**http://spacescience.nasa.gov/missions**
Cassini/Huygens	Mission scheduled to arrive at Saturn in 2004	http://saturn.jpl.nasa.gov/index.cfm
Chandra X-Ray Observatory	Latest discoveries, educational activities, and other information from the Chandra X-Ray Observatory	http://chandra.harvard.edu
Far Ultraviolet Spectroscopic Explorer (FUSE)	Ultraviolet observatory in space	http://fuse.pha.jhu.edu
Galileo	Mission orbiting Jupiter	http://www.jpl.nasa.gov/galileo
Hubble Space Telescope	Latest discoveries, educational activities, and other information from the Hubble Space Telescope	http://hubble.stsci.edu
Mars Exploration Program	Information on current and planned Mars missions	http://mars.jpl.nasa.gov
Microwave Anisotropy Probe (MAP)	Mission to study the cosmic microwave background	http://map.gsfc.nasa.gov
Space Infrared Telescope Facility (SIRTF)	Infrared observatory scheduled for launch in 2002	http://sirtf.caltech.edu
Stratospheric Observatory for Infrared Astronomy (SOFIA)	Airborne observatory scheduled to begin flights in 2002	http://sofia.arc.nasa.gov

KEY OBSERVATORY SITES

The following table lists the Web pages leading to major ground-based observatories.

Site	*Description*	*Web Address*
World's Largest Optical Telescopes	**Direct links to most of the world's major optical observatories**	**http://www.seds.org/billa/bigeyes.html**
Arecibo Observatory (Puerto Rico)	World's largest single-dish radio telescope	http://www.naic.edu
Cerro Tololo Inter-American Observatory	Links to major observatories on site in Cerro Tololo, Chile	http://www.ctio.noao.edu
European Southern Observatory	Links to European telescope projects in Chile, including the Very Large Telescope	http://www.eso.org
Mauna Kea Observatories	Links to major observatories in Hawaii, including Keck, Gemini, Subaru, CFHT, and others	http://www.ifa.hawaii.edu/mko
Mt. Palomar Observatory	Powerful telescope near San Diego	http://www.astro.caltech.edu/palomarpublic
National Optical Astronomy Observatory	Home page for United States national observatories in Arizona, Hawaii, and Chile	http://www.noao.edu
National Radio Astronomy Observatory	Home page for United States national radio observatories, including the Very Large Array (VLA)	http://www.nrao.edu

MORE ASTRONOMICAL WEB SITES

The following Web sites are some of the authors' favorites among many other non-commercial resources for astronomy.

Site	*Description*	*Web Address*
The Astronomy Place	**Don't forget to start here for all your astronomical Web surfing.**	**http://www.astronomyplace.com**
American Association of Variable Star Observers (AAVSO)	One of the largest organizations of amateur astronomers in the world. Check this site if you are interested in serious amateur astronomy.	http://www.aavso.org
Astronomical Society of the Pacific	An organization for both professional astronomers and the general public, devoted largely to astronomy education.	http://www.astrosociety.org
Astronomy Picture of the Day	An archive of beautiful pictures, updated daily.	http://antwrp.gsfc.nasa.gov/apod
AstroWeb	Listing of major resources for astronomy on the Web.	http://www.stsci.edu/astroweb/astronomy.html
Canadian Space Agency	Home page for Canada's space program.	http://www.space.gc.ca
European Space Agency (ESA)	Home page for this international agency.	http://www.esa.int
The Extrasolar Planets Encyclopedia	Information about the search for and discoveries of extrasolar planets.	http://cfa-www.harvard.edu/planets
NASA Home Page	Learn almost anything you want about NASA.	http://www.nasa.gov
NASA Science News	Read the latest news from NASA; has option to subscribe to e-mail notices of news releases.	http://science.nasa.gov
The Nine Planets (University of Arizona)	A multimedia tour of the solar system.	http://www.nineplanets.org
The Planetary Society	Has more than 100,000 members who are interested in planetary exploration and the search for life in the universe.	http://planetary.org
The SETI Institute	Devoted to the search for other civilizations.	http://www.seti.org
Voyage Scale Model Solar System	Take a virtual tour of the Voyage Scale Model Solar System.	http://www.voyageonline.org

I | The 88 Constellations

CONSTELLATION NAMES (ENGLISH EQUIVALENT IN PARENTHESES)

Andromeda (The Chained Princess)
Antlia (The Air Pump)
Apus (The Bird of Paradise)
Aquarius (The Water Bearer)
Aquila (The Eagle)
Ara (The Altar)
Aries (The Ram)
Auriga (The Charioteer)
Boötes (The Herdsman)
Caelum (The Chisel)
Camelopardalis (The Giraffe)
Cancer (The Crab)
Canes Venatici (The Hunting Dogs)
Canis Major (The Great Dog)
Canis Minor (The Little Dog)
Capricornus (The Sea Goat)
Carina (The Keel)
Cassiopeia (The Queen)
Centaurus (The Centaur)
Cepheus (The King)
Cetus (The Whale)
Chamaeleon (The Chameleon)
Circinus (The Drawing Compass)
Columba (The Dove)
Coma Berenices (Berenice's Hair)
Corona Australis (The Southern Crown)
Corona Borealis (The Northern Crown)
Corvus (The Crow)
Crater (The Cup)
Crux (The Southern Cross)
Cygnus (The Swan)
Delphinus (The Dolphin)
Dorado (The Goldfish)
Draco (The Dragon)
Equuleus (The Little Horse)
Eridanus (The River)

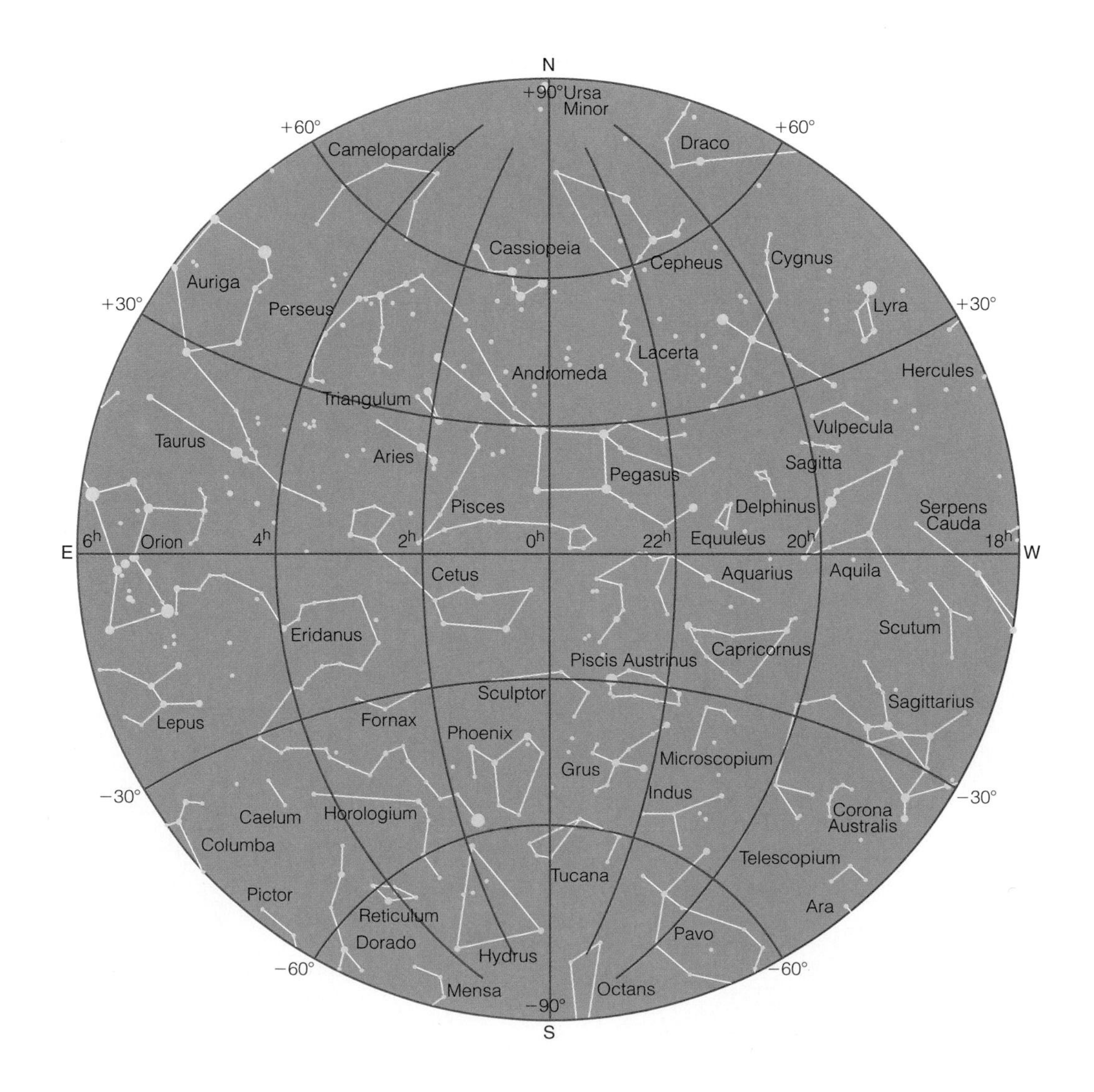

Fornax (The Furnace)
Gemini (The Twins)
Grus (The Crane)
Hercules
Horologium (The Clock)
Hydra (The Sea Serpent)
Hydrus (The Water Snake)
Indus (The Indian)
Lacerta (The Lizard)
Leo (The Lion)
Leo Minor (The Little Lion)
Lepus (The Hare)
Libra (The Scales)
Lupus (The Wolf)
Lynx (The Lynx)
Lyra (The Lyre)
Mensa (The Table)
Microscopium (The Microscope)
Monoceros (The Unicorn)
Musca (The Fly)
Norma (The Level)
Octans (The Octant)
Ophiuchus (The Serpent Bearer)
Orion (The Hunter)
Pavo (The Peacock)
Pegasus (The Winged Horse)
Perseus (The Hero)
Phoenix (The Phoenix)
Pictor (The Painter's Easel)
Pisces (The Fish)
Piscis Austrinus (The Southern Fish)
Puppis (The Stern)
Pyxis (The Compass)
Reticulum (The Reticle)
Sagitta (The Arrow)
Sagittarius (The Archer)
Scorpius (The Scorpion)
Sculptor (The Sculptor)
Scutum (The Shield)
Serpens (The Serpent)
Sextans (The Sextant)
Taurus (The Bull)
Telescopium (The Telescope)
Triangulum (The Triangle)
Triangulum Australe (Southern Triangle)
Tucana (The Toucan)
Ursa Major (The Great Bear)
Ursa Minor (The Little Bear)
Vela (The Sail)
Virgo (The Virgin)
Volans (The Flying Fish)
Vulpecula (The Fox)

CONSTELLATION LOCATIONS

Each of the charts on these pages shows half of the celestial sphere in projection, so you can use them to learn the approximate locations of the constellations. The grid lines are marked by right ascension and declination.

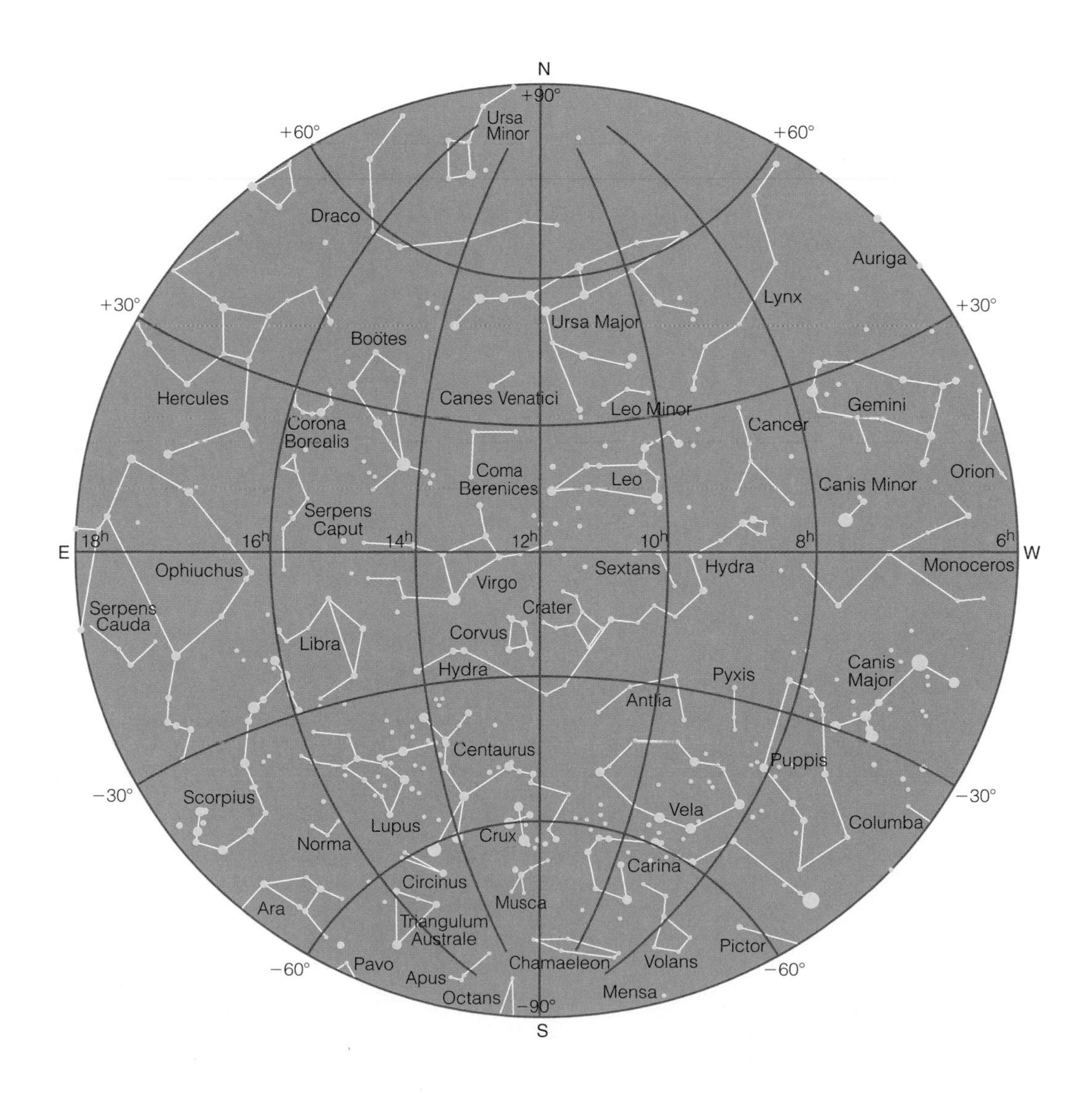

J | Star Charts

How to use the star charts:

Check the times and dates under each chart to find the best one for you. Take it outdoors within an hour or so of the time listed for your date. Bring a dim flashlight to help you read it.

On each chart, the round outside edge represents the horizon all around you. Compass directions around the horizon are marked in yellow. Turn the chart around so the edge marked with the direction you're facing (for example, north, southeast) is down. The stars above this horizon now match the stars you are facing. Ignore the rest until you turn to look in a different direction.

The center of the chart represents the sky overhead, so a star plotted on the chart halfway from the edge to the center can be found in the sky halfway from the horizon to straight up.

The charts are drawn for 40°N latitude (for example, Denver, New York, Madrid). If you live far south of there, stars in the southern part of your sky will appear higher than on the chart and stars in the north will be lower. If you live far north of there, the reverse is true.

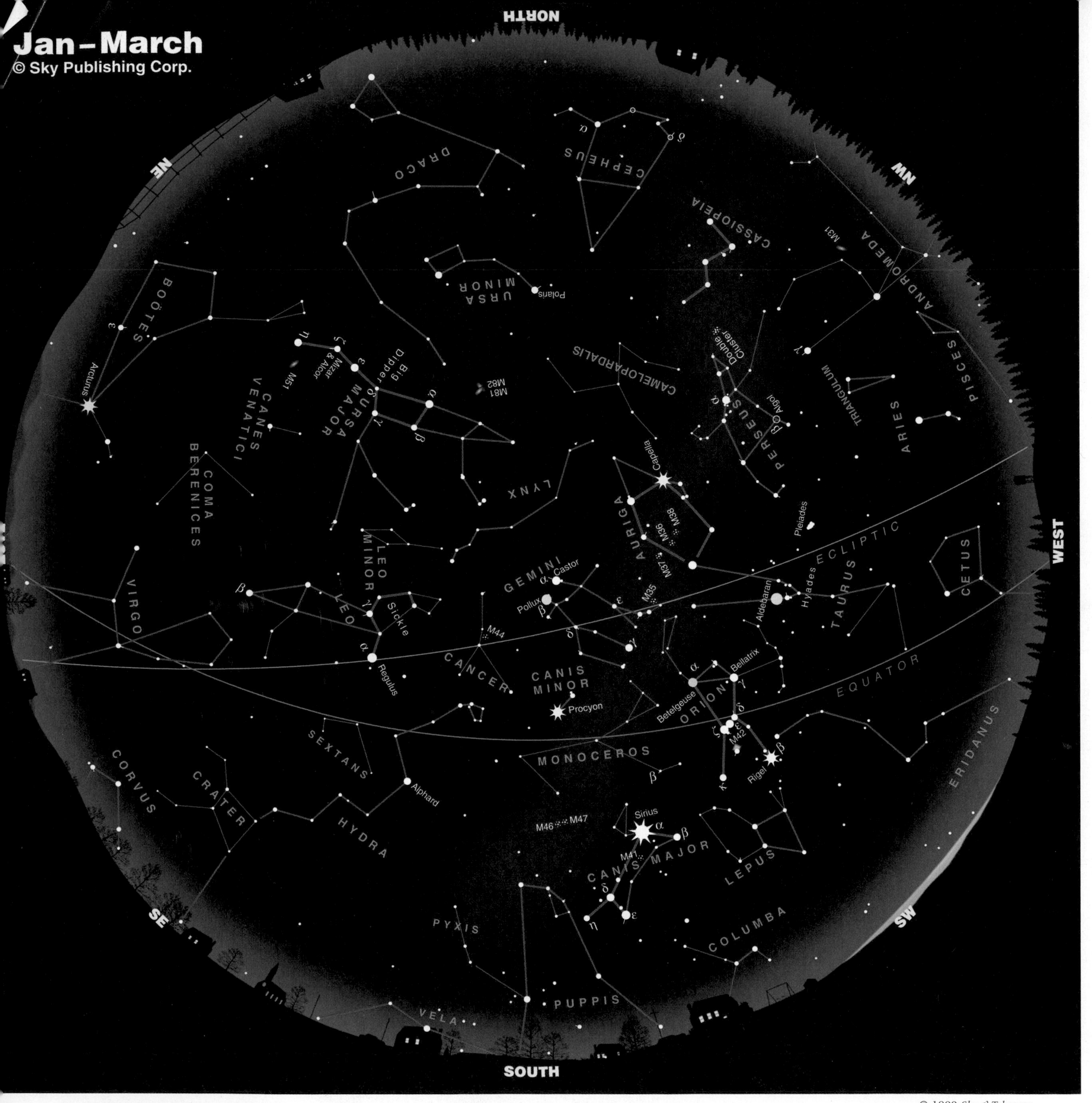

Use this chart January, February, and March.

Early January — 1 A.M.
Late January — Midnight

Early February — 11 P.M.
Late February — 10 P.M.

Early March — 9 P.M.
Late March — Dusk

Use this chart April, May, and June.

Early April — 3 A.M.*
Late April — 2 A.M.*

Early May — 1 A.M.*
Late May — Midnight*

Early June — 11 P.M.*
Late June — Dusk

*Daylight Saving Time

Use this chart July, August, and September.

Early July — 1 A.M.*
Late July — Midnight*

Early August — 11 P.M.*
Late August — 10 P.M.*

Early September — 9 P.M.*
Late September — Dusk

*Daylight Saving Time

Use this chart October, November, and December.

Early October — 1 A.M.*	Early November — 10 P.M.	Early December — 8 P.M.
Late October — Midnight*	Late November — 9 P.M.	Late December — 7 P.M.

*Daylight Saving Time

Glossary

Note: This glossary includes all the bold terms from this book, plus many other common astronomical terms.

21-cm line A spectral line from atomic hydrogen with wavelength 21 cm (in the radio portion of the spectrum).

absolute magnitude A measure of an object's luminosity; defined to be the apparent magnitude the object would have if it were located exactly 10 parsecs (32.6 light-years) away.

absolute zero The coldest possible temperature, which is 0 K.

absorption (of light) The process by which matter absorbs radiative energy.

absorption-line spectrum A spectrum that contains absorption lines.

accelerating universe One of the four general models that may describe the way the expansion of the universe changes with time. In this model, a repulsive force (*see* dark energy) causes the rate of expansion to accelerate with time.

acceleration The rate at which an object's velocity changes. Its standard units are m/s^2.

acceleration of gravity The acceleration of a falling object. On Earth, the acceleration of gravity, designated by g, is 9.8 m/s^2.

accretion The process by which small objects gather together to make larger objects.

accretion disk A rapidly rotating disk of material that gradually falls inward as it orbits a starlike object (e.g., white dwarf, neutron star, or black hole).

active galactic nuclei The unusually luminous centers of some galaxies, thought to be powered by accretion onto supermassive black holes. Quasars are the brightest type of active galactic nuclei; radio galaxies also contain active galactic nuclei.

active galaxy A term sometimes used to describe a galaxy that contains an active galactic nucleus.

adaptive optics A technique in which a ground-based telescope's mirror flexes rapidly to compensate for the bending of starlight caused by atmospheric turbulence.

altitude (above horizon) The angular distance between the horizon and an object in the sky.

amino acids The building blocks of proteins.

analemma The figure-8 path traced by the Sun over the course of a year when viewed at the same place and the same time each day; represents the discrepancies between apparent and mean solar time.

Andromeda Galaxy (M 31; the Great Galaxy in Andromeda) The nearest large spiral galaxy to the Milky Way.

angular momentum Momentum attributable to rotation or revolution. The angular momentum of an object moving in a circle of radius r is the product m × v × r.

angular resolution (of a telescope) The smallest angular separation that two pointlike objects can have and still be seen as distinct points of light (rather than as a single point of light).

angular size (or angular distance) A measure of the angle formed by extending imaginary lines outward from our eyes to span an object (or between two objects).

annihilation *See* matter–antimatter annihilation

annular solar eclipse A solar eclipse during which the Moon is directly in front of the Sun but its angular size is not large enough to fully block the Sun; thus, a ring (or annulus) of sunlight is still visible around the Moon's disk.

Antarctic Circle The circle on the Earth with latitude 66.5°S.

antielectron *See* positron

antimatter Refers to any particle with the same mass as a particle of ordinary matter but whose other basic properties, such as electrical charge, are precisely opposite.

aphelion The point at which an object orbiting the Sun is farthest from the Sun.

apogee The point at which an object orbiting Earth is farthest from the Earth.

apparent brightness The amount of light reaching us per unit area from a luminous object; often measured in units of $watts/m^2$.

apparent magnitude A measure of the apparent brightness of an object in the sky, based on the ancient system developed by Hipparchus.

apparent retrograde motion Refers to the apparent motion of a planet, as viewed from Earth, during the period of a few weeks or months when it moves westward relative to the stars in our sky.

apparent solar time Time measured by the actual position of the Sun in your local sky; defined so that noon is when the Sun is on the meridian.

arcminutes (or minutes of arc) One arcminute is $\frac{1}{60}$ of 1°.

arcseconds (or seconds of arc) One arcsecond is $\frac{1}{60}$ of an arcminute, or $\frac{1}{3,600}$ of 1°.

Arctic Circle The circle on the Earth with latitude 66.5°N.

asteroid A relatively small and rocky object that orbits a star; asteroids are sometimes called minor planets because they are similar to planets but smaller.

asteroid belt The region of our solar system between the orbits of Mars and Jupiter in which asteroids are heavily concentrated.

astrobiology The study of life on Earth and beyond; emphasizes research into questions of the origin of life, the conditions under which life can survive, and the search for life beyond Earth.

astronomical unit (AU) The average distance (semimajor axis) of Earth from the Sun, which is about 150 million km.

atmospheric pressure The surface pressure resulting from the overlying weight of an atmosphere.

atomic mass number The combined number of protons and neutrons in an atom.

atomic number The number of protons in an atom.

atoms Consist of a nucleus made from protons and neutrons surrounded by a cloud of electrons.

aurora Dancing lights in the sky caused by charged particles entering our atmosphere; called the aurora borealis in the Northern Hemisphere and the aurora australis in the Southern Hemisphere.

autumnal equinox *See* fall equinox

azimuth (usually called direction in this book) Direction around the horizon from due north, measured clockwise in degrees. E.g., the azimuth of due north is 0°, due east is 90°, due south is 180°, and due west is 270°.

bar The standard unit of pressure, approximately equal to the Earth's atmospheric pressure at sea level.

baryonic matter Refers to ordinary matter made from atoms (because the nuclei of atoms contain protons and neutrons, which are both baryons).

baryons Particles, including protons and neutrons, that are made from three quarks.

basalt A type of volcanic rock that makes a low-viscosity lava when molten.

Big Bang The event that gave birth to the universe.

Big Crunch If gravity ever reverses the universal expansion, the universe will someday begin to collapse and presumably end in a Big Crunch.

Big Rip The hypothesized end of the universe if dark energy accelerates the expansion so much that galaxies, stars, and planets are eventually ripped apart.

binary star system A star system that contains two stars.

biosphere Refers to the "layer" of life on Earth.

BL Lac objects The name given to a class of active galactic nuclei that probably represent the centers of radio galaxies whose jets happen to be pointed directly at us.

blackbody radiation *See* thermal radiation

black hole A bottomless pit in spacetime. Nothing can escape from within a black hole, and we can never again detect or observe an object that falls into a black hole.

black smokers Structures around seafloor volcanic vents on Earth that support a wide variety of life.

blueshift A Doppler shift in which spectral features are shifted to shorter wavelengths, which occurs when an object is moving toward the observer.

bound orbits Orbits on which an object travels repeatedly around another object; bound orbits are elliptical in shape.

brown dwarf An object too small to become an ordinary star because electron degeneracy pressure halts its gravitational collapse before fusion becomes self-sustaining; brown dwarfs have mass less than $0.08M_{Sun}$.

bubble (interstellar) The surface of a bubble is an expanding shell of hot, ionized gas driven by stellar winds or supernovae; inside the bubble, the gas is very hot and has very low density.

bulge (of a spiral galaxy) The central portion of a spiral galaxy that is roughly spherical (or football shaped) and bulges above and below the plane of the galactic disk.

Cambrian explosion The dramatic diversification of life on Earth that occurred between about 540 and 500 million years ago.

carbonate rock A carbon-rich rock, such as limestone, that forms underwater from chemical reactions between sediments and carbon dioxide. On Earth, most of the outgassed carbon dioxide currently resides in carbonate rocks.

carbonate–dioxide cycle The process that cycles carbon dioxide between Earth's atmosphere and surface rocks.

Cassini division A large, dark gap in Saturn's rings, visible through small telescopes on Earth.

CCD (charge coupled device) A type of electronic light detector that has largely replaced photographic film in astronomical research.

celestial coordinates The coordinates of right ascension and declination that fix an object's position on the celestial sphere.

celestial equator (CE) The extension of the Earth's equator onto the celestial sphere.

celestial navigation Navigation on the surface of the Earth accomplished by observations of the Sun and stars.

celestial sphere The imaginary sphere on which objects in the sky appear to reside when observed from Earth.

Celsius (temperature scale) The temperature scale commonly used in daily activity internationally. Defined so that, on Earth's surface, water freezes at 0°C and boils at 100°C.

central dominant galaxy A giant elliptical galaxy found at the center of a dense cluster of galaxies, apparently formed by the merger of several individual galaxies.

Cepheid *See* Cepheid variable

Cepheid variable A particularly luminous type of pulsating variable star that follows a period–luminosity relation and hence is very useful for measuring cosmic distances.

Chandrasekhar limit *See* white dwarf limit

charged particle belts Zones in which ions and electrons accumulate and encircle a planet.

chemical enrichment The process by which the abundance of heavy elements (heavier than helium) in the interstellar medium gradually increases over time as these elements are produced by stars and released into space.

chromosphere The layer of the Sun's atmosphere below the corona; most of the Sun's ultraviolet light is emitted from this region, in which the temperature is about 10,000 K.

circumpolar star A star that always remains above the horizon as seen from a particular latitude.

climate Describes the long-term average of weather.

close binary A binary star system in which the two stars are very close together.

closed universe The universe is closed if its average density is greater than the critical density, in which case spacetime must curve back on itself to the point where its overall shape is analogous to that of the surface of a sphere. In the absence of a repulsive force (*see* cosmological constant), a closed universe would someday stop expanding and begin to contract.

cluster of galaxies A collection of a few dozen or more galaxies bound together by gravity; smaller collections of galaxies are simply called groups.

cluster of stars A group of anywhere from several hundred to a million or so stars; star clusters come in two types—open clusters and globular clusters.

CNO cycle The cycle of reactions by which intermediate- and high-mass stars fuse hydrogen into helium.

coasting universe One of the four general models that may describe the way the expansion of the universe changes with time. In this model, the mass density of the universe is smaller than the critical density, so that the collective gravity of all matter cannot halt the expansion. In the absence of a repulsive force (dark energy), such a universe would keep expanding forever with little change in its rate of expansion.

coma (of a comet) The dusty atmosphere of a comet created by sublimation of ices in the nucleus when the comet is near the Sun.

comet A relatively small, icy object that orbits a star.

comparative planetology The study of the solar system by examining and understanding the similarities and differences among worlds.

condensation The formation of solid or liquid particles from a cloud of gas.

conduction (of energy) The process by which thermal energy is transferred by direct contact from warm material to cooler material.

conjunction (of a planet with the Sun) When a planet and the Sun line up in the sky.

conservation of angular momentum (law of) The principle that, in the absence of net torque (twisting force), the total angular momentum of a system remains constant.

conservation of energy (law of) The principle that energy (including mass-energy) can be neither created nor destroyed, but can only change from one form to another.

conservation of momentum (law of) The principle that, in the absence of net force, the total momentum of a system remains constant.

constellation A region of the sky; 88 official constellations cover the celestial sphere.

convection The energy transport process in which warm material expands and rises, while cooler material contracts and falls.

convection cell An individual small region of convecting material.

convection zone (of a star) A region in which energy is transported outward by convection.

core (of a planet) The dense central region of a planet that has undergone differentiation.

core (of a star) The central region of a star, in which nuclear fusion can occur.

Coriolis effect Causes air or objects moving on a rotating planet to deviate from straight-line trajectories.

corona (solar) The tenuous uppermost layer of the Sun's atmosphere; most of the Sun's X-rays are emitted from this region, in which the temperature is about 1 million K.

coronal holes Regions of the corona that barely show up in X-ray images because they are nearly devoid of hot coronal gas.

cosmic microwave background The remnant radiation from the Big Bang, which we detect using radio telescopes sensitive to microwaves (which are short-wavelength radio waves).

cosmic rays Particles such as electrons, protons, and atomic nuclei that zip through interstellar space at close to the speed of light.

cosmological constant The name given to a term in Einstein's equations of general relativity. If it is not zero, then it represents a repulsive force or a type of energy (sometimes called dark energy) that might cause the expansion of the universe to accelerate with time.

cosmological horizon The boundary of our observable universe, which is where the lookback time is equal to the age of the universe. Beyond this boundary in spacetime, we cannot see anything at all.

cosmological principle The idea that matter in the universe is evenly distributed on very large scales, and the universe therefore has no center or edges.

critical density The precise average density for the entire universe that marks the dividing line between a recollapsing universe and one that will expand forever.

critical universe One of the four general models that may describe the way the expansion of the universe changes with time. In this model, the mass density of the universe equals the critical density. The universe will never collapse, but in the absence of a repulsive force (dark energy) it will expand more and more slowly as time progresses.

crust (of a planet) The low-density surface layer of a planet that has undergone differentiation.

cycles per second Units of frequency for a wave; describes the number of peaks (or troughs) of a wave that pass by a given point each second. Equivalent to hertz.

dark energy The name given to the as-yet-unknown form of energy that appears to be causing the expansion of the universe to accelerate. Sometimes described as a repulsive force, or by the names *quintessence* or *cosmological constant*.

dark matter Matter that we infer to exist from its gravitational effects but from which we have not detected any light; dark matter apparently dominates the total mass of the universe.

daylight saving time Standard time plus 1 hour, so that the Sun appears on the meridian around 1 P.M. rather than around noon.

declination (dec) Analogous to latitude, but on the celestial sphere; it is the angular north-south distance between the celestial equator and a location on the celestial sphere.

degeneracy pressure A type of pressure unrelated to an object's temperature, which arises when electrons (electron degeneracy pressure) or neutrons (neutron degeneracy pressure) are packed extremely tightly (so that rules the quantum laws known as the exclusion and uncertainty principles come into play).

degenerate object An object in which degeneracy pressure is the primary pressure pushing back against gravity, such as a brown dwarf, white dwarf, or neutron star.

deuterium A form of hydrogen in which the nucleus contains a proton and a neutron, rather than only a proton (as is the case for most hydrogen nuclei).

differential rotation Describes the rotation of an object in which the equator rotates at a different rate than the poles.

differentiation The process in which gravity separates materials according to density, with high-density materials sinking and low-density materials rising.

diffraction grating A finely etched surface that can split light into a spectrum.

diffraction limit The angular resolution that a telescope could achieve if it were limited only by the interference of light waves; it is smaller (i.e., better angular resolution) for larger telescopes.

dimension (mathematical) Describes the number of independent directions in which movement is possible; e.g., the surface of the Earth is two-dimensional because only two independent directions of motion are possible (north-south and east-west).

direction (in local sky) One of the two coordinates (the other is altitude) needed to pinpoint an object in the local sky. It is the direction, such as north, south, east, or west, in which you must face to see the object. *See also* azimuth

disk component (of a galaxy) The portion of a spiral galaxy that looks like a disk and contains an interstellar medium with cool gas and dust; stars of many ages are found in the disk component.

disk population (of stars in a galaxy) Refers to stars that orbit within the disk of a spiral galaxy. Sometimes called Population I.

Doppler effect (shift) The effect that shifts the wavelengths of spectral features in objects that are moving toward or away from the observer.

dust (or dust grains) Tiny solid flecks of material; in astronomy, we often discuss interplanetary dust (found within a star system) or interstellar dust (found between the stars in a galaxy). *See also* interstellar dust grains

dust tail (of a comet) One of two tails seen when a comet passes near the Sun (the other is the plasma tail); composed of small solid particles pushed away from the Sun by the radiation pressure of sunlight.

dwarf elliptical galaxy A small elliptical galaxy with less than about a billion stars.

eccentricity A measure of how much an ellipse deviates from a perfect circle; defined as the center-to-focus distance divided by the length of the semimajor axis.

eclipse Occurs when one astronomical object casts a shadow on another or crosses our line of sight to the other object.

eclipse seasons Periods during which lunar and solar eclipses can occur because the nodes of the Moon's orbit are nearly aligned with the Earth and Sun.

eclipsing binary A binary star system in which the two stars happen to be orbiting in the plane of our line of sight, so that each star will periodically eclipse the other.

ecliptic The Sun's apparent annual path among the constellations.

ecliptic plane The plane of the Earth's orbit around the Sun.

ejecta (from an impact) Debris ejected by the blast of an impact.

electromagnetic force One of the four fundamental forces; it is the force that dominates atomic and molecular interactions.

electromagnetic radiation Another name for light of all types, from radio waves through gamma rays.

electromagnetic spectrum The complete spectrum of light, including radio waves, infrared, visible light, ultraviolet light, X-rays, and gamma rays.

electromagnetic wave A synonym for light, which consists of waves of electric and magnetic fields.

electron degeneracy pressure Degeneracy pressure exerted by electrons, as in brown dwarfs and white dwarfs.

electrons Fundamental particles with negative electric charge; the distribution of electrons in an atom gives the atom its size.

electron-volt (eV) A unit of energy equivalent to 1.60×10^{-19} joule.

electroweak era The era of the universe during which only three forces operated (gravity, strong force, and electroweak force), lasting from 10^{-38} second to 10^{-10} second after the Big Bang.

electroweak force The force that exists at high energies when the electromagnetic force and the weak force exist as a single force.

element (chemical) A substance made from individual atoms of a particular atomic number.

ellipse A type of oval that happens to be the shape of bound orbits. An ellipse can be drawn by moving a pencil along a string whose ends are tied to two tacks; the locations of the tacks are the foci (singular, focus) of the ellipse.

elliptical galaxies Galaxies that appear rounded in shape, often longer in one direction, like a football. They have no disks and contain very little cool gas and dust compared to spiral galaxies, though they often contain very hot, ionized gas.

elongation (greatest) For Mercury or Venus, the point at which it appears farthest from the Sun in our sky.

emission (of light) The process by which matter emits energy in the form of light.

emission-line spectrum A spectrum that contains emission lines.

emission nebula Another name for an ionization nebula. *See also* ionization nebula

energy Broadly speaking, energy is what can make matter move. The three basic types of energy are kinetic, potential, and radiative.

equation of time Describes the discrepancies between apparent and mean solar time.

equivalence principle The fundamental starting point for general relativity, which states that the effects of gravity are exactly equivalent to the effects of acceleration.

era of atoms The era of the universe lasting from about 380,000 years to about 1 billion years after the Big Bang, during which it was cool enough for neutral atoms to form.

era of galaxies The present era of the universe, which began with the formation of galaxies when the universe was about 1 billion years old.

era of nuclei The era of the universe lasting from about 3 minutes to about 380,000 years after the Big Bang, during which matter in the universe was fully ionized and opaque to light. The cosmic microwave background represents light released at the end of this era.

era of nucleosynthesis The era of the universe lasting from about 0.001 second to about 3 minutes after the Big Bang, by the end of which virtually all of the neutrons and about one-seventh of the protons in the universe had fused into helium.

erosion The wearing down or building up of geological features by wind, water, ice, and other phenomena of planetary weather.

eruption The process of releasing hot lava on the planet's surface.

escape velocity The speed necessary for an object to completely escape the gravity of a large body such as a moon, planet, or star.

evaporation The process by which atoms or molecules escape into the gas phase from a liquid.

event horizon The boundary that marks the "point of no return" between a black hole and the outside universe; events that occur within the event horizon can have no influence on our observable universe.

excited state (of an atom) Any arrangement of electrons in an atom that has more energy than the ground state.

exposure time The amount of time for which light is collected to make a single image.

extrasolar planet A planet orbiting a star other than our Sun.

Fahrenheit (temperature scale) The temperature scale commonly used in daily activity in the United States. Defined so that, on Earth's surface, water freezes at 32°F and boils at 212°F.

fall equinox (autumnal equinox) Refers both to the point in Virgo on the celestial sphere where the ecliptic crosses the celestial equator and to the moment in time when the Sun appears at that point each year (around September 21).

fault (geological) A place where rocks slip sideways relative to one another.

feedback relationships Processes in which one property amplifies (positive feedback) or counteracts (negative feedback) the behavior of properties.

Fermi's paradox The question posed by Enrico Fermi about extraterrestrial intelligence—"So where is everybody?"—which asks why we have not observed other civilizations even though simple arguments would suggest that some ought to have spread throughout the galaxy by now.

field An abstract concept used to describe how a particle would interact with a force. For example, the idea of a gravitational field describes how a particle would react to the local strength of gravity, and the idea of an electromagnetic field describes how a charged particle would respond to forces from other charged particles.

filter (for light) A material that transmits only particular wavelengths of light.

fireball A particularly bright meteor.

flare star A small, spectral type M star that displays particularly strong flares on its surface.

flat (or Euclidean) geometry Refers to any case in which the rules of geometry for a flat plane hold, such as that the shortest distance between two points is a straight line.

flat universe A universe in which the overall geometry of spacetime is flat (Euclidean), as would be the case if the density of the universe is equal to the critical density.

focal plane The place where an image created by a lens or mirror is in focus.

focus (of a lens or mirror) The point at which rays of light that were initially parallel (such as light from a distant star) converge.

force Anything that can cause a change in momentum.

frame of reference (in relativity) Two (or more) objects share the same frame of reference if they are not moving relative to each other.

free-fall Refers to conditions in which an object is falling without resistance; objects are weightless when in free-fall.

free-float frame A frame of reference in which all objects are weightless and hence float freely.

frequency Describes the rate at which peaks of a wave pass by a point; measured in units of 1/s, often called cycles per second or hertz.

frost line The boundary in the solar nebula beyond which ices could condense; only metals and rocks could condense within the frost line.

fundamental forces There are four known fundamental forces in nature: gravity, the electromagnetic force, the strong force, and the weak force.

fundamental particles Subatomic particles that cannot be divided into anything smaller.

galactic cannibalism The term sometimes used to describe the process by which large galaxies merge with other galaxies in collisions. Central dominant galaxies are products of galactic cannibalism.

galactic disk (of a spiral galaxy) *See* disk component

galactic wind A wind of low-density but extremely hot gas flowing out from a starburst galaxy, created by the combined energy of many supernovae.

galaxy A huge collection of anywhere from a few hundred million to more than a trillion stars, all bound together by gravity.

galaxy cluster *See* cluster of galaxies

galaxy evolution The formation and development of galaxies.

Galilean moons The four moons of Jupiter that were discovered by Galileo: Io, Europa, Ganymede, and Callisto.

gamma-ray burst A sudden burst of gamma rays from deep space; such bursts apparently come from distant galaxies, but their precise mechanism is unknown.

gamma rays Light with very short wavelengths (and hence high frequencies)—shorter than those of X-rays.

gap moons Tiny moons located within a gap in a planet's ring system. The gravity of a gap moon helps clear the gap.

gas phase The phase of matter in which atoms or molecules can move essentially independently of one another.

gas pressure Describes the force (per unit area) pushing on any object due to surrounding gas. *See also* pressure

genetic code The "language" that living cells use to read the instructions chemically encoded in DNA.

geocentric universe (ancient belief in) The idea that the Earth is the center of the entire universe.

geological processes The four basic geological processes are impact cratering, volcanism, tectonics, and erosion.

geology The study of surface features (on a moon, planet, or asteroid) and the processes that create them.

giant molecular cloud A very large cloud of cold, dense interstellar gas, typically containing up to a million solar masses worth of material. *See also* molecular clouds

giants (luminosity class III) Stars that appear just below the supergiants on the H–R diagram because they are somewhat smaller in radius and lower in luminosity.

global positioning system (GPS) A system of navigation by satellites orbiting the Earth.

global wind patterns (or global circulation) Wind patterns that remain fixed on a global scale, determined by the combination of surface heating and the planet's rotation.

globular cluster A spherically shaped cluster of up to a million or more stars; globular clusters are found primarily in the halos of galaxies and contain only very old stars.

grand unified theory (GUT) A theory that unifies three of the four fundamental forces—the strong force, the weak force, and the electromagnetic force (but not gravity)—in a single model.

granulation (on the Sun) The bubbling pattern visible in the photosphere, produced by the underlying convection.

gravitation (law of) *See* universal law of gravitation

gravitational constant The experimentally measured constant G that appears in the law of universal gravitation;

$$G = 6.67 \times 10^{-11} \frac{m^3}{kg \times s^2}$$

gravitational contraction The process in which gravity causes an object to contract, thereby converting gravitational potential energy into thermal energy.

gravitational encounter Occurs when two (or more) objects pass near enough so that each can feel the effects of the other's gravity and can therefore exchange energy.

gravitational equilibrium Describes a state of balance in which the force of gravity pulling inward is precisely counteracted by pressure pushing outward. Also known as hydrostatic equilibrium.

gravitational lensing The magnification or distortion (into arcs, rings, or multiple images) of an image caused by light bending through a gravitational field, as predicted by Einstein's general theory of relativity.

gravitational redshift A redshift caused by the fact that time runs slow in gravitational fields.

gravitational time dilation The slowing of time that occurs in a gravitational field, as predicted by Einstein's general theory of relativity.

gravitational waves Predicted by Einstein's general theory of relativity, these waves travel at the speed of light and transmit distortions of space through the universe. Although not yet observed directly, we have strong indirect evidence that they exist.

gravitationally bound system Any system of objects, such as a star system or a galaxy, that is held together by gravity.

gravity One of the four fundamental forces; it is the force that dominates on large scales.

grazing incidence (in telescopes) Reflections in which light grazes a mirror surface and is deflected at a small angle; commonly used to focus high-energy ultraviolet light and X-rays.

great circle A circle on the surface of a sphere whose center is at the center of the sphere.

Great Red Spot A large, high-pressure storm on Jupiter.

greenhouse effect The process by which greenhouse gases in an atmosphere make a planet's surface temperature warmer than it would be in the absence of an atmosphere.

greenhouse gases Gases, such as carbon dioxide, water vapor, and methane, that are particularly good absorbers of infrared light but are transparent to visible light.

Gregorian calendar Our modern calendar, introduced by Pope Gregory in 1582.

ground state (of an atom) The lowest possible energy state of the electrons in an atom.

group (of galaxies) A few to a few dozen galaxies bound together by gravity. *See also* cluster of galaxies

GUT era The era of the universe during which only two forces operated (gravity and the grand-unified-theory or GUT force), lasting from 10^{-43} second to 10^{-38} second after the Big Bang.

GUT force The proposed force that exists at very high energies when the strong force, the weak force, and the electromagnetic force (but not gravity) all act as one.

H II region Another name for an ionization nebula. *See* ionization nebula

habitable zone The region around a star in which planets could potentially have surface temperatures at which liquid water could exist.

half-life The time it takes for half of the nuclei in a given quantity of a radioactive substance to decay.

halo (of a galaxy) The spherical region surrounding the disk of a spiral galaxy.

Hawking radiation Radiation predicted to arise from the evaporation of black holes.

heavy elements In astronomy, heavy elements generally refers to all elements except hydrogen and helium.

helium-capture reactions Fusion reactions that fuse a helium nucleus into some other nucleus; such reactions can fuse carbon into oxygen, oxygen into neon, neon into magnesium, and so on.

helium flash The event that marks the sudden onset of helium fusion in the previously inert helium core of a low-mass star.

helium fusion The fusion of three helium nuclei into one carbon nucleus; also called the triple-alpha reaction.

hertz (Hz) The standard unit of frequency for light waves; equivalent to units of 1/s.

Hertzsprung–Russell (H–R) diagram A graph plotting individual stars as points, with stellar luminosity on the vertical axis and spectral type (or surface temperature) on the horizontal axis.

high-mass stars Stars born with masses above about $8M_{Sun}$; these stars will end their lives by exploding as supernovae.

horizon A boundary that divides what we can see from what we cannot see.

horizontal branch The horizontal line of stars that represents helium-burning stars on an H–R diagram for a cluster of stars.

horoscope A predictive chart made by an astrologer; in scientific studies, horoscopes have never been found to have any validity as predictive tools.

hot spot (geological) A place within a plate of the lithosphere where a localized plume of hot mantle material rises.

hour angle (HA) The angle or time (measured in hours) since an object was last on the meridian in the local sky. Defined to be 0 hours for objects that are on the meridian.

Hubble's constant A number that expresses the current rate of expansion of the universe; designated H_0, in this book it is stated in units of kilometers per second per million light-years.

Hubble's law Mathematically expresses the idea that more distant galaxies move away from us faster; its formula is $v = H_0 \times d$, where v is a galaxy's speed away from us, d is its distance, and H_0 is Hubble's constant.

hydrogen compounds Compounds that contain hydrogen and were common in the solar nebula, such as water (H_2O), ammonia (NH_3), and methane (CH_4).

hydrogen-shell burning Hydrogen fusion that occurs in a shell surrounding a stellar core.

hydrosphere Refers to the "layer" of water on the Earth consisting of oceans, lakes, rivers, ice caps, and other liquid water and ice.

hydrostatic equilibrium *See* gravitational equilibrium

hyperbola The precise mathematical shape of one type of unbound orbit (the other is a parabola) allowed under the force of gravity; at great distances from the attracting object, a hyperbolic path looks like a straight line.

hypernova A term sometimes used to describe a supernova (explosion) of a star so massive that it leaves a black hole behind.

hyperspace Any space with more than three dimensions.

hypothesis A tentative model proposed to explain some set of observed facts, but which has not yet been rigorously tested and confirmed.

ices (in solar system theory) Materials that are solid only at low temperatures, such as the hydrogen compounds water, ammonia, and methane.

image A picture of an object made by focusing light.

impact The collision of a small body (such as an asteroid or comet) with a larger object (such as a planet or moon).

impact crater A bowl-shaped depression left by the impact of an object that strikes a planetary surface (as opposed to burning up in the atmosphere).

impact cratering The excavation of bowl-shaped depressions (impact craters) by asteroids or comets striking a planet's surface.

inflation (of the universe) A sudden and dramatic expansion of the universe thought to have occurred at the end of the GUT era.

infrared light Light with wavelengths that fall in the portion of the electromagnetic spectrum between radio waves and visible light.

inner solar system Generally considered to encompass the region of our solar system out to about the orbit of Mars.

intensity (of light) A measure of the amount of energy coming from light of specific wavelength in the spectrum of an object.

interferometry A telescopic technique in which two or more telescopes are used in tandem to produce much better angular resolution than the telescopes could achieve individually.

interstellar cloud A cloud of gas and dust between the stars.

interstellar dust grains Tiny solid flecks of carbon and silicon minerals found in cool interstellar clouds; they resemble particles of smoke and form in the winds of red giant stars.

interstellar medium Refers to gas and dust that fills the space between stars in a galaxy.

interstellar ramjet A hypothesized type of spaceship that uses a giant scoop to sweep up interstellar gas for use in a nuclear fusion engine.

intracluster medium Hot, X-ray-emitting gas found between the galaxies within a cluster of galaxies.

inverse square law Any quantity that decreases with the square of the distance between two objects is said to follow an inverse square law.

inverse square law for light The formula that relates apparent brightness, luminosity, and distance:

$$\text{apparent brightness} = \frac{\text{luminosity}}{4\pi \times (\text{distance})^2}$$

Io torus A donut-shaped charged-particle belt around Jupiter that approximately traces Io's orbit.

ionization The process of stripping an electron from an atom.

ionization nebula A colorful, wispy cloud of gas that glows because neighboring hot stars irradiate it with ultraviolet photons that can ionize hydrogen atoms.

ionosphere A portion of the thermosphere in which ions are particularly common (due to ionization by X-rays from the Sun).

ions Atoms with a positive or negative electrical charge.

irregular galaxies Galaxies that look neither spiral nor elliptical.

isotopes Each different isotope of an element has the same number of protons but a different number of neutrons.

jets High-speed streams of gas ejected from an object into space.

joule The international unit of energy, equivalent to about $\frac{1}{4{,}000}$ of a Calorie.

jovian planets Giant gaseous planets similar in overall composition to Jupiter.

Kelvin (temperature scale) The most commonly used temperature scale in science, defined such that absolute zero is 0 K and water freezes at 273.15 K.

Kepler's first law States that the orbit of each planet about the Sun is an ellipse with the Sun at one focus.

Kepler's laws of planetary motion Three laws discovered by Kepler that describe the motion of the planets around the Sun.

Kepler's second law States that, as a planet moves around its orbit, it sweeps out equal areas in equal times. This tells us that a planet moves faster when it is closer to the Sun (near perihelion) than when it is farther from the Sun (near aphelion) in its orbit.

Kepler's third law States that the square of a planet's orbital period is proportional to the cube of its average distance from the Sun (semimajor axis), which tells us that more distant planets move more slowly in their orbits. In its original form, written $p^2 = a^3$. *See also* Newton's version of Kepler's third law

kinetic energy Energy of motion, given by the formula $\frac{1}{2}mv^2$.

Kirchhoff's laws A set of rules that summarizes the conditions under which objects produce thermal, absorption line, or emission line spectra. In brief: (1) An opaque object produces thermal radiation. (2) An absorption line spectrum occurs when thermal radiation passes through a thin gas that is cooler than the object emitting the thermal radiation. (3) An emission line spectrum occurs when we view a cloud of gas that is warmer than any background source of light.

Kuiper belt The comet-rich region of our solar system that spans distances of about 30–100 AU from the Sun; Kuiper belt comets have orbits that lie fairly close to the plane of planetary orbits and travel around the Sun in the same direction as the planets.

Large Magellanic Cloud One of two small, irregular galaxies (the other is the Small Magellanic Cloud) located about 150,000 light-years away; it probably orbits the Milky Way Galaxy.

large-scale structure (of the universe) Generally refers to structure of the universe on size scales larger than that of clusters of galaxies.

latitude The angular north-south distance between the Earth's equator and a location on the Earth's surface.

leap year A calendar year with 366 rather than 365 days; our current calendar (the Gregorian calendar) has a leap year every 4 years (by adding February 29) except in century years that are not divisible by 400.

lenticular galaxies Galaxies that look lens-shaped when seen edge-on, resembling spiral galaxies without arms. They tend to have less cool gas than normal spiral galaxies but more gas than elliptical galaxies.

life track A track drawn on an H–R diagram to represent the changes in a star's surface temperature and luminosity during its life; also called an evolutionary track.

light-collecting area (of a telescope) The area of the primary mirror or lens that collects light in a telescope.

light curve A graph of an object's intensity against time.

light pollution Human-made light that hinders astronomical observations.

light-year The distance that light can travel in 1 year, which is 9.46 trillion km.

liquid phase The phase of matter in which atoms or molecules are held together but move relatively freely.

lithosphere The relatively rigid outer layer of a planet; generally encompasses the crust and the uppermost portion of the mantle.

Local Bubble (interstellar) The bubble of hot gas in which our Sun and other nearby stars apparently reside. *See also* bubble (interstellar)

Local Group The group of more than 30 galaxies to which the Milky Way Galaxy belongs.

local sidereal time (LST) Sidereal time for a particular location, defined according to the position of the spring equinox in the local sky. More formally, the local sidereal time at any moment is defined to be the hour angle of the spring equinox.

local sky The sky as viewed from a particular location on Earth (or another solid object). Objects in the local sky are pinpointed by the coordinates of altitude and direction (or azimuth).

Local Supercluster The supercluster of galaxies to which the Local Group belongs.

longitude The angular east-west distance between the prime meridian (which passes through Greenwich) and a location on the Earth's surface.

lookback time Refers to the amount of time since the light we see from a distant object was emitted. I.e., if an object has a lookback time of 400 million years, we are seeing it as it looked 400 million years ago.

low-mass stars Stars born with masses less than about $2M_{Sun}$; these stars end their lives by ejecting a planetary nebula and becoming a white dwarf.

luminosity The total power output of an object, usually measured in watts or in units of solar luminosities ($L_{Sun} = 3.8 \times 10^{26}$ watts).

luminosity class Describes the region of the H–R diagram in which a star falls. Luminosity class I represents supergiants, III represents giants, and V represents main-sequence stars; luminosity classes II and IV are intermediate to the others.

lunar eclipse Occurs when the Moon passes through the Earth's shadow, which can occur only at full moon; may be total, partial, or penumbral.

lunar maria The regions of the Moon that look smooth from Earth and actually are impact basins.

lunar phase Describes the appearance of the Moon as seen from Earth.

MACHOs Stands for *massive compact halo objects* and represents one possible form of dark matter in which the dark objects are relatively large, like planets or brown dwarfs.

magma Underground molten rock.

magnetic braking The process by which a star's rotation slows as its magnetic field transfers its angular momentum to the surrounding nebula.

magnetic field Describes the region surrounding a magnet in which it can affect other magnets or charged particles in its vicinity.

magnetic-field lines Lines that represent how the needles on a series of compasses would point if they were laid out in a magnetic field.

magnetosphere The region surrounding a planet in which charged particles are trapped by the planet's magnetic field.

magnitude system A system of describing stellar brightness by using numbers, called magnitudes, based on an ancient Greek way of describing the brightnesses of stars in the sky. This system uses apparent magnitude to describe a star's apparent brightness and absolute magnitude to describe a star's luminosity.

main sequence (luminosity class V) The prominent line of points running from the upper left to the lower right on an H–R diagram; main-sequence stars shine by fusing hydrogen in their cores.

main-sequence fitting A method for measuring the distance to a cluster of stars by comparing the apparent brightness of the cluster's main sequence with the standard main sequence.

main-sequence lifetime The length of time for which a star of a particular mass can shine by fusing hydrogen into helium in its core.

main-sequence turnoff A method for measuring the age of a cluster of stars from the point on its H–R diagram where its stars turn off from the main sequence; the age of the cluster is equal to the main-sequence lifetime of stars at the main-sequence turnoff point.

mantle (of a planet) The rocky layer that lies between a planet's core and crust.

Martian meteorite This term is used to describe meteorites found on Earth that are thought to have originated on Mars.

mass A measure of the amount of matter in an object.

mass-energy The potential energy of mass, which has an amount $E = mc^2$.

mass exchange (in close binary star systems) The process in which tidal forces cause matter to spill from one star to a companion star in a close binary system.

mass extinction An event in which a large fraction of the species living on Earth go extinct, such as the event in which the dinosaurs died out about 65 million years ago.

mass increase (in relativity) Refers to the effect in which an object moving past you seems to have a mass greater than its rest mass.

mass-to-light ratio The mass of an object divided by its luminosity, usually stated in units of solar masses per solar luminosity. Objects with high mass-to-light ratios must contain substantial quantities of dark matter.

massive-star supernova A supernova that occurs when a massive star dies, initiated by the catastrophic collapse of its iron core; often called a Type II supernova.

matter–antimatter annihilation Occurs when a particle of matter and a particle of antimatter meet and convert all of their mass-energy to photons.

mean solar time Time measured by the average position of the Sun in your local sky over the course of the year.

meridian A half-circle extending from your horizon (altitude 0°) due south, through your zenith, to your horizon due north.

metallic hydrogen Hydrogen that is so compressed that the hydrogen atoms all share electrons and thereby take on properties of metals, such as conducting electricity. Occurs only under very high pressure conditions, such as that found deep within Jupiter.

metals (in solar system theory) Elements, such as nickel, iron, and aluminum, that condense at fairly high temperatures.

meteor A flash of light caused when a particle from space burns up in our atmosphere.

meteor shower A period during which many more meteors than usual can be seen.

meteorite A rock from space that lands on Earth.

meteoroid A name sometimes given to the small particles in space that can create meteors if they happen to enter Earth's atmosphere.

Metonic cycle The 19-year period, discovered by the Babylonian astronomer Meton, over which the lunar phases occur on the same dates.

microwaves Light with wavelengths in the range of micrometers to millimeters. Microwaves are generally considered to be a subset of the radio wave portion of the electromagnetic spectrum.

mid-ocean ridges (on Earth) Long ridges of undersea volcanoes, along which mantle material erupts onto the ocean floor and pushes apart the existing seafloor on either side. These ridges are essentially the source of new seafloor crust, which then makes its way along the ocean bottom for millions of years before returning to the mantle at a subduction zone.

Milky Way Used both as the name of our galaxy and to refer to the band of light we see in the sky when we look into the plane of the Milky Way Galaxy.

millisecond pulsars Pulsars with rotation periods of a few thousandths of a second.

model (scientific) A representation of some aspect of nature that can be used to explain and predict real phenomena without invoking myth, magic, or the supernatural.

molecular bands The tightly bunched lines in an object's spectrum that are produced by molecules.

molecular clouds Cool, dense interstellar clouds in which the low temperatures allow hydrogen atoms to pair up into hydrogen molecules (H_2).

molecular dissociation The process by which a molecule splits into its component atoms.

molecule Technically the smallest unit of a chemical element or compound; in this text, the term refers only to combinations of two or more atoms held together by chemical bonds.

momentum The product of an object's mass and velocity.

moon An object that orbits a planet.

mutations Errors in the copying process when a living cell replicates itself.

natural selection The process by which mutations that make an organism better able to survive get passed on to future generations.

neap tides The lower-than-average tides on Earth that occur at first- and third-quarter moon, when the tidal forces from the Sun and Moon oppose one another.

nebula A cloud of gas in space, usually one that is glowing.

nebular capture The process by which icy planetesimals capture hydrogen and helium gas to form jovian planets.

nebular theory The detailed theory that describes how our solar system formed from a cloud of interstellar gas and dust.

net force The overall force to which an object responds; the net force is equal to the rate of change in the object's momentum, or equivalently to the object's mass 3 acceleration.

neutrino A type of fundamental particle that has extremely low mass and responds only to the weak force; neutrinos are leptons and come in three types—electron neutrinos, mu neutrinos, and tau neutrinos.

neutron degeneracy pressure Degeneracy pressure exerted by neutrons, as in neutron stars.

neutron star The compact corpse of a high-mass star left over after a supernova; typically contains a mass comparable to the mass of the Sun in a volume just a few kilometers in radius.

neutrons Particles with no electrical charge found in atomic nuclei, built from three quarks.

newton The standard unit of force in the metric system:

$$1 \text{ newton} = 1\frac{\text{kg} \times \text{m}}{\text{s}^2}$$

Newton's first law of motion States that, in the absence of a net force, an object moves with constant velocity.

Newton's laws of motion Three basic laws that describe how objects respond to forces.

Newton's second law of motion States how a net force affects an object's motion. Specifically: force 5 rate of change in momentum, or force = mass × acceleration.

Newton's third law of motion States that, for any force, there is always an equal and opposite reaction force.

Newton's universal law of gravitation *See* universal law of gravitation

Newton's version of Kepler's third law This generalization of Kepler's third law can be used to calculate the masses of orbiting objects from measurements of orbital period and distance. Usually written as:

$$p^2 = \frac{4\pi^2}{G(M_1 + M_2)}a^3$$

nodes (of Moon's orbit) The two points in the Moon's orbit where it crosses the ecliptic plane.

nonbaryonic matter Refers to exotic matter that is not part of the normal composition of atoms, such as neutrinos or the hypothetical WIMPs.

nonscience As defined in this book, nonscience is any way of searching for knowledge that makes no claim to follow the scientific method, such as seeking knowledge through intuition, tradition, or faith.

north celestial pole (NCP) The point on the celestial sphere directly above the Earth's North Pole.

nova The dramatic brightening of a star that lasts for a few weeks and then subsides; occurs when a burst of hydrogen fusion ignites in a shell on the surface of an accreting white dwarf in a binary star system.

nuclear fission The process in which a larger nucleus splits into two (or more) smaller particles.

nuclear fusion The process in which two (or more) smaller nuclei slam together and make one larger nucleus.

nucleus (of an atom) The compact center of an atom made from protons and neutrons.

nucleus (of a comet) The solid portion of a comet, and the only portion that exists when the comet is far from the Sun.

observable universe The portion of the entire universe that, at least in principle, can be seen from Earth.

Occam's razor A principle often used in science, holding that scientists should prefer the simpler of two models that agree equally well with observations. Named after the medieval scholar William of Occam (1285–1349).

Olbers' paradox Asks the question of how the night sky can be dark if the universe is infinite and full of stars.

Oort cloud A huge, spherical region centered on the Sun, extending perhaps halfway to the nearest stars, in which trillions of comets orbit the Sun with random inclinations, orbital directions, and eccentricities.

opaque (material) Describes a material that absorbs light.

open cluster A cluster of up to several thousand stars; open clusters are found only in the disks of galaxies and often contain young stars.

open universe The universe is open if its average density is less than the critical density, in which case spacetime has an overall shape analogous to the surface of a saddle.

opposition The point at which a planet appears opposite the Sun in our sky.

orbital resonance Describes any situation in which one object's orbital period is a simple ratio of another object's period, such as $\frac{1}{2}$, $\frac{1}{4}$, or $\frac{5}{3}$. In such cases, the two objects periodically line up with each other, and the extra gravitational attractions at these times can affect the objects' orbits.

outer solar system Generally considered to encompass the region of our solar system beginning at about the orbit of Jupiter.

outgassing The process of releasing gases from a planetary interior, usually through volcanic eruptions.

oxidation Refers to chemical reactions, often with the surface of a planet, that remove oxygen from the atmosphere.

ozone The molecule O_3, which is a particularly good absorber of ultraviolet light.

ozone depletion Refers to the declining levels of atmospheric ozone found worldwide on Earth, especially in Antarctica, in recent years.

ozone hole A place where the concentration of ozone in the stratosphere is dramatically lower than is the norm.

parabola The precise mathematical shape of a special type of unbound orbit allowed under the force of gravity; if an object in a parabolic orbit loses only a tiny amount of energy, it will become bound.

paradigm (in science) Refers to general patterns of thought that tend to shape scientific beliefs during a particular time period.

paradox A situation that, at least at first, seems to violate common sense or contradict itself. Resolving paradoxes often leads to deeper understanding.

parallax The apparent shifting of an object against the background, due to viewing it from different positions. *See also* stellar parallax

parallax angle Half of a star's annual back-and-forth shift due to stellar parallax; related to the star's distance according to the formula

$$\text{distance in parsecs} = \frac{1}{p}$$

where p is the parallax angle in arcseconds.

parsec (pc) Approximately equal to 3.26 light-years; it is the distance to an object with a parallax angle of 1 arcsecond.

partial lunar eclipse A lunar eclipse in which the Moon becomes only partially covered by the Earth's umbral shadow.

partial solar eclipse A solar eclipse during which the Sun becomes only partially blocked by the disk of the Moon.

particle accelerator A machine designed to accelerate subatomic particles to high speeds in order to create new particles or to test fundamental theories of physics.

particle era The era of the universe lasting from 10^{-10} second to 0.001 second after the Big Bang, during which subatomic particles were continually created and destroyed and ending when matter annihilated antimatter.

peculiar velocity (of a galaxy) The component of a galaxy's velocity relative to the Milky Way that deviates from the velocity expected by Hubble's law.

penumbra The lighter, outlying regions of a shadow.

penumbral (lunar) eclipse A lunar eclipse in which the Moon passes only within the Earth's penumbral shadow and does not fall within the umbra.

perigee The point at which an object orbiting the Earth is nearest to the Earth.

perihelion The point at which an object orbiting the Sun is closest to the Sun.

period-luminosity relation The relation that describes how the luminosity of a Cepheid variable star is related to the period between peaks in its brightness; the longer the period, the more luminous the star.

phase (of matter) Describes the way in which atoms or molecules are held together; the common phases are solid, liquid, and gas.

photon An individual particle of light, characterized by a wavelength and a frequency.

photosphere The visible surface of the Sun, where the temperature averages just under 6,000 K.

pixel An individual "picture element" on a CCD.

Planck era The era of the universe prior to the Planck time.

Planck time The time when the universe was 10^{-43} second old, before which random energy fluctuations were so large that our current theories are powerless to describe what might have been happening.

Planck's constant A universal constant, abbreviated h, with value $h = 6.626 \times 10^{-34}$ joule $\times$ s.

planet An object that orbits a star and that, while much smaller than a star, is relatively large in size; there is no "official" minimum size for a planet, but the nine planets in our solar system all are at least 2,000 km in diameter.

planetary nebula The glowing cloud of gas ejected from a low-mass star at the end of its life.

planetesimals The building blocks of planets, formed by accretion in the solar nebula.

plasma A gas consisting of ions and electrons.

plasma tail (of a comet) One of two tails seen when a comet passes near the Sun (the other is the dust tail); composed of ionized gas blown away from the Sun by the solar wind.

plate tectonics The geological process in which plates are moved around by stresses in a planet's mantle.

plates (on a planet) Pieces of a lithosphere that apparently float upon the denser mantle below.

Population I (of stars in a galaxy) *See* disk population

Population II (of stars in a galaxy) *See* spheroidal population

positron The antimatter equivalent of an electron. It is identical to an electron in virtually all respects, except it has a positive rather than a negative electrical charge.

potential energy Energy stored for later conversion into kinetic energy; includes gravitational potential energy, electrical potential energy, and chemical potential energy.

power The rate of energy usage, usually measured in watts (1 watt = 1 joule/s).

precession The gradual wobble of the axis of a rotating object around a vertical line.

pressure Describes the force (per unit area) pushing on an object. In astronomy, we are generally interested in pressure applied by surrounding gas (or plasma). Ordinarily, such pressure is related to the temperature of the gas (*see* thermal pressure). In objects such as white dwarfs and neutron stars, pressure may arise from a quantum effect (*see* degeneracy pressure). Light can also exert pressure. (*See* radiation pressure.)

primary mirror The large, light-collecting mirror of a reflecting telescope.

prime focus (of a reflecting telescope) The first point at which light focuses after bouncing off the primary mirror; located in front of the primary mirror.

prime meridian The meridian of longitude that passes through Greenwich, England, defined to be longitude 0°.

primitive meteorites Meteorites that formed at the same time as the solar system itself, about 4.6 billion years ago. They are sometimes known as chondrites, a technical name that relates to their structure.

processed meteorites Meteorites that apparently once were part of a larger object that "processed" the original material of the solar nebula into another form. They are sometimes known as achondrites.

protogalactic cloud A huge, collapsing cloud of intergalactic gas from which an individual galaxy formed.

proton–proton chain The chain of reactions by which low-mass stars (including the Sun) fuse hydrogen into helium.

protons Particles found in atomic nuclei with positive electrical charge, built from three quarks.

protoplanetary disk A disk of material surrounding a young star (or protostar) that may eventually form planets.

protostar A forming star that has not yet reached the point where sustained fusion can occur in its core.

protostellar disk A disk of material surrounding a protostar; essentially the same as a protoplanetary disk, but may not necessarily lead to planet formation.

protostellar wind The relatively strong wind from a protostar.

protosun The central object in the forming solar system that eventually became the Sun.

pseudoscience Something that purports to be science or may appear to be scientific but that does not adhere to the testing and verification requirements of the scientific method.

pulsar A neutron star from which we see rapid pulses of radiation as it rotates.

pulsating variable stars Stars that alternately grow brighter and dimmer as their outer layers expand and contract in size.

quantum mechanics The branch of physics that deals with the very small, including molecules, atoms, and fundamental particles.

quarks The building blocks of protons and neutrons, quarks are one of the two basic types of fermions (leptons are the other).

quasar The brightest type of active galactic nucleus.

radar ranging A method of measuring distances within the solar system by bouncing radio waves off planets.

radial motion The component of an object's motion directed toward or away from us.

radial velocity The portion of any object's total velocity that is directed toward or away from us. This part of the velocity is the only part that we can measure with the Doppler effect.

radiation pressure Pressure exerted by photons of light.

radiation zone (of a star) A region of the interior in which energy is transported primarily by radiative diffusion.

radiative diffusion The process by which photons gradually migrate from a hot region (such as the solar core) to a cooler region (such as the solar surface).

radiative energy Energy carried by light; the energy of a photon is Planck's constant times its frequency, or $h \times f$.

radio galaxy A galaxy that emits unusually large quantities of radio waves; thought to contain an active galactic nucleus powered by a supermassive black hole.

radio lobes The huge regions of radio emission found on either side of radio galaxies. The lobes apparently contain plasma ejected by powerful jets from the galactic center.

radio waves Light with very long wavelengths (and hence low frequencies)—longer than those of infrared light.

radioactive element (or radioactive isotope) A substance whose nucleus tends to undergo spontaneous change.

radioactive decay The event in which a radioactive nucleus undergoes change, such as a change in its number of protons or neutrons or both.

radiometric dating The process of determining the age of a rock (i.e., the time since it solidified) by comparing the present amount of a radioactive substance to the amount of its decay product.

recession velocity (of a galaxy) The speed at which a distant galaxy is moving away from us due to the expansion of the universe.

recollapsing universe One of the four general models that may describe the way the expansion of the universe changes with time. In this model, the collective gravity of all its matter eventually halts and reverses the expansion. The galaxies will come crashing back together, and the universe will end in a fiery Big Crunch.

red giant A giant star that is red in color.

red-giant winds The relatively dense but slow winds from red giant stars.

redshift (Doppler) A Doppler shift in which spectral features are shifted to longer wavelengths, caused when an object is moving away from the observer.

reflecting telescope A telescope that uses mirrors to focus light.

reflection (of light) The process by which matter changes the direction of light.

reflection nebula A nebula that we see as a result of starlight reflected from interstellar dust grains. Reflection nebulae tend to have blue and black tints.

refracting telescope A telescope that uses lenses to focus light.

resonance *See* orbital resonance

rest wavelength The wavelength of a spectral feature in the absence of any Doppler shift or gravitational redshift.

retrograde motion Motion that is backward compared to the norm; e.g., we see Mars in apparent retrograde motion during the periods of time when it moves westward, rather than the more common eastward, relative to the stars.

revolution The orbital motion of one object around another.

right ascension (RA) Analogous to longitude, but on the celestial sphere; it is the angular east-west distance between the vernal equinox and a location on the celestial sphere.

rings (planetary) Consist of numerous small particles orbiting a planet within its Roche zone.

rocks (in solar system theory) Material common on the surface of the Earth, such as silicon-based minerals, that are solid at temperatures and pressures found on Earth but typically melt or vaporize at temperatures of 500–1,300 K.

rotation The spinning of an object around its axis.

rotation curve A graph that plots rotational (or orbital) velocity against distance from the center for any object or set of objects.

runaway greenhouse effect A positive feedback cycle in which heating caused by the greenhouse effect causes more greenhouse gases to enter the atmosphere, which further enhances the greenhouse effect.

saddle-shaped (or hyperbolic) geometry Refers to any case in which the rules of geometry for a saddle-shaped surface hold, such as that two lines that begin parallel eventually diverge.

Sagittarius Dwarf A small, dwarf elliptical galaxy that is currently passing through the disk of the Milky Way Galaxy.

saros cycle The period over which the basic pattern of eclipses repeats, which is about 18 years $11\frac{1}{3}$ days.

satellite Any object orbiting another object.

scattered light Light that is reflected into random directions.

Schwarzschild radius A measure of the size of the event horizon of a black hole.

science The search for knowledge that can be used to explain or predict natural phenomena in a way that can be confirmed by rigorous observations or experiments.

scientific method An organized approach to explaining observed facts through science.

scientific theory A model of some aspect of nature that has been rigorously tested and has passed all tests to date.

secondary mirror A small mirror in a reflecting telescope, used to reflect light gathered by the primary mirror toward an eyepiece or instrument.

sedimentary rock A rock that formed from sediments created and deposited by erosional processes.

seismic waves Earthquake-induced vibrations that propagate through a planet.

semimajor axis Half the distance across the long axis of an ellipse; in this text, it is usually referred to as the average distance of an orbiting object, abbreviated a in the formula for Kepler's third law.

SETI (search for extraterrestrial intelligence) The name given to observing projects designed to search for signs of intelligent life beyond Earth.

Seyfert galaxies The name given to a class of galaxies found relatively nearby and that have nuclei much like those of quasars, except that they are less luminous.

shepherd moons Tiny moons within a planet's ring system that help force particles into a narrow ring. A variation on gap moons.

shield volcano A shallow-sloped volcano made from the flow of low-viscosity basaltic lava.

shock wave A wave of pressure generated by gas moving faster than the speed of sound.

sidereal day The time of 23 hours 56 minutes 4.09 seconds between successive appearances of any particular star on the meridian; essentially the true rotation period of the Earth.

sidereal month About $27\frac{1}{4}$ days, the time required for the Moon to orbit the Earth once (as measured against the stars).

sidereal period (of a planet) A planet's actual orbital period around the Sun.

sidereal time Time measured according to the position of stars in the sky rather than the position of the Sun in the sky. See also local sidereal time

sidereal year The time required for the Earth to complete exactly one orbit as measured against the stars; about 20 minutes longer than the tropical year on which our calendar is based.

silicate rock A silicon-rich rock.

singularity The place at the center of a black hole where, in principle, gravity crushes all matter to an infinitely tiny and dense point.

Small Magellanic Cloud One of two small, irregular galaxies (the other is the Large Magellanic Cloud) located about 150,000 light-years away; it probably orbits the Milky Way Galaxy.

solar activity Refers to short-lived phenomena on the Sun, including the emergence and disappearance of individual sunspots, prominences, and flares; sometimes called solar weather.

solar day Twenty-four hours, which is the average time between appearances of the Sun on the meridian.

solar eclipse Occurs when the Moon's shadow falls on the Earth, which can occur only at new moon; may be total, partial, or annular.

solar flares Huge and sudden releases of energy on the solar surface, probably caused when energy stored in magnetic fields is suddenly released.

solar luminosity The luminosity of the Sun, which is approximately 4×10^{26} watts.

solar maximum The time during each sunspot cycle at which the number of sunspots is the greatest.

solar minimum The time during each sunspot cycle at which the number of sunspots is the smallest.

solar nebula The piece of interstellar cloud from which our own solar system formed.

solar neutrino problem Refers to the disagreement between the predicted and observed number of neutrinos coming from the Sun.

solar prominences Vaulted loops of hot gas that rise above the Sun's surface and follow magnetic-field lines.

solar sail A large, highly reflective (and thin, to minimize mass) piece of material that can "sail" through space using pressure exerted by sunlight.

solar system (or star system) Consists of a star (sometimes more than one star) and all the objects that orbit it.

solar wind A stream of charged particles ejected from the Sun.

solid phase The phase of matter in which atoms or molecules are held rigidly in place.

sound wave A wave of alternately rising and falling pressure.

south celestial pole (SCP) The point on the celestial sphere directly above the Earth's South Pole.

spacetime The inseparable, four-dimensional combination of space and time.

spectral lines Bright or dark lines that appear in an object's spectrum, which we can see when we pass the object's light through a prismlike device that spreads out the light like a rainbow.

spectral resolution Describes the degree of detail that can be seen in a spectrum; the higher the spectral resolution, the more detail we can see.

spectral type A way of classifying a star by the lines that appear in its spectrum; it is related to surface temperature. The basic spectral types are designated by a letter (OBAFGKM, with O for the hottest stars and M for the coolest) and are subdivided with numbers from 0 through 9.

spectroscopic binary A binary star system whose binary nature is revealed because we detect the spectral lines of one or both stars alternately becoming blueshifted and redshifted as the stars orbit each other.

spectroscopy (in astronomical research) The process of obtaining spectra from astronomical objects.

spectrum (of light) *See* electromagnetic spectrum

speed The rate at which an object moves. Its units are distance divided by time, such as m/s or km/hr.

speed of light The speed at which light travels, which is about 300,000 km/s.

spherical geometry Refers to any case in which the rules of geometry for the surface of a sphere hold, such as that lines that begin parallel eventually meet.

spheroidal component (of a galaxy) The portion of any galaxy that is spherical (or football-like) in shape and contains very little cool gas; generally contains only very old stars. Elliptical galaxies have only a spheroidal component, while spiral galaxies also have a disk component.

spheroidal galaxy Another name for an elliptical galaxy.

spheroidal population (of stars in a galaxy) Refers to stars that orbit within the spheroidal component of a galaxy. Thus, elliptical galaxies have only a spheroidal population (they lack a disk population), while spiral galaxies have spheroidal population stars in their bulges and halos. Sometimes called Population II.

spiral arms The bright, prominent arms, usually in a spiral pattern, found in most spiral galaxies.

spiral density waves Gravitationally driven waves of enhanced density that move through a spiral galaxy and are responsible for maintaining its spiral arms.

spiral galaxies Galaxies that look like flat, white disks with yellowish bulges at their centers. The disks are filled with cool gas and dust, interspersed with hotter ionized gas, and usually display beautiful spiral arms.

spring equinox (vernal equinox) Refers both to the point in Pisces on the celestial sphere where the ecliptic crosses the celestial equator and to the moment in time when the Sun appears at that point each year (around March 21).

spring tides The higher-than-average tides on Earth that occur at new and full moon, when the tidal forces from the Sun and Moon both act along the same line.

standard candle An object for which we have some means of knowing its true luminosity, so that we can use its apparent brightness to determine its distance with the luminosity–distance formula.

standard model (of physics) The current theoretical model that describes the fundamental particles and forces in nature.

standard time Time measured according to the internationally recognized time zones.

star A large, glowing ball of gas that generates energy through nuclear fusion in its core. The term star is sometimes applied to objects that are in the process of becoming true stars (e.g., protostars) and to the remains of stars that have died (e.g., neutron stars).

star cluster *See* cluster of stars

starburst galaxy A galaxy in which stars are forming at an unusually high rate.

steady state theory A now-discredited theory that held that the universe had no beginning and looks about the same at all times.

stellar evolution The formation and development of stars.

stellar parallax The apparent shift in the position of a nearby star (relative to distant objects) that occurs as we view the star from different positions in the Earth's orbit of the Sun each year.

stellar wind A stream of charged particles ejected from the surface of a star.

stratosphere An intermediate-altitude layer of the atmosphere that is warmed by the absorption of ultraviolet light from the Sun.

stratovolcano A steep-sided volcano made from viscous lavas that can't flow very far before solidifying.

stromatolites Rocks that appear to be fossils of ancient bacterial "colonies."

strong force One of the four fundamental forces; it is the force that holds atomic nuclei together.

subduction (of tectonic plates) The process in which one plate slides under another.

subduction zones Places where one plate slides under another.

subgiant A star that is between being a main-sequence star and being a giant; subgiants have inert helium cores and hydrogen-burning shells.

sublimation The process by which atoms or molecules escape into the gas phase from a solid.

summer solstice Refers both to the point on the celestial sphere where the ecliptic is farthest north of the celestial equator and to the moment in time when the Sun appears at that point each year (around June 21).

sunspot cycle The period of about 11 years over which the number of sunspots on the Sun rises and falls.

sunspots Blotches on the surface of the Sun that appear darker than surrounding regions.

superbubble Essentially a giant interstellar bubble, formed when the shock waves of many individual bubbles merge to form a single, giant shock wave.

supercluster Superclusters consist of many clusters of galaxies, groups of galaxies, and individual galaxies and are the largest known structures in the universe.

supergiants (luminosity class I) The very large and very bright stars that appear at the top of an H–R diagram.

supermassive black hole Giant black hole, with a mass millions to billions of times that of our Sun, thought to reside in the centers of many galaxies and to power active galactic nuclei.

supernova The explosion of a star.

Supernova 1987A A supernova witnessed on Earth in 1987; it was the nearest supernova seen in nearly 400 years and helped astronomers refine theories of supernovae.

supernova remnant A glowing, expanding cloud of debris from a supernova explosion.

synchronous rotation Describes the rotation of an object that always shows the same face to an object that it is orbiting because its rotation period and orbital period are equal.

synodic month (or lunar month) The time required for a complete cycle of lunar phases, which averages about $29\frac{1}{2}$ days.

synodic period (of a planet) The time between successive alignments of a planet and the Sun in our sky; measured from opposition to opposition for a planet beyond Earth's orbit, or from superior conjunction to superior conjunction for Mercury and Venus.

tangential motion The component of an object's motion directed across our line of sight.

tangential velocity The portion of any object's total velocity that is directed across (perpendicular to) our line-of-sight. This part of the velocity cannot be measured with the Doppler effect. It can be measured only by observing the object's gradual motion across our sky.

tectonics The disruption of a planet's surface by internal stresses.

temperature A measure of the average kinetic energy of particles in a substance.

terrestrial planets Rocky planets similar in overall composition to Earth.

theories of relativity (special and general) Einstein's theories that describe the nature of space, time, and gravity.

thermal energy Represents the collective kinetic energy, as measured by temperature, of the many individual particles moving within a substance.

thermal escape The process in which atoms or molecules in a planet's exosphere move fast enough to escape into space.

thermal pressure The ordinary pressure in a gas arising from motions of particles that can be attributed to the object's temperature.

thermal radiation The spectrum of radiation produced by an opaque object that depends only on the object's temperature; sometimes called "blackbody radiation."

thermosphere A high, hot X-ray-absorbing layer of an atmosphere, just below the exosphere.

tidal force A force that is caused when the gravity pulling on one side of an object is larger than that on the other side, causing the object to stretch.

tidal friction Friction within an object that is caused by a tidal force.

tidal heating A source of internal heating created by tidal friction. It is particularly important for satellites with eccentric orbits such as Io and Europa.

time dilation Refers to the effect in which you observe time running slower in reference frames moving relative to you.

torque A twisting force that can cause a change in an object's angular momentum.

total lunar eclipse A lunar eclipse in which the Moon becomes fully covered by the Earth's umbral shadow.

total solar eclipse A solar eclipse during which the Sun becomes fully blocked by the disk of the Moon.

totality (eclipse) The portion of either a total lunar eclipse during which the Moon is fully within the Earth's umbral shadow or a total solar eclipse during which the Sun's disk is fully blocked by the Moon.

transmission (of light) The process in which light passes through matter without being absorbed.

transparent (material) Describes a material that transmits light.

triple-alpha reaction *See* helium fusion

Trojan asteroids Asteroids found within two stable zones that share Jupiter's orbit but lie 60° ahead of and behind Jupiter.

tropic of Cancer The circle on the Earth with latitude 23.5°N. It is the northernmost latitude at which the Sun ever passes directly overhead (at noon on the summer solstice).

tropic of Capricorn The circle on the Earth with latitude 23.5°S. It is the southernmost latitude at which the Sun ever passes directly overhead (at noon on the winter solstice).

tropical year The time from one spring equinox to the next, on which our calendar is based.

troposphere The lowest atmospheric layer, in which convection and weather occur.

turbulence Rapid and random motion.

Type I/II supernovae A way of classifying supernovae based on the spectral lines we observe from them. Type Ia supernovae are thought to be caused by what we call white dwarf supernovae in this book, while Type Ib, Ic, and Type II supernovae are thought to be massive star supernovae.

ultraviolet light Light with wavelengths that fall in the portion of the electromagnetic spectrum between visible light and X-rays.

umbra The dark central region of a shadow.

unbound orbits Orbits on which an object comes in toward a large body only once, never to return; unbound orbits may be parabolic or hyperbolic in shape.

universal law of gravitation The law expressing the force of gravity (F_g) between two objects, given by the formula

$$F_g = G\frac{M_1 M_2}{d^2}$$

$$\left(G = 6.67 \times 10^{-11}\frac{m^3}{kg \times s^2}\right).$$

universal time (UT) Standard time in Greenwich (or anywhere on the prime meridian).

universe The sum total of all matter and energy.

velocity The combination of speed and direction of motion; it can be stated as a speed in a particular direction, such as 100 km/hr due north.

vernal equinox *See* spring equinox

visible light The light our eyes can see, ranging in wavelength from about 400 to 700 nm.

visual binary A binary star system in which we can resolve both stars through a telescope.

voids Huge volumes of space between superclusters that appear to contain very little matter.

volcanism The eruption of molten rock, or lava, from a planet's interior onto its surface.

wavelength The distance between adjacent peaks (or troughs) of a wave.

weak force One of the four fundamental forces; it is the force that mediates nuclear reactions; also the only force besides gravity felt by weakly interacting particles.

weakly interacting particles Particles, such as neutrinos and WIMPs, that respond only to the weak force and gravity; that is, they do not feel the strong force or the electromagnetic force.

weather Describes the ever-varying combination of winds, clouds, temperature, and pressure in a planet's troposphere.

weight The net force that an object applies to its surroundings; in the case of a stationary body on the surface of the Earth, weight = mass × acceleration of gravity.

weightless A weight of zero, as occurs during free-fall.

white-dwarf limit (also called the Chandrasekhar limit) The maximum possible mass for a white dwarf, which is about $1.4M_{Sun}$.

white dwarf supernova A supernova that occurs when an accreting white dwarf reaches the white-dwarf limit, ignites runaway carbon fusion, and explodes like a bomb; often called a Type Ia supernova.

white dwarfs The hot, compact corpses of low-mass stars, typically with a mass similar to the Sun compressed to a volume the size of the Earth.

WIMPs Stands for weakly interacting massive particles and represents a possible form of dark matter consisting of subatomic particles that are dark because they do not respond to the electromagnetic force.

winter solstice Refers both to the point on the celestial sphere where the ecliptic is farthest south of the celestial equator and to the moment in time when the Sun appears at that point each year (around December 21).

wormholes The name given to hypothetical tunnels through hyperspace that might connect two distant places in our universe.

X-rays Light with wavelengths that fall in the portion of the electromagnetic spectrum between ultraviolet light and gamma rays.

X-ray binary A binary star system that emits substantial amounts of X-rays, thought to be from an accretion disk around a neutron star or black hole.

X-ray burster An object that emits a burst of X-rays every few hours to every few days; each burst lasts a few seconds and is thought to be caused by helium fusion on the surface of an accreting neutron star in a binary system.

Zeeman effect The splitting of spectral lines by a magnetic field.

zenith The point directly overhead, which has an altitude of 90°.

zodiac The constellations on the celestial sphere through which the ecliptic passes.

Credits and Acknowlegements

Chapter 1 CO NASA Earth Observing System **01.01** NASA/Goddard Institute for Space Studies **01.03** Jerry Lodriguss/Astropix LLC **01.06a** Stan Maddock **01.07** NASA/Goddard Institute for Space Studies **01.08** Akira Fujii **01.10** Blakeley Kim (Impact), Corel Corporation (sea anemone), NASA Earth Observing System (Earth), Corel Corporation (pyramid) **01.13** NASA Earth Observing System (motion of stars) **01.14** NASA Earth Observing System

Chapter 2 CO David Nunuk **02.01** Gordon Garradd **02.06** Richard Tauber Photography **02.14** Akira Fujii **02.19** Akira Fujii/Akira Fujii **02.20** Pearson Education/Benjamin Cummings Publishing Company **02.22a** Akira Fujii **02.22b** Dennis diCicco **02.22c** Akira Fujii **02.23a,b, c** Akira Fujii **02.24** Akira Fujii 02.26 Akira Fujii **02.27b** Photograph by Tunc Tezel

Chapter 3 CO Margaret Curtis **03.01 (margin)** Giraudon/Art Resource, N.Y. (Copernicus) **03.02** Michael Yamashita/CORBIS, NY **03.02 (margin)** Hulton Archive/Getty Images (Brahe) **03.03** N. Pecnik/Visuals Unlimited **03.03 (margin)** Erich Lessing/Art Resource, N.Y. (Kepler) **03.04** Kenneth Garrett Photography **03.05** William E. Woolam **03.05 (margin)** Bettmann/Corbis/Bettmann (Galileo) **03.06** Margaret Curtis **03.07** Richard A. Cooke, III / Stone/Getty Images **03.08** Loren McIntyre/Woodfin Camp & Associates **03.09** Jeff Henry/Peter Arnold, Inc. **03.10** Courtesy of Oliver Strewe **03.11a** Courtesy of Carl Sagan Productions, Inc. From *Cosmos* (Random House) **03.11b** Courtesy of Carl Sagan Productions, Inc. From *Cosmos* (Random House) **03.12** Corbis/Bettmann **03.14** The Granger Collection, New York **03.19** Anthony Ayiomamitis

Chapter 4 CO Hubble Heritage Team **04.01(margin)** Corbis/Bettmann (Newton) **04.05a** NASA/Goddard Institute for Space Studies **04.05b** Duomo/Corbis/Bettmann **04.05c** NASA/Goddard Institute for Space Studies **04.08(left)** Alvis Upitis/Image Bank/Getty Images **04.08 (right)** Fred Dana/Corbis/Bettmann **04.08a** Getty Images, Inc.**04.13** U.S. Department of Energy **04.16** NASA/Goddard Institute for Space Studies **04.21 (left and right)** Bill Bachmann/Jeff Gnass Photography

Chapter 5 CO NOAO/AURA/NSF **05.01** Runk/Schoenberger/Grant Heilman Photography, Inc. **05.14b** Yerkes Observatory, University of Chicago **05.15a** National Optical Astronomy Observatories **05.16** Russ Underwood/C.A.R.A. / W. M. Keck Observatory **05.17** David Parker, 1997/Science Library. The Arecibo Observatory is part of the National Astronomy and Ionosphere Center, which is operated by Cornell University under a cooperative agreement with the National Science Foundation **05.18a** NASA/CXC/SAO **05.19** NASA/Jet Propulsion Laboratory **05.20** Richard J. Wainscoat **05.21** NASA/Johnson Space Center **05.22a** NASA/Goddard Space Flight Center **05.22b** SciTech Web Team **05.22c** NASA/CXC/SAO **05.22d** SciTech Web Team **05.22e** NASA/Jet Propulsion Laboratory **05.22f** NASA Earth Observing System 05.23a CFHT Corporation 05.23b CFHT Corporation **05.24** Joel Gordon Photography

Chapter 6 CO NASA/Jet Propulsion Laboratory **06.02a** Big Bear Solar Observatory New Jersey Institute of Technology and NASA Marshall Space Flight Center **06.02b** NASA/Marshall Space Flight Center **06.03a** NASA/Marshall Space Flight Center **06.03b** From the Voyage scale model solar system, developed by Challenger Center for Space Science Education, the Smithsonian Institution, and NASA. Image created by ARC Science Simulations, ©2001 **06.04a** NASA/Marshall Space Flight Center **06.04b** From the Voyage scale model solar system, developed by Challenger Center for Space Science Education, the Smithsonian Institution, and NASA. © 2001 David P. Anderson, Southern Methodist University **06.05b** From the Voyage scale model solar system, developed by Challenger Center for Space Science Education, the Smithsonian Institution, and NASA. Image created by ARC Science Simulations, © 2001 **06.05b** NASA Earth Observing System **06.06a** NASA Earth Observing System **06.07** From the Voyage scale model solar system, developed by Challenger Center for Space Science Education, the Smithsonian Institution, and NASA. Image created by ARC Science Simulations, © 2001 **06.08** From the Voyage scale model solar system, developed by Challenger Center for Space Science Education, the Smithsonian Institution, and NASA. Image created by ARC Science Simulations, © 2001 **06.09** From the Voyage scale model solar system, developed by Challenger Center for Space Science Education, the Smithsonian Institution, and NASA. Image created by ARC Science Simulations, © 2001 **06.10** From the Voyage scale model solar system, developed by Challenger Center for Space Science Education, the Smithsonian Institution, and NASA. Image created by ARC Science Simulations, © 2001 **06.11** Dr. R. Albrecht **06.13** APL and NASA **06.14** Niescja Turner and Carter Emmart **06.16** NASA/Jet Propulsion Laboratory **06.17** Anglo-Australian Observatory, photograph by David Malin **06.18** NASA/Jet Propulsion Laboratory **06.19a** Courtesy of Alfred Schultz and Helen Hart **06.19b** Courtesy of John Bally and Nathan Smith, CASA, University of Colorado **06.20** NASA/Jet Propulsion Laboratory **06.22** Courtesy of Robert Haag Meteorites **06.24** NASA Earth Observing System **06.25a** NASA/Jet Propulsion Laboratory **06.25b** NASA/Jet Propulsion Laboratory **06.26** Don Davis

Chapter 7 07.01 NASA/Jet Propulsion Laboratory (Mercury, Venus and Earth), Akira Fujii (Moon (globe), NASA/Jet Propulsion Laboratory (Mars) **07.03** Roger Ressmeyer/CORBIS. NY **07.03 (inset)** Viking Project, NASA, NSSDC **07.04 (inset, left and right)** Viking Project, NASA, NSSDC (image from Viking I in 1976), Mar Global Surveyor, MLS, NASA ("face of Mars") **07.05** Jules Bucher/Photo Researchers, Inc. **07.06** Courtesy of Daniel Hershman **07.07** Don Davis **07.08** Courtesy of Brad Snowder **07.09** Paul Chesley/Stone/Getty Images (eruption of active volcano) **07.10** U.S. Geological Survey, Denver **07.11a** NASA Earth Observing System **07.11b** NASA/Jet Propulsion Laboratory **07.12a** Gene Ahrens/Bruce Coleman Inc. **07.12b** J. Messerschmidt/Bruce Coleman Inc. **07.12c** Craig Aurness/CORBIS. NY **07.12d** C. C. Lockwood/D. Donne Bryant Stock Photography **07.15** Lunar and Planetary Institute **07.16a** NASA/Goddard Institute for Space Studies **07.16b** NASA/Goddard Space Flight Center **07.17a** NASA/Jet Propulsion Laboratory **07.17b** NASA/Jet Propulsion Laboratory **07.17c** Courtesy of Mark Robinson/Northwestern University **07.18** Courtesy of Mark Robinson **07.19** J.Bell (Cornell), M. Wolff (Space Science Inst.), Hubble Heritage Team (STScl/AURA) **07.20** Mars Global Surveyor/NSSDC **07.21a** NASA/Goddard Institute for Space Studies **07.21b** NASA/Goddard Institute for Space Studies **07.22** NASA/Jet Propulsion Laboratory **07.23a** EROS Data Center, U.S. Geological Survey **07.23b** NASA/Jet Propulsion Laboratory **07.23c** Dr. David E. Smith, NASA, and MOLA Science Team **07.23d** R. P. Irwin III and G. A. Franz **07.24a** NASA Earth Observing System **07.24b** NASA Earth Observing System **07.24c** NASA Earth Observing System **07.25** NASA/Jet Propulsion Laboratory **07.26** NASA/Jet Propulsion Laboratory **07.28** NASA, the Magellan project. **07.29a** NASA/Jet Propulsion Laboratory **07.29b** NASA/Jet Propulsion Laboratory **07.29c** NASA/Jet Propulsion Laboratory **07.29d** NASA/Jet Propulsion Laboratory **07.30** NASA/Jet Propulsion Laboratory **07.32** Digital image by Dr. Peter W. Sloss

Chapter 8 08.08b NASA Earth Observing System **08.14** PIRL/Lunar & Planetary Laboratory (Univ. of Arizona) and NASA; NASA/JPL; University of Arizona and NASA **08.20** Courtesy of Calvin Hamilton **08.27b** NASA/JPL **08.29** NASA/JPL/Space Science Institute/NASA/Jet Propulsion Laboratory

Chapter 9 CO Photodisc/Getty Images **09.01** Photo made by Eleanor F. Helin/JPL **09.02** NASA Earth Observing System **09.02a** NASA/Jet Propulsion Laboratory **09.02b** NASA/Jet Propulsion Laboratory **09.02c** APL/NASA **09.02d** APL/NASA **09.05** Jonathan Blair/Corbis Los Angeles **09.06** Courtesy of Robert Haag Meteorites **09.07a** Peter Ceravolo **09.07b** Tony & Daphne Hallas **09.08b** Astuo Kuboniwa, March 9, 1997, 19:25:00.out, BISTAR Astronomical Observatory, Japan. **09.08b(inset)** Halley Multicolour Camera Team, Giotto, ESA, © MPAE **09.09c** © Vic and Jen Winters/Icstars Astronomy **09.12b** Courtesy Eliot Young **09.13a** Courtesy of Hal Weaver and T. E. Smith **09.13b** Courtesy of Paul Schenk **09.14a** MSSO, ANU/Science Library/Photo

Researchers, Inc. **09.14b** HST Comet Team **09.15** Courtesy of Kirk Johnson **09.16** Virgil L. Sharpton **09.17** Quade Paul/fiVth.com **09.18** Tass/Sovfoto/Eastfoto

Chapter 10 CO SOHO, ESA, NASA **10.01** Corel Corporation **10.08b** Ambio-Royal Swedish Academy of Sciences **10.09** National Optical Astronomy Observatories **10.10** Brookhaven National Laboratory **10.11** Lawrence Berkeley National Laboratory **10.12a(left and right)** Ambio-Royal Swedish Academy of Sciences **10.12b** National Solar Observatory **10.14b** From NASA's TRACE mission **10.15** From NASA's TRACE Mission **10.16** Courtesy of B. Haish and G. Slater **10.17** NASA/Goddard Space Flight Center

Chapter 11 11.03 NASA/Jet Propulsion Laboratory **11.04** NASA/Jet Propulsion Laboratory **11.07(center)** Lowell Observatory **11.11** Anglo-Australian Observatory, photograph by David Malin **11.12** NASA/Jet Propulsion Laboratory

Chapter 12 CO European Southern Observatory **12.01** Anglo-Australian Observatory, photograph by David Malin **12.02** NASA Earth Observing System **12.02b** IPAC **12.04a** C. Burrows and J. Morse (STScI), J. Hester (Arizona State University) **12.04b** C. Burrows and J. Morse (STScI), J. Hester (Arizona State University) **12.09a** Nordic Optical telescope, La Palma **12.09b** Andrew Fruchter and the ERO Team (Sylvia Baggett/STScI, Richard Hook/ST.ECF, Zoltan Levay/STScI) **12.09c** NASA and the Hubble Heritage Team (STScI/AURA) **12.09d** NASA/Jet Propulsion Laboratory **12.18** European Southern Observatory **12.19a** Anglo-Australian Observatory, photograph by David Malin **12.19b** Anglo-Australian Observatory, photograph by David Malin

Chapter 13 CO NASA/Jet Propulsion Laboratory **13.01** NASA/Jet Propulsion Laboratory **13.04b** M. Shara, B. Williams, and D. Zurek (STScI); R. Gilmozzi (ESO): D. Prialnik (Tel Aviv Univ.) **13.06** NASA/Jet Propulsion Laboratory **13.08** European Southern Observatory **13.16** Anglo-Australian Observatory, photograph by David Malin

Chapter 14 CO Dr. Axel Mellinger **14.01a** NASA/Jet Propulsion Laboratory **14.01b** NASA/Jet Propulsion Laboratory **14.03a** Anglo-Australian Observatory, photograph by David Malin **14.03b** J. Hester and P. Scowen **14.03c** John Bally, University of Colorado **14.03d** NASA/Jet Propulsion Laboratory **14.03e** Snowden Hodges **14.03f** Anglo-Australian Observatory, photograph by David Malin **14.04** NASA/Jet Propulsion Laboratory **14.05** Anglo-Australian Observatory, photograph by David Malin **14.06** NSA/CXC/SAO **14.07a** Tony & Daphne Hallas **14.07b** NASA **14.09** John Bally, University of Colorado **14.12** Anglo-Australian Observatory, photograph by David Malin **14.14** Anglo-Australian Observatory, photograph by David Malin **14.15** Anglo-Australian Observatory, photograph by David Malin **14.16** N. Sconville (Caltech) and T. Rector (NOAO) **14.17** J. Trauger **14.18a** NASA/Jet Propulsion Laboratory **14.18b** NASA/Jet Propulsion Laboratory **14.18c** NASA/Jet Propulsion Laboratory **14.18d** NASA/Jet Propulsion Laboratory **14.19** Quade Paul14.20a E. Kopan (IPAC/Caltech) **14.20b** NRAO/VLA, F. Zadeh et al14.20c NRAO/AUI (D. A. Roberts, F. Yusef-Zadeh, W. Goss) **14.20d** European Southern Observatory14.22 NASA/Jet Propulsion Laboratory

Chapter15 CO NASA/Jet Propulsion Laboratory **15.01** © Sky Publishing Corporation, reproduced with permission **15.02b** Hubble Heritage Team (AURA/STScI) **15.03** Anglo-Australian Observatory, photograph by David Malin **15.04** Anglo-Australian Observatory, photograph by David Malin and Steve Lee **15.05** Anglo-Australian Observatory, photograph by David Malin **15.06** Anglo-Australian Observatory, photograph by David Malin **15.07(SBc)** David and Christine Smith/Adam Block/NOAO/AURA/NSF **15.07(SBb)** Daryl Seibel/Adam Block/NOAO/AURA/NSF **15.07 (SBa)** Bob Birket and John Evelan/Adam Block/NOAO/AURA/NSF **15.07 (SBO)** Courtesy of Zsolt Frei and James E. Gunn **15.07 (EO)** Courtesy Zsolt Frei and James E. Gunn **15.07 (E5)** Courtesy Zsolt Frei and James E. Gunn **15.07 (SO)** Courtesy Zsolt Frei and James E. Gunn **15.07 (Sa)** Rick Barry/Adam Block/NOAO/AURA/NSF **15.07 (Sb)** Michael Chase/Adam Block/NOAO/AURA/NSF **15.07 (Sc)** Tom Boemer and David Young/Adam Block/NOAO/AURA/NSF **15.08** Hubble Heritage Team (AURA/STScI/NASA) **15.09** N. Benitez (JHU), T. Broadhurst (The Hebrew University), H. Ford (Jhu), M. Clampin (STScI), G. Harting (STScI), G. Illingworth (UCO/Lick Observatory), the ACS Science Team and ESA **15.13** The Observatories of The Carnegie Institution of Washington **15.19** Robert Williams and the HDF Team (STScI) and NASA **15.20** Brad Whitmore (STScI)15.21 Neg./Transparency no. [___]. (Photo by Frank Summers. Courtesy Dept. of Library Services, American Museum of Natural History. **15.22** Courtesy of Dr. Michael J. West/University of Hawaii **15.23a** NASA/SAO/G. Fabbiano et al. **15.23b** Copyright © Subaru Telescope, National Astronomical Observatory of Japan. All rights reserved **15.24** Hubble heritage Team (AURA/STScI/NASA) **15.25** John Bahcall/Institute for Advanced Study and NASA **15.26** Dr. William Keel/ National Radio Astronomy Observatory **15.27** John Biretta **15.28** NASA/Jet Propulsion Laboratory

Chapter 16 CO A. Fruchter, the ERO Team (STScI, ST.ECF) **16.06** Courtesy of Caltech **16.07** NASA Earth Observing System **16.08** W. N. Colley and E. Turner (Princeton University), J. A. Tyson (Bell Labs, Lucent Technologies) **16.10** A. Fruchter, the ERO Team (STScI, ST.ECF) **16.11 (inset photos)** Charles Alcock **16.12** Michael Strauss, Princeton University **16.13a** Harvard-Smithsonian Center for Astrophysics **16.13b**, c Emiko-Rose Koike/fiVth.com **16.14** Andrey Kravtsov

Chapter 17 CO E. Bunn/University of Richmond **17.05** Roger Ressmeyer/CORBIS. NY **17.08** E. Bunn/University of Richmond **17.16(left)** John Kieffer/Peter Arnold, Inc. **17.16(right)** Joel Gordon/Joel Gordon Photography

Chapter 18 CO Seth Shostak **18.02** Jeff Greenberg/Visuals Unlimited **18.04a** Biological Photo Service **18.04b** S. M. Awramik, University of California/Biological Photo Service **18.04c** S. M. Awramik, University of California/Biological Photo Service **18.05** Nih R. Feldman/Visuals Unlimited **18.07a** Woods Hole Oceanographic Institution **18.07b** George Steinmetz Photography **18.08** Roger Ressmeyer/Corbis Los Angeles18.09b F. M. Menger and Kurt Gabrielson **18.11** T. E. Adams/Visuals Unlimited **18.12** NASA/Jet Propulsion Laboratory **18.13** NASA/Johnson Space Center **18.14a** NASA/Jet Propulsion Laboratory **18.14b** R. L. Folk and F. L. Lynch **18.17** Yeshe Fenner/Space Telescope Institute **18.18** Seth Shostak **18.20** Seth Shostak **18.22a** NASA/Jet Propulsion Laboratory **18.22b** NASA/Jet Propulsion Laboratory **18.23** NASA/Jet Propulsion Laboratory **18.24** NASA/Jet Propulsion Laboratory

Index

Page references in italics refer to figures. Page references preceded by a "t" refer to tables. Page references followed by an "n" refer to footnotes.